Solid

Rectangular Solid (Parallele piped)
Surface Area $S = 2\,lw + 2\,wh + 2\,lh$
Volume $V = lwh$

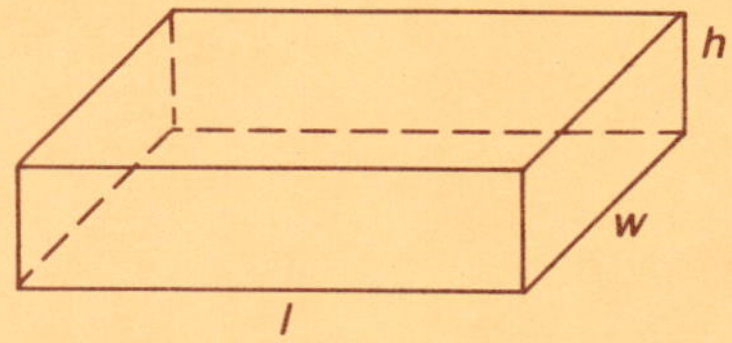

Right Circular Cylinder
Surface Area $S = 2\pi rh + 2\pi r^2$
Volume $V = \pi r^2 h$

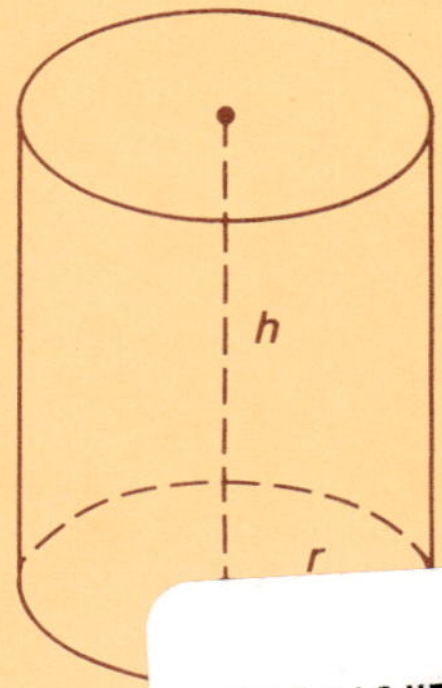

Right Circular Cone
Surface Area $S = \pi rl + \pi r^2$
Volume $V = \dfrac{1}{3}\pi r^2 h$

Sphere
Surface Area $S = 4\pi r^2$
Volume $V = \dfrac{4}{3}\pi r^3$

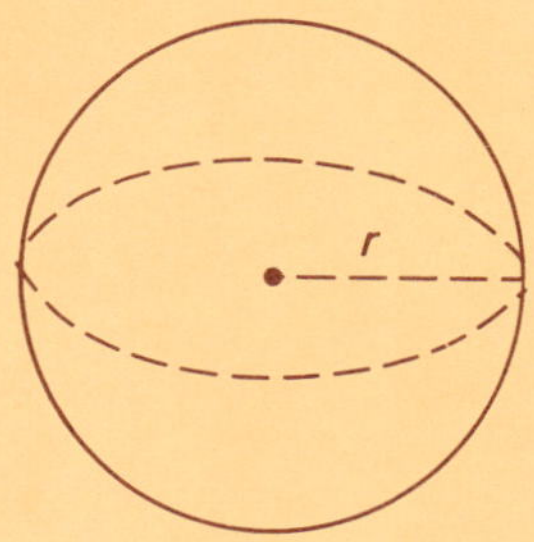

Right Pyramid
Volume $V = \dfrac{1}{3}Bh$
(B is the area of the base)

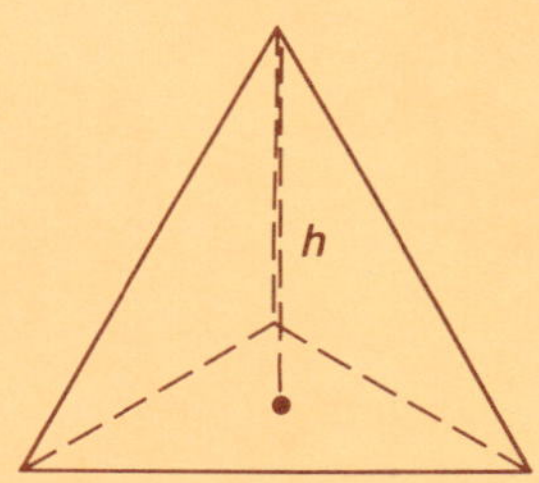

ESSENTIALS OF TECHNICAL
MATHEMATICS

ESSENTIALS OF TECHNICAL MATHEMATICS

The technical achievement of the St. Louis Arch is vividly demonstrated by the perfect union of the two arching shafts. A deviation of only $\frac{1}{64}$th of an inch in the bases would have caused the 630 foot structures not to meet.

Harry L. Nustad • Terry H. Wesner

Henry Ford Community College

Wm. C. Brown Company Publishers
Dubuque, Iowa

wcb group

Wm. C. Brown Chairman of the Board
Mark C. Falb Executive Vice-President

wcb

Wm. C. Brown Publishers
College Division

Lawrence E. Cremer President
James L. Romig Vice-President, Product Development
David Wm. Smith Vice-President, Marketing
David A. Corona Vice-President, Production and Design
E. F. Jogerst Vice-President, Cost Analyst
Marcia H. Stout Marketing Manager
Marilyn A. Phelps Manager of Design
William A. Moss Production Editorial Manager
Mary M. Heller Visual Research Manager

Book Team

Robert B. Stern Senior Editor
James M. McNeil Senior Designer
Julie A. Kennedy Senior Production Editor
Faye M. Schilling Visual Research Editor
Mavis M. Oeth Permissions Editor

Cover photo by Arteago Photos

Library of Congress Catalog Card Number: 82–072522

ISBN: 0–697–08551–1
2–08551–01

Printed in the United States of America

To our wives Dene and Mary Ann

Contents

Preface

Essentials of Technical Mathematics is a beginning level text for technical mathematics. We assume that students using this book have only an arithmetic background.

We have tried to make the text as readable as possible by presenting the material in a manner similar to that which the instructor might use in the classroom. Notes to the students are used throughout the text to highlight important ideas and to point out possible errors that the student might make in that particular area of study. Each idea is supported by real-life applications relative to that concept.

Essentials of Technical Mathematics provides a large selection of drill and practice problems followed by numerous applied problems dealing with the new ideas. There are 1,300 examples in the text to provide sufficient illustrations of each concept so that the students can become acquainted with the various problems they will encounter in the problem sets.

There is a total of 6,600 problems in the text of which over 1,400 are applications to such areas as electronics, mechanics, architecture, manufacturing engineering technology, fluid power, building and construction, civil technology, physics, data processing, chemistry, and drafting. The correct emphasis on the use of approximate numbers in applications has been followed throughout the text.

Within the exercise sets there are "trial problems" denoted by a • whose solution is shown in its entirety in the student study guide that accompanies the text.

Chapter summaries are provided to highlight the important ideas within the chapter. This is followed by a chapter review problem set, in which problems are presented in the same order as the sections of the chapter. The chapter review problems are designed to show the student if he or she needs further work on a particular section. A chapter test is provided and is designed to show the student if he or she can identify the problems without knowing the section from which they are drawn.

The text has been designed so that the instructor may skip several of the beginning chapters if the students are already familiar with that material. Most of the chapters have been designed so that the instructor may pick the chapters and sections he or she wants to teach without having to cover the topics in the precise order of the book. For example, an instructor can undertake the study of the trigonometry chapters after a basic algebra review and the study of angles and triangles in the geometry chapters.

The text starts out with a strong review of arithmetic followed by a separate chapter on ratio, proportion, and percent. Then chapter 4 covers using the hand-held calculator with the various operations and chaining operations and the special keys that the technical student would be most likely to use in his or her work. We have also included a section on estimating in this chapter. The use of the calculator is then urged throughout the text. For example, the next chapter on the metric system and approximate numbers encourages the student to use the calculator.

The text then presents the uses of algebra followed by the uses of geometry for the technician. The geometry consists of the important theorems and postulates that will be most useful to the student. No attempt is made to develop theorem-proof abilities in the student; rather, emphasis is placed upon the practical applications of geometry. Then there are chapters on logarithms, graphing, and systems of equations. The final three chapters are devoted to a study of the basic concepts of trigonometry.

Acknowledgments

We wish to express our heartfelt thanks and grateful appreciation for the many comments and suggestions given to us during the preparation of this text. In particular, we wish to thank Philip Mahler, Richard Matthei, and Richard Rowe for their excellent effort in reviewing each stage of the book and supplying us with numerous valuable comments, suggestions, and constructive criticisms.

We also wish to thank the members of the Technical Division of Henry Ford Community College. A special thanks to Michael Joannides, Enrico Puzzuoli, J. Stuart Spencer, and Leslie Spiers for their assistance in developing an excellent set of technical applications.

The authors would personally like to thank Professor John Seaton of Prince Georges Community College and Roberta Stern for working all the problems in this textbook. Their hard work is most appreciated.

In addition we wish to thank Nancy Leach for her superb help in preparing the manuscript and instructor's manual.

A special thank you goes out to Harold Elliott for introducing us to William C. Brown Company, Publishers.

List of Reviewers

1. **Natalie G. Sanders**
   Edison State Community College
2. **August J. Bethem**
   California State College, California, PA
3. **Andrew Wargo**
   Bucks County Community College
4. **Richard Wagenknecht**
   St. Paul Vocational Technical Institute
5. **Morris L. Shoss**
   San Antonio College
6. **John Seaton**
   Prince George Community College

ESSENTIALS OF TECHNICAL MATHEMATICS

1
Fundamental Operations

A technician using electronic equipment to test metal stress and fatigue.
(Photograph courtesy of Ford Motor Company)

1.1 Addition and Subtraction of Whole Numbers

The most basic use of our number system is for counting. We use 1, 2, 3, 4, 5, and so on as symbols to represent the **counting** or **natural numbers.** If we also include 0 in this group of numbers, we have the **whole numbers.** Our system of numbers uses ten symbols, called **digits** from the Latin word for "finger." These symbols are 0, 1, 2, 3, 4, 5, 6, 7, 8, and 9. We can use these symbols to write a number of any size. The symbols were invented to stand for the number of objects being counted. However, if counting continued to one more than nine, there was originally no symbol to represent the number. The number zero was the last invented, and, eventually, it was put together with the number 1 to get 10, the symbol used to represent ten objects. From this beginning, eleven was invented to represent 10 + 1, denoted by 11; twelve was established to represent 10 + 2, denoted by 12; and so on.

The digits that make up these succeeding numbers take on **place value**. When the number 28 came into being, for example, it came to mean two "groups" of ten objects plus eight single objects, or

$$(2 \cdot 10) + (8 \cdot 1) = 20 + 8 = 28$$

note

The raised dot between the 2 and the 10 and between the 8 and the 1 denotes the operation of multiplication. Other ways to denote multiplication are with $\times$ or parentheses: 8×1 or (8)(1), for example.

Place value is determined by position. In the number 28, the 8 is in the position called the *units* place and the 2 is in the position called the *tens* place. Therefore, we call the 8 the *units digit* and the 2 the *tens digit*. It is the use of this place value of digits that allows us to write a number of any magnitude.

When we count to 99, we have

$$(9 \cdot 10) + (9 \cdot 1) = 90 + 9 = 99$$

and if we add one more object to this, we have ten "groups" of ten objects, which is called one hundred, denoted by 100. The 1 is in the *hundreds* place and is called the *hundreds digit*. Thus, the number 436 means

$$(4 \cdot 100) + (3 \cdot 10) + (6 \cdot 1) = 400 + 30 + 6 = 436$$

As the numbers get greater and greater we continue adding symbols to the left of the previous digits. In our number system, starting at the right, the first place is the units place; the next place to the left is the tens place; the third is the hundreds place; the fourth, the *thousands* place; fifth, the *ten-thousands* place; sixth, the *hundred-thousands* place; and so on. In the number 2,456,389, we have the following:

MILLIONS	HUNDRED-THOUSANDS	TEN-THOUSANDS	THOUSANDS	HUNDREDS	TENS	UNITS
2	4	5	6	3	8	9

Now consider the first of the operations on whole numbers, **addition**. When two or more numbers are added, each number is called an **addend** and the result is called the **sum** of the numbers. The symbol used to indicate addition is the **plus sign**, denoted by +. *When we perform addition, we must be careful to add only like quantities.*

To add numbers with two, three, four, or more digits in them, we usually arrange them in a column, placing units digits under units digits, tens under tens, hundreds under hundreds, and so on, thereby enabling us to add like quantities. For example, to find the sum of

$$23 + 456 + 736 + 4{,}058$$

we would write the numbers in column form:

$$\begin{array}{r} 23 \\ 456 \\ 736 \\ \underline{4{,}058} \end{array}$$

To perform the addition, we first add the units digits:

$$3 + 6 + 6 + 8 = 23$$

and place the 3 in the units column. Since the 2 represents the tens digit, which is to be ''carried'' to the next column, the tens column, we place the 2 above the tens column. Thus,

$$\begin{array}{r} {\scriptstyle 2} \\ 23 \\ 456 \\ 736 \\ \underline{4{,}058} \\ 3 \end{array}$$

Now we add the tens digits:

$$2 + 2 + 5 + 3 + 5 = 17$$

Place the 7 in the tens column and ''carry'' the 1 above the hundreds column to get

$$\begin{array}{r} {\scriptstyle 1\,2} \\ 23 \\ 456 \\ 736 \\ \underline{4{,}058} \\ 73 \end{array}$$

Now add the hundreds column:

$$1 + 4 + 7 + 0 = 12$$

and place the 2 in the hundreds column, carrying the 1 above the thousands column to get

$$\begin{array}{r} {\scriptstyle 1\,1\,2} \\ 23 \\ 456 \\ 736 \\ \underline{4{,}058} \\ 273 \end{array}$$

Finally, add the 1 and 4 to get 5.

$$\begin{array}{r} {}^{1\,1\,2} \\ 23 \\ 456 \\ 736 \\ \underline{4{,}058} \\ 5{,}273 \end{array}$$

Thus, we have

$$23 + 456 + 736 + 4{,}058 = 5{,}273$$

or

$$\begin{aligned} &(5 \cdot 1{,}000) + (2 \cdot 100) + (7 \cdot 10) + (3 \cdot 1) \\ &= 5{,}000 + 200 + 70 + 3 = 5{,}273 \end{aligned}$$

In order to be sure your answer is correct, adopt a method for checking it. Probably the best way is to add the columns again, this time reversing the direction in which you add. To illustrate,

$$\begin{array}{r} {}^{1\,\;1\,1} \\ 3{,}426 \\ 723 \\ 402 \\ \underline{95} \\ 4{,}646 \end{array} \qquad \textbf{Check} \qquad \begin{array}{r} {}^{1\,\;1\,1} \\ 95 \\ 402 \\ 723 \\ \underline{3{,}426} \\ 4{,}646 \end{array}$$

Example If one pipe is 6 ft 7 in long and a second pipe is 4 ft 9 in long, what is their combined length? Add the numbers in column form:

$$\begin{array}{rr} 6 \text{ ft} & 7 \text{ in} \\ \underline{4 \text{ ft}} & \underline{9 \text{ in}} \\ 10 \text{ ft} & 16 \text{ in} \end{array}$$

We observe that the 16 in can be written in a more useful form as 1 ft 4 in. When we do this, we write the 4 in and carry the 1 ft from the inch column over to the foot column:

$$\begin{array}{rr} {}^{1\text{ ft}} & \\ 6 \text{ ft} & 7 \text{in} \\ \underline{4 \text{ ft}} & \underline{9 \text{in}} \\ 11 \text{ ft} & 4 \text{in} \end{array}$$

We assume that students are familiar with some common units of measure and their symbols, such as yard—yd, foot—ft, inch—in, hour—h, minute—min, second—s, gallon—gal, quart—qt, ounce—oz, and so on. In chapter 5 we discuss units of measure and their symbols.

Example Find the perimeter of the figure shown in figure 1.1.

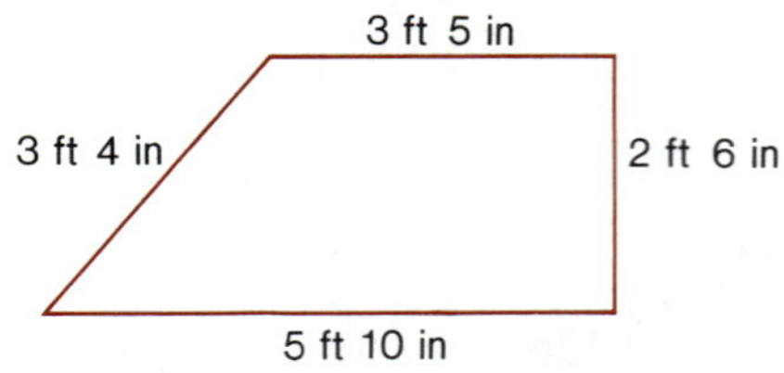

Figure 1.1

The perimeter of this geometric figure or any geometric figure in a plane is the total distance around the figure. Therefore, we have:

$$\begin{array}{rr} 3\text{ ft} & 5\text{ in} \\ 2\text{ ft} & 6\text{ in} \\ 5\text{ ft} & 10\text{ in} \\ \underline{3\text{ ft}} & \underline{4\text{ in}} \\ 13\text{ ft} & 25\text{ in} \end{array}$$

The 25 in is also 2 ft 1 in, so we write the 1 in and carry the 2 ft from the inch column to the foot column. Now we have:

$$\begin{array}{rr} \overset{2\text{ ft}}{3\text{ ft}} & 5\text{ in} \\ 2\text{ ft} & 6\text{ in} \\ 5\text{ ft} & 10\text{ in} \\ \underline{3\text{ ft}} & \underline{4\text{ in}} \\ 15\text{ ft} & 1\text{ in} \end{array}$$

Therefore, the perimeter of this geometric figure is 15 ft 1 in.

The inverse or opposite of adding whole numbers is the operation of subtracting them. When one number is subtracted from another, we call the result the **difference**. The symbol used to indicate subtraction is the **minus sign**, denoted by $-$. To indicate the subtraction of 6 from 18, we write

$$18 - 6$$

which is read "18 subtract 6." In the example

$$18 - 6 = 12$$

we call 18 the **minuend** and 6 the **subtrahend**; 12 is the difference.

As with addition, we place the digits—units, tens, hundreds, and so on—in columns when we subtract whole numbers. For example, write 597 − 248 in columns:

$$\begin{array}{rl} 597 & \text{(minuend)} \\ -\ \underline{248} & \text{(subtrahend)} \end{array}$$

and subtract. Consider the units digits. We first subtract the 8 from the 7. However, since the 8 is greater than the 7, it is necessary to use **borrowing**, which is the opposite of carrying. That is, we "borrow" 1 ten from the tens digit in the number 597 and make the tens digit 8:

$$\begin{array}{r} 5\overset{8}{\cancel{9}}7 \\ -\ \underline{248} \end{array}$$

Now add the borrowed 10 to the units digit (7) to get 17 and subtract the 8 from 17 to get a difference in the units column of 9:

$$\begin{array}{r} 5\overset{8}{\cancel{9}}{}^{1}7 \\ -\ \underline{248} \\ 9 \end{array}$$

Now proceed to the tens column and subtract 4 from 8:

$$\begin{array}{r} 5\overset{8}{\cancel{9}}{}^{1}7 \\ -\ \underline{248} \\ 49 \end{array}$$

Subtracting the hundreds digits we obtain a difference of 3. Thus, we have

$$\begin{array}{r} 5\overset{8}{\cancel{9}}{}^{1}7 \\ -\ \underline{248} \\ 349 \end{array}$$

The difference 597 − 248 is, therefore, 349. To check the subtraction, add the difference, 349, and the subtrahend, 248; the sum of these two numbers must equal the minuend, 597. Since

$$349 + 248 = 597$$

you know that your answer is correct.

Examples

1. Subtract 479 from 836.

Solution

$$\begin{array}{r} \overset{7}{\cancel{8}}{}^{1}\overset{2}{\cancel{3}}{}^{1}6 \\ -\ \underline{479} \\ 357 \end{array}$$

Check

$$\begin{array}{r} \overset{1}{3}\overset{1}{5}7 \\ \underline{479} \\ 836 \end{array}$$

2. Subtract 563 from 902.

Solution

$$\begin{array}{r} \overset{8}{\cancel{9}}{}^{1}\overset{9}{\cancel{0}}{}^{1}2 \\ -\ \underline{563} \\ 339 \end{array}$$

Check

$$\begin{array}{r} \overset{1}{3}\overset{1}{3}9 \\ \underline{563} \\ 902 \end{array}$$

3. A board 8 ft 7 in long has a piece 3 ft 11 in cut from it. What is the length of the remaining piece of board?

$$\begin{array}{r} 8\text{ ft }\ \ 7\text{ in} \\ -\ \underline{3\text{ ft } 11\text{ in}} \end{array}$$

Since we cannot subtract the 11 in from the 7 in, we must borrow 1 ft from the foot column; this measure comes to the inch column as 12 in. We now subtract the 11 in from the 19 in (7 in + 12 in) and we have:

$$\begin{array}{r} \overset{7}{\not{8}}\text{ ft }\ \overset{19}{\not{7}}\text{ in} \\ -\ \underline{3\text{ ft } 11\text{ in}} \\ 4\text{ ft }\ \ 8\text{ in} \end{array}$$

Therefore, the remaining length of board is 4 ft 8 in.

When a series of numbers involving only addition and subtraction are written horizontally, the operations should be taken in order from left to right as they occur. For example, in

$9 - 3 + 4 + 3 - 6 - 1 + 4$

Subtract	$9 - 3 = 6$	$6 + 4 + 3 - 6 - 1 + 4$
then add	$6 + 4 = 10$	$10 + 3 - 6 - 1 + 4$
and add	$10 + 3 = 13$	$13 - 6 - 1 + 4$
then subtract	$13 - 6 = 7$	$7 - 1 + 4$
and subtract	$7 - 1 = 6$	$6 + 4$
and add	$6 + 4 = 10$	10

so $\quad 9 - 3 + 4 + 3 - 6 - 1 + 4 = 10$

If any quantity is enclosed with grouping symbols such as parentheses, ( ), brackets, [ ], or braces, { }, we treat the quantity within as a single number. Thus, in

$9 - (3 + 1) + (6 - 2) - (5 - 1)$

we perform operations within parentheses first to get

$9 - 4 + 4 - 4$

which gives

$5 + 4 - 4 = 9 - 4 = 5$

Therefore,

$9 - (3 + 1) + (6 - 2) - (5 - 1) = 5$

Examples Perform the indicated operations.

$$
\begin{aligned}
1.\ 8 - 3 + 2 - 5 - 1 &= 5 + 2 - 5 - 1 \\
&= 7 - 5 - 1 \\
&= 2 - 1 \\
&= 1
\end{aligned}
$$

$$
\begin{aligned}
2.\ (6 + 1) - (8 - 5) + 7 + (9 - 6) &= 7 - 3 + 7 + 3 \\
&= 4 + 7 + 3 \\
&= 11 + 3 \\
&= 14
\end{aligned}
$$

Exercise Set 1.1

Find the sum of the following problems. *Do not use a calculator.*

1.	2.	3.	4.	5.
387	893	864	746	867
246	747	246	554	598

6.	7.	8.	9.	10.
707	916	153	791	763
723	597	467	398	854
281	307	395	476	636

11.	12.	13.	14.	15.
183	371	602	915	498
763	298	193	598	413
543	702	529	943	529
288	699	977	372	639

16. 736 + 1,234 + 34 + 607
17. 14,769 + 5 + 704 + 59 + 998
18. 15,707 + 394 + 21 + 1,234 + 63
19. 89 + 6 + 76 + 1,436,705 + 36,451
20. 500 + 7,364 + 82 + 91,674 + 2

Find the difference in the following problems.

21.	22.	23.	24.
376	707	639	907
− 123	− 519	− 299	− 389

25.	26.	27.	28.
1,234	7,947	9,008	6,793
− 1,217	− 1,217	− 3,214	− 5,009

29.	30.	31.	32.
3,473	1,093	7,920	9,769
− 963	− 576	− 776	− 8,936

Perform the indicated operations.

•33. $43 + (9 - 6) - (7 + 3) - (8 - 3)$

34. $64 - (9 - 8) + (8 - 2) - (3 + 5)$

35. $92 + (6 - 3) - (9 - 2) - (7 - 5)$

36. $34 - (18 - 7) - (7 + 9) + (36 - 17)$

37. $42 - 6 + 7 - 8 + 14 - 36$

38. $17 + 3 - 11 - 4 + 19 - 8$

39. $5 + 14 - 3 - 9 - 2 + 14$

40. $63 - 21 - 14 + 15 + 9 - 8 - 16$

Find the sum in the following problems.

41. 4 ft 3 in 2 ft 5 in	42. 5 ft 7 in 3 ft 4 in	•43. 9 ft 7 in 5 ft 6 in
44. 3 ft 9 in 10 ft 11 in	•45. 10 h 15 min 8 h 50 min	46. 3 h 37 min 2 h 52 min
47. 14 h 7 min 2 h 57 min	48. 18 h 42 min 4 h 18 min	49. 6 h 37 min 2 h 23 min
50. 14 gal 3 qt 2 gal 2 qt	51. 5 gal 1 qt 4 gal 3 qt	52. 7 gal 2 qt 4 gal 2 qt
53. 9 gal 3 qt 5 gal 3 qt	54. 7 gal 3 qt 2 gal 2 qt	

Find the difference in the following problems.

55. 5 ft 10 in − 2 ft 4 in	56. 6 ft 7 in − 3 ft 4 in	57. 9 ft 5 in − 4 ft 6 in
58. 10 ft 4 in − 3 ft 4 in	59. 10 h 45 min − 8 h 12 min	60. 5 h 37 min − 2 h 18 min
61. 16 h 16 min − 7 h 42 min	62. 8 h 11 min − 4 h 47 min	63. 14 gal 3 qt − 2 gal 2 qt
64. 8 gal 3 qt − 5 gal 1 qt	•65. 9 gal 1 qt − 3 gal 3 qt	66. 7 gal 2 qt − 3 gal 3 qt

Find the perimeters of the given figures.

67.

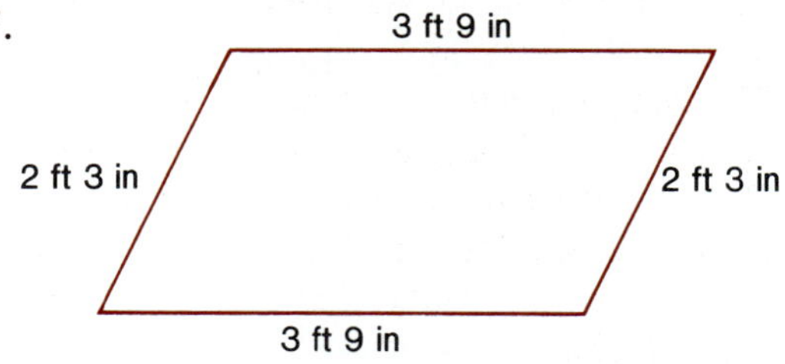

68.

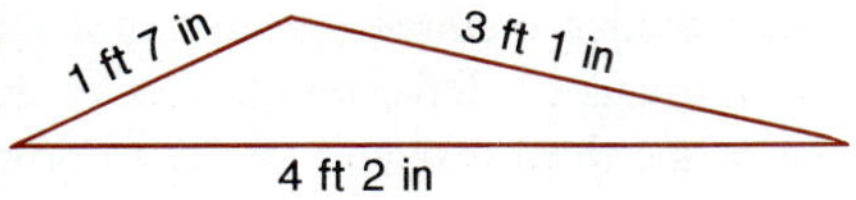

69.

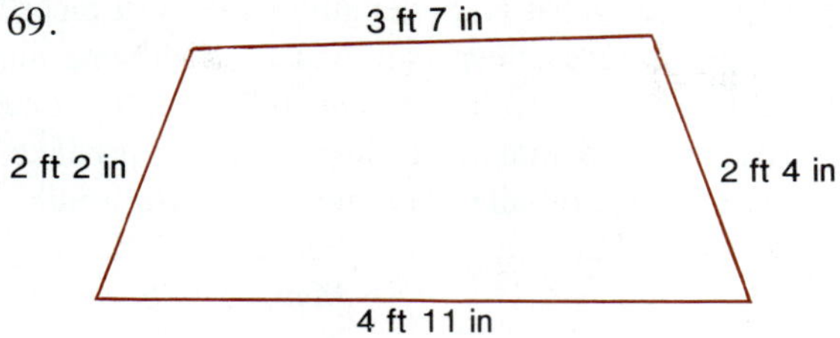

70.

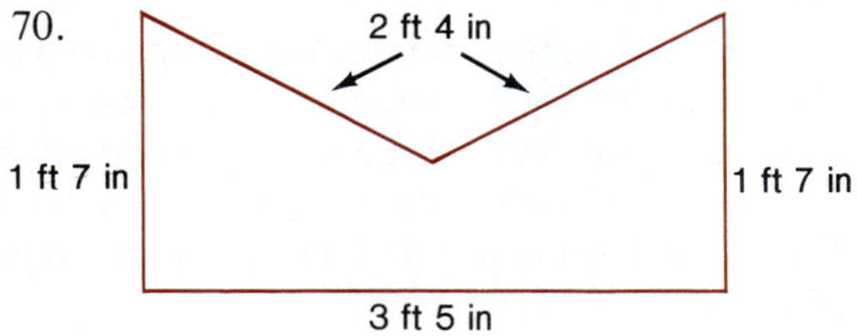

71.

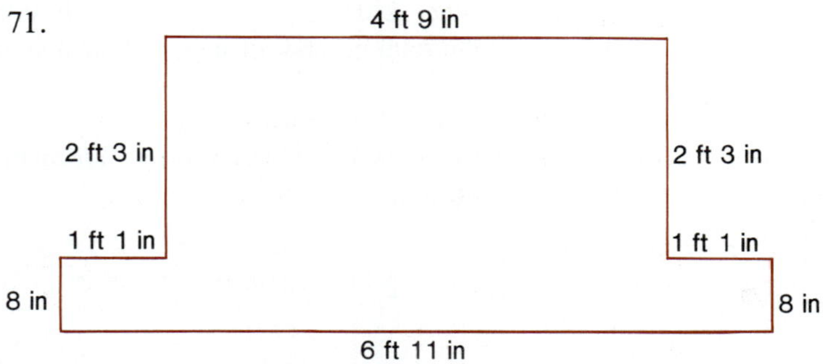

72. The water meter readings for six consecutive months were 536 gal, 367 gal, 703 gal, 901 gal, 699 gal, and 823 gal. How many gallons were used in the six months?
73. The electric meter readings for two consecutive months were 30,273 kilowatt-hours (kwh) and 33,091 kilowatt-hours (kwh). How many kilowatt-hours were used in the month?
74. The belts in a pulley system measure 36 ft, 32 ft, 24 ft, 63 ft, 91 ft, and 8 ft. How many feet of belts are there in the system?
75. The total resistance of a series-connected circuit of two or more series-connected resistors is equal to the sum of the individual resistances. Find the total resistance of a series-connected circuit whose resistors are 68 ohms, 623 ohms, 2 ohms, and 829 ohms.

76. The total resistance of three resistors in a series-connected circuit is 903 ohms. If two of the resistors have resistances of 36 ohms and 372 ohms, what is the resistance of the third resistor? (Refer to problem 75.)
77. A two-by-four 24 ft long is cut into three pieces. Two of the pieces are 8 ft and 10 ft long. What is the length of the third piece?
78. An electronics salesperson had the following number of sales during a 5-month period: 321, 274, 76, 95, and 142. Find the total number of sales.
79. The Bradford Manufacturing Company produced the following number of subchassis for a missile plant during a given year:

January 876	February 732	March 1,086
April 524	May 1,586	June 905
July 654	August 232	September 1,473
October 1,843	November 1,956	December 1,301

Find the total number constructed by the company during this year.
80. An electronics supply house has 432 resistors of a certain type. If 36 are sold during the first week, 72 during the second week, 29 during the third week, and 58 during the fourth week, how many were left at the end of the month?
81. The total number of new students in a technical program for a given four-semester period was:

First semester 2,236	Second semester 1,934
Third semester 2,077	Fourth semester 1,876

What was the total number of new technical students during this period?

1.2 Multiplication and Division of Whole Numbers

Multiplication can be thought of as a shortened form of *repeated* addition. For instance, if we wish to add four 8s, then

$$8 + 8 + 8 + 8 = 32$$

Another way of saying this repeated addition is:

$$4 \times 8 = 32$$

where the symbol $\times$ means multiply or times; thus, the statement can be read as "4 times 8," and 4×8 means using 8 as an addend four times. The number 32 is called the **product** of 4 times 8. In the multiplication of two numbers, the number that is to be taken a given number of times—in our example, the 8—is called the **multiplicand** and the other number is called the **multiplier**—in our example, the 4.

Using the previous multiplication, note that

$$4 \times 8 = 8 \times 4$$

since

$$4 + 4 + 4 + 4 + 4 + 4 + 4 + 4 = 32$$

which demonstrates that

$$8 \times 4 = 32$$

This fact tells us that *the order in which we multiply numbers together will not change the answer;* this is called the **commutative law of multiplication.** This same idea is true for addition. That is, *the order in which we add numbers will not change the answer,* and this fact is called the **commutative law of addition.**

Consider the product of

$$19 \times 5 \times 2$$

If we recognize the product

$$5 \times 2 = 10$$

and then

$$19 \times 10 = 190$$

the computation becomes easier. Using parentheses to indicate which operation to perform first, we have:

$$(19 \times 5) \times 2 = 19 \times (5 \times 2) = 190$$

This example indicates that *when we are multiplying, we may group our numbers any way we wish and our answer will not be affected.* This idea of grouping is called the **associative law of multiplication.** It also applies when we are doing addition. In that case, it is called the **associative law of addition.** These laws do not work for subtraction or division.

As we noted in the previous section, multiplication may be indicated in other ways that tend to make the problem appear less complicated. One commonly used method is to place a raised dot between the numbers:

$$5 \times 6 = 5 \cdot 6$$

This method prevents the confusion that may develop when the cross, or $\times$, is used in other ways, as seen later in the text. Another way to indicate multiplication is to enclose the numbers in parentheses without any sign between them, as

$$5 \times 6 = (5)(6)$$

So

$$5 \times 6 = 5 \cdot 6 = (5)(6) = 30$$

The numbers in a multiplication problem are referred to as the **factors** *of the product.* The numbers 5 and 6 are the factors of 30 in the previous illustration. For another example, the factors of 36 might be 4 and 9, 12 and 3, 6 and 6, or 36 and 1 since

$$4 \cdot 9 = 12 \cdot 3 = 6 \cdot 6 = 36 \cdot 1 = 36$$

Notice that 36 also has factors of 2, 2, 3, and 3 since

$$2 \cdot 2 \cdot 3 \cdot 3 = 4 \cdot 9 = 36$$

Often, identical factors appear more than once in an indicated product, as in

$$2 \cdot 2 \cdot 3 \cdot 3 = 36$$

In this situation, the numbers 2 and 3 are factors of the product two times. It becomes useful in many situations like this to write the indicated product of two or more identical factors in a shortened form.

Consider the indicated products

$4 \cdot 4 \cdot 4 = 64$

and

$3 \cdot 3 \cdot 3 \cdot 3 = 81$

A more convenient way of writing $4 \cdot 4 \cdot 4$ is 4^3, which is read "4 to the third power" or "4 cubed." We call the number 4 the **base** of the expression and the small number 3 to the upper right the **exponent.** Thus,

$4 \cdot 4 \cdot 4 = 4^3 = 64$

In like fashion,

$3 \cdot 3 \cdot 3 \cdot 3$

may be written

3^4

where 3 is the base and 4 the exponent, and the expression is read, "3 to the fourth power." Then

$3 \cdot 3 \cdot 3 \cdot 3 = 3^4 = 81$

Notice that *the exponent tells how many times the base is used as a factor in an indicated product*. We call this form of a product the **exponential form.** That is, the exponential form of the product $3 \cdot 3 \cdot 3 \cdot 3$ is 3^4.

Because the number 2 is used as an exponent quite often, it is given a special name. We read 3^2 as "3 to the second power" or more commonly "3 squared." Similarly, with the exponent 3 we read 2^3 as "2 to the third power" or more commonly, "2 cubed." Thus, in the previous example,

$2 \cdot 2 \cdot 3 \cdot 3 = 36$

could be written in exponential form by

$2^2 \cdot 3^2 = 36$

Examples Write in exponential form.

1. $5 \cdot 5 \cdot 5 = 5^3$ (read "5 cubed")
2. $2 \cdot 2 \cdot 2 \cdot 2 \cdot 2 = 2^5$ (read "2 to the fifth power")

---

note

If no exponent is written, then it is understood that the exponent is 1. That is, 5 is the same as 5^1.

To multiply numbers containing more than one digit, place one number above the other so that the units digits are in line. Normally, the greater number is placed above the smaller, as with the indicated product

$534 \cdot 25$

which we can write

$$\begin{array}{r} 534 \\ \times \underline{25} \end{array}$$

Begin by multiplying the units digits

$$5 \cdot 4 = 20$$

in which case the units digit in the product, 0, is placed beneath the units digits of the two numbers and the 2, standing for 20, or 2 tens, is carried over to the tens digit of the multiplicand.

$$\begin{array}{r} \overset{2}{5}34 \\ \times \underline{25} \\ 0 \end{array}$$

Now multiply 5 and 3 tens to get 15 tens, which, when added to the 2 tens carried gives 17 tens, or 1 hundred and 7 tens. The 7 is placed in the tens column and the 1 hundred is carried to the hundreds column.

$$\begin{array}{r} \overset{1\,2}{5\,3}4 \\ \times \underline{25} \\ 70 \end{array}$$

Now multiply 5 and 5 hundreds to get 25 hundreds, which, added to the 1 hundred carried, gives 26 hundreds. Since this represents 2 thousands and 6 hundreds, we write the 6 in the hundreds place and the 2 in the thousands place to get

$$\begin{array}{r} \overset{1\,2}{5\,3}4 \\ \times \underline{25} \\ 2670 \end{array}$$

which gives a partial product of 2670, the result of multiplying by 5.

Next, we multiply by the 2 in the multiplier, which is actually 20 since it is in the tens place and represents 2 tens. Therefore, we multiply 20 and 4 to get 80 or 8 tens. Place the 8 in the tens place on the next line down:

$$\begin{array}{r} 534 \\ \times \underline{25} \\ 2670 \\ 8 \end{array}$$

Now, when we multiply the tens digit of the multiplier, 2, times the tens digit of the multiplicand, 3, we are multiplying

$$20 \times 30 = 600$$

so place the 6 in the hundreds place.

$$\begin{array}{r} 534 \\ \times \underline{25} \\ 2670 \\ 68 \end{array}$$

Finally, multiply the 2 in the multiplier times the 5 of the multiplicand to get

$$20 \times 500 = 10{,}000$$

Place the 1 in the ten-thousands place, and fill in the space with the 0 in the thousands place to get the second partial product. We then add the partial products to get 13,350.

```
     534
×     25
    2670
   1068
  13,350
```

If the multiplier contains a zero, the multiplication of this digit need not be shown since *the product of zero and any number is zero.*

Examples

1. Multiply 3,978 by 306.

```
     3,978                 3,978
  ×    306              ×    306
     23868      or         23868    (multiply by 6)
     0000                11934      (multiply by 300)
   11934               1,217,268
 1,217,268
```

To check your multiplication, reverse the multiplier and the multiplicand. Since the order in which numbers are multiplied does not change the product, you should get the same answer.

2. Multiply 243 by 196.

```
Solution        243      Check      196
             ×  196              ×  243
               1458                 588
              2187                 784
              243                 392
             47,628              47,628
```

The process of multiplying numbers containing more than one digit uses a very important law of mathematics. We can perform the multiplication

$4 \cdot 26$

in a way other than using column form:

```
     26
×     4
    104
```

Since $26 = 20 + 6$, then $4 \cdot 26$ can be written

$4 \cdot (20 + 6)$

and we note that

$$\begin{aligned} 4 \cdot (20 + 6) &= (4 \cdot 20) + (4 \cdot 6) \\ &= 80 + 24 \\ &= 104 \end{aligned}$$

We have performed the multiplication in a horizontal form, achieving the same correct result. The principle used here is called the **distributive law of multiplication over addition.** In words, *the distributive law states that if the sum of two numbers is multiplied by a third number, then we multiply this third number times each of the numbers that we are adding, and add these products to find the final result.*

Examples Multiply using the distributive law.

1. $5 \cdot 32 = 5 \cdot (30 + 2)$
   $= (5 \cdot 30) + (5 \cdot 2)$
   $= 150 + 10$
   $= 160$
2. $6 \cdot 48 = 6 \cdot (40 + 8)$
   $= (6 \cdot 40) + (6 \cdot 8)$
   $= 240 + 48$
   $= 288$

We put this law to much greater use later in the text. In preparation for this, you are asked, in exercise set 1.2, to perform some multiplications using the distributive law.

The process of dividing two whole numbers is the inverse operation of multiplying two numbers. Thus, if we wish to divide 24 by 6, the resulting answer, called the **quotient,** is found by asking "What number multiplied by 6 will produce 24?" The quotient of

$$24 \div 6 = 4$$

since $4 \times 6 = 24$. We denote division in three ways:

$$24 \div 6 \qquad 24/6 \qquad \frac{24}{6}$$

The $\div$ is the symbol for "is divided by" and the second and third forms are fractions. In this example, 24 is called the **dividend** and 6 is called the **divisor.**

Now consider dividing numbers containing several digits. To do the problem

$$43{,}263 \div 8$$

we use a process called long division:

```
     5,407     remainder 7
  8)43,263
    40
     32
     32
       63
       56
        7
```

The quotient could also be written as $5{,}407\frac{7}{8}$, which is a more useful way of stating it. To check our answer, we multiply the quotient times the divisor and add the remainder as follows:

$$\begin{array}{r} 5{,}407 \\ \times \quad 8 \\ \hline 43{,}256 \\ + \quad 7 \\ \hline 43{,}263 \end{array}$$

Example Divide 324,677 by 45.

Solution

$$\begin{array}{r} 7{,}215 \quad \text{remainder } 2 \\ 45\overline{)324{,}677} \\ \underline{315} \\ 96 \\ \underline{90} \\ 67 \\ \underline{45} \\ 227 \\ \underline{225} \\ 2 \end{array}$$

The quotient is $7{,}215\frac{2}{45}$.

Example Light travels at a rate of 186,000 miles per second (mi/s). How long does it take light to travel 5,022,000 mi? How far will light travel in 1 minute (min)?

Since we have the total distance traveled and the distance light travels in 1 s, we can find the number of seconds by dividing 5,022,000 by 186,000 as follows:

$$\begin{array}{r} 27 \\ 186{,}000\overline{)5{,}022{,}000} \\ \underline{3\ 720\ 00} \\ 1\ 302\ 000 \\ \underline{1\ 302\ 000} \end{array}$$

Therefore, it takes light 27 s to travel 5,022,000 mi.

If light travels 186,000 mi in 1 s and we wish to know how far it will travel in 1 min, which is 60 s, we can determine this by multiplying as follows:

$$\begin{array}{r} 186{,}000 \\ \times \quad 60 \\ \hline 11{,}160{,}000 \end{array}$$

Therefore, in 1 min light travels 11,160,000 mi.

Exercise Set 1.2

Write the following in expanded form and determine the answer.

Example $6^3 = 6 \cdot 6 \cdot 6 = 216$

1. 2^3	•2. 3^4	3. 4^3	4. $(25)^2$
5. 5^4	6. 2^8	•7. $(4)^2$	8. $(7)^3$
9. $3^3 \cdot 5^2$	10. $2^4 \cdot 3^2 \cdot 4^2$	11. 0^3	12. $2^3 \cdot 0^2 \cdot 3^3$

Write the following in exponential form.

Example $8 \cdot 8 \cdot 8 = 8^3$

13. $3 \cdot 3 \cdot 3$
14. $5 \cdot 5 \cdot 5 \cdot 5 \cdot 5$
15. $7 \cdot 7 \cdot 7 \cdot 7 \cdot 7 \cdot 7 \cdot 7$
16. $2 \cdot 2 \cdot 2 \cdot 2 \cdot 2 \cdot 2 \cdot 2$
17. $3 \cdot 3 \cdot 2 \cdot 2 \cdot 2 \cdot 2$
18. $5 \cdot 4 \cdot 4 \cdot 4 \cdot 6 \cdot 6$
•19. $9 \cdot 9 \cdot 7 \cdot 7 \cdot 7 \cdot 5 \cdot 5$
20. $4 \cdot 4 \cdot 3 \cdot 3 \cdot 3 \cdot 2 \cdot 2 \cdot 2 \cdot 2 \cdot 5 \cdot 5 \cdot 5$

Perform the following multiplications using the distributive law.

Example
$$\begin{aligned} 4 \cdot 18 &= 4 \cdot (10 + 8) \\ &= (4 \cdot 10) + (4 \cdot 8) \\ &= 40 + 32 \\ &= 72 \end{aligned}$$

•21. $8 \cdot 19$	22. $6 \cdot 24$	23. $5 \cdot 32$
24. $7 \cdot 27$	25. $9 \cdot 22$	26. $12 \cdot 17$
27. $14 \cdot 19$	28. $16 \cdot 25$	29. $26 \cdot 23$

Multiply the following.

30. $\begin{array}{r} 3{,}242 \\ \underline{36} \end{array}$	31. $\begin{array}{r} 5{,}107 \\ \underline{98} \end{array}$	32. $\begin{array}{r} 7{,}421 \\ \underline{72} \end{array}$	33. $\begin{array}{r} 7{,}932 \\ \underline{53} \end{array}$
34. $\begin{array}{r} 7{,}786 \\ \underline{223} \end{array}$	35. $\begin{array}{r} 9{,}008 \\ \underline{149} \end{array}$	36. $\begin{array}{r} 9{,}612 \\ \underline{877} \end{array}$	37. $\begin{array}{r} 3{,}920 \\ \underline{501} \end{array}$

38.	39.	40.	41.
6,666	7,767	5,362	7,091
<u>2,004</u>	<u>2,222</u>	<u>8,212</u>	<u>5,328</u>

Divide the following.

42. 734,214 ÷ 6	43. 24,363 ÷ 3	44. 50,004 ÷ 9
45. 32,736 ÷ 25	46. 57,423 ÷ 36	47. 60,716 ÷ 86
48. 69,506 ÷ 89	49. 72,186 ÷ 142	50. 88,888 ÷ 916
51. 7,777,777 ÷ 103	•52. 784,367 ÷ 921	

53. Sound travels at the approximate rate of 1,080 feet per second (ft/s).
    a. How far will it travel in 8 s?
    b. How much farther than 1 mi will it travel in 6 s?
54. Light travels at a rate of 186,000 mi/s. How long will it take to travel 1,674,000 mi?
•55. How long does it take light from the sun to reach earth if the sun is approximately 93,000,000 mi away? (Refer to problem 54.)
56. How far will light travel in 2 h? (Refer to problem 54.)
57. One way to find the average of a set of numbers is to add the numbers and divide by the number of numbers in the set. If the voltage being transmitted from one point to another point is 360 volts, 478 volts, 732 volts, and 275 volts, when measured at hour intervals, what was the average voltage during the length of time over which the measurements were made?
58. In certain electrical circuits, the product of the resistance in ohms and the current in amperes (amp) is equal to the voltage in volts across the resistor. A particular heating element has a resistance of 11 ohms per unit of length. If the heating element is 24 ft long and the current through the wire is 3 amps, what is the voltage across the element?
•59. The product of the resistance and the current squared is equal to the power in watts dissipated by the resistor in certain electric circuits. If three resistors of 6, 8, and 12 ohms are placed in series, what power is dissipated in each resistor when the circuit current is 4 amps?
60. When equal resistances are connected in parallel, the resistance of the circuit is equal to the value of one resistance divided by the number of resistors connected in parallel. If four resistors of 7,200 ohms each are connected in parallel, what is their combined resistance?
61. When capacitors of equal capacitance are connected in series, the capacitance of the combination is equal to the value of one capacitance divided by the number of capacitors. If five capacitors of 355 microfarads are connected in series, what is their combined capacitance?
62. In cents, at a cost of 32 cents per pound (cents/lb), find the cost of 132 castings, each weighing 17 lb.
63. One hundred and twenty-seven tons of coal was purchased for $8,001. What was the price per ton?

64. The distance from New York to San Francisco is about 2,606 mi. A passenger train leaves New York bound for San Francisco, at 7:00 A.M. traveling at 90 mi/h. How long will it take to make the trip? (*Hint:* Distance = rate × time can be changed to time = distance ÷ rate.)
65. Between Chicago and Detroit, a distance of 282 mi, a driver averages 47 mi/h. How long will it take him to make the trip?
66. A trip of 369 mi takes 9 h to complete. What was the average rate of speed?
67. A gas station owner bought 10,250 gal of gas at $1.50 per gal. What is his total bill?
68. If a car gets 27 mi/gal, how far can you travel on 14 gal?

1.3 The Number Line

Let us now consider an important aid in understanding mathematics called the **number line.** *The number line is a line on which we visually represent our system of numbers* (figure 1.2).

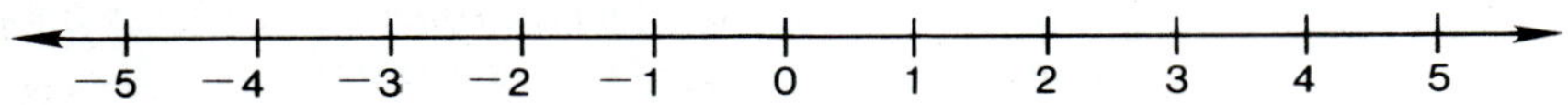

Figure 1.2

We see that as we move to the right of the number 0, we have the natural numbers 1, 2, 3, 4, and so on. If we move to the left of the number 0, we have the negatives, or opposites of the natural numbers. Negative numbers are used to represent values less than zero. We may use negative numbers to represent degrees of temperature below zero, altitudes below sea level, losses in money, and so on. If we combine the natural numbers, the negatives of the natural numbers, and zero, we have the group of numbers called the **integers.** There are many more numbers on the number line other than the integers, and we discuss them later in this section. Several of them appear on the number line in figure 1.3.

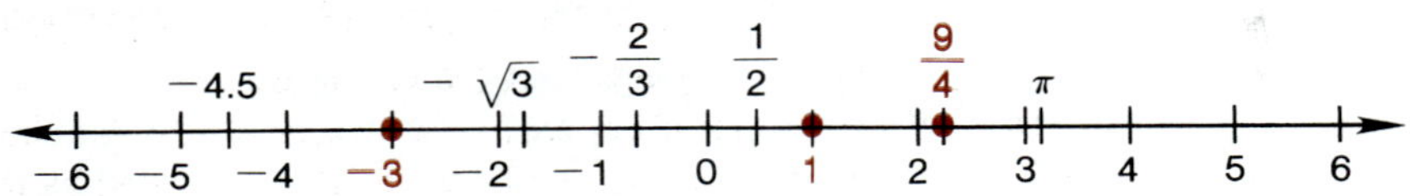

Figure 1.3

The number that is associated with each point on the line is called the **coordinate** of the point. The point that is associated with each number is called the **graph** of that number. In figure 1.3, the numbers 1, $\frac{9}{4}$, and -3 are the *coordinates* of the points indicated on the line by solid circles, and the solid circles are the *graphs* of the numbers 1, $\frac{9}{4}$, and -3.

The **origin,** 0, on the number line is of particular interest because if we move to the right of the origin we are dealing with *positive numbers,* and if we move to the left of the origin we are dealing with *negative numbers*. The origin itself is neither positive nor negative.

The direction we move on the number line is also important. If we move to the right, we are moving in a positive direction and our numbers are *increasing*. If we move to the left, we are moving in a negative direction and our numbers are *decreasing* (figure 1.4).

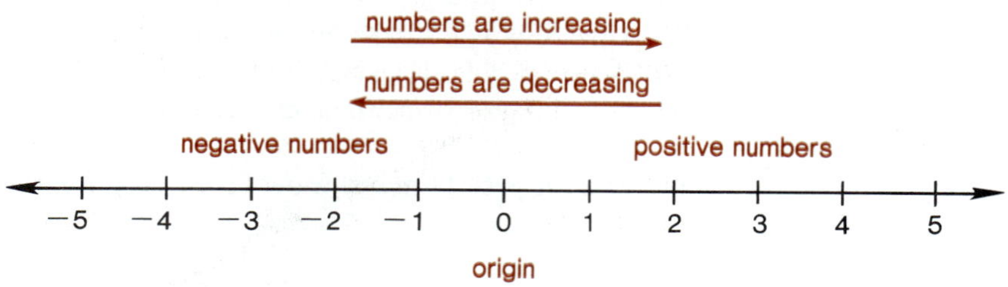

Figure 1.4

If we pick two arbitrary points on the number line and represent them by a and b, where a and b represent some *unspecified* numbers, we observe that there is an *order* relationship between a and b (figure 1.5).

Figure 1.5

The point associated with a is to the left of the point associated with b. Because of the position of the two points on the number line, we say a is *less than* b, which in symbols is $a < b$. We might also indicate that b is to the right of a by saying b is *greater than* a, which in symbols is $b > a$. The symbols $<$ (less than) and $>$ (greater than) are inequality symbols called *strict inequalities;* they denote an order relationship between numbers.

Examples

Insert the proper inequality symbol ($<$ or $>$) between the following numbers.

1. 5 7 answer: $5 < 7$
2. 8 3 answer: $8 > 3$
3. −5 −3 answer: $-5 < -3$
4. −9 −12 answer: $-9 > -12$
5. 0 −4 answer: $0 > -4$
6. 0 3 answer: $0 < 3$

There are two other inequality symbols called *weak inequalities*. These are *less than or equal to*, $\leqslant$, and *greater than or equal to*, $\geqslant$. The weak inequality symbol $\geqslant$, greater than or equal to, denotes that one number could either be greater than a second number or equal to the second number. For example, we could say that $4 \geqslant 3$ since 4 is greater than 3 or $4 \geqslant 4$ since 4 is equal to 4. The other weak inequality symbol ≤, less than or equal to, denotes that one number could either be less than a second number or equal to the second number. For example, we could say $3 \leq 4$ since 3 is less than 4, or $4 \leq 4$ since 4 is equal to 4.

As we study the number line, we observe a very useful property called *symmetry*. Our numbers are symmetrical on either side of the origin. That is, if we go four units to the right of 0 we come to the number 4, and if we go four units to the left of the origin we come to the opposite of 4, which is -4. Each of these numbers is four units away from the origin. This idea of how far a given number is from the origin is called the **absolute value** of the number. *The absolute value of a number is thought of as the undirected distance that the number is from the origin, or simply the* **numerical value** *of the number without any sign.* The symbol for absolute value is $| \; |$.

Example Find the indicated absolute values.

1. $|-7| = 7$
2. $|3| = 3$
3. $|0| = 0$
4. $\left|\frac{2}{3}\right| = \frac{2}{3}$
5. $|-3.7| = 3.7$

Notice that in drawing the number line, we place an arrowhead at each end. These indicate that the number line continues in both directions forever. Graphically, the arrowheads indicate that the number system contains an unlimited amount of numbers.

To develop the ideas of addition, subtraction, multiplication, and division, we have dealt with only some of the numbers that make up our complete number system, called the **real number system.** The real numbers are composed of two major groups called **rational numbers** and **irrational numbers.** The rational numbers are any numbers that can be expressed as the quotient of two integers. You may think of these as anything that can be written as a fraction. By writing an integer over 1, we see that the integers are part of the rationals. The irrationals are numbers that cannot be expressed as the quotient of two integers. The irrationals are discussed in chapter 10.

In conclusion, all the properties that we study relative to whole numbers or integers extend to the rationals, irrationals, and thence to the real numbers.

Exercise Set 1.3

Plot the graph of the following numbers, using a different number line for each problem.

Example $-3, -2, 1, 1\frac{1}{2}, 3$

answer:

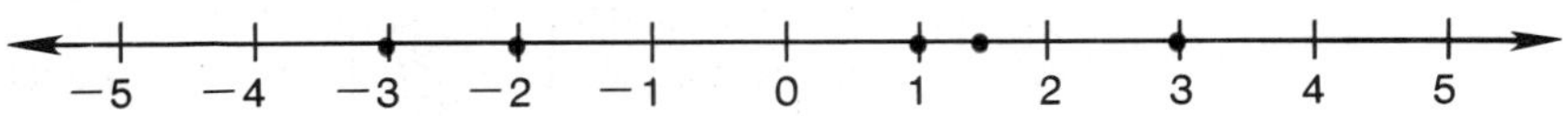

•1. $-3, -1, \frac{1}{2}, 3, 4$ 2. $-2, -1, \frac{3}{4}, 2, 4$ 3. $-5, -2, 0, 1, 3\frac{1}{2}$

4. $-1, 0, \frac{2}{3}, 1, 3$ 5. $-4, -2, 1, 2\frac{1}{2}, 4$ 6. $-3, -1, \frac{1}{2}, 1, 2$

Find the coordinates of the solid circles on the following number lines.

Example

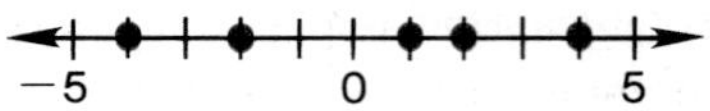

answer: $-4, -2, 1, 2, 4$

7. 

•8. 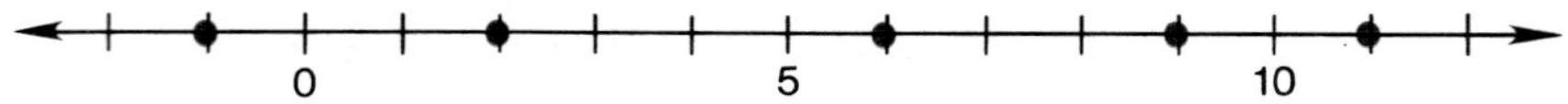

9. 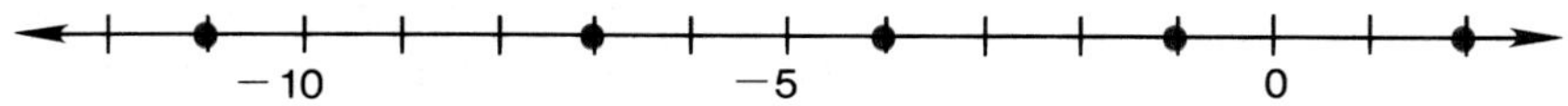

10. 

11. 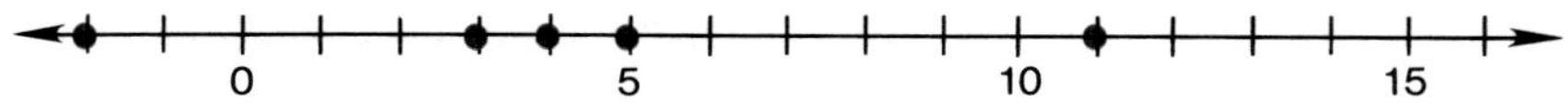

12. 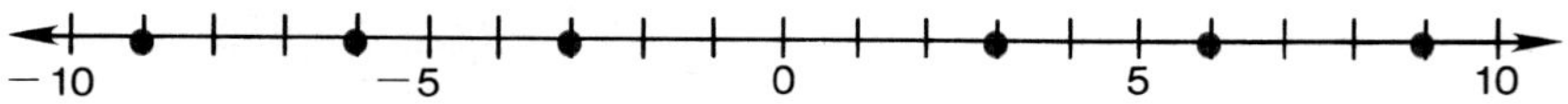

Write the absolute value of the following numbers.

13. -4 14. 0 15. 2
16. 8 17. -5 18. -7
19. 4 20. $\frac{2}{3}$ •21. $-\frac{1}{2}$
22. $-\frac{3}{4}$ 23. $1\frac{1}{2}$ 24. $-2\frac{3}{4}$
25. $\frac{5}{8}$ 26. -2 27. 6

Insert the proper inequality symbol ($<$ or $>$) between the following numbers.

Example 6 9 -4 -7 $|-6|$ $|3|$

answer: $6 < 9$ answer: $-4 > -7$ answer: $|-6| > |3|$

28. 5 7 •29. 4 8 30. 6 3
31. 9 2 32. -2 -4 33. -3 -8

34. -9 $\quad -6$	•35. -10 $\quad -5$	36. 0 $\quad 2$
37. 0 $\quad 4$	38. -3 $\quad 0$	39. 0 $\quad -6$
40. $\lvert -2\rvert$ $\quad \lvert -4\rvert$	41. $\lvert 5\rvert$ $\quad \lvert -7\rvert$	42. $\lvert -3\rvert$ $\quad \lvert 4\rvert$
43. $\lvert 0\rvert$ $\quad \lvert -2\rvert$	44. $\lvert -3\rvert$ $\quad \lvert 0\rvert$	•45. $\lvert -8\rvert$ $\quad \lvert -5\rvert$
46. $\lvert -9\rvert$ $\quad \lvert 7\rvert$	47. $\lvert -6\rvert$ $\quad \lvert -2\rvert$	

1.4 Addition and Subtraction of Integers

Now let us extend to the integers what we know about adding and subtracting whole numbers. We saw on the number line that integers consist of positive integers, which are natural numbers, negative integers, which are the negatives of the natural numbers, and zero, which is neither negative nor positive.

When we perform the operations of addition and subtraction on the integers, we refer to the process as operations with **signed numbers.** We use the symbol for subtraction, the minus sign $(-)$, to indicate a negative number, and the symbol for addition, the plus sign $(+)$, to indicate a positive number. For a positive number, we need use the plus sign only if we wish to emphasize the fact that the number is positive since when there is no sign, it is understood to be positive.

We now consider the whole numbers as signed numbers and examine the operation of addition. Consider the following example. $2 + 7$ can be written as $(+2) + (+7)$ if we wish to emphasize the fact that 2 and 7 are positive numbers. When we perform the addition, we have $(+2) + (+7) = (+9)$. We see that we added the 2 and the 7 together to get 9 and that the answer, $(+9)$, has the same sign as the 2 and the 7.

Examples

Add the following.

1. $(+4) + (+5) = (+9)$
2. $(+3) + (+8) = (+11)$
3. $(+6) + (+2) = (+8)$

Although we are familiar with the operation of adding two positive numbers, we can summarize the procedure by saying: When we add two positive numbers, we combine (add) them numerically and prefix the answer with their common sign, $+$.

note

The numerical value of a number is also called the absolute value of the number.

Let us now examine the addition of two negative numbers. To understand the operation of adding two negative numbers, let us consider a common everyday situation where this occurs. The money you withdraw from a savings account can be considered a negative amount or loss to your savings. If you make two withdrawals, the result is an even greater loss to your savings. For example, if you withdraw \$10, represented by \$$(-10)$, from the account, and you then withdraw \$15, represented by \$$(-15)$, your total withdrawal will be \$$(-10)$ + \$$(-15)$ = \$$(-25)$. We can summarize this process by saying: When we add two negative numbers, we combine them numerically and prefix the answer with their common sign, $-$.

Examples

Add the following.

1. $(-2) + (-7) = (-9)$
2. $(-9) + (-4) = (-13)$
3. $(-20) + (-30) = (-50)$
4. $(-6) + (-11) = (-17)$

We observe from the examples that, when we add signed numbers where the signs are the same, the procedure is the same. Therefore, *when we add two signed numbers and their signs are the same, we combine them numerically and prefix the answer with their common sign.*

Examples

Add the following.

1. $(+2) + (+13) = (+15)$
2. $(-4) + (-6) = (-10)$
3. $(-11) + (-12) = (-23)$
4. $(+5) + (+5) = (+10)$
5. $(-6) + (-8) = (-14)$
6. $(-9) + (-9) = (-18)$

note

Observe that every time we write a signed number, we write it inside parentheses. This precaution ensures that the sign of the number does not get lost during the problem, and that the sign of the number does not become confused with the operation symbol for addition or subtraction.

To consider the addition of two numbers with different signs, we again refer to the savings example. If you make a deposit to and a withdrawal from your savings account, depositing more money than you withdraw, you will have a net gain to your savings. For example, if you deposit \$40, represented by \$(+40), and withdraw \$25, represented by \$(−25), the net result to the savings account will be \$(+40) + \$(−25) = \$(+15). Note that the answer is the numerical difference between the 40 and the 25, prefixed by the sign of the numerically greater value, (+40). Consider a second example, where you deposit \$25, \$(+25), and withdraw \$40, \$(−40). The net result to your savings this time is a loss of \$15, shown by \$(+25) + \$(−40) = \$(−15). Again note that the answer is the numerical difference between the two numbers, prefixed by the sign of the numerically greater number (−40).

We can now summarize the procedure for adding numbers of different signs: *To find the sum of a positive and negative number, we find the numerical difference of the numbers and prefix the answer with the sign from the number with the greater numerical value.*

Examples Add the following.

1. $(-6) + (+10) = (+4)$
2. $(-8) + (+5) = (-3)$
3. $(+12) + (-7) = (+5)$
4. $(-6) + (+9) = (+3)$
5. $(+14) + (-26) = (-12)$
6. $(+10) + (-16) = (-6)$

To summarize the procedure for adding signed numbers:

1. If the signs are the same, combine them numerically and prefix the answer by their common sign.
2. If the signs are different, find the numerical difference of the numbers and prefix the answer by the sign of the numerically greater number.

To subtract integers, we observe that 10 − 6 is a number, 4, which, when added to 6, gives 10. Consider the following examples:

Examples

1. $\begin{array}{r} (+7) \\ -\ \underline{(+4)} \end{array}$ To find the answer we ask what number added to $(+4)$ gives $(+7)$, and the answer is $(+3)$. Therefore, $\begin{array}{r} (+4) \\ +\ \underline{(?)} \\ (+7) \end{array}$

   $$(+7) - (+4) = (+3)$$

2. $\begin{array}{r} (+4) \\ -\ \underline{(+7)} \end{array}$ To find the answer we ask what number added to $(+7)$ gives $(+4)$, and the answer is (-3). Therefore, $\begin{array}{r} (+7) \\ +\ \underline{(?)} \\ (+4) \end{array}$

   $$(+4) - (+7) = (-3)$$

3. $\begin{array}{r} (-7) \\ -\ \underline{(+4)} \end{array}$ To find the answer we ask what number added to $(+4)$ gives (-7), and the answer is (-11). Therefore, $\begin{array}{r} (+4) \\ +\ \underline{(?)} \\ (-7) \end{array}$

   $$(-7) - (+4) = (-11)$$

4. $\begin{array}{r} (+4) \\ -\ \underline{(-7)} \end{array}$ To find the answer we ask what number added to (-7) gives $(+4)$, and the answer is $(+11)$. Therefore, $\begin{array}{r} (-7) \\ +\ \underline{(?)} \\ (+4) \end{array}$

   $$(+4) - (-7) = (+11)$$

5. $\begin{array}{r} (-7) \\ -\ \underline{(-4)} \end{array}$ To find the answer we ask what number added to (-4) gives (-7), and the answer is (-3). Therefore, $\begin{array}{r} (-4) \\ +\ \underline{(?)} \\ (-7) \end{array}$

   $$(-7) - (-4) = (-3)$$

6. $\begin{array}{r} (-4) \\ -\ \underline{(-7)} \end{array}$ To find the answer we ask what number added to (-7) gives (-4), and the answer is $(+3)$. Therefore, $\begin{array}{r} (-7) \\ +\ \underline{(?)} \\ (-4) \end{array}$

   $$(-4) - (-7) = (+3)$$

From these examples, note that we obtain the same results if we change the operation from subtraction to addition and change the sign of the number we are subtracting.

Examples

1. $\begin{array}{r} (+7) \\ -\ \underline{(+4)} \end{array}$ $\xrightarrow[\text{and add}]{\text{change the sign}}$ $\begin{array}{r} (+7) \\ +\ \underline{(-4)} \\ (+3) \end{array}$

2. $\begin{array}{r} (+4) \\ -\ \underline{(+7)} \end{array}$ $\xrightarrow[\text{and add}]{\text{change the sign}}$ $\begin{array}{r} (+4) \\ +\ \underline{(-7)} \\ (-3) \end{array}$

3. $\begin{array}{r} (-7) \\ -\ \underline{(+4)} \end{array}$ $\xrightarrow[\text{and add}]{\text{change the sign}}$ $\begin{array}{r} (-7) \\ +\ \underline{(-4)} \\ (-11) \end{array}$

4. $\begin{array}{r} (+4) \\ -\ \underline{(-7)} \end{array}$ $\xrightarrow[\text{and add}]{\text{change the sign}}$ $\begin{array}{r} (+4) \\ +\ \underline{(+7)} \\ (+11) \end{array}$

5. $\begin{array}{r} (-7) \\ -\ \underline{(-4)} \end{array}$ $\xrightarrow[\text{and add}]{\text{change the sign}}$ $\begin{array}{r} (-7) \\ +\ \underline{(+4)} \\ (-3) \end{array}$

6. $\begin{array}{r} (-4) \\ -\ \underline{(-7)} \end{array}$ $\xrightarrow[\text{and add}]{\text{change the sign}}$ $\begin{array}{r} (-4) \\ +\ \underline{(+7)} \\ (+3) \end{array}$

To summarize the procedure for subtracting signed numbers:

Step 1. Change the operation from subtraction to addition.

Step 2. Change the sign of the number that is being subtracted.

Step 3. Perform the addition using the rules for adding signed numbers.

Examples

Subtract the following.

1. $(+8) - (+5) = (+8) + (-5) = (+3)$
2. $(+4) - (+11) = (+4) + (-11) = (-7)$
3. $(-9) - (+5) = (-9) + (-5) = (-14)$
4. $(+6) - (-8) = (+6) + (+8) = (+14)$
5. $(-12) - (-6) = (-12) + (+6) = (-6)$
6. $(-5) - (-14) = (-5) + (+14) = (+9)$
7. $0 - (+6) = 0 + (-6) = (-6)$
8. $0 - (-16) = 0 + (+16) = (+16)$

Example If a person borrows $1,800 and has $700 in savings, what is her net worth? We represent the borrowed amount as $(−1,800) and her savings as $(+700). Her net worth is then represented by $(−1,800) + $(+700) = $(−1,100).

Exercise Set 1.4

Find each sum or difference.

Example $(-7) + (+3) - (+12) =$

We perform the addition and subtraction in order from left to right; and we have

$$(-7) + (+3) - (+12) =$$
$$(-4) - (+12) =$$
$$(-4) + (-12) = (-16)$$

1. $(+5) + (+7)$
2. $(+3) + (-5)$
3. $(+7) + (-2)$
4. $(-14) + (-10)$
5. $(-10) + 0$
6. $(+13) + 0$
•7. $(-8) + 3$
8. $(+12) + (-8)$
9. $4 - 5$
10. $(-6) - 2$
11. $4 - (-2)$
12. $(-3) - (-7)$
•13. $(-8) - (-4)$
14. $4 - (-8)$
15. $10 + (-5) + (-2)$
16. $4 - 9$
17. $3 + (-4) + 1$
18. $(-3) + (-3) + (-2)$
19. $8 - (-4)$
20. $(-12) - (-16)$
21. $8 - (-6)$
22. $14 - 4$
23. $(-6) + 0$
24. $9 - 11$
25. $(-4) - 0$
26. $15 - (-3)$
27. $12 + (-3)$
28. $7 - 0$
29. $6 + (-10)$
30. $0 - 14$
31. $0 + (-8)$
32. $0 - (-7)$
33. $(-25) - 0$
34. $0 - (-6)$
35. $(+11) - (11)$
36. $0 - (-39)$
37. $(-1) - (+18)$
38. $(-37) - (+26)$
39. $(-12) - (-10) - (+8) + (+24)$
40. $(-30) + 14 - 8 - 20$
41. $(-25) + 4 - 32 + 28 + 3$
42. $(24) - (-12) - (+12) + (-13)$
43. $(-2) - (+3) + (-4) - (-5) + (-6)$
44. $5 - (-3) + (-4) + 0 + (7) - (+7)$
45. $(-15) - (13) - (-7) - 32 + (-2) - (-20) + (-4)$
•46. $(-17) - (11) + (-12) - (-5) + (18) + (3) - (-16)$
47. $(18) + (-21) - (-73) + (-4) + (31) - (-45) - (14)$
48. $(-312) - (-415) + 247 + (-763) - (658) - (-1,342)$
49. $(239) - (-305) - (247) + (643) + (123) - (394)$

For problems 50–54, replace the question mark with one of the following symbols to make a true statement: $<$, $=$, $>$.

50. $(7 + 5) ? (8 - 3)$
•51. $-(7 - 9) ? (10 - 12)$
52. $-(4 + 6) ? (-3 - 7)$
53. $-(-7 + 8) ? (10 - 8)$
54. $(9 - 3) ? -(2 + 5)$

•55. How many degrees rise or fall is there when the temperature changes from:
   a. 3°C to 34°C
   b. (−8)°C to 47°C
   c. (−37)°C to (−46)°C
   d. (−78)°C to (−21)°C
   e. 5°C to (−49)°C
   f. 86°C to 35°C

56. A certain point in a circuit is (−145) volts with respect to the ground, while another point is (+219) volts with respect to the ground. What is the potential difference between the two points?

•57. A man owes two other men $343 and $205, respectively, and a third man owes him $176. In terms of positive and negative quantities, how does the man stand monetarily?

58. The temperature at 7 P.M. on a given day is 64°F. If the hourly recorded temperature *changes* starting at 8 P.M. are (−5)°F, (−3)°F, (−1)°F, (+2)°F, (+6)°F, (+3)°F, (−4)°F, (−5)°F, (−2)°F, (−2)°F, and (+8)°F, what is the temperature at:
   a. midnight?
   b. 10 P.M.?
   c. 3 A.M.?
   d. 6 A.M.?
   e. the highest temperature reading?
   f. the lowest temperature reading?

59. The value of a certain stock on a given day at the New York Stock Exchange is $23.35. The changes in the value of the stock over a 9-day period are as follows: 21¢ gain, 13¢ loss, 7¢ loss, 5¢ gain, 18¢ gain, 36¢ loss, 2¢ gain, 44¢ gain. What is the value of the stock:
   a. on the 5th day?
   b. on the last day?
   c. at its lowest value?
   d. at its highest value?

60. If a person has $78 just after paying off a debt of $23, how much money did he have before paying off the debt? Write a statement involving the subtraction of integers to show your answer.

61. A temperature of (−18)°C is increased by 25°C. What is the resulting temperature?

62. A temperature of (−6)°C is decreased by 32°C. What is the resulting temperature?

1.5 Multiplication and Division of Integers

We are already familiar with the idea that the product of two positive numbers is positive. This fact can be seen by considering multiplication as repeated addition. For example, $3 \cdot 2 = 2 + 2 + 2 = 6$. We can compute the product $3 \cdot (-2)$ by computing $(-2) + (-2) + (-2) = (-6)$. Since we have already seen that the order in which we multiply numbers will not affect the answer, we can say that $(-2) \cdot 3 = 3 \cdot (-2) = (-6)$. This discussion leads us to the idea that *when we multiply numbers with opposite signs, the answer will be negative.*

Examples Multiply the following.

1. $(-2) \cdot 3 = (-6)$
2. $(-4) \cdot 4 = (-16)$
3. $5(-2) = (-10)$
4. $7(-3) = (-21)$

A further illustration of the results of multiplying a positive number and a negative number is found in the following pattern:

$$3 \cdot 3 = 9$$
$$2 \cdot 3 = 6$$
$$1 \cdot 3 = 3$$
$$0 \cdot 3 = 0$$
$$(-1) \cdot 3 = (-3)$$
$$(-2) \cdot 3 = (-6)$$
$$(-3) \cdot 3 = (-9)$$

We observe from this pattern that the answers decrease each time by 3. It logically follows, therefore, that *the product of a negative number and a positive number is a negative number.*

We will observe another pattern when we consider the following:

$$3(-3) = (-9)$$
$$2(-3) = (-6)$$
$$1(-3) = (-3)$$
$$0(-3) = 0$$
$$(-1)(-3) = 3$$
$$(-2)(-3) = 6$$
$$(-3)(-3) = 9$$

Note from this pattern that the answers increase by 3 each time, and it logically follows that *the product of two negative numbers is a positive number.*

Examples Multiply the following.

1. $(-2)(-4) = (+8)$
2. $(-3)(-7) = (+21)$
3. $(-6)(-5) = (+30)$
4. $(-12)(-4) = (+48)$
5. $(-5)(-5) = (+25)$

Since the operation of dividing two numbers is the inverse operation of multiplication, the same principles are true for division. The quotient of two positive numbers, $36 \div 4$ or $\frac{36}{4}$, is the number that, multiplied by 4, gives 36. Therefore,

$$36 \div 4 = \frac{36}{4} = 9$$

since

$$9 \cdot 4 = 36$$

Thus, the quotient of two positive numbers is positive.

The quotient of two negative numbers may be determined in the same way. To find the quotient of $(-20) \div (-5) = \frac{(-20)}{(-5)}$, we wish to find the number that, multiplied by (-5), gives (-20). That number is 4, since $(-5)(+4) = (-20)$. Therefore, $\frac{(-20)}{(-5)} = (+4)$, and we observe that *the quotient of two negative numbers is a positive number.*

Examples

Divide the following.

1. $\frac{(-14)}{(-7)} = (+2)$
2. $(-36) \div (-6) = (+6)$
3. $\frac{(-18)}{(-3)} = (+6)$
4. $\frac{(-48)}{(-12)} = (+4)$

Finally, to divide a positive number by a negative number or a negative number by a positive number, consider the following divisions:

$$(-14) \div (2) = \frac{(-14)}{2} = (-7)$$

since

$$(-7)(2) = (-14)$$

and

$$(24) \div (-6) = \frac{24}{(-6)} = (-4)$$

because

$$(-4)(-6) = 24$$

We find that *the quotient of a positive number and a negative number is always a negative number.*

Examples

Divide the following.

1. $\frac{(-18)}{(+6)} = (-3)$
2. $(-24) \div (+3) = (-8)$
3. $\frac{(+15)}{(-5)} = (-3)$
4. $\frac{(+42)}{(-7)} = (-6)$

We can summarize the rules of multiplication and division of signed numbers as follows:

1. When we multiply or divide numbers having like signs, the answer will be positive.
2. When we multiply or divide numbers having different signs, the answer will be negative.

Using these rules, we can determine the sign of our answer when we multiply three or more signed numbers. Consider the following examples.

1. $(-1)(2)(3)(4) = (-2)(3)(4) = (-6)(4) = (-24)$
2. $(-1)(2)(-3)(4) = (-2)(-3)(4) = (6)(4) = 24$
3. $(-1)(2)(-3)(-4) = (-2)(-3)(-4) = (6)(-4) = (-24)$
4. $(-1)(-2)(-3)(-4) = (2)(-3)(-4) = (-6)(-4) = 24$

We observe that *if the numbers being multiplied include an odd number of negative numbers, the answer will be negative*. See examples 1 and 3. *If the numbers being multiplied include an even number of negative numbers, the answer will be positive.* See examples 2 and 4.

Examples

1. $(-7)(-2)(5) = (14)(5) = 70$ (even number of negative numbers)
2. $(-6)(2)(-4) = (-12)(-4) = 48$ (even number of negative numbers)
3. $(-3)(5)(4) = (-15)(4) = (-60)$ (odd number of negative numbers)

We can extend our rule about even and odd numbers of negative numbers to the idea of division. Consider the following examples.

1. $\frac{(-1)(12)}{(2)(3)} = \frac{(-12)}{(6)} = (-2)$ (odd number of negative numbers)
2. $\frac{(-1)(12)}{(2)(-3)} = \frac{(-12)}{(-6)} = 2$ (even number of negative numbers)

3. $\dfrac{(-1)(12)}{(-2)(-3)} = \dfrac{(-12)}{(6)} = (-2)$ (odd number of negative numbers)

4. $\dfrac{(-1)(-12)}{(-2)(-3)} = \dfrac{(12)}{(6)} = 2$ (even number of negative numbers)

To summarize, the rule for three or more signed numbers in a multiplication or division problem is as follows: *If the numbers being multiplied or divided include an odd number of negative numbers, the answer will be negative; otherwise, it will be positive.*

note

The rule concerning multiplication or division of three of more signed numbers applies only when the operations are strictly multiplication and/or division. The problem $\dfrac{(-8) + (-4)}{(-2)}$, which involves addition and division, has an odd number of negative numbers, but the solution is $\dfrac{(-8) + (-4)}{(-2)} = \dfrac{(-12)}{(-2)} = (+6)$. We cannot apply the rule here because we are not performing only multiplication and/or division.

The number 0 is the only number that cannot be used as a divisor. To see why zero is excluded as a divisor, recall that a division problem is checked by multiplying the divisor times the quotient to get the dividend. If this idea is applied with zero as a divisor, we observe the following. Suppose there is a number x, such that $3 \div 0 = x$. Then $x \cdot 0$ will have to be equal to 3 for the answer to check, but this product is 0 regardless of the value of x. Therefore, we cannot find an answer for this problem and we say that the answer is *undefined*. If we try to divide zero by zero and again call our answer x, we have $0 \div 0 = x$. When we check our work, $0 \cdot x = 0$, we see that any value for x will work. In this situation, we say that the answer is *indeterminate*. We therefore decide that *division by zero is not allowed.*

It is important to note that although division by zero is not allowed, the division of zero by some other number is allowed. We can see that since $0 \cdot (-4) = 0$, then $0 \div (-4) = 0$. Thus, *zero divided by any number other than zero is always 0.*

Examples

Perform the division, if possible.

1. $\dfrac{0}{5} = 0$

2. $\dfrac{2}{0} = \text{undefined}$

3. $\dfrac{-7}{0} = \text{undefined}$

4. $\frac{0}{-7} = 0$

5. $\frac{0}{0} =$ indeterminate

In section 1.2, we introduced the idea of exponents. Exponents show when a number is multiplied times itself two or more times; that is, an exponent represents how many times a number is used as a factor in the problem. Now let us relate the idea of exponents to signed numbers.

Examples

1. $2^4 = 2 \cdot 2 \cdot 2 \cdot 2 = 16$ (The exponent 4 indicates how many times the base 2 is used as a factor.)
2. $(+2)^4 = (+2)(+2)(+2)(+2) = (+16)$
3. $(-2)^4 = (-2)(-2)(-2)(-2) = (+16)$ (even number of negative numbers)
4. $(-2)^3 = (-2)(-2)(-2) = (-8)$ (odd number of negative numbers)

Remember that it is good practice to place signed numbers inside parentheses. With this in mind, we can see that there is a definite difference between $(-2)^4$ and -2^4. In the first case, the parentheses denote that the negative number is to be taken four times; that is, $(-2)^4 = (-2)(-2)(-2)(-2) = (+16)$. In the second case, since there are no parentheses around the number, it is not a negative number to be raised to a power, but rather the negative of the answer gotten when 2 is raised to the fourth power; that is, $-2^4 = -(2^4) = -(2 \cdot 2 \cdot 2 \cdot 2) = -(16) = (-16)$.

Examples

Perform the indicated multiplication.

1. $(-3)^3 = (-3)(-3)(-3) = (-27)$
2. $-3^3 = -(3 \cdot 3 \cdot 3) = (-27)$
3. $(-3)^4 = (-3)(-3)(-3)(-3) = (+81)$
4. $-3^4 = -(3 \cdot 3 \cdot 3 \cdot 3) = (-81)$

We can summarize this idea as follows: *A negative number raised to an even power will have a positive answer and a negative number raised to an odd power will have a negative answer.*

note

This rule applies only to a negative number raised to a power, which is denoted by a negative number inside parentheses and an exponent outside the parentheses.

Exercise Set 1.5

Perform the indicated operations, if possible.

1. $(-3) \cdot (-5)$
2. $0 \cdot (-6)$
3. $4 \cdot (-7)$
4. $(-8) \cdot 3$
5. $4 \cdot (-3) \cdot 5$
6. $(-2)(2)(-2)$
7. $4 \cdot (-9) \cdot 0$
8. $(-3)(-2)(-8)$
•9. $(-1)(-4)(5)$
10. $(-5)(2)(4)(3)$
11. $7 \cdot (-1)(-3)(-5)$
12. $2 \cdot (-3)(-1)(2)(-2)(3)$
•13. $(2)(3)(-4)(-1)(5)(-1)(-1)$
14. $(-4)(-3)(7)(-1)(-2)(4)(6)$
15. $\dfrac{(-14)}{(-7)}$
16. $\dfrac{(-4)}{0}$
17. $\dfrac{(-15)}{5}$
•18. $\dfrac{32}{(-4)}$
19. $\dfrac{0}{(-9)}$
20. $\dfrac{18}{3}$
21. $\dfrac{(-22)}{(-11)}$
22. $\dfrac{(-4)(-3)}{(-6)}$
•23. $\dfrac{-(18)(2)}{(-4)}$
24. $\dfrac{(16)(2)}{(-8)}$
25. $18 \div (-3)$
26. $(-16) \div 2$
27. 2^4
•28. $(-4)^2$
29. $(-5)^4$
30. $(-3)^3$
31. $(-2)^5$
32. $(-1)^{31}$
33. $(-6)^2$
34. -4^3
•35. $-(-2)^2$
36. $(3)^4 - (3)^2$
37. $(21 - 17)^3$
•38. $(-13)^2 - (15)^2$
39. $(-1)^3(-2)^3$
40. $(5)(-2)^6(-1)^3$
41. $\dfrac{0}{0}$
42. $(-2)(0)(3)(-4)$
43. $\dfrac{-8}{0}$
44. $0(-8)$

Replace the comma with the proper symbol to make the statement true: ($<$, $=$, $>$).

•45. $(-7)(-8 + 5), (3)(-2 - 1)$
46. $9 \cdot (-4 + 7), (21 + 6)$
47. $(-3)(-2)(-1), (-4)(8)(0)$
48. $(-3)(-8)(3), (3)(12)(-2)$
49. $32 \cdot (-1)(1), (2)(-2)(8)(1)$

Find two integers such that their product is the first number and their sum is the second number.

Example 4, -4 answer: Since $(-2)(-2) = 4$ and $(-2) + (-2) = -4$, then -2 and -2 are the integers.

-27, -6 answer: Since $(-9)(3) = -27$ and $(-9) + (3) = -6$, then -9 and 3 are the integers.

50. -30, 1	51. -16, 0	52. 20, -9	53. -11, 10
54. 0, -7	55. -72, -21	56. 12, 7	57. 48, 16
58. 35, 12	59. 4, -4	60. -8, 7	61. -9, 0
62. -12, 1	63. -15, 2	64. -18, 3	65. 25, 10

Example A football player carried the ball 8 times, making the following yardages: gain of 6 yd, loss of 3 yd, loss of 4 yd, gain of 4 yd, gain of 3 yd, loss of 1 yd, loss of 2 yd, gain of 5 yd. Show his gains and losses by positive and negative integers. What was his average gain or loss per carry?

$(+6)$ yd, (-3) yd, (-4) yd, $(+4)$ yd, $(+3)$ yd, (-1) yd, (-2) yd, $(+5)$ yd.

To find an average, add together all of the values and divide by the total number of values:

$$\frac{(+6)\text{ yd} + (-3)\text{ yd} + (-4)\text{ yd} + (+4)\text{ yd} + (+3)\text{ yd} + (-1)\text{ yd} + (-2)\text{ yd} + (+5)\text{ yd}}{8}$$

$$= \frac{(+8)\text{ yd}}{8} = (+1)\text{ yd}$$

66. The temperature at 1 P.M. for 7 consecutive days in January was 5°C, -8°C, -7°C, -1°C, 10°C, -6°C, and 0°C. What was the average temperature for the seven days?

•67. Assume the stock market showed the following gains and losses during 6 consecutive hours of trading on a given day, and determine the average gain or loss during that 6-hour period: gain 36 points, loss 23 points, loss 72 points, gain 25 points, loss 31 points, loss 21 points.

68. Over a 5-day period, the price of a particular stock suffered losses of \$3 the first two days and \$2 the last three days. If the stock originally sold for \$88, what was its price after the 5-day period?

69. A woman acquires a debt of \$6 each day for 5 days. If we represent the \$6 debt by (-6), write a statement of the change in her assets after 5 days. What is the change?

Chapter Summary

1. We can add and subtract only **like quantities.**
2. When we add numbers or multiply numbers, the order in which we perform the operations will not affect the answer.
3. If a quantity is enclosed in a grouping symbol, we treat the quantity within as a single number.
4. When we add numbers or multiply numbers, we may group the numbers any way we wish, and the answer will not be affected.
5. The numbers in an indicated multiplication are referred to as the **factors** of the **product.**
6. The **exponent** tells how many times the **base** is used as a factor in an indicated product.
7. The product of zero and any number is zero.
8. The **distributive law** allows us to distribute an indicated multiplication over an indicated sum.
9. The **number line** is a line on which our system of numbers can be visually represented.
10. The symbols that denote an **order relationship** between numbers are: $<$ (less than), $>$ (greater than), $\leqslant$ (less than or equal to), and $\geqslant$ (greater than or equal to).
11. The **absolute value** of a number is the undirected distance that the number is from the **origin,** 0, or simply the **numerical value** of the number without any sign. The symbol for absolute value is $|\ |$.
12. To add signed numbers:
    a. If the signs are the same, combine them numerically and prefix the answer by their common sign.
    b. If the signs are different, find the numerical difference of the numbers and prefix the answer by the sign of the numerically greater number.
13. To subtract signed numbers:
    a. Change the operation from subtraction to addition.
    b. Change the sign of the number that was being subtracted.
    c. Perform the addition, using the rules for adding signed numbers.
14. When multiplying or dividing with signed numbers, the following rules apply:
    a. When the signs are the same, the answer will be positive.
    b. When the signs are different, the answer will be negative.
15. Division by zero is not allowed.

Chapter Review

[1.1] Find the sum.

1. $\begin{array}{r} 375 \\ 438 \\ \underline{967} \end{array}$

2. $\begin{array}{r} 943 \\ 607 \\ 896 \\ \underline{214} \end{array}$

3. $\begin{array}{r} 493 \\ 576 \\ 847 \\ 968 \\ \underline{302} \end{array}$

4. $36 + 2 + 1{,}376 + 27{,}403$

5. $50{,}007 + 246 + 71 + 6 + 1{,}234$

Find the difference in the following numbers.

6. $\begin{array}{r} 7{,}936 \\ -\ \underline{279} \end{array}$ 7. $\begin{array}{r} 6{,}009 \\ -\ \underline{3{,}924} \end{array}$ 8. $\begin{array}{r} 5{,}936 \\ -\ \underline{2{,}749} \end{array}$

9. 3,694 − 248

10. 8,270 − 6,937

Perform the indicated operations.

11. 23 − (8 − 2) + (6 − 4) − (2 + 8)
12. 74 − (1 + 7) + (14 − 6) − (36 − 24)

[1.2] Find the indicated product in the following problems.

13. $\begin{array}{r} 374 \\ \times\ \underline{46} \end{array}$ 14. $\begin{array}{r} 5{,}794 \\ \times\ \underline{703} \end{array}$ 15. $\begin{array}{r} 1{,}473 \\ \times\ \underline{6{,}293} \end{array}$

Find the indicated quotient of the following numbers.

16. 46,345 ÷ 8 17. 23,746 ÷ 32 18. 33,333 ÷ 123

[1.3] Find the absolute value of the following numbers.

19. -11 20. $-\frac{3}{4}$ 21. 4

[1.4] Find the sum or difference.

22. $(-1) + (-3)$
23. $6 - 3$
24. $7 - 13$
25. $(-4) + (5)$
26. $(-8) + (2)$
27. $7 - (-8)$
28. $(-8) - (-4)$
29. $(-3) - (6)$
30. $0 + (-3) - (-7) + 3 - (+4)$
31. $4 - 3 + 7 - 8 + 12 - (-3)$

[1.5] Find the product or quotient.

32. $3 \cdot (-7)$
33. $(-4) \cdot (-3)$
34. $(-8) \cdot (3) \cdot (-1)$
35. $8 \cdot (-9) \cdot (-1) \cdot (-2)$
36. $(-4) \cdot (3) \cdot (-5) \cdot 0$
37. $\frac{-14}{2}$
38. $\frac{-8}{-4}$
39. $24 \div (-4)$
40. $\frac{7}{0}$
41. $\frac{0}{-8}$

42. Two integral factors of 36 whose sum is -13 are ________________.

43. $3 \cdot (-5)$ is ($=$ or $\neq$) $(-5) \cdot 3$? ________________.

44. A business person suffers successive financial losses of \$3,000, \$2,560, and \$3,300 on three business transactions. If a loss is denoted by a negative number and the person originally had \$52,000:
    a. Write a statement, using negative numbers to represent assets after the losses.
    b. Find the total assets after the losses.
45. At 7 A.M. on February 23, the temperature was $-17°$. At noon that same day the reading was 23°. How much did the temperature rise from 7 A.M. to noon?
46. The temperature readings during a 5-hour period were 63°, 72°, 80°, 75°, and 69°.
    a. Represent by positive and negative integers how much rise (+) and fall (−) there was from hour to hour.
    b. Was the total rise greater than, equal to, or less than the total fall? By how much? (Represent by a positive or negative integer.)
    c. If the sixth hour showed a $-11°$ change, what was the temperature? Write a statement involving a negative integer representing this answer.

Chapter Test

Perform the indicated operations.

1. $(5)(-2)(3) =$
2. $8 - (-4) =$
3. $(2)(-5)(0)(4) =$
4. $-9 - 4 =$
5. $-12 - (-14) =$
6. $(-4)(-5) =$
7. $\dfrac{0}{-3} =$
8. $27 - (2 - 4) + 7 =$
9. $(287)(314) =$
10. $\dfrac{-4 + 4}{3 - 3} =$
11. $(817) + (1{,}013) + (487) + (23) =$
12. $407{,}124 \div 387 =$
13. $\dfrac{-18}{-3} =$
14. $\dfrac{21}{-7} =$
15. $8{,}172 + 9{,}301 + 1.034 =$
16. $6 - (-8) + 4 =$
17. $(-2) - (-6) + 5 =$
18. $18{,}106 - 10{,}947 =$
19. $(-2)(-5)(-3) =$
20. $16 - (7 - 3) + (8 - 4) =$
21. $(-8)(2)(-4) =$
22. $1{,}407 \times 303 =$
23. $\dfrac{(-3)(-8)}{(-2)(3)} =$
24. $\dfrac{14}{0} =$
25. $0 + (-5) - (-7) - 3 =$
26. $16 - (5 - 8) - 18 =$
27. $(-14) - 14 + 7 =$
28. If a person makes \$4 per hour straight time and \$6 per hour overtime, what is the person's total pay if he worked 40 h straight time and 11 h overtime?

2
Operations with Numbers

Technicians using inspection equipment to gauge parts to 0.0002 inch.
(Photograph by Kenneth Walters, courtesy of Henry Ford Community College)

2.1 Fundamental Principle of Fractions

A fraction is part of a whole. It can be thought of as a certain number of equal parts of the whole. For example, if we wish to consider the fraction $\frac{3}{4}$ we have the idea that something has been divided into four equal parts, and we are discussing three of them (figure 2.1).

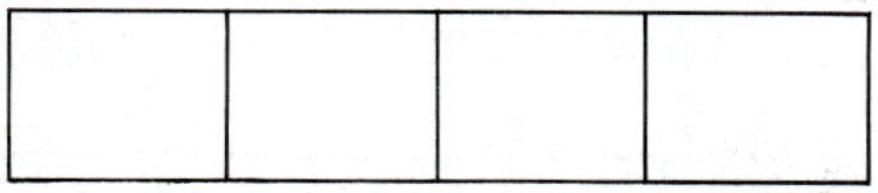

a whole divided into 4 equal parts

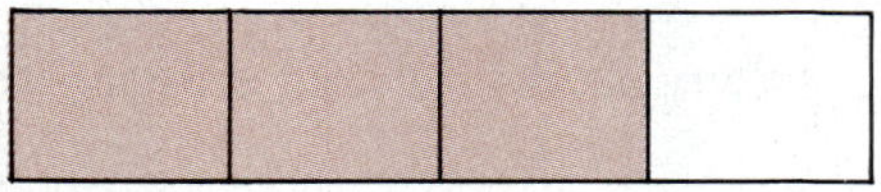

3 of the 4 equal parts or $\frac{3}{4}$ of the whole

Figure 2.1

In the fraction $\frac{3}{4}$, the number on top, 3, is called the **numerator** of the fraction, and the number on the bottom, 4, is called the **denominator** of the fraction. The fraction $\frac{3}{4}$ is called a **proper fraction.** *A proper fraction is a fraction whose numerator is less than its denominator.* Some other examples of proper fractions are:

$$\frac{1}{2}, \frac{5}{8}, \frac{7}{16}, \frac{2}{3}, \frac{4}{9}, \frac{11}{16}$$

An **improper fraction** *is a fraction whose numerator is greater than or equal to its denominator.* The following are examples of improper fractions:

$$\frac{3}{2}, \frac{5}{3}, \frac{12}{12}, \frac{16}{5}, \frac{47}{9}$$

A fraction is also a division problem, where we divide the denominator into the numerator. For example, $\frac{6}{5}$ may be read as "6 divided by 5." Since the numerator in an improper fraction is greater than or equal to its denominator, the improper fraction represents a number that is greater than or equal to 1.

Consider the improper fraction $\frac{37}{4}$. If we carry out the indicated division of 4 divided into 37, we find that 4 divides into 37 nine times with a remainder of 1. To write this remainder, we place it over the original divisor. Thus, the fraction becomes $\frac{37}{4} = 9\frac{1}{4}$. The whole part, 9, represents the number of times the denominator divided into the numerator, and the fractional part, $\frac{1}{4}$, represents the remainder, 1, over the original denominator, 4.

We call $9\frac{1}{4}$ the **mixed number** form of the improper fraction $\frac{37}{4}$. We observe from our example that if we wish *to change an improper fraction to a mixed number, we divide the denominator into the numerator*. If it divides exactly, we have a whole number. If it does not divide exactly, we have a mixed number, where the number of times the denominator divided into the numerator is the whole part and the remainder over the original denominator is the fractional part.

Examples Change the improper fractions to mixed numbers.

1. $\frac{16}{5}$ Carry out the indicated division $\begin{array}{r} 3 \\ 5\overline{)16} \\ \underline{15} \\ 1 \end{array}$ to get a quotient of 3 with a remainder of 1, which you write as $3\frac{1}{5}$. Therefore, $\frac{16}{5} = 3\frac{1}{5}$.

2. $\frac{47}{9}$ Dividing $\begin{array}{r} 5 \\ 9\overline{)47} \\ \underline{45} \\ 2 \end{array}$, we get a remainder of 2.

   Therefore, $\frac{47}{9} = 5\frac{2}{9}$.

3. $\frac{12}{12}$ In this problem the numerator and denominator are equal. We recall that when we divide a number by itself (except for 0), the answer is 1. Therefore, $\frac{12}{12}$ is equal to the whole number 1.

To change a mixed number to an improper fraction, multiply the whole number times the denominator and add this to the numerator, placing the sum over the original denominator.

Examples Change the mixed numbers to improper fractions.

1. $7\frac{3}{8} = \frac{(7 \cdot 8) + 3}{8} = \frac{56 + 3}{8} = \frac{59}{8}$
2. $4\frac{5}{12} = \frac{(4 \cdot 12) + 5}{12} = \frac{48 + 5}{12} = \frac{53}{12}$
3. $5\frac{3}{4} = \frac{(5 \cdot 4) + 3}{4} = \frac{20 + 3}{4} = \frac{23}{4}$

Before we deal with the fundamental principle of fractions, we need to study a special type of number called a *prime number*.

The whole numbers can be separated as follows:

1. **Prime numbers** are whole numbers greater than 1 that are exactly divisible only by themselves, and by 1. For example,

   2, 3, 5, 7, 11, 13

   are all prime numbers since each is divisible only by 1 and by itself.
2. Any whole number, other than 0 and 1, that is *not* prime is called a **composite number.** Thus, a composite number is divisible by numbers other than itself and 1. For example,

   4, 9, 28, 42

   are composite numbers since 4 is divisible by 2; 9 is divisible by 3; 28 is divisible by 2, 4, 7, and 14; and 42 is divisible by 2, 3, 6, 7, 14, 21.

Any composite number can be expressed as a product of prime numbers. We call this characteristic the **prime factor form** of the composite number. To illustrate, the prime factor form of the composite number 12 is

$$2 \cdot 2 \cdot 3 = 2^2 \cdot 3$$

since $2^2 \cdot 3 = 12$, and 2 and 3 are both prime numbers.

To determine the prime factors of a number, divide by the least prime number, 2, if possible, and continue to do so until a remainder is obtained that is not divisible by 2. Then divide this quotient by 3, if possible. Continue this process with increasing prime numbers until the quotient obtained is a prime number. The prime factor form of the composite number is the indicated product of the divisors and the final quotient. For example, to determine the prime factor form of 196, we proceed as follows:

$$\begin{array}{r|l} 2 & \underline{196} \\ 2 & \underline{\ \ 98} \\ 7 & \underline{\ \ 49} \\ & \ \ \ \ 7 \end{array}$$

Then, $196 = 2 \cdot 2 \cdot 7 \cdot 7 = 2^2 \cdot 7^2$ is prime factor form.

Examples

Write in prime factor form.

1. 60

   Solution
   $$\begin{array}{r|l} 2 & \underline{60} \\ 2 & \underline{30} \\ 3 & \underline{15} \\ & \ \ 5 \end{array}$$

   Then, $60 = 2 \cdot 2 \cdot 3 \cdot 5 = 2^2 \cdot 3 \cdot 5$

2. 250

Solution
$$\begin{array}{r|l} 2 & 250 \\ \hline 5 & 125 \\ \hline 5 & 25 \\ \hline & 5 \end{array}$$

Then, $250 = 2 \cdot 5 \cdot 5 \cdot 5 = 2 \cdot 5^3$

It will be useful to remember the following:

1. A number is divisible by 2 if the last digit is 0, 2, 4, 6, or 8.
2. A number is divisible by 3 if the sum of the digits of the number is divisible by 3. For example, 198 is divisible by 3 since $1 + 9 + 8 = 18$ and 18 is divisible by 3.
3. A number is divisible by 5 if the last digit of the number is 0 or 5.

There are also methods for determining the divisibility of a number by the prime numbers 7 and 11, but we will not consider them here since they are too complex. Use trial and error for these other prime numbers.

With the aid of prime numbers, we are now ready to introduce and discuss the **fundamental principle of fractions.** This principle says that *we can multiply or divide the numerator and the denominator of a fraction by the same nonzero number, and the resulting fraction will be equivalent (equal) to the original one*. We can write this principle with mathematical symbols as follows:

$$\frac{a}{b} = \frac{a \cdot c}{b \cdot c} \quad \text{where } b \text{ and } c \text{ are not equal to } 0$$

Recall that we can multiply any number by 1 and thus return to what we started with. To illustrate:

$$\frac{2}{3} \cdot 1 = \frac{2}{3}$$

Since we know that $\frac{2}{2} = 1$, then $\frac{2}{3} \cdot 1 = \frac{2}{3} \cdot \frac{2}{2} = \frac{(2 \cdot 2)}{(3 \cdot 2)} = \frac{4}{6}$. That is, $\frac{2}{3} = \frac{4}{6}$. We can say that the fraction $\frac{2}{3}$ is equivalent to $\frac{4}{6}$.

From this example, we see that the fundamental principle of fractions is used to "build up" fractions into equivalent fractions with denominators that are multiples of the given denominator. Consider the fraction

$$\frac{1}{4}$$

which we might wish to build up to an equivalent fraction with 12 as the denominator.

Thus, we want

$$\frac{1}{4} = \frac{?}{12}$$

If we note that

$$12 = 4 \cdot 3$$

it becomes apparent that an application of the fundamental principle of fractions will dictate that we must multiply the numerator, 1, by 3 also, in order to obtain a fraction that is equal to the original. Therefore

$$\frac{1}{4} = \frac{1 \cdot 3}{4 \cdot 3} = \frac{3}{12}$$

and we determine that the numerator of the new fraction must be 3.

Examples

Use the fundamental principle of fractions to determine the missing numerators.

1. $\frac{5}{7} = \frac{?}{42}$

   Since $42 = 7 \cdot 6$, multiply 5 by 6 to get the desired numerator, 30:

   $$\frac{5}{7} = \frac{5 \cdot 6}{7 \cdot 6} = \frac{30}{42}$$

2. $\frac{-14}{3} = \frac{?}{21}$

   Since $21 = 3 \cdot 7$, the missing numerator is $-14 \cdot 7 = -98$. That is,

   $$\frac{-14}{3} = \frac{-14 \cdot 7}{3 \cdot 7} = \frac{-98}{21}$$

3. $\frac{5}{8} = \frac{?}{56}$

   Since $56 = 8 \cdot 7$, multiply $5 \cdot 7$ to get the desired numerator, 35:

   $$\frac{5}{8} = \frac{5 \cdot 7}{8 \cdot 7} = \frac{35}{56}$$

4. $\frac{17}{3} = \frac{?}{12}$

   Since $12 = 3 \cdot 4$, multiply $17 \cdot 4$ to get the desired numerator, 68:

   $$\frac{17}{3} = \frac{17 \cdot 4}{3 \cdot 4} = \frac{68}{12}$$

---

note

To determine what number to multiply the numerator by, we divide the original denominator into the new denominator and multiply the numerator of the original fraction by this quotient. This result gives us the numerator of the equivalent fraction.

This important principle is also the rule by which we can reduce any fraction to its **lowest terms.** *When reduced to lowest terms, the numerator and denominator have no common factor other than 1.* For example, $\frac{5}{6}$ is reduced to lowest terms since the only factor the numerator and denominator have in common is 1. That is, 1 is the only number that can divide evenly into both the numerator and denominator. On the other hand, $\frac{12}{16}$ is *not* reduced to lowest terms since the numerator and denominator have a common factor of 4. Therefore, we can write

$$\frac{12}{16} = \frac{3 \cdot 4}{4 \cdot 4}$$

If we apply the fundamental principle of fractions and divide both numerator and denominator by 4, we get

$$\frac{12}{16} = \frac{3 \cdot 4}{4 \cdot 4} = \frac{3}{4}$$

Since 3 and 4 have no common factors other than 1, we say that $\frac{3}{4}$ is reduced to lowest terms.

Examples

Reduce to lowest terms using the fundamental principle of fractions.

1. $\frac{16}{24} = \frac{2 \cdot 8}{3 \cdot 8} = \frac{2}{3}$
2. $\frac{-25}{45} = \frac{-5 \cdot 5}{9 \cdot 5} = \frac{-5}{9}$
3. $\frac{150}{192} = \frac{75 \cdot 2}{96 \cdot 2} = \frac{75}{96} = \frac{25 \cdot 3}{32 \cdot 3} = \frac{25}{32}$

In these examples, the process of reducing the fraction is done by observing the **greatest common factor** of both numerator and denominator and dividing it out. The *greatest common factor is the largest number that will divide evenly into the numerator and denominator*. Note from previous example 3 that sometimes the greatest common factor is not immediately observed, and several steps of reduction must be performed. A more positive approach to reducing fractions is to put the numerator and denominator in prime factor form and then perform any possible reductions.

Examples

1. $\frac{16}{24} = \frac{2 \cdot \not{2} \cdot \not{2} \cdot \not{2}}{3 \cdot \not{2} \cdot \not{2} \cdot \not{2}} = \frac{2}{3}$
2. $\frac{-25}{45} = \frac{-5 \cdot \not{5}}{3 \cdot 3 \cdot \not{5}} = -\frac{5}{9}$
3. $\frac{150}{192} = \frac{\not{2} \cdot \not{3} \cdot 5 \cdot 5}{\not{2} \cdot 2 \cdot 2 \cdot 2 \cdot 2 \cdot 2 \cdot \not{3}} = \frac{25}{32}$

If you reduce by the inspection method, always inspect the result to be sure it is reduced as low as possible.

In any case, the process of reducing a fraction to lowest terms is one of factoring the numerator and denominator so there is a factor, or factors, common to both. Therefore, we can reduce a fraction by observing the largest number (the greatest common factor) that will divide evenly into the numerator and denominator. We can also reduce using the prime factorization of the numerator and denominator and dividing out the common primes.

Exercise Set 2.1

Change to a whole or mixed number.

Example Change $\frac{37}{8}$ to a mixed number: $\frac{37}{8} = 37 \div 8 = 4\frac{5}{8}$

1. $\frac{18}{7}$
2. $\frac{23}{5}$
3. $\frac{11}{4}$
4. $\frac{49}{6}$
5. $\frac{67}{12}$
6. $\frac{29}{15}$
7. $\frac{19}{4}$
8. $\frac{43}{21}$
9. $\frac{65}{6}$
10. $\frac{84}{11}$
11. $\frac{15}{2}$
12. $\frac{41}{3}$
13. $\frac{53}{7}$
14. $\frac{44}{9}$
15. $\frac{80}{13}$
16. $\frac{64}{8}$
17. $\frac{72}{8}$
18. $\frac{45}{8}$
19. $\frac{58}{8}$
20. $\frac{13}{4}$
21. $\frac{21}{6}$
22. $\frac{33}{16}$
23. $\frac{30}{4}$
24. $\frac{48}{16}$
25. $\frac{40}{16}$
26. $\frac{40}{32}$
27. $\frac{64}{32}$
28. $\frac{73}{32}$
29. $\frac{80}{64}$
30. $\frac{192}{64}$

Change to improper fractions.

Examples 1. Change $3\frac{3}{4}$ to an improper fraction:

$$3\frac{3}{4} = \frac{(3 \cdot 4) + 3}{4} = \frac{12 + 3}{4} = \frac{15}{4}$$

2. Change 6 to eighths:

$$6 = 6 \cdot \frac{8}{8} = \frac{6 \cdot 8}{8} = \frac{48}{8}$$

31. $2\frac{1}{2}$
32. $5\frac{1}{4}$
33. $6\frac{2}{3}$
34. $7\frac{1}{8}$
35. $2\frac{15}{16}$
36. $7\frac{1}{7}$
•37. $3\frac{7}{32}$
38. $2\frac{5}{64}$
39. $5\frac{5}{21}$
40. $4\frac{2}{11}$
41. $6\frac{5}{6}$
42. $10\frac{3}{5}$
43. $4\frac{14}{17}$
44. $12\frac{1}{3}$
45. $4\frac{5}{23}$
46. 4 to sixteenths
47. 8 to thirty-seconds
48. 3 to eighths
49. 5 to twelfths
50. 7 to fourths
51. $3\frac{3}{4}$
52. $2\frac{5}{8}$
53. $4\frac{7}{8}$
54. $2\frac{11}{12}$
55. $9\frac{1}{4}$
56. $2\frac{15}{16}$
57. $5\frac{17}{32}$
58. $4\frac{13}{16}$
59. $90\frac{1}{4}$
60. $100\frac{1}{2}$
61. $58\frac{2}{3}$
62. $3\frac{23}{64}$

Write in prime factor form.

63. 8
64. 26
65. 64
66. 56
67. 65
68. 38
69. 20
70. 106
71. 205
72. 206
73. 195
•74. 385
75. 102
76. 285
77. 2,310

Apply the fundamental principle of fractions to get equivalent fractions.

78. $\frac{3}{5} = \frac{?}{25}$
79. $\frac{5}{8} = \frac{?}{24}$
80. $\frac{2}{3} = \frac{?}{18}$

•81. $\frac{13}{23} = \frac{?}{138}$

82. $\frac{-13}{18} = \frac{?}{126}$

83. $\frac{5}{6} = \frac{?}{42}$

84. $\frac{8}{7} = \frac{?}{56}$

85. $\frac{12}{7} = \frac{?}{49}$

86. $\frac{-14}{15} = \frac{?}{120}$

87. $\frac{19}{9} = \frac{?}{108}$

88. $\frac{4}{8} = \frac{?}{2}$

89. $\frac{6}{10} = \frac{?}{5}$

90. $\frac{16}{20} = \frac{?}{5}$

91. $\frac{45}{63} = \frac{?}{7}$

92. $\frac{72}{24} = \frac{?}{3}$

93. $\frac{27}{78} = \frac{?}{26}$

94. $\frac{33}{77} = \frac{3}{?}$

95. $\frac{35}{63} = \frac{5}{?}$

96. $\frac{60}{72} = \frac{5}{?}$

•97. $\frac{81}{111} = \frac{27}{?}$

Apply the fundamental principle of fractions to reduce the following to lowest terms.

98. $\frac{10}{12}$

•99. $\frac{-16}{48}$

100. $\frac{18}{30}$

101. $\frac{21}{-42}$

102. $\frac{15}{33}$

103. $-\frac{12}{64}$

104. $\frac{7}{28}$

105. $\frac{-56}{252}$

106. $\frac{160}{192}$

107. $-\frac{140}{252}$

108. $\frac{252}{288}$

109. $\frac{288}{432}$

110. $\frac{20}{15}$

111. $\frac{36}{15}$

112. $\frac{60}{32}$

Convert the given fractions to equivalent fractions having the required higher denominators.

Example $\frac{3}{8}$ to thirty-seconds: $\frac{3}{8} = \frac{?}{32} = \frac{3 \cdot 4}{8 \cdot 4} = \frac{12}{32}$

113. $\frac{5}{8}$ to sixty-fourths

114. $\frac{2}{3}$ to twelfths

115. $\frac{9}{16}$ to thirty-seconds

116. $\frac{11}{18}$ to one hundred forty-fourths

117. $\frac{5}{4}$ to sixteenths

118. $\frac{7}{8}$ to thirty-seconds

•119. $\frac{3}{4}$ to sixtieths

120. $\frac{13}{8}$ to thirty-seconds

121. $\frac{23}{16}$ to sixty-fourths

122. Ohms resistance law says $R = \frac{E}{I}$, where R is the resistance in ohms, E the voltage drop across the resistor in volts, and I the current through the resistor in amps.

a. If $E = 3$ volts and $I = 15$ amp, then $R = \frac{3}{15}$ ohms. Reduce to lowest terms.

b. If $E = 16$ volts and $I = 5$ amp, then $R = \frac{16}{5}$ ohms. Write R as a mixed number.

c. If $E = 8$ volts and $I = 2$ amp, then $R = \frac{8}{2}$ ohms. What kind of a number represents R?

123. The fraction $\frac{3}{4}$ is often referred to as the *ratio* of 3 to 4. Find the ratio of 286 volts to 12 volts, reduced to lowest terms. What kind of number is this?

124. Find the ratio of 32 amp to 128 amp. What kind of a number is this? (Refer to problem 123.)

125. The outside diameter of a lock washer is found to be $\frac{5}{16}$ in. To use this measurement, a technician must have its equivalent in sixty-fourths. What measurement will he use?

126. The diametral pitch, P, of a standard involute-tooth gear is given by $P = \frac{N}{D}$, where N is the number of teeth and D the pitch diameter. What is the diametral pitch of a gear having 38 teeth and a pitch diameter of 6 in? State your answer as an improper fraction reduced to lowest terms and as a mixed number.

127. The efficiency of a motor is given by efficiency = output/input. What is the efficiency of a motor expressed as a fraction reduced to lowest terms if the motor receives 2,500 watts and delivers 2,250 watts?

2.2 Sums and Differences of Fractions

To be able to perform the operations of addition and subtraction on fractions, we must keep in mind an important principle for adding and subtracting any type of number: *Only like quantities can be added or subtracted.* In the case of fractions, like quantities are called **like fractions,** and *like fractions are those that have the same denominator*.

To find the sum of two or more fractions having the same denominator, consider the following example:

$$\frac{3}{8} + \frac{5}{8} + \frac{7}{8}$$

Since this problem means "3 eighths add 5 eighths add 7 eighths," then an equivalent statement is "15 eighths," and we can conclude that

$$\frac{3}{8} + \frac{5}{8} + \frac{7}{8} = \frac{15}{8}$$

Therefore, we see that

$$\frac{3}{8} + \frac{5}{8} + \frac{7}{8} = \frac{3 + 5 + 7}{8} = \frac{15}{8}$$

note

When adding fractions, the denominators *must* be the same—that is, we add like quantities. The operation of addition (or subtraction) is done only in the numerator; the denominator remains the same.

Summary: To add or subtract like fractions:

1. Add or subtract the numerators of the given fractions.
2. Write the answer found in step 1 over their common denominator.
3. Reduce the resulting fraction to lowest terms.

Examples

Perform the indicated addition or subtraction.

1. $\frac{3}{7} + \frac{2}{7} = \frac{3 + 2}{7} = \frac{5}{7}$
2. $\frac{1}{5} + \frac{4}{5} = \frac{1 + 4}{5} = \frac{5}{5} = 1$
3. $\frac{3}{8} + \frac{1}{8} = \frac{3 + 1}{8} = \frac{4}{8} = \frac{1}{2}$
4. $\frac{7}{16} - \frac{5}{16} = \frac{7 - 5}{16} = \frac{2}{16} = \frac{1}{8}$
5. $\frac{31}{32} - \frac{15}{32} = \frac{31 - 15}{32} = \frac{16}{32} = \frac{1}{2}$
6. $\frac{7}{8} - \frac{1}{8} + \frac{3}{8} = \frac{7 - 1 + 3}{8} = \frac{6 + 3}{8} = \frac{9}{8}$
7. $\frac{13}{16} - \frac{9}{16} - \frac{2}{16} = \frac{13 - 9 - 2}{16} = \frac{4 - 2}{16} = \frac{2}{16} = \frac{1}{8}$

Now consider the sum

$$\frac{5}{8} + \frac{3}{5}$$

We cannot use the previous property since the denominators are not the same. One or both of the fractions will have to be replaced by equivalent fractions, called **higher terms,** which have the same denominator. To replace such fractions, we can use one of two possible approaches. The first approach can be stated mathematically as follows:

$$\frac{a}{b} + \frac{c}{d} = \frac{(a \cdot d) + (b \cdot c)}{b \cdot d}$$

That is, we multiply the numerator of the first fraction times the denominator of the second fraction. To this product, we add the product of the first denominator times the second numerator. Then we place this sum over the product of the two denominators. To illustrate, consider the previous sum, $\frac{5}{8} + \frac{3}{5}$, where

$$a = 5,\ b = 8,\ c = 3,\ d = 5$$

By this principle,

$$\begin{aligned} \frac{5}{8} + \frac{3}{5} &= \frac{(5 \cdot 5) + (8 \cdot 3)}{8 \cdot 5} \\ &= \frac{25 + 24}{40} \\ &= \frac{49}{40} \end{aligned}$$

Note that this result indicates that since

$$\frac{25 + 24}{40} = \frac{25}{40} + \frac{24}{40}$$

then

$$\frac{25}{40} = \frac{5}{8} \quad \text{and} \quad \frac{24}{40} = \frac{3}{5}$$

and the fractions have been made "higher," with the **common denominator** of 40.

Example Find the sum.

$$\begin{aligned} \frac{7}{8} + \frac{5}{6} &= \frac{(7 \cdot 6) + (8 \cdot 5)}{8 \cdot 6} \\ &= \frac{42 + 40}{48} \\ &= \frac{82}{48} = \frac{41}{24} \end{aligned}$$

The concept of the common denominator leads us to the second approach for changing fractions into higher terms. Notice that in the previous example, $\frac{7}{8}$ and $\frac{5}{6}$ could be built to higher terms with a denominator of 24. That is,

$$\frac{7}{8} = \frac{7 \cdot 3}{8 \cdot 3} = \frac{21}{24} \quad \text{and} \quad \frac{5}{6} = \frac{5 \cdot 4}{6 \cdot 4} = \frac{20}{24}$$

Then

$$\frac{7}{8} + \frac{5}{6} = \frac{21}{24} + \frac{20}{24} = \frac{21 + 20}{24} = \frac{41}{24}$$

We call 24 the **least common multiple (LCM),** or **least common denominator (LCD),** of 6 and 8 since *it is the smallest number that can be divided evenly by 6 and 8.*

Inspection will often indicate which LCM is necessary, but where the LCM is not obvious, the following procedure should be used:

1. Express each number in prime factor form. If a prime appears more than once as a factor in any number, write it in exponential form.
2. From the numbers involved, list all the *different* prime factors.
3. Raise each factor involved to the greatest power of that factor found in any of the numbers. (Remember that if there is no exponent it means the exponent is 1.)

Examples

Find the LCMs.

1. 24 and 18

   First, put 24 and 18 in prime factor form.

   $24 = 2 \cdot 2 \cdot 2 \cdot 3 = 2^3 \cdot 3$
   $18 = 2 \cdot 3 \cdot 3 = 2 \cdot 3^2$

   Second, write down all the different primes from the first step:

   2 and 3

   Third, raise each of these primes to their greatest power in step 1:

   $2^3 \cdot 3^2$

   Our LCM is $2^3 \cdot 3^2 = 8 \cdot 9 = 72$.

2. 6, 8, 14

   Step 1. $6 = 2 \cdot 3$
   $8 = 2 \cdot 2 \cdot 2 = 2^3$
   $14 = 2 \cdot 7$

   Step 2. 2, 3, and 7 are the different primes.

   Step 3. $2^3 \cdot 3 \cdot 7$

   Our LCM is $2^3 \cdot 3 \cdot 7 = 8 \cdot 3 \cdot 7 = 24 \cdot 7 = 168$.

3. 9, 20, 30

Step 1. $9 = 3 \cdot 3 = 3^2$

$20 = 2 \cdot 2 \cdot 5 = 2^2 \cdot 5$

$30 = 2 \cdot 3 \cdot 5$

Step 2. The different primes are 2, 3, and 5.

Step 3. $2^2 \cdot 3^2 \cdot 5$

Our LCM is $2^2 \cdot 3^2 \cdot 5 = 4 \cdot 9 \cdot 5 = 36 \cdot 5 = 180$.

To find the sum of $\frac{5}{24} + \frac{7}{18}$, we proceed as follows.

1. The LCM of 24 and 18 is 72.
2. $\frac{5}{24} = \frac{5 \cdot 3}{24 \cdot 3} = \frac{15}{72}$

   $\frac{7}{18} = \frac{7 \cdot 4}{18 \cdot 4} = \frac{28}{72}$
3. $\frac{5}{24} + \frac{7}{18} = \frac{15}{72} + \frac{28}{72} + \frac{15 + 28}{72} = \frac{43}{72}$

note

Once we find the LCM of 24 and 18, we use the fundamental principle of fractions to build the fractions to higher terms. They now have the same denominator, and we are able to add the like fractions.

Summary: To add or subtract *unlike* fractions:

1. Find the LCM of all the fractions involved.
2. Change each fraction by the fundamental principle of fractions to an equivalent fraction having as its denominator the LCM of step 1.
3. Perform the addition and subtraction of the like fractions.
4. Reduce the results to lowest terms.

Examples

Perform the indicated addition or subtraction.

1. $\frac{3}{4} + \frac{2}{3} + \frac{5}{8}$

Step 1. The LCM of 4, 3, and 8 is 24.

Step 2. $\frac{3}{4} = \frac{3 \cdot 6}{4 \cdot 6} = \frac{18}{24}$

$\frac{2}{3} = \frac{2 \cdot 8}{3 \cdot 8} = \frac{16}{24}$

$\frac{5}{8} = \frac{5 \cdot 3}{8 \cdot 3} = \frac{15}{24}$

Step 3. $\frac{3}{4} + \frac{2}{3} + \frac{5}{8} = \frac{18}{24} + \frac{16}{24} + \frac{15}{24} = \frac{18 + 16 + 15}{24} = \frac{49}{24}$

2. $\frac{3}{20} + \frac{5}{24} + \frac{7}{30}$

Step 1. The LCM of 20, 24, and 30 is 120.

Step 2. $\frac{3}{20} = \frac{3 \cdot 6}{20 \cdot 6} = \frac{18}{120}$

$\frac{5}{24} = \frac{5 \cdot 5}{24 \cdot 5} = \frac{25}{120}$

$\frac{7}{30} = \frac{7 \cdot 4}{30 \cdot 4} = \frac{28}{120}$

Step 3. $\frac{3}{20} + \frac{5}{24} + \frac{7}{30} = \frac{18}{120} + \frac{25}{120} + \frac{28}{120} = \frac{18 + 25 + 28}{120}$

$= \frac{71}{120}$

3. $4\frac{1}{5} + 3\frac{3}{4} + 6\frac{1}{4}$

Change each mixed fraction to an improper fraction:

$4\frac{1}{5} = \frac{21}{5}$, $3\frac{3}{4} = \frac{15}{4}$, $6\frac{1}{4} = \frac{25}{4}$

Step 1. The LCM of 4 and 5 is 20.

Step 2. $\frac{21}{5} = \frac{21 \cdot 4}{5 \cdot 4} = \frac{84}{20}$

$\frac{15}{4} = \frac{15 \cdot 5}{4 \cdot 5} = \frac{75}{20}$

$\frac{25}{4} = \frac{25 \cdot 5}{4 \cdot 5} = \frac{125}{20}$

Step 3. $4\frac{1}{5} + 3\frac{3}{4} + 6\frac{1}{4} = \frac{21}{5} + \frac{15}{4} + \frac{25}{4} = \frac{84}{20} + \frac{75}{20} + \frac{125}{20}$

$= \frac{84 + 75 + 125}{20} = \frac{284}{20}$

Step 4. $\frac{284}{20} = \frac{71 \cdot \not{4}}{5 \cdot \not{4}} = \frac{71}{5}$ or $14\frac{1}{5}$

4. $\frac{7}{8} - \frac{1}{3}$

Step 1. The LCM of 8 and 3 is 24.

Step 2. $\frac{7}{8} = \frac{7 \cdot 3}{8 \cdot 3} = \frac{21}{24}$

$\frac{1}{3} = \frac{1 \cdot 8}{3 \cdot 8} = \frac{8}{24}$

Step 3. $\frac{7}{8} - \frac{1}{3} = \frac{21}{24} - \frac{8}{24} = \frac{21 - 8}{24} = \frac{13}{24}$

5. $3\frac{7}{8} - 2\frac{3}{4}$

Change each of the mixed numbers to an improper fraction:

$3\frac{7}{8} = \frac{31}{8}, \quad 2\frac{3}{4} = \frac{11}{4}$

Step 1. The LCM of 8 and 4 is 8.

Step 2. $\frac{31}{8}$ already has the LCM.

$\frac{11}{4} = \frac{11 \cdot 2}{4 \cdot 2} = \frac{22}{8}$

Step 3. $3\frac{7}{8} - 2\frac{3}{4} = \frac{31}{8} - \frac{11}{4} = \frac{31}{8} - \frac{22}{8} = \frac{31 - 22}{8}$

$= \frac{9}{8}$ or $1\frac{1}{8}$

6. $\frac{3}{4} - \frac{1}{8} + \frac{1}{3}$

Step 1. The LCM of 4, 8, and 3 is 24.

Step 2. $\frac{3}{4} = \frac{3 \cdot 6}{4 \cdot 6} = \frac{18}{24}$

$\frac{1}{8} = \frac{1 \cdot 3}{8 \cdot 3} = \frac{3}{24}$

$\frac{1}{3} = \frac{1 \cdot 8}{3 \cdot 8} = \frac{8}{24}$

Step 3. $\frac{3}{4} - \frac{1}{8} + \frac{1}{3} = \frac{18}{24} - \frac{3}{24} + \frac{8}{24} = \frac{18 - 3 + 8}{24}$

$= \frac{15 + 8}{24} = \frac{23}{24}$

7. $\frac{7}{12} - \frac{1}{6} - \frac{2}{15}$

Step 1. The LCM of 12, 6, and 15 is 60.

Step 2. $\frac{7}{12} = \frac{7 \cdot 5}{12 \cdot 5} = \frac{35}{60}$

$\frac{1}{6} = \frac{1 \cdot 10}{6 \cdot 10} = \frac{10}{60}$

$\frac{2}{15} = \frac{2 \cdot 4}{15 \cdot 4} = \frac{8}{60}$

Step 3. $\frac{7}{12} - \frac{1}{6} - \frac{2}{15} = \frac{35}{60} - \frac{10}{60} - \frac{8}{60} = \frac{35 - 10 - 8}{60}$

$= \frac{25 - 8}{60} = \frac{17}{60}$

Exercise Set 2.2

Find the LCM of the following sets of numbers.

1. 3, 8, 10
2. 9, 15, 21
3. 6, 14, 18
4. 5, 10, 12
5. 16, 24, 36
6. 12, 16, 24
7. 5, 7, 11
8. 10, 20, 30
9. 6, 9, 12
10. 10, 14, 18
11. 10, 15, 20
12. 10, 15, 24
13. 14, 21, 18
14. 8, 12, 18

•15. 22, 24, 30

Perform the indicated operations of addition and subtraction.

16. $\frac{1}{3} + \frac{1}{3}$
17. $\frac{2}{5} + \frac{3}{10}$
18. $\frac{1}{3} + \frac{1}{4}$
19. $\frac{5}{6} - \frac{1}{6}$
20. $\frac{4}{5} - \frac{2}{10}$

•21. $\frac{5}{6} - \frac{3}{8}$

22. $\frac{2}{7} + \frac{2}{3} + \frac{5}{7}$
23. $\frac{5}{8} + \frac{3}{4} + \frac{1}{2}$
24. $1 + \frac{5}{8}$
25. $3 + \frac{5}{6}$
26. $4 - \frac{3}{5}$
27. $1 + \frac{3}{8} - \frac{1}{6}$
28. $\frac{2}{3} + \frac{3}{4}$
29. $\frac{3}{5} + \frac{7}{15}$
30. $\frac{5}{6} - \frac{1}{3}$
31. $\frac{1}{2} + \frac{1}{5} + \frac{1}{10}$
32. $\frac{3}{8} - \frac{1}{12}$
33. $\frac{7}{24} - \frac{3}{16}$
34. $3\frac{1}{2} + 4\frac{3}{4} - 2\frac{1}{8}$
35. $3 + 2\frac{3}{4} - 4\frac{1}{2}$
36. $17\frac{1}{2} + 12\frac{3}{4}$
37. $8\frac{3}{16} - 4\frac{5}{8}$
38. $34\frac{1}{3} + 17\frac{5}{6}$
39. $\frac{9}{16} + \frac{5}{18} - \frac{2}{15}$

40. $\frac{7}{15} + \frac{5}{6} - \frac{3}{4}$

41. $\frac{7}{54} + \frac{19}{45}$

42. $\frac{19}{45} - \frac{28}{75} + \frac{4}{15}$

43. $5\frac{5}{6} + 2\frac{1}{4} - 3\frac{7}{8}$

44. $4\frac{5}{16} + 6\frac{4}{21} - 8\frac{9}{35}$

•45. $2\frac{7}{16} - 9\frac{5}{18} - 7\frac{4}{15}$

46. $87\frac{1}{2} + 33\frac{7}{15} - 26\frac{5}{6}$

•47. By how much must the diameter of a $\frac{3}{4}$-in shaft be reduced to bring it down to $\frac{39}{64}$ in?

48. A steel casting $\frac{7}{8}$ in thick is finished by taking a cut $\frac{3}{64}$ in deep. What is the final thickness?

49. What is the difference in thickness between a $\frac{9}{16}$-in thick steel plate and a $\frac{7}{8}$-in thick steel plate?

50. From a bar of brass $16\frac{1}{2}$ in long, the following length pieces are cut: $1\frac{3}{8}$ in, $3\frac{1}{2}$ in, and $4\frac{1}{16}$ in. What is the final length of the bar, allowing $\frac{1}{16}$ in for each cut?

51. Allowing $\frac{1}{8}$ in for each cut, what must be the length of a bar of brass to provide four pieces of the following lengths: $1\frac{5}{6}$ in, $5\frac{3}{8}$ in, $\frac{11}{32}$ in, $3\frac{1}{4}$ in.

52. A planer takes a $\frac{5}{32}$-in cut on a plate that is $\frac{9}{4}$ in thick. What is the final thickness of the plate?

53. A shaft that is supposed to measure $\frac{5}{8}$ in in diameter measures $\frac{49}{64}$ in. How much is its oversize?

54. Find the inside diameter of the pipe shown in the following cross section:

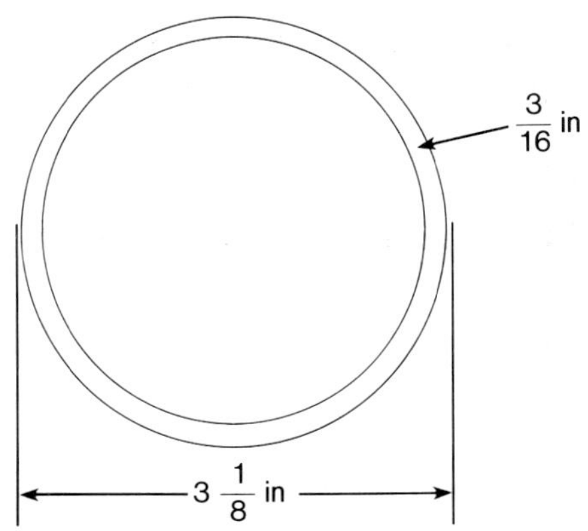

55. A triangular metal plate has the measurements given in the accompanying figure. Find the perimeter of the plate—that is, the distance around the plate.

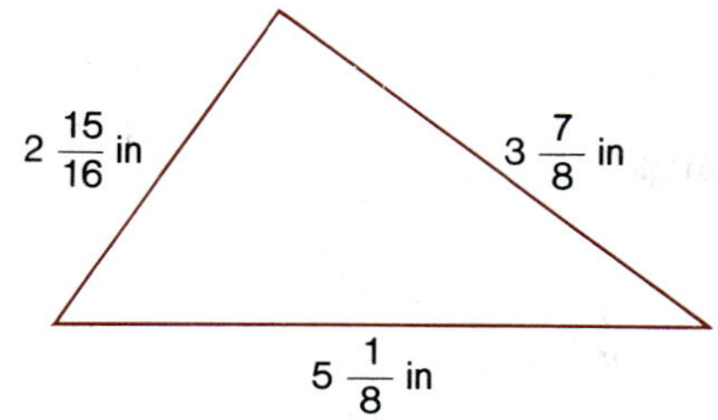

56. A lens formula is $\frac{1}{q} = \frac{1}{f} - \frac{1}{p}$. How much is q if $f = 150$ and $p = 50$?

57. By Kirchhoff's current law, the total current through a parallel circuit is equal to the sum of the currents in all branches. That is, $I_t = I_1 + I_2 + I_3 + \cdots$. What is I_t if $I_1 = 1\frac{1}{2}$ amp, $I_2 = \frac{7}{4}$ amp, and $I_3 = 2\frac{2}{5}$ amp?

58. In the accompanying figure, find the length of side a if the perimeter is $10\frac{3}{4}$ ft.

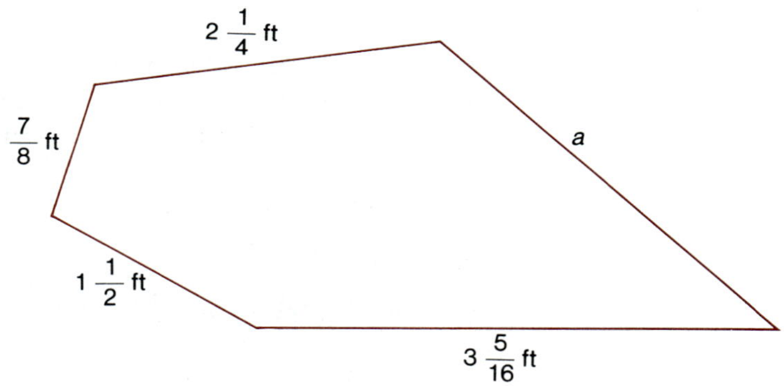

2.3 Products and Quotients of Fractions

When we first studied the idea of multiplication, we observed that it was a shortened form of repeated addition. Recall that if we wish to multiply

$$3 \cdot \frac{1}{4}$$

we could represent this product as a repeated addition problem:

$$\frac{1}{4} + \frac{1}{4} + \frac{1}{4} = \frac{3}{4}$$

since $3 \cdot \frac{1}{4}$ means to add $\frac{1}{4}$ three times. We notice, however, that the product $\frac{3}{4}$ could be obtained by

$$3 \cdot \frac{1}{4} = \frac{3 \cdot 1}{4} = \frac{3}{4}$$

(Multiplying the whole number times the numerator of the fraction.)

In like fashion, $2 \cdot \frac{3}{8}$ means

$$\frac{3}{8} + \frac{3}{8} = \frac{3 + 3}{8} = \frac{2 \cdot 3}{8} = \frac{6}{8} = \frac{3}{4}$$

Therefore, *to multiply a whole number times a fraction, you multiply the whole number times the numerator and put the product over the denominator.*

We can also look at the previous example in another way. Since $2 = \frac{2}{1}$, then

$$2 \cdot \frac{3}{8} = \frac{2}{1} \cdot \frac{3}{8}$$

and we can obtain the same result by

$$\frac{2 \cdot 3}{1 \cdot 8} = \frac{6}{8} = \frac{3}{4}$$

note

The same result is achieved when we state the whole number as a fraction with 1 in the denominator, and then proceed to multiply the numerators together and put this product over the product of the denominators. We can then use this to determine the product

$$\frac{2}{3} \cdot \frac{5}{8} = \frac{2 \cdot 5}{3 \cdot 8} = \frac{10}{24} = \frac{5}{12}$$

In general, if we want to multiply two fractions:

1. Multiply the numerators of the fractions together to get the numerator of the product.
2. Multiply the denominators of the fractions together to get the denominator of the product.

Examples

Find the product.

1. $\frac{3}{4} \cdot \frac{5}{7} = \frac{3 \cdot 5}{4 \cdot 7} = \frac{15}{28}$
2. $\frac{5}{6} \cdot \frac{3}{8} = \frac{5 \cdot 3}{6 \cdot 8} = \frac{15}{48} = \frac{5}{16}$

In example 2, $\frac{15}{48}$ has not yet been reduced to lowest terms since there is still a common factor, 3, by which it can be reduced. This 3 can be divided into the numerator and denominator before the terms are multiplied. That is,

$$\frac{5}{6}\cdot\frac{3}{8}=\frac{5\cdot\overset{1}{\cancel{3}}}{\underset{2}{\cancel{6}}\cdot 8}=\frac{5\cdot 1}{2\cdot 8}=\frac{5}{16}$$

We see from this example, that *if we are multiplying fractions, we should divide out any factors that are common to both numerator and denominator and then perform the multiplication with the remaining factors.* If we reduce before we perform the multiplication, our product will be reduced to lowest terms.

Examples Multiply the following.

1. $\frac{3}{4}\cdot\frac{2}{9}=\frac{\overset{1}{\cancel{3}}}{\underset{2}{\cancel{4}}}\cdot\frac{\overset{1}{\cancel{2}}}{\underset{3}{\cancel{9}}}=\frac{1\cdot 1}{2\cdot 3}=\frac{1}{6}$

note

When the 3 and the 2 were divided out of the numerators, the number 1 was left.

2. $\frac{25}{36}\cdot\frac{42}{55}\cdot\frac{24}{70}$

Recall that we can reduce a fraction by inspection or by putting the numerator and denominator in prime factor form. In this example, to be sure that all the common factors are divided out, we use prime factor form.

$$\begin{aligned}\frac{25}{36}\cdot\frac{42}{55}\cdot\frac{24}{70}&=\frac{\cancel{5}\cdot\cancel{5}}{\cancel{2}\cdot\cancel{2}\cdot\cancel{3}\cdot\cancel{3}}\cdot\frac{\cancel{2}\cdot\cancel{3}\cdot\cancel{7}}{\cancel{5}\cdot 11}\cdot\frac{\cancel{2}\cdot\cancel{2}\cdot 2\cdot\cancel{3}}{\cancel{2}\cdot\cancel{5}\cdot\cancel{7}}\\&=\frac{1}{1}\cdot\frac{1}{11}\cdot\frac{2}{1}\\&=\frac{1\cdot 1\cdot 2}{1\cdot 11\cdot 1}=\frac{2}{11}\end{aligned}$$

3. $\frac{48}{63}\cdot\frac{42}{54}\cdot\frac{36}{32}$ in prime factor form is

$$\begin{aligned}\frac{48}{63}\cdot\frac{42}{54}\cdot\frac{36}{32}&=\frac{\cancel{2}\cdot\cancel{2}\cdot\cancel{2}\cdot\cancel{2}\cdot\cancel{3}}{\cancel{3}\cdot\cancel{3}\cdot\cancel{7}}\cdot\frac{\cancel{2}\cdot\cancel{3}\cdot\cancel{7}}{\cancel{2}\cdot\cancel{3}\cdot\cancel{3}\cdot 3}\cdot\frac{\cancel{2}\cdot 2\cdot\cancel{3}\cdot\cancel{3}}{\cancel{2}\cdot\cancel{2}\cdot\cancel{2}\cdot\cancel{2}\cdot\cancel{2}}\\&=\frac{2}{3}\end{aligned}$$

To multiply mixed numbers, we change them first to improper fractions and then proceed with the multiplication.

Examples

Multiply the following.

1. $\left(3\frac{4}{7}\right)\left(6\frac{3}{10}\right)$

Step 1. First, we change the mixed numbers to improper fractions:

$$3\frac{4}{7} = \frac{(3 \cdot 7) + 4}{7} = \frac{21 + 4}{7} = \frac{25}{7}$$

$$6\frac{3}{10} = \frac{(6 \cdot 10) + 3}{10} = \frac{60 + 3}{10} = \frac{63}{10}$$

Step 2. Then we proceed with the multiplication:

$$\left(3\frac{4}{7}\right)\left(6\frac{3}{10}\right) = \frac{\overset{5}{\cancel{25}}}{\underset{1}{\cancel{7}}} \cdot \frac{\overset{9}{\cancel{63}}}{\underset{2}{\cancel{10}}} = \frac{45}{2} \quad \text{or} \quad 22\frac{1}{2}$$

2. $\left(3\frac{3}{4}\right)\left(2\frac{2}{5}\right)$

Step 1. Changing to improper fractions we have:

$$3\frac{3}{4} = \frac{(3 \cdot 4) + 3}{4} = \frac{12 + 3}{4} = \frac{15}{4}$$

$$2\frac{2}{5} = \frac{(2 \cdot 5) + 2}{5} = \frac{10 + 2}{5} = \frac{12}{5}$$

Step 2. Multiplying we have

$$\left(3\frac{3}{4}\right)\left(2\frac{2}{5}\right) = \frac{\overset{3}{\cancel{15}}}{\underset{1}{\cancel{4}}} \cdot \frac{\overset{3}{\cancel{12}}}{\underset{1}{\cancel{5}}} = \frac{9}{1} = 9$$

Consider the quotient $\frac{6}{2}$. We know this to be 3. But from previous discussion,

$$\frac{6}{2} = 6 \cdot \frac{1}{2} = 3$$

This statement says that "6 *divided by* 2" is equal to "6 *times* $\frac{1}{2}$." We call $\frac{1}{2}$ the **reciprocal** of 2. It is easy to see that the reciprocal of a number is determined by "inverting" the number—turning it upside down. That is, since

$$2 = \frac{2}{1}$$

then the reciprocal of $\frac{2}{1}$ is $\frac{1}{2}$. Extending this same idea to fractions, we can determine that

$$\frac{7}{8} \div \frac{3}{4} = \frac{\frac{7}{8}}{\frac{3}{4}} = \frac{7}{\cancel{8}_2} \cdot \frac{\cancel{4}^1}{3} = \frac{7 \cdot 1}{2 \cdot 3} = \frac{7}{6}$$

note

Instead of dividing by the number $\frac{3}{4}$, we multiplied by its reciprocal. We observe that the reciprocal of the fraction is $\frac{4}{3}$, and that the reciprocal of a fraction is found by inverting the fraction; that is, turning the fraction upside down.

In general, *to divide one fraction by another fraction: Invert the fraction that is the divisor, and multiply this times the numerator.*

Examples

Find the quotient.

1. $\frac{7}{8} \div \frac{6}{7} = \frac{7}{8} \cdot \frac{7}{6} = \frac{7 \cdot 7}{8 \cdot 6} = \frac{49}{48}$ or $1\frac{1}{48}$

2. $\frac{\frac{4}{5}}{\frac{3}{7}} = \frac{4}{5} \cdot \frac{7}{3} = \frac{4 \cdot 7}{5 \cdot 3} = \frac{28}{15}$ or $1\frac{13}{15}$

3. $3\frac{1}{4} \div 5\frac{2}{3}$

We change the mixed numbers to improper fractions:

$$3\frac{1}{4} \div 5\frac{2}{3} = \frac{13}{4} \div \frac{17}{3}$$

$$= \frac{13}{4} \cdot \frac{3}{17} = \frac{13 \cdot 3}{4 \cdot 17} = \frac{39}{68}$$

4. $\frac{7}{8} \div 3 = \frac{7}{8} \div \frac{3}{1} = \frac{7}{8} \cdot \frac{1}{3} = \frac{7 \cdot 1}{8 \cdot 3} = \frac{7}{24}$

5. $\frac{\frac{9}{10}}{4} = \frac{\frac{9}{10}}{\frac{4}{1}} = \frac{9}{10} \cdot \frac{1}{4} = \frac{9 \cdot 1}{10 \cdot 4} = \frac{9}{40}$

Please note that division by zero is still undefined for fractions as it was for integers. That is,

$$\frac{4}{5} \div 0 = \text{undefined}$$

since we know of no number that, multiplied by zero, gives a product other than zero. However,

$$0 \div \frac{7}{8} = 0 \cdot \frac{8}{7} = 0$$

Therefore, we can divide zero by a fraction and get zero as an answer.

Exercise Set 2.3

Find the following products and quotients and reduce to lowest terms.

•1. $\frac{5}{6} \cdot \frac{3}{5}$

2. $\frac{2}{3} \cdot \frac{5}{6}$

3. $\frac{7}{8} \cdot \frac{7}{12}$

4. $\frac{7}{5} \cdot \frac{3}{2}$

5. $\frac{7}{9} \cdot \frac{3}{4}$

6. $\frac{3}{4} \cdot 6$

7. $\frac{3}{7} \div \frac{4}{5}$

8. $\frac{12}{25} \div \frac{8}{15}$

•9. $\frac{6}{7} \div 3$

10. $4 \div \frac{3}{8}$

11. $\frac{15}{17} \div \frac{3}{5}$

12. $4 \div \frac{7}{2}$

•13. $\frac{4}{5} \cdot \frac{2}{3} \cdot \frac{3}{8}$

14. $\frac{9}{8} \cdot \frac{2}{3} \cdot \frac{3}{8}$

15. $17 \div 2\frac{1}{3}$

16. $\frac{12}{5} \cdot \frac{5}{16} \cdot \frac{4}{3}$

17. $\frac{3}{13} \cdot \frac{7}{15} \cdot \frac{11}{14}$

18. $12 \cdot 1\frac{5}{6}$

19. $30 \cdot 2\frac{9}{16}$

20. $1\frac{1}{5} \cdot 2\frac{1}{2}$

21. $4\frac{4}{5} \cdot 2\frac{1}{2}$

22. $7\frac{1}{3} \cdot 2\frac{4}{7}$

23. $2\frac{3}{4} \cdot 3\frac{3}{7} \cdot 5\frac{3}{5}$

24. $7\frac{1}{2} \div 5\frac{1}{4}$

•25. $5\frac{2}{3} \div 4\frac{1}{7}$

26. $\frac{8}{\frac{2}{3}}$

27. $\frac{17}{\frac{3}{4}}$

28. $\dfrac{\frac{11}{12}}{2}$

•29. $\dfrac{\frac{16}{17}}{8}$

30. $\dfrac{\frac{7}{8}}{\frac{4}{3}}$

31. $\dfrac{\frac{15}{64}}{\frac{45}{8}}$

32. $\frac{8}{3} \cdot \frac{4}{7} \div \frac{15}{4}$

•33. $\frac{8}{3} \div \frac{15}{14} \cdot \frac{4}{7}$

34. $\frac{8}{3} \div \frac{4}{7} \cdot \frac{15}{4}$

35. $\frac{8}{3} \cdot \frac{15}{14} \div \frac{4}{7}$

36. What is the total length of 25 pieces of steel, each $5\frac{1}{2}$ in long?

37. The volume of a rectangular block is found by multiplying the length times the width times the height.

   a. What is the volume in cubic inches of a rectangular block of wood $15\frac{1}{3}$ in long, $4\frac{1}{4}$ in wide, and $2\frac{7}{8}$ in high?

   b. What is the volume in cubic inches of a block of steel $8\frac{1}{2}$ in long, $2\frac{1}{8}$ in wide, and $1\frac{3}{4}$ in thick?

38. A space $61\frac{1}{2}$ in long is divided into 14 equal parts. What is the length of each part?

•39. The diagonal of a square can be approximated by dividing the length of the side by $\frac{707}{1{,}000}$. How long is the diagonal of a $2\frac{1}{2}$-in square?

40. How many holes spaced $1\frac{7}{16}$ in center to center can be drilled in an angle iron $22\frac{5}{8}$ in long, allowing $1\frac{1}{4}$ in end distances?

41. Find the length of a piece of stock required to make 10 taper keys, each $7\frac{1}{2}$ in long, allowing $\frac{1}{8}$-in waste per cut.

42. How many binding posts $1\frac{5}{6}$ in long can be cut from a piece of brass rod $29\frac{3}{4}$ in long if we allow $\frac{1}{8}$-in waste per cut?

43. How many washers $\frac{5}{32}$ in thick can be made from a piece of stock $36\frac{3}{4}$ in long, allowing $\frac{1}{16}$-in waste per cut?

44. If we wish to cut from a sheet of aluminum that is 24 × 45 in, small rectangular pieces that are $2\frac{2}{3} \times 3\frac{3}{4}$ in, how many such pieces can be cut from the sheet?

45. A standard dry cell has a voltage of $1\frac{1}{2}$ volts. Find the total voltage of 7 dry cells connected in series. (*Note:* total voltage = voltage per battery × number of batteries)

46. A portable tape recorder has two speeds. One is $2\frac{7}{8}$ in per second and the other is twice as fast. What is the speed of the second?

•47. How many $\frac{3}{4}$'s are there in 16?

48. What happens to the value of a fraction if:
    a. Its numerator is multiplied by 3?
    b. Its denominator is multiplied by 3?
    c. Both its numerator and denominator are multiplied by 3?

2.4 Decimal Fractions

Decimal fractions are those fractions which have denominators that are powers of ten. That is, the denominators are 10, 100, 1,000, 10,000, and so on. Recall that in a number such as 23, the digits 2 and 3 have place value as follows:

$$23 = (2 \cdot 10) + (3 \cdot 1)$$

Now consider the same two digits with a dot, called the **decimal point,** in front of them, .23. We call this number a **decimal fraction,** or just plain **decimal.** (It is standard procedure to place a zero to the left of the decimal point if the decimal is less than 1. The zero helps the reader see the decimal point and emphasizes the fact that we are dealing with a decimal fraction.) In this new form, we have

$$0.23 = \left(2 \cdot \frac{1}{10}\right) + \left(3 \cdot \frac{1}{100}\right) = \frac{2}{10} + \frac{3}{100} = \frac{20}{100} + \frac{3}{100}$$

which can be written simply as

$$\frac{23}{100}$$

and read "twenty-three hundredths." The decimal point is placed so that the number of digits to the right of it indicates the number of zeros in the fraction's denominator. If there is one digit to the right of the decimal point, the denominator is 10, read *tenths;* if there are two digits to the right of the decimal point, the denominator is 100, read *hundredths;* if three digits to the right of the decimal point, the denominator is 1,000, read *thousandths;* and so on.

To read a decimal fraction: Read the whole number (if any); next read "and" for the decimal point; then read the portion after the decimal point as a whole number; and finally read the name of the decimal place of the last digit on the right.

Examples

1. 0.37 is read "thirty-seven hundredths."
2. 5.0008 is read "five and eight ten-thousandths."
3. 456.039 is read "four hundred fifty-six and thirty-nine thousandths."
4. 654,321.123456 is read "six hundred fifty-four thousand, three hundred twenty-one and one hundred twenty-three thousand, four hundred fifty-six millionths":

HUNDRED-THOUSANDS	TEN-THOUSANDS	THOUSANDS	HUNDREDS	TENS	UNITS		TENTHS	HUNDREDTHS	THOUSANDTHS	TEN-THOUSANDTHS	HUNDRED-THOUSANDTHS	MILLIONTHS
6	5	4	3	2	1	.	1	2	3	4	5	6

To convert common fractions to decimal fractions if the denominator is a power of ten (10, 100, 1,000, 10,000, and so on), write the numerator and place the decimal point so that there are as many decimal places as there are zeros in the denominator.

Examples

1. Write $\frac{35}{100}$ as a decimal.

   Since there are two zeros in the denominator, there must be two digits to the right of the decimal point. Therefore, $\frac{35}{100} = 0.35$.

2. $\frac{259}{1{,}000} = 0.259$
3. $\frac{36}{10{,}000} = 0.0036$
4. $\frac{4{,}123}{1{,}000} = 4.123$

It is a good idea to become familiar with several of the common fractions and their decimal equivalent:

Fraction	Decimal Equivalent
$\frac{1}{4}$	0.25
$\frac{1}{2}$	0.50
$\frac{3}{4}$	0.75
$\frac{1}{8}$	0.125
$\frac{3}{8}$	0.375
$\frac{5}{8}$	0.625
$\frac{7}{8}$	0.875

However, there are a far greater number for which we do not bother to remember the decimal equivalent. Since a fraction is an indicated division, we can determine its decimal equivalent by performing the division. For example, with a fraction such as $\frac{7}{8}$, we divide 7 by 8. Since 8 does not go into 7 evenly, we place the decimal point and additional zeros after the 7, and carry out the division as follows:

$$8\overline{)7.000}\quad = 0.875$$

Thus, $\frac{7}{8} = 0.875$. With a fraction like $\frac{5}{6}$, we find that no matter how many zeros are added after the decimal point, the quotient will not end—there is always a remainder of 2. In a case like this, we must decide how far to carry out the answer. If we want to state the decimal equivalent to the nearest thousandth, which means three places to the right of the decimal point, we get

$$\frac{5}{6} = 0.833\frac{1}{3}$$

and since $\frac{1}{3}$ is less than $\frac{1}{2}$, it is correct to say

$$\frac{5}{6} \approx 0.833$$

accurate to the nearest thousandth. The symbol $\approx$ means ''is approximately equal to''; we use it whenever the quotient is not exact.

The following rules are used to round decimals off to a particular decimal place:

1. If the digit to the right of the position to be rounded is greater than or equal to 5, add 1 to the given position.
2. If the digit to the right of the position to be rounded is less than 5, leave the given position as it is.

Examples

1. Convert the fraction to a decimal rounded to the nearest thousandth.

   a. $\frac{3}{8} = 8\overline{)3.000}$ (quotient 0.375) so $\frac{3}{8} = 0.375$ (no rounding necessary)

   b. $\frac{2}{7} = 7\overline{)2.000}$ (quotient $0.285\frac{5}{7}$) so $\frac{2}{7} \approx 0.286$ (rounded to the nearest thousandth)

2. Round to the nearest hundredth.

   a. $0.235 \approx 0.24$
   b. $0.372 \approx 0.37$
   c. $2.365 \approx 2.37$
   d. $4.075 \approx 4.08$

To change a decimal fraction to a common fraction, consider the number following the decimal point as the numerator of the fraction; the denominator will be the number 1 with as many zeros following as there are digits after the decimal point. We then reduce the resulting fraction.

Examples

Change to a fraction and reduce.

1. $0.6 = \frac{6}{10} = \frac{3}{5}$
2. $0.45 = \frac{45}{100} = \frac{9}{20}$
3. $0.125 = \frac{125}{1,000} = \frac{1}{8}$
4. $0.33 = \frac{33}{100}$
5. $0.02 = \frac{2}{100} = \frac{1}{50}$

In examples 1, 2, 3, and 5 note that the resulting fraction is not completely reduced since the numerator and denominator contain factors common to both. We apply the fundamental principle of fractions to reduce them.

To change the decimal

$$0.16\frac{2}{3}$$

to its fractional form, we first write it as the common fraction

$$\frac{16\frac{2}{3}}{100}$$

Now we apply the fundamental principle of fractions again, and multiply the numerator and denominator by 3, which is the denominator of the fraction $\frac{2}{3}$. This procedure enables us to eliminate the fraction in the numerator as follows:

$$\frac{16\frac{2}{3} \cdot 3}{100 \cdot 3} = \frac{\frac{50}{3} \cdot \frac{3}{1}}{100 \cdot 3} = \frac{50}{300} = \frac{1}{6}$$

Examples

Change the decimal to a fraction and reduce.

1. $0.33\frac{1}{3} = \frac{33\frac{1}{3}}{100}$

$$= \frac{\frac{100}{3}}{100} \cdot \frac{3}{3} \qquad \text{Multiply by } \frac{3}{3}$$

$$= \frac{100}{300}$$

$$= \frac{1}{3}$$

2. $0.22\frac{2}{9} = \frac{22\frac{2}{9}}{100}$

$$= \frac{\frac{200}{9}}{100} \cdot \frac{9}{9} \qquad \text{Multiply by } \frac{9}{9}$$

$$= \frac{200}{900}$$

$$= \frac{2}{9}$$

3. $0.8\frac{1}{3} = \dfrac{8\frac{1}{3}}{10}$

$= \dfrac{\frac{25}{3}}{10} \cdot \dfrac{3}{3}$ Multiply by $\dfrac{3}{3}$

$= \dfrac{25}{30}$

$= \dfrac{5}{6}$

If we want to change a mixed number to a decimal number or a decimal number to a mixed number, we need consider only the fractional part—the integer part remains unchanged. To change $3\frac{1}{4}$ to a decimal number, for example, we simply change the $\frac{1}{4}$ to its decimal equivalent. That is, $\frac{1}{4} = 0.25$—therefore, $3\frac{1}{4} = 3.25$.

Examples

1. Change the mixed number to a decimal number.

a. $5\frac{7}{8}$

Since $\frac{7}{8} = 0.875$

Then, $5\frac{7}{8} = 5.875$

b. $6\frac{1}{5}$

Since $\frac{1}{5} = 0.2$

Then, $6\frac{1}{5} = 6.2$

2. Change the decimal number to a mixed number.

a. 4.45

$$\text{Since } 0.45 = \frac{45}{100} = \frac{9}{20}$$

$$\text{Then, } 4.45 = 4\frac{9}{20}$$

b. 7.375

$$\text{Since } 0.375 = \frac{375}{1{,}000} = \frac{3}{8}$$

$$\text{Then, } 7.375 = 7\frac{3}{8}$$

Exercise Set 2.4

Write in decimal form.

1. Three hundred fifty-three thousandths
2. Ten and fifteen ten-thousandths
3. Forty and seven tenths
4. Five hundred sixty-seven hundred-thousandths
5. One thousand two and thirty thousandths

6. $5\frac{7}{10}$ 7. $23\frac{15}{1{,}000}$ 8. $25\frac{3}{10{,}000}$

9. $73\frac{1}{100}$ 10. $\frac{267}{10{,}000}$

Write the following in words.

11. 0.36 12. 0.736 13. 2.007
14. 450.09 15. 0.00703 16. 2,673.19
17. 50,007.0306 18. 36.000002 19. 45.2763

Change the following to decimals (round to three decimal places).

20. $\frac{3}{4}$ 21. $\frac{3}{20}$ 22. $\frac{5}{8}$ 23. $\frac{7}{16}$

24. $\frac{9}{32}$ 25. $\frac{13}{20}$ 26. $\frac{36}{64}$ 27. $\frac{57}{75}$

28. $\frac{4}{9}$ 29. $\frac{7}{40}$ 30. $\frac{5}{11}$ 31. $\frac{8}{13}$

32. $\frac{2}{7}$ 33. $\frac{7}{8}$ 34. $1\frac{3}{5}$ 35. $7\frac{1}{6}$

36. $5\frac{5}{12}$ 37. $2\frac{3}{10}$ 38. $11\frac{17}{100}$ •39. $5\frac{9}{14}$

Convert to fractions (reduced to lowest terms).

40. 0.16	41. 0.375	42. 2.125	•43. 0.0625
44. 0.128	45. 0.6875	46. 3.5625	47. 0.71875
48. 0.250	49. 0.75	50. $0.66\frac{2}{3}$	51. $0.12\frac{1}{2}$
•52. $0.22\frac{2}{9}$	53. 7.06	54. 4.44	55. 6.85
56. 9.125	57. 11.1875	58. 15.245	

2.5 Operations with Decimal Numbers

To add two or more decimals, we write them so that the decimal points are under one another. Then we proceed as in the addition of whole numbers, and place the decimal point in the sum directly under the other decimal points.

Examples

Add the following.

1. To add 5.67, 32.046, 251.7367, and 0.92. We arrange the numbers in columns, with the decimal points aligned, and then add:

$$\begin{array}{r} 5.67 \\ 32.046 \\ 251.7367 \\ +\ \underline{0.92} \\ 290.3727 \end{array}$$

2. 63.4 + 75 + 0.056 + 19.18

$$\begin{array}{r} 63.4 \\ 75. \\ 0.056 \\ +\ \underline{19.18} \\ 157.636 \end{array}$$

To find the difference between two decimal numbers, we write the numbers so that the decimal points are under one another, and then proceed as in the subtraction of whole numbers. We place the decimal point in the difference directly under the other decimal points.

Examples

1. Subtract the following.

   a. 53.67 − 22.07

$$\begin{array}{rl} 53.67 & \text{minuend} \\ -\ \underline{22.07} & \text{subtrahend} \\ 31.60 & \text{difference} \end{array}$$

b. Subtract 18.7 from 39.62.

$$\begin{array}{r} 39.62 \\ -\ \underline{18.7} \\ 20.92 \end{array}$$

2. What is the perimeter of figure 2.2?
(Recall that the perimeter is the total distance around the figure.)

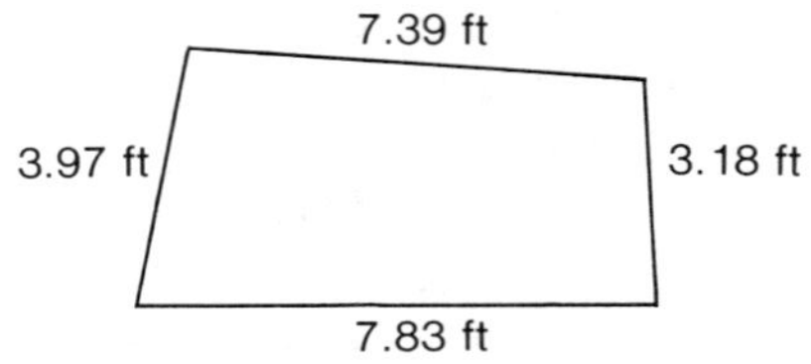

Figure 2.2

answer:

$$\begin{array}{r} 3.97 \text{ ft} \\ 7.39 \text{ ft} \\ 3.18 \text{ ft} \\ +\ \underline{7.83} \text{ ft} \\ 22.37 \text{ ft} \end{array}$$

To find the product of two decimal numbers, we first multiply as in whole numbers. Next, we add up the number of decimal places there are in the numbers multiplied. We use this sum to find the position of the decimal point in the product. Beginning at the right of the product, we count off to the left as many decimal places as there are in the sum, and insert the decimal point. This may give a product with a far greater accuracy than is desired by the technician, so the result will most often be rounded off to be of more practical use.

Examples

Multiply the following.

1. (2.36)(0.403)

note

There will be five decimal places in the product since there are two in the multiplicand and three in the multiplier.

$$\begin{array}{rl} 2.36 & \text{multiplicand} \\ \underline{0.403} & \text{multiplier} \\ 708 & \\ \underline{944} & \\ 0.95108 & \text{product} \end{array}$$

2. (18.14)(106.4)

$$\begin{array}{r} 18.14 \\ \underline{106.4} \\ 7256 \\ 10884 \\ \underline{1814} \\ 1{,}930.096 \end{array}$$

---

note

We need not write the problem in column form with the decimal points in the same column, since the arrangement of the numbers being multiplied does not help determine our answer.

To divide two decimal numbers, it is important that the divisor be made into a whole number. Therefore, it is necessary for us to move the decimal point in the divisor such that it is to the right of the last nonzero number in the divisor. In order to keep the problem equivalent to the original problem, we move the decimal point in the dividend in the same direction, the same number of places.

Examples

Prepare the decimal problem for division.

1. $360.5 \div 1.03$

   We move the decimal point two places to the right in the divisor, thus changing 1.03 into 103. Also, we move the decimal point in 360.5 two places to the right, making it 36,050. Our new problem, in which the divisor is a whole number, is:

   $36{,}050 \div 103$

   which is equivalent to our original problem.

note

When we moved the decimal point in 360.5 two places to the right, a zero had to be inserted to mark the position of the decimal point.

2. $30.814 \div 2.17$

   We move the decimal point two places to the right in each number to get $3{,}081.4 \div 217$.

note

The procedure of moving the decimal point the same number of places in both the numerator and denominator is an application of the fundamental principle of fractions. In the examples, we multiplied both numbers by 100 so the decimal point can be moved two places to the right.

Once we have made the divisor into a whole number, we are ready to proceed with the division in the same fashion as we did with whole numbers.

Examples

Divide the following.

1. 360.5 ÷ 1.03

   We change the divisor into a whole number and move the decimal point in the dividend the same number of places.

```
          .
103)36,050.
```

   Next, we put the problem into long division format, marking the decimal point in the answer directly above the decimal point in the dividend. We now proceed with the division:

```
        350.
103)36,050.
    30 9
     5 15
     5 15
        0
```

   The answer is 350.

2. 30.814 ÷ 2.17

   3,081.4 ÷ 217

```
       14.2
217)3,081.4
    2 17
      911
      868
       434
       434
         0
```

When we are performing division, we may want to carry out the answer to a certain place value. In that case, we carry the division out to one place past the desired value, and then apply the rounding procedure.

Examples

1. Divide and round to two decimal places (that is, to the nearest hundredth).

   12.4 ÷ 0.37

   1,240. ÷ 37.

```
         33.513
  37)1,240.000
     1 11
       130
       111
        190
        185
          50
          37
          130
          111
```

   We now round 33.513 to the hundredth position; thus,
   $12.4 \div 0.37 \approx 33.51$.

2. If a car travels 462.3 mi and uses 15.8 gal of gas, how many miles per gallon did the car achieve? (Round to the nearest tenth.)
   To determine the fuel economy, we divide the total number of miles traveled by the amount of gasoline used:

   $$\frac{462.3 \text{ mi}}{15.8 \text{ gal}}$$

   Carrying out the division between the decimal numbers to the hundredths position:

```
          29.25
  158)4,623.00
      3 16
      1 463
      1 422
         410
         316
          940
          790
          150
```

   Rounding back to the tenths we get 29.3 mi/gal.

Exercise Set 2.5

Find the indicated sum or difference in the following.

•1. $6.8 + 0.354 + 2.78 + 7.083 + 2.002$
2. $4.76 + 0.573 + 3.57 + 40.09 + 13$
3. $8.0007 + 360.01 + 25.72 + 6.362 + 140.2$
4. $7.0001 + 8 + 7.067 + 803.1 + 5.25$
5. $10.03 + 3.113 + 0.3342 + 0.0763 + 0.005$
6. $27.376 - 14.007$
7. $367.0076 - 210.02$
8. $836 - 0.367$
•9. $1.07 - 0.00036$
10. $4{,}563.2 - 274.063$
11. $19.07 + 2.736 - 8.007 - 4.1$
12. $0.00436 + 2.63 + 0.127 - 4.36$
13. $14 - 0.07 + 7.3 - 8.0073$
14. $363.45 - 43.606 - 120.6 + 42$
•15. $0.73 + 9.631 - 42.6 - 0.079 + 57.0008$

Multiply the following.

16. $0.375 \cdot 100$
17. $42.63 \cdot 1{,}000$
18. $246.37 \cdot 10$
19. $5.6076 \cdot 10{,}000$
20. $42.6 \cdot 73$
21. $7.006 \cdot 1.36$
22. $703.6 \cdot 1.7$
•23. $56.37 \cdot 0.0076$
24. $346.137 \cdot 0.07743$
25. $30.0303 \cdot 0.030303$

Divide the following.

26. $0.84 \div 0.7$ (to nearest hundredth)
•27. $0.0364 \div 0.35$ (to nearest thousandth)
28. $89 \div 0.26$ (to nearest hundredth)
29. $4.53972 \div 8.03$ (to nearest hundredth)
30. $362.4763 \div 2.342$ (to nearest thousandth)
•31. $7{,}076.374 \div 60.05$ (to nearest hundredth)
32. $0.00073 \div 0.0126$ (to nearest thousandth)
33. $3.63473 \div 0.00237$ (to nearest hundredth)
34. $803.4 \div 56$ (to nearest thousandth)
35. $25.7 \div 6.47$ (to nearest hundredth)

In problems 36–45 round all calculations to three decimal places.

36. A shaft that should be $1\frac{3}{16}$ in in diameter measures 1.317 in. How much is its oversize?

•37. A round piece of work is $1\frac{5}{8}$ in in diameter. How deep a cut should be taken to bring the diameter down to 1.487 in?

38. Find the thickness of the pipe in the accompanying figure if its inside diameter is 1.067 in and its outside diameter is 1.44 in.

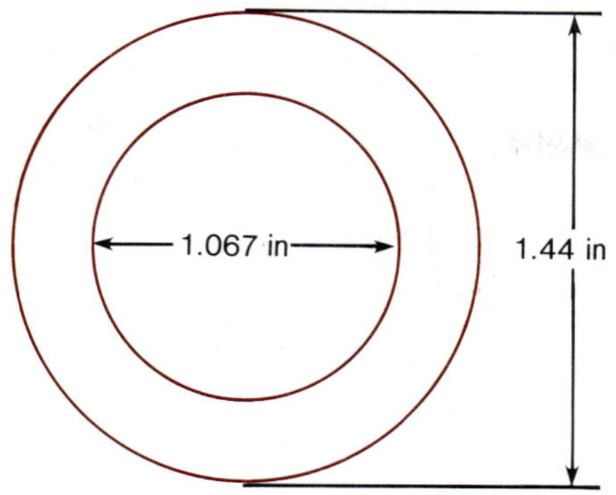

39. Find the depth of the tooth in the accompanying figure.

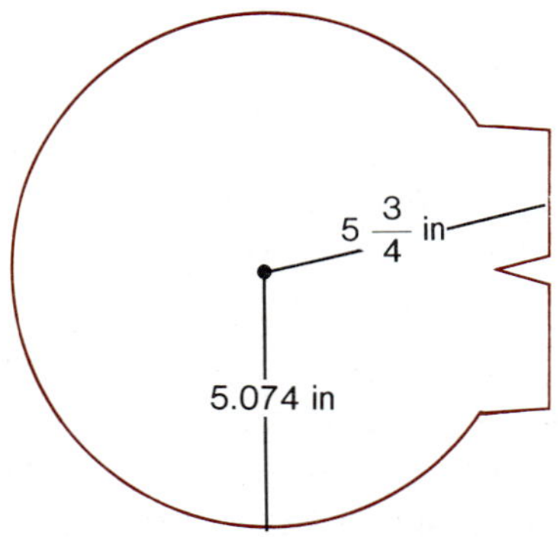

40. Allowing $\frac{1}{2}$ in for each cut, what must be the length of a bar of brass to provide four pieces of the following lengths: 1.406 in, 4.30 in, 0.763 in, and 3.25 in?
41. A planer takes a cut of 0.064 in on a plate that is 3.53 in thick. What is the final thickness of the plate?
42. A precision instrument reads 119.43 volts on a supply line. A later reading with the same instrument shows 119.025 volts. What is the voltage drop on the line?
43. A set of resistors is arranged in series. The resistors have the following values: $R_1 = 2.03$ ohms, $R_2 = 14.6$ ohms, $R_3 = 142.7$ ohms, $R_4 = 1{,}920.76$ ohms. What is the total resistance of the circuit? (Add the resistances.)
44. When capacitors are connected in parallel, the total equivalent capacitance, C, is equal to the sum of the individual capacitors. Find the total equivalent capacitance if the following capacitors are connected in parallel: $C_1 = 0.0007\mu f$, $C_2 = 0.03\mu f$, $C_3 = 0.13\mu f$, $C_4 = 0.008\mu f$, $C_5 = 0.025\mu f$. (μf is read microfarads.)
45. A battery is made up of seven cells connected in series. If each cell has a terminal voltage of 3.22 volts, what is the sum of the terminal voltage of the battery?
46. Find the total length of 32 pieces of steel, each 5.125 in long.

47. From a piece of pipe 73.02 in long we want 14 pieces of equal length. How long will each piece be, correct to the nearest hundredth?
48. Cast iron weighs 450 lb/ft^3, wrought iron weighs 480 lb/ft^3, and steel weighs 490 lb/ft^3.
    a. What is the weight of each type of metal per cubic inch if 1 cubic foot = 1,728 cubic inches?
    b. What is the weight of a wrought iron rod that contains 52.75 cubic inches?
    c. What part of a cubic foot is contained in a steel shaft weighing 205.85 lb? (State your answer as a decimal to the nearest thousandth.)
    d. What is the weight of 210 iron castings if each contains 35.5 cubic inches?
49. If a cubic foot of water weighs 62.5 lb, how many pounds of water are there in a tank containing 506.3 cubic feet?
50. How many gallons of water will the tank in problem 49 contain if 231 cubic inches = 1 gal?
51. A piece of electronic equipment consumes 4.8 kilowatt-hours (kwh) of energy in one day. How many kilowatt-hours of energy will be consumed by the end of a month of 31 days?
52. How many individual cells does an 800-volt dry cell battery have if each cell has a terminal voltage of 2.5 volts?
53. A special type of tube produces a current output that is 150,000 times greater than the input. What output current can be expected if the input light produces a current of 0.003 milliamp?
54. The wavelength of a radio wave in meters is obtained by

    $$\frac{300{,}000}{\text{frequency in kilohertz}}$$

    Find the wavelength in meters if the frequency is 3,400 kilohertz, correct to the nearest hundredth of a meter.

2.6 Order of Operations

When we are performing several different types of arithmetic operations within an expression, there must be a rule establishing the order in which the operations will be performed. To show that such a rule is necessary, consider the following numerical expression,

$$3 + 4 \cdot 5 - 3$$

More than one answer is possible for this expression, depending on the order in which we perform the operations. To illustrate,

$$3 + 4 \cdot 5 - 3 = 7 \cdot 2 = 14$$

if we add and subtract as indicated *before* we multiply. However,

$$3 + 4 \cdot 5 - 3 = 3 + 20 - 3 = 20^*$$

if we multiply before we add or subtract. A third possibility is

$$3 + 4 \cdot 5 - 3 = 3 + 4 \cdot 2 = 3 + 8 = 11$$

*This is the correct answer.

if we subtract, then multiply, and finally add. To standardize the answer, the following set of rules was established, defining the order of operations, or priorities:

1. Perform any operations within a grouping symbol such as ( ), [ ], { }, |absolute value|, $\sqrt{\text{radicals}}$, and so on, and those in the numerator or denominator of a fraction. Within a grouping symbol, we perform the operations according to the established priorities.
2. Perform indicated powers and roots.
3. Perform multiplication and division.
4. Perform addition and subtraction.

note

Within any priority every operation is of equal importance (that is, multiplication is no more important than division in priority 3). Therefore, when performing a particular priority, start at the left and proceed to the right, doing only those operations within the priority that we are performing, as we come to them, without skipping around.

To illustrate this order, consider the numerical expression

$$6 + 5(7 - 3) - 2^2$$

First, we evaluate and carry out the operation within the grouping symbol (parentheses) to get

$$6 + 5(4) - 2^2$$

Then we perform the indicated power:

$$6 + 5(4) - 4$$

Next, we carry out the multiplication:

$$6 + 20 - 4$$

The last step is to perform the addition and subtraction in order from left to right:

$$26 - 4 = 22$$

Examples

Perform the indicated operations in proper order and simplify.

1. $3 + 4(6 - 2) - \dfrac{6 + 3}{2^2 - 1}$

   First we evaluate within the grouping symbol to get

   $$3 + 4(4) - \frac{6 + 3}{2^2 - 1}$$

   Then we perform the operations in the fraction to get

   $$3 + 4(4) - \frac{9}{4 - 1}$$

   which equals

   $$3 + 4(4) - \frac{9}{3}$$

Now we multiply and divide, in order, to get

$$3 + 16 - \frac{9}{3} = 3 + 16 - 3$$

and finally we add and subtract, in order:

$$19 - 3 = 16$$

2. $$\frac{3(2+4)}{4-2} - \frac{4+6}{5} = \frac{3(6)}{4-2} - \frac{4+6}{5} = \frac{18}{2} - \frac{10}{5} = 9 - 2 = 7$$

3. $$3^2 - \frac{4(8-3)}{10} = 3^2 - \frac{4(5)}{10} = 9 - \frac{20}{10} = 9 - 2 = 7$$

Exercise Set 2.6

Perform the indicated operations and simplify.

1. $\frac{4+2}{3} + 2$
2. $-6 \cdot 7 + 8$
3. $6 + 5 \cdot 4$
4. $\frac{1}{5} \cdot 5 + 6$
5. $3 + 8 \cdot 0$
6. $\frac{12}{3} + 4$
7. $-2 + 10 \cdot \frac{1}{5}$
8. $4(3 - 2)(2 + 1)$
9. $0(5 + 2) + 3$
10. $\frac{24 \cdot 3}{9} - 6$
11. $\frac{6(9-2)}{4} + \frac{4}{3-1}$
12. $(7 + 6) + (3 - 2)$
13. $(15 + 8) - (7 + 3)$
14. $(32 + 2) - (5 - 4)$
15. $(24 - 6) \div 3$
16. $(37 - 4) \div 11$

•17. $\frac{42+8}{8+2}$

18. $12 \cdot 4 + 2$
19. $15 \cdot 3^2 - 14$
20. $(8 - 3)(5 + 3)$
21. $3(6 - 2)(7 + 1)$
22. $12 + 3 \cdot 16 \div 4^2 - 2$
23. $10 - 3 \cdot 4 \div 6 - 5$
24. $8 - (12 + 3) - 4 \cdot 3$
25. $4(2 - 5)^2 - 2(3 - 4)$
26. $6(-8 + 10) - 5(4 - 7)$

•27. $\frac{5(3-5)}{2} - \frac{27}{-3}$

28. $\frac{3(8-6)}{2} - \frac{8}{-2}$

29. The surface of a flat ring whose inside diameter is 2 in and whose outside diameter is 3 in is found by $\frac{22}{7} \cdot 3^2 - \frac{22}{7} \cdot 2^2$. Find the area.

30. The area of a ring section whose inside diameter is 18 in and whose outside diameter is 26 in is found by $3.14 \cdot \frac{26 + 18}{2} \cdot \frac{26 - 18}{2}$. Find the surface area.

31. To find the pitch diameter, D, of a gear with 36 teeth and an outside diameter of 8 in we get $D = \frac{36(8)}{38 + 3}$. Find D.

•32. The length of the threaded portion of a pipe that has threads that are perfect at top and bottom, with an outside diameter of 2.875 in and an $\frac{1}{8}$-in pitch is determined by Brigg's formula: $A = (0.8 \cdot 2.875 + 4.8) \cdot \frac{1}{8}$. Find A.

33. A Murray loop is used to determine the point at which a telephone line is grounded. The unknown distance to the point of the ground, x, for a length of the loop of 32.0 mi and resistances of 222 and 384 ohms is given by

$$x = \frac{384}{222 + 384} \cdot 32.0.$$ Find x.

34. The velocity in feet per second, V, of sound (traveling through air) for an air temperature of 62° Fahrenheit is given by $V = 1{,}090 + 1.14(62 - 32)$. Find V.

35. The total resistance of an electrical circuit connected in parallel, with resistances of 5 ohms, 10 ohms, and 11.5 ohms can be found by

$$R_t = \frac{5 \cdot 10 \cdot 11.5}{(5 \cdot 10) + (5 \cdot 11.5) + (10 \cdot 11.5)}$$

Find R_t.

36. The electromotive force (emf) of a storage battery is 6.3 volts, its internal resistance is 0.015 ohm, and it is used to drive a dynamotor that requires 300 watts. What is the voltage across the battery terminals, E_b, if

$$E_b = 6.3 - 0.015 \cdot \frac{300}{0.82}?$$

37. To convert 74° Fahrenheit to Celsius temperature, we use:

$$C = \frac{5}{9}(74 - 32).$$

Find C.

Chapter Summary

1. A **fraction** is part of a whole.
2. A **proper fraction** is a fraction whose numerator is less than its denominator.
3. An **improper fraction** is a fraction whose numerator is greater than or equal to the denominator.
4. A **mixed number** is an improper fraction stated as the sum of a whole number and a fraction.
5. The **fundamental principle of fractions** says the same nonzero number can be multiplied times the numerator and the denominator of a fraction to get an equivalent fraction.
6. A fraction is **reduced to lowest terms** when the numerator and denominator have no common factors other than 1.
7. The **least common multiple** (or **least common denominator**) of two or more numbers is the smallest number exactly divisible by the given numbers.
8. A **prime number** is a whole number greater than 1 that is exactly divisible only by itself and by 1.
9. The **prime factor form** of a composite number is that number stated as a product of prime numbers.
10. To find the sum or difference of decimal fractions, place them in columns so that the decimal points are under one another in a column and add or subtract as in whole numbers, placing the decimal point in the sum or difference directly under the others.
11. To multiply decimal fractions, multiply as with whole numbers and point off in the product as many decimal places, starting at the right, as there are decimal places in the multiplier and multiplicand together.
12. To divide two decimal fractions, move the decimal point in the divisor the number of places necessary to make it a whole number. Move the decimal point in the dividend the same number of places to the right. Then divide in the usual manner, placing the decimal point in the quotient above the decimal point in the dividend.
13. **Order of Operations**
    a. Perform operations within grouping symbols and in the numerator/denominator of a fraction.
    b. Perform indicated powers and roots.
    c. Perform multiplication and division.
    d. Perform addition and subtraction.

Chapter Review

[2.1] Using the fundamental principle of fractions, reduce to lowest terms.

1. $\frac{15}{24}$ 2. $\frac{39}{72}$ 3. $\frac{76}{48}$

Apply the fundamental principle of fractions, and find the missing term.

4. $\frac{4}{13} = \frac{?}{52}$ 5. $\frac{5}{8} = \frac{?}{96}$

[2.2] Find the indicated sum or difference.

6. $\frac{1}{5} + \frac{3}{7}$

7. $\frac{5}{6} - \frac{2}{3}$

8. $\frac{8}{9} + \frac{3}{4} - \frac{1}{3}$

9. $\frac{9}{16} + \frac{7}{10} - \frac{4}{15}$

10. $1\frac{1}{2} + 2\frac{2}{3}$

11. $4\frac{5}{6} - 3\frac{3}{4}$

12. $2\frac{1}{2} + 3\frac{3}{5} - 1\frac{7}{8}$

13. $4\frac{1}{5} - 3\frac{1}{3} + 6\frac{3}{4}$

[2.3] Find the indicated product or quotient.

14. $\frac{5}{7} \cdot \frac{3}{5}$

15. $\frac{6}{7} \div \frac{4}{9}$

16. $\frac{8}{3} \cdot \frac{6}{5}$

17. $\frac{9}{8} \div \frac{15}{4}$

18. $\frac{6}{7} \cdot \frac{3}{4} \cdot \frac{1}{5}$

19. $\frac{5}{6} \cdot \frac{7}{8} \div \frac{3}{4}$

20. $8\frac{2}{3} \div 2\frac{3}{4}$

21. $\frac{4}{5} \div 6$

22. $8 \div \frac{6}{5}$

23. $\frac{4}{3} \div \frac{8}{9}$

[2.4] Write as a decimal.

24. Six hundred twenty-five and thirty-one thousandths

25. $4\frac{5}{10{,}000}$

Write in words.

26. 0.0394

27. 45.47

Change to a decimal. Round to three decimal places if necessary.

28. $\frac{7}{20}$

29. $\frac{3}{16}$

30. $\frac{52}{65}$

Change to a fraction reduced to lowest terms.

31. 0.46

32. 0.365

33. 0.5075

[2.5] Perform the indicated operations.

34. $5.03 + 0.006 + 400.1 + 25 + 0.7364$

35. $2.73 - 0.0091$

36. $3.62 \cdot 0.004$

37. $53.06 \div 0.007$ (to nearest hundred)

38. At 16¢ ($0.16) per pound, find the cost of 150 castings, each weighing 7.4 lb.

39. In the accompanying figure the actual inside diameter of a ''1''-in pipe is 1.04 in. The actual outside diameter is 1.315 in. Find the thickness of the pipe.

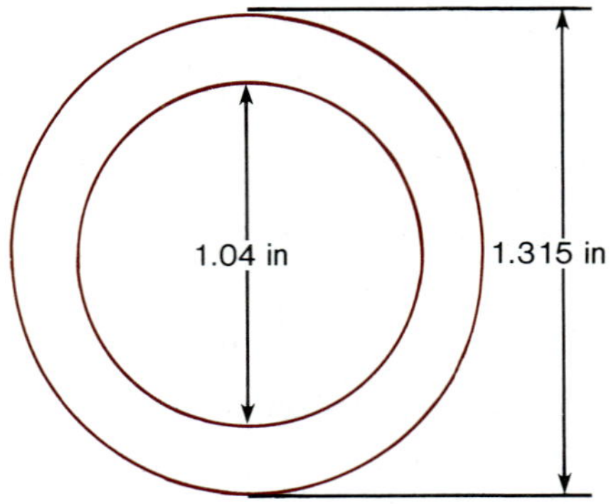

40. What must be the length of a bar of brass to provide pieces of length $1\frac{3}{8}$ in, $3\frac{3}{4}$ in, and $4\frac{5}{16}$ in, allowing $\frac{1}{8}$ in for each cut?

41. Find the length a in the bolt in the accompanying figure.

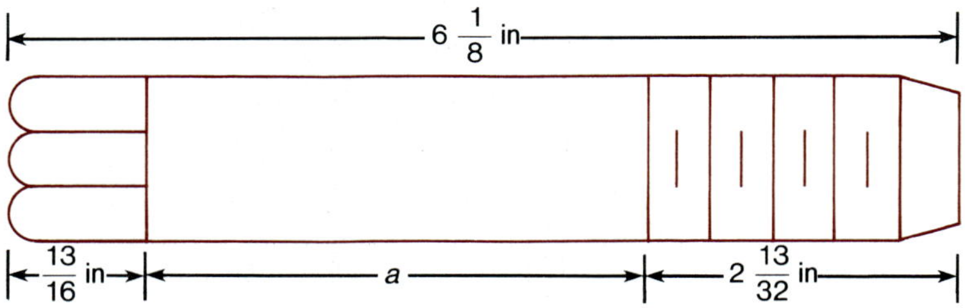

42. If two cells are connected in series-opposing with a voltmeter and one cell has a terminal voltage of $\frac{9}{10}$ volts and the other has a terminal voltage of $\frac{3}{4}$ volts, what value will the voltmeter indicate? (*Hint:* voltmeter reading equals the difference between terminal voltages.)

43. The accompanying figure depicts two cells connected in series-aiding with a voltmeter.

If one cell has a terminal voltage of $\frac{7}{8}$ volts and the other cell has a terminal voltage of $\frac{11}{12}$ volts, what value will the voltmeter indicate? (*Hint:* voltmeter reading is the sum of the terminal voltages.)

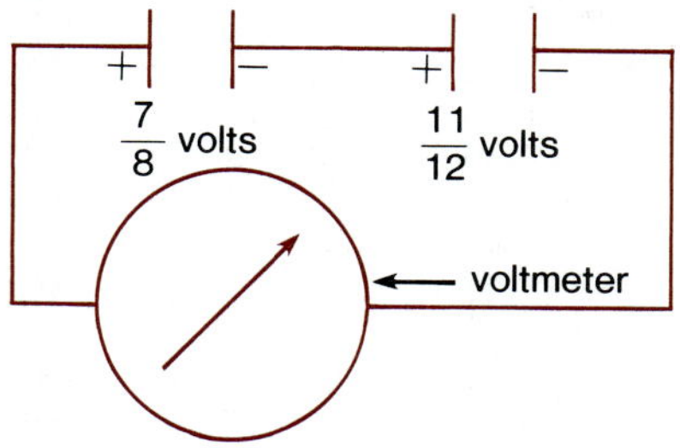

44. The total resistance, or equivalent resistance, of a two-branch parallel circuit, as illustrated in the accompanying figure, is equal to the product of the resistance values divided by their sum. Find the total resistance if the two resistances are 20.5 ohms and 30 ohms.

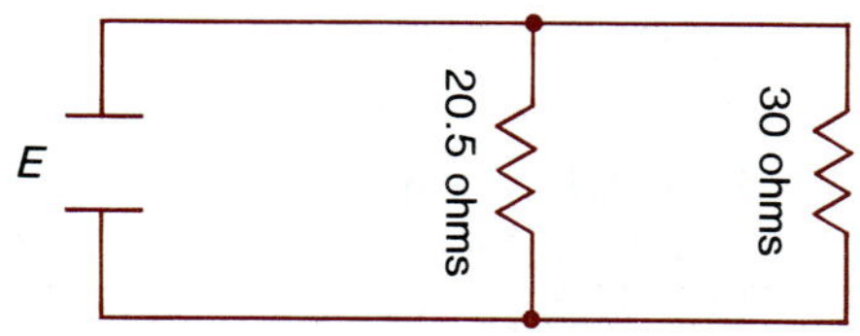

45. If marble weighs 168.5 lb/ft^3, how much does 7.3 ft^3 of marble weigh?

46. What is the inside diameter of a piece of tubing if the outside diameter is 0.672 in and the wall thickness is $\frac{9}{64}$ in? (correct to nearest thousandth)

[2.6] Perform the indicated operations and simplify.

47. $3 + 4 \cdot 5$

48. $5(3 - 2)(4 + 1)$

49. $\dfrac{15(9 - 6)}{5} + \dfrac{14}{5 + 2}$

50. $(25 - 1) \div 6$

51. $8 \cdot 2^3 - 15$

52. $4(7 - 3)(4 + 5)$

53. $18 - 6 \cdot 5 \div 2 - 4$

54. $11 - (8 + 5) - 5 \cdot 3$

55. $5(3 - 6)^2 - 2(4 - 6)$

56. $5(-7 + 12) - 4(3 - 8)$

Chapter Test

Perform the indicated operations and simplify.

1. $\frac{2}{3} + \frac{3}{4} - \frac{5}{6}$
2. $\frac{24}{35} \cdot \frac{15}{18}$
3. $1\frac{1}{2} + 3 + 2\frac{1}{4}$
4. $\frac{15}{38} \div 5$
5. $\frac{8}{3} \div \frac{5}{6}$
6. $9 \div \frac{1}{2}$
7. $7.04 + 13.6$
8. $18.9 - 6.041$
9. $27.3 \cdot 1.04$
10. $58.5 \div 1.3$
11. $7 + 4 \cdot 3$
12. $\frac{2}{5} - \frac{1}{7}$
13. $4 - 2\frac{5}{9}$
14. $6(4 - 3)(5 + 2)$
15. $3\frac{2}{3} + 2\frac{5}{8}$
16. $4 \cdot 2^3 - 18 + 3 \cdot 5$
17. $4\frac{3}{8} \div 2\frac{3}{4}$
18. $0.04 \cdot 0.031$
19. $\frac{18(14 - 4)}{5} + \frac{24}{7 + 5}$
20. $3\frac{1}{4} + 4\frac{3}{8} - 1\frac{5}{6}$
21. $51{,}783 \div 1.23$
22. $\frac{5}{6} \cdot \frac{4}{3} \cdot \frac{9}{15}$
23. $4(-6 - 3) - 5(8 - 11)$
24. $\frac{6}{7} \cdot \frac{1}{5} \div \frac{4}{3}$
25. $\left(8\frac{3}{4}\right) \cdot \left(1\frac{1}{5}\right)$
26. $2(4 - 7)^2 - 3(5 - 8)$
27. A person sold 1,200 lb of newspaper for \$30. What was the price per pound?
28. How many shims 0.014 in thick are needed to shim a space 0.182 in?
29. A two-by-four, 18 ft long, is cut into two pieces such that one piece is $10\frac{5}{8}$ ft long. What is the length of the other piece?

3
Ratio, Proportion, and Percent

A technician using a universal tester to perform tensile tests.
(Photograph by Kenneth Walters, courtesy of Henry Ford Community College)

3.1 Ratio

The relationship of one number to another can be stated in many ways. In chapter 2 we studied arithmetic fractions whose numerator and denominator indicated one such relationship—given the fraction $\frac{a}{b}$, the numerator a represents a equal parts of a whole that is divided into b equal parts. We interpreted this fraction as the quotient of a divided by b. Now we consider a similar relationship between two numbers, called a **ratio.** A ratio is the comparison of two numbers by division. *The ratio of one number to another number is determined when one number is divided by another*. That is, the ratio of a to b is

$$a/b, \quad \frac{a}{b}, \quad a \div b, \quad \text{or } a{:}b$$

note

We read "$a{:}b$" by "the ratio of a to b," where a and b are called the *terms* of the ratio.

Examples

1. The ratio of 3 to 4 is given by $\frac{3}{4}$ or 3:4.
2. The ratio of 15 to 9 is given by $\frac{15}{9}$ or 15:9. Simplifying this fraction by reducing it to lowest terms gives $\frac{5}{3}$ or 5:3. Thus, the ratio of 15 to 9 is the same as the ratio 5:3. Ratios are always reduced to their lowest terms. To say the ratio is "15:9" is not acceptable form.
3. The ratio of $\frac{5}{6}$ to 10 is given by $\frac{\frac{5}{6}}{10}$ or $\frac{5}{6}{:}10$. Again, the ratio can be reduced to lower terms:

   $$\frac{5}{6} \div 10 = \frac{5}{6} \cdot \frac{1}{10} = \frac{1}{12}$$

   and so the ratio $\frac{5}{6}{:}10$ becomes 1:12.
4. The ratio of $2\frac{1}{2}$ to $3\frac{1}{4}$ is given by

   $$\frac{2\frac{1}{2}}{3\frac{1}{4}} \quad \text{or} \quad 2\frac{1}{2}{:}3\frac{1}{4}$$

   Now $2\frac{1}{2} = \frac{5}{2}$ and $3\frac{1}{4} = \frac{13}{4}$ so we have

   $$\frac{5}{2} \div \frac{13}{4} = \frac{5}{2} \cdot \frac{4}{13} = \frac{10}{13}$$

   Therefore, the ratio $2\frac{1}{2}{:}3\frac{1}{4}$ becomes 10:13.

Ratios are used to indicate relationships in many areas of the physical and technical world.

1. Maps are printed to scale—for example, 20 mi to 1 in.
2. Pressure is measured by the ratio of force per unit of area—for example, 14.7 lb/in^2.

note
When the units are not the same, the ratio represents a rate.

3. The auto mechanic interprets engine specifications by compression ratio—9 to 1 or 9:1, for example.
4. The machinist is concerned with gear ratio—such as 2 to 1 or 2:1.
5. Density is measured in mass per unit of volume—for example, $10 g/mm^3$.

note
This ratio represents a *rate*.

6. The carpenter talks of the *pitch ratio* of a roof—that is, the ratio of the *rise* of a rafter to the *span* of the roof.

When comparing two measurable quantities by ratio, it is not necessary that they have the same unit of measure. If the units are not the same, each unit must be included in expressions of the ratio.

Examples

1. 3 ft to 4 in is given by 3 ft:4 in.
2. 400 mi to 8 h is given by 400 mi:8 h.
3. 50 mi to 1 in is given by 50 mi:1 in.
4. 10 lb to 3 oz is given by 10 lb:3 oz.

If the quantities have the same unit of measure, the ratio then becomes a number with no unit designation required.

Examples

1. 8 in to 2 in

$$8\text{ in}:2\text{ in} = \frac{8\text{ in}}{2\text{ in}} = \frac{4}{1} = 4:1$$

2. 45 min to 60 min

$$45\text{ min}:60\text{ min} = \frac{45\text{ min}}{60\text{ min}} = \frac{3}{4} = 3:4$$

3. 64 kg to 12 kg

$$64\text{ kg}:12\text{ kg} = \frac{64\text{ kg}}{12\text{ kg}} = \frac{16}{3} = 16:3$$

4. 32 h to 72 h

$$32\text{ h}:72\text{ h} = \frac{32\text{ h}}{72\text{ h}} = \frac{4}{9} = 4:9$$

5. 80 teeth to 30 teeth

$$80\text{ teeth}:30\text{ teeth} = \frac{80\text{ teeth}}{30\text{ teeth}} = \frac{8}{3} = 8:3$$

If the compared quantities are not in the same unit of measure but *can be stated* in the same unit, we should do so; this ratio, too, then becomes a number.

Examples

1. 3 ft to 4 in
   Since 3 ft = 36 in,
   then, $36\text{ in}:4\text{ in} = \frac{36\text{ in}}{4\text{ in}} = \frac{9}{1} = 9:1.$
2. 6 days to 3 weeks
   Since 3 weeks = 21 days,
   then, $6\text{ days}:21\text{ days} = \frac{6\text{ days}}{21\text{ days}} = \frac{2}{7} = 2:7.$
3. 30 s to 7 min
   Since $7\text{ min} = 7 \cdot 60\text{ s} = 420\text{ s}$
   then, $30\text{ s}:420\text{ s} = \frac{30\text{ s}}{420\text{ s}} = \frac{1}{14} = 1:14.$
4. 3 mi to 1,000 ft
   Now 1 mi = 5,280 ft,
   then, $3\text{ mi} = 3 \cdot 5{,}280\text{ ft} = 15{,}840\text{ ft}$
   and
   we have $15{,}840\text{ ft}:1{,}000\text{ ft} = \frac{15{,}840\text{ ft}}{1{,}000\text{ ft}} = \frac{396}{25} = 396:25.$
5. 35 cents to 4 dollars
   Since there are 400 cents in 4 dollars,
   then, $35\text{ cents}:400\text{ cents} = \frac{35\text{ cents}}{400\text{ cents}} = \frac{7}{80} = 7:80.$

---

note

When changing to a common unit of measure, it is easiest to change to the *smaller* unit of measure, as we did in the preceding examples. Changing to the larger unit of measure usually produces fractions, which are more difficult to reduce.

Sometimes we cannot (or don't wish to) state the compared quantities in the same unit of measure. There are situations when the comparison is more meaningful when stated with different units of measure.

Examples

Express the following as a ratio and reduce to lowest terms.

1. 350 mi to 7 h

   $$350 \text{ mi}:7 \text{ h} = \frac{350 \text{ mi}}{7 \text{ h}} = \frac{50 \text{ mi}}{1 \text{ h}}$$

   We state the ratio by 50 miles per hour.

2. 240 mi to 4 in

   $$240 \text{ mi}:4 \text{ in} = \frac{240 \text{ mi}}{4 \text{ in}} = \frac{60 \text{ mi}}{1 \text{ in}}$$

   We state the ratio by 60 mi/in (scale for a map).

3. The pressure P exerted against a surface is the ratio of the force F applied perpendicular to the area A of the surface. Thus, $P = \frac{F}{A}$.

   If $F = 28$ lb and the area $A = 4\text{ in}^2$, then $P = \frac{28 \text{ lb}}{4 \text{ in}^2} = \frac{7 \text{ lb}}{1 \text{ in}^2}$, which we state by $P = 7\text{ lb/in}^2$.

4. An automobile travels 400 mi on 25 gal of gasoline. We have the ratio $400 \text{ mi}:25 \text{ gal} = \frac{400 \text{ mi}}{25 \text{ gal}} = \frac{16 \text{ mi}}{1 \text{ gal}}$, which we state by 16 mi/gal.

Exercise Set 3.1

In problems 1 through 16, express the given ratios in two ways, reduced to lowest terms: $\frac{a}{b}$ and $a:b$.

Example 15 to 9 becomes $15:9 = \frac{15}{9} = \frac{5}{3} = 5:3$.

answer: $\frac{5}{3}$ and 5:3

1. 12 to 7
2. 8 to 19
3. 16 to 6
4. 24 to 9
5. 15 to 5
6. 7 to 42
7. 32 to 60
8. 90 to 40
9. $\frac{1}{2}$ to 6
10. 8 to $\frac{3}{4}$
11. $2\frac{1}{2}$ to 10
12. 12 to $1\frac{3}{4}$

13. $\frac{5}{6}$ to $\frac{2}{3}$

14. $\frac{7}{5}$ to $\frac{3}{4}$

•15. $3\frac{1}{4}$ to $2\frac{2}{3}$

16. $5\frac{5}{6}$ to 3

17. 4.2 to 2.1

18. 2.4 to 6.0

•19. 1.02 to 2.38

20. 3.06 to 2.55

In problems 21 through 38, find the indicated ratios reduced to lowest terms.

21. 6 in to 14 in
22. 4 ft to 18 ft
23. 25 cm to 10 cm
24. 35 lb to 5 lb
25. 36 km to 24 km
26. 48 lb to 16 lb
•27. 15 in to 3 ft
28. 10 ft to 4 yd
29. \$3 to 35¢
30. 5 days to 15 weeks
31. 30 min to 13 h
32. 16 lb to 8 oz
•33. 4 lb to 24 ft^3
34. 50°C to 5 in^3
35. 16 g to 2 cm^3
36. 300 mi to 10 gal
37. 1,020 mi to 17 h
38. 105 kg to 35 m^3

39. The output in horsepower (hp) is the useful energy delivered by an engine and the input in horsepower is the amount of energy delivered to an engine. The *mechanical efficiency* (ME) of the engine is given by the ratio:

$$ME = \frac{\text{output}}{\text{input}}$$

Find the mechanical efficiency of an engine rated to deliver 425 hp that actually delivers only 375 hp.

40. The pitch of a roof is the ratio of the rise of a rafter to the span of the roof.

$$\text{pitch} = \frac{\text{rise of rafter}}{\text{span of roof}}$$

Find the pitch if the roof rises 7 ft in a span of 21 ft.

41. The smaller of two belted pulleys makes 240 revolutions per minute, and the larger one makes 100 revolutions per minute. What is the ratio of the speed of the larger pulley to that of the smaller pulley? Of the smaller pulley to the larger pulley?

42. The *mechanical advantage* (MA) of a machine is given by the ratio:

$$MA = \frac{\text{resistance to effort (output)}}{\text{input effort}}$$

If a machine has an effort of 30 lb that results in a resistance of 120 lb, what is the mechanical advantage of the machine?

43. The mechanical advantage of a hydraulic press can be represented by the ratio

$$MA = \frac{\text{area of the large piston}}{\text{area of the small piston}}$$

If the large piston has an area of 32 cm^2 and the small piston has an area of 12 cm^2, find the mechanical advantage of the press.

44. Tool steel may be worked at a cutting speed of 20 ft/min in a lathe, and cast iron may be worked at a cutting speed of 45 ft/min. What is the ratio of the cutting speed of tool steel to cast iron?
45. An automobile engine is rated at 350 hp. When tested, the engine produces only 325 hp. What is the mechanical efficiency of the engine? (Refer to problem 39.)
46. An electric motor uses 10 volts of electricity to produce an equivalent output of 8 volts. What is the efficiency of the motor?
•47. A particular stock costing $63 paid an earnings of $6. What is the cost:earning ratio?
48. A room is 24 ft long and 18 ft wide. What is the ratio of its length to its width?
49. A cement block weighs 120 lb and a steel block weighs 1,860 lb. What is the ratio of the weight of the steel block to the weight of the cement block?
50. A mathematics class contains 22 male students and 10 female students. What is the ratio of the male students to the female students?
51. A doctor earns $60,000 in a given year and pays income taxes of $8,400. What is the ratio of taxes to income?
52. The stress caused by a heavy load is defined by the following ratio:

$$\text{stress} = \frac{\text{distorting force } (F)}{\text{area } (A)},$$

measured in lb/in^2. Find the stress of a force of a 4,400-lb load on an area of 1,200 in^2.
53. Power is defined by the ratio of the work done (F_x) to the time taken (t). Find the power in ft-lb/s if 42 ft-lb of work is done in 6 s.
54. The magnification (M) of an object by a lens is given by the ratio $M = \dfrac{q}{p}$, where q = the image distance and p = the object distance from the lens. Find the magnification of an object whose image distance is 27 ft and object distance is 12 ft.
55. The pitch of a roof is given by

$$\text{pitch} = \frac{\text{rise}}{2 \times \text{run}}$$

or

$$\text{pitch} = \frac{\text{rise}}{\text{span}}$$

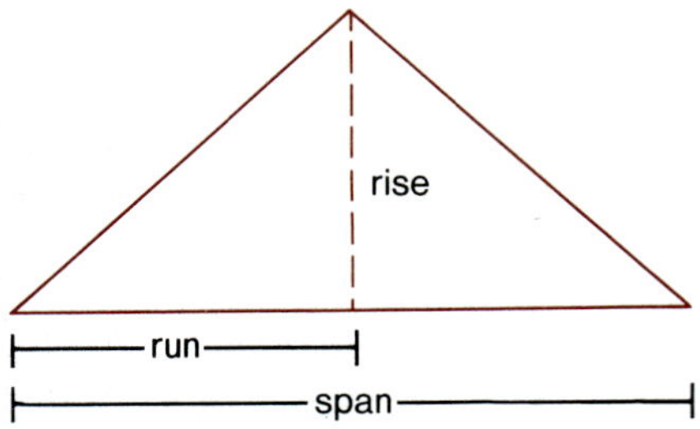

What is the pitch of a shed roof, illustrated in the accompanying figure, having a rise of 3 m and a run of 8 m?
56. What is the pitch of a shed roof having a run of 8 ft 0 in and a rise of 2 ft 6 in?
57. A building having a gable roof has a rise of 8 ft 3 in and a span of 24 ft 0 in. What is the pitch of the roof?
58. What is the pitch of a gable roof having a span of 28 ft 0 in and a rise of 14 ft 0 in?

3.2 Proportion

A proportion establishes a relationship between two ratios. By definition, a **proportion** is a statement of equality of two ratios. Thus, given the ratios a to b and c to d, the statement that

$$\frac{a}{b} = \frac{c}{d} \quad \text{or} \quad a:b = c:d$$

is called a proportion. We read the statement

$$a:b = c:d$$

by "a is to b as c is to d."

The numbers a, b, c, and d are called the **terms** of the proportion. For identification purposes, a, b, c, and d are called the first, second, third, and fourth terms of the proportion, respectively.

Example

Consider the proportion

$$\frac{3}{4} = \frac{6}{8} \quad \text{or} \quad 3:4 = 6:8$$

The 3 is the first term, 4 the second term, 6 the third term, and 8 the fourth term of the proportion.

Usually, however, the terms of a proportion are identified in another way by their placement in the proportion. In $a:b = c:d$, the "outside" two terms, a and d, are called the **extremes** of the proportion, and the "inside" two terms, b and c, are called the **means** of the proportion.

Examples

1. In the proportion $3:4 = 6:8$, the extremes of the proportion are 3 and 8, and the means of the proportion are 4 and 6.
2. In the proportion $\frac{5}{6} = \frac{15}{18}$, the extremes of the proportion are 5 and 18, and the means of the proportion are 6 and 15.

Now consider the proportion $\frac{3}{4} = \frac{6}{8}$ or $3:4 = 6:8$. Note that $3 \cdot 8 = 4 \cdot 6$ since both products are 24. This is an important property of a proportion and is used to solve for one of the terms of the proportion if we know the other three terms. That is, if

$$\frac{a}{b} = \frac{c}{d} \quad \text{or} \quad a:b = c:d$$

then

$$a \cdot d = b \cdot c$$

That is, this property of a proportion can be stated as: in a proportion the product of the extremes equals the product of the means. Thus, using the previous example, $3 \cdot 8 = 4 \cdot 6$ and then

$$3 = \frac{4 \cdot 6}{8} \qquad \text{and} \qquad 8 = \frac{4 \cdot 6}{3}$$

We note that either extreme can be found by multiplying the means and dividing the product by the other extreme. The fact that $4 = \frac{3 \cdot 8}{6}$ and $6 = \frac{3 \cdot 8}{4}$ means that either mean can be found by multiplying the extremes and dividing the product by the other mean. Therefore, if one ratio is known and one part of the second ratio is known, we can find the unknown part of the second ratio by using the property of proportions.

Examples

Find the unknown part of the given proportion.

1. $\frac{x}{8} = \frac{16}{64}$ The extremes of the proportion are x and 64, and the means are 8 and 16. Thus, $64 \cdot x = 8 \cdot 16$, and so $x = \frac{8 \cdot 16}{64} = \frac{128}{64} = 2$. Therefore, $x = 2$.

2. $\frac{49}{y} = \frac{35}{5}$ The extremes are 49 and 5, and the means are y and 35. Therefore, $35 \cdot y = 5 \cdot 49$, so $y = \frac{5 \cdot 49}{35} = 7$.

3. $15{:}6 = z{:}2$ The extremes of the proportion are 15 and 2, and the means are 6 and z. Therefore, $6 \cdot z = 15 \cdot 2$, and so $z = \frac{15 \cdot 2}{6} = 5$.

4. $18{:}R = 16{:}25$ The extremes are 18 and 25; the means are R and 16. Thus, $16 \cdot R = 18 \cdot 25$, so $R = \frac{18 \cdot 25}{16} = \frac{9 \cdot 25}{8} = \frac{225}{8} = 28\frac{1}{8}$.

note

In each of these examples we are finding the number that makes the statement a proportion. Ratios and proportions are used to solve a variety of problems in the technical field and elsewhere.

Examples

1. Two gears are in the ratio of 4:5. If the smaller gear has 32 teeth, how many teeth are there in the larger gear?
   Let x = the number of teeth in the larger gear. Then, 4 is to 5 as 32 is to x, and we have the proportion

   $$\frac{4}{5} = \frac{32}{x} \quad \text{or} \quad 4:5 = 32:x$$

   Then

   $$4 \cdot x = 5 \cdot 32$$

   $$x = \frac{5 \cdot 32}{4} = 40 \text{ teeth}$$

2. On a map, 1 in represents 6 mi. How many inches are needed to represent 28 mi?
   Let x equal the number of inches representing 28 mi. Then, 1 in is to 6 mi as x in is to 28 mi and we have the proportion

   $$\frac{1 \text{ in}}{6 \text{ mi}} = \frac{x \text{ in}}{28 \text{ mi}} \quad \text{or} \quad 1 \text{ in}:6 \text{ mi} = x \text{ in}:28 \text{ mi}$$

   so

   $$6 \cdot x = 1 \cdot 28$$

   $$x = \frac{1 \cdot 28}{6} = \frac{28}{6} = \frac{14}{3} = 4\frac{2}{3}$$

   Therefore, 28 mi is represented by $4\frac{2}{3}$ in on the map.

   note

   In example 2, the same units of measure are in the numerators of the ratios and the same units of measure are in the denominators. It is important to remember to set up a proportion like this when solving for an unknown.

3. A 75-lb bag of fertilizer covers 3,000 ft^2 of lawn. How many pounds will be needed to fertilize a lawn that has an area of 1,750 ft^2?
   Let x equal the number of pounds of fertilizer needed to cover 1,750 ft^2. Then, 75 lb is to 3,000 ft^2 as x lb is to 1,752 ft^2, and we have the proportion

   $$\frac{75 \text{ lb}}{3{,}000 \text{ ft}^2} = \frac{x \text{ lb}}{1{,}750 \text{ ft}^2}$$

   Then, $3{,}000 \cdot x = 75 \cdot 1{,}750$, and

   $$x = \frac{75 \cdot 1{,}750}{3{,}000} = \frac{175}{4} = 43\frac{3}{4}$$

   So 1,750 ft^2 of grass will require $43\frac{3}{4}$ pounds of fertilizer.

Exercise Set 3.2

Find the value of the unknown that makes the statement a proportion.

•1. $\frac{9}{x} = \frac{36}{5}$

2. $\frac{y}{7} = \frac{30}{42}$

3. $\frac{5}{9} = \frac{p}{20}$

4. $\frac{14}{10} = \frac{21}{z}$

5. $6:15 = x:8$

6. $R:12 = 15:100$

•7. $1\frac{1}{2}:a = 4\frac{3}{4}:2$

8. $\frac{3}{4}:4 = \frac{1}{2}:b$

9. $1.2:x = 3.6:9$

10. $4.5:3 = y:2$

11. $\frac{3\frac{1}{4}}{5} = \frac{2\frac{1}{2}}{a}$

12. $\frac{2.4}{4.2} = \frac{b}{2.1}$

Solve the following problems by first choosing a letter to represent the unknown and then setting up the proper proportion.

•13. A person earns $180 per week. How many weeks must he work to earn $1,260?

•14. A 15-in gear meshes with a 9-in gear. How many teeth must the smaller gear have if the larger gear has 90 teeth?

15. An automobile uses 8 liters of gasoline to travel 84 km. How many liters are needed to travel 1,428 km?

16. If 24 g of water yield 4 g of hydrogen, how many grams of hydrogen will there be in 216 g of water?

17. The operating instructions for a gasoline chain saw call for a 16:1 fuel-to-oil mixture. How many *pints* of oil are needed to mix with 88 gal of fuel?

18. The power-to-weight ratio of a given engine is 5:3. What is the weight of the engine if it produces 650 hp?

19. If a 20-lb casting costs $1.50, how much will a 42-lb casting cost?

20. A copper wire 300 ft long has a resistance of 1,024 ohms. What is the resistance of 2,000 ft of copper wire?

•21. A 12-in shaft is to have a taper of 0.310 in in 5 in. Find the taper of the entire length of the shaft.

22. In a hydraulic press the force on the output piston is to the force on the input piston as the area of the output piston is to the area of the input piston. That is,

$$\frac{F_o}{F_i} = \frac{A_o}{A_i} \quad \text{or} \quad F_o:F_i = A_o:A_i$$

Find the area of the input piston if the F_o is 15.2 lb, F_i is 6.5 lb, and A_o is 10.4 in^2.

23. If the ratio of the wins to the losses of the Detroit Tigers in a given season is 6:5, how many games did they lose if they won 90 games?

•24. A rectangular picture that is 10 in long and 8 in wide is to be enlarged to 36 in wide. What will be the length of the enlargement?

•25. The corresponding sides of the accompanying triangles are in proportion. Find the dimensions of the missing sides x and y.

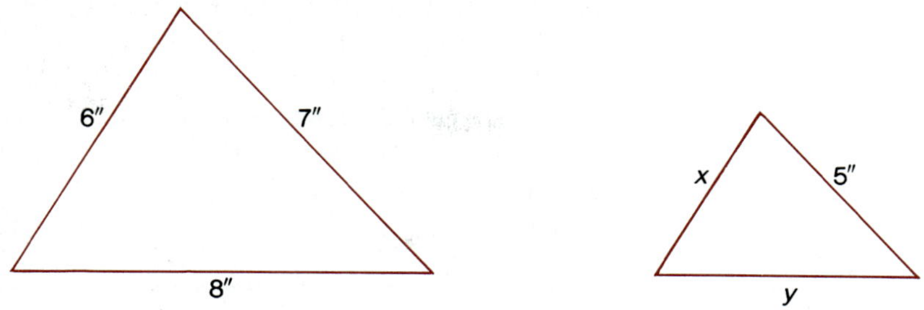

Hint: The corresponding sides are 6 in and x, 7 in and 5 in, 8 in and y.

26. A machine operator can produce 14 parts in 20 min. How long will it take her to produce 224 parts?
27. A punch machine can make 72 holes in 4 min. How many holes can the machine make in 3 h?
28. A roof rises $6\frac{1}{2}$ ft in a rafter span of 9 ft. At this rate, what will be the rise in a 15-ft span?
29. John can type 3 pages of an English paper in 15 min. How long will it take him to type 54 pages? (State your answer in hours and minutes.)

•30. On a drafting scale, $\frac{1}{8}$ in represents 1 ft. What length will a measurement of $2\frac{5}{8}$ in on the scale represent?

31. An automobile engine uses $\frac{3}{4}$ qt of oil in 900 mi. How much oil will it take in 3,000 miles?
32. How many pounds of nails are required for 4,520 sq ft of metal lath if 8 lb are used for each 1,000 sq ft?
33. A plasterer uses $1\frac{1}{4}$ cu yd of sand for each 100 sq ft. How much sand is needed for 5,375 sq ft?

3.3 Variation

If two quantities are so related that, as one quantity changes the other quantity also changes, then the relationship is a **variation.** Variation between quantities is the basis for many formulas used in scientific and technical fields. Involved in variation are two kinds of quantities: *variables* and *constants*. A **variable** is a quantity that can take on more than one value in a given discussion. A **constant** is a quantity that takes on only one value in a discussion.

Standard procedure in mathematics is to let the letters near the beginning of the alphabet—a, b, and c—be used as constants, while letters near the end of the alphabet—x, y, and z—are used as variables. Another letter often used as a constant is the letter k. In our discussions on variation, k will be used extensively as the constant.

Consider the variables x and y and the constant k. To say the ratio of the two variables remains constant is to say that $y/x = k$. Then, if we multiply both sides of the equation by x, solving for y, we get $y = kx$. This typé of a relationship is known as *direct variation*. The equation is read "y varies directly with x" or "y is directly proportional to x," and k is called the **proportionality constant,** or **constant of variation. Direct variation** occurs when the two variable quantities increase or decrease at the same time. That is, if an increase in the value of one quantity produces an increase in the value of the other and a decrease in the value of one quantity produces a decrease in the value of the other, then k is positive.

Examples

1. Consider the distance an automobile covers while traveling at a constant rate of speed. The distance traveled, d, is equal to the rate, r, times the time traveled, t. That is, $d = rt$. When $r = 40$ mi/h, then $d = 40t$. As t increases, d increases.

t	1	2	3	4	5
d	40	80	120	160	200

So we say, "d varies directly as t" or "d is directly proportional to t," and 40 is the constant of variation k.

2. Consider the perimeter, p, of a square having side s. Now, $p = 4s$, and 4 is the constant variation k. Then "p varies directly with s."

note

Dividing both sides of $p = 4s$ by 4, we get

$$s = \frac{p}{4} = \frac{1}{4}p$$

The proportionality constant k is $\frac{1}{4}$ while "s varies directly with p."

3. The area of a circle, A, is given by $A = 3.14r^2$, where the constant of variation k is 3.14 and the "area A varies directly as the square of the radius r." As r increases or decreases, A increases or decreases.

Given the direct variation $y = kx$ and a pair of corresponding values of x and y, we can determine the value of the constant of variation k. If we then substitute this value for k in the equation, we can find the value of y, given any specific value for x, and vice versa.

Examples

1. Let y vary directly with x, and suppose $x = 4$ when $y = 20$. Since $y = kx$, then, by substitution, $20 = k \cdot 4$. Dividing both sides by 4 yields $k = \frac{20}{4} = 5$. Thus, $y = 5x$, and when $x = 7$, $y = 5 \cdot 7 = 35$.
2. The formula for the velocity v in feet per second of a free-falling body when it starts from rest is given by $v = gt$, where g is a constant and t is the time of the fall. If $v = 64$ units when $t = 2$ s, find the constant g and then find v when $t = 6$ s. Now, $64 = g \cdot 2$, and dividing both sides by 2 yields $g = 32$. So, $v = 32 \cdot t$, and when $t = 6$, then $v = 32(6) = 192$ ft/s.

A second kind of variation of two quantities occurs if, when one quantity increases, the other quantity decreases, or as one quantity decreases, the other quantity increases. Such a variation is called **inverse variation** and is expressed by the equation:

$$y = \frac{k}{x} \qquad \text{or} \qquad xy = k$$

which is read "y varies inversely as x" or "y is inversely proportional to x."

Examples

1. Boyle's law for gases states that the volume V of a gas varies inversely to the pressure P, provided the mass and temperature are maintained at a constant. Thus,

$$V = \frac{k}{P} \qquad \text{or} \qquad P = \frac{k}{V},$$

and given $V = 12\ \text{ft}^3$, when $P = 100\ \text{lb/in}^2$, then $12 = \frac{k}{100}$ and $k = 12 \cdot 100 = 1{,}200$. Therefore,

$$V = \frac{1{,}200}{P} \qquad \text{or} \qquad P = \frac{1{,}200}{V}$$

The volume increases as the pressure decreases, and decreasing volume increases the pressure. Now find V when $P = 50\ \text{lb/in}^2$.

$$V = \frac{1{,}200}{50} = 24\ \text{ft}^3$$

2. The speed s of a gear in a mechanical system of gears varies inversely as the size of the gear T. Thus,

$$s = \frac{k}{T} \quad \text{or} \quad T = \frac{k}{s}$$

Suppose $T = 25$ teeth when $s = 150$ rev/min (rpm). Then $150 = \frac{k}{25}$, and multiplying both sides by 25 we get $k = 150 \cdot 25 = 3{,}750$.

Now find s when $T = 75$ teeth.

$$s = \frac{3{,}750}{75} = 50 \text{ rpm}$$

As the number of teeth in the gear increases, the speed of the gear decreases.

A third variation relates one variable quantity to two or more other variable quantities. When a quantity varies directly as the product of two or more quantities, the variation is called a **joint variation.**

Examples

1. The formula for the area of a triangle, A, is given by $A = \frac{1}{2}bh$, where b = the length of the base and h = the altitude of the triangle. The constant of variation, k, is $\frac{1}{2}$, and we say "A varies directly as b and h" or "A varies directly as the product of b and h."
2. The diametral pitch, P, of a gear is given by $P = \frac{kN}{D}$, where N is the number of teeth in the gear and D is the pitch diameter. The pitch P varies directly as the product of the number of teeth and the reciprocal of the pitch diameter, which is then a joint variation. We say the "pitch of the gear varies directly as the number of teeth and inversely as the pitch diameter." Then
   a. As N increases, P increases; as N decreases, P decreases.
   b. As D increases, P decreases; as D decreases, P increases.
3. Coulomb's law says the magnitude F of the force between two charges of electricity is $F = k \cdot \frac{q_A \cdot q_B}{r^2}$, where q_A and q_B are the two charges that are a distance r apart. Then the force F varies directly as the product of q_A and q_B, and the reciprocal of the square of r, and we have a joint variation.

In all of these examples, k can be determined if the values of the variables are given. Then, any one of the variables can be evaluated, knowing the value of the others.

Exercise Set 3.3

In exercises 1 through 20, express the given statements as equations, using the constant of variation k.

1. s varies directly with t.
2. P varies directly with s.
3. I is directly proportional to R.
4. A varies directly as the square of r.
•5. V varies directly as the cube of r.
6. A is directly proportional to the product of l and w.
•7. V varies directly with h and the square of r.
8. P varies inversely with T.
•9. E varies inversely with I.
10. V varies directly with Q and inversely with C.
•11. N varies directly with the cube of s and inversely with the square of p.
12. R varies directly with the square of v and inversely with g.
13. v varies directly with the square root of the product of a and b.
14. Q varies directly with the product of r and s and inversely with t.
•15. V is directly proportional to x and inversely proportional to the product of y and the square of z.
16. The resistance R to the flow of electricity in a conductor varies inversely as the diameter d of the wire.
17. The momentum M of a body varies directly as the product of its mass m and its velocity v.
18. The force of impact F varies directly as the product of the mass m and the velocity v and inversely as the product of the acceleration of gravity g and time t.
19. The force due to centripetal force F of an automobile traveling around a curve is directly proportional to the product of the weight m and the square of the velocity v and inversely proportional to the radius r of the curve.
20. The wavelength L of a sound wave varies directly as the velocity v of the wave and inversely as the frequency of vibrations n.

In exercises 21 through 30, find the proportionality constant (or constant of variation).

•21. y varies directly as x, and $y = 20$ when $x = 5$.
22. p varies directly as s, and $p = 42$ when $s = 7$.
•23. A varies directly as the square of r, and $A = 22$ when $r = 7$.
24. s varies inversely as t, and $s = 15$ when $t = 4$.
25. V is directly proportional to the cube of r, and $V = 81$ when $r = 3$.
26. I is directly proportional to the product of E and R, and $I = 24$ when $E = 3$ and $R = 5$.
•27. S varies inversely with T, and $S = 48$ when $T = 2.5$.
28. n varies inversely with the square root of p, and $n = 14$ when $p = 9$.
•29. p varies directly with m and inversely with v, and $p = 18$ when $m = 4$ and $v = 3$.
30. a is directly proportional to b^2 and inversely proportional to c, and $a = 42$ when $b = 4$ and $c = 3$.

In exercises 31 through 40 find the value of the indicated variable after evaluating the constant of variation.

•31. x varies directly with y; if $x = 36$ when $y = 4$, find x when $y = 13$.
32. w varies inversely with l; if $w = 6$ when $l = 9$, find w when $l = 8$.
33. Find R_1 when $R_2 = 12$, if R_1 varies inversely with R_2 and $R_1 = 48$ when $R_2 = 4$.
34. Find f when $p = 3$ and $q = 6$, if f varies directly with the product of p and q and $f = 54$ when $p = 2$ and $q = 3$.
•35. If y varies inversely with the square of z, find y when $z = 3$, if $y = 12$ when $z = 2$.
36. Find A when $r = 3$, if A is directly proportional to the square of r and $A = 48$ when $r = 4$.
37. If v varies directly with s and inversely with t, find v when $s = 8$ and $t = 6$, if $v = 20$ when $s = 4$ and $t = 2$.
38. If T varies directly with the square of s and inversely with the cube of m, find T when $s = 2$ and $m = 3$, if $T = 14$ when $s = 3$ and $m = 6$.
•39. If a varies directly with b and inversely with c^2, find b when $a = 4$ and $c = 9$, if $a = 10$ when $b = 1$ and $c = 2$.
40. Find T when $p = 40$ and $V = 2$ if p varies directly with T and inversely with V, and $p = 30$ when $T = 6$ and $V = 8$.
41. Hooke's law states that the amount of tension (force) F applied to a wire is directly proportional to the elongation of the wire s. If $s = 0.1$ in when $F = 20$ lb, find F when $s = 0.6$ in.
42. The volume of sales V varies inversely with the unit price p of a commodity. If $V = 4{,}000$ units when $p = \$1.50$, find p when $V = 3{,}000$ units.
43. If the current I flowing through an electrical circuit varies inversely with the resistance R, and $I = 5$ amp when $R = 4$ ohms, find I when $R = 6$ ohms.
44. The current I flowing into a circuit varies directly with the voltage E. When the ammeter reading is $I = 10$ amps, $E = 150$ volts. Find I when $E = 100$ volts.
45. The velocity v of a sound wave varies directly with the product of the frequency n and the wavelength l. If $v = 12$ ft/s when $n = 2$ and $l = 5$, find v when $n = 6$ and $l = 8$.
46. The intensity of illumination on a surface E in footcandles varies inversely with the square of the distance d of the light source from the surface. If $E = 6.4$ footcandles when $d = 5$ ft, find E when $d = 8$ ft.
•47. The power p of a lens, measured in diopters, varies inversely with the focal length f. If $p = 4$ when $f = \frac{1}{4}$, find p when $f = 3.20$.
48. The field intensity H of a magnetic field varies directly with the force F acting upon it and inversely with the strength of the pole m. If $H = 3$ oersteds when $F = 750$ dynes and $m = 200$, find H when $F = 500$ dynes and $m = 175$.
49. The ideal gas law states that the pressure p of a gas varies directly with the absolute temperature T of the gas and inversely with the volume V of the gas. State this relationship using C as the constant of variation. Solve for V.

50. The electrical resistance of a wire R is directly proportional to the length of the wire L and inversely proportional to the square of the diameter of the wire d. State this equation using the proportionality constant k. Solve for L.
51. The simple interest I earned by a savings deposit in a given time t varies jointly as the principal p and the interest r. If I = \$180 when p = \$1,000 and r = 0.06, find I when p = \$650 and r = 0.06, in the same amount of time.
52. The kinetic energy KE (energy due to motion) of an object varies directly as the square of its velocity v. Given KE = 5,000 g cm^2/s^2 when v = 15 cm/s, find KE when v = 25 cm/s.

3.4 Percent

In this section we consider the concept of percent. The word *percent* means "parts per hundred" (*centum* means "100" in Latin), and we define **percent** as the number of parts per hundred. On the basis of this definition, percent can be represented by a fraction with a denominator of 100. Also, percent is designated by the symbol %. Thus, 5% represents 5 parts per 100, and we write $5\% = \frac{5}{100}$. Recall from the study of fractions and decimals that $\frac{5}{100}$ is read "five hundredths," and can also be written as the decimal 0.05. Thus, $5\% = \frac{5}{100} = 0.05$. There are three ways to show hundredths: by fraction, by decimal, and by percent. Consider the following examples.

Examples

Write the following fractions as (a) a common fraction, (b) a decimal, and (c) a percent.

1. Thirty-five hundredths

   (a) $\frac{35}{100}$ (b) 0.35 (c) 35%

2. Four hundredths

   (a) $\frac{4}{100}$ (b) 0.04 (c) 4%

3. One hundred seventy-five hundredths

   (a) $\frac{175}{100}$ (b) 1.75 (c) 175%

From these examples note that percents can be converted to decimals, decimals to percents, percents to fractions, and fractions to percents.

Examples

1. Change 25% to a decimal.

$$25\% = \frac{25}{100} = 0.25$$

2. Change 0.73 to a percent.

$$0.73 = \frac{73}{100} = 73\%$$

3. Change 250% to a decimal.

$$250\% = \frac{250}{100} = 2.50$$

4. Change 3.25 to a percent.

$$3.25 = 3\frac{25}{100} = \frac{325}{100} = 325\%$$

5. Change 0.4% to a decimal.

$$0.4\% = \frac{0.4}{100} = \frac{0.4}{100} \cdot \frac{10}{10} = \frac{4}{1{,}000} = 0.004$$

6. Change 0.052 to a percent.

$$0.052 = \frac{52}{1{,}000} = \frac{5.2}{100} = 5.2\%$$

The following rules apply in converting a percent to a decimal and a decimal to a percent.

1. To change a percent to a decimal, move the decimal point *two places to the left* and remove the % symbol.
2. To change a decimal to a percent, move the decimal point *two places to the right* and add the % symbol.

Examples

Convert the following percents to decimals.

1. 47% = .47. = 0.47
2. 475% = 4.75. = 4.75
3. 0.7% = .00.7 = 0.007
4. $\frac{1}{2}\%$ = 0.5% = .00.5 = 0.005
5. $5\frac{3}{4}\%$ = 5.75% = .05.75 = 0.0575

Convert the following decimals to percents.

6. $0.78 = 0.78 = 78\%$
7. $3.14 = 3.14 = 314\%$
8. $0.009 = 0.009 = 0.9\%$
9. $0.076 = 0.076 = 7.6\%$
10. $0.0025 = 0.0025 = 0.25\% = \frac{1}{4}\%$

Now consider the following examples of changing a percent to a fraction, and a fraction to a percent.

Examples

1. Change 32% to a fraction.

$$32\% = \frac{32}{100} = \frac{8}{25}$$

2. Change 125% to a fraction.

$$125\% = \frac{125}{100} = \frac{5}{4}$$

3. Change 0.25% to a fraction.

$$0.25\% = \frac{0.25}{100} \cdot \frac{100}{100} = \frac{25}{10{,}000} = \frac{1}{400}$$

4. Change $\frac{7}{8}$ to a percent.

$$\frac{7}{8} = 0.875 = 87.5\%$$

5. Change $\frac{6}{5}$ to a percent.

$$\frac{6}{5} = 1.20 = 120\%$$

6. Change $\frac{3}{25}$ to a percent.

$$\frac{3}{25} = 0.12 = 12\%$$

7. Change $\frac{7}{6}$ to a percent.

$$\frac{7}{6} = 1.1666 = 116.7\%$$ (rounded to the nearest tenth of a percent)

We can see the following rules hold:

1. To change a percent to a fraction, write the number over 100 and reduce the fraction to lowest terms.

   note

   If the numerator is a decimal number, move the decimal point to the right the number of places necessary to make it a whole number and move the decimal point in the denominator this same number of places. (This means the numerator and denominator have been multiplied by the same power of 10 as in the previous example 3.)

2. To change from a fraction to a percent, first change the fraction to a decimal by dividing the numerator by the denominator. Then change the decimal to a percent.

Examples

Convert the following percents to fractions.

1. $35\% = \dfrac{35}{100} = \dfrac{7}{20}$

2. $160\% = \dfrac{160}{100} = \dfrac{8}{5}$

3. $0.45\% = \dfrac{0.45}{100} = \dfrac{45}{10{,}000} = \dfrac{9}{2{,}000}$

4. $\dfrac{3}{8}\% = \dfrac{\frac{3}{8}}{100} = \dfrac{3}{8} \div 100 = \dfrac{3}{8} \cdot \dfrac{1}{100} = \dfrac{3}{800}$

   note

   After placing $\frac{3}{8}$ over 100, we performed the indicated division to get the fraction.

5. $16\frac{1}{2}\% = \dfrac{16\frac{1}{2}}{100} = \dfrac{16.5}{100} = \dfrac{165}{1{,}000} = \dfrac{33}{200}$

Exercise Set 3.4

In exercises 1 through 10 change the given percent to an equivalent decimal.

1. 52%
2. 7%
•3. 325%
4. 736%
•5. 0.8%
6. 0.094%
7. 8.6%
8. 20.3%
•9. $\frac{3}{4}\%$
10. $10\frac{5}{8}\%$

In exercises 11 through 20 change the given decimal to an equivalent percent.

•11. 0.37	12. 0.01	13. 4.73	14. 32.3
15. 0.008	16. 0.0048	•17. 15	18. 0.0009
19. 0.00056	20. 0.0275		

In exercises 21 through 30 change the given percent to an equivalent fraction reduced to lowest terms.

21. 56%	22. 6%	•23. 155%	24. 360%
•25. 0.9%	26. 0.082%	•27. $\frac{1}{4}\%$	28. $9\frac{2}{5}\%$
29. 0.75%	30. 0.37%		

In exercises 31 through 40 change the given fraction to an equivalent percent, rounded off to the nearest tenth of a percent where necessary.

•31. $\frac{4}{5}$	32. $\frac{6}{10}$	33. $\frac{5}{6}$	34. $\frac{11}{10}$	35. $\frac{4}{35}$
36. $\frac{17}{20}$	37. $\frac{10}{3}$	38. $\frac{7}{9}$	39. $\frac{3}{25}$	•40. $\frac{15}{7}$

3.5 Applications of Percent

Consider a bucket that will hold 4 gal of a liquid. Suppose the bucket contains 3 gal of water. Since $\frac{3}{4} = \frac{75}{100} = 0.75 = 75\%$, we can say:

The bucket is $\frac{3}{4}$ full.

The bucket is 0.75 full.

The bucket is 75% full.

In this discussion three important numerical quantities are involved:

1. 3, which is called the *amount* and is denoted by A.
2. 4, representing the *whole* quantity, which is called the *base* and is denoted by B.
3. 75, which represents the *percent* and is denoted by P.

Consider the proportion

$$\frac{3}{4} = \frac{75}{100}$$

formed by these quantities. Using the names given to the numbers of this proportion, we get the proportion

$$\frac{\text{amount } (A)}{\text{base } (B)} = \frac{\text{percent } (P)}{100}$$

which is called the **percent proportion.**

note

The fourth term of the percent proportion is *always* the constant 100; the other three terms—A, B, and P—are variable quantities.

In working all percent problems we must know two of the three variable quantities. The third (unknown) quantity is found using the percent proportion. In order to do this, we must be able to identify the amount (A), base (B), and the percent (P) in the given problem. This identification can be made in the following way:

1. The percent P is the number written with the word *percent* or the symbol %.

   Examples 1. 15 is 50% of what number? P is 50.
   2. 75% of 35 is what number? P is 75.
   3. 16 is what percent of 48? P is the unknown.

2. The base B is the number that represents 100% or the "whole" amount. This number *usually* follows the words *percent of*.

   Examples 1. What number is 52% of 60? B is 60 and P is 52.
   2. 75 is 40% of what number? B is the unknown quantity and P is 40.
   3. 26 is what percent of 13? B is 13 and P is the unknown quantity.

3. The amount A is the remaining number after the percent P and the base B have been identified, and usually represents part of the whole, B. A is *usually* before or after the word *is*.

   Examples 1. 14 is what percent of 16? Since B is 16 and P is the unknown quantity, then A is 14.
   2. 72% of 14 is what amount? Since P is 72 and B is 14, then A is the unknown quantity.
   3. 600 is 35% of what number? Since P is 35 and B is the unknown quantity, then A is 600.

Now let us identify all three quantities in the following examples.

Examples

1. What percent of 14 is 28? P is the unknown; A is 28; B is 14.
2. 46% of 23 is what number? P is 46; B is 23; A is the unknown quantity.
3. 33 is 135% of what number? P is 135; B is the unknown quantity; A is 33.

Thus, there are three basic types of percent problems, each involving the use of the percent proportion $\frac{A}{B} = \frac{P}{100}$. They are:

1. Given A and B, using $\frac{A}{B} = \frac{P}{100}$, solve for P.

2. Given A and P, using $\frac{A}{B} = \frac{P}{100}$, solve for B.

3. Given B and P, using $\frac{A}{B} = \frac{P}{100}$, solve for A.

Recall that to solve for an unknown quantity in a proportion, we use the property: The product of the extremes equals the product of the means. Then, since in the percent proportion A and 100 are the extremes and B and P are the means, we *always* use the relationship $100 \cdot A = B \cdot P$.

Examples

1. 12 is what percent of 48?

   P is the unknown quantity, B is 48, and A is 12. Therefore, $\frac{12}{48} = \frac{P}{100}$, and reducing the left side (which should always be done when possible), we get $\frac{1}{4} = \frac{P}{100}$. Then $P = \frac{1 \cdot 100}{4} = 25$. Therefore, 12 is 25% of 48.
2. 15% of what number is 36?

   P is 15, B is the unknown quantity, and A is 36. Then $\frac{36}{B} = \frac{15}{100}$, and reducing on the right side, $\frac{36}{B} = \frac{3}{20}$, so $3 \cdot B = 36 \cdot 20$. Thus, $B = \frac{36 \cdot 20}{3} = 240$. Therefore, 15% of 240 is 36.
3. 40% of 54 is what number?

   P is 40, B is 54, and A is the unknown quantity. Then $\frac{A}{54} = \frac{40}{100}$, and reducing on the right side, $\frac{A}{54} = \frac{2}{5}$, so $A = \frac{54 \cdot 2}{5} = \frac{108}{5} = 21.6$. Therefore, 40% of 54 is 21.6.

When solving a percent word problem:

1. Restate the problem into one of the three types of percent problems.
2. Identify the quantities, P, B, and A.
3. Solve the resulting proportion for the unknown quantity.

Example

1. A painter spent 3 h of an 8-h day painting the front of a house. What percent of the workday did he spend painting the front of the house? This problem is asking: 3 is what percent of 8? P is the unknown quantity, B is 8, and A is 3. So $\frac{3}{8} = \frac{P}{100}$ and $8 \cdot P = 3 \cdot 100$. Then $P = \frac{3 \cdot 100}{8} = 37.5$. So $P = 37.5\%$.
2. A clothing salesperson receives a commission of 7.5% of her total sales. In a given week she sells $8,500 worth of clothes. How much commission does she receive? This problem is asking: 7.5% of $8,500 is what amount? P is 7.5, B is 8,500, and A is the unknown quantity. Thus, $\frac{A}{8{,}500} = \frac{7.5}{100}$, and $100 \cdot A = 8{,}500 \cdot 7.5$, so
$$A = \frac{8{,}500 \cdot 7.5}{100} = 85 \cdot 7.5 = 637.5$$
Therefore, $A = \$637.50$.
3. Francine paid $3.32 in sales tax when she purchased some tools. If the sales tax is 4%, what was the cost of the tools? The problem can be stated: 4% of what number is $3.32? P is 4, B is the unknown quantity, and A is $3.32. So $\frac{3.32}{B} = \frac{4}{100}$, and $4 \cdot B = 3.32 \cdot 100$. Then
$$B = \frac{3.32 \times 100}{4} = 3.32 \cdot 25 = 83.$$ Therefore, $B = \$83$.

Exercise Set 3.5

Answer the following questions.

•1. What is 24% of 30?
2. 87% of 41 is what number?
•3. 16 is 40 percent of what number?
4. 20 is 50 percent of what number?
5. What percent of 50 is 35?
6. 3 is what percent of 24?
7. 64 is 25% of what number?
8. 5% of 450 is what number?
•9. What percent of 14 is 20?
10. 73 is 100% of what number?
•11. 2.7% of 80 is what number?
12. 30 is what percent of 90?
13. What percent of 100 is 79?
14. 135% of 70 is what number?
15. 3.6 is what percent of 14.4?

16. What is $66\frac{2}{3}\%$ of 72?
17. 60 is 120% of what number?
18. 13 is what percent of 42?
19. 24.4% of what number is 20?
20. What percent of 16 is 96?

In exercises 21 through 29 restate each of the following problems as one of the three types of percent problems. Set up a percent proportion and solve for the unknown.

21. 230 people were given eye exams. If 46 of the people were found to have poor eyesight, what percent of the people had poor eyesight?
22. On a mathematics test containing 52 problems, John had 73% correct. How many correct answers did John have on the test?
•23. A parts salesperson's commission is 16% of his total sales. If he received a commission of $268, what were his total sales?
24. If the sales tax is $262.28 on a purchase of $8,756, what percent is the sales tax rate?
•25. Deming, New Mexico, claims to have the purest drinking water in the world. If its water is 99.9% pure, how much of 360 liters of water is pure?
26. The price of steak rose 9% during the past year. If this was an increase of 36¢ per pound, what was the old price?
27. A 4,000-lb automobile contains 450 lb of rubber. What percent of the total weight of the auto is rubber?
28. A motor receiving 10 hp delivers 8.2 hp of work. What percent of the input is the output?
29. The inspectors in a factory rejected 33 pieces as imperfect. If this represents $2\frac{1}{2}\%$ of the daily production, how many pieces were produced?

Solve exercises 30 through 68 using the percent proportion.

30. A mechanic receives an increase of $12.60 per week. If this represents an increase of 12% of her salary, what was her old salary? What is her new salary?
31. A standard formula for type metal calls for $77\frac{1}{2}\%$ lead, $6\frac{1}{2}\%$ tin, and 16% antimony. If there are 350 lb of the type metal, how many pounds of each metal are there?
32. The indicated horsepower of a steam engine is 8.6 hp and the effective hp is 7.4. What percent of the indicated horsepower is the effective horsepower?
•33. A motor has 82% efficiency and delivers $9\frac{1}{2}$ hp. What horsepower does it receive? (82% efficiency means the output is 82% of the input.)
34. When water freezes, its volume expands by 9%. How many cubic meters of water are necessary to make 9.2 cu m of ice?
35. An automobile engine actually burns 89% of the fuel it uses. If it uses 36 gal, how many gallons does it burn?

36. Metal expands when heated. A particular metal will expand approximately 0.02% of its length when exposed to the heat of a normal day. How much will a 36-ft long beam expand during the day?
37. The plumber took 350 ft of pipe along on a job. There were 70 ft left when he completed the job. What percent of the pipe was used?
38. A rough casting that weighs 45 lb loses 8 lb in a machining operation. What percent of the weight is lost?
39. A 440-lb bronze casting is 21.3% zinc. How many pounds of the casting are zinc?
40. The ratio of antifreeze to water in a radiator is 2:7. What percent of the content of the radiator is antifreeze?
41. A contractor estimates that on a $35,000 contract, 51% is paid for labor, 9% is paid for supervision, and the remainder goes for materials. How much does he pay for each item?
42. A woman's weekly pay is $250.40. If 12.3% of this is paid in social security taxes, how much does she pay in social security taxes?
43. Jason sells an automobile for $4,260, losing 15% of the original price of the auto. How much did he pay for the automobile?
44. There are 1,760 yards in a mile. What percent of a mile is 100 yards?
•45. An auto mechanic receives a wage increase of $16.80. If this represents 12% of his present weekly wages, what are his present weekly wages? What are his new wages?
46. The total price of an article is 104% of the selling price. If a snowblower sells for $269, what is the total price you would pay for the snowblower?
47. Lucy spends 26% of her salary for room and board. If she spends $120 per week on room and board, what is her weekly salary?
48. A toaster costs a dealer $9.50 and his profit is to be 20% of the selling price. Find the selling price. (*Hint:* $9.50 is what percent of the selling price?)
49. A 12% tax is levied on all public transportation. If the price of an airplane ticket without tax is $288, what is the total cost of the ticket?
50. The sales tax on an automobile is $239.04. If the sales tax is 4% of the selling price, what is the total price of the auto, including the sales tax?

Example It is estimated that 3,450 board feet (bd ft) of roof boards are needed to roof a house. If an additional 15% is allowed for waste, find the number of board feet needed.

Now $P = 15$ and $B = 3{,}450$ bd ft. We want A where A is the amount of waste. Then $\frac{A}{3{,}450} = \frac{15}{100}$ and

$$100 \cdot A = 15 \cdot 3{,}450$$

$$A = \frac{15 \cdot 3{,}450}{100}$$

$$= 517 \cdot 5$$

Then $3{,}450 + 517.5 = 3{,}967.5$ bd ft are needed altogether.

51. The total area to be covered with insulation board is 1,475 sq ft. If it is estimated that an extra 11% of the area should be added for waste, how many square feet of insulation should be ordered?
52. It is estimated that about 70 lb of flashing material for drip caps is needed for a job. If 15% of the estimate is added to allow for waste, how many pounds should be ordered altogether?
53. A carpenter's helper estimates she needs 450 bd ft of flooring. How many board feet should the carpenter order if she adds 22.5% to this amount?
54. A certain job is estimated to require 9,450 face bricks. How many bricks are needed if 13% of the estimate is added for waste?
55. A contractor estimates the cost of building a house at $63,475. If the contractor adds 12% of the cost for profit, what is the exact amount of his bid? (*Note:* bid = cost + profit)
56. What is the bid made if estimated costs are $5,349, and the profit to be made is $15\frac{3}{4}\%$ of the cost?
57. To an estimate of $95,473 a contractor adds 3% of the cost for incidentals, and 10% of the estimate and incidentals for profit. What is her total bid, correct to the nearest dollar?
58. A general contractor's bid shows the following net costs and profits (percent of the costs) added by subcontractors:

Heating	10,723.41	+ 7%
Plumbing	19,432.15	+ 8%
Masonry	25,142.05	+13%
Electrical	9,100.50	+10%

The general contractor estimates additional costs of $92,563.53 and adds 17% of the total for profit. What is the general contractor's total bid?

59. A housing contractor receives a profit of $35,600 on a job. If this amount is 15% of the total charge, what does the home owner pay?
60. If $1,073.50 is allowed for excavating in constructing a house, what is the cost of the house if the excavation amount is $3\frac{1}{4}\%$ of the total cost?
61. If simple interest (I) = principal (P) · rate (R) · time (T), where R is the rate for 1 year, what is the yearly interest payment if a bank lends a contractor $157,450 at a rate of 18% per year?
62. A contractor borrows $75,000 on a building loan at 17.5% yearly interest. What does the loan cost him for one year?
63. If the amount (A) = principal (P) + interest (I), what amount must a contractor pay back to the bank after 1 year if she borrows $55,400 at 18.5% yearly interest?
64. A contractor gives a bank a note for $35,550 at a rate of 4.5% for 1 month. How much interest is charged for 3 months? How much total must he pay the bank after 3 months?
65. The weight of a wood pattern used to make a casting is 8% of the 450-lb casting. How much does the pattern weigh?

66. A patternmaker buys 575 bd ft of lumber. If $9\frac{1}{2}\%$ of this amount is eventually scrapped, how many board feet remain for making patterns?

67. If a net profit of \$1,060 is made on a job run in a machine shop and the contract price is \$8,480, what percent of profit has been realized?

68. The finished diameter of a shaft turned in an engine lathe measures 3.242 in. The blueprint calls for it to be 3.247 in in diameter. What is the machinist's percent of error?

Chapter Summary

1. The **ratio** of one number to another is that number divided by the other.
2. The "ratio of a to b" can be written $\frac{a}{b}$; $a \div b$; $a:b$.
3. a and b are called the **terms** of the ratio.
4. A **proportion** is a statement that two ratios are equal.
5. The proportion $\frac{a}{b} = \frac{c}{d}$ can be written $a:b = c:d$, which is read "a is to b as c is to d."
6. In the proportion $\frac{a}{b} = \frac{c}{d}$:
   a. a, b, c, and d are called the **first, second, third,** and **fourth** terms of the proportion.
   b. a and d are called the **extremes** of the proportion, and b and c the **means** of the proportion.
7. In a proportion, the product of the extremes equals the product of the means.
8. y varies *directly* as x, written $y = kx$, where k is a constant, if, when x increases y also increases, and when x decreases y also decreases. We say "y is **directly proportional** to x."
9. y is **inversely proportional** to x is written $y = \frac{k}{x}$, where k is a constant, if, when x increases y decreases, and when x decreases y increases.
10. k is the **constant of variation,** or **proportionality constant.**
11. A **joint variation** occurs when one quantity varies directly as the product of two or more quantities.
12. **Percent** means the number of parts per hundred.
13. To change percent to a decimal, move the decimal point *two* places to the *left* and remove the % symbol.
14. To change from a decimal to a percent, move the decimal point *two* places to the *right* and annex the % symbol.
15. The percent proportion is given by $\frac{A}{B} = \frac{P}{100}$, where A is the amount, B is the base, and P is the percent.

Chapter Review

[3.1] Express the given ratio in two ways, reduced to lowest terms.

1. 10 to 15
2. 18 to 6
3. 45 to 60
4. 16 to 40
5. $\frac{3}{4}$ to 8
6. $1\frac{1}{3}$ to $3\frac{1}{2}$
7. 1.05 to 2
8. 4.2 to 1.4

[3.1] Find the indicated ratio, reduced to lowest terms.

9. 20¢ to 35¢
10. 16 lb to 6 lb
11. 24 g to 36 g
12. 3 ft to 9 in
13. 10 oz to 4 lb
14. 250 mi to 5 gal

[3.2] Solve the given proportions for the unknown quantities.

15. $\frac{x}{4} = \frac{20}{24}$
16. $\frac{10}{12} = \frac{y}{48}$
17. $\frac{9}{8} = \frac{30}{a}$
18. $R{:}14 = 1{:}21$
19. $\frac{4}{5}{:}b = 2{:}\frac{5}{6}$
20. $3.2{:}1.2 = 4{:}t$

[3.3] Express the given statements as equations, using the constant k.

21. x varies directly as p.
22. y varies inversely as z.
23. a is directly proportional to the square of b.
24. R is inversely proportional to the square of d.
25. E is directly proportional to the product of I and R.
26. x is directly proportional to y and varies inversely as z.

[3.3] In exercises 27 through 30 find the value of the indicated variable after evaluating the proportionality constant.

27. p varies directly as y. If $p = 10$ when $y = 2$, find p when $y = 3$.
28. If a varies inversely as b, and $a = 4$ when $b = 7$, find a when $b = 14$.
29. If I is inversely proportional to the square of d and if $I = 4$ when $d = 3$, find d when $I = 9$.
30. If x is directly proportional to y and inversely proportional to z, find x when $y = 3$ and $z = 4$, if $x = 3$ when $y = 2$ and $z = 10$.
31. Mr. John weighs 175 pounds and his daughter weighs 75 pounds. Compare Mr. John's weight with his daughter's weight, using a ratio.
32. An economy car averages 28 mi to the gallon of gasoline. A heavier car averages 18 mi to the gallon of gasoline. Find the ratio of the mileage of the heavier car to the mileage of the economy car.
33. Five kg of sugar costs $2.60. How many kilograms of sugar can be purchased for $13.00?
34. 660 bricks are required for 720 cm of a wall. How many bricks will be required for 1,800 cm?
35. If $\frac{5}{6}$ cm on a map represents 12 km, how many centimeters represent 30 km?

[3.4] Solve the following.

36. Convert the following percents to decimals.
    a. 46%
    b. 145%
    c. 0.6%
    d. 0.35%

37. Convert the following decimals to percents.
   a. 0.73 b. 0.047 c. 5.75 d. 0.0015
38. Convert the following percents to fractions.
   a. 36% b. 205% c. 0.12% d. 0.015%
39. Convert the following fractions to percents.
   a. $\frac{4}{5}$ b. $\frac{7}{25}$ c. $\frac{3}{16}$ d. $\frac{7}{8}$

[3.5] Solve the following.

40. 16% of 32 is what number?
41. 12 is what percent of 48?
42. 20% of what number is 35?
43. 77.5 is 31% of what number?
44. 300 is what percent of 45?
45. What is 245% of $450?
46. $87\frac{1}{2}$% of what weight is 13 lb?
47. What percent of a foot is $\frac{5}{6}$ of an inch?
48. Find 23.23% of 0.0056.
49. 1,080 is 125% of what number?
50. 500 g of a solution contains 75 g of a drug. Find what percent of the solution is the drug.
51. A certain ore yields $3\frac{3}{4}$% of iron. How many pounds of iron are there in $1\frac{1}{2}$ tons of ore? (1 ton = 2,000 lb.)
52. Inspectors in a ball bearing factory rejected 42 ball bearings as imperfect. If this number represents $2\frac{1}{2}$% of the day's production, how many bearings are produced in 1 day?
53. Alumel is a thermocouple material used in making thermoelectric pyrometers for metallurgical temperature measurement and control up to 3,000°F. If alumel is 94% nickel, 3% manganese, 2% aluminum, and 1% silicon, how much of each is there in 32.5 oz of alumel?
54. In 95 lb of solder, $11\frac{1}{2}$ lb of lead and $83\frac{1}{2}$ lb of tin are used. What percent of each is used?
55. The selling price of a lawnmower is $139. If there is a 10% sale price discount on the selling price of the mower, what is the sale price? If there is an additional 5% discount on the sale price for paying cash, what is the cash price?

Chapter Test

1. Reduce the ratio 52 to 39 to lowest terms.
2. Write the ratio of 1 h to 10 min.
3. Compare 70¢ to $2 by ratio.
4. Find the ratio of $1\frac{3}{4}$ to $3\frac{1}{2}$.

5. A building that is 26 ft high casts a $6\frac{1}{2}$-ft shadow. Find the ratio of the height of the building to the length of its shadow.
6. Solve the proportion $\frac{x}{18} = \frac{24}{30}$ for x.
7. A painter uses 3 gal of paint in doing 2 rooms. How many gallons would he need to paint 20 rooms?
8. A car burns $2\frac{1}{2}$ qt of oil on an 1,800-mi trip. How many quarts of oil can the owner expect to use on a 12,000-mi trip?
9. If y varies directly as x and $y = 6$ when $x = 4$, find y when $x = 14$.
10. If the resistance (R) in an electric circuit varies directly as the voltage E, and $R = 4.2$ when $E = 60$, find E when $R = 6.6$.
11. A ball starts from rest and rolls down an incline of uniform slope. The distance s that the ball moves varies directly as the square of the time t. If the ball rolls 12 ft in 8 s, find the distance it will roll in 12 s. Find the time it will take to roll 27 ft.
12. Change each from a percent to a decimal.

    a. 13% b. 420% c. 0.05% d. $6\frac{3}{5}\%$
13. Change each decimal to a percent.

    a. 0.56 b. 0.072 c. 6 d. 0.0014
14. Change each percent to a fraction, reduced to lowest terms.

    a. 64% b. 140% c. 0.5% d. 0.125%
15. Change each fraction to a percent.

    a. $\frac{4}{5}$ b. $\frac{17}{25}$ c. $\frac{38}{50}$ d. $2\frac{3}{8}$
16. 32 is what percent of 960?
17. 15% of what number is 120?
18. 350% of 7,800 is what number?
19. What percent of a foot is $\frac{5}{8}$ of an inch?
20. $12\frac{1}{2}\%$ of what amount of money is \$104?
21. A finished casting weighs 51.2 lb. This amount is $97\frac{1}{2}\%$ of the original weight. What is the original weight?
22. The efficiency of a motor is given by

$$\frac{\text{efficiency}}{100} = \frac{\text{power output}}{\text{power input}}$$

    If the power output is 630 watts and the power input is 810 watts, find the efficiency of the motor to the nearest percent.
23. A loudspeaker is 44% efficient. If 60 watts of power is delivered to the speaker, what is the actual output?
24. If the power output of a motor is 500 watts and its efficiency is 63%, what is the power input?
25. Of a total production of 3,015 ball bearings manufactured on a particular day, 110 are found to be defective. What percent of the total was defective?

4
Hand-Held Calculators

Computers enable technicians to make highly accurate mathematical models of proposed equipment and subject them to forces simulating real-life conditions. *(Photograph courtesy of Ford Motor Company)*

4.1 Introduction

Until the early 1970s technicians and scientists used the slide rule for calculations involving multiplication, division, squaring and cubing, for finding the square root, finding the cube root, and so on. Since then, however, with the development of the electronic microchip, the hand-held electronic calculator has become readily available to everyone and has replaced the slide rule. The calculator is easier to use and has *much* greater accuracy than the slide rule.

The era of modern computing—which led to the calculator—began in the early 1800s. Charles Babbage, an English mathematician, is considered the "father of modern computing." In 1872, E. D. Barbour patented the first printing calculator. In 1926, a company owned by William S. Burroughs produced 1 million mechanical calculators. It was not until 1963, however, that the price of calculators was reduced enough to make electronic calculators practical. Since that time, the price of single-chip portable calculators has further decreased to such a point that owning one is desirable, feasible, and practical for the average individual, and of particular importance to the technician.

There is a wide variety of calculators on the market, with a wide range of operations. Since it is impossible here to discuss all the various calculators and their operations, we will direct our attention to the types most generally used by students and discuss the features common to most calculators. We will discuss the use of the calculator in performing arithmetic operations, squares, square roots, reciprocals, percent, powers of a numerical base, and calculator memory. Other special features and variations will be noted.

All calculators come with a manual that can be used to learn how to manipulate it. For further direction in the use of any specific model, refer to the accompanying manual.

4.2 The Keyboard

When a number is entered, or a result calculated, the number or result shows on the **display,** or **read out,** at the top of the calculator. Below the display is the **keyboard,** containing the keys and their special functions. Our discussion here is based on an 8-digit display even though many calculators display 10 or more digits. The operation, however, is essentially the same, and if the result contains more digits than the display is capable of showing, the result will be rounded off. (Some calculators will *truncate*—that is, discard—digits beyond a certain place.)

When entries or calculations involve numbers that are too large or too small—that is, numbers that have too many digits for the display—the numbers cannot be entered. The use of scientific notation, discussed in section 4.8, increases the digital capacity of the calculator.

Some operations do not have defined results. For example, it is not possible to: (1) divide by zero or (2) take the square root of a negative number. If such an entry, or calculation, is attempted, the calculator indicates an error by displaying E, a flashing display of **E**'s, or some other type of display. Note which of these your calculator uses.

The **on/off** switch, generally located just below the display, turns the power supply to the calculator **on** or **off.** This power is supplied by batteries or an electrical wall outlet. Some calculators have a special display that indicates when batteries need replacement or recharging. The user should become acquainted with this and any other special displays the calculator may use. The **power-on** condition is generally indicated by the illumination of a digit, usually **0,** on the display.

The keyboard keys can be divided into four groups: control, data entry, basic function, and special. We first describe the keys contained in each of these groups and the function of each group. We discuss separately the operations performed by the keys.

4.3 The Control Keys

The **control keys** are used to clear, or erase, data from the calculator display. The control keys are the following:

[C] clear key
[CD] clear display key
[CE] clear entry key

The [C] key is used to clear the calculator and display of all information being calculated before a new calculation is started. This key sets the calculator display to **0** for the start of a new entry.

note
The [C] key *will not* clear the calculator of any information stored in the memory.

Example

To clear a previous calculation,

press [C]
and read **0** on the display.

Note that most calculators clear all previously entered data, except for that in the memory, also after the [=] key is pressed. But, unlike with the [C] key, the last calculation remains on the display and may be used further by entering an operation and other data. If another number is entered after the [=] key, the calculation on the display is eliminated.

The [CE] or [CD] key is used to clear the last entry, or last calculated result. Its use has no effect on any other part of a calculation. This key is most often used to erase a wrong entry that has just been entered. Refer to your manual for additional information on the use of the [CE] or [CD] key.

Example

Correcting the mistakes made in attempting to enter [1] [2] [+] [1] [=] to add 12 and 1.

1. [1] [2] [3]	[3] entered by mistake
2. [CD] or [CE]	1, 2, 3, replaced by **0** on the display
3. [1] [2] [+] [4]	[4] entered by mistake
4. [CD] or [CE]	4 replaced by **0** on the display
5. [1]	0 replaced by **1** on the display
6. [=]	**13** read on the display

4.4 Data Entry Keys

The data entry keys consist of the numbered keys, [0], [1], . . . , [9]; the decimal point key, [•]; the pi key, [π]; the change of sign key, [+/−]; and the enter exponent key, [EE].

The digit keys [0], [1], [2], . . . , [9] are used to enter the digits 0 through 9 into the display and into the calculator's processor, where the actual calculations are performed. These keys are also used to enter an exponent of 10 when scientific notation is used. This topic is discussed in section 4.8. The numbers are entered into the display in the order they are pressed.

Example

To enter 4,763, press [4] [7] [6] [3] and read **4763** on the display.

The **decimal point key,** [•], is used to enter a decimal point. If the number being entered is a decimal fraction, pressing the [•] key will place the decimal point in the proper location.

note

Each time a number is entered, the decimal point appears to the number's right unless it is specifically placed elsewhere. This is the **assumed decimal point.**

Examples

1. To enter 37.46, press [3] [7] [•] [4] [6] and read **37.46** on the display.
2. To enter 123, press [1] [2] [3] and read **123.** on the display. Note that the assumed decimal point appears.

The **pi key,** [π], enters the value of π, which is approximately equal to 3.1415927, into the display. This key is used in computing the areas of circles and the volumes of certain geometric figures, such as the sphere, right circular cylinder, and the cone. We use this key after discussing the operation of multiplication on the calculator.

The **change sign key,** [+/−], changes the sign of the number shown on the display. Every number, when entered, is assumed to be a positive number. This key is also used to change the sign of the exponent when scientific notation is used. To enter a negative number, first enter the number and then press the change sign key.

note

If your calculator does not have a [+/−] key, but has a **minus key,** [−], you must enter the minus sign *before* entering the number.

Examples

1. To enter −4.976, press [4] [•] [9] [7] [6] [+/−] and read **−4.976** on the display.
2. To enter $2.936 \cdot 10^{-8}$, press [2] [•] [9] [3] [6] [EE] [8] [+/−] and read **2.936 −08** on the display.

The **enter exponent key,** [EE], instructs the calculator that the next number is to be entered in scientific notation. To indicate a power of ten, depress the [EE] key and then the exponent to be entered.

Example

To enter $3.56 \cdot 10^7$, press [3] [•] [5] [6] [EE] [7] and, on the display, read **3.56 07** which means the original number, $3.56 \cdot 10^7$.

We use this key more extensively in section 4.8 where we discuss scientific notation.

4.5 Basic Function Keys

The basic function keys are the arithmetic keys that perform the operations of addition, subtraction, multiplication, and division, together with the equals key. These are the keys that are used most on the calculator. Performing the arithmetic calculations with the calculator is simple and direct. The numbers and functions are entered in the same sequence as the expression is written, with some variations that are discussed later.

The **equals key,** [=], is used to tell the calculator to complete the calculation of all previously entered data and functions to provide the desired calculation result. This key is used to obtain intermediate results as well as final results.

The **add key,** [+], is used to add the following entry to the previous entry, or calculated result.

Examples

1. To enter 3 + 5, press [3] [+] [5] [=] and read **8** on the display.
2. To enter 3.56 + 3.073, press [3] [•] [5] [6] [+] [3] [•] [0] [7] [3] [=] and read **6.633** on the display.
3. To enter 9.99999 + 0.00001, press [9] [•] [9] [9] [9] [9] [9] [+] [•] [0] [0] [0] [0] [1] [=] and read **10** on the display.
4. To enter (−4.6) + 5.7, press [4] [•] [6] [+/−] [+] [5] [•] [7] [=] and read **1.1** on the display.
5. To enter 7.4 + (−8.03), press [7] [•] [4] [+] [8] [•] [0] [3] [+/−] [=] and read **−0.63** on the display.
6. To enter (−47) + (−7.47), press [4] [7] [+/−] [+] [7] [•] [4] [7] [+/−] [=] and read **−54.47** on the display.

The **subtract key,** [−], subtracts the next entry from the previous entry, or calculated result.

Examples

1. To subtract 38.56 − 21.09, press [3] [8] [•] [5] [6] [−] [2] [1] [•] [0] [9] [=] and read **17.47** on the display.
2. To subtract 42.06 − (−16.3) press [4] [2] [•] [0] [6] [−] [1] [6] [•] [3] [+/−] [=] and read **58.36** on the display.
3. To subtract (−4.07) − 10.6, press [4] [•] [0] [7] [+/−] [−] [1] [0] [•] [6] [=] and read **−14.67** on the display.
4. To subtract −14.07 − (−61.3) press [1] [4] [•] [0] [7] [+/−] [−] [6] [1] [•] [3] [+/−] [=] and read **47.23** on the display.

When a sequence of additions and/or subtractions are performed on the calculator, the operations are performed in exactly the same order as the problem is written from left to right. All operations are performed on the number shown in the display.

Examples

Perform the indicated operations:

1. 24.3 + 3.06 + 5.1 + 231.73
   Press [2] [4] [•] [3] [+] [3] [•] [0] [6] [+] [5] [•] [1] [+] [2] [3] [1] [•] [7] [3] [=] and read **264.19** on the display.
2. 21.43 − 3.4 − 6 − 0.002
   Press [2] [1] [•] [4] [3] [−] [3] [•] [4] [−] [6] [−] [0] [•] [0] [0] [2] [=] and read **12.028** on the display.
3. 43.67 + 21.04 − 17.8 − 1.6 + 2
   Press [4] [3] [•] [6] [7] [+] [2] [1] [•] [0] [4] [−] [1] [7] [•] [8] [−] [1] [•] [6] [+] [2] [=] and read **47.31** on the display.

Exercise Set 4.1

Perform the indicated operations.

1. 3 + 4 + 5
2. 15 + 25 + 32
3. 14.7 + 2.37 + 5.06
4. 421.9 + 0.008 + 14.97
5. 19.23 + 426.73 + 1.0793
6. 2.7 + 0.7364 + 82 + 724.63
7. 42.07 − 12.4
8. 234.76 − 15.724
9. 51.6 − 0.083 − 14.72
10. 17.43 − 0.0093 − 8.076 − 1.7
11. 14.076 + 24.7 − 16.943

12. 473.7 − 27.075 − 26.27 + 1.73
•13. 47.73 + 2,437.6 − 16.7 − 187.43 + 1.00897
14. 67.3 − 1.0779 + 472.9 − 15.89 + 167.69
15. 4,173.4 + 703.92 − 1,073.927 − 62.14 + 143.092

The **multiply key,** [×], is used to multiply the previous entry, or calculated result, by the next entry.

Examples

1. To multiply 5.396 × 2.92, press [5] [•] [3] [9] [6] [×] [2] [•] [9] [2] [=] and read **15.75632** on the display.
2. To multiply 23 × (−14.7), press [2] [3] [×] [1] [4] [•] [7] [+/−] [=] and read **−338.1** on the display.
3. To multiply 14.03 × 16.2 × 1.473, press [1] [4] [•] [0] [3] [×] [1] [6] [•] [2] [×] [1] [•] [4] [7] [3] [=] and read **334.79228** on the display.

Exercise Set 4.2

Perform the indicated multiplications.

1. 24 · 145
2. 247 · 1,073
•3. 24.67 · 9.705
4. 57.089 · 2,946.67
5. 375.09 · (−28.5)
6. −354.45 · 3,862.9
7. −34.765 · (−29.48)
8. 3.6 · 89.4 · 243.07
9. 0.0293 · 57.4 · 78.05 · 35
10. 35.3 · 0.89034 · 1.6 · 24.35
11. 3.56 · (−2.89) · (−253.04)
12. −0.456 · (−34.6) · (−283.1)
•13. 23 · (375.4) · (−2.9) · 29.457
14. −3.4 · (−78.43) · (−2.3) · (−5.049)

The **divide key,** [÷], is used to divide the previous entry, or calculated result, by the next entry.

Examples

1. To divide 5.023 by 1.36, press [5] [•] [0] [2] [3] [÷] [1] [•] [3] [6] [=] and read **3.6933824** on the display.
2. 42.73 ÷ 0.0076
   Press [4] [2] [•] [7] [3] [÷] [•] [0] [0] [7] [6] [=] and read **5622.3684** on the display.
3. $\dfrac{409.3}{-6.73}$
   Press [4] [0] [9] [•] [3] [÷] [6] [•] [7] [3] [+/−] [=] and read **−60.817236** on the display.

Exercise Set 4.3

Divide the following.

1. $\dfrac{234}{6.89}$	2. $\dfrac{408.6}{34.89}$
3. $\dfrac{56.29}{496.7}$	4. $\dfrac{2{,}938.67}{0.0075}$
5. $\dfrac{-23.45}{8.38}$	6. $\dfrac{396.57}{-26.498}$
7. −4.78 ÷ −0.685	8. −2,934.57 ÷ −35.67
•9. 3.0284 ÷ −62.34	10. −0.3854 ÷ 4.64
11. 294.5843 ÷ −38.95	12. −0.00475 ÷ −0.048

4.6 Special Function Keys

These keys operate only on the quantity displayed. They *do not* complete any previous calculation instructions.

The **reciprocal key,** [1/x] , is used to find the reciprocal of the number that is displayed. This key divides any number into 1 without the user having to press the [÷] and [=] keys.

Example

To find $\dfrac{1}{25}$, press [2] [5] [1/x] and read **0.04** on the display.

If the calculator does not have a [1/x] key, or is not programmed, the problem must be worked out as a division problem 1 ÷ x.

Example

To find $\dfrac{1}{25}$ without a reciprocal key, press [1] [÷] [2] [5] [=] and read **0.04** on the display.

note

In general, when performing such operations on the calculator, if the result shows eight digits on the display, the calculation may, or may not, be rounded off. If fewer than eight digits, the result is usually exact.

The **square key,** [x^2], is used to find the square of the number on the display, without using the [=] key.

Example To find $(25)^2$, press [2] [5] [x^2] and read **625** on the display.

If the calculator does not have a [x^2] key, or is not programmed for squares, the calculation must be worked as a repeated multiplication problem.

Example To find $(35)^2$, press [3] [5] [×] [3] [5] [=] and read **1225** on the display.

The **square root key,** [$\sqrt{x}$], is used to find the square root of the number on the display. Finding the square root of a number is the reverse of squaring the number. Given x is some number, the symbol for the square root of x is $\sqrt{x}$. If x can be stated as a product of two equal numbers, one of the numbers is the square root of x. That is, $\sqrt{x} = n$ if $n \cdot n = x$.

Examples

1. To find $\sqrt{25}$, since $25 = 5 \cdot 5$, then $\sqrt{25} = 5$.
2. To find $\sqrt{144}$, since $144 = 12 \cdot 12$, then $\sqrt{144} = 12$.

Suppose we wish to determine $\sqrt{5}$. We do not know an exact number, n, such that $n \cdot n = 5$. Such a number is called an *irrational number* and is discussed in chapter 1. To find $\sqrt{5}$ on the calculator, press [5] [$\sqrt{x}$] [=] and read **2.236068** on the display. This is an approximate number that has been rounded off to six decimal places.

Examples

1. To find $\sqrt{256}$, press [2] [5] [6] [$\sqrt{x}$] and read **16** on the display.
2. To find $\sqrt{23}$, press [2] [3] [$\sqrt{x}$] and read **4.7958315** on the display.
3. To find $\sqrt{44.3}$, press [4] [4] [•] [3] [$\sqrt{x}$] and read **6.6558245** on the display.
4. Consider $\sqrt{-4}$. Press [4] [+/−] [$\sqrt{x}$] and read an error notation since the square root of a negative number is not defined.

The **y to the power x key,** $\boxed{y^x}$, instructs the calculator to raise a number to a given power. That is, the $\boxed{y^x}$ key uses the number as a factor x times in an indicated product. The first quantity entered is the value assigned to y.

Examples

Find the given power.

1. 6^3 Press $\boxed{6}$ $\boxed{y^x}$ $\boxed{3}$ $\boxed{=}$ and read **216** on the display.
2. $(1.26)^4$ Press $\boxed{1}$ $\boxed{\cdot}$ $\boxed{2}$ $\boxed{6}$ $\boxed{y^x}$ $\boxed{4}$ $\boxed{=}$ and read **2.5204738** on the display.
3. $(0.005)^5$ Press $\boxed{\cdot}$ $\boxed{0}$ $\boxed{0}$ $\boxed{5}$ $\boxed{y^x}$ $\boxed{5}$ $\boxed{=}$ and read **3.125 −12** on the display.

note

The result is $3.125 \cdot 10^{-12}$, or 0.000000000003125.

The **xth root of y key,** $\boxed{\sqrt[x]{y}}$, is used to find the root of a number. We can extend the idea of square root to the xth root of y, denoted by $\sqrt[x]{y}$, where x is called the index of the radical, y is called the radicand, and $\sqrt[x]{y}$ is called radical notation. To generalize, $\sqrt[x]{y} = n$ if $\overbrace{n \cdot n \cdot n \cdots n}^{x \text{ factors}} = y$.

Examples

1. To find $\sqrt[5]{32}$, press $\boxed{3}$ $\boxed{2}$ $\boxed{\sqrt[x]{y}}$ $\boxed{5}$ $\boxed{=}$ and read **2** on the display, because $\overbrace{2 \cdot 2 \cdot 2 \cdot 2 \cdot 2}^{5 \text{ factors}} = 32$.
2. To find $\sqrt[4]{81}$, press $\boxed{8}$ $\boxed{1}$ $\boxed{\sqrt[x]{y}}$ $\boxed{4}$ $\boxed{=}$ and read **3** on the display, because $\overbrace{3 \cdot 3 \cdot 3 \cdot 3}^{4 \text{ factors}} = 81$.
3. To find $\sqrt[3]{64}$, press $\boxed{6}$ $\boxed{4}$ $\boxed{\sqrt[x]{y}}$ $\boxed{3}$ $\boxed{=}$ and read **4** on the display, because $\overbrace{4 \cdot 4 \cdot 4}^{3 \text{ factors}} = 64$.
4. Consider $\sqrt[3]{7}$. The answer to this problem is *irrational* since there is no exact number n such that $\overbrace{n \cdot n \cdot n}^{3} = 7$; the answer in the display is an approximate number. To find $\sqrt[3]{7}$, we press $\boxed{7}$ $\boxed{\sqrt[x]{y}}$ $\boxed{3}$ $\boxed{=}$ and read **1.9129312** on the display.

Exercise Set 4.4

Perform the indicated operations.

•1. $\frac{1}{5}$	2. $\frac{1}{6}$	3. $\frac{1}{45}$	4. $\frac{1}{469}$
5. $\frac{1}{4.7}$	6. $\frac{1}{78.47}$	7. $\frac{1}{0.00456}$	8. $(56)^2$
9. $(285)^2$	10. $(4.6)^2$	11. $(34.89)^2$	12. $(0.0298)^2$
•13. $(-24.67)^2$	14. $(-0.237)^2$	15. $\sqrt{441}$	16. $\sqrt{1,024}$
17. $\sqrt{4.32}$	18. $\sqrt{0.0374}$	19. $\sqrt{3,924.5}$	20. $\sqrt{35.47}$
•21. $\sqrt{5.6789}$	22. 6^4	23. $(25)^3$	24. $(2.9)^4$
25. $(0.34)^3$	26. $(3.23)^5$	27. $(-27)^4$	28. $(-6.23)^5$
29. $\sqrt[6]{64}$	30. $\sqrt[3]{125}$	31. $\sqrt[3]{.008}$	32. $\sqrt[5]{-32}$
33. $\sqrt[3]{729}$	34. $\sqrt[4]{81}$	35. $\sqrt[8]{256}$	36. $\sqrt[3]{11}$
37. $\sqrt[4]{20}$	38. $\sqrt[3]{-18.1}$	•39. $\sqrt[5]{.0073}$	40. $\sqrt[4]{1.4}$
41. $\sqrt[6]{3.14}$	42. $\sqrt[5]{31.9}$		

The **percent key,** [%] , is used in several ways to compute problems involving percent.

Examples

1. To find a given percent of a number:
   3.6% of 425 or 3.6% · 425, press [4] [2] [5] [×] [3] [•] [6] [%] [=] and read **15.3** on the display.
2. To convert a fraction to a percent:
   Change $\frac{3}{8}$ to a percent. Press [3] [÷] [8] [%] [=] and read **37.5** on the display.
3. To calculate the cost of an article, including tax:
   To find \$400 + 4% tax, press [4] [0] [0] [+] [4] [%] [=] and read **416** on the display.
4. To calculate the price of an article less a discount:
   To find \$150 less 15% discount, press [1] [5] [0] [−] [1] [5] [%] [=] and read **127.5** on the display; the display figure means the price is \$127.50.

Examples

1. To find 7% of 72 you must
   a. Multiply 7 by 0.01 to get 0.07.
   b. Multiply 0.07 · 72 by pressing [•] [0] [7] [×] [7] [2] [=] and read **5.04** on the display.

2. To convert $\frac{5}{8}$ to a percent
   a. Perform the indicated division by pressing [5] [÷] [8] [=] and read **0.625** on the display.
   b. Multiply this result by 100 by pressing [×] [1] [0] [0] [=] and read **62.5** on the display. Thus, $\frac{5}{8} = 62.5\%$.
3. To find \$560 + 4% of \$560 in tax
   a. Multiply 4 by 0.01 to get 0.04.
   b. Multiply 560 by 0.04 by pressing [•] [0] [4] [×] [5] [6] [0] [=] and read **22.4** on the display. (\$22.40)
   c. Add \$560 to \$22.40 by pressing [5] [6] [0] [+] [2] [2] [•] [4] [=] and read **582.4** on the display. (\$582.40)

Exercise Set 4.5

Convert the following to percents.

1. $\frac{7}{8}$
2. $\frac{34}{89}$
•3. $\frac{0.34}{0.12}$
4. $\frac{145}{789}$
5. $\frac{0.564}{0.45}$

Calculate the following.

6. 6.7% of 342
•7. 43.1% of 48
8. 9.23% of 567
9. 0.56% of 78.9
10. 134% of 625
11. 721.9% of 1.34
12. 0.124% of 6.14
•13. \$240 + 5% tax on \$240
14. \$34.56 + 4.5% tax on \$34.56
•15. \$14.78 less 3.7% of \$14.78
16. \$300 less 4.9% of \$300
•17. \$1,300 + 5.6% of \$1,300 less 1.3% of that amount
18. \$405.78 + 1.9% of \$405.78 less 9.8% of that amount

4.7 Mixed Operations

When evaluating expressions containing a sequence of addition, subtraction, multiplication, and division operations, it is sometimes necessary to change the order in which the operations are performed in order to conform to the capabilities of the calculator. Where the order of the expression is such that the sequence of operations is performed in exactly the same order as the problem is written from left to right, there is no particular problem.

Examples

Perform the indicated operations.

1. $3 \cdot 5 + 4$
   Press [3] [×] [5] [+] [4] [=] and read **19** on the display.
2. $5 \cdot 4 \div 2 - 6 + 3$
   Press [5] [×] [4] [÷] [2] [−] [6] [+] [3] [=] and read **7** on the display.
3. $(-18.4) \div 4 + 1.07$
   Press [1] [8] [•] [4] [+/−] [÷] [4] [+] [1] [•] [0] [7] [=] and read **−3.53** on the display.

Notice, in these illustrations, that the order of operations to be performed is such that the operations of multiplication and division are performed *before* the operations of addition and/or subtraction. The hierarchy of operations requires that this order be observed—that is, multiplication and/or division must be performed in order from left to right *before* addition and/or subtraction. Some calculators observe this hierarchy. Others do not. In the latter case, if a problem is such that the operations are not listed in hierarchical order, then adjustments must be made according to the capabilities of the calculator. It should be noted that where parentheses are involved, the operations indicated therein are *always* performed first according to the hierarchy of operations.

Examples

Perform the indicated operations.

1. $3 + 4 \cdot 5$
   Press [4] [×] [5] [+] [3] [=] and read **23** on the display.
2. $(8 + 3) \cdot 6$
   Press [8] [+] [3] [=] [×] [6] [=] and read **66** on the display.

note

In example 2 the parentheses indicate that the operation inside must be performed first so the *order will not be changed*. Also, the [=] key between [3] and [×] is optional on some calculators.

3. $2 \cdot 5 + 6 \cdot 3$
   If the calculator contains the **parentheses keys,** [(] and [)], this problem can be treated as though it is written $(2 \cdot 5) + (6 \cdot 3)$. Press [(] [2] [×] [5] [)] [+] [(] [6] [×] [3] [)] [=] and read **28** on the display. If there are no parentheses keys, the **storage key,** [sto], and the **recall key,** [rcl] can be used. Press [2] [×] [5] [=] [sto] [6] [×] [3] [=] [+] [rcl] [=] and read **28** on the display.

note

The [=] key between [3] and [+] is optional. Note too that some calculators have the key [M] or [M+] for the storage key.

4. (6.4 − 3.8)(14.6 + 8.2) (*Note:* Product of a difference and a sum)
   a. If a calculator contains parentheses keys, press [(] [6] [•] [4] [−] [3] [•] [8] [)] [×] [(] [1] [4] [•] [6] [+] [8] [•] [2] [)] [=] and read **59.28** on the display.
   b. Without the parentheses, and using the memory of the calculator, press [6] [•] [4] [−] [3] [•] [8] [=] [sto] [1] [4] [•] [6] [+] [8] [•] [2] [=] [×] [rcl] [=] and read **59.28** on the display.

5. $\dfrac{24}{3 + 7}$
   a. Using the [sto] and [rcl] keys, press [3] [+] [7] [=] [sto] [2] [4] [÷] [rcl] [=] and read **2.4** on the display.
   b. Using the reciprocal key, [1/x] , press [3] [+] [7] [=] [1/x] [×] [2] [4] [=] and read **2.4** on the display.

6. $\dfrac{8 + 16}{6}$

   Normal order of operations is indicated in this problem, so press [8] [+] [1] [6] [=] [÷] [6] [=] and read **4** on the display.

7. $\dfrac{6 \cdot 21}{2 \cdot 7}$
   a. This problem can be rewritten: $\dfrac{6 \cdot 21}{2 \cdot 7} = \dfrac{(6 \cdot 21)}{(2 \cdot 7)} = (6 \cdot 21)$ $\div\ (2 \cdot 7)$. So, using parentheses, press [(] [6] [×] [2] [1] [)] [÷] [(] [2] [×] [7] [)] [=] and read **9** on the display.
   b. Using storage, press [2] [×] [7] [=] [sto] [6] [×] [2] [1] [÷] [rcl] [=] and read **9** on the display.
   c. Since $\dfrac{6 \cdot 21}{2 \cdot 7}$ can also be written as $6 \cdot 21 \cdot \dfrac{1}{2} \cdot \dfrac{1}{7}$, we can perform this calculation by pressing [6] [×] [2] [1] [÷] [2] [÷] [7] [=] and read **9** on the display.

note

[6] [×] [2] [1] [÷] [2] [×] [7] [=] is equal to $6 \cdot 21 \cdot \dfrac{1}{2} \cdot 7$, or $\dfrac{6 \cdot 21 \cdot 7}{2}$, which is **441** and that is wrong.

8. $\dfrac{(8.1)(21.4)(0.71)}{(46.7)(0.082)}$

   Using storage, press [4] [6] [•] [7] [×] [•] [0] [8] [2] [=] [sto] [8] [•] [1] [×] [2] [1] [•] [4] [×] [•] [7] [1] [÷] [rcl] [=] and read **32.13856** on the display.

When performing mixed operation calculations, the user must experiment to determine which of the above-mentioned procedures the calculator is capable of following.

Exercise Set 4.6

Perform the indicated operations on a calculator.

1. $4 + 5 \cdot 6$
2. $5 - 3 \cdot 5$
•3. $6.2 + 7.8 \cdot 4.3$
4. $6.93 - 2.07 \cdot 3.91$
5. $5 \cdot 7 + 6$
6. $8.3 \cdot 7.06 - 5.7$
•7. $\dfrac{62}{14 \cdot 73}$
8. $\dfrac{5.07}{3.16 \cdot 9.8}$
9. $5.9 \div 4.6 \cdot 7$
10. $0.07 \div 3.71 \cdot 42.9$
11. $\dfrac{15 + 14}{36}$
12. $\dfrac{8.9 + 7.8}{1.4}$
•13. $\dfrac{18.36 - 14.2}{10.3}$
14. $7 \cdot 8 \div 4 - 3 + 5$
15. $6.04 \div 4.0 \cdot 7.62 + 0.08 - 2.11$
16. $(1.2 + 6.3) \cdot 4.2$
•17. $(5.02 - 2.13) \div 6.2$
18. $8 \cdot 7 - 9 \cdot 5$
•19. $(8.07 - 2.76)(3.3 + 4.7)$
20. $7 \cdot (2 + 6) \div 4$
21. $4.167 \cdot 8.502 + 6.01 \cdot 5.763$
22. $\dfrac{5.7 + 6.8}{3.1}$
23. $\dfrac{8.073 - 6.71}{2.04}$
24. $\dfrac{7.3 - 9.6}{8.5 + 1.67}$
25. $\dfrac{46}{4 + 123}$
26. $\dfrac{8}{15 - 42}$
•27. $\dfrac{6.73}{2.4 + 7 + 6.3}$
28. $\dfrac{7 \cdot 32}{6 \cdot 5}$
29. $\dfrac{(-7.3) \cdot (-1.43)}{(-2.6)}$ (observe the sign of the result)
30. $\dfrac{(72.7)(0.73)}{(-5.72)(14.3)}$ (observe the sign of the result)
31. $\left(\dfrac{4.076}{8.1}\right) - 5.89 + 8.06 - 0.0028$
32. $\dfrac{(7.9)(16.84)(15.023)}{(18.62)(8.793)}$
33. $\left(\dfrac{18.7}{16.2}\right)\left(\dfrac{10.1}{3.6}\right) + 22.097 - 8.735$
34. If we key $6 + 4 \div 3 - 7$ into a calculator which obeys the hierarchy of operations, which of the following problems is being solved?

    a. $\dfrac{6 + 4}{3 - 7}$ b. $\dfrac{6 + 4}{3} - 7$ c. $6 + \dfrac{4}{3} - 7$ d. $6 + \dfrac{4}{3 - 7}$

    Suppose the calculator in this problem can do only one operation at a time. Which of the problems would it solve?

35. Analyze the problem $8 \div 3 + 1 - 2 \div 7$ in the same way as in exercise 34. Consider

a. $\dfrac{8}{3 + 1} - \dfrac{2}{7}$ b. $\dfrac{8}{3} + 1 - \dfrac{2}{7}$ c. $\dfrac{8}{3} + \dfrac{1 - 2}{7}$

d. $\dfrac{\frac{8}{3} + 1 - 2}{7}$

Which of these problems is being solved using the hierarchy? Which is being solved doing one operation at a time?

36. The sequence [3] [+] [4] [×] [5] [=] may be used to solve one of the following problems. Which one?

a. $3 + 4 \cdot 5$ b. $(3 + 4) \cdot 5$

37. The sequence [5] [+] [6] [÷] [2] [+] [1] [3] [=] may be used to solve one of the following problems. Which one?

a. $(5 + 6) \div (2 + 13)$ b. $5 + (6 \div 2) + 13$

c. $(5 + 6) \div 2 + 13$

38. The area of a trapezoid is given by $A = \dfrac{1}{2} h(b + c)$, where h is the height of the trapezoid, and b and c are the parallel sides (bases). Using a calculator, find A when $h = 4.6$, and the bases are $b = 7.6$ and $c = 16.2$.

39. The total area of a surface of a right circular cylinder with closed ends is given by $A = \pi r^2(h + 2)$. Find the area when r, the radius of the base, is 1.02 in and h, the height of the cylinder, is 5.6 in.

40. Given the formula $E = 1200 \cdot R \cdot (t_1 - t_2)$, find E when $R = 0.000021$, $t_1 = 212$, and $t_2 = 32$.

•41. The area of the surface of a washer (ring) is given by $A = A_1 - A_2$, where $A_1 = \pi r_1^2$ and $A_2 = \pi r_2^2$, A_1 being the area of the outside circle with radius r_1, and A_2 is the area of the inside (hollow) circle with radius r_2. Find the area when $r_1 = 5.6$ cm and $r_2 = 2.07$ cm.

42. The volume of a cylindrical pipe is given by $V = V_1 - V_2$, where $V_1 = \pi r_1^2 h$ and $V_2 = \pi r_2^2 h$. Find V when $r_1 = 6$ cm, $r_2 = 4.2$ cm, and $h = 15$ cm.

43. A diagonal of a square is the line joining two opposite vertices (corners) of the square. The length of the side of a square, s, in relationship to the length of the diagonal, d, is given by $s = \dfrac{d\sqrt{2}}{2}$. Find s when $d = 22.4$ mm.

44. The area of an equilateral triangle, a triangle having three equal sides, is given by $A = \dfrac{h^2\sqrt{3}}{3}$. Find the area, A, when the altitude, h, is 5.1 in.

45. The resistance in a parallel circuit, R_t, is given by $R_t = \dfrac{R_1R_2}{R_1 + R_2}$, where R_1 and R_2 are the resistances in ohms of parallel lines. Find R_t when $R_1 = 1.2$ ohms and $R_2 = 2.6$ ohms.

4.8 Scientific Notation

In technical fields, very large and very small numbers are often used. To work with such numbers on the calculator, we must often enter them in scientific notation form. We define the scientific notation form of a number N to be the product

$$a \times 10^n$$

a is a number greater than or equal to 1 and less than 10; n is an integer. To achieve this form of N, we use the following procedure:

1. Move the decimal point to a position immediately following the first nonzero digit in N.
2. Count the number of places the decimal point has been moved. This is the power n to which 10 is raised.
3. If (a) the decimal point is moved to the *left*, n is *positive;* if (b) the decimal point is moved to the *right*, n is *negative*.

Examples

Express the following in scientific notation.

1. 250

   $250 = 2.50. \times 10^2 = 2.5 \times 10^2$
2. 45,000,000

   $45{,}000{,}000 = 4.5000000. \times 10^7 = 4.5 \times 10^7$
3. 5

   $5 = 5 \times 10^0$ (Note: $10^0 = 1$)
4. 0.000152

   $0.000152 = 0.0001.52 \times 10^{-4} = 1.52 \times 10^{-4}$

---

note

To write a negative number in scientific notation, use the same procedure as for a positive number, except place a negative sign, $-$, in front of a.

Example

To write -0.0234 in scientific notation, we proceed as follows:

$-0.0234 = -0.02.34 \times 10^{-2} = -2.34 \times 10^{-2}$

Sometimes it is necessary to convert a number in scientific notation to its standard form. To do this, we apply the rules in reverse. Therefore,

1. When the power of ten is *positive,* the decimal point is moved to the *right* n places.
2. When the power of ten is *negative,* the decimal point is moved to the *left* n places.

Examples

Express the following in standard form.

1. 1.45×10^4

   Since the exponent of 10 is positive 4, we move the decimal point 4 places to the *right* to get

   $$1.45 \times 10^4 = 1.4500 = 14{,}500$$

2. 5.23×10^{-3}

   The *negative* exponent, -3, tells us to move the decimal point 3 places to the *left* to get

   $$5.23 \times 10^{-3} = 0.005.23 = 0.00523$$

note

In examples 1 and 2, it was necessary to add zeros to locate the decimal point in the proper position.

3. -4.07×10^{-2}

   With a negative exponent, -2, we move the decimal point 2 places to the left to get

   $$-4.07 \times 10^{-2} = -0.04.07 = -0.0407$$

note

The negative sign preceding the number is just carried along into the standard form.

Exercise Set 4.7

Express the following in scientific notation.

1. 255	2. 65,000,000	•3. 12,345
4. 14,800	5. 155,000	6. 14.36
7. 855.076	8. 1,570.7	9. 1,007,600
10. 6,000,736	11. 0.00012	12. 0.0863
13. 0.0000081	14. 0.0000147	15. 0.0007
16. 0.12079	17. 0.000000000094	18. -456
19. $-4{,}500$	20. -0.00087	21. $-5{,}850{,}000$
22. -0.0567	23. -45.78	24. $-34{,}000{,}000$
•25. -0.00000002985		

Convert the following numbers in scientific notation to their standard form.

26. 2.07×10^3	•27. 4.99×10^7	28. 5.061×10^5
29. 7.23×10^0	30. 1.073×10^4	31. 4.2×10^{-3}
32. 7.611×10^{-7}	33. 1.47×10^{-6}	34. 5.0×10^{-2}
•35. 7.89×10^{-4}	36. -2.3×10^5	37. -4.82×10^{-9}
38. -2.61×10^2	39. -4.92×10^{-6}	40. -9.3×10^8

As previously indicated, on the calculator the EE key is used to write a number in scientific notation. In order to use this key it is necessary, however, to know the power of ten required to write the number in this form.

Examples

Express the following numbers in scientific notation on the calculator.

1. 527
   Since $527 = 5.27 \times 10^2$, press [5] [•] [2] [7] [EE] [2] and read **5.27 02** on the display.
2. 65,000,000
   Since $65{,}000{,}000 = 6.5 \times 10^7$, press [6] [•] [5] [EE] [7] and read **6.5 07** on the display.
3. 0.000723
   Since $0.000723 = 7.23 \times 10^{-4}$, press [7] [•] [2] [3] [EE] [4] [+/−] and read **7.23 −04** on the display.
4. −0.00456
   Since $-0.00456 = -4.56 \times 10^{-3}$, press [4] [•] [5] [6] [+/−] [EE] [3] [+/−] to get **−4.56 −03** on the display.

Exercise Set 4.8

Express the following numbers in scientific notation on the calculator.

1. 5,076
2. 23
•3. 4,700,000
4. 6,720
5. 8,123,000
6. 0.00147
7. 0.763
8. 0.0000901
•9. 0.00023
10. 0.06893
11. −56,000
12. −0.00000732

Express the following values in scientific notation on the calculator.

13. The mass of the moon is 8,060,000,000,000,000,000,000 tons.
14. General Motors reported a gross earnings of $7,230,000,000 in the year 1979.
15. 1 cu in equals 0.0005787 cu ft.
16. 43,560 sq ft equals 1 acre.
17. 1 g equals 0.0022046 lb.
18. 46,656 cu in equals 1 cu yd.
19. 1 kg equals 0.001102 ton.
20. A particular gauge of copper wire has a resistance of 0.00006203 ohms/lb.
21. A light-year, the distance light travels in 1 year, is 5,872,000,000 mi.
22. 3,785.3 milliliters equals 1 gal.

When computing with very large or very small measurements on the calculator, it is sometimes necessary to use scientific notation. First, we must express the number in scientific notation on the display, and then we multiply or divide as necessary. If the calculator is unable to express the number in standard form because the number of places is too great, the result will be stated in scientific notation on the display.

Examples

Perform the indicated operations on the calculator using scientific notation.

1. 52,000,000 × 45
   Press [5] [•] [2] [EE] [7] [×] [4] [5] [=] and read **2.34 09** on the display.
2. 0.00000000763 × 103
   Press [7] [•] [6] [3] [EE] [9] [+/−] [×] [1] [0] [3] [=] and read **7.8589 −07** on the display.
3. 62,004,000 × 5,800
   Press [6] [•] [2] [0] [0] [4] [EE] [7]
   [×] [5] [•] [8] [EE] [3] [=] and read **3.596232 11** on the display.
4. 5,700 ÷ 0.00083
   Press [5] [•] [7] [EE] [3] [÷] [8] [•] [3] [EE] [4] [+/−] [=]
   and read **6.867,4699 06** on the display.
5. 0.000123 ÷ 0.0813
   Press [1] [•] [2] [3] [EE] [4] [+/−]
   [÷] [8] [•] [1] [3] [EE] [2] [+/−] [=] and read
   1.5129151 −03 on the display.

Exercise Set 4.9

Multiply or divide the following using scientific notation.

1. $5.23 \times 10^9 \times 1.073 \times 10^6$
2. $5.12 \times 10^6 \times 6.2 \times 10^4$
3. $8.473 \times 10^3 \times 8.4 \times 10^{-4}$
4. $1.673 \times 10^{-4} \times 7.5 \times 10^{-6}$
5. $6.23 \times 10^2 \div 5.73 \times 10^4$
6. $1.47 \times 10^3 \div 8.03 \times 10^1$
7. •$7.82 \times 10^{-2} \div 5.6 \times 10^5$
8. $7.23 \times 10^5 \div 6.075 \times 10^{-9}$
9. 0.876 × 21.46
10. 0.000476 × 0.0053
11. 0.000000089 × 0.145
12. 473 × 0.0000579
13. •0.00053 × 0.634 × 0.0000000146
14. 243 × 0.00073 ÷ 65,000
15. 6,184,000 ÷ 0.0079 × 52.1
16. The amount of stress on a piece of metal is given by

$$\text{stress} = \frac{\text{force producing the stress}}{\text{area of the surface}}$$

    Find the stress on a piece of metal where the force is 452.6 kg and the area is 0.000763 sq cm.
17. If the volume of a right circular cylinder is given by $V = \pi r^2 h$, find the volume of a cylinder that has radius $r = 0.0073$ m and height $h = 12$ m.
18. A unit of light intensity is a lumen. If 1 lumen equals 0.001496 watts, how many lumens are there in 4,760,000 watts?

4.9 Estimating

It is often useful to be able to estimate the result of a particular computation on the calculator. Estimating can be done using basic rules for rounding off numbers. Note, however, that the precision of the data that is to be computed will greatly affect the precision of the result.

When rounding off a number to a given decimal place, we use the following rules. Look at the digit immediately to the right of the digit to be rounded off:

1. If it is less than 5, drop it and all following digits. The digit to be rounded *remains the same*.
2. If it is equal to or greater than 5, drop it and all following digits. *Increase by one* the digit to be rounded.

There are two reasons why we might want to be able to estimate a result. First, we can compare the approximation of the value of a computation with the actual result to check the legitimacy of the latter.

note

Simple errors are often made on the calculator by pressing the wrong button.

Second, an estimation helps us place the decimal point in the resulting product.

For example, to estimate the product of the following:

$3.4 \cdot 4.6 \cdot 2.5$

we first round off to the nearest unit to get

$3 \cdot 5 \cdot 3$

A fair estimation of the product then is:

$3 \cdot 5 \cdot 3 = 45$

note

Since an estimation yields approximate results, the symbol $\approx$, which means "is approximately equal to," is used.

Thus, we know that

1. $3.4 \cdot 4.6 \cdot 2.5 \approx 45$.
2. The accurate result will have *two digits* to the left of the decimal point and the remaining digits to the right.

After calculation we find that the actual result is $3.4 \cdot 4.6 \cdot 2.5 = 39.1$, and since there are two digits to the left of the decimal point, we can feel reasonably certain that the computation is correct.

To estimate the indicated product

$4.03 \cdot 53.6 \cdot 8.563$

we make two such maneuvers. First, rounding off to the nearest unit,

$4.03 \cdot 53.6 \cdot 8.563 \approx 4 \cdot 54 \cdot 9$

Since

$4 \cdot 54 \cdot 9 = 36 \cdot 54$

rounding off to the nearest ten we get

$36 \cdot 54 \approx 40 \cdot 50 = 2{,}000$

and so we estimate

$4.03 \cdot 53.6 \cdot 8.563 \approx 2{,}000$

Using the calculator, we obtain $4.03 \cdot 53.6 \cdot 8.563 = 1{,}849.6765$, which is reasonable by the estimation. The estimation has determined that the result of the computation will contain four digits to the left of the decimal point and that the product will be in thousands.

Finally, to combine the operations of multiplication and division, as technicians are often asked to do, consider the problem

$$\frac{0.734 \cdot 4.213 \cdot 50.4}{2.73 \cdot 5.06}$$

Rounding off to the nearest unit we get

$$\frac{0.734 \cdot 4.213 \cdot 50.4}{2.73 \cdot 5.06} \approx \frac{1 \cdot 4 \cdot 50}{3 \cdot 5} = \frac{200}{15}$$

and since

$$\frac{200}{15} \approx \frac{200}{20} = 10,$$

the result, by implication, will be in tens. That is, there will be two digits to the left of the decimal point. Using the calculator,

$$\frac{0.734 \cdot 4.213 \cdot 50.4}{2.73 \cdot 5.06} = 11.282483$$

which appears reasonable.

In general, to estimate a given computation use the following steps:

1. Round each factor to the nearest unit if the numbers lie between 0 and 100.
2. If the resulting estimation involves computation that one cannot mentally perform, round to the nearest ten and proceed as before.
3. Continue this procedure to hundreds, thousands, and so on, as necessary.

Examples

Estimate the following products and quotients.

1. $$\begin{aligned} 3.07 \cdot 5{,}232 \cdot 62.6 &\approx 3 \cdot 5{,}000 \cdot 63 \\ &\approx 3 \cdot 60 \cdot 5{,}000 \\ &\approx 180 \cdot 5{,}000 \\ &\approx 200 \cdot 5{,}000 \\ &\approx 1{,}000{,}000 \end{aligned}$$

2. $$\begin{aligned} \frac{54.3 \cdot 72.6}{576} &\approx \frac{50 \cdot 70}{600} \\ &\approx \frac{3{,}500}{600} \\ &\approx \frac{35}{6} \\ &\approx 6 \end{aligned}$$

note

To get from step 2 to step 3 in example 2, divide the numerator and denominator by 100.

Exercise Set 4.10

Estimate the answers in the following indicated operations. (The symbol ≈ should be used.)

•1. $5.8 \cdot 6.92$

2. $3.9 \cdot 5.67 \cdot 2.081$

3. $42.1 \cdot 1.763 \cdot 2.914$

4. $52.7 \cdot 7.5 \cdot 8.0$

•5. $123.6 \cdot 7.43 \cdot 9.92 \cdot 1.567$

6. $\dfrac{83.6}{7.076}$

7. $\dfrac{603.7}{25.16}$

8. $\dfrac{10.085}{1.893}$

9. $\dfrac{5.79 \cdot 6.43}{9.25}$

10. $\dfrac{7.89 \cdot 6.1}{11.83}$

•11. $\dfrac{7.9 \cdot 8.0 \cdot 6.22}{4.374 \cdot 2.56}$

12. $\dfrac{9.6 \cdot 15.06 \cdot 3.1}{5.36 \cdot 2.934}$

13. $\dfrac{1.06 \cdot 3.112 \cdot 5.9 \cdot 6.56}{14.02 \cdot 2.91}$

14. $\dfrac{4.1 \cdot 3.7 \cdot 5.6 \cdot 9.7}{15.2 \cdot 2.77 \cdot 3.9}$

15. $\dfrac{8.6 \cdot 7.5 \cdot 2.1 \cdot 1.3 \cdot 4.5}{9.3 \cdot 8.5 \cdot 5.03}$

16. $\dfrac{24.7 \cdot 30.8 \cdot 504.3}{54.326 \cdot 78.6}$

•17. $\dfrac{0.793 \cdot 476.23 \cdot 78.92 \cdot 6.43}{1.96 \cdot 295.4 \cdot 65.493}$

4.10 RPN Calculators

RPN stands for "reverse Polish notation." The "Polish" is a reference to the great Polish mathematician Jan Lukasiewicz (wū-ka-shā′-vich), whose initial work and suggestions led Hewlett-Packard Company to the creation and patent of this type of calculator. Hewlett-Packard does not have the only calculator that uses RPN, and they were probably not the first to use the stacking and RPN data entry system. In the mid-1960s, both Friden and Monroe had calculators using this type of data entry system. Currently, five companies use the RPN system on their calculators.

In 1951 Lukasiewicz, in a book on logic, recommended a parenthesis-free algebraic notation in which the operation symbols, such as +, −, ÷, ·, preceded the numbers on which operations are to be performed. For example, to multiply 5 and 6 in Lukasiewicz' notation the order is · 5 6. Placing the operation symbol after the numbers (5 6 ·) gives reverse Polish notation (RPN).

Most calculators on the market today use some form of an **AES** (algebraic entry system) where the operation symbol comes between the two numbers on which the operation is to be performed. We used this type of system in the preceding sections of this chapter. Now we examine some of the differences between AES and RPN and discuss how to use an RPN calculator.

The two main differences between AES and RPN calculators lie in the areas of data entry and data storage. First, we examine the differences in data entry.

Consider the problem 5 · 6 − 4. In an AES calculator we press: [5] [×] [6] [−] [4] [=] and read **26** on the display. The operation symbol comes between the numbers on which the operations are performed. In an RPN calculator we press: [5] [Enter↑] [6] [×] [4] [−] and read **26** on the display. The operation symbol comes after the numbers on which the operations are performed, and the [Enter↑] key is needed to insert the first of a pair of numbers into the calculator.

The second major difference between the two calculators is in the internal storage of the data on which the operations are to be performed. In an AES calculator there are generally two internal storage registers, but there are three registers if the calculator has operational hierarchy and four internal registers if the calculator has two-level parentheses. The RPN calculator has four internal registers for data, although some of the less expensive models may have only three.

To demonstrate the usefulness of the internal data registers, consider the following problem: $20 - 3 \cdot 4$. On an AES calculator, if we press [2] [0] [−] [3] [×] [4] [=], we get **68** unless the calculator has operational hierarchy or parentheses. Recall, however, that multiplication is performed before addition; therefore, the correct answer should have been **8.** On an RPN calculator the problem would be entered as follows: [2] [0] [Enter↑] [3] [Enter↑] [4] [×] [−] and read on the display as **8.** An RPN calculator uses a stacking process to hold data in the internal registers until the operation is performed, and in many cases, a complex problem can be handled without having the user store or write down intermediate answers that will be used later.

Note that this discussion is in no way intended to be complete. It is intended, instead, to highlight the major differences between AES and RPN calculators.

We shall now look at how to solve problems using an RPN calculator. We assume the student is familiar with the calculator's on/off switch, display, and the location of all the keys on the calculator. Now examine the uses of each of these keys.

1. Basic Function Keys
   a. The **add key,** [+], is used to add the two previous entries or calculated results. To add 7 and 9, press [7] [Enter↑] [9] [+] and read **16** on the display.
   b. The **subtract key,** [−], subtracts the two previous entries or calculated results. To subtract 18 from 27, press [2] [7] [Enter↑] [1] [8] [−] and read **9** on the display.
   c. The **multiply key,** [×], is used to multiply the two previous entries or calculated results. To multiply 15 and 11, press [1] [5] [Enter↑] [1] [1] [×] and read **165** on the display.
   d. The **divide key,** [÷], is used to divide the two previous entries or calculated results. To divide 140 by 16, press [1] [4] [0] [Enter↑] [1] [6] [÷] and read **8.75** on the display.

Exercise Set 4.11

Work the problems in exercise sets 4.1, 4.2, and 4.3.

2. Special Function Keys
   a. The **reciprocal key,** [$1/x$], is used to find the reciprocal of the number that is displayed. To find the reciprocal of 0.2, press [•] [2] [$1/x$] and read **5** on the display.
   b. The **square key,** [x^2], is used to find the square of the number that is displayed. To find the square of 6.5, press [6] [•] [5] [x^2] and read **42.25** on the display.

c. The **square root key,** [$\sqrt{x}$], is used to find the square root of the number that is displayed. To find the square root of 81, press [8] [1] [$\sqrt{x}$] and read **9** on the display.

d. The ***y* to the power *x* key,** [y^x], instructs the calculator to raise a number to a given power. To find 3 raised to the fourth power, press [3] [Enter ↑] [4] [y^x] and read **81** on the display.

e. The ***x*th root of *y* key,** [$\sqrt[x]{y}$], is used to find the *x*th root of a number. To find the fifth root of 243, press [2] [4] [3] [Enter ↑] [5] [$\sqrt[x]{y}$] and read **3** on the display.

Exercise Set 4.12

Work the problems in exercise set 4.4.

f. The **percent key,** [%] , is used to compute percentages. To find 7.5% interest on $1,200, press [1] [2] [0] [0] [Enter ↑] [7] [•] [5] [%] and read **90** on the display.

Exercise Set 4.13

Work the problems in exercise set 4.5.

3. Mixed Operations
The following are a few examples to demonstrate how more complex problems are handled with an RPN calculator.

a. $(9 + 7) \cdot 15$
Press [9] [Enter ↑] [7] [+] [1] [5] [×] and read **240** on the display.

b. $\dfrac{8 + 4}{10}$
Press [8] [Enter ↑] [4] [+] [1] [0] [÷] and read **1.2** on the display.

c. $(5 + 7) \cdot (6 + 3)$
Press [5] [Enter ↑] [7] [+] [6] [Enter ↑] [3] [+] [×] and read **108** on the display.

d. $\dfrac{(6 \cdot 5) + (3 \cdot 4)}{(5 \cdot 7)}$
Press [6] [Enter ↑] [5] [×] [3] [Enter ↑] [4] [×] [+] [5] [Enter ↑] [7] [×] [÷] and read **1.2** on the display.

e. $\sqrt{\dfrac{(18 \cdot 7) + 18}{6 + 10}}$
Press [1] [8] [Enter ↑] [7] [×] [1] [8] [+] [6] [Enter ↑] [1] [0] [+] [÷] [$\sqrt{x}$] and read **3** on the display.

Exercise Set 4.14

Work the problems in exercise set 4.6.

Chapter Summary

1. The results of all calculations on the calculator are entered on the **display,** or **read out.**
2. The **control keys** consist of the
   a. clear key [C]
   b. clear entry key [CD] or [CE]
3. The **data entry keys** consist of the
   a. digit keys [0], [1], [2], ⋯, [9]
   b. decimal point key [•]
   c. change sign key [+/−]
   d. enter exponent key [EE]
4. The **basic function keys** consist of the
   a. arithmetic keys [+], [−], [×], [÷]
   b. equals key [=]
5. The **special function keys** consist of the
   a. reciprocal key [$1/x$]
   b. square key [x^2]
   c. square root key [$\sqrt{x}$]
   d. y to the power x key [y^x]
   e. percent key [%]
   f. xth root of y [$\sqrt[x]{y}$]
6. The capabilities of the calculator must be noted by the user when performing mixed operations—does it recognize, for example, the hierarchy of operations or not?
7. A number written in scientific notation is of the form

   $$a \times 10^n$$

   where a is the number 1 or a number between 1 and 10, and n is an integer.
8. The **enter exponent key,** [EE], is used in writing a number in scientific notation on the calculator.
9. **Estimating** the result of a computation helps determine the legitimacy of results of the manipulations performed and helps place the decimal point.

Chapter Review

[4.5] Perform the indicated operations on a calculator.

1. $3.4 \cdot 1.6 \div 8.4 - 3.1$
2. $8.4 \div 9.3 - 6.6 + 6.9$
3. $\dfrac{45.3 - 15.8}{2.6 + 7.8}$
4. $\dfrac{(-4.35)(-2.7)}{(5.8)(-1.3)}$
5. $\dfrac{4.78 + 8.12 - 0.89}{4.67 - 1.98}$
6. $\dfrac{0.0084}{34} - 0.058 + 5.92$

[4.6] Find the following powers and roots.

7. $(45.8)^2$
8. $(0.00567)^3$
9. $(45{,}000)^4$
10. $\sqrt{46.9}$
11. $\sqrt{0.0234}$

[4.6] Calculate the following.

12. 34.5% of 789
13. 1.07% of 9.8
14. 243% of 109.5
15. \$890 plus 4.5% tax on \$890
16. \$1,050 less 3.8% of \$1,050

Convert the following to percents.

17. $\frac{47}{83}$ 18. $\frac{0.56}{67}$ 19. $\frac{234}{156}$

[4.8] Perform the following calculations using scientific notation and give the answer in scientific notation.

20. $34{,}000 \times 5{,}600$
21. $2{,}800{,}000 \times 9{,}080{,}000$
22. 0.00056×0.0000078
23. $3{,}000{,}000 \times 0.0000456$
24. $\frac{45{,}000}{0.00123}$
25. $\frac{0.0000709}{4{,}500}$

[4.9] Estimate the following products and quotients.

26. $4.56 \cdot 8.12 \cdot 89.1 \cdot 34.67$
27. $458 \cdot 802.1 \cdot 45.82$
28. $\frac{406.7 \cdot 25.9}{2.34 \cdot 502}$
29. $\frac{63.1 \cdot 0.89}{492.5 \cdot 1.49}$
30. $\frac{0.78 \cdot 36.5 \cdot 9.92}{319.3 \cdot 76.7}$

Chapter Test

Perform the indicated operations using a calculator.

1. $5.4 \cdot 3.06 \cdot 52.1$
2. $\frac{5.06 \cdot 31.4}{4.7}$
3. $\frac{5.17 + 0.367}{15 - 20.2}$

4. $\sqrt{47.6}$
5. $\sqrt{0.236}$
6. $(72.43)^2$
7. $(0.0073)^3$
8. $52 + 3.6 \div 4.72 - 6$
9. $(73.7 + 8.4) \cdot (63.7 - 0.073)$
10. $0.769 \cdot 32.4 + 18 \div 13.2$
11. Convert the fraction $\frac{42}{23}$ to a percent.
12. Calculate 15.6% of 373.4.
13. Calculate \$15.34 plus 4% tax on \$15.34.

Write the following in scientific notation.

14. 56,000,000
15. 0.00732

Write the following in standard form.

16. 1.43×10^3
17. 5.63×10^{-4}

Estimate the following products and quotients.

18. $5.7 \cdot 10.5 \cdot 7.8 \cdot 1.2$
19. $\frac{42.3 \cdot 6.5}{18.6 \cdot 5.0}$
20. Evaluate $0.000534 \cdot 0.000000124$ on your calculator, using scientific notation.

5

Measurement and Approximate Numbers

A technician turning a diameter on a South Bend lathe.
(Photograph by Kenneth Walters, courtesy of Henry Ford Community College)

5.1 Basic Units of Measurement

There are seven basic units of measure: length, mass or weight, time, temperature, electric current, light intensity, and molecular substance. Mass is measured by the quantity of matter in a body, as opposed to weight, which is measured by the pull of gravity on the body. The weight of an object differs according to the distance the object is from the earth. At great distances from the earth, for example, the weight of an object will be close to zero. The weight of an object also changes depending on its location on earth—an object will weigh slightly more at sea level, for instance, than at the top of a mountain. The mass of an object does not change regardless of its location since mass is independent of gravity.

All other measurements are expressed in terms of these basic units or on units derived from them. Some examples of derived units of measurement are:

1. Volume (in cubic units)—as a product of length × length × length.
2. Velocity (for example, feet per second)—as the quotient of length divided by unit of time, or length per unit of time.
3. Density (for example, kilogram per cubic meter)—obtained by dividing the unit of mass by the cube of a length unit.

5.2 British Units of Measurement

There are two basic systems of measure: the **International System of Units** (SI, or metric) and the **British system.** The United States currently uses the British system, but most major industrial firms will have completed a voluntary conversion to the International system by the mid-1980s. The standard units of length in the British system are the *inch, foot, yard,* and *mile*. The basic unit of weight in this system is the *pound*. Other units of weight are the *ounce* and *ton*. The basic unit of time is the *second*. The British system is often called the *foot-pound-second* (FPS) system of measurement, where the *foot* is the standard unit of length. In table 5.1, we have a partial table of relations within the British system of measure.

Numbers that express units of measurement are called **denominate** numbers. In using such numbers it is often necessary to change from one unit of measurement to another. When this change is performed within a given system of measurement, such as SI to SI or British to British, we call the procedure **reduction.** When the change is from one system of measurement to a different system of measurement, we call the procedure **conversion.** We shall first look at reduction in the British system using table 5.1.

Where the relation between two units is known, it is easy to reduce from one unit of measurement to its equivalent. To illustrate, we can easily reduce a value of 4 yd to the equivalent value of 12 ft when we know there are 3 feet per yard. However, reducing complex combinations of units to equivalent values requires more manipulation. We need a procedure for making this reduction because there are many units and combinations of units in common technical use.

The procedure for reduction involves multiplication of the original number and units of measure with related units in fractional form until the desired value and unit is obtained. The units are treated as quantities and are "cancelled out" until all that

Table 5.1 British System of Measures

Units of Length

12 inches	=	1 foot
3 feet	=	1 yard
$16\frac{1}{2}$ feet	=	1 rod
5,280 feet	=	1 mile

Units of Area

144 in^2	=	1 ft^2
9 ft^2	=	1 yd^2
160 rd^2	=	1 acre
640 acres	=	1 mi^2

Dry Measure

2 pints	=	1 quart
8 quarts	=	1 peck
4 pecks	=	1 bushel

Energy

1 BTU	=	778 foot-pound
1 kilowatt-hour	=	3,413 BTU
1 BTU	=	252 calories

Power

1 horsepower (hp)	=	550 ft-lb/s
1 horsepower (hp)	=	0.75 kilowatt
1 kilowatt	=	3,413 BTU/h

Units of Mass or Weight

16 ounces	=	1 pound
2,000 pounds	=	1 short ton
2,240 pounds	=	1 long ton

Units of Volume

1,728 in^3	=	1 ft^3
27 ft^3	=	1 yd^3

Liquid Measure

2 cups	=	1 pint
2 pints	=	1 quart
4 quarts	=	1 gallon
$31\frac{1}{2}$ gallons	=	1 barrel
231 in^3	=	1 gallon

Symbolism

Unit	*Symbol*
inch	in
foot	ft
yard	yd
rod	rd
mile	mi
cup	c
pint	pt
quart	qt
gallon	gal
peck	pk
bushel	bu
ounce	oz
pound	lb
short ton	t
long ton	T
hour	h
minute	min
second	s
week	wk
year	yr

remains is the unit that is desired. When stating a number and units in fractional (or ratio) form, it is important to be aware that any single relationship can be stated in two different ways. To illustrate, consider the relationship from table 5.1.

$$16 \text{ oz} = 1 \text{ lb}$$

If we wish to state this as "ounces per pound," we get the fraction

$$\frac{16 \text{ oz}}{1 \text{ lb}} \qquad \textit{Note: } \frac{16 \text{ oz}}{1 \text{ lb}} = 1$$

If, however, we need the relationship for "pounds per ounce," we state it as

$$\frac{1 \text{ lb}}{16 \text{ oz}} \qquad \textit{Note: } \frac{1 \text{ lb}}{16 \text{ oz}} = 1$$

The form that we use in a reduction depends on the units that are desired and the units given.

Let us consider the following example. To reduce 4 yd to inches we use the relationships 3 ft = 1 yd and 12 in = 1 ft and multiply as follows:

$$\left(\frac{4 \cancel{\text{yd}}}{1}\right)\left(\frac{3 \cancel{\text{ft}}}{1 \cancel{\text{yd}}}\right)\left(\frac{12 \text{ in}}{1 \cancel{\text{ft}}}\right)$$

The yard and foot units have been "cancelled out" leaving only the inch unit, which was desired. Thus,

$$\begin{aligned} 4 \text{ yd} &= 4 \cdot 3 \cdot 12 \text{ in} \\ &= 144 \text{ in} \end{aligned}$$

The units of each ratio are selected so that at least one preceding unit is cancelled.

Examples

1. Reduce 16,000 oz to tons.

Solution

a. Use 1 lb = 16 oz
   1 ton = 2,000 lb

b. $\left(\frac{16{,}000 \cancel{\text{oz}}}{1}\right)\left(\frac{1 \cancel{\text{lb}}}{16 \cancel{\text{oz}}}\right)\left(\frac{1 \text{ ton}}{2{,}000 \cancel{\text{lb}}}\right)$

c. $16{,}000 \text{ oz} = \left(\frac{16{,}000}{1}\right)\left(\frac{1}{16}\right)\left(\frac{1}{2{,}000}\right) \text{ tons}$

$$= \frac{1}{2} \text{ ton}$$

2. Reduce 6 gal to pints.

Solution

a. Use 1 gal = 4 qt
   1 qt = 2 pt

b. $\left(\frac{6 \cancel{\text{gal}}}{1}\right)\left(\frac{4 \cancel{\text{qt}}}{1 \cancel{\text{gal}}}\right)\left(\frac{2 \text{ pt}}{1 \cancel{\text{qt}}}\right)$

c. $\begin{aligned} 6 \text{ gal} &= 6 \cdot 4 \cdot 2 \text{ pt} \\ &= 48 \text{ pt} \end{aligned}$

3. Reduce 55 mi/h to feet per second.

Solution

a. Use 1 mi = 5,280 ft
1 h = 60 min
1 min = 60 s

b. $\left(\frac{55 \text{ mi}}{1 \text{ h}}\right)\left(\frac{5{,}280 \text{ ft}}{1 \text{ mi}}\right)\left(\frac{1 \text{ h}}{60 \text{ min}}\right)\left(\frac{1 \text{ min}}{60 \text{ s}}\right)$

c. $$55 \frac{\text{mi}}{\text{h}} = (55)\,(5{,}280)\left(\frac{1}{60}\right)\left(\frac{1}{60}\right)\frac{\text{ft}}{\text{s}}$$
$$= \frac{242}{3}\frac{\text{ft}}{\text{s}}$$
$$= 80\frac{2}{3}\frac{\text{ft}}{\text{s}}$$

4. Reduce 2 yd^3 (read "cubic yards") to in^3 (read "cubic inches").

Solution

a. Use 1 yd = 36 in

b. $$\left(\frac{2 \text{ yd}^3}{1}\right)\left(\frac{36 \text{ in}}{1 \text{ yd}}\right)^3 = \frac{2 \text{ yd}^3}{1} \cdot \frac{(36)^3 \text{ in}^3}{1 \text{ yd}^3}$$

note

Since we are dealing with cubic yards, we use the relationship between yards and inches raised to the third power.

c. $$2 \text{ yd}^3 = 2 \cdot (36)^3 \text{ in}^3$$
$$= 2 \cdot 46{,}656 \text{ in}^3$$
$$= 93{,}312 \text{ in}^3$$

5. Reduce 20 lb/yd^2 to ounces per square foot.

Solution

a. Use 1 yd = 3 ft
1 lb = 16 oz

b. $$\left(\frac{20 \text{ lb}}{1 \text{ yd}^2}\right)\left(\frac{16 \text{ oz}}{1 \text{ lb}}\right)\left(\frac{1 \text{ yd}}{3 \text{ ft}}\right)^2 = \frac{20 \text{ lb}}{1 \text{ yd}^2} \cdot \frac{16 \text{ oz}}{1 \text{ lb}} \cdot \frac{1 \text{ yd}^2}{9 \text{ ft}^2}$$

c. $$20 \text{ lb/yd}^2 = (20)(16)\left(\frac{1}{9}\right) \text{ oz/ft}^2$$
$$= \frac{320}{9} \text{ oz/ft}^2$$
$$= 35\frac{5}{9} \text{ oz/ft}^2$$

note

In example 5 that in order to reduce square yards to square feet we used the ratio of the ''yards per foot'' and squared the result to get square feet.

Exercise Set 5.1

Reduce the following, using the calculator where necessary. Round the answer to three decimal places.

1. 36.7 oz to pounds
2. 4,360 lb to short tons
•3. 5.3 t (short tons) to long tons
4. 43.6 pk to bushels
5. 8.39 gal to pints
6. 8.2 bu to pints (dry measure)
•7. 386 qt to barrels
8. 9.36 yd^2 to square feet
9. $5\frac{3}{8}$ ft^2 to square inches
10. 12 mi^2 to acres
11. 4 mi to rods
12. 30 t/h to pounds per second
•13. 60 lb/ft^2 to ounces per square inch
14. 10.6 lb/in^2 to short tons per square yard
15. 18 ft^3/h to cubic yards per second (Round to 6 decimal places.)
16. Find the equivalent of 2 hp in BTU per minute.
17. A section of pipe weighs 20 oz/in of length. Reduce this to short tons per mile.
18. The density of water is about 62.4 lb/ft^3. Convert this to ounces per cubic inch.
•19. A pumping station consumes 100 kwh of energy each day, working 6 hours per day. Find the rate of energy consumption in foot-pounds per minute. (Round to the nearest hundred.)
20. A septic tank can be emptied by a drain pipe at the rate of 2.5 barrels/min. Find the rate of drain in quarts per second.
21. A grain combine can harvest the grain in a field at the rate of 5.65 acres/h. Find the rate of harvest in square rods per second.
22. A person consumes 2,000 calories per day; how many foot-pounds per hour in energy is this?
23. A jackhammer exerts 5,000 lb/in^2 when in use; how many short tons per square foot is this?
•24. A painter can paint 250 ft^2/h. What is this rate in square inches per second?

5.3 SI Units of Measurement

The standard unit of measure of length in the SI system is the *meter* and the standard unit of mass is the *kilogram*. The unit of time is the same as in the British system, the *second*. In 1960 the *International System of Units (SI)* was established using the metric system as its basis, and when we refer to the metric system we are using SI units. The metric system, like our number system, is based on the number 10; as a result, calculations and reductions are much easier with the metric system than with the British system. In the metric system we have several base units of measure that we combine with the proper prefix to form the desired unit of measure. You should observe that the same prefixes are used no matter what the base unit of measurement. We shall use table 5.2 for SI units of measure

Table 5.2 SI Units of Measure

Units of Length

10 decimeters	= 1 meter
100 centimeters	= 1 meter
1,000 millimeters	= 1 meter
1,000 meters	= 1 kilometer

Units of Mass or Weight

10 decigrams	= 1 gram
100 centigrams	= 1 gram
1,000 milligrams	= 1 gram
1,000 grams	= 1 kilogram

Liquid Measure

10 deciliters	= 1 liter
100 centiliters	= 1 liter
1,000 milliliters	= 1 liter
1,000 liters	= 1 kiloliter

Symbolism

Unit	*Symbol*
meter	m
gram	g
liter	l

Metric Prefixes

Prefix	*Symbol*	*Meaning*
mega	M	million
kilo	k	thousand
hecto	h	hundred
deka	dk	ten
deci	d	tenth
centi	c	hundredth
milli	m	thousandth
micro	μ	millionth

As with the British system, we can make necessary reductions by multiplying units, starting with the original and working to the desired unit through other related units. To illustrate, to change

5 m to centimeters

we read from the table that

100 cm = 1 m

thus we have

$$\left(\frac{5\ \not{m}}{1}\right)\left(\frac{100\text{ cm}}{1\ \not{m}}\right) = (5)(100)\text{ cm}$$

$$= 500\text{ cm}$$

Examples

1. Reduce 500 ml to kiloliters.

Solution

a. Use 1 l = 1,000 ml
   1 kl = 1,000 l

b. $$\left(\frac{500\,\cancel{\text{ml}}}{1}\right)\left(\frac{1\,\cancel{\text{l}}}{1{,}000\,\cancel{\text{ml}}}\right)\left(\frac{1\text{ kl}}{1{,}000\,\cancel{\text{l}}}\right)$$

c. $$\begin{aligned}500\text{ ml} &= (500)\left(\frac{1}{1{,}000}\right)\left(\frac{1}{1{,}000}\right)\text{ kl}\\ &= \frac{1}{2{,}000}\text{ kl}\\ &= 0.0005\text{ kl}\end{aligned}$$

2. Reduce 200 cm^3/min to cubic millimeters per second.

Solution

a. Use 1 min = 60 s
   1 m = 100 cm
   1 m = 1,000 mm

b. $$\left(\frac{200\text{ cm}^3}{1\text{ min}}\right)\left(\frac{1\text{ min}}{60\text{ s}}\right)\left(\frac{1\text{ m}}{100\text{ cm}}\right)^3\left(\frac{1{,}000\text{ mm}}{1\text{ m}}\right)^3$$

c. $$\begin{aligned}&200\text{ cm}^3/\text{min}\\ &= (200)\left(\frac{1}{60}\right)\left(\frac{1}{100}\right)^3(1{,}000)^3\text{ mm}^3/\text{s}\\ &= \frac{10{,}000}{3}\text{ mm}^3/\text{s}\\ &= 3{,}333\frac{1}{3}\text{ mm}^3/\text{s}\end{aligned}$$

note

If you are reducing in the metric system within a base unit, the following table will be of help in moving the decimal point.

↑ For each place you move up the scale, move the decimal point one place to the left.	kilo hecto deka meter, liter, gram deci centi milli	↓ For each place you move down the scale, move the decimal point one place to the right.

Example

1. Reduce 6 hg to centigrams.
   a. When moving from hectograms to centigrams, move four places in the scale (hecto to deka, deka to gram, gram to deci, and deci to centi).
   b. Move the decimal point four places to the right.
   6.0000. hg = 60,000 cg
2. Reduce 400 ml to dekaliters.
   a. Move four places up in the scale.
   b. Move the decimal point four places to the left.
   .0400. ml = 0.04 dkl

note

When the decimal point in 400 was moved four places to the left, a zero had to be inserted to mark the position of the decimal point.

Exercise Set 5.2

Reduce the following, using the calculator where necessary.

•1. 36.2 cm to meters
2. 5 km to centimeters
•3. 80,300 mg to kilograms
4. 523.6 cg to grams
5. 43.07 kg to grams
6. 0.00321 kl to liters
•7. 4,200 ml to liters
8. 42.6 cm^2 to square meters
•9. 4.92 cm^2 to square decimeters
10. 49.3 m^3 to cubic centimeters
11. 82,500 m^3 to cubic kilometers
12. 0.00314 kg/cm^2 to centigrams per square millimeter
13. 0.23 km/min to meters per second
14. 30 kg/m^2 to grams per square centimeter
•15. 925 g/cm^3 to kilograms per cubic meter
16. An auto is traveling at a rate of 42.3 km/h. What is its rate of speed in meters per minute?
17. At a meal, Sigrud consumed 632 g of meat in 5 min. How many milligrams per minute is this?
18. John can run 525 m/min. What rate is this in centimeters per second?
19. An automobile engine takes approximately 4.63 liters of oil. How many milliliters is this? How many kiloliters is this?
20. A storage tank fills at the rate of 573 cm^3/min. What is this rate in cubic meters per second?

5.4 Conversions

In the preceding sections of this chapter we learned how to reduce units of measurement to other units in the same system. Now we will learn how to change units from one system to another; this procedure is called **conversion.** To help us do conversion we use table 5.3, which gives approximate relations between some of the more important specific units in the two systems, British and metric.

Table 5.3 Approximate Relations Between the British and Metric Systems

Units of Length	*Units of Area*
1 inch = 2.54 centimeters*	1 in^2 = 6.452 cm^2
1 foot = 30.48 centimeters	1 m^2 = 10.76 ft^2
1 meter = 39.37 inches	
1 mile = 1.609 kilometers	*Measures of Weight or Mass*
1 kilometer = 0.621 mile	1 pound = 453.6 grams
	1 ounce = 28.35 grams
Measures of Volume	1 kilogram = 2.205 pounds
1 in^3 = 16.39 cm^3	
1 ft^3 = 28.32 liters	
1 pint = 473.2 cm^3	
1 liter = 1.056 quarts	

**Note:* This is exact by definition.

The procedure by which we convert from one system to another is the same as the one we used to reduce from one unit of measure to another within the same system.

Examples

1. Convert 300 gal to liters. (Round to the units.)

Solution $\left(\frac{300\ \cancel{gal}}{1}\right)\left(\frac{4\ \cancel{qt}}{1\ \cancel{gal}}\right)\left(\frac{1\ l}{1.056\ \cancel{qt}}\right)$

$$\begin{aligned} 300 \text{ gal} &= (300)(4)\left(\frac{1}{1.056}\right) \text{ liters} \\ &= \left(\frac{1{,}200}{1.056}\right) \text{ liters} \\ &\approx 1{,}136 \text{ liters} \end{aligned}$$

note

The symbol $\approx$ means "approximately equal to."

2. Convert 0.055 km/min to inches per second. (Round to the tenths.)

Solution $\left(\dfrac{0.055\ \cancel{km}}{1\ \cancel{min}}\right)\left(\dfrac{1\ \cancel{min}}{60\ s}\right)\left(\dfrac{1{,}000\ \cancel{m}}{1\ \cancel{km}}\right)\left(\dfrac{39.37\ in}{1\ \cancel{m}}\right)$

$$\begin{aligned} 0.055 \text{ km/min} &= (0.055)\left(\frac{1}{60}\right)(1{,}000)(39.37) \text{ in/s} \\ &= \frac{(0.055)\cdot(1{,}000)\cdot(39.37)}{60} \text{ in/s} \\ &\approx 36.1 \text{ in/s} \end{aligned}$$

3. Convert 9.43 oz/in² to grams per square centimeter. (Round to the hundredths.)

Solution $\left(\dfrac{9.43\ \cancel{oz}}{\cancel{in^2}}\right)\left(\dfrac{28.35\ g}{1\ \cancel{oz}}\right)\left(\dfrac{1\ \cancel{in^2}}{6.452\ cm^2}\right)$

$$\begin{aligned} 9.43 \text{ oz/in}^2 &= (9.43)(28.35)\left(\frac{1}{6.452}\right) \text{ g/cm}^2 \\ &= \frac{(9.43)\cdot(28.35)}{6.452} \text{ g/cm}^2 \\ &\approx 41.44 \text{ g/cm}^2 \end{aligned}$$

4. Convert 16.4 in³/h to cubic centimeters per second. (Round to the ten-thousandths.)

Solution $\left(\dfrac{16.4\ \cancel{in^3}}{1\ \cancel{h}}\right)\left(\dfrac{16.39\ cm^3}{1\ \cancel{in^3}}\right)\left(\dfrac{1\ \cancel{h}}{60\ \cancel{min}}\right)\left(\dfrac{1\ \cancel{min}}{60\ s}\right)$

$$\begin{aligned} 16.4 \text{ in}^3\text{/h} &= (16.4)(16.39)\left(\frac{1}{60}\right)\left(\frac{1}{60}\right) \text{ cm}^3\text{/s} \\ &= \frac{(16.4)\cdot(16.39)}{(60)\cdot(60)} \text{ cm}^3\text{/s} \\ &\approx 0.0747 \text{ cm}^3\text{/s} \end{aligned}$$

Exercise Set 5.3

Convert the following. Leave your answer rounded to the hundredths.

•1. 2,450 in to meters
2. 525 gal to liters
3. 5.62 kg to pounds
4. 6.73 km to miles
•5. 3.62 ft³ to liters
6. 3,760 oz to kilograms
•7. 378 in² to square meters
8. 0.036 km to inches
9. 3.5 mi² to square centimeters
10. 528 in² to square centimeters

11. 42.6 ft/s to centimeters per minute
12. 20.3 ft^3/h to cubic centimeters per second
•13. 0.021 kg/cm^2 to pounds per square inch
14. 0.0037 km/min to inches per second
•15. 2.6 t/yd^3 to grams per cubic centimeter
16. 5.37 oz/in^2 to grams per square centimeter
17. 0.453 in^3/oz to cubic centimeters per gram
18. 360 mph (miles per hour) to kilometers per minute
19. 42 km/s to miles per hour
20. 0.075 t/ft^3 to kilograms per cubic meter
21. The shipping weight of a radio set is 22 lb. What is this weight in kilograms?
22. An imported transistor radio has dimensions $8 \cdot 4 \cdot 3.5$ cm. Find its dimensions in inches.
•23. A quarter-inch wrench is to be used on a fitting. This fitting takes what size wrench in centimeters?
24. The wavelength of radio waves at a frequency of 300 megacycles is 10 cm. What is this wavelength in inches?
•25. The diameter of No. 10 wire is 0.102 in. What is its radius in centimeters?
26. The capacitance of a certain power line is 0.00802 microfarads per mile. What is this capacitance in microfarads per kilometer?
27. A certain grade of twisted pair transmission line has a loss of 6.4 decibels per 100 feet. What is the loss in decibels per meter?
28. If 1,000 cycles = 1 kilocycle, then 1,500 kilocycles per second equals how many cycles per minute?
29. A certain resistance wire has a resistance of 3 ohms/ft. What is the resistance of 6 ft of this wire in ohms per meter?

5.5 Exact vs. Approximate Numbers

A large portion of the data that a technician uses is measured. Since measurements can never be exact, the technician is constantly working with approximate numbers. Exact numbers are those numbers obtained by counting, by definition, or by computation involving exact numbers. For example,

523 pages in a book
4 wheels on an automobile
15 pulleys in a system
5 resistors in an electrical system
8 pistons in an automobile

are exact numbers. There are exactly 12 in in 1 ft by definition. All other numbers are approximate. For example,

$3\frac{1}{2}$ in—the length of the piston
John's weight—176 lb
3.5 ohms resistance in a resistor
5.26 lb/in^2 pressure

are all approximate numbers. When we construct a device to measure something, there will always be physical limitations in the construction of the device as well as limitations in the technician's ability to correctly read the measurement. Thus every measured value has built-in limitations and we must take into account these limitations when using numerical values made from measurements. When we said above that

John weighs 176 lb

there exists a possible error of plus or minus 0.5 lb (± 0.5 lb) because the instrument used to measure weight could be read accurately only to the nearest pound. That is, John could weigh anywhere from 175.5 lb to 176.5 lb and still have an approximate weight of 176 lb. In the same way

1. 3.5 ohms resistance has a possible error of ± 0.05 (3.45 to 3.55).
2. 5.26 lb/in^2 pressure has a possible error of ± 0.005 (5.255 to 5.265).
3. 42,000 as the population of a given city has a possible error of ± 500 (41,500 to 42,500).

The plus or minus notation is called the **tolerance** of the number. It can be found by placing a 5 after the last reliable digit in the number and then subtracting the original number from the one just formed.

Example

Find the tolerance of 3.5 ohms resistance.

1. Place a 5 at the end of the number: $3.5\underline{5}$
2. Subtract the original number from the one just formed: 3.55 − 3.5 = 0.05
3. The tolerance is plus or minus this difference (± 0.05 ohm).

Therefore, in notation indicating tolerance, 3.5 ohms means 3.5 ohms ± 0.05 ohm.

5.6 Significant Digits

The degree of accuracy of an approximate number depends on the number of reliable digits, called **significant digits** or **significant figures,** in the number. The following are the rules by which we determine if a digit is significant.

Nonzero digits are always significant. To illustrate:

5.23 is correct to three significant digits.
412.15 is correct to five significant digits.
0.95263 is correct to five significant digits.

When zeros are found in a number, their position determines whether or not they are significant. First, zeros found *at the end of a whole number are usually not considered significant*. These zeros are considered to be placeholders unless there is definite information to the contrary. For example:

42,000 is correct to two significant digits.
40,320 is correct to four significant digits.
59,003,600 is correct to six significant digits.

note

If a zero at the end of a whole number is significant, it should be overscored with a tilde (˜) to indicate that it is significant. 42,0$\tilde{0}$0 indicates that the zeros in the hundreds and tens columns are significant and the number has four significant digits.

It is important to observe from the above examples that zeros found *between the first and last significant digits are considered significant*. For example, the 0 between the 4 and 3 in 40,320 and the two zeros between the 9 and 3 of 59,003,600 are considered significant.

Zeros found at the beginning of a decimal fraction (not a mixed decimal) are not significant. Their sole purpose is to serve as placeholders in the decimal. Thus,

0.0734 is correct to three significant digits.
0.0009045 is correct to four significant digits.
0.00205 is correct to three significant digits.

note

If we change 9¢ to $.09 or 4.5 cm to 0.0045 m, we see that the zeros at the beginning of a decimal number to the right of the decimal point are used only to locate the decimal point.

Finally, zeros found *at the end of a decimal number are considered significant*. For example:

45.20 is correct to four significant digits.
0.0076400 is correct to five significant digits.
9.00300 is correct to six significant digits.

note

If a number is written in scientific notation, then all the digits are significant.

$40{,}320 = 4.032 \times 10^4$ is correct to four significant digits.
$.00205 = 2.05 \times 10^{-3}$ is correct to three significant digits.
$45.20 = 4.520 \times 10^1$ is correct to four significant digits.

Summary of rules:

1. All nonzero digits are significant.
2. Zeros between the first and last significant digits are significant.
3. Zeros at the end of a decimal number are significant.
4. Zeros at the end of a whole number are usually not considered significant.
5. Zeros at the beginning of a decimal fraction are not significant.

Often data that is collected may have a greater number of significant digits than is necessary. When this occurs, we **round off** the numbers, which is accomplished by dropping a significant digit (or digits) at the right end of the number. Recall the rules of rounding from chapter 2 and observe the following examples:

5.623 becomes 5.62 when rounded to three significant digits.

Thus, when the digit dropped *is less than 5,* the remaining digits are not changed.

82.056 becomes 82.06 when rounded to four significant digits.

When the digit dropped *is greater than or equal to 5,* increase the digit to its left by one.

9.8765 becomes 9.877 when rounded to four significant digits.

Examples

1. 8.34465 becomes 8.34 when rounded to three significant digits.
2. 0.4536 becomes 0.5 when rounded to one significant digit.
3. 2,740 becomes 2,700 when rounded to two significant digits.
4. 185,499 becomes 185,000 when rounded to three significant digits.

---

note

Observe from examples 3 and 4 that when rounding whole numbers, the dropped digits are replaced with zeros to hold the positions of the rest of the digits.

The **accuracy** of a number is determined by counting the number of significant digits. That is,

11.6 has an accuracy of three significant digits.
0.09 has an accuracy of one significant digit.
5,300 has an accuracy of two significant digits.
67 has an accuracy of two significant digits.
0.00870 has an accuracy of three significant digits.

When comparing the accuracy of two or more approximate numbers, the number having the most significant digits is the most accurate. Thus, 5,324 is more accurate than 10.8 since there are four significant digits in 5,324 and only three significant digits in 10.8.

The **precision** of a number is determined by the position of the right-most significant digit. To illustrate, the following numbers are precise as indicated:

11.6 is precise to the nearest tenth.
0.09 is precise to the nearest hundredth.
5,300 is precise to the nearest hundred.

note

5,300 *may be* correct to the nearest unit, but unless we are told this we must assume that it is not.

67 is precise to the nearest unit.
0.00870 is precise to the nearest hundred-thousandth.

note

When comparing the precision of two or more measurements, be sure that all the measurements are in the same unit of measure. For example, 5 mm is more precise than 1.2 m, because if both measurements were expressed in terms of mm, we would have 1.2 m = 1,200 mm and 5 mm. The 5 mm is precise to the units, whereas the 1,200 mm is precise to the hundreds.

When comparing the precision of two or more approximate numbers, the number with the significant digit farthest to the right, relative to the decimal point, is the most precise. That is, 10.8 is more precise than 5,324 since 10.8 is correct to the nearest tenth and 5,324 is correct to the nearest unit. The number 5,324 is more accurate than 10.8, but 10.8 is more precise than 5,324.

The accuracy of a product or quotient of two numbers is equal to the accuracy of the *least accurate* of the factors or of the divisor and dividend.

Example Find the product of the approximate numbers 2.36 and 0.035.

```
   2.36
  0.035
  -----
   1180
   708
-------
0.08260
```

Since 2.36 has three significant digits and 0.035 has two significant digits, then the answer will have only two significant digits. Therefore,

$$(2.36)(0.035) \approx 0.083$$

Example Find the quotient 2,380 ÷ 32.

```
       74.375
32) 2,380.000
    2 24
      140
      128
       120
        96
       240
       224
        160
        160
```

Since 32 is the least accurate number, with two significant digits, the quotient will have two significant digits. Therefore,

$$2{,}380 \div 32 \approx 74.$$

The precision of a sum or difference is equal to the precision of the *least precise* of the addends or of the subtrahend and minuend.

Example

Find the sum of 32.403, 18.16, and 2.2047.

$$\begin{array}{r} 32.403 \\ 18.16 \\ \underline{2.2047} \\ 52.7677 \end{array}$$

Since the sum must be rounded off to the same precision as the least precise (18.16), then the sum is ≈ 52.77.

Example

Find the difference 48.1 − 12.382.

$$\begin{array}{r} 48.1 \\ -\ \underline{12.382} \\ 35.718 \end{array}$$

Rounding the difference to the least precise number (48.1) makes the difference ≈ 35.7.

When multiplying or dividing *exact numbers,* the rules to follow are:

1. The product or quotient of two exact numbers is exact.
2. The quotient of two exact numbers can be stated to as high a degree of accuracy as desired if the two numbers do not divide exactly.
3. The product or quotient of an exact and an approximate number gives an approximate number with the same number of significant digits as in the given approximate number.

Example

If we have the exact number 187 and the approximate number 1.4, our results when adding or subtracting these numbers would be precise to the tenths, and our results when multiplying or dividing these numbers would be accurate to two significant digits.

1. $187 + 1.4 = 188.4$
2. $187 - 1.4 = 185.6$
3. $(187) \cdot (1.4) \approx 260$
4. $(187) \div (1.4) \approx 130$

The result of extracting the *root of an approximate number* will have the same number of significant digits as the number.

Examples

Find the square root of the following approximate numbers.

1. $\sqrt{7} \approx 3$
2. $\sqrt{7.0} \approx 2.6$
3. $\sqrt{7.00} \approx 2.65$
4. $\sqrt{16} = 4.0$

Rules for combined operations with approximate numbers:

1. All computations should be done according to the order of operations.
2. When all operations of a given priority have been completed, all the results from those operations should be rounded according to the rules of approximate numbers.

Examples

Perform the indicated operations on the given approximate numbers.

1. $3.17 + (1.86)(6.42)$
   $\approx 3.17 + 11.9$
   ≈ 15.1
2. $\sqrt{6.8} + (2.3)^2 - 1.47$
   $\approx 2.6 + 5.3 - 1.47$
   ≈ 6.4
3. $(6.41 + 82.1)(107 + 1043)$
   $\approx (88.5)(1{,}15\tilde{0})$
   $\approx 102{,}000$

note

In example 3, a tilde (~) is used above the zero in $1{,}15\tilde{0}$ to indicate that the zero is significant.

Exercise Set 5.4

State the accuracy and precision of the following significant digits.

Example 3.63 answer: Accurate to three significant digits, precise to the hundredths

1. 0.425	2. 0.0763	3. 1.008
4. 53,000	•5. 50,630	6. 12.360
7. 0.81023	8. 9.000760	•9. 0.000012
10. 0.1240	11. 1,234.0	12. 60,007.02

Round the following to the number of significant digits indicated in parentheses.

Example 5.0637 (3)   answer: 5.06

13. 0.4736 (2)	14. 5.7563 (3)	•15. 9,007.076 (5)
16. 0.0445 (2)	17. 5.66783 (3)	18. 0.00312 (1)
19. 73.1532 (3)	20. 0.9234 (1)	•21. 535.000 (2)
22. 6.7342 (4)	23. 10.946 (3)	24. 15.0076 (4)

Indicate whether each of the following is an exact or an approximate number.

25. 4 hubcaps on an automobile
26. 4.8 s to run 50 yd
•27. 36 in in 1 yd
28. 8,000 mi is the diameter of the earth.
29. Garden City has a population of 40,000 people.
30. 32 students in our mathematics class
31. 3 amp of current in an electric circuit
32. The diameter of a brass rod is 1.325 in.
33. The gear has 32 teeth.
34. The number of quarts in a gallon is 4.
•35. A plane is traveling 275 mph.

List the numbers in each of the following groups in order of (1) accuracy and (2) precision, starting with the most accurate or most precise.

Example 1.87; 2,345; 45.836   answer: (1) 45.836; 2,345; 1.87
(2) 45.836; 1.87; 2,345

36. 3.023; 24.7; 82,362
37. 8,600,000; 102,000,100,000; 82,730
38. 243.6; 82.000; 0.20; 540
•39. 6.031; 0.0056; 89.7; 0.00073
40. 0.0108; 43.36; 22,000,000

Write in tolerance notation as determined by the number of significant digits:

Example 3.89   answer: $3.895 - 3.89 = 0.005$
Therefore, 3.89 ± 0.005

41. 32	42. 0.014	43. 425.6
44. 8,000	45. 190	46. 0.000832
•47. 5,600,000	48. 0.000002030	

Compute the following, stating your answer to the proper accuracy or precision.

Example (14.2)(2.7) answer:

$$\begin{array}{r} 14.2 \\ \underline{2.7} \\ 994 \\ \underline{284} \\ 38.34 \end{array}$$

$(14.2)(2.7) \approx 38$

14.2 is accurate to three significant digits, and 2.7 is accurate to two significant digits; therefore, the answer will be accurate to two significant digits.

49. (0.3245)(1.02)
50. $82.361 \div 5$
51. (3.1)(81.07)
52. $8.814 \div 72.6$
•53. (0.000256)(421.6)
54. $78.302 \div 5.2$
55. $400 \div 6.032$
56. (902)(2.003)
57. $\pi \div 8.07$
58. $(\pi)(0.0793)$
59. $452.62 + 37.746 + 105.5007$
60. $18.2 + 82.883 + 0.08 + 47.607$
61. $8.090 - 5.8634$
62. $0.00535 - 0.003819$
63. $0.09952 + 0.132 + 0.8976$
64. $400.002 + 0.0070 + 6.73$
65. $75.00 - 2.4$
66. $82.36 - 1.7892107$
67. $500.3 + 2.0076 - 89.076$
68. $22.309 - 0.008 + 1.00009$
69. $0.8700 + 0.0009306 - 0.7000$
70. $86.0 - 75.673 - 7.12345$
71. $\sqrt{3.48}$
72. $\sqrt{5.61}$
73. $\sqrt{9.0}$
74. $\sqrt{25}$
75. $15.12 + (6.87)(1.45)$
76. $1.893 + (7.41)(18.42)$
77. $\sqrt{4.9} + (4.7)^2 - 8.63$
78. $\sqrt{64} + (5.3)^2 - 9.67$
79. $(5.83 + 9.46)(11.91 - 8.37)$
80. $(9.98 - 6.32)(4.59 + 5.62)$

Solve the following, stating the answer as precisely or accurately as the measurements dictate.

81. The area of a parallelogram is determined by multiplying the length of the base by the altitude. What is the area of a parallelogram if the base is 5.36 in and the altitude is 3.06 in?
82. What is the largest possible area in problem 81 and the smallest possible area? (Consider the measurements having tolerance 0.005.)
83. What must be the length of a bolt to pass through a metal plate having a thickness of 1.027 in and a bracket having a thickness of 0.735 in if we put a lock washer having a thickness of 0.27 in and a nut having a thickness of 0.634 in on it?
84. What range of values can a technician expect from a 2,200-ohm resistor with a tolerance of 110 ohms?
85. Three resistors measure 48.7 ohms, 76.35 ohms, and 100.6 ohms. What is their total resistance?
86. On a chassis drawing the length of the four sides are: 21.06, 7.32, 4.3, and 22.076 in. What is the perimeter (sum of the sides) of the chassis?
87. What is the voltage (E) value produced by a current (I) 0.81 amp through a 672-ohm resistor (R)? ($E = IR$)

88. A 6-volt light bulb draws 0.25 amp when cold and 0.125 when hot. What is the difference between its hot and cold resistance values? (*Hint:* use $R = \frac{E}{I}$ and refer to problem 87.)

89. Find the perimeter of the following figure.

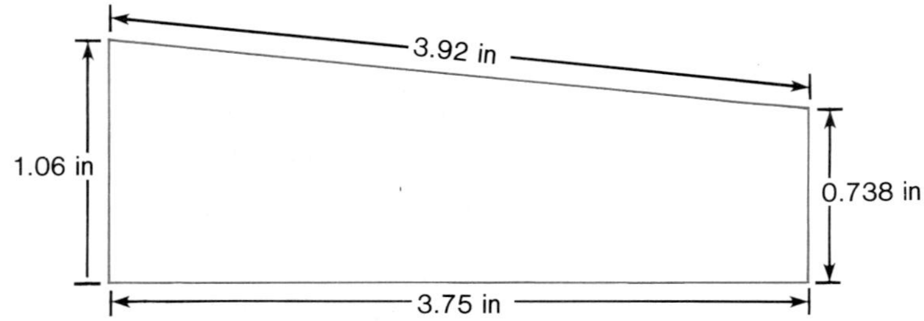

Chapter Summary

1. The **British system** of measure has standard units of measure of *foot-pound-second*.
2. The **SI system** of measure has standard units of measure of *meter-kilogram-second*.
3. Units of measure are **reduced** within each system of measure.
4. Units of measure are **converted** from one system of measure to another.
5. **Significant digits** in a number are determined by the following principles:
   a. Nonzero digits are significant.
   b. Zeros found at the end of a whole number are *not* significant.
   c. Zeros found between the first and last significant digits are significant.
   d. Zeros found at the beginning of a decimal number are *not* significant.
   e. Zeros found at the end of a decimal number are significant.
6. When **rounding off** a number to a given number of significant digits, we follow these principles:
   a. If the first digit dropped is less than 5, no change is made in the remaining part.
   b. If the first digit dropped is 5 or more, the digit to its immediate left is increased by 1.
7. The **tolerance** of an approximate number is stated by plus or minus one-half the unit in the right-most significant digit.
8. The **precision** of an approximate number is determined by the position of the right-most significant digit.
9. The **accuracy** of an approximate number is determined by the number of significant digits.

Chapter Review

[5.1, 5.2] Reduce the following. Round your answer to three decimal places.

1. 5,360 lb to short tons
2. 5.63 gal to pints
3. 7.36 yd^2 to ft^2
4. 8 mi to rods
5. 23.9 cm to meters

6. 736 cg to grams
7. 5,300 ml to liters
8. 39.5 cm^2 to m^2
9. 43 lb/s to ounces per minute
10. 9.36 cm^2 to square decimeters
11. 0.37 km/min to meters per second
12. 22 ft^3/h to cubic yards per minute
13. 10 kg/m^2 to grams per square centimeter
14. 2.63 t/in^2 to pounds per square foot
15. 25 mm^3 to cubic centimeters

[5.3] Convert the following, using the calculator where necessary. Round your answers to the hundredths.

16. 2.36 mi to kilometers
17. 5,670 oz to kilograms
18. 4.3 mi^2 to square kilometers
19. 5.6 ft^3 to liters
20. 836 in^2 to square centimeters
21. 82 ft/s to centimeters per minute
22. 5 cm^3/s to cubic feet per hour
23. 7.6 oz/in^2 to grams per square centimeter
24. 560 mi/h to kilometers per minute
25. 0.083 t/ft^3 to kilograms per cubic meter
26. If the shipping weight of a flywheel is 53.2 kg, what is its weight in pounds?
27. A 3/8-in wrench is used in a fitting. This fitting takes what size of wrench in centimeters?
28. The diameter of No. 12 wire is 0.936 cm. What is its radius in inches?
29. The high-frequency resistance of No. 10 copper wire at a frequency of 100 megacycles is 0.00321 ohms/cm. What is the resistance of 3 ft of this wire at the same frequency?
30. 1,600,000 cycles per minute converts to how many kilocycles per hour if 1,000 cycles = 1 kilocycle?
31. The owner's manual for a certain foreign car says to inflate the tires to 2.4 kg/cm^2. How many pounds per square inch is this?
32. If an automobile's fuel tank holds 19.9 gal, how much does it hold in liters?

[5.4, 5.5, 5.6] The following are accurate to how many significant digits?

33. 0.364
34. 0.0045
35. 402.3
36. 430,000
37. 50.600
38. 5.006

Round off to the indicated number of significant digits.

39. 4.037 to 3 significant digits
40. 0.00735 to 2 significant digits
41. 50.374 to 4 significant digits
42. 463,425 to 5 significant digits

Write in tolerance notation:

43. 72.3

44. 82,300

45. 0.00093

In each of the following sets of numbers, determine the most and least (a) precise and (b) accurate.

46. 52; 0.367; 22,400
47. 702.6; 0.000873; 79,000

Perform the indicated operations on the following approximate numbers.

48. $9.080 - 4.8544$
49. $14.13 + 7.001 + 5.913$
50. $(5.142)(87.935)$
51. $\sqrt{14.2}$
52. $87.61 - (3.87)(9.431)$

Solve the following, stating your answer as precisely or accurately as indicated by the measurements.

53. Bell metal weighs 14.2 times as much as mahogany. If a pattern of mahogany weighs 9.25 lb, what will the bell metal casting weigh?
54. A shaft with diameter of 2.434 in is to be turned down on a lathe by making a cut 0.21 in deep. What will the finished diameter of the shaft be?
55. What would be the total thickness of a 4-ply sheet of plywood made with plies of the following thicknesses: 0.1234 in, 0.210 in, 0.109 in, and 0.3092 in? Allow 0.0147 in of bonding material between each ply.
56. Given a neon bulb with voltage (E) of 75 volts and a current flow (I) of 0.015 amp, what is the resistance (R) if $R = \dfrac{E}{I}$?

Chapter Test

Perform the indicated reduction or conversion. Round your answers to the hundredths.

1. 453 cl to liters
2. 3 gal to cups
3. 3 yd^2 to square inches
4. 16 oz to centiliters
5. 30 mi/h to kilometers per hour
6. 10.2 ft^2 to square yards
7. 9.8 lb/ft^3 to grams per cubic centimeter

Perform the indicated operations on the following approximate numbers.

8. $(7.8)(0.345)$
9. $(14.7)(0.23)$
10. $481.7 + 394.96 + 2.471$
11. $14.86 - 3.9841$

Round off to the indicated number of significant digits.

12. 5.951 to 2 significant digits
13. 14.4999 to 3 significant digits
14. 31.008 to 3 significant digits

Solve the following, stating your answer as precisely or accurately as indicated by the measurements.

15. The edge of a cube is measured at 1.87 in. Find the volume. ($v = e^3$)
16. The dimensions of a room are given as 12 ft wide and 16.5 ft long. How many square yards of carpet would be needed to cover the floor? ($A = l \cdot w$)
17. A batch of pistons weighs 183 lb. If there are 43 pistons, what is the approximate weight of each one?

6
Algebraic Expressions

A technician checking performance and durability tests under controlled conditions using sophisticated computerized test facilities and equipment.
(Photograph courtesy of Ford Motor Company)

6.1 Algebraic Expressions

We define a **variable** *to be a symbol that may take on different values in a given discussion;* that is, *it holds a place for a number*. A variable is capable of taking on any one of the different values that it represents. In the relationship:

$$I = \frac{115}{R}$$

I and R are variables since they both can assume varying numerical values to make the statement true.

A **constant** *is a symbol that does not change its value*. In the relationship:

$$I = \frac{115}{R}$$

115 is a constant. A number is a constant. If a symbol represents only one value during a given discussion, that symbol is a constant.

With the use of variables and constants we can adopt a new convention for indicating the product of two or more variables, or the product of a constant and one or more variables. From now on, the symbol

$$3d$$

will mean "3 times d." That is, when a constant and a variable are written next to each other we will assume the indicated operation to be multiplication. The same will apply to two or more variables written with no indicated operation between them.

$$\frac{1}{2}at$$

will mean

$$\frac{1}{2} \cdot a \cdot t$$

Any meaningful collection of variables, constants, grouping symbols, and signs of operation is called an **algebraic expression.** Examples of algebraic expressions are:

$$\frac{1}{2}at^2, \qquad \frac{RS}{R + S}, \qquad lwh$$

In an algebraic expression, the symbols that are separated by a plus or minus sign are called the **terms** *of the expression*. Thus, in the expression,

$$a^2 + 2ab + b^2$$

a^2, $2ab$, and b^2 are the terms.

note

When several terms are grouped together, we consider the group to be a single term.

Example

$(x^2 + y^2)$ is one term.

$(a^2 + 2ab + b^2) + c$ is two terms.

$\frac{RS}{R + S}$ is one term since the fraction bar acts as a grouping symbol as well as a division symbol. That is, $\frac{RS}{R + S}$ means

$$(RS) \div (R + S).$$

In the expression $3at$, 3 is called the **numerical coefficient.** The numerical coefficient tells how many at's are in the problem.

Since we often talk about the numerical coefficients of an expression, we eliminate the word *numerical* and just say "coefficient" when we discuss numerical coefficients. If no numerical coefficient appears in an expression, the coefficient is *understood* to be 1.

Example

In the algebraic expression $6x + 3y + z$, 6 is the coefficient of x, 3 is the coefficient of y, and 1 is understood to be the coefficient of z.

A special kind of algebraic expression is a polynomial. A **polynomial** is an algebraic expression with the following characteristics: *it has real number coefficients; all variables are raised only to natural number exponents; and the operations involving variables are limited to addition, subtraction, and multiplication.* A polynomial containing just one term is called a **monomial;** one containing two terms is called a **binomial;** and a polynomial containing three terms is called a **trinomial.** A polynomial containing two or more terms is also a **multinomial.** No special names are given to polynomials containing more than three terms.

Examples

1. $\frac{4}{x + 2}$ and $3\sqrt{x} + x$ are not polynomials since they contain variables in the denominator or under the radical sign.
2. x, $4x$, 3, and $5x^2y$ are monomials.
3. $3x + 1$, $x + y$, and $81W^2 - 9T^2$ are binomials.
4. $5x^3 + 2y - 1$ and $z^2 + 9z - 10$ are trinomials.

Many problems that technicians encounter are stated verbally and need to be translated into an algebraic statement. While there is no standard procedure for changing a verbal statement into an algebraic statement, the following guidelines should be of use:

1. Read the statement carefully.
2. Note what information is given.
3. Determine any prior knowledge that may be useful.
4. Let some letter represent one of the unknown quantities.
5. Put any other unknown quantities in terms of the first unknown.
6. Make sure in steps 4 and 5 that the unknown quantities represent numerical quantities.
7. Using the given conditions in the statement and the unknowns from steps 4 and 5, write an algebraic expression.

Examples

1. The product of a and b: $a \cdot b$
2. The sum of a and 4: $a + 4$
3. x decreased by 9: $x - 9$
4. y divided by 3: $\frac{y}{3}$
5. A number plus 6: $n + 6$
6. Two times a number, decreased by 5: $2n - 5$
7. A number divided by 3, increased by 2: $\frac{n}{3} + 2$

Exercise Set 6.1

Specify the number of terms in each expression.

1. $3x + 4y$
2. $5xyz$
3. $4x^2 + 3x - 1$
4. $x^3 - 4x + 7$
5. $\frac{6x}{5}$
6. $\frac{x}{3}$
•7. $8xy + \frac{5y}{2} - 6x$
8. $\frac{15x^2y}{8}$
•9. $5x^3 + (3x^2 - 4)$
10. IR
11. $2(fL)$
12. $R + jx$
13. $\frac{R}{Pf} + z + R$
14. $\frac{L}{RC} + ei + IR$

Given the algebraic expression $8x^2y + 5xyz - 9x^2y^2 - 5y^3z$, answer the following.

15. What is the numerical coefficient of each term?
16. How many terms are there in the expression?

Write an algebraic expression for each of the following.

17. The sum of a and b
18. 3 times a, subtracted from b
•19. 7 less than x
20. 5 more than y
•21. The sum of x and y, divided by z
22. x times the sum of y and z
23. a decreased by 5
24. a decreased by b
•25. $\frac{1}{2}$ of x decreased by 2 times x
26. A number decreased by 12
27. A number increased by 4
28. 3 times a number, increased by 1
29. A number divided by 5
30. 2 times the sum of a number and 4
•31. A number minus 6, divided by 11

6.2 Evaluating Algebraic Expressions

An extremely important process studied in algebra is that of finding the **numerical evaluation** of a given algebraic expression when we are given a number replacement for the variable or variables. The technician frequently needs to know the numerical value of an expression. For example, to find the cutting speed of a piece of cast iron turned in a lathe, the following formula may be used:

$$\pi \cdot d \cdot r$$

where d represents the diameter expressed in feet and r is the number of revolutions per minute. For example, when $d = \frac{5}{6}$ in and $r = 180$, the cutting speed becomes approximately

$$\left(\frac{22}{7}\right)\left(\frac{5}{6}\right)\left(\frac{1}{12}\right)(180) = \frac{275}{7}$$

$$= 39\frac{2}{7} \text{ ft/min}$$

note

When replacing the letters with the numbers they are representing, it is good procedure to put each of the numbers inside parentheses.

Example $x^2 + 2x - 7$, when $x = 4$ is $(\ \ )^2 + 2(\ \ ) - 7$ without the x; substituting the 4 in place of the x gives $(4)^2 + 2(4) - 7$. Using the order of operations we have

$$16 + 2(4) - 7$$
$$16 + 8 - 7$$
$$24 - 7$$
$$17$$

Therefore, the formula $x^2 + 2x - 7$, evaluated when $x = 4$, is 17.

note

$$(-2)^2 = (-2)(-2) = 4 \text{ but } -2^2 = -(2 \cdot 2) = -(4) = -4$$

Algebraic expressions that are most frequently used by technicians are **formulas.** An example of a formula in electricity is

$$I = \frac{E}{R}$$

where I represents the current in a certain part of a circuit, E is the potential difference in voltage across that part of the circuit, and R is the resistance of that part of the circuit. Thus, $\frac{E}{R}$ is an algebraic expression that is a symbol for I. To find I we must evaluate $\frac{E}{R}$. To illustrate, given $E = 110$ volts and $R = 44$ ohms, then

$$I = \frac{E}{R} = \frac{(110)}{(44)} = \frac{10}{4} = 2\frac{1}{2}$$

and since I is measured in amp,

$$I = 2\frac{1}{2} \text{ amps}$$

In some formulas two or more measurements of the same unit may be given. It is customary to label these by using **subscripts.** For example, we might label three different measurements of resistance in an electrical circuit as: R_1, R_2, R_3. Here, 1, 2, and 3 are the *subscripts* and are always written to the lower right of the letter. These symbols are read "R sub-one," "R sub-two," and "R sub-three."

Note the difference between subscripts and exponents. Subscripts merely help us to distinguish between distinct units of measure; while exponents indicate the number of times a given quantity is used as a factor in an indicated product. Also, exponents are written to the upper right of the quantity and subscripts are written to the lower right of the quantity.

Examples

1. In

$$E_t = E_p + E_g$$

E_p and E_g use the subscripts p and g to denote two different measurements of voltage. E_t represents another measurement of voltage, the total of the other two.

2. In

$$\frac{1}{R_t} = \frac{1}{R_1} + \frac{1}{R_2} + \frac{1}{R_3}$$

R_1, R_2, and R_3 represent three different measurements of resistance, and R_t represents the total resistance.

Exercise Set 6.2

Evaluate the following expressions if $a = 2$, $b = 3$, $c = -2$, $d = -3$.

1. $2a + b - c$
2. $(a + b)$
•3. $3a - 2b - (c + d)$
4. $a - 3(c + b)$
5. $(3a + 2b)(a - c)$
6. $2ab(c + d)$
•7. $ac - bd$
8. $3c - 2(3a + b)$
9. $7a - d(6b + c)$
10. $(3a - 5c)(2b - 4d)$
•11. $(5c - 3a)(4d - 2b)$
12. $(5c - d)4a$
13. $5a + 7b - 3c(a - d)$
14. $(4a + b) - (3a - b)(c + 2d)$
15. $a^2 - c^2$
16. $b^2 + 2d^2$
17. $3ab - 4c^2 + d$
18. $(c + d)^2$
19. $(c - d)^2$
20. $a^2b^2 + c^2d^2$
21. $3ac - 2a^2c^2$
22. $ab - ac$
23. $(ab)^2 - (ac)^2$
24. $(c - d)^2(a + b)$
25. $a^3b - c^3d$
26. $c^2 - d^3$
27. $3d^2 - 2c^3$
28. $(3d - 5c)^3$

In problems 29–41, round your answer to two decimal places.

•29. Find the HP in $HP = \dfrac{PLAN}{33{,}000}$ when

a. $P = 75$, $L = 3$, $A = 10$, $N = 160$
b. $P = 110$, $L = 11$, $A = 24$, $N = 120$

30. Evaluate HP in $HP = \dfrac{D^2 \cdot N}{2.5}$ when

a. $D = 4$, $N = 16$
b. $D = 3\frac{1}{2}$, $N = 5$

31. In $I = \dfrac{E}{R}$ evaluate I when

a. $E = 230$, $R = 46$
b. $E = 220$, $R = 33$

32. In $C_t = \dfrac{C_1 \cdot C_2}{C_1 + C_2}$, find C_t when

a. $C_1 = 4$, $C_2 = 6$
b. $C_1 = 250$, $C_2 = 350$

•33. Evaluate $\dfrac{I^2R - 120E^2}{R}$ when

a. $E = 5$, $I = 12$, $R = 10$
b. $E = 12$, $I = 8$, $R = 16$

34. Find R_t when $\dfrac{1}{R_t} = \dfrac{1}{R_1} + \dfrac{1}{R_2} + \dfrac{1}{R_3}$ and

a. $R_1 = 2$, $R_2 = 6$, $R_3 = 10$
b. $R_1 = 3$, $R_2 = 8$, $R_3 = 18$

35. Find V when $V = \dfrac{4}{3}\pi R^3$ where $R = 4$ and we use 3.14 for π.

36. Evaluate $E_g + E_p + E_r$ when
a. $E_g = 5$, $E_p = 8$, $E_r = 3$
b. $E_g = \dfrac{3}{4}$, $E_p = 2$, $E_r = 3$
c. $E_g = \dfrac{5}{6}$, $E_p = 3$, $E_r = \dfrac{1}{2}$

37. Solve for X_L in $X_L = 2\pi rh$ using 3.14 for π and
a. $h = 500$, $r = 4.75$
b. $h = 1{,}600{,}000$, $r = 0.0045$

38. Find the value of R_t in $R_t = \dfrac{R_1 \cdot R_2}{R_1 + R_2}$ when

a. $R_1 = 8$, $R_2 = 12$
b. $R_1 = 72$, $R_2 = 360$
c. $R_1 = 5\frac{1}{2}$, $R_2 = 26\frac{1}{2}$
d. $R_1 = 4.25$, $R_2 = 18.25$
e. $R_1 = 16{,}000$, $R_2 = 80{,}000$

39. Solve for R_x in $R_x = \dfrac{R_aR_b + R_bR_c + R_aR_c}{R_a}$ when

a. $R_a = 8$, $R_b = 5$, $R_c = 24$
b. $R_a = 12$, $R_b = 4$, $R_c = 9$
c. $R_a = 30$, $R_b = 40$, $R_c = 50$

40. Solve for R_x in $R_x = \dfrac{R_bR_c}{R_a + R_b + R_c}$ when

a. $R_a = 6$, $R_b = 8$, $R_c = 10$
b. $R_a = 40$, $R_b = 50$, $R_c = 60$

•41. Evaluate $\frac{1}{3}\pi r^2 h$ using 3.14 for π and

   a. $r = 3.5,\ h = 9$
   b. $r = 6,\ h = 5.6$

In problems 42–61, round to the accuracy of the measured data.

42. The formula used to convert the measured diagonal to diagonal-pyramid hardness (DPH) is given by $DPH = \frac{1.854L}{d^2}$, where L = applied load in kilograms and d = diagonal length of square impression. Find DPH if $L = 502$ and $d = 11.5$.
43. The energy E of x-rays is given by $E = h \cdot v$, where $h = 6.62 \times 10^{-27}$ erg-seconds and v = frequency of the radiation. Find E when $v = 4.25$.
44. The horsepower required to move a given volume at a known pressure is given by $HP = GPM \cdot P \cdot 0.00058$. Find HP when $GPM = 55$ gal/min and $P = 230$ lb/in^2.
45. The horsepower required by a hydraulic pump is given by $\text{HP} = \frac{GPM \cdot \text{psi}}{1{,}714}$. Find HP when the $GPM = 10.0$ gal/min and the psi $= 3{,}0\bar{0}0$ lb/in^2.
46. The required ratio of gearing (R) of a milling machine is given by

$$R = (A - N) \cdot \left(\frac{40}{A}\right),$$

where N = required number of divisions, A = approximate number of divisions. Find R when $A = 28\bar{0}$ and $N = 271$.
47. The cutting speed of a milling cutter (in feet per minute of a point on the circumference of the cutter) is determined by $rpm = \frac{4CS}{D}$, where CS = cutting speed of the metal and D = diameter of the cutter. Find the rpm (revolutions per minute) when $CS = 45$ ft/min and $D = 2$ in.
48. The tap drill size (TDS) of a drill needed to drill threads in a nut is given by $TDS = D - \left(\frac{1}{N}\right)$, where D = diameter of the nut and N = number of threads per inch. Find the TDS with a 0.5 in tap and 10 threads per inch.
49. The circular pitch (CP) of a spur gear is given by $CP = \frac{PD}{0.3183N}$, where PD = pitch diameter and N = number of teeth. Find CP when $PD = 1.25$ in and $N = 12$.
50. In a gear system, the velocity (V) of the driving gear is defined by $V = \frac{vn}{N}$, where v = velocity of the follower gear, n = number of teeth of follower gear, and N = number of teeth of driving gear. Find V when $v = 90.0$ revolutions per minute, $n = 30$ teeth, $N = 65$ teeth.

51. The stiffness of a metal is measured by $E = \dfrac{\Delta G}{\Delta \varepsilon}$, where E = modulus of stiffness, ΔG = change in stress required to produce $\Delta \varepsilon$ = change in strain. Find E when $\Delta G = 0.0367$ and $\Delta \varepsilon = 1.312$.

52. A piece of steel 2.5 in in diameter is to be turned in a lathe. What number of revolutions per minute (rpm) is necessary to give a cutting speed (CS) of 80.0 ft/min if $rpm = \dfrac{4CS}{d}$?

53. It is necessary to drag a box $60\tilde{0}$ ft across a level lot in 3.00 min. The force required to pull the box is $2{,}00\tilde{0}$ lb. What is the horsepower needed to do this if horsepower is defined by $HP = \dfrac{L \cdot W}{33{,}000 \cdot t}$, where L = length to be moved, W = force exerted through distance L, and t = time in minutes required to move the box through L?

54. What is the cutting speed (CS) of a shaper if the length of the stroke (L) is 8.00 in and the machine is operating at $3\tilde{0}$ strokes per minute (N)? Use the formula $CS = 0.14NL$.

55. A pulley 12.00 in in diameter that is running at $32\tilde{0}$ rpm is connected by a belt to a pulley 9.00 in in diameter. How many revolutions per minute will the smaller pulley make if $s = \dfrac{SD}{d}$, where s = speed of smaller pulley and d = diameter of smaller pulley, S = speed of larger pulley, and D = diameter of larger pulley?

56. Find the thermal stress (S) existing between the inside and outside of a heated metal that is defined by $S = (\delta E)(\Delta T)$, where the coefficient of linear expansion δ is 6.50×10^{-6}, modulus of elasticity (E) is 30.0×10^{6} lb/in^2 and the difference in temperature (ΔT) is $80\tilde{0}°$. [(ΔT) is read "delta T" and δE is read "delta E."]

57. The power in any part of an electrical circuit is given by the formula $P = \dfrac{E^2}{R}$, where P is the power in watts, E the voltage across that part, and R is the resistance of that part. Find the power expended when (a) $E = 23\tilde{0}$ volts and $R = 10\tilde{0}$ ohms; (b) $E = 35.00$ volts and $R = 6.00$ ohms.

58. In problem 57, what happens to the power P if the resistance R remains constant and the voltage E is (a) doubled, (b) tripled, (c) halved?

59. In problem 57, what happens to the power P when the voltage E remains constant and the resistance R is (a) doubled, (b) tripled, (c) halved?

60. The power in any part of a given circuit is also given by the formula $P = I^2R$, where P is the power in watts, I the current in amperes through that part, and R the resistance in ohms of that part. Find the power expended when (a) $I = 25.00$ amps and $R = 5.000$ ohms; (b) $I = 0.250$ amp and $R = 1{,}00\tilde{0}$ ohms.

61. In problem 60, when R remains constant, what happens to P when the current I is (a) doubled, (b) tripled, (c) halved?

Calculator Problems

Using the rules of approximate numbers, evaluate the following expressions if $a = 1.87$, $b = 3.14$, $c = -2.19$, $d = -0.33$.

62. $3a - 2b + c$
63. $5b - c - d$
64. $(a + b)^2$
65. $(2a - c)^2$
66. $(c + d)^2$
•67. $(2a - b)(3a + c)$
68. $a^2 - c^2$
69. $a^2 + d^2$
70. $c^2 - d^2$
71. $ab - cd$
72. $a^2b + c^2d$
73. $3a(5c - d)$
74. $(5a - c)(2b + d)$
75. $(4c - 3d)^2$

6.3 Algebraic Addition and Subtraction

Since algebraic expressions (and polynomials) represent a single number when the variables are replaced by numbers, the ideas and properties that apply to operations with numbers also apply to such expressions. Hence, when two or more terms of an expression contain the same variable factor, such as R in $2R + 5R$, the principles of addition of like quantities permits us to write $2R + 5R = (2 + 5)R = 7R$.

Terms in which the variable part is the same are called **like terms,** *or* **similar terms.**

note

To have similar terms, the variables must be the same and their respective exponents must be the same.

Example

x^2y and xy^2 both contain the same variables but are not similar terms because the powers of the respective variables are not the same.

It is *only similar terms* that we can add or subtract. Therefore, $3R + 6I$ could not be written more simply since we do not have similar terms.

Examples

Write as a single term if possible.

1. $5e + 2e + 3e = (5 + 2 + 3)e = 10e$
2. $4i + 3i + 5r = (4 + 3)i + 5r = 7i + 5r$
3. $3x^2 + 2x^2 + 3x + x = (3 + 2)x^2 + (3 + 1)x = 5x^2 + 4x$

In the preceding examples we simplified the expressions by combining the like terms.

Consider the sum

$$(3x^2 + 2x + 5) + (4x^2 + 3x + 6)$$

Since we cannot perform any of the addition within the parentheses, we remove them, giving:

$$3x^2 + 2x + 5 + 4x^2 + 3x + 6$$

Recall that the order in which we add numbers will not affect the answer (this property is called the *commutative law of addition*); hence, we can rewrite the problem as follows:

$$3x^2 + 4x^2 + 2x + 3x + 5 + 6$$

Recall, too, that we can group the numbers we are adding any way we wish without affecting our answer (this property is called the *associative law of addition*). Grouping together the similar terms, we have:

$$\begin{aligned}(3x^2 + 4x^2) + (2x + 3x) + (5 + 6) &= (3 + 4)x^2 + (2 + 3)x + (5 + 6) \\ &= 7x^2 + 5x + 11\end{aligned}$$

Thus, we have written the sum of these two algebraic expressions in simpler form by applying properties that we learned in arithmetic.

Since each algebraic expression represents a number, the difference of two terms in an expression can be written as an indicated sum. For example,

$$5E - 3E = 5E + (-3)E = [5 + (-3)]E = 2E$$

To find the difference between two algebraic expressions such as

$$(5x^2 + 3x + 2) - (2x^2 - 3x + 4)$$

we first note that the indicated operations within parentheses cannot be performed since they are not like terms. We then employ the definition of subtraction to get

$$(5x^2 + 3x + 2) - (2x^2 - 3x + 4) = (5x^2 + 3x + 2) + [-(2x^2 - 3x + 4)]$$

and now the right side becomes

$$5x^2 + 3x + 2 + (-2)x^2 + 3x + (-4)$$

note

If, for the expression $-(2x^2 - 3x + 4)$, we read the minus sign as "the opposite of," it becomes clear that we want the opposite of every term grouped together by the parentheses when we remove the parentheses. In general, if we have a grouping symbol preceded by a minus sign, this means that we want the opposite of each term in the group when we remove the grouping symbol.

Commuting and associating, we get

$$[5x^2 + (-2)x^2] + [(3x) + (3x)] + [2 + (-4)]$$

Then, adding the like terms, we get

$$[5 + (-2)]x^2 + [3 + 3]x + (-2)$$

which equals

$$3x^2 + 6x - 2$$

Examples

Simplify by finding the indicated sums and/or differences.

1. $6x + 5x - 3x = 6x + 5x + (-3)x$
$= [6 + 5 + (-3)]x$
$= 8x$

2. $(3x^2 - 4x + 1) + (5x^2 + 2x - 3)$
$= 3x^2 + (-4)x + 1 + 5x^2 + 2x + (-3)$
$= 3x^2 + 5x^2 + (-4)x + 2x + 1 + (-3)$
$= (3 + 5)x^2 + [(-4) + 2]x + [1 + (-3)]$
$= 8x^2 + (-2)x + (-2)$
$= 8x^2 - 2x - 2$

3. $(8R^2 - 2R + 3) - (6R^2 + 6R - 1)$
$= 8R^2 + (-2)R + 3 + (-6)R^2 + (-6)R + 1$
$= 8R^2 + (-6)R^2 + (-2)R + (-6)R + 3 + 1$
$= [8 + (-6)]R^2 + [(-2) + (-6)]R + [3 + 1]$
$= (2)R^2 + (-8)R + (4)$
$= 2R^2 - 8R + 4$

In section 6.2 we established some guidelines for writing an algebraic expression. We now extend those guidelines to show that we can translate a verbal problem into an equation that we will eventually solve for the unknown quantities.

The following are guidelines that should be useful in translating a verbal problem into an equation.

1. Read the problem carefully.
2. Note what information is given and what information you are asked to find.
3. Determine any prior knowledge that may be useful in setting up an equation.
4. Let some letter represent one of the unknown quantities.
5. Put any other unknown quantities in terms of the first unknown.
6. Make sure in steps 4 and 5 that the unknown quantities represent numerical quantities.
7. Using the given conditions in the problem and the unknowns from steps 4 and 5, write an equation.

Examples

1. One number is five times another number and their sum is 36.

Lesser number	*Greater number*	*Equation*
x	$5x$	$x + 5x = 36$

2. The length of a rectangle is three times as long as its width and its perimeter is 40.

Width	*Length*	*Equation*
x	$3x$	$2(x) + 2(3x) = 40$

note

In example 2 we need the prior knowledge that the perimeter of a rectangle is given by formula $P = 2w + 2l$.

3. The sum of three consecutive integers is 36.

First	*Second*	*Third*	*Equation*
x	$x + 1$	$x + 2$	$(x) + (x + 1) + (x + 2) = 36$

note

In example 3, we need the prior knowledge that if x is an integer, then if we add 1 to it, we get the next consecutive integer and if we add 2 to the first, we get the third consecutive integer.

4. Machine A is able to produce 12 more parts in 1 hour than machine B. In 8 hours their combined output is 896 parts.

Machine A	*Machine B*	*Equation*
$x + 12$	x	$8[(x + 12) + x] = 896$

Exercise Set 6.3

Simplify the following by finding the indicated sum and/or difference.

1. $3x + 4x + 7x$
2. $2i^2r - 4i^2r + 6i^2r$
•3. $4ab + 11ab - 10ab - 8ab$
4. $d^2 + d - 3d^2 + d^4 + 4d^2$
•5. $5x + x^2 - x + 6x^2$
6. $5x^2y - 3xy + 5y + 6xy - x^2y$
7. $a^2b - b^3 - ab^2 + 2a^3 - 5ab^2$
8. $3a + b + 2a - 5c - b - 2x^2 + 8a$
9. $x + 2x^2 - 5 + x^3 - 2x - 2x^2$
10. $3i + 8i - 6i + 9i$
11. $3i + 8I - 6i - 17I$
12. $28ei - 73ei + ei + 11ei - 9ei$
13. $(2x^2y - xy^2 + 7xy) + (xy^2 - 5x^2y + 8xy)$
14. $(5x^2 - y^2) - (6x^2 - 3y^2) - (8x^2 + 2y^2)$
•15. $(8a^3 - 2a^2b + 4ab^2 - 6b^3) - (4a^3 - 3a^2b - 2ab^2 - b^3)$
16. $(13R - 24wL) + (46wL - 16R - 26Z)$
17. $(48A + 3B) - (-22A - 6B)$
18. $(3.061I^2R - 0.062iI + 32W) + (0.72iI - 3I^2R)$
19. $(8eI + 9I^2R) - (13eI - 14IR)$
20. $(18E + 31R) - (23E - 14IR)$
21. $(E - 3I + 2) - (E + 5I - 8)$
22. $(2R + 6Z - 10K) - (8X + 3Z - 6R + 4)$

Write an expression for problems 23–41.

•23. The sum of two currents is 80 amps. If the greater current is I amps, express the smaller current in terms of I.

24. The sum of two pressures is 232 lb/in^2. If the lesser of the two pressures is P lb/in^2, express the greater pressure in terms of P.

25. The difference between two voltages is 25 volts. If the lesser voltage is E volts, express the greater voltage in terms of E.

26. The difference between the volumes of two containers is 432 in^3. If the volume of the greater container is V in^3, express the volume of the lesser in terms of V.

27. There are four voltages. The second voltage is two times the first; the third is one-half of the second; and the fourth is four times the third. Let E represent the first voltage in volts. Express the others in terms of E and find the sum of all four voltages in terms of E.

•28. There are three resistances in which the second is two-thirds of the first and the third is six times the second. If R represents the first resistance in ohms, express the others in terms of R and find their sum.

29. Write an expression that represents a resistance of $25R$ ohms more than $10R$ ohms less $13R$ ohms. Combine and simplify the expression.

30. Write an expression for the sum of $16I$ amperes and $13I$ amperes taken away from $43I$ amperes. Combine and simplify the expression.

31. A capacitance C exceeds another capacitance C_1 by 180 microfarads. Express the capacitance C_1 in terms of C.

32. Write an expression for the amount that $(3IR - 2IE)$ is greater than $(IR - 3IE)$. Simplify by removing the parentheses and combining.

33. The second side of a triangle is three-fourths as long as the first side, and the third side is twice as long as the second side. Write an expression for the perimeter in terms of the first side S. Find the perimeter in terms of S.

34. The radius of a given circle is r. The radius of a second circle is three times as long as the first, and the radius of a third circle is $1\frac{1}{4}$ the length of the first radius. If the circumference of a circle is given by $2\pi r$, write an expression for the sum of the circumferences of these circles in terms of r. Simplify.

35. The length of a rectangle exceeds its width by 36 ft. Write an expression for the perimeter in terms of the length l and simplify.

36. A piston weighs one-half as much as its connecting rod. Given an 8-cylinder engine: (a) Write an expression for the sum of the weights of the connecting rods in terms of the weight of the pistons, P; (b) write an expression for the sum of the weights of the pistons in terms of the weight of the connecting rods, C. Simplify each expression.

37. The tensile strength of copper wire is approximately twice that of aluminum wire. Write an expression for the tensile strength of aluminum wire in terms of the tensile strength of a copper wire, C.

38. A steel rod is cut into three pieces so that the longest piece is four times as long as the shortest piece, and the shortest piece is two-thirds as long as the other piece. Write an expression for the sum of the three pieces in terms of the middle-length piece, M. Simplify.
39. Three pipes are to fill a tank. Water flows through the first pipe at four-fifths the rate it flows through the second pipe. If water flows through the third pipe at twice the rate of the first pipe, write an expression for the rate the water flows through all three pipes in terms of the rate of the second pipe, R. Simplify.
40. A tank is filled by two pipes at the same time it is being drained by a third pipe. If the fluid flows through the first of the filling pipes at three times the rate it flows through the second filling pipe, and the emptying pipe flows at three-fifths the rate of the first pipe, write an expression for the rate the tank is filling and emptying in terms of the rate of flow through the second filling pipe, P. Simplify.
•41. A 50-lb bar is made from silver and tin. The bar contains $2\frac{1}{2}$ times more tin than it does silver. Write an expression for the weight of the bar in terms of the weight of the silver, s.

Calculator Problems

Using the rules of approximate numbers, simplify the following by finding the indicated sum and/or difference.

42. $1.417A + 13.804A - 0.017A$
43. $2.96x - 3.8y - 8.412y - 0.0056x$
44. $0.6E - 0.09R - 0.08E + 1.79R$
45. $3.24R + 127.47R + 96.0R - 5.6R$
46. $(0.25x - 0.035y) + (0.0875x + 1.023y)$
47. $(248.4A + 0.275B) + (87.9A - 12.43B)$
48. $(7.5R - 18.04S) - (3.75R - 9.09S)$
•49. $(35.004y - 0.014z) - (19.101y - 1.514z)$
50. $(0.002x^2 + 9.18x - 11.004) + (11.17x^2 - 0.14x + 6.17)$
51. $(8.4x^2 - 1.85x + 0.20) + (5.26x^2 + 74.66x - 7.5)$
52. $(9.73x^2 + 700.8x - 0.069) - (46.9x^2 + 0.63x - 0.0528)$
53. $(1.236x^2 - 0.035x + 1.375) - (0.045x^2 - 88.80x - 5.007)$
•54. $(8.85x^2 - 0.50x + 13.055) - (4.499x^2 - 1.07x + 1.067)$

6.4 Exponents

In chapter 1 we discussed some basic laws of exponents as related to numbers. Since variables are symbols for numbers, let us now apply these laws to them. Recall that the expression x^4 is called the **exponential form** of the product

$$x \cdot x \cdot x \cdot x$$

and that x is the **base** and 4 is the **exponent.** x^4 is read "the fourth power of x." Consider the following example.

$$x^2 \cdot x^3 = (x \cdot x) \cdot (x \cdot x \cdot x) = x \cdot x \cdot x \cdot x \cdot x$$
$$= x^5$$

which leads to the conclusion that

$$x^2 \cdot x^3 = x^{2+3} = x^5$$

In words this law says,

> To multiply factors having *like bases,* add the exponents to get the exponent of the given base in the product. In symbols this law is $a^m \cdot a^n = a^{m+n}$.

note

If no exponent is associated with a numeral or a variable, the exponent is understood to be 1. That is, $a^1 = a$.

Examples

Find the product.

1. $x^3 \cdot x^5 = x^{3+5} = x^8$
2. $3^2 \cdot 3^4 = 3^{2+4} = 3^6 = 729$
3. $e^2 \cdot e^3 \cdot e^4 = e^{2+3+4} = e^9$
4. $a^2 \cdot a \cdot a^3 = a^{2+1+3} = a^6$

Now consider

$$x^6/x^2 = \frac{x \cdot x \cdot x \cdot x \cdot x \cdot x}{x \cdot x}$$

$$= \frac{\cancel{x} \cdot \cancel{x} \cdot x \cdot x \cdot x \cdot x}{\cancel{x} \cdot \cancel{x}}$$

By reducing the numerator and denominator of the fraction by two factors of x each, we get

$$\frac{x^6}{x^2} = x \cdot x \cdot x \cdot x = x^4$$

Thus

$$\frac{x^6}{x^2} = x^{6-2} = x^4$$

A rule for dividing like bases is as follows:

> To divide quantities having like bases, subtract the exponent of the denominator from the exponent of the numerator to get the exponent of the given base in the quotient. In symbols, this law is $\frac{a^m}{a^n} = a^m \div a^n = a^{m-n}$

note

For the moment, we will restrict the exponent of the numerator to be greater than the exponent of the denominator.

Examples Find the quotient.

1. $\dfrac{x^7}{x^2} = x^{7-2} = x^5$

2. $\dfrac{5^4}{5} = 5^{4-1} = 5^3 = 125$

3. $\dfrac{i^3 \cdot i^2}{i^4} = i^{3+2-4} = i$

It is important to note that these first two laws apply *only* to factors *having the same base*.

Two additional laws of exponents that can be derived from the first law are illustrated by the following.

$$\begin{aligned}(xy)^3 &= (x \cdot y)(x \cdot y)(x \cdot y) && \text{commuting and associating}\\ &= (x \cdot x \cdot x) \cdot (y \cdot y \cdot y)\\ &= x^3 \cdot y^3\\ &= x^3y^3\end{aligned}$$

When a group of factors is raised to a power, raise each of the factors in the group to that power. That is, $(ab)^n = a^nb^n$.

Examples Find the product.

1. $(e \cdot i)^4 = e^4 \cdot i^4$
2. $(3 \cdot 4)^3 = 3^3 \cdot 4^3 = 1{,}728$
3. $(5IR)^2 = 5^2 \cdot I^2 \cdot R^2 = 25I^2R^2$

note

The quantity $(a + b)^3$ is not equal to $a^3 + b^3$ because a and b are terms, not factors as the rule specified. If we consider $(a + b)$ to be a single term, then by the definition of exponents we have:

$$(a + b)^3 = (a + b)(a + b)(a + b)$$

We will see how to multiply this in section 6.5.

$$(x^4)^3 = x^4 \cdot x^4 \cdot x^4 = x^{\overbrace{4+4+4}^{3}} = x^{4 \cdot 3} = x^{12}$$

We observe that

$$(x^4)^3 = x^{4 \cdot 3} = x^{12}$$

which indicates that when raising a number to a power and then raising that to another power, we multiply the exponents to get the power of the given base. That is,

$$(a^m)^n = a^{mn}$$

Examples

Find the product.

1. $(y^3)^2 = y^{3 \cdot 2} = y^6$
2. $(4^2)^5 = 4^{2 \cdot 5} = 4^{10}$
3. $(E^4)^5 = E^{4 \cdot 5} = E^{20}$

Further Examples

1. $a^{2x} \cdot a^{3x} = a^{2x+3x} = a^{5x}$
2. $x^{2a-1} \cdot x^{3a+2} = x^{(2a-1)+(3a+2)} = x^{2a-1+3a+2} = x^{5a+1}$
3. $(x^a)^{2a} = x^{a \cdot 2a} = x^{2a^2}$
4. $x^{5n} \div x^{3n} = x^{5n-3n} = x^{2n}$
5. $(a^2b^x)^x = (a^2)^x(b^x)^x = a^{2x}b^{x \cdot x} = a^{2x}b^{x^2}$
6. $(3x^2y)(2x^2y) = (3 \cdot 2)(x^2x^2)(yy) = 6x^4y^2$
7. $5a^2 \cdot a + 2a \cdot a^2 = 5a^3 + 2a^3 = 7a^3$
8. $4x(x^2)^2 + (2x)^3x = 4x \cdot x^4 + 2^3x^3 \cdot x = 4x^5 + 8x^4$

Exercise Set 6.4

Perform the indicated multiplications.

1. $6 \cdot 6 \cdot 6 \cdot 6$
2. $x \cdot x \cdot x \cdot x$
3. $I \cdot I \cdot I \cdot I \cdot I$
4. $L \cdot L \cdot w \cdot w \cdot h$
5. $IR \cdot IR \cdot IR$
6. $I \cdot I \cdot E \cdot E \cdot R \cdot R \cdot R$
•7. $4^3 \cdot 4^2$
8. $x^3 \cdot x^4 \cdot x$
9. $I^2 \cdot I \cdot I^3$
10. $V^2 \cdot V \cdot V \cdot V^4$
11. $(-5y)(3y)$
12. $(-5x^2)(4x^3)$
13. $(4ab)(5ab)$
14. $(-3IR)(3IR)(-4IR)$
•15. $(xy)(xyz)(-xz)$
16. $(-5ab^2c)(-2a^3c^2)(-3bc)$
•17. $(5i^3)(-3i)(-2i^2)(-i)$
18. $(4P)(-2P^2)(-P^3)(3P)$
•19. $x \cdot x \cdot (x + y) \cdot y$
20. $i \cdot i + 1 \cdot 1 \cdot 1$
21. $(xy)^3$
22. $\dfrac{3^4}{3^2}$
23. $\dfrac{R^8}{R^4}$
24. $\dfrac{a^5}{a^3}$
25. $\dfrac{y^5 \cdot y}{y^2}$
26. $\dfrac{a^4a^2}{a^5}$
27. $(2ab)^2$
28. $(xy^2)^3$
29. $(-5x^2y^3)^2$
•30. $(-3cd^2)^3$
31. $-(5i^2R)^3$
32. $\left(-\dfrac{1}{2}i^2v\right)(-5a^n)$
33. $(3a^n)(-5a^n)$
34. $(10^2)^3$
35. $(10^z)^4$
36. $x \cdot x + 3 \cdot x \cdot x$
37. $x^{3a} \cdot x^{2a}$
•38. $x^{a+1} \cdot x^{2a+1}$

39. $(2a^2bc^3)^a$

40. $(3x^2)^2 + 2x \cdot x^3$

41. $(5x)^2 - 10x \cdot x$

•42. $(2x^2)^2 \cdot x^3 + (3x^2)^3$

43. $(x^3)^2 \cdot x^2 - (2x^2)^3$

44. $(x^3)^a$

45. $(x^{2a})^{3a}$

46. $(a^{5b})^{4b}$

47. $(2a^nb^m)^p$

48. $\dfrac{a^7b^5}{a^4b^2}$

49. $\dfrac{x^9y^7}{x^4y}$

50. $\dfrac{3^4a^5b^3}{3^2a^4b}$

51. $\dfrac{5^3a^7b^3}{5^2a^3b}$

Calculator Problems Using the rules of approximate numbers, perform the indicated multiplications.

52. $(2.34)^2$

53. $(-3.1)^3$

54. $(-2.8)^2$

55. $(5.75)^3$

56. $(2.37a^2)(3.14a)$

57. $(0.03x^5)(1.07x^4)$

58. $(1.93a^2b^3)(14.7ab^5)$

59. $(5.7a^3b^9)(6.3a^7b^2)$

60. $(11.2a^4b^2)^2$

61. $(0.31x^5y^7)^2$

62. $(-6.36xy^{11})^2$

63. $(12.1xyz^4)^3$

6.5 Algebraic Products

We now consider the same properties of exponents when applied to variables in polynomials. Recall from the previous section to multiply the monomials

$$5e \cdot 4i$$

we apply the commutative and associative laws of multiplication and write this as a product of the numerical coefficients times the product of the variables. That is,

$$5e \cdot 4i = (5 \cdot 4) \cdot (e \cdot i) = 20ei$$

Note that the product of e and i can be written *only* as the indicated product ei since these do not represent like bases. To find the product of

$$3x^2 \cdot 5x$$

we apply the same laws plus the laws of exponents to get

$$3x^2 \cdot 5x = (3 \cdot 5) \cdot (x^2 \cdot x) = 15 \cdot x^3 = 15x^3$$

Examples Find the indicated products.

1. $4x \cdot 3xy = (4 \cdot 3) \cdot (x \cdot x) \cdot y = 12x^2y$
2. $8V^3 \cdot 4V^3 \cdot 3V = (8 \cdot 4 \cdot 3) \cdot (V^3 \cdot V^3 \cdot V) = 96V^7$
3. $(-2e^2) \cdot (3ie) = (-2 \cdot 3) \cdot (e^2 \cdot e) \cdot i = -6e^3i$

When multiplying a monomial times a binomial or a trinomial, the *distributive law* becomes the basis for determining the product. For example, to multiply

$$3x^2y(x^2 + xy + y^2)$$

we multiply the monomial $3x^2y$ times each term in the trinomial to get

$$(3x^2y \cdot x^2) + (3x^2y \cdot xy) + (3x^2y \cdot y^2)$$

which yields

$$3x^2y(x^2 + xy + y^2) = 3x^4y + 3x^3y^2 + 3x^2y^3$$

In each indicated product you should observe that we multiplied like bases by using the law of exponents. For example,

$$3x^2y \cdot x^2 = 3 \cdot (x^2 \cdot x^2) \cdot y = 3x^4y$$

Examples

Find the product.

1. $5y(2y + 3) = 5y \cdot 2y + 5y \cdot 3$
 $= 10y^2 + 15y$
2. $x^3(x^2 + xy - y^2) = x^3 \cdot x^2 + x^3 \cdot xy - x^3 \cdot y^2$
 $= x^5 + x^4y - x^3y^2$
3. $5(2x - 1) + 3(x + 4) = 10x - 5 + 3x + 12$
 $= 13x + 7$
4. $6(2r - 3) - 4(r - 8) = 12r - 18 - 4r + 32$
 $= 8r + 14$

The product of two binomials requires the use of the distributive law twice. That is, in the product

$$(2L + P)(3L - 2P)$$

we apply the law once by multiplying each term of the first binomial times the second binomial to get

$$2L(3L - 2P) + P(3L - 2P)$$

Then we apply it again as we did previously when multiplying a monomial times a binomial and the product becomes

$$(2L \cdot 3L) - (2L \cdot 2P) + (P \cdot 3L) - (P \cdot 2P)$$

And so

$$(2L + P)(3L - 2P) = 6L^2 - 4LP + 3LP - 2P^2$$

Adding like terms gives

$$6L^2 - LP - 2P^2$$

Notice in this product that *each term of one binomial factor is multiplied by each term of the other binomial factor*. This procedure is completely consistent with multiplying whole numbers and integers, where each digit of one number is multiplied times each digit of the other number.

Examples Find the product.

1. $$\begin{aligned}(e + 3)(e - 4) &= e(e - 4) + 3(e - 4)\\ &= (e \cdot e) - (e \cdot 4) + (3 \cdot e) - (3 \cdot 4)\\ &= e^2 - 4e + 3e - 12\\ &= e^2 - e - 12\end{aligned}$$
2. $$\begin{aligned}(2i + 3)(5i - 2) &= 2i(5i - 2) + 3(5i - 2)\\ &= (2i \cdot 5i) - (2i \cdot 2) + (3 \cdot 5i) - (3 \cdot 2)\\ &= 10i^2 - 4i + 15i - 6\\ &= 10i^2 + 11i - 6\end{aligned}$$

With continued practice in such multiplication you should observe that a general pattern will emerge that is easily adaptable to the product of any two binomials containing similar terms. That is, the first term of the product is the product of the first terms of the binomials:

$$6x^2$$
$$(2x + 1)(3x - 3)$$

The second term is the sum of the products of the two "outer" terms and the two "inner" terms:

$$3x$$
$$(2x + 1)(3x - 3)$$
$$-6x$$

Thus, the second term is $-6x + 3x = -3x$. The last term in the product is the product of the last terms in the binomials.

$$-3$$
$$(2x + 1)(3x - 3)$$

Thus,

$$(2x + 1)(3x - 3) = 6x^2 - 3x - 3$$

The ability to perform this process easily will lead to the multiplication becoming a mental process and thus saving time.

note

In the product $(a + b)(2x - y)$ we cannot apply this method because we do not have similar terms:

$$\begin{aligned}(a + b)(2x - y) &= a(2x - y) + b(2x - y)\\ &= 2ax - ay + 2bx - by\end{aligned}$$

Two special products warrant discussion since they appear so often, and you should be able to put down the products without computation. First, consider the product

$$(x + 6)^2 = (x + 6)(x + 6)$$

which becomes

$$x^2 + 6x + 6x + 36$$

When we apply the distributive law to combine the second and third terms we get

$$x^2 + 12x + 36$$

This is called the "square of a binomial" and has certain characteristics. Inspection shows that in

$$(x + 6)^2 = x^2 + 12x + 36$$

the three terms of the product can be obtained in the following manner:

1. The first term of the product is the *square of the first term* of the binomial $[(x)^2 = x^2]$.
2. The second term of the product is *two times the product of the two terms of the binomial* $(2 \cdot x \cdot 6 = 12x)$.
3. The third term of the product is the *square of the second term* of the binomial $[(6)^2 = 36]$.

If we apply the above to

$$(x - 7)^2$$

we get

$$x^2 + [2 \cdot x \cdot (-7)] + (-7)^2$$

and so

$$\begin{aligned}(x - 7)^2 &= x^2 + (-14x) + 49\\ &= x^2 - 14x + 49\end{aligned}$$

In general, for numbers x and y,

$$(x + y)^2 = x^2 + 2xy + y^2$$

Examples Find the product.

1. $$\begin{aligned}(2x + 3)^2 &= (2x)^2 + (2 \cdot 2x \cdot 3) + (3)^2\\ &= 4x^2 + 12x + 9\end{aligned}$$
2. $$\begin{aligned}(5I - 4E)^2 &= (5I)^2 + [2 \cdot 5I \cdot (-4E)] + (-4E)^2\\ &= 25I^2 + (-40IE) + 16E^2\\ &= 25I^2 - 40IE + 16E^2\end{aligned}$$

A second special product comes from the product

$$(x + 3)(x - 3)$$

which becomes

$$x^2 + 3x - 3x - 9$$

when we apply the distributive law. Combining the middle terms we get

$$x^2 - 9$$

Thus,

$$(x + 3)(x - 3) = x^2 - 9$$

Special characteristics are evident in this product, too; namely, the product may be obtained by

1. Squaring the first term of the factors.
2. Subtracting the square of the second term of the factors.

This special product is referred to as the "product of the sum and the difference of the same two numbers" and *always* equals the *square of the first term minus the square of the second term*. In general, for numbers a and b,

$$(a + b)(a - b) = a^2 - b^2$$

To generalize the rule of multiplying two polynomials together: *We multiply each of the terms in the first polynomial times each of the terms in the second polynomial.*

Examples

Find the product.

1. $(x + 7)(x - 7) = (x)^2 - (7)^2 = x^2 - 49$
2. $(2i + 3)(2i - 3) = (2i)^2 - (3)^2 = 4i^2 - 9$
3. $(3R - 2E)(3R + 2E) = (3R)^2 - (2E)^2 = 9R^2 - 4E^2$
4. $(2a^2 + 3a + 2)(a - 2)$
   $= (2a^2 \cdot a) - (2a^2 \cdot 2) + (3a \cdot a) - (3a \cdot 2) + (2 \cdot a) - (2 \cdot 2)$
   $= 2a^3 - 4a^2 + 3a^2 - 6a + 2a - 4$
   $= 2a^3 - a^2 - 4a - 4$
5. $(a + b)(a - 2b)(a - b)$
   $= [(a + b) \cdot (a - 2b)](a - b)$
   $= [(a \cdot a) - (a \cdot 2b) + (ab) - (b \cdot 2b)](a - b)$
   $= [a^2 - 2ab + ab - 2b^2](a - b)$
   $= [a^2 - ab - 2b^2](a - b)$
   $= (a^2 \cdot a) - (a^2 \cdot b) - (ab \cdot a) + (ab \cdot b) - (2b^2 \cdot a) + (2b^2 \cdot b)$
   $= a^3 - a^2b - a^2b + ab^2 - 2ab^2 + 2b^3$
   $= a^3 - 2a^2b - ab^2 + 2b^3$
6. $(a + b)^3 = (a + b)(a + b)(a + b)$
   $= [(a + b)(a + b)](a + b)$
   $= [(a \cdot a) + (a \cdot b) + (a \cdot b) + (b \cdot b)](a + b)$
   $= [a^2 + ab + ab + b^2](a + b)$
   $= [a^2 + 2ab + b^2](a + b)$
   $= (a^2 \cdot a) + (a^2 \cdot b) + (2ab \cdot a) + (2ab \cdot b) + (b^2 \cdot a) + (b^2 \cdot b)$
   $= a^3 + a^2b + 2a^2b + 2ab^2 + ab^2 + b^3$
   $= a^3 + 3a^2b + 3ab^2 + b^3$

Exercise Set 6.5

Write the product of the following polynomials and simplify.

1. $2ab(a^2 - bc + c^2)$
2. $6R(4i + 7zI)$
3. $3R(5I^2 - 7i^2)$
4. $-ei(e^4 - e^2i^2 - i^4)$
•5. $-5ab^2(3a^2 - ab + 4b^2)$
6. $6x^2(4x^2 - 2x + 3)$
7. $3ir(i^2 - 2ir - r^2)$
8. $(2x)(x - y + 5)(5y)$
9. $5(2e - 3) + 3(e + 4)$
10. $6(2r - 3) - 4(r - 8)$
11. $8(4E - 3e) - 3(E - 9e)$
12. $I(3R - 6) - 3I(R + 2)$
13. $2R(i^2 - 3I^2) + R(6I^2 - 8i)$
14. $2w(M + 3L) - 2(6L - 5M)$
15. $6\left(\frac{E}{3} - \frac{e}{2} + \frac{R}{6}\right)$
16. $8\left(\frac{I^2R}{4} + \frac{i^2r}{2}\right)$
•17. $(x + 3)(x + 4)$
18. $(P + 5)(P - 3)$
19. $(y - 9)(y - 4)$
20. $(z + 7)(z - 11)$
21. $(V + 1)(V + 1)$
22. $(r - 1)(r - 1)$
23. $(R - 3)^2$
24. $(R + 2)(R - 2)$
25. $(F + 3)(F - 3)$
26. $(3x + 2)(x - 4)$
27. $(3a - 5)(2a - 7)$
28. $(3 - 2y)(2 - y)$
29. $(7 + 2x)(2x - 7)$
30. $(4r + 3)(r - 12)$
31. $(3k + w)(k - 6w)$
32. $(E - 6IR)(5E + 4IR)$
•33. $(E + 6e)^2$
34. $(2R + 3r)^2$
35. $(2e + 3i)(2e - 3i)$
36. $(4L - w)(4L + w)$
•37. $(a + 4b)(a^2 - 2ab + b^2)$
38. $(x - 2y)(2x^2 - 3xy + y^2)$
39. $(I + 4)(6I^2 - 3I + 7)$
40. $(R - r)(R^2 - 2Rr + r^2)$
•41. $(x^2 - 2x - 3)(x^2 + x + 4)$
42. $(a^2 - 3a + 6)(a^2 + 2a - 5)$
43. $(V - 6)(V - 2)(V + 1)$
44. $(2F - 1)(F + 2)(2F + 1)$
45. $(R - r)(R + r)(2R - 3r)$
46. $(a + b)^3$
47. $(a - b)^3$
48. $(2a + b)^3$
•49. $(a - 2b)^3$

Calculator Problems

Using the rules of approximate numbers, write the product of the following polynomials and simplify.

50. $3.7a(2.1a^2 - 6.4a + 7.1)$
•51. $1.06R^2(1.1R^3 - 3.96R^2 + 2.11R)$
52. $(-5.17)x^2y^2(-4.87x^3y + 8.86x^4y^4)$
53. $(-7.21)a^2b^3(7.1ab^4 - 5.73a^2 - 0.24ab^5)$
54. $(x + 7.59)(x - 13.27)$
55. $(a - 3.19)(a - 1.63)$
56. $(x - 3.57)(x - 4.18)$
57. $(x - 4.57)^2$
58. $(y + 6.38)^2$

6.6 Algebraic Quotients

We are now going to look at the indicated quotient of two polynomials as a fraction. Recall from section 2.3 that division by zero is undefined and, therefore, we must restrict the values that the variables can assume so that no divisor, or denominator, can be equal to zero. For example, in the indicated quotient

$$\frac{3x^3 - 9x^2 + 15x}{3x}$$

$x \neq 0$ since, if we allow x to be zero, we have $3 \cdot 0 = 0$ in the denominator and the result is undefined. To perform this division, recall a fundamental principle of fractions:

$$\frac{a}{c} + \frac{b}{c} = \frac{a + b}{c}$$

We can turn this rule around, and we have

$$\frac{a + b}{c} = \frac{a}{c} + \frac{b}{c}$$

which is the procedure we use to divide a polynomial of more than one term by a monomial. To carry out this division, we divide every term of the polynomial by the monomial, as follows

$$\frac{3x^3}{3x} - \frac{9x^2}{3x} + \frac{15x}{3x}$$

Using the law of exponents for dividing like bases, we divide the coefficients and the variables to get

$$\begin{aligned} \frac{3}{3} \cdot \frac{x^3}{x} - \frac{9}{3} \cdot \frac{x^2}{x} + \frac{15}{3} \cdot \frac{x}{x} &= x^{3-1} - 3x^{2-1} + 5 \\ &= x^2 - 3x + 5 \end{aligned}$$

Examples

Find the indicated quotient.

1. $$\begin{aligned} \frac{6x^4 - 6x^3 + 12x^2}{6x} &= \frac{6x^4}{6x} - \frac{6x^3}{6x} + \frac{12x^2}{6x} \\ &= x^3 - x^2 + 2x \end{aligned}$$

2. $$\begin{aligned} \frac{8e^4 + 4e^2 - 12e}{4e} &= \frac{8e^4}{4e} + \frac{4e^2}{4e} - \frac{12e}{4e} \\ &= 2e^3 + e - 3 \end{aligned}$$

Consider the quotient where the divisor is not a monomial.

$$\frac{y^2 - y - 2}{y - 2}$$

This problem involves the division of two multinomials. We handle this just like the long division of numbers. We set it up in the form

$$y - 2 \overline{\smash{)}\, y^2 - y - 2}$$

note

The divisor and dividend must be arranged in descending powers of one variable with zeros inserted to hold the position of any missing term.

First, we divide y of the divisor into y^2 of the dividend to get

$$\begin{array}{r} y \\ y - 2\,\overline{)\,y^2 - y - 2} \end{array}$$

Now we multiply y of the quotient times the entire divisor and write the product below the first two terms of the dividend:

$$\begin{array}{rl} & \phantom{y^2 - {}}y \\ y - 2\,\big) & \overline{y^2 - y - 2} \\ & \underline{y^2 - 2y} \end{array}$$

We subtract the result, yielding

$$\begin{array}{rl} & \phantom{y^2 - {}}y \\ y - 2\,\big) & \overline{y^2 - y - 2} \\ & \underline{-\,(y^2 - 2y)} \\ & \phantom{y^2 - 2{}}y \end{array} \qquad \text{or} \qquad \begin{array}{rl} & \phantom{y^2 - {}}y \\ y - 2\,\big) & \overline{y^2 - y - 2} \\ & \underline{-\,y^2 + 2y} \\ & \phantom{y^2 + 2{}}y \end{array}$$

Now we bring down the next term of the dividend, -2, to form an expression with the remainder y:

$$\begin{array}{rl} & \phantom{y^2 - {}}y \\ y - 2\,\big) & \overline{y^2 - y - 2} \\ & \underline{y^2 - 2y} \\ & \phantom{y^2 - {}}y - 2 \end{array}$$

We divide the y of the divisor into the y of this remainder expression and place the result, 1, in the quotient:

$$\begin{array}{rl} & \phantom{y^2 - {}}y + 1 \\ y - 2\,\big) & \overline{y^2 - y - 2} \\ & \underline{y^2 - 2y} \\ & \phantom{y^2 - {}}y - 2 \end{array}$$

We multiply the 1 times the entire divisor, placing this product below the expression $y - 2$. Subtracting, we get a remainder of zero, and so the quotient we want is $y + 1$:

$$\begin{array}{rl} & \phantom{y^2 - {}}y + 1 \\ y - 2\,\big) & \overline{y^2 - y - 2} \\ & \underline{y^2 - 2y} \\ & \phantom{y^2 - {}}y - 2 \\ & \phantom{y^2 - {}}\underline{y - 2} \\ & \phantom{y^2 - y - {}}0 \end{array}$$

Therefore,

$$\frac{y^2 - y - 2}{y - 2} = y + 1$$

Notice that y cannot take on the value of 2 in this case since, if it did, we would have $2 - 2 = 0$ in the denominator and the quotient would be undefined.

It is important for you to note that one of the true "pitfalls" in this operation lies in *subtracting* the product of a term in the quotient times the divisor from the terms of the dividend. In the previous example, when we subtract

$$\begin{array}{r}
y \phantom{{}-y-2} \\
y - 2\,\overline{)\,y^2 - y - 2} \\
(-)\ \underline{y^2 - 2y} \phantom{{}-2}
\end{array}$$

we do so by *adding* the opposite of the subtrahend. Recall that when a grouping symbol is prefixed by a minus sign, we change the sign of each of the enclosed terms when we remove the grouping symbol.

$$(y^2 - y) - (y^2 - 2y)$$

becomes

$$y^2 - y + (-y^2) + 2y$$

which equals

$$[y^2 + (-y^2)] + [2y - y] = y$$

and so the remainder is y. Most errors in this type of problem are committed at this point.

Examples Find the indicated quotient.

1. $\dfrac{x^2 + 3x - 4}{x + 4}$; $\begin{array}{r}
x - 1 \\
x + 4\,\overline{)\,x^2 + 3x - 4} \\
\underline{x^2 + 4x} \phantom{{}-4} \\
-x - 4 \\
\underline{-x - 4} \\
0
\end{array}$

$= x - 1$

2. $\dfrac{2x^2 + x - 3}{x - 1}$; $\begin{array}{r}
2x + 3 \\
x - 1\,\overline{)\,2x^2 + x - 3} \\
\underline{2x^2 - 2x} \phantom{{}-3} \\
3x - 3 \\
\underline{3x - 3} \\
0
\end{array}$

$= 2x + 3$

If, when we perform the indicated division, we still have a remainder after "bringing down" all the terms of the dividend, we proceed as follows:

$$\frac{e^2 + 5e + 6}{e - 2}\ ; \quad e - 2\,\overline{\big)\,e^2 + 5e + 6}^{\;e + 7}$$

$$\begin{array}{r} \underline{e^2 - 2e} \\ 7e + 6 \\ \underline{7e - 14} \\ 20 \end{array}$$

$$= e + 7 + \frac{20}{e - 2}$$

In the problem

$$\frac{x^3 - x + 2}{x - 3}$$

note that no term in the dividend contains x^2. The division will be easier to perform if the term $0x^2$ is inserted as a placeholder so that all powers of the variable x are present in descending order. Thus, we have

$$\frac{x^3 + 0x^2 - x + 2}{x - 3}$$

and the value of the dividend has not been changed since we have added $0x^2$, which is zero. Therefore, to perform the division we get

$$x - 3\,\overline{\big)\,x^3 + 0x^2 - x + 2}^{\;x^2 + 3x + 8}$$

$$\begin{array}{r} \underline{x^3 - 3x^2} \\ 3x^2 - x \\ \underline{3x^2 - 9x} \\ 8x + 2 \\ \underline{8x - 24} \\ 26 \end{array}$$

and so

$$\frac{x^3 - x + 2}{x - 3} = x^2 + 3x + 8 + \frac{26}{x - 3}$$

Exercise Set 6.6

Perform the indicated divisions.

1. $\dfrac{8x^3}{2x}$
2. $\dfrac{x^3y^2}{xy}$
3. $\dfrac{-15x^5}{3x^2}$
4. $\dfrac{-x^4y^2z}{xy^2z}$

•5. $\dfrac{22a^2b^3c}{11ab^2c}$

6. $\dfrac{-65x^4y^2z}{13xy}$

7. $\dfrac{-28E^3I}{-7EI}$

8. $\dfrac{35e^3i^2}{-5e^2i}$

9. $\dfrac{3(a - b)^2}{a - b}$

10. $\dfrac{-24l^2w^3h^4}{8lwh^3}$

11. $\dfrac{4x - 6}{2}$

12. $\dfrac{6x - 9}{3}$

•13. $\dfrac{bx^2 - bx}{bx}$

14. $\dfrac{a^3 - 3a^2 + 2a}{a}$

15. $\dfrac{12x^3 - 8x^2 + 3x}{4x}$

16. $\dfrac{15a^3 - 9a^2 + 12a - 6}{3}$

17. $(12x^4 - 16x^3 - 8x + 4) \div 4$

18. $(16R^3 - 4R^2 - 20R) \div 4$

19. $\dfrac{24\theta^2 - 12\theta}{-6}$

20. $\dfrac{13R - I^2R^2 + I^2R}{I^2R}$

•21. $\dfrac{x^2y - xy^2 - 2xy^3}{xy}$

22. $\dfrac{14a^2b^3 - 21a^2b^2 - 28ab}{7ab}$

23. $\dfrac{R(i - 1) - E(i - 1)}{i - 1}$

24. $\dfrac{30x^3y^4 + 21x^2y^2 - 18x^2y^4}{3x^2y^2}$

•25. $\dfrac{I(e - ir) - i(e - ir)}{e - ir}$

26. $\dfrac{a^2 - 7a + 10}{a - 2}$

•27. $\dfrac{x^2 + 8x + 15}{x + 5}$

28. $\dfrac{e^2 + 5e + 6}{e + 3}$

29. $\dfrac{E^2 - E - 72}{E + 8}$

30. $(R^2 - 11R + 30) \div (R - 5)$

31. $(a^2 + 6a + 9) \div (a + 3)$

32. $(4a^2 + 4a + 1) \div (2a + 1)$

33. $(9a^2 - 24a + 16) \div (3a - 4)$

34. $(27a^3 - 1) \div (3a - 1)$

35. $(x^3 - 8) \div (x - 2)$

36. $(x^4 - 16) \div (x - 2)$

37. $(x^2 + 2xy + y^2) \div (x + y)$

38. $8x^2 - 22xy - 21y^2) \div (4x + 3y)$

39. $(15a^2 + 8ab - 12b^2) \div (3a - 2b)$

40. $(9a^2 - 2ab - 6b^2) \div (a + b)$

41. $\dfrac{x^3 + 4x^2 + 7x + 6}{x + 2}$

42. $\dfrac{2F^3 - 3F^2 - 13F + 12}{F - 5}$

•43. $\dfrac{I^3 + 6I^2 + 7I - 8}{I - 1}$

44. $\dfrac{6V^4 - V^3 - 8V^2 - 8V - 19}{2V - 3}$

45. $(15a^2 + 28a - 32) \div (-5a + 4)$

46. A contractor uses the formula $x^2 + 6x + 8$ to represent the square footage of a room. If he decides that the length of the room will be represented by $x + 4$, what will the width of the room be in terms of x? If he wants the room to be 12 ft long, what will be its width? What will be the square footage of the room?

47. An electrician uses the formula $x^2 + 5x + 6$ to determine the amount of wire to order when wiring a house. If the formula comes from multiplying the number of rooms times the number of outlets and she knows the number of rooms to be $x + 2$, find the number of outlets in terms of x. If, in a particular house, there are 6 rooms, how many outlets will there be? How much wire should she order?

Chapter Summary

1. A **variable is** a symbol that may take on different values in a given discussion.
2. A **constant** is the symbol that does not change its value.
3. An **algebraic expression** is a meaningful collection of variables, constants, grouping symbols, and signs of operation.
4. The symbols separated by plus or minus signs in an algebraic expression are the **terms** of the expression.
5. In the expression $8x$, 8 is called the **numerical coefficient.**
6. A *polynomial* is a special kind of algebraic expression. A *monomial* is a polynomial containing one term; a *binomial* contains two terms; a *trinomial* contains three terms; a *multinomial* contains two or more terms.
7. In an algebraic expression we can combine only **like terms.**
8. In the expression x^6, x is called the **base** and 6 the **exponent.**
9. We can *add, subtract, multiply,* and *divide* algebraic expressions.
10. Different measures of the same thing are denoted by **subscripts.**

Chapter Review

[6.1] Name each polynomial by the number of terms:

1. $4x + 3x^2 + 2$ 2. $6k^3$ 3. $7xy + 3$ 4. $EI - R + R^2$

[6.2] Evaluate the following.

5. Evaluate $HP = \dfrac{PLANK}{33{,}000}$ when $P = 40$, $L = 25$, $A = 2$, $N = 60$, $K = 11$.

6. Evaluate $\dfrac{1}{R} + \dfrac{1}{R_1} + \dfrac{1}{R_2}$ when $R = 7$, $R_1 = 6$, $R_2 = 8$.

7. Evaluate z^2 when $z^2 = R^2 + x^2$ given $R = 4$ and $x = 8$.

8. Evaluate R when $R = \dfrac{P \cdot L}{D^2}$ given

   a. $P = 6, L = 8, D = 4$ b. $P = 7, L = 3, D = \dfrac{2}{3}$

9. Evaluate V_2 when $V_2 = \dfrac{P_1V_1}{P_2}$, given $P_1 = 780.0$, $V_1 = 80.0$, $P_2 = 60.0$.

[6.3] Find the indicated sums and/or differences.

10. $3E - E + 6E - 2E$
11. $5I^2 + 6I - I + 3I^2$
12. $(4P + 3R) + (6R - 2P)$
13. $(8V_0 - 3V_1) - (6V_0 - V_1)$
14. $(5E - 4R + 1) - (6I + 3E - 2R)$

[6.4] Perform the indicated multiplication.

15. $x^4 \cdot x^3$
16. $x^2 \cdot x^3 \cdot x^5$
17. $(-3x^2)(2x^3)$
18. $(-2cd^2)^4$
19. $(3a^2b^3c^5)^4$
20. $x \cdot x^2 + 4x^2 \cdot x$
21. $a^{x+1} \cdot a^{2x+3}$
22. $a^{3x} \cdot a^{2x}$
23. $(x^{3a})^{4a}$
24. $(2a^rb^s)^t$

[6.5] Find the indicated products.

25. $4x(3x^2 - 2x + 4)$
26. $6R^2(2ER - 4R + 6E^2)$
27. $(3x + 4)(2x - 3)$
28. $(7e - 3i)(2e + i)$
29. $(2P - 3)^2$
30. $(3R + 4)(3R - 4)$
31. $(E - 3)(E^2 + 2Ee - e^2)$
32. $(x^2 + 2x - 3)(x^2 - 4x + 1)$
33. $(V + 1)(2V - 3)(3V - 2)$

[6.6] Find the indicated quotients.

34. $\dfrac{24x^3}{-3x}$

35. $\dfrac{2E^2 - 3E + 5E^3}{E}$

36. $(8I^2 - 2I - 3) \div (2I - 1)$

37. $\dfrac{20n^3 - 19n^2 - 13n + 12}{4n - 3}$

Chapter Test

Perform the indicated operations and simplify.

1. $(3x^2 - 2x + 4) - (x^2 - 5x + 7)$
2. $(3a^2b)(2ab^4)(a^5b^2)$
3. $(3x^2y - 2xy + 5xy^2) - (x^2y - 3xy + 2x^2y^2)$
4. $(2a - b)^2$
5. $3x(2x - y) - 2x(x + y)$
6. $(3x)(4x + 3y)(2x)$
7. $(3x - 2y)(x + 4y)$
8. $3a^2b^3(9a^4b + 2ab - 7ab^3)$
9. $(2x^5y^2)(-4x^9y)$
10. $\dfrac{2^3a^2b^4}{4ab^2}$

11. $(x^2 - x - 6) \div (x + 2)$
12. $(2R - 3S)(2R + 3S)$
13. $(x^2 - 8x + 13) \div (x - 2)$
14. $5x^5y^4(2x^2y^2 - 4xy + 3)$
15. $(3x + y)^2$
16. $\dfrac{-18x^5y^4}{3^2x^2y^2}$
17. $(x^2 + 5 - 10x + 6x^3) \div (2x - 1)$
18. $3(2a - b) - 4(a - 2b)$
19. $(2x^2 + x + 1)(x^2 - x - 4)$
20. $(5x^2 + 3x - 4) - (2x^2 - 5x - 9)$
21. $(a + b)(a - b)$
22. $\dfrac{15a^9b^{11}c^4}{5a^5b^4c^3}$
23. $(2x^3 - 3x^2 - 13x + 12) \div (x - 3)$
24. $(x - y)^3$
25. $(2a + b)(a - b)(a + b)$
26. $\dfrac{5(2x - y)^3}{(2x - y)^2}$
27. $(5a - b)(3a + b)(2a - 4b)$

Evaluate the following if $a = -5$, $b = 3$, $c = 4$, $d = -3$.

28. $(a - 2d)^2$
29. $(2a - 3b) - (5c + d)$
30. $(3a - b)(3a + b)$

7
Linear Equations and Inequalities

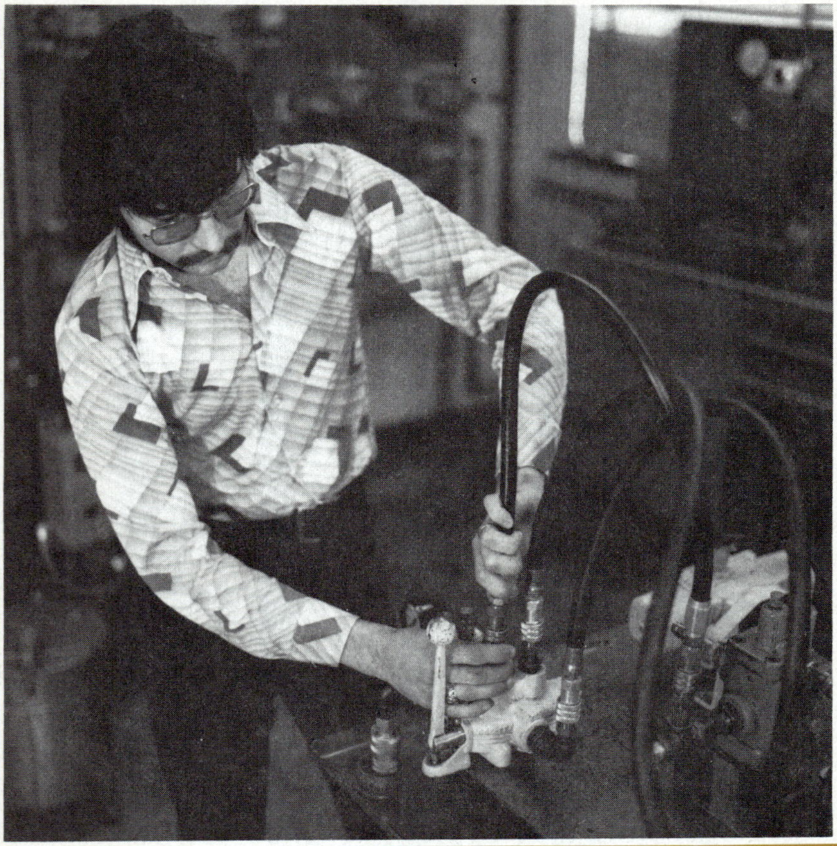

A technician connecting a circuit to observe pressure and flow differentials.
(Photograph by Kenneth Walters, courtesy of Henry Ford Community College)

7.1 Identities and Conditional Equations

A **mathematical statement** is a declarative sentence that can be labeled true or false. Consider the statements $2 + 3 = 5$ and $3 + 4 = 8$. By inspection we can see that the first mathematical sentence is true while the second one is false. Other mathematical sentences such as $2 + x = 8$ and $4 - x = 7$ cannot be judged as true or false. Such sentences are called **open sentences.** The truth or falsity of the sentence is "open" since we do not know the value that the variable represents. In mathematics, sentences involving equality relationships are called **equations.**

A replacement value for the variable that forms a true statement is called a **root,** or **solution,** of that open sentence. We say that a root of the given equation *satisfies* that equation.

Example

If we replace x by 6 in the equation $2 + x = 8$, the equation is true (that is, $2 + 6 = 8$). Then 6 is a root of the equation.

In the preceding example, there is just one root, 6, since this is the only replacement value for x for which the equation is true. Equations that are not true for all permissible number replacements of the variable are called **conditional equations.** If the equation is true for every permissible value of the variable, it is called an **identity.** For example,

$$2(x + 3) = 2x + 6$$

is true for any real number replacement for x and is thus called an *identity*. Properties such as

$$a + (b + c) = (a + b) + c \qquad \text{and} \qquad a \cdot b = b \cdot a$$

are other examples of identities.

In this chapter, we are primarily concerned with *first-degree conditional equations* called **linear equations.** A first-degree equation means that the exponent of the variable we are solving for is 1 and that the solution will be one value.

Example

$2 + x = 8$ and $4 - x = 7$ are first-degree equations since only one replacement value for the variable will satisfy the equation. By inspection we see that $2 + (6) = 8$, so 6 is the solution of the first equation, and $4 - (-3) = 7$, so -3 is the solution of the second equation.

---

note

Certain equations do not have a solution—for example, $x = x + 1$. There is no number that is equal to one more than itself.

Exercise Set 7.1

Find the solution by inspection. If the equation is an identity, indicate that.

Examples 1. $x + 5 = 11$. We can ask: what number added to 5 gives 11? That number is 6.

2. $3x = 15$. We ask ourselves: 3 times what number gives 15? That number is 5.

3. $4 + x = x + 4$. We observe that this problem is an application of the commutative law of addition and is true for all numbers; therefore it is an identity.

1. $x + 3 = 5$
2. $7 + x = 15$
•3. $4 = x - 7$
4. $11 = 3 - x$
5. $18 = x - 4$
6. $5x = x(3 + 2)$
•7. $8x = 32$
8. $I + 8 = 8 + I$
9. $y - 3 = -3 + y$
10. $6E = 28$
•11. $5(y - 2) = 5y - 10$
12. $y - 20 = 3$
•13. $\frac{5}{3}x = 20$
14. $4x = -3$
15. $\frac{1}{3}i = \frac{i}{3}$
16. $8x = 18$
17. $5(E + 3) = 5E + 15$
18. $y \cdot \left(\frac{1}{y}\right) = 1$
19. $\frac{3}{2}V = 21$
20. $R + 9 = 23$

Determine whether the given statement is true or false after the variable in each equation is replaced with the given number:

Examples 1. $3x + 3 = 6$; 1
Replacing x with 1 we get $3(1) + 3 = 6$, so $6 = 6$ and the statement is *true*.

2. $2x - 1 = 3$; 4
Replacing x with 4 we get $2(4) - 1 \neq 3$ so the statement is *false*.

21. $4 + x = 8$; 4
22. $3 - x = 4$; -1
•23. $x + 7 = 10$; 3
24. $3x - 2 = 4$; 2
25. $4 + \frac{x}{3} = 6$; 6
26. $5x - 2 = x + 1$; -1
27. $3x + 2 = 2x - 3$; 2
28. $\frac{1}{R} + 5 = 4$; -1
29. $\frac{2}{5} = 2 - \frac{8}{I}$; 4
30. $3I + 7 + 2I = I - 5$; -3
31. $4(E + 6) = 2E$; -12
32. $4e - 6 = \frac{1}{e}$; 1

33. $8P + 6 = 2P - 6; -2$

34. $\frac{4}{5}V + 2 = 10; 10$

•35. $7W - 3 = 2W + 2; -2$

Example Given the statement $A = L \cdot W$, if $A = 48$ and $L = 6$, find W.

Solution Substituting 48 in place of A and 6 in place of L we have $48 = 6W$. By inspection, we ask ourselves what number times 6 gives 48? That number is 8, so $W = 8$.

36. Given the formula $E = IR$.
   a. If $E = 24$ and $I = 6$, find R.
   b. If $E = 0.14$ and $R = 2$, find I.

37. Given the formula $I = \frac{E}{R}$.
   a. If $I = 20$ and $E = 4$, find R.
   b. If $I = 1.5$ and $R = 0.6$, find E.

38. Given the formula $V = LWH$.
   a. If $V = 36$, $L = 6$, $W = 2$, find H.
   b. If $V = 240$, $L = 20$, $H = 5$, find W.
   c. If $V = 7$, $W = 0.5$, $H = 4$, find L.

39. Given the statement $z = 5t - 3$.
   a. If $z = 2$, find t.
   b. If $z = 27$, find t.
   c. If $z = 4.5$, find t.

40. Given the algebraic statement of variation $Wt = k$.
   a. If $W = 16$ and $k = 24$, find t.
   b. If $t = 3$ and $k = 16$, find W.

41. Given $R_1 + R_2 + R_3 = 165$.
   a. If $R_1 = 46$ and $R_2 = 35$, find R_3.
   b. If $R_1 = 102$ and $R_3 = 16$, find R_2.

42. Given the relationship $R = \frac{4}{5}T$.
   a. Find T if $R = 16$.
   b. Find T if $R = \frac{25}{12}$.

43. Given $2N_1 - N_2 = 24$.
   a. Find N_1 when $N_2 = 8$.
   b. Find N_2 when $N_1 = 7$.

44. Given input = output/efficiency.
   a. Find output when input = 6.25 and efficiency = 0.80.
   b. Find efficiency when input = 8 and output = 6.

45. Given the formula for the horsepower of an electrical motor $\text{HP} = \frac{IE}{750}$.
   a. Find the amperes I when HP = 20 and volts $E = 15$.
   b. Find the volts E when HP = 25 and amperes $I = 150$.

7.2 Solving Equations—I

In section 7.1, the first-degree conditional equations we considered were such that inspection allowed us to determine the solution. Unfortunately, most equations cannot be solved by inspection, and a procedure must be developed so we can determine the roots.

To develop this procedure, let us examine the equation

$$6(x + 1) = 4x + 10$$

whose solution is not evident by inspection. A series of steps to find the solution might be as follows:

$$\begin{aligned} 6(x + 1) &= 4x + 10 \\ 6x + 6 &= 4x + 10 \\ 2x + 6 &= 10 \\ 2x &= 4 \\ x &= 2 \end{aligned}$$

In all of these equations the root of the equation is 2, and for that reason they are said to be **equivalent equations.** This series of equations demonstrates the procedure for solving for an unknown. The process consists of forming equivalent equations until we get to the point of having the unknown on one side of the equation and the solution on the other side. The operations that we will be using are as follows:

1. We can add or subtract the same number on both sides of the equation.
2. We can multiply or divide both sides of the equation by the same nonzero number.

Using these two operations and making sure that what we do to one side of the equation we do the exact same thing to the other side, we form equivalent equations.

There are four basic steps to solving a first-degree conditional equation. With the two basic operations in mind, we shall now apply these steps to the equation

$$6(x + 1) = 4x + 10$$

Step 1 *Simplify the equation.* In simplifying an equation we carry out all indicated operations. That is, we perform all indicated addition, subtraction, multiplication, and division. We remove all grouping symbols and, in general, do anything to the problem that would make it easier to deal with.

In our example, step 1 involves carrying out the indicated multiplication on the left-hand side:

$$\begin{aligned} 6(x + 1) &= 4x + 10 \\ 6x + 6 &= 4x + 10 \end{aligned}$$

Step 2 *Have all terms with the unknown on one side of the equation.*

note

A good habit to get into in step 2 is always to have all the unknowns on the side that originally contained the greater number of unknowns to begin with. This procedure will ensure a positive coefficient for the unknown.

In our example, the greater number of x's is on the left-hand side. Therefore, we want to remove the x's from the right-hand side; we can do this by subtracting $4x$ from *both* sides of the equation:

$$6x + 6 \underline{- 4x} = 4x + 10 \underline{- 4x}$$

The result is

$$2x + 6 = 10$$

Step 3 *Remove anything that is being added to or subtracted from the unknown.* In our example, 6 is being added to the $2x$, and we remove it by subtracting 6 from *both* sides of the equation.

$$2x + 6 \underline{- 6} = 10 \underline{- 6}$$

The result is

$$2x = 4$$

Step 4 *Divide each side of the equation by the coefficient of the unknown.* In our example, 2 is the coefficient of x. That is, 2 is being multiplied times x; therefore, to remove it, we divide both sides of the equation by 2.

$$\frac{2x}{2} = \frac{4}{2}$$

The result is the solution

$$x = 2$$

In general, to remove a coefficient we divide both sides of the equation by the coefficient. However, in the case where the coefficient is a fraction, we recall that when we divide by a fraction we invert and multiply. Therefore, if the coefficient is a fraction, we will multiply both sides of the equation by the reciprocal of the coefficient. For example

$$\frac{2}{3}x = 12$$

$$\frac{3}{2} \cdot \frac{2}{3}x = \frac{12}{1} \cdot \frac{3}{2}$$

$$x = 18$$

You should observe that in steps 2, 3, and 4, to remove something, we first determine what the number is doing relative to the unknown and then do the inverse operation. In our example, in step 2, for instance, we wanted to remove a positive $4x$ and did so by subtracting $4x$ from both sides. In step 3, 6 was being added to the unknown and we removed it by subtracting 6 from both sides. Finally, in step 4, the 2 was multiplied times the unknown, and we removed it by dividing both sides by 2.

To check our work, we substitute the solution in place of the unknown in the original equation and if we get a true statement, we say the solution "satisfies" the equation.

Example

In the equation $6(x + 1) = 4x + 10$, we found that $x = 2$. Substituting 2 in place of x in the original equation gives

$$\begin{aligned} 6[(2) + 1] &= 4(2) + 10 \\ 6[3] &= 8 + 10 \\ 18 &= 18 \end{aligned}$$

Thus, 2 satisfies the equation.

To summarize, the primary goal in solving an equation is to isolate the unknown on one side of the equation and everything else on the other side. (If the unknown were on both sides of the equation, the answer for the unknown would be in terms of itself and would not be an answer.) To solve for the unknown we form equivalent equations, making sure, after we have simplified, that whatever we do to one side of the equation we do the exact same operation to the other side. Finally, having found a solution, we substitute it back into the original equation to see that it forms a true statement.

Examples

Find the solution and check.

1. $6y + 5 - 7y = 10 - 2y + 3$

Solution Combine like terms on both sides to get

$$5 - y = 13 - 2y$$

Add $2y$ to both sides to get

$$\begin{aligned} 5 - y + 2y &= 13 - 2y + 2y \\ 5 + y &= 13 \end{aligned}$$

Subtract 5 from both sides to get

$$\begin{aligned} 5 + y - 5 &= 13 - 5 \\ y &= 8 \end{aligned}$$

The solution is 8.

Check

$$\begin{aligned} 6(8) + 5 - 7(8) &= 10 - 2(8) + 3 \\ 48 + 5 - 56 &= 10 - 16 + 3 \\ 53 - 56 &= -6 + 3 \\ -3 &= -3 \end{aligned}$$

2. $\dfrac{4x + x}{3} = 10$

Solution Add $4x$ and x to get $\dfrac{5x}{3} = 10$, where $\dfrac{5}{3}$ is the coefficient of x.

Multiply both sides by $\dfrac{3}{5}$ to get

$$\frac{3}{5} \cdot \frac{5x}{3} = \frac{3}{5} \cdot \frac{10}{1}$$

$$x = 6$$

c. The solution is 6.

Check

$$\frac{4(6) + (6)}{3} = 10$$

$$\frac{24 + 6}{3} = 10$$

$$\frac{30}{3} = 10$$

$$10 = 10$$

You should observe from the last two examples, that although there are four basic steps in solving a linear equation, all four steps are not always used in each problem.

Exercise Set 7.2

Find the solutions of the following equations.

Example $5x + 2(x - 1) = 4 - 3x$

Step 1 $5x + 2x - 2 = 4 - 3x$

$7x - 2 = 4 - 3x$

Step 2 $10x - 2 = 4$

Step 3 $10x = 6$

Step 4 $x = \dfrac{6}{10}$

$x = \dfrac{3}{5}$

1. $2x = 4$
2. $5x = 15$
3. $3x = 11$
4. $7x = 12$
5. $5x = -10$
6. $-3x = 12$
7. $-2x = 8$
8. $\dfrac{x}{2} = 8$
9. $\dfrac{x}{4} = 24$
10. $\dfrac{3x}{2} = 8$

11. $\dfrac{5x}{3} = 18$

12. $x + 7 = 11$

13. $x - 4 = 9$

14. $x + 5 = -6$

15. $x - 12 = -11$

16. $x + 5 = 5$

17. $x - 4 = -4$

18. $3x + 1 = 10$

•19. $5x - 2 = 13$

20. $4x + 7 = 7$

21. $6x + 2 = 2$

22. $5x + 2x = x + 6$

•23. $2x + (3x - 1) = 4 - x$

24. $2x + 3x - 6x = 4x - 8$

25. $\dfrac{x}{2} + 7 = 14$

26. $5 - \dfrac{3x}{5} = 11$

•27. $\dfrac{9x}{11} = \dfrac{3}{4}$

28. $\dfrac{5x + 2x}{6} = 10$

29. $3(2x - 1) = 4x + 3$

30. $5(7x - 3) = 30x + 11$

31. $12x - 8 = 5x + 2$

32. $9x + 4 = x - 11$

33. $2 + 5x - 16 = 6x - 4$

34. $3(2x + 5) = 4(x - 3)$

•35. $8 - 2(3x + 4) = 5x - 16$

36. $(3x + 2) - (2x - 5) = 7$

37. $(7x - 6) - (4 - 3x) = 27$

38. $2(x - 4) - 3(5 - 2x) = 16$

39. $3(2x + 3) = 5 - 4(x - 2)$

40. $6(3x - 2) = 7(x - 3) - 2$

Calculator Problems

Using the rules of approximate numbers, find the solutions of the following equations.

41. $3.7x = 1.9$

42. $5.6x = 23.41$

43. $-1.87x = 3.49$

44. $-2.43x = -6.01$

45. $1.01x + 3.27 = 5.19$

46. $8.89x - 6.17 = 11.28$

•47. $10.1x + 9.07 = 8.14$

48. $7.37x - 6.18 = 3.01x - 11.2$

49. $5.48x - 17.1 = 9.63x + 12$

50. $17.0x - 6.41 = 0.41x - 0.44$

7.3 Solving Equations—II

Equations that contain one or more terms that are fractions are called **fractional equations.** Such an equation is

$$\frac{x}{3} + 4 = \frac{5}{4}$$

We simplify a fractional equation by eliminating the fraction from the problem. This elimination can be done by finding the least common denominator (LCD) of *all* the denominators of the fractions in the problem and then multiplying *both* sides of the equation by the LCD. In this example, the LCD of 3 and 4 is 12. Now we multiply both sides of the equation by 12. That is,

$$12\left(\frac{x}{3} + 4\right) = 12 \cdot \frac{5}{4}$$

$$\left(12 \cdot \frac{x}{3}\right) + (12 \cdot 4) = 12 \cdot \frac{5}{4}$$

$$4x + 48 = 15$$

We now can proceed through the rest of the steps to determine the solution:

$$4x + 48 = 15$$
$$4x + 48 \underline{-48} = 15 \underline{-48}$$
$$4x = -33$$
$$\frac{4x}{4} = \frac{-33}{4}$$
$$x = \frac{-33}{4}$$

Examples

1. $\dfrac{x}{5} - \dfrac{2x}{3} = 1$

   The LCD of 5 and 3 is 15.

   $$15\left(\frac{x}{5} - \frac{2x}{3}\right) = 15 \cdot 1$$

   note

   Although the right side of the equation contains no fractions, we must still multiply that side by the LCD, thereby multiplying both sides of the equation by the exact same number.

   $$15 \cdot \frac{x}{5} - 15 \cdot \frac{2x}{3} = 15 \cdot 1$$
   $$3x - 10x = 15$$
   $$-7x = 15$$
   $$x = -\frac{15}{7}$$

2. $\dfrac{2x + 1}{4} = \dfrac{3}{8}$

   The LCD is 8.

   $$8 \cdot \frac{2x + 1}{4} = 8 \cdot \frac{3}{8}$$
   $$2(2x + 1) = 3$$

   note

   After dividing the 4 into the 8, the 2 is multiplied times the entire numerator because the numerator is a group.

   $$4x + 2 = 3$$
   $$4x = 1$$
   $$x = \frac{1}{4}$$

3. $\frac{y}{4} + \frac{3}{5} = \frac{y + 1}{10}$

The LCD is 20.

$$
\begin{aligned}
20\left(\frac{y}{4} + \frac{3}{5}\right) &= 20\left(\frac{y + 1}{10}\right) \\
20 \cdot \frac{y}{4} + 20 \cdot \frac{3}{5} &= 20\left(\frac{y + 1}{10}\right) \\
5y + 12 &= 2(y + 1) \\
5y + 12 &= 2y + 2 \\
3y + 12 &= 2 \\
3y &= -10 \\
y &= \frac{-10}{3}
\end{aligned}
$$

4. $\frac{R}{3} - \frac{R - 1}{6} = 2$

The LCD is 6.

$$
\begin{aligned}
6\left(\frac{R}{3} - \frac{R - 1}{6}\right) &= 6 \cdot 2 \\
6 \cdot \frac{R}{3} - 6\left(\frac{R - 1}{6}\right) &= 6 \cdot 2 \\
2R - (R - 1) &= 12 \\
2R - R + 1 &= 12
\end{aligned}
$$

note

A common error is not distributing the negative sign to both terms inside the parentheses.

$$
\begin{aligned}
R + 1 &= 12 \\
R &= 11
\end{aligned}
$$

The following are additional examples involving parentheses.

Examples

1. $2x(x - 1) = (2x + 1)(x - 2)$

First we simplify by carrying out the indicated multiplication:

$$2x^2 - 2x = 2x^2 - 4x + x - 2$$

Then we add like terms:

$$2x^2 - 2x = 2x^2 - 3x - 2$$

We have not yet studied how to solve an equation where the unknown is raised to the second power. Therefore, in order to continue in this problem, we must eliminate all terms containing x to the second power.

Since the same number of x^2's appears on both sides of the equation, we can eliminate them by subtracting $2x^2$ from both sides.

$$2x^2 - 2x - 2x^2 = 2x^2 - 3x - 2 - 2x^2$$
$$-2x = -3x - 2$$

Now we can proceed with the problem.

$$-2x + 3x = -3x - 2 + 3x$$
$$x = -2$$

2. $3x(x + 2) - (3x - 1)(x + 1) = 4$

It is important to enclose the product of the two binomials in parentheses first, thereby avoiding a sign error.

$$3x(x + 2) - [(3x - 1)(x + 1)] = 4$$
$$3x^2 + 6x - [3x^2 + 3x - x - 1] = 4$$
$$3x^2 + 6x - [3x^2 + 2x - 1] = 4$$

Now we remove the parentheses and change the signs of all the terms within:

$$3x^2 + 6x - 3x^2 - 2x + 1 = 4$$

Then we combine like terms to get

$$4x + 1 = 4$$
$$4x = 3$$
$$x = \frac{3}{4}$$

3. $$\frac{5x}{2} - \frac{4x}{3} = x + \frac{1}{4}$$
$$12\left(\frac{5x}{2} - \frac{4x}{3}\right) = 12\left(x + \frac{1}{4}\right)$$
$$12 \cdot \frac{5x}{2} - 12 \cdot \frac{4x}{3} = 12x + 12 \cdot \frac{1}{4}$$
$$30x - 16x = 12x + 3$$
$$14x = 12x + 3$$
$$2x = 3$$
$$x = \frac{3}{2}$$

Exercise Set 7.3

Find the solution.

Example

$$\frac{2x}{3} - \frac{x}{12} = 1$$
$$12\left(\frac{2x}{3} - \frac{x}{12}\right) = 1 \cdot 12$$
$$8x - x = 12$$
$$7x = 12$$
$$x = \frac{12}{7}$$

1. $6x = 18$
2. $3 = x - 4$
•3. $\frac{V}{6} = 9$
4. $2e + 1 = 3$
5. $2q + 5 = 11$
6. $5 + 3R = R + 4$
7. $8I - 6 = 3I + 9$
8. $3 + 2x - 1 = 12 - 3x$
9. $x + 1 + 3x = 5$
10. $36E - 8 + 4E - 9 = 2 + 10E$
11. $x - 2 = x + 3 - 2x$
12. $2i + 3i + i = i + 9 + 2i$
•13. $3x + 12x + 17 = x - 4x + 107$
14. $4e - 3 = 5e - 16 - 21e + 39$
15. $4x - 3 + 3x - 4 = 6x - 8$
16. $\frac{1}{R} = \frac{4}{5}$
17. $\frac{R}{2} = \frac{9}{31}$
18. $\frac{5x}{2} = 20$
19. $\frac{2R}{3} = -4$
20. $\frac{5x - x}{6} = 4$
21. $\frac{3x + x}{5} = -2$
22. $\frac{i}{2} + \frac{i}{9} = 11$
•23. $\frac{x}{4} - \frac{x}{5} = 1$
24. $6 - \frac{y}{2} = y$
•25. $\frac{2x}{3} - 3 = x - 3$
26. $\frac{E}{12} = \frac{E}{4} - 48$
•27. $\frac{x - 3}{6} = \frac{2x + 2}{3}$
28. $\frac{2x - 1}{3} = \frac{3x + 1}{4}$
29. $\frac{2x - 1}{5} = \frac{x + 1}{2}$
30. $\frac{30 - e}{4} = \frac{50 - e}{8}$

Example $6(x - 2) - (x + 1) = 2$

$$6x - 12 - x - 1 = 2$$
$$5x - 13 = 2$$
$$5x = 15$$
$$x = 3$$

31. $2(x + 5) = 16$
32. $6 = 2(2x - 1)$
33. $x - (8 - x) = 2$
34. $2x - (3 - x) = 0$
35. $3(7 - 2i) = 30 - 7(i + 1)$
36. $6r - 2(2r + 5) = 6(5 + r)$
37. $(x - 2)(x - 4) = (x + 2)(x - 3)$
38. $x^2 - (x - 1)^2 = 19$
39. $R^2 - (R + 1)^2 = -21$
40. $(2x + 1)(x - 3) = (x - 2)(x + 1) + x^2$
41. $(I - 2)^2 = I^2 - 8$
42. $(E - 1)^2 = E^2 - 15$

•43. $4 - (x - 3)(x + 2) = 10 - x^2$

44. $6 + 3r - r^2 = 4 - (r - 2)(r + 3)$

In problems 45–51, round to the accuracy of the measured data.

45. The total creep of a metal (E_p) at time t is given by $E_p = E_o + V_o t$ where $E_o =$ original creep, $t =$ time, and $V_o =$ the original volume. Find V_o when $E_p = 16.0$, $E_o = 9.0$, and $t = 3.0$.
46. The diametral pitch (DP) of a spur gear is measured by $DP = \dfrac{N + 2}{OD}$, where $N =$ number of teeth and $OD =$ outside diameter. Find OD when $N = 16$ and $DP = 0.60$.
47. If the resistance of annealed copper wire is given by $R = \dfrac{PL}{d^2}$, where $P =$ specific resistance of the material, $d =$ diameter of the wire at a particular temperature, and $L =$ length of the wire, find P when $L = 450$, $d = 30.0$, and $R = 5.0$.
48. The number of teeth in a driver gear (t_2) of a gear system is defined by $t_2 = \dfrac{(t_1)(r_1)}{r_2}$, where $t_1 =$ number of teeth of driven gear, $r_1 =$ revolutions of the driven gear, and $r_2 =$ revolutions of the driven gear. Find t_1 when $r_1 = 120.0$ rpm, $r_2 = 80.0$ rpm, and $t_2 = 24$ teeth.
49. To convert Celsius temperature to Fahrenheit we use $\text{F} = \dfrac{9}{5}\text{C} + 32$. Find C when (a) $\text{F} = 18.0°$, (b) $\text{F} = -27.0°$, (c) $\text{F} = 2.0°$.

50. The Stefan-Boltzmann Law in metallurgy, which is the temperature scale of radiation pyrometers, is given by $W = KT^4$. Find K when (a) $W = 36.0$, $T = 2.00$; (b) $W = 243$, $T = 3.0$.

•51. The total conductance in a parallel circuit is equal to the sum of the conductances. The conductance is equal to $\frac{1}{R}$, where R represents resistance. Thus, for three resistors R_1, R_2, R_3, the total conductance (C_t) is defined by

$$C_t = \frac{1}{R_1} + \frac{1}{R_2} + \frac{1}{R_3}.$$

Find C_t when $R_1 = 6.0$, $R_2 = 7.0$, $R_3 = 9.0$.

7.4 Solving Literal Equations

When an equation contains two or more variables, it is called a **literal equation.** The kind of literal equation that is most familiar to us is the formula. A **formula** is a rule expressing some scientific relationship in the form of an equation involving letters, symbols, and constant terms. To illustrate, the power P expended in an electrical circuit is equal to the product of its current I squared times the resistance R of the circuit. Stated as a formula, this is

$$P = I^2R$$

and is a literal equation involving the variables P, I, and R.

It is often desirable to solve for one quantity contained in the literal equation in terms of the other quantities. To do this we employ all the procedures previously learned for finding the solution of a linear equation. In the formula $P = I^2R$, P is stated in terms of I and R. Let us now solve for R in terms of I and P. Since I^2 is the coefficient of R and we wish to remove it, we divide both sides by I^2 to get

$$\frac{P}{I^2} = \frac{I^2R}{I^2}$$

$$\frac{P}{I^2} = R$$

$$R = \frac{P}{I^2}$$

Consider the equation

$$a = \frac{x}{b} + c$$

To solve for x in terms of a, b, and c, we form a series of equivalent equations until we have x isolated on one side and everything else on the other side. We proceed as follows:

Step 1 Since this is a fractional equation, we multiply both sides by the denominator b.

$$b \cdot a = b\left(\frac{x}{b} + c\right) \qquad b \neq 0$$

$$b \cdot a = b \cdot \frac{x}{b} + b \cdot c$$

$$ab = x + bc$$

Step 2 All the terms containing x are on one side of the equation.

Step 3 Since bc is added to x, we subtract bc from both sides.

$$ab - bc = x + bc - bc$$
$$ab - bc = x$$

Hence

$$x = ab - bc$$

note

We stated in the first step that $b \neq 0$. We did this for two reasons. First, according to our procedure, we can multiply both sides of an equation only by a nonzero number. Second, the b in the denominator of the original problem cannot be zero, or the problem would be undefined. We discuss restrictions of this type in a later chapter.

Examples

Solve for the indicated variable.

1. $A = E + IR$ for R.

   Subtract E from both sides.

   $$A - E = E + IR - E$$
   $$A - E = IR$$

   Divide both sides by I.

   $$\frac{A - E}{I} = \frac{IR}{I}$$
   $$\frac{A - E}{I} = R$$

   Therefore, $R = \dfrac{A - E}{I}$.

2. $5T - S = T - 3R$ for T.

   Subtract T from both sides.

   $$5T - S - T = T - 3R - T$$
   $$4T - S = -3R$$

   Add S to both sides.

   $$4T - S + S = -3R + S$$
   $$4T = S - 3R$$

   Divide both sides by 4.

   $$\frac{4T}{4} = \frac{S - 3R}{4}$$

   Hence, $T = \dfrac{S - 3R}{4}$.

To solve a literal equation (or formula) involving parentheses we proceed as follows. Consider the equation

$$A = \frac{h}{2}(b + c)$$

To solve for b in terms of A, h, and c, we first eliminate the denominator by multiplying by 2.

$$2A = 2 \cdot \frac{h}{2}(b + c)$$

$$2A = h(b + c)$$

We now perform the indicated multiplication to get

$$2A = hb + hc$$

We then subtract hc from both sides:

$$2A - hc = hb + hc - hc$$

leaving

$$2A - hc = hb$$

Finally, dividing by h we have

$$\frac{2A - hc}{h} = \frac{hb}{h}$$

$$\frac{2A - hc}{h} = b$$

Therefore,

$$b = \frac{2A - hc}{h}$$

In like fashion we can solve for c in terms of A, b, and h or h in terms of A, b, and c.

Example $4x - 3(x - a) = 8a$ for x.

Multiply on the left side.

$$4x - 3x + 3a = 8a$$

Combine terms on the left side.

$$x + 3a = 8a$$

Subtract $3a$ from both sides.

$$x + 3a - 3a = 8a - 3a$$

$$x = 5a$$

Finally, let us solve the equation $xy + y = 1$ for y. First we shall look at an equation that is similar to this:

$$3y + y = 1$$

Recall exactly how the $3y$ and the y are added. According to the rules of algebra, if we have similar terms, we add together the coefficients and this sum becomes the new coefficient. That is

$$3y + y = (3 + 1)y = 4y$$

And in the problem we have

$$4y = 1$$

$$y = \frac{1}{4}$$

We now apply this idea to the original problem:

$$xy + y = 1$$

We want to know how many y's we have. Proceeding as we did in the previous example, we add together the coefficients x and 1 as follows:

$$xy + 1y = (x + 1)y$$

Therefore, in our problem, we have

$$xy + y = 1$$

$$(x + 1)y = 1$$

and $(x + 1)$ is the coefficient of y. We then divide both sides by $x + 1$:

$$\frac{(x + 1)y}{(x + 1)} = \frac{1}{(x + 1)}$$

$$y = \frac{1}{x + 1}$$

This technique is actually the distributive law and the procedure is called *factoring*. Factoring is taken up more completely in the next chapter.

Examples

Solve for the indicated variable.

1. $4r + e = ir$ for r.

   Subtract $4r$ from both sides:

   $$4r + e - 4r = ir - 4r$$

   $$e = ir - 4r$$

   We now have all the terms containing r on one side and everything else is on the other. To find how many r's are in the problem, we combine all the coefficients of r, yielding

   $$e = (i - 4)r$$

   Then we divide by $i - 4$.

   $$\frac{e}{(i - 4)} = \frac{(i - 4)r}{(i - 4)}$$

   $$\frac{e}{i - 4} = r$$

   Hence, $r = \dfrac{e}{i - 4}$.

2. $\frac{1}{a} + \frac{1}{x} = \frac{1}{b}$ for x.

Multiply by the product of the denominators a, x, and b:

$$axb\left(\frac{1}{a} + \frac{1}{x}\right) = \frac{1}{b}(axb)$$

$$axb \cdot \frac{1}{a} + axb \cdot \frac{1}{x} = \frac{1}{b} \cdot axb$$

$$bx + ab = ax$$

Next, we get all the x's on one side by subtracting bx.

$$bx + ab - bx = ax - bx$$

$$ab = ax - bx$$

Then we combine the coefficients of x, giving

$$ab = (a - b)x$$

Finally, we divide by the coefficient, $a - b$,

$$\frac{ab}{(a - b)} = \frac{(a - b)x}{(a - b)}$$

$$\frac{ab}{a - b} = x$$

Therefore,

$$x = \frac{ab}{a - b}$$

Exercise Set 7.4

Solve the following for the indicated letter.

•1. $2xy = 3$ for x

2. $3x - 5b = x$ for x

•3. $5abx + 6a = abx$ for x

4. $ay + c^2 = 0$ for y

5. $ax = a^2b - ax$ for x

6. $\frac{ay}{c} + b = 0$ for y

7. $by - c = a$ for y

8. $7x - 8a - 3x = 3a + 2x$ for x

•9. $9y - 12a = 3y + 9a - y$ for y

10. $3x - 2a = 5x + 3b - 3x$ for x

Solve each formula for the indicated letter.

11. $pV = K$ for p

12. $E = IR$ for I

13. $V = lwh$ for h

14. $E = mc^2$ for m

15. $s = \frac{1}{2}at^2$ for a

16. $v^2 = 2gh$ for g

•17. $M = \frac{1}{6}gh^2$ for g

18. $W = FR$ for R

19. $P_1V_1 = P_2V_2$ for V_2

20. $v = k + gt$ for g

21. $P = 2L + 2W$ for L

22. $P_1 = \dfrac{P_2V_2}{V_1}$ for V_1

23. $I = \dfrac{E}{R + r}$ for E

24. $\dfrac{E}{e} = \dfrac{R + r}{r}$ for e

25. $\mathrm{F} = \dfrac{9}{5}\mathrm{C} + 32$ for C

26. $A = 2\pi rh + 2\pi r^2$ for h

27. $F = \dfrac{Wv^2}{gr}$ for g

28. $I = \dfrac{E_g - E_t}{R}$ for R

•29. $a = \dfrac{v - v_o}{t}$ for v

30. $x = v_o t + \dfrac{1}{2}at^2$ for a

Solve the following equations for x or y.

•31. $cx = b - x$

32. $cx + b = ax$

33. $3x - 4(x - 2a) = 3a$

34. $a(a - x) = c(b - x)$

•35. $(y - 2)(a + 3) = a$

36. $(y - 1)(b + 3) = 2b$

•37. $\dfrac{a - b}{b} + \dfrac{3}{2b} = \dfrac{2}{x}$

38. $\dfrac{x}{a} - \dfrac{a}{b} = \dfrac{b}{c}$

•39. $\dfrac{2x + 4a}{2a} + \dfrac{3x + 4a}{2a} = a$

Solve each formula for the indicated letter.

40. $A = \dfrac{h}{2}(b + c)$ for c

41. $S = 2r(r + h)$ for h

42. $S = 3(d + D)$ for d

43. $A = 2r(h + r)$ for h

44. $s = \dfrac{N(a + L)}{2}$ for a

•45. $\dfrac{1}{R} = \dfrac{1}{R_1} + \dfrac{1}{R_2}$ for R

46. $\dfrac{1}{C} = \dfrac{1}{C_1} + \dfrac{1}{C_2}$ for C_2

47. $L_t = L_o + L_o at$ for L_o

48. $A = \dfrac{B}{1 + B}$ for B

In problems 49, 52, and 54, round to the accuracy of the measured data.

49. To convert from a Fahrenheit (F) temperature reading to Celsius (C) temperature we use the formula $\mathrm{C} = \dfrac{5}{9}(\mathrm{F} - 32)$. Find F when C is (a) -10.0, (b) 25.0, (c) 9.0.

50. The area of a trapezoid is given by $A = \dfrac{h}{2}(b_1 + b_2)$, where A = area, h = altitude, and b_1, b_2 = the bases. Solve for h.

•51. When two impedances Z_1 and Z_2 are connected in parallel, the resultant joint impedance Z_t is determined by

$$Z_t = \frac{Z_1Z_2}{Z_1 + Z_2}$$

Solve for (a) Z_1, (b) Z_2.

52. In problem 51, what is Z_1 when $Z_t = 3.0$ and $Z_2 = 6.0$?

53. A lens formula is $\frac{1}{f} = \frac{1}{p} + \frac{1}{q}$. Solve for (a) f, (b) q.

54. In problem 53, if the focal length $f = -4\tilde{0}$ and the object distance $p = 4\tilde{0}$, what is the image distance q?

55. If $\frac{x}{a} + \frac{x}{b} + \frac{x}{c} = 1$, solve for b.

56. A pulley formula is given by $P = \frac{W(R - r)}{2R}$. Solve for r.

57. In problem 56, solve for R.

58. If a source voltage consists of n cells in series, each with an electromotive force of E volts and an internal resistance of r ohms, the current flowing through a load of R ohms is given by $I = \frac{nE}{R + nr}$ amperes. Solve for n.

59. The area of a trapezoid is given by $A = \frac{h(b_1 + b_2)}{2}$, where h is the altitude, and b_1 and b_2 are the parallel bases. Solve for b_1.

60. The overall changes in length of a specimen L is the difference in the length at any particular measurement I and the original length I_o. That is, $L = I - I_o$. Solve for I_o in terms of L and I.

61. The actual area of a pipe needed to handle a given flow is determined by $A = \frac{GPM \cdot (0.3208)}{V}$, where V = velocity in feet per second and GPM = gallons per minute.
    a. Solve for V in terms of A and GPM.
    b. Solve for GPM in terms of A and V.

62. Electrical horsepower is measured by $HP = \frac{VI}{746}$, where V = voltage and I = current. Solve for V in terms of HP and I.

63. Torricelli's theorem for determining the velocity at an outlet of a liquid discharging into free atmosphere is given by $V^2 = 2gh$, where V = velocity in feet per second, g = acceleration of gravity, and h = height in feet. Solve for h in terms of V and g.

64. The required ratio of gearing R for milling machines is given by

$$R = (A - N) \cdot \left(\frac{40}{A}\right)$$

    a. Solve for N in terms of R and A.
    b. Solve for A in terms of R and N.

65. The number of strokes (N) required to shape a piece of metal with a given speed (cs) and a stroke length (L) is defined by $N = \frac{cs \cdot 7}{L}$.
    a. Solve for cs in terms of N and L.
    b. Solve for L in terms of N and cs.

66. The outside diameter (OD) of a spur gear with a given number of teeth (N) and pitch diameter (PD) is given by $OD = \dfrac{N + 2}{\dfrac{N}{PD}}$.

   a. Solve for PD in terms of OD and N.
   b. Solve for N in terms of OD and PD.

67. The resistance of a wire is proportional to its length. That is, $\dfrac{R_1}{L_1} = \dfrac{R_2}{L_2}$.

   a. Solve for L_1 in terms of R_1, R_2, L_2.
   b. Solve for R_2 in terms of R_1, L_1, L_2.

68. A useful formula in trigonometry when dealing with triangles is

$$a^2 = b^2 + c^2 - 2bc(\cos A).$$

Solve for $\cos A$ in terms of a, b, and c.

7.5 Verbal Problems

Many problems that technicians encounter are stated verbally and need to be translated into an equation. In chapter 6 we saw how to write a verbal statement as an algebraic expression, and later as an equation.

We are now ready to combine our abilities of writing an equation and solving an equation in order to solve a verbal problem. Although there is no standard procedure for solving a verbal problem, the following guidelines should be useful.

1. Read the problem carefully.
2. Note what information is given and what information you are asked to find.
3. Determine any prior knowledge that may be useful in setting up an equation.
4. Let some letter represent one of the unknown quantities.
5. Put any other unknown quantities in terms of the first unknown.
6. Make sure in steps 4 and 5 that the unknown quantities represent numerical quantities—that is, a number, not a quality.
7. Using the given conditions in the problem and the unknowns from 4 and 5, write an equation.
8. Solve the equation for the unknown and relate this answer to any other unknowns in the problem.
9. Check your results in the original statement of the problem.

Examples

1. One whole number is five times another whole number and their sum is 36. Find the whole numbers.

Lesser number	*Greater number*	*Equation*
x	$5x$	$x + 5x = 36$

Solution
$$\begin{aligned} x + 5x &= 36 \\ 6x &= 36 \\ x &= 6 \end{aligned}$$

Therefore, the lesser number is 6 and the greater number is five times the lesser number ($5x$) and is $5(6) = 30$.

2. The length of a rectangle is three times as long as its width, and its perimeter is 40. Find the dimensions.

note

In example 2 we need the prior knowledge that the perimeter of a rectangle is given by the formula $P = 2w + 2l$.

Width	*Length*	*Equation*
x	$3x$	$2(x) + 2(3x) = 40$

Solution
$$\begin{aligned} 2(x) + 2(3x) &= 40 \\ 2x + 6x &= 40 \\ 8x &= 40 \\ x &= 5 \end{aligned}$$

Therefore, the width, x, is 5 and the length is $3x = 3(5) = 15$.

3. The sum of three consecutive integers is 36. Find the integers.

note

For this problem we need the prior knowledge that consecutive integers differ by 1. Therefore, we add 1 to the first to get the second, and we add 2 to the first to get the third, and so on.

First	*Second*	*Third*	*Equation*
x	$x + 1$	$x + 2$	$x + (x + 1) + (x + 2) = 36$

Solution
$$\begin{aligned} x + (x + 1) + (x + 2) &= 36 \\ x + x + 1 + x + 2 &= 36 \\ 3x + 3 &= 36 \\ 3x &= 33 \\ x &= 11 \end{aligned}$$

Hence, the first integer is $x = 11$, the second integer is $x + 1 = (11) + 1 = 12$, and the third integer is $x + 2 = (11) + 2 = 13$.

4. Machine A is able to produce 12 more parts in 1 hour than machine B. In 8 hours, their combined output is 896 parts. Determine the number of parts produced in 1 h by each machine.

Machine A	*Machine B*	*Equation*
$x + 12$	x	$8\,[(x + 12) + x] = 896$

Solution
$$\begin{aligned} 8\,[(x + 12) + x] &= 896 \\ 8\,[x + 12 + x] &= 896 \\ 8\,[2x + 12] &= 896 \\ 16x + 96 &= 896 \\ 16x &= 800 \\ x &= 50 \end{aligned}$$

Therefore, machine B produces $x = 50$ parts per hour and machine A produces $x + 12 = (50) + 12 = 62$ parts per hour.

5. A woman has \$10,000, part of which she invests at 11% and the rest at 8%. If her total income from the two investments is \$980, how much does she have invested at each rate?

note

For this problem we need the prior knowledge that interest = principle · rate · time. In the examples and homework problems, time will be 1 yr and the rate will be yearly.

11% Investment	*8% Investment*	*Equation*
x	$10{,}000 - x$	$x(0.11) + (10{,}000 - x)(0.08) = 980$

note

If we have \$10,000 to invest and we invest x dollars at 11%, then the amount left to invest at 8% will be the total minus what we had already invested, $10{,}000 - x$.

Solution
$$\begin{aligned} x(0.11) + (10{,}000 - x)0.08 &= 980 \\ 0.11x + 800 - 0.08x &= 980 \\ 0.03x + 800 &= 980 \\ 0.03x &= 180 \\ x &= 6{,}000 \end{aligned}$$

Hence, she has $x =$ \$6,000 at 11% and $10{,}000 - x = 10{,}000 - 6{,}000 =$ \$4,000 at 8%.

6. If one man can do a job in 10 hours and a second man can do the same job in 12 hours, how long will it take for both working together to do the job?

If we let x equal the number of hours required for both to do the job, then the first man can do $\left(\frac{1}{10} \text{ of the job per hour}\right)(x \text{ hours}) = \frac{x}{10}$ of the total job, and the second man can do $\left(\frac{1}{12} \text{ of the job per hour}\right)$ $(x \text{ hours}) = \frac{x}{12}$ of the total job.

First man	*Second man*	*Equation*
$\frac{x}{10}$	$\frac{x}{12}$	$\frac{x}{10} + \frac{x}{12} = 1$

The 1 in the equation represents one complete job.

Solution

$$\frac{x}{10} + \frac{x}{12} = 1$$

$$60\left(\frac{x}{10} + \frac{x}{12}\right) = 1 \cdot 60$$

$$60 \cdot \frac{x}{10} + 60 \cdot \frac{x}{12} = 1 \cdot 60$$

$$6x + 5x = 60$$

$$11x = 60$$

$$x = \frac{60}{11}$$

Therefore, it will take $5\frac{5}{11}$ hours for the two men working together to complete the job.

7. What quantities of silver 65% pure and 45% pure must be mixed together to give 100 g of silver 50% pure?

note

If we know the sum of two numbers, we can let x represent one of the numbers, and the total minus x can represent the other number.

Number of grams of 65%	*Number of grams of 45%*	*Equation*
x	$100 - x$	$(0.65)x + (0.45)(100 - x) = (0.50)100$

Solution

$$(0.65)x + (0.45)(100 - x) = (0.50)100$$

$$0.65x + 45 - 0.45x = 50$$

$$0.20x + 45 = 50$$

$$0.20x = 5$$

$$x = 25$$

Therefore, we need 25 g of 65% pure silver and 100 g − 25 g = 75 g of 45% pure silver.

Exercise Set 7.5

For these problems, write an equation and solve for the unknown quantities.

•1. One number is 18 more than a second number. If their sum is 62, find the two numbers.

2. One number is 9 less than another number. If their sum is 47, find the numbers.

•3. The difference of two numbers is 17. Find the numbers if their sum is 87.

4. If 3 times a number is then increased by 11 and the answer is 47, what is the number?

•5. If a number is divided by 4, and then that result is increased by 6, the answer is 13. Find the number.

6. If a number is decreased by 14, and then that result is divided by 5, the answer is 15. Find the number.

7. Nine times a number is decreased by 4, leaving 59. What is the number?
8. One-third of a number is 8 less than one-half of the number. Find the number.
•9. The difference between one-half of a number and one-third of a number is 9. Find the number.
10. The sum of three consecutive even integers is 72. Find the integers.
•11. The sum of three consecutive odd integers is 51. Find the integers.
12. One number is 4 times a second number and their sum is 65. Find the numbers.
•13. One number is 9 times a second number and their sum is 120. Find the numbers.
14. The sum of three numbers is 43. The first is three times the second and the third is 6 less than the first. Find the numbers.
•15. The sum of three numbers is 63. The first is twice the second and the third is three times the first. Find the numbers.
16. One number is 7 more than another number. Find the two numbers if three times the larger exceeds four times the smaller by 5.
•17. One number is 4 more than another number. Find the two numbers if two times the larger number is 7 less than 5 times the smaller.
18. The length of a rectangle is 9 ft more than its width. The perimeter of the rectangle is 58 ft. Find the dimensions. (Prior knowledge: Perimeter = 2 times the length plus 2 times the width.)
19. The width of a rectangle is 3 ft less than its length. The perimeter of the rectangle is 70 ft. Find the dimensions. (See problem 18.)
20. The width of a rectangle is $\frac{1}{3}$ of its length. If the perimeter is 96 ft, find the dimensions.
21. A 12-ft board is cut into two pieces so that one piece is 4 ft longer than the other. How long is each piece?
22. A 24-ft rope is cut into two pieces so that one piece is twice as long as the other. How long is each piece?
23. The sum of two currents is 80 amps. If the greater current is 24 amps more than the lesser current, find their values.
24. A 50-ft extension cord is cut into two pieces such that one piece is 12 ft longer than the other piece. How long is each piece?
25. The sum of the number of teeth on two gears is 64, and their difference is 12. How many teeth are on each gear?
26. Two gears have a total of 83 teeth, and one gear has 15 less teeth than the other. How many teeth are on each gear?
27. Two electrical voltages have a total of 126 V. If one voltage is 32 volts more than the other, find the voltages.
28. The sum of two voltages is 85 volts and their difference is 32 volts. Find the voltages.
29. The sum of two resistors in a series is 24 ohms and their difference is 14 ohms. How many ohms are in each resistor?
30. One resistor exceeds another resistor by 25 ohms, and their sum is 67 ohms. How many ohms are in each resistor?

•31. Phil has \$20,000, part of which he invests at 8% interest and the rest at 6%. If his total income from the two investments is \$1,460, how much did he invest at each rate?

32. Nancy has \$18,000. She invests part of her money at $7\frac{1}{2}\%$ interest and the rest at 9%. If her income from the two investments is \$1,560, how much did she invest at each rate?

33. Rich has \$18,000, part of which he invests at 10% interest and the rest at 8%. If his income from each investment is the same, what does he have invested at each rate?

34. Dick has \$30,000, part of which he invests at 9% interest and the rest at 7%. If his income from the 7% investment is \$820 more than that from the 9% investment, how much was invested at each rate?

•35. Lynne made two investments totaling \$25,000. On one investment she made an 18% profit, but on the other investment she took an 11% loss. If her net gain is \$2,180, how much was each investment?

36. Larry made two investments totaling \$21,000. One investment made him a 13% profit, but on the other investment he took a 9% loss. If his net loss is \$196, how much was in each investment?

37. Dana has \$14,000 invested at 7% and is going to invest an additional amount at 11% so that her total investment will make 9%. How much does she need to invest at 11% to achieve this?

38. A man can do a job in 5 hours and his son can do the job in 9 hours. How long will it take them to do the job together?

•39. In problem 38, if the man had a second son who could do the job in 10 hours, how long would it take all three working together to do the job?

40. An auto mechanic has two bottles of battery acid. One contains a 10% solution and the other a 4% solution. How many cubic centimeters of each solution must be used to make 120 cm^3 of a solution that is 6% acid?

41. A metallurgist wishes to form 2,000 kg of an alloy that is 80% copper. This alloy is to be obtained by fusing an alloy that is 68% copper with one that is 83% copper. How many kilograms of each alloy must be used?

42. If a jeweler wishes to form 12 oz of 75% pure gold from sources that are 60% and 80% pure gold, how much of each substance must he mix together?

43. A chemist wishes to make 1,000 liters of a 3.5% acid solution by mixing a 2.5% solution with a 25% solution. How many liters of each solution is necessary?

44. A pharmacist is able to fill 200 3-grain and 2-grain capsules using 500 grains of a certain drug. How many capsules of each kind does she fill?

45. A solution that is 38% silver nitrate is to be mixed with a solution that is 3% silver nitrate to obtain 100 cl of solution that is 5% silver nitrate. How many centiliters of each solution should be used in the mixture?

46. A druggist has two solutions, one 60% hydrogen peroxide and the other 30% hydrogen peroxide. How many liters of each should be mixed to obtain 30 liters of a solution that is 40% hydrogen peroxide?

7.6 Solving Inequalities

In chapter 1 we studied the meaning of the inequality symbols representing less than ($<$), greater than ($>$), less than or equal to ($\leqslant$), and greater than or equal to ($\geqslant$). These symbols define the *sense* or *order* of an inequality. The following are some examples of how we use these symbols:

1. If we want to state symbolically that 4 is less than 7, we write $4 < 7$.
2. If we wish to denote that the variable x is representing 5 or any number greater than 5, we write $x \geqslant 5$.

   note

   $x \geqslant 5$ represents any real number that is greater than or equal to 5, and not just any integer greater than or equal to 5.

3. If we wish to denote that the variable T is representing any number less than 3, and not 3 itself, we write $T < 3$.

When we replace the equal sign in a linear equation with one of these inequality symbols, we form a **linear inequality.** The following are examples of linear inequalities.

$$2x \geqslant 6$$

$$2x + 5x - 1 < 3x + 2$$

$$\frac{3x}{4} - 1 > 8$$

$$5(2x + 1) \leqslant 7x - 4x + 3$$

A major difference between the linear equation and the linear inequality is the solution. The solution of a linear equation has at most one value, whereas the solution of a linear inequality consists of an unlimited number of values. Consider the inequality $2x \geqslant 6$. We can see by inspection that if we substitute 3, $3\frac{1}{2}$, 4, or 5 for x, the inequality will be true. In fact, if we substitute any number greater than or equal to 3 into the inequality, the inequality will be true. Thus, the solution has an unlimited number of solutions and we state this fact as $x \geqslant 3$.

Another way to indicate the solution of an inequality is by graphing. To graph the solution to $x \geqslant 3$, we simply draw a number line (as we did in chapter 1), place a solid circle at 3 on the number line, and draw an arrow extending from this to the right.

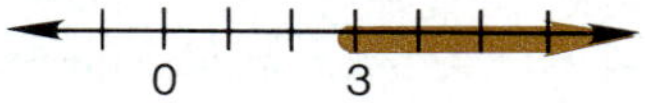

The solid line indicates that *all* numbers greater than or equal to 3 are part of the graph.

Examples

Graph the following linear inequalities.

1. $x < 4$ Here, x is representing all real numbers less than 4, but not 4 itself. To denote the fact that x cannot be equal to 4, put a hollow circle at 4.

2. $x \geqslant -3$ The greater than or equal to symbol, $\geqslant$, indicates that the point -3 is contained, and a solid circle is placed at -3.

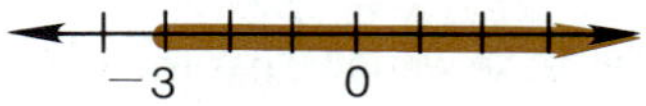

note

In graphing inequalities, a strict inequality ($<$ or $>$) is represented by a hollow circle at the number. A weak inequality ($\leqslant$ or $\geqslant$) is represented by a solid circle at the number.

3. $-3 \leqslant x < 4$ This statement is called a **compound inequality** and is read "-3 is less than or equal to x *and* x is less than 4." We place a solid circle at -3 to denote that -3 is included and place a hollow circle at 4 to show that 4 is not included. Then we draw a line segment between the two circles.

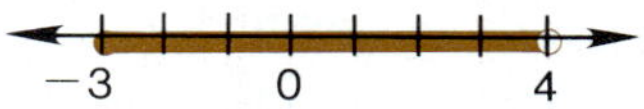

The properties that we use to solve inequalities are similar to those that we used to solve linear equations. They are as follows:

1. The same number can be added to or subtracted from *both* sides of the inequality. We can multiply or divide *both* sides of the inequality by the same positive number and still maintain the direction of the inequality symbol.
2. We can multiply or divide *both* sides of an inequality by the same *negative* number provided that we *reverse* the direction of the inequality symbol.

To demonstrate these operations, consider the inequality $6 < 12$.

Adding a number: If we add 4 to both sides, we have

$$6 + 4 < 12 + 4$$
$$10 < 16$$

which is still a correct statement.

Subtracting a number: If we subtract 5 from both sides of the inequality, we have

$$6 - 5 < 12 - 5$$
$$1 < 7$$

which is still a correct statement.

Multiplying by a positive number: If we multiply 4 times both sides of the inequality, we have

$$6 \cdot 4 < 12 \cdot 4$$
$$24 < 48$$

which is still a correct statement.

Dividing by a positive number: If we divide both sides of the inequality by 3, we have

$$\frac{6}{3} < \frac{12}{3}$$
$$2 < 4$$

which is still a correct statement.

Multiplying by a negative number: If we multiply -2 times both sides of the inequality, we have

$$6(-2) < 12(-2)$$
$$-12 < -24$$

We realize at this point that our inequality is no longer correct. For our inequality to be correct, we need to *reverse the direction of the inequality symbol,* thereby reversing the sense of the inequality. Thus, we have

$$-12 > -24$$

which is a correct statement.

Dividing by a negative number: If we divide both sides of the inequality by -3, we must reverse the direction of the inequality symbol to get:

$$6 < 12$$
$$\frac{6}{-3} > \frac{12}{-3} \quad \text{reversing the direction of the inequality symbol}$$
$$-2 > -4$$

which is a correct statement, where the sense of the inequality has been reversed.

To summarize, we observe that these operations are the same as those for linear equations with one exception: *Whenever we multiply or divide both sides of an inequality by a negative number, we must reverse the direction of the inequality symbol.*

Now let us solve some inequalities. The procedure for solving an inequality follows the same four steps used to solve a linear equation.

Examples

Find the solution.

1. $2x + 5x - 1 < 4x + 2$

Step 1 *Simplify* the problem by carrying out the indicated addition on the left side.

$$2x + 5x - 1 < 4x + 2$$
$$7x - 1 < 4x + 2$$

Step 2 *All the terms containing the unknown, x, must be on one side of the inequality*. Therefore, we subtract $4x$ from both sides of the inequality.

$$7x - 1 < 4x + 2$$
$$7x - 4x - 1 < 4x - 4x + 2$$
$$3x - 1 < 2$$

note

A negative coefficient of the unknown can be avoided if we form equivalent inequalities where the unknown appears only on the side of the inequality that has the greater coefficient of the unknown.

Step 3 *All the terms not involving the unknown must be on the other side of the inequality*. Therefore, we add 1 to both sides of the inequality.

$$3x - 1 < 2$$
$$3x - 1 + 1 < 2 + 1$$
$$3x < 3$$

Step 4 *Form an equivalent inequality where the coefficient of the unknown is 1*. Hence, we divide both sides of the inequality by 3.

$$3x < 3$$
$$\frac{3x}{3} < \frac{3}{3}$$
$$x < 1$$

The solution can be left in this form or graphical form.

note

In step 4 we must be careful to observe whether we are multiplying or dividing by a positive or negative number so that we will form the correct inequality.

2. $-2x \leqslant 4$ The only operation that needs to be performed to solve the inequality is to divide by -2. Since we are dividing by a negative number, we must remember to *reverse* the direction of the inequality symbol.

$$-2x \leqslant 4$$

$$\frac{-2x}{-2} \geqslant \frac{4}{-2}$$

$$x \geqslant -2$$ or

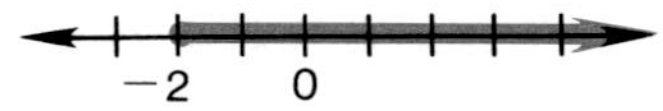

3.

$$5(2x + 1) \leqslant 7x - 4x + 3$$
$$10x + 5 \leqslant 3x + 3$$

Simplify by multiplying on the left side and combining like terms on the right side.

$$10x - 3x + 5 \leqslant 3x - 3x + 3$$

Subtract $3x$ from both sides.

$$7x + 5 \leqslant 3$$
$$7x + 5 - 5 \leqslant 3 - 5$$

Subtract 5 from both sides.

$$7x \leqslant -2$$

$$\frac{7x}{7} \leqslant \frac{-2}{7}$$

Divide both sides by 7.

$$x \leqslant \frac{-2}{7}$$ or

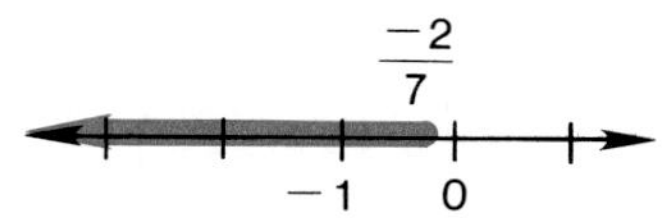

4. $-3 \leqslant 2x + 1 < 5$ When solving a compound inequality, the solution must be such that the unknown appears only in the middle of the inequality. We can still use all of our properties, if we apply them to all three parts, and we must reverse the direction of *all* the inequality symbols when multiplying or dividing by a negative number.

$$-3 \leqslant 2x + 1 < 5$$
$$-3 - 1 \leqslant 2x + 1 - 1 < 5 - 1$$

Subtract 1 from all three parts.

$$-4 \leqslant 2x < 4$$

$$\frac{-4}{2} \leqslant \frac{2x}{2} < \frac{4}{2}$$

Divide all three parts by 2.

$$-2 \leqslant x < 2$$ or

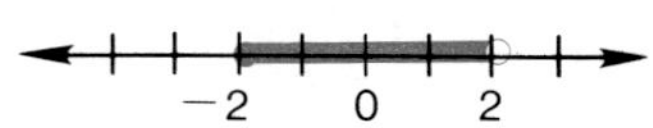

5. $-6 < -3x - 3 \leq 9$

$$-6 + 3 < -3x - 3 + 3 \leq 9 + 3$$ Add 3 to all three parts.

$$-3 < -3x \leq 12$$

$$\frac{-3}{-3} > \frac{-3x}{-3} \geq \frac{12}{-3}$$ Divide all three parts by -3, reversing the direction of *both* inequality symbols.

$$1 > x \geq -4$$ or

Exercise Set 7.6

Graph the following.

Examples 1. $x \geq -1$

2. $-2 < x \leq 3$

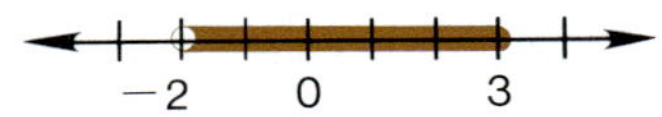

1. $x > 2$
2. $x < -2$
3. $x \geq 1$
4. $x \geq 4$
5. $x < 0$
6. $x < -2$
7. $x > 0$
8. $x \leq 3$
9. $x \leq -4$
10. $-2 < x < 0$
11. $-1 < x < 2$
12. $-3 \leq x \leq 4$
13. $0 \leq x \leq 5$
14. $1 \leq x < 4$
15. $-1 < x \leq 3$

In the following problems, solve for x. Leave your answer in both inequality form and graphical form.

Examples 1. $4x + 1 \geq 2x - 3$

$$4x - 2x + 1 \geq 2x - 2x - 3$$
$$2x + 1 \geq -3$$
$$2x + 1 - 1 \geq -3 - 1$$
$$2x \geq -4$$
$$\frac{2x}{2} \geq \frac{-4}{2}$$
$$x \geq -2$$ or

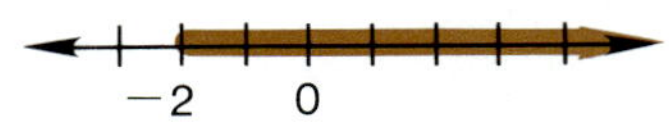

2. $-5 < 2x + 1 \leqslant 3$

$$-5 - 1 < 2x + 1 - 1 \leqslant 3 - 1$$
$$-6 < 2x \leqslant 2$$
$$\frac{-6}{2} < \frac{2x}{2} \leqslant \frac{2}{2}$$
$$-3 < x \leqslant 1$$

or

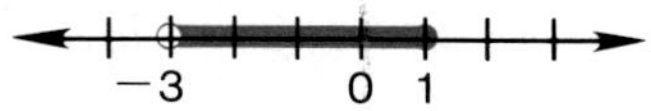

16. $4x > 10$
17. $2x \leqslant 5$
•18. $3x \geqslant 15$
19. $5x < 30$
•20. $\frac{3}{4}x < 9$
21. $\frac{2}{3}x \geqslant 12$
22. $-4x < 12$
23. $-3x \leqslant 27$
•24. $-6x > 18$
25. $-2x < 10$
26. $4x + 3x \geqslant 2x + 7$
27. $8x - 2x > 4x - 5$
•28. $3x + 2x < x + 6$
29. $\frac{3x}{2} > 8$
30. $\frac{4x}{3} > 12$
31. $x + 7 \leqslant 9$
32. $x + 4 \geqslant -12$
33. $x - 5 < 6$
34. $x - 12 < -9$
35. $2x + (3x - 1) > 5 - x$
36. $2(3x + 1) < 7$
37. $3(2x - 1) \geqslant 4x + 3$
38. $4(5x - 3) \leqslant 25x + 11$
39. $12x - 8 > 5x + 2$
40. $9x + 4 > x - 11$
41. $2 + 5x - 16 < 6x - 4$
42. $3(2x + 5) > 4(x - 3)$
43. $8 - 2(3x + 4) > 5x - 16$
44. $(3x + 2) - (2x - 5) > 7$
45. $(7x - 6) - (4 - 3x) \leqslant 27$
•46. $2(x - 4) - 16 \leqslant 3(5 - 2x)$
47. $3(1 - 2x) \geqslant 2(4 - 4x)$
48. $4(5 - x) > 7(2 - x)$
49. $3(2x + 3) \geqslant 5 - 4(x - 2)$
50. $6(3x - 2) \leqslant 7(x - 3) - 2$
51. $-1 < 2x + 3 < 4$
52. $-3 < 3x - 4 < 6$
53. $-2 \leqslant 5x + 2 \leqslant 3$
54. $0 \leqslant 7x - 1 \leqslant 7$
55. $-5 < 4x + 3 \leqslant 8$
56. $-2 < -x \leqslant 3$
57. $-1 \leqslant -x < 4$
58. $-4 \leqslant 2 - x < 3$
59. $-3 < 4 - x \leqslant 5$
60. $1 < 3 - 4x < 6$
61. $0 \leqslant 1 - 3x < 7$
62. $-4 \leqslant 3 - 2x \leqslant 0$

Write an inequality to represent statements 63–67.

Example A student must score below 65 to fail the examination.
If x is the student's score, then the inequality is

$$x < 65$$

63. The student must score at least 72 on the final to pass the course.
64. The temperature today will be less than 38.
65. A company needs to order at least 8 new lift trucks.
66. A company will hire at least 2 new employees, but not more than 7.
67. The selling price (P) must be at least twice the cost (C).

Write an inequality for problems 68–76, using the given information, and solve.

Example Three times a number added to 5 is greater than 14. Find all numbers that satisfy this condition.
If x = the number, then the equation is:

$$3x + 5 > 14$$
$$3x + 5 - 5 > 14 - 5$$
$$3x > 9$$
$$\frac{3x}{3} > \frac{9}{3}$$
$$x > 3$$

The solution is all numbers greater than 3.

68. When 7 is subtracted from 2 times a number, the result is greater than or equal to 9. Find all numbers that satisfy this condition.
69. Five times a number minus 11 is less than 19. Find all numbers that satisfy this condition.
70. If one-third of a number is added to 23, the result is greater than 30. Find all numbers that satisfy this condition.
71. A student has scores of 7, 6, and 8 on three quizzes. What must she score on the fourth quiz to have an average of 7 or higher?
72. A student has scores of 72, 67, and 81 on three exams. If an average of 70 is required to pass the course, what must he score on the fourth test to pass?
73. Two times a number plus 4 is greater than 6 but less than 14. Find all numbers that satisfy these conditions.
74. Four times a number minus 7 is greater than 17 but less than 25. Find all numbers that satisfy these conditions.
75. The perimeter of a square must be greater than 20 in but less than 108 in. Find all numbers that satisfy these conditions.
76. The perimeter of a rectangle must be less than 100 ft. If the length is known to be 30 ft, find all numbers that could represent the width. (*Note:* The width of a real rectangle must be a positive number.)

Chapter Summary

1. A **mathematical statement** is a declarative sentence that can be labeled true or false.
2. An equation that is true for every permissible value of the variable is called an **identical equation** or an **identity.**
3. An equation that is not true for all permissible values of the variable is called a **conditional equation.**
4. A replacement value for the variable that forms a true statement (satisfies that equation) is called a **root,** or **solution,** of that equation.
5. A **linear equation** is an equation where the variable that we are solving for is raised to the first power.
6. A **literal equation** is an equation that contains two or more variables.
7. When we replace the equals sign in a linear equation with an inequality symbol, we form a **linear inequality.**
8. Whenever we multiply or divide both sides of an inequality by a negative number, we must reverse the direction of the inequality symbol.

Chapter Review

[7.1] Determine whether the given statement is true or false when the variable in each equation is replaced with the given number.

1. $x + 7 = 11;\ \ 4$
2. $x - 5 = 12;\ \ 17$
3. $2x + 1 = 9;\ \ 2$
4. $3x - 4 = 14;\ \ 3$
5. $5x - 1 = 21;\ \ 5$
6. $4x + 7 = 27;\ \ 5$
7. $\frac{x}{2} + 5 = 12;\ \ 14$
8. $6 - \frac{x}{3} = 2;\ \ 12$

[7.2, 7.3] Find the solutions of the following equations.

9. $3x = 9$
10. $4x = 12$
11. $-2x = 14$
12. $-3x = 21$
13. $\frac{x}{3} = 4$
14. $\frac{x}{2} = 7$
15. $\frac{3x}{5} = 9$
16. $\frac{2x}{7} = 6$
17. $x + 5 = 12$
18. $x - 4 = 17$
19. $2x + 5 = 11$
20. $3x - 8 = 6$
21. $x + (2x - 1) = 6$
22. $x + 3x = 5 + 7$
23. $3(2x - 1) = 4x - 2$
24. $5(2\mathrm{x} + 3) = 2x - 7$
25. $(3x - 2) - (4x - 1) = 3x$
26. $3(x - 1) - 2(x + 1) = 4$
27. $\frac{x}{2} - \frac{1}{3} = 4$
28. $\frac{x}{4} + \frac{2}{3} = 6$
29. $\frac{2x}{5} - \frac{x}{2} = 1$
30. $\frac{3x}{4} + \frac{x}{2} = 2$
31. $\frac{x - 2}{6} + \frac{x + 3}{6} = 3$
32. $\frac{2x - 3}{4} - \frac{x + 2}{4} = 1$

33. $\frac{x + 2}{3} + \frac{x + 1}{2} = x$

34. $\frac{2x - 1}{3} = \frac{2x + 1}{4}$

35. $(x - 2)(x - 3) = x^2 + 4$

36. $(x - 1)^2 = x^2 + 3x + 2$

[7.4] Solve the following for x.

37. $3ax = 4$

38. $2x - 3a = 7$

39. $4x + 2y = 3x + y$

40. $3ax + 2by = 4bx$

41. $2(x + y) = 3x - y$

42. $3a - 2bx = 9a + bx$

43. $5abx - 3b = 7b + x$

44. $2ax + 2b = 3b + ax$

45. $2abx + 6a = abx$

46. $a(2x - y) = 3ax + y$

47. $a(x + 2y) = b(2x - y)$

48. $2a(2x + y) = b(x + y)$

[7.5] Write an equation for problems 49–53 and solve for the unknown quantities.

49. The difference of two numbers is 23. Find the two numbers if their sum is 105.
50. If a number is divided by 9 and then that result is increased by 7, the answer is 11. Find the number.
51. The difference between one-third of a number and one-fifth of the number is 6. Find the number.
52. John invested part of \$20,000 at 8% and the rest at 7%. If his income from the 8% investment is \$250 more than that from the 7% investment, how much was invested at each rate?
53. Marina made two investments totaling \$25,000. On one investment she made a 12% profit but on the other she took a 19% loss. If her net loss is \$1,030, how much was in each investment?

[7.6] Solve for x.

54. $3x > 12$

55. $5x \leqslant 15$

56. $-2x < 14$

57. $-4x > 16$

58. $2x + 1 < 5$

59. $7x - 4 > 11$

60. $3x + 7 < 5x - 2$

61. $9x + 13 \geqslant 4x + 7$

62. $6(2x - 1) \leqslant 3x - 4$

Chapter Test

Solve for x.

1. $3x - 2 = 5x + 6$

2. $7x - 4 = 3x + 11$

3. $3x - 2 = x - 7$

4. $6x - 5a = 2b + 1$

5. $-2x - 5 = 5x + 3$

6. $5x - 17 = 12$

7. $5bx - 7b = 3bx - 2b$

8. $ax = bx + a$

9. $3(2x - 1) = 7x - 4$

10. $-3x - 7 \geqslant -4x - 5$

11. $4x - 7 = 7x - 3$

12. $2ax - 6a = 3bx + b$

13. $5x + 12 = 3x - 9$

14. $4x + 14 = 7x + 2$

15. $bx = 3ax - 4b$

16. $13x + 7 \geqslant 6x - 9$

17. $\frac{3x}{4} - 5 = 1$

18. $7 + 3(5x - 2) = 14$

19. $2x - 3 = 5x + 11$

20. $2(3x - 4) \geqslant 5(x - 1)$

21. $7a(2x - b) = 3ax + b$

22. $\frac{3x}{2} - \frac{11}{3} + \frac{x}{2} = 1$

23. $\frac{2x}{3} - 4 < 5x - 1$

24. $(x - 1)^2 = x^2 + 2x + 4$

25. Henry invested part of \$14,000 at 10% and the rest at 6%. If his total income from the investments is \$1,080, how much does he have invested at each rate?

26. One number is 7 less than a second number. Find the two numbers if 3 times the smaller is 2 more than 2 times the larger.

8
Factoring

A technician using a shadow-graph to inspect a part.
(Photograph by Kenneth Walters, courtesy of Henry Ford Community College)

8.1 Common Factors

To find the solution of an equation that is not linear, we need to study a technique called **factoring a polynomial.** Factoring polynomials is also useful in dealing with algebraic fractions, since, as we have seen with arithmetic fractions, we must have the fractions in factored form in order to reduce or to find a common denominator.

The first type of factoring that we will examine is finding the **greatest common factor.** Recall the statement of the distributive law:

$$a(b + c) = ab + ac$$

The left side is called the factored form of $ab + ac$. Let us now use the distributive law to write $3x + 6$ in a factored form. First, $3x + 6$ can be written as

$$\underline{3} \cdot x + \underline{3} \cdot 2$$

By applying the distributive law, we have

$$3x + 6 = \underline{\underline{3}} \cdot x + \underline{\underline{3}} \cdot 2 = 3(x + 2)$$

$3(x + 2)$ is the completely factored form of $3x + 6$.

We observe that this type of factoring, as its name implies, involves looking for numbers or variables that are *common to all* of the original terms. In our example, 3 was common to all of the original terms and we were able to factor it out.

note

When we factor something out, we are dividing it out of all the original terms.

Examples

Factor the polynomials.

1. $10x^2 + 15y = \underline{\underline{5}} \cdot 2x^2 + \underline{\underline{5}} \cdot 3y$
   $= 5(2x^2 + 3y)$
2. $12S - 42R = \underline{\underline{6}} \cdot 2S - \underline{\underline{6}} \cdot 7R$
   $= 6(2S - 7R)$
3. $18xy + 12xz = \underline{\underline{6x}} \cdot 3y + \underline{\underline{6x}} \cdot 2z$
   $= 6x(3y + 2z)$

In these examples, we determined the greatest common factor by inspection. In some problems, however, this may not be possible, and the following procedure will be necessary. Let us factor the polynomial $12x^3y + 30x^2y^3$.

Step 1 Factor each term such that it is the product of primes and variables to powers.

$12x^3y$	$30x^2y^3$
$2^2 \cdot 3 \cdot x^3 \cdot y$	$2 \cdot 3 \cdot 5 \cdot x^2 \cdot y^3$

Step 2 Write down all the prime numbers and variables that are common to *every* term.

$$2 \cdot 3 \cdot x \cdot y$$

note

We do not have 5 as part of our greatest common factor since it does not appear in *all* of the terms.

Step 3 Take the numbers and variables in step 2 and raise them to the *lowest* power to which they were raised in any of the terms.

$$2^1 \cdot 3^1 \cdot x^2 \cdot y^1 = 6x^2y$$

This is the greatest common factor (GCF).

Step 4 We can now write the polynomial in its factored form.

$$12x^3y + 30x^2y^3$$
$$= \underline{6x^2y} \cdot 2x + \underline{6x^2y} \cdot 5y^2$$
$$= 6x^2y(2x + 5y^2)$$

note

If you do not see by inspection what the greatest common factor is multiplied by in each term, simply divide the GCF into each term. The answer tells what the GCF is multiplied by. In our example we could have divided each term by $6x^2y$ as follows:

$$\frac{12x^3y}{6x^2y} = 2x \quad \text{and} \quad \frac{30x^2y^3}{6x^2y} = 5y^2$$

and our factorization would have been the same as in step 4:

$$6x^2y(2x + 5y^2)$$

Examples

Factor the polynomials.

1. $2r^4s - 4r^2s^2 - 8r^3s^3 = 2 \cdot r^4 \cdot s - 2^2 \cdot r^2 \cdot s^2 - 2^3 \cdot r^3 \cdot s^3$
   Note that 2, r, and s are common factors and that the lowest respective powers are $2^1r^2s^1$. Therefore, the completely factored form is $2r^2s(r^2 - 2s - 4rs^2)$.
2. $72a^2b^5 - 84a^3b^4 + 48a^4b^2$
   $= 2^3 \cdot 3^2 \cdot a^2 \cdot b^5 - 2^2 \cdot 3 \cdot 7 \cdot a^3 \cdot b^4 + 2^4 \cdot 3 \cdot a^4 \cdot b^2$
   Each term has 2, 3, a, and b in common, and the GCF is $2^2 \cdot 3 \cdot a^2 \cdot b^2 = 12a^2b^2$. The completely factored form is
   $12a^2b^2(6b^3 - 7ab^2 + 4a^2)$.
3. $3x^3y^2 + 15x^2y^4 + 3xy^2 = 3 \cdot x^3 \cdot y^2 + 3 \cdot 5 \cdot x^2 \cdot y^4 + 3 \cdot x \cdot y^2$
   We have 3, x, and y in common and the GCF is $3xy^2$. The factored form is $3xy^2(x^2 + 5xy^2 + 1)$.

   note

   In example 3, the last term in the factored form is 1. This occurs when a term and the GCF are the same—that is, whenever we are able to factor all the numbers and variables out of a given term.

4. $3x^2 + 3x^3 - 6x^2y = 3 \cdot x^2 + 3 \cdot x^3 - 2 \cdot 3 \cdot x^2 \cdot y$
   3 and x are common factors in each term and the GCF is $3x^2$. The completely factored form is $3x^2(1 + x - 2y)$.

Notice in example 4 that $3x^2 + 3x^3 - 6x^2y$ can also be factored to

$$3x(x + x^2 - 2xy) \text{ or } 6x\left(\frac{1}{2}x + \frac{1}{2}x^2 - xy\right)$$

Thus, a given polynomial may be factored in many ways, unless restrictions are placed on the procedure. Since we want to factor each polynomial in a *unique* manner that will not permit such variations in the results, we want to adopt the following criteria for a completely factored polynomial. A polynomial is considered to be in **completely factored form** when it satisfies these criteria (which ensure its uniqueness):

1. The algebraic expression is written as a product of polynomials with integral coefficients.
2. No polynomial (other than a monomial) in factored form contains another polynomial factor with integral coefficients.

We see that $3x^2(1 + x - 2y)$ is the completely factored form of the expression $3x^2 + 3x^3 - 6x^2y$ since all coefficients are integers, and, except for the monomial factor, $3x^2$, the remaining factor contains no other factor with integral coefficients.

In general, whenever we factor a monomial out of a polynomial, we will factor the monomial so that it has a positive coefficient. You should realize that we can also factor out the opposite, or negative, of this common factor. In our previous example, for instance, we could have factored out $-3x^2$ and the completely factored form would have been

$$-3x^2(-1 - x + 2y)$$

Observe that the only change in the answer when we factor out the opposite of the common factor is that this changes the signs of all the terms inside the parentheses.

Examples

Factor the polynomial two ways: first, so that the common factor has a positive coefficient, and second, so that the common factor has a negative coefficient.

1. $36x^3 - 54x^2 + 30x$
   a. $6x(6x^2 - 9x + 5)$
   b. $-6x(-6x^2 + 9x - 5)$
2. $6a^3b^3 - 2a^2b^5 + 4ab^4$
   a. $2ab^3(3a^2 - ab^2 + 2b)$
   b. $-2ab^3(-3a^2 + ab^2 - 2b)$
3. $3x + 5y - 7z$

   Observe that the greatest common factor between the three terms is only the number 1.

   a. $1(3x + 5y - 7z)$
   b. $-1(-3x - 5y + 7z)$

A final consideration when looking for common factors comes from recalling the fact that when something is within a grouping symbol, we treat the quantity as just one number. Therefore, if we have a quantity common to all of the terms, it can be factored out of the polynomial.

Examples

Factor completely.

1. $x(a - 2b) + y(a - 2b)$

   The quantity $(a - 2b)$ is common to both terms. We can then factor the common quantity out of each term and place the remaining values from each term in a second parentheses:

   $$x\underline{(a - 2b)} + y\underline{(a - 2b)}$$
   $$(a - 2b) \cdot (x + y)$$

   Common factor   *Remaining values*

2. $3a^2(x + 3y) - 6b(x + 3y)$

   3 and $(x + 3y)$ are common and we have:

   $$\underline{3} \cdot a^2\underline{(x + 3y)} - 2 \cdot \underline{3} \cdot b \cdot \underline{(x + 3y)}$$
   $$3(x + 3y) \cdot (a^2 - 2b)$$

   Common factor   *Remaining values*

3. $4x^2(a - 4b) + 12xy(a - 4b)$

   4, x, and $(a - 4b)$ are common and we have:

   $$\underline{4} \cdot \underline{x} \cdot x\,\underline{(a - 4b)} + 3 \cdot \underline{4} \cdot \underline{x} \cdot y\underline{(a - 4b)}$$
   $$4x(a - 4b) \cdot (x + 3y)$$

   Common factor   *Remaining values*

It is important that we remember to look for common factors *first* when attempting to determine the completely factored form of any polynomial. If we fail to do this, then the answer may not be in a completely factored form or we may not see how to factor the problem by an appropriate rule.

Exercise Set 8.1

Write in completely factored form.

Example $9x + 6y + 3 = 3 \cdot 3x + 3 \cdot 2y + 3 \cdot 1$
$3(3x + 2y + 1)$

1. $2y + 6$
2. $3E - 12$
3. $4x^2 + 8y$
4. $8y^2 + 10x^2$
5. $3x^2y + 15z$
6. $5R^2 + 10RI - 20I$
7. $7e - 14i + 21r$
8. $8E - 12I + 16R$
9. $15xy - 18z + 3x^2$
10. $18I - 27I + 3$
11. $42xy - 51y^2 + 7$
12. $15a^2 - 27b^2 + 12ab$
13. $8x - 10y + 12z - 18w$
14. $15L^2 - 21W^2 + 36H$

•15. $20E^2 - 60ER + 45I^2$

Supply the missing factors in problems 16–25.

Example $-3i - 6e = -3(\quad)$
Since $-3i - 6e = -3 \cdot i + -3 \cdot 2e$, then
$-3i - 6e = -3(i + 2e)$

16. $-2e + 2 = -2(\quad)$
17. $-6x - 9 = -3(\quad)$
18. $-5I + 10R = 5(\quad)$
19. $6x - 8z - 12w = 2(\quad)$
20. $-4E^3 - 6EV + 18EV^2 - 24V^3 = -2(\quad)$
21. $-12L + 15W - 6H = -3(\quad)$
22. $-9ac + 18a^2c - 27c^2 = -9(\quad)$
23. $-16R^2 - 32R + 48E = 16(\quad)$
24. $5 - 10r^2 + 15e - 25r = 5(\quad)$
•25. $-6w + 12s - 6w^2 - 6w^3 - 18 = -6(\quad)$

Write in completely factored form.

26. $4x^2 + 8x$
27. $3x^2y + 6xy$
28. $8x^3 + 4x^2$
29. $2R^4 - 6R^2$
30. $3x^2 - 3xy + 3x$
31. $2x^3 - x^2 + x$
32. $24e^2 + 12e - 6e^3$
•33. $15er + 18er^2 - 3e^2r$
34. $2x^4 - 6x^2 + 8x$
35. $xy^2 + xyz + xy$
36. $3IR^2 - 6I^2R^2 + 12IR$
37. $2L^3 - 18L + 2L^2$
38. $V^2 + V^3 - V^4 + 2V$
39. $5p^2 + 10p - 15p^3$
40. $16x^3y - 3x^2y^2 + 24x^2y^3$

Supply the missing factors in problems 41–50.

41. $-xy - xz = -x(\quad)$
42. $-R^2 - RE = -R(\quad)$
43. $-ei - e^2i = ei(\quad)$
44. $-3r + ir^3 = -r(\quad)$
45. $-x + x^2 - x^3 = -x(\quad)$
46. $-V_1 + 2V_1V_2 + V_1V^2 = V_1(\quad)$
•47. $-xyz + x^2yz - xy^2z + xyz^2 = -xyz(\quad)$
48. $-4x^2 + 8x - 12x^3 = -4x(\quad)$
49. $-10e^2r^2 + 15er - 20r^3e^3 = 5er(\quad)$
50. $-24R_1R_2 - 16R_1 + 32R_1^2 = -8R_1(\quad)$

Factor the polynomials in problems 51–60 in two ways: first, so that the common factor has a positive coefficient, and second, so that the common factor has a negative coefficient.

51. $4x^3 - 8x^2$
52. $12x^2y^3 + 6xy^2$
53. $8a^3 - 32a^2b$
54. $9y^5 - 15y^3$
55. $24a^3 - 12a + 6a^2$
56. $24a^3b^2 - 3a^2b^2 + 12a^2b^5$
57. $7R^2 + 14R - 21R^3$
•58. $3x^2y - 6xy^4 + 15x^3y^2$
59. $7x^2 + 9y^2$
60. $21a^3b - 10c^4$

Write in completely factored form.

61. $x(a + b) + y(a + b)$

62. $3a(x - y) + b(x - y)$

•63. $15x(2a + b) + 10y(2a + b)$

64. $21R(L + 2N) - 35S(L + 2N)$

65. $3xy(a + 4b) + 6xy(a + 4b)$

66. $4RS(2P + q) - 8RT(2P + q)$

•67. $8a(b + 6) - (b + 6)$

68. $2x(y + 32) - (y + 32)$

69. $12x^2y(a - 2b) + 8xy^2(a - 2b)$

70. $15a^3b^2(x + y) + 30a^2b^5(x + y)$

Solve the equations in problems 71–81 for the indicated variable.

Example Solve the equation $RT - 2T = 6$ for T, and evaluate the equation for T when $R = 8$

Solution $(R - 2)T = 6$ Factor out T

$$\frac{(R - 2)T}{(R - 2)} = \frac{6}{R - 2}$$ Divide both sides by the coefficient $(R - 2)$

answer: $$T = \frac{6}{R - 2}$$

Evaluate T when $R = 8$

$$T = \frac{6}{(8) - 2} = \frac{6}{6} = 1$$

•71. Solve for R when $2RW - R = 4W$. Evaluate R when $W = 13$.

72. Solve for K when $KL - 5 = 2K$. Evaluate K when $L = 10$.

•73. Solve for L when $LP^2 - LP = L + 6$. Evaluate for L when $P = 2$.

74. Solve for I when $IR^2 = 3R - 4I + 6$. Evaluate I when $R = 3$.

75. Solve $\frac{Q + 3}{Q} = R$ for Q. Evaluate Q when $R = -3$.

76. Solve $\frac{8E - 3I}{2E} = 2I$ for E. Evaluate E when $I = -4$.

77. The pitch diameter D of a gear with outside diameter D_o and number of teeth N is given by $D(N + 2) = D_oN$. Solve for N.

78. The total surface area of a right circular cone is given by $A = \pi rs + \pi r^2$; factor the right side. (π is the Greek letter pi.)

79. The equation of a rocket fired vertically upward into the air is given by $S = 560t - 16t^2$, where the rocket is S feet from the ground after t seconds. Factor the right side.

80. In engineering, the equation for deflection of a beam is given by

$$Y = \frac{2wx^4}{48EI} - \frac{3lwx^3}{48EI} - \frac{l^3wx}{48EI}.$$

Factor the right side.

81. The total surface area of a piston is given by $2\pi rh + 2\pi r^2$, where r is the radius of the base and h is the height of the piston. Factor this expression.

8.2 The Difference of Two Squares

In section 6.5 we saw that the product of $(a + b)(a - b)$ was $a^2 - b^2$. We refer to the indicated product $(a + b)(a - b)$ as the "*product* of the *sum* and *difference* of the same two numbers." Notice that in one factor we *add* the numbers and in the other we find the *difference* between these same numbers. The product will *always* be the "*difference of the squares* of the two numbers."

To factor the difference of two squares, we reverse the formula from section 6.5 and have $a^2 - b^2 = (a + b)(a - b)$. Verbally we have:

(first term)^2 − (second term)^2 factors into
(first term + second term) · (first term − second term)

In order to use this factoring technique, we must be able to recognize **perfect squares.** For example, 16, x^2, $25a^2$, $9y^4$ are all perfect squares since

$$16 = 4 \cdot 4 = (4)^2$$
$$x^2 = x \cdot x = (x)^2$$
$$25a^2 = 5a \cdot 5a = (5a)^2$$
$$9y^4 = 3y^2 \cdot 3y^2 = (3y^2)^2$$

We see from the examples that a perfect square is any quantity that can be written as the product of two identical quantities.

The procedure for factoring the difference of two squares is as follows:

Step 1 Identify the fact that we have a perfect square minus another perfect square.

Step 2 Rewrite the problem as a first term squared minus a second term squared.

(first term)^2 − (second term)^2

Step 3 Factor the problem into the quantity of the first term plus the second term times the quantity of the first term minus the second term.

(first term + second term)(first term − second term)

Examples

Write in completely factored form.

1. $x^2 - 9 = (x)^2 - (3)^2$
   $= (x + 3)(x - 3)$
2. $4P^2 - 25V^2 = (2P)^2 - (5V)^2$
   $= (2P + 5V)(2P - 5V)$
3. $R^4 - 16E^2 = (R^2)^2 - (4E)^2$
   $= (R^2 + 4E)(R^2 - 4E)$
4. $81a^4b^6 - 4c^4 = (9a^2b^3)^2 - (2c^2)^2$
   $= (9a^2b^3 + 2c^2)(9a^2b^3 - 2c^2)$

The first step in any factoring problem is to look for common factors. Often an expression that does not appear to be factorable becomes so by taking out the common factor. Consider

$$3x^2 - 27$$

which contains the common factor 3. Thus, it becomes

$$3(x^2 - 9)$$

Now we recognize the quantity inside the parentheses as the difference of two squares, and the completely factored form is:

$$3(x + 3)(x - 3)$$

Another thing that we must be watching for is that when we have applied a factoring rule to a problem, *we must inspect all parts of the answer to make sure that nothing will factor further*. Consider $x^4 - 81$, which factors as the difference of two squares into

$$(x^2)^2 - (9)^2 = (x^2 + 9)(x^2 - 9)$$

Upon further inspection we observe that the second factor is also the difference of two squares; thus, the completely factored form will be

$$(x^2 + 9)[(x)^2 - (3)^2] = (x^2 + 9)(x + 3)(x - 3)$$

note

In the above example $x^2 + 9$ is called the sum of two squares, and it will not factor in the system of real numbers.

Examples

Write in completely factored form.

1. $2x^2 - 18y^2 = 2(x^2 - 9y^2)$
$= 2[(x)^2 - (3y)^2]$
$= 2(x + 3y)(x - 3y)$
2. $3w^2 - 48 = 3(w^2 - 16)$
$= 3[(w)^2 - (4)^2]$
$= 3(w + 4)(w - 4)$
3. $5a^4 - 45a^2b^2 = 5a^2(a^2 - 9b^2)$
$= 5a^2[(a)^2 - (3b)^2]$
$= 5a^2(a + 3b)(a - 3b)$
4. $a^4 - 16 = (a^2)^2 - (4)^2$
$= (a^2 + 4)(a^2 - 4)$
$= (a^2 + 4)[(a)^2 - (2)^2]$
$= (a^2 + 4)(a + 2)(a - 2)$
5. $2x^4 - 32 = 2(x^4 - 16)$
$= 2[(x^2)^2 - (4)^2]$
$= 2(x^2 + 4)(x^2 - 4)$
$= 2(x^2 + 4)[(x)^2 - (2)^2]$
$= 2(x^2 + 4)(x + 2)(x - 2)$

note

A common error in examples 1, 2, 3, and 5 is factoring out something that is common but forgetting to include it as a factor in the final answer.

Exercise Set 8.2

Write in completely factored form.

1. $x^2 - 1$
2. $x^2 - 25$
3. $a^2 - 4$
4. $R^2 - I^2$
•5. $9 - E^2$
6. $49 - R^2$
7. $1 - k^2$
8. $4a^2 - 25b^2$
9. $16i^2 - r^2$
10. $144x^2 - 121y^2$
•11. $x^4 - 1$
12. $r^4 - 9$
13. $i^2 - 16r^2$
14. $P^4 - 81$
15. $R^2 - 4I^2$
16. $x^4 - y^4$
17. $25 - W^4$
18. $8x^2 - 32y^2$
•19. $3a^2 - 27b^2$
20. $5r^2 - 75e^2$
21. $20 - 5b^2$
22. $50 - 2x^2$
•23. $x^2y^2 - 4z^2$
24. $r^2i^2 - 25e^2$
•25. $R^4a^4 - 16E^4$
26. $128P^2 - 2V^2$
27. $16e^4 - 1$
28. $49i^2 - 64r^4$
29. $125p^2 - 20v^2$
30. $98x^2y^2 - 50p^2c^2$
31. In engineering, the equation of transverse shearing stress in a rectangular beam is given in two forms: (a) $T = \frac{V}{8I}(h^2 - 4v_1^2)$; (b) $T = \frac{3V}{2A}\left(1 - \frac{4v_1^2}{h^2}\right)$. Factor the right-hand side in each part.
32. In engineering, the equation for the moment of inertia of a square of size a on a side with a square hole in its center of size b on a side is $\frac{1}{12}a^4 - \frac{1}{12}b^4$. Factor this equation.

8.3 Determining When a Trinomial Will Factor

In order to learn how to factor a trinomial of the form $ax^2 + bx + c$, we must first review some ideas of arithmetic.

The first step in our procedure is to find two integers, if they exist, that satisfy some given conditions in the trinomial. For example, we might wish to find two integers, called m and n, such that their product is 48 and their sum is 16. In many problems we will be able to determine the two integers by inspection, but for those problems where the m and n values are not obvious, we use the following procedure.

Find a pair of integers, m and n, whose product is 48 and whose sum is 16. That is, $m \cdot n = 48$ and $m + n = 16$.

Step 1 Make a table of two columns. The heading of the first column will be the numerical value of the product of m and n, and the heading of the second column will be the sign of this product.

48	+

Step 2 Take the natural numbers, $\{1,2,3,4,\cdots\}$, and divide them into the product of m and n. Those that divide in evenly we write as a factored form with the quotient times the divisor.

48	+
$48 \cdot 1$	
$24 \cdot 2$	
$16 \cdot 3$	
$12 \cdot 4$	
$8 \cdot 6$	
$6 \cdot 8$	
$4 \cdot 12$	
$3 \cdot 16$	
$2 \cdot 24$	
$1 \cdot 48$	

We note that the top five factorizations are the same as the bottom five; therefore, we need only perform this procedure until the factors repeat.

$8 \cdot 6$
The factors repeat.
$6 \cdot 8$

Hence, we have

48	+
$48 \cdot 1$	
$24 \cdot 2$	
$16 \cdot 3$	
$12 \cdot 4$	
$8 \cdot 6$	

Step 3 Use the sign in the second column as an operation symbol, and perform the operation on the factors in the first column.

48	+
$48 \cdot 1$	$48 + 1 = 49$
$24 \cdot 2$	$24 + 2 = 26$
$16 \cdot 3$	$16 + 3 = 19$
$12 \cdot 4$	$12 + 4 = 16$
$8 \cdot 6$	$8 + 6 = 14$

Step 4 From the second column, find the value that is numerically equivalent to the $m + n$ value in the beginning of the problem. That is, $m + n = 16$, and $12 + 4 = 16$ satisfies this condition in our problem.

Step 5 Give the two numbers found in step 4 the signs that will yield the correct product and sum. In our example, the product was positive; therefore, the two factors have the same sign, and since their sum is positive, both numbers must be positive. Our answer, then, is 4 and 12, since $4 \cdot 12 = 48$ and $4 + 12 = 16$.

Examples

Find two integers, if they exist, whose product is the first number and whose sum is the second number.

1. 24 and -11

Step 1

24	+

Step 2

24	+
$24 \cdot 1$	
$12 \cdot 2$	
$8 \cdot 3$	
$6 \cdot 4$	
$4 \cdot 6 \leftarrow$ repeats	

Steps 3 and 4

24	+
$24 \cdot 1$	$24 + 1 = 25$
$12 \cdot 2$	$12 + 2 = 14$
$8 \cdot 3$	$8 + 3 = 11$
$6 \cdot 4$	$6 + 4 = 10$

Step 5 The product was positive and the sum was negative; therefore, both numbers are negative: $(-3)(-8) = 24$ and $(-3) + (-8) = -11$.

answer: -3 and -8

2. -36, 5

Step 1

36	−

Step 2

36	−
$36 \cdot 1$	
$18 \cdot 2$	
$12 \cdot 3$	
$9 \cdot 4$	
$6 \cdot 6$	
$4 \cdot 9 \leftarrow$ repeats	

Steps 3 and 4

36	−
$36 \cdot 1$	35
$18 \cdot 2$	16
$12 \cdot 3$	9
$9 \cdot 4$	5
$6 \cdot 6$	0

Step 5 The product is negative; therefore, one number is positive and the other is negative. The sum gets its sign from the sign of the numerically larger. Therefore, the numerically larger, 9, is positive and the other, 4, is negative: $(9)(-4) = -36$ and $9 + (-4) = 5$.

answer: 9 and -4

3. -12, -1

Step 1

12	−

Step 2

12	−
$12 \cdot 1$	
$6 \cdot 2$	
$4 \cdot 3$	
$3 \cdot 4 \leftarrow$ repeats	

Steps 3 and 4

12	−
$12 \cdot 1$	11
$6 \cdot 2$	4
$4 \cdot 3$	1

Step 5 The product is negative; therefore, one number is positive and the other is negative. The sum gets its sign from the sign of the numerically larger. Therefore, the numerically larger, 4, is negative and the other, 3, is positive: $(-4) \cdot (3) = -12$ and $(-4) + 3 = -1$.

answer: -4 and 3

4. 24, 12

Step 1

24	+

Step 2

24	+
$24 \cdot 1$	
$12 \cdot 2$	
$8 \cdot 3$	
$6 \cdot 4$	
$4 \cdot 6 \leftarrow$ repeats	

Steps 3 and 4

24	+
$24 \cdot 1$	25
$12 \cdot 2$	14
$8 \cdot 3$	11
$6 \cdot 4$	10

Step 5 Since 12 does not appear in the second column, there are no two integers that satisfy the given conditions. Hence, there is no solution.

note

Remember that we use this procedure only when the two integers cannot be determined by inspection.

Let us now go on to factor trinomials of the form $ax^2 + bx + c$. This is called the **standard form of the trinomial:** We have a single variable and the terms of the polynomial are arranged in descending powers of that variable. The a, b, and c in our standard form represent integral constants. For example, $4x^2 - 11x + 6$ is a trinomial in standard form, where $a = 4$, $b = -11$, and $c = 6$. We are going to use our m and n procedure to determine if this trinomial will factor. *The trinomial will factor if and only if there are two integers, m and n, whose product is equal to $a \cdot c$ and whose sum is equal to b.* In our specific trinomial, we want $m \cdot n$ to be equal to $a \cdot c = 4 \cdot 6 = 24$ and $m + n$ to be equal to b, which is -11. Therefore, this trinomial will factor if we find m and n values satisfying the condition of $m \cdot n = 24$ and $m + n = -11$. We saw in the preceding example 1 that the m and n values that satisfy these conditions are -3 and -8. Since we were able to find these m and n values, we are assured that this trinomial will factor. If we could not find m and n values, as in example 4, we would say that the trinomial will not factor.

Examples

Determine if the trinomial will factor.

1. $4x^2 + 8x + 3$ $\qquad a = 4, b = 8, c = 3$
   $a \cdot c = 4 \cdot 3 = 12 = m \cdot n$
   $b = 8 = m + n$
   The trinomial will factor since there exist values for m and n satisfying the given conditions; $6 \cdot 2 = 12$ and $6 + 2 = 8$.
2. $12x^2 - 17x + 6$ $\qquad a = 12, b = -17, c = 6$
   $a \cdot c = 12 \cdot 6 = 72 = m \cdot n$
   $b = -17 = m + n$
   The trinomial will factor since we can find values for m and n; $(-8)(-9) = 72$ and $(-8) + (-9) = -17$.

3. $x^2 - 7x - 18$ $\quad a = 1, b = -7, c = -18$
$a \cdot c = (1)(-18) = -18 = m \cdot n$
$b = -7 = m + n$
The trinomial will factor since there exist values for m and n; $(2)(-9) = -18$ and $2 + (-9) = -7$.

4. $4x^2 - 9 = 4x^2 + 0x - 9$ $\quad a = 4, b = 0, c = -9$
$a \cdot c = 4(-9) = -36 = m \cdot n$
$b = 0 = m + n$
This will factor since there exist values for m and n; $(6)(-6) = -36$ and $6 + (-6) = 0$.

note

Example 4 is the difference between two squares, but if we write it as a trinomial in standard form, our m and n procedure will also determine if it factors.

5. $3x^2 + 5x - 4$ $\quad a = 3, b = 5, c = -4$
$a \cdot c = 3(-4) = -12 = m \cdot n$
$b = 5 = m + n$
Since no two integers satisfy these conditions, we say that this trinomial will not factor.

Exercise Set 8.3

Find two integers, m and n, that multiply together to equal the first number and add together to equal the second number. If m and n do not exist, put DNE as your answer.

1. 12, 7	2. $-4, -3$	3. 12, 8	4. $-24, 2$
•5. 48, 16	6. 48, 19	7. 48, 26	8. 24, 25
•9. $-30, 1$	10. $-6, -1$	11. $7, -8$	12. $-24, -10$
13. 144, 24	14. 144, 25	15. 144, 26	16. 144, 30
17. $-24, 5$	18. 24, 14	19. $-8, 7$	20. $24, -10$
21. $-24, 10$	22. 34, 19	23. 24, 13	24. $4, -4$
25. $4, -5$	•26. $-72, -21$	27. 24, 15	28. $20, -9$
29. 4, 5	30. 25, 10	31. 16, 8	32. 36, 12
33. 49, 14	•34. $-30, -17$	35. 12, 13	36. 30, 31
37. $30, -17$	38. $-30, -11$	39. 18, 11	40. $-72, 21$
41. $-30, 7$	42. $-30, 8$	43. $-72, 34$	44. $18, -9$
45. 35, 12	46. 72, 27	47. $-18, -17$	48. $72, -18$
49. $-72, -6$	50. $-18, 3$		

Determine, by finding m and n, if the following polynomials will factor.

•51. $3x^2 + 4x + 1$	52. $4i^2 - 5i + 1$	53. $3e^2 + 7e + 2$
54. $x^2 + 10x + 25$	55. $x^2 + 5x + 7$	56. $2b^2 - 9b + 10$
57. $5i^2 - 2i - 3$	58. $2r^2 + 13r + 18$	59. $4x^2 + 20x + 21$
60. $x^2 + 16x + 64$	•61. $9y^2 - 21y - 8$	62. $10v^2 + 7v - 6$
63. $10L^2 + 9L + 2$	64. $x^2 - 81$	65. $2i^2 + 15i + 7$

66. $2x^2 - x - 4$	67. $4x^2 - 49$	68. $2x^2 - 5x - 12$
69. $12x^2 - 13x + 3$	70. $5x^2 - 4x - 1$	71. $3x^2 + 4x + 5$
72. $9z^2 - 30z + 25$	73. $14x^2 - x - 3$	74. $5x^2 + 11x - 12$
75. $4x^2 - 17x + 4$	76. $12x^2 + 16x + 5$	77. $3x^2 - 7x + 9$
78. $12x^2 - 37x + 28$	79. $6x^2 + 23x + 15$	80. $12x^2 + 5x - 28$

8.4 Factoring Trinomials

In the previous section we learned how to determine if a trinomial is *factorable*. To actually perform the factoring, let us now combine that procedure with the ideas of factoring out what is common. Consider the product $(2x + 1)(2x + 3)$. By multiplying these two quantities together we get a trinomial:

$$\begin{aligned}(2x + 3)(2x + 1) &= 4x^2 + 2x + 6x + 3 \\ &= 4x^2 + 8x + 3\end{aligned}$$

To completely factor the trinomial $4x^2 + 8x + 3$ entails reversing this procedure to get $(2x + 3)$ and $(2x + 1)$.

The trinomial is factorable if we can find two integers such that $a \cdot c = 4 \cdot 3 = 12 = m \cdot n$ and $b = 8 = m + n$. The two values that satisfy these conditions for this example are 2 and 6. In being able to find these two integers, we are sure that this trinomial will factor. If we observe the multiplication process in our example, we see that m and n appear as the coefficients of the middle terms that are to be combined for our final answer:

$$\begin{aligned}(2x + 1)(2x + 3) &= 4x^2 + \underline{\underline{2x + 6x}} + 3 \\ &= 4x^2 + 8x + 3\end{aligned}$$

This is precisely what we are to do with the m and n values. We replace the middle term in the trinomial with these values. In our example, m and n are 2 and 6 and we replace the 8 with them.

$$4x^2 + \overbrace{2x + 6x}^{8x} + 3$$

Our next step is to group the first two terms and the last two terms:

$$(4x^2 + 2x) + (6x + 3)$$

At this point we factor out what is common in each pair. We see that the first two terms contain the common factor $2x$ and the last two terms contain the common factor 3:

$$2x(2x + 1) + 3(2x + 1)$$

When we reach this point, what is inside the parentheses in each term will be the same. Since the quantity $(2x + 1)$ represents just one number, and this number is common to both terms, we can factor it out:

$$2x(2x + 1) + 3(2x + 1)$$

Common to both terms

Having factored out what is common, what is left in each term is placed in a second pair of parentheses.

$$2x(2x + 1) + 3(2x + 1)$$
$$(2x + 1) \cdot (2x + 3)$$

Common factor *Remaining values*

The trinomial is factored.

A summary of the steps is as follows:

Step 1 Determine if the trinomial is factorable by finding m and n. If m and n do not exist, we conclude that the trinomial will not factor.

Step 2 Replace the middle term in x by the sum of mx and nx.

Step 3 Place parentheses about the first and second terms and about the third and fourth terms. Factor out what is common to each pair.

Step 4 Factor out the common quantity of each term and place the remaining values from each term in a second pair of parentheses.

Examples

1. $6x^2 + 17x + 12$

Step 1 $a \cdot c = 6 \cdot 12 = 72 = m \cdot n$
$b = 17 = m + n$
m and n are 8 and 9

Step 2 $6x^2 + \overbrace{8x + 9x}^{17x} + 12$

Step 3 $(6x^2 + 8x) + (9x + 12)$
$2x(3x + 4) + 3(3x + 4)$

Step 4 $(3x + 4)(2x + 3)$

note

The order in which we place m and n into the problem will not affect the outcome. For example,

Step 2 (alternate) $6x^2 + \overbrace{9x + 8x}^{17x} + 12$
Step 3 $(6x^2 + 9x) + (8x + 12)$
$3x(2x + 3) + 4(2x + 3)$
Step 4 $(2x + 3)(3x + 4)$

The outcome in step 4 is the same regardless of the order of m and n in the problem.

2. $6x^2 + 5x + 1$

Step 1 $a \cdot c = 6 \cdot 1 = 6 = m \cdot n$
$b = 5 = m + n$
m and n are 2 and 3.

Step 2 $6x^2 + \overbrace{2x + 3x}^{5x} + 1$

Step 3 $(6x^2 + 2x) + (3x + 1)$
$2x(3x + 1) + 1(3x + 1)$
In the last two terms, we observe that the greatest common factor is simply 1 or -1.

Step 4 $(3x + 1)(2x + 1)$

3. $4x^2 - 5x + 1$

Step 1 $a \cdot c = 4 \cdot 1 = 4 = m \cdot n$
$b = -5 = m + n$
m and n are -4 and -1.

Step 2 $4x^2 \overbrace{- 4x - x}^{-5x} + 1$

Step 3 $(4x^2 - 4x) + (-x + 1)$
$4x(x - 1) - 1(x - 1)$
We again see that the last two terms have only 1 or -1 as a common factor. This time we choose to factor out the -1 so that we will have the same quantity inside each parentheses.

note
In the second step, if the third term is preceded by a minus sign, we usually factor out the negative factor.

Step 4 $(x - 1)(4x - 1)$

4. $12x^2 - 11x + 2$

Step 1 $a \cdot c = 12 \cdot 2 = 24 = m \cdot n$
$b = -11 = m + n$
m and n are -3 and -8.

Step 2 $12x^2 \overbrace{- 3x - 8x}^{-11x} + 2$

Step 3 $(12x^2 - 3x) + (-8x + 2)$
$3x(4x - 1) - 2(4x - 1)$
In the last two terms, we have 2 or -2 as a common factor. We factor out the -2 so that we will have the same quantity appear inside the parentheses.

Step 4 $(4x - 1)(3x - 2)$

5. $6x^2 + 7x + 4$

Step 1 $a \cdot c = 6 \cdot 4 = 24 = m \cdot n$
$b = 7 = m + n$
These m and n values are not obvious by inspection; therefore, we use the procedure to determine them.

24	+
$24 \cdot 1$	$24 + 1 = 25$
$12 \cdot 2$	$12 + 2 = 14$
$8 \cdot 3$	$8 + 3 = 11$
$6 \cdot 4$	$6 + 4 = 10$
$4 \cdot 6 \leftarrow$ repeats	

We observe that a sum of 7 was never obtained in the second column; therefore, $6x^2 + 7x + 4$ will not factor.

6. $24x^2 - 9x - 15 = 3(8x^2 - 3x - 5)$

note
Recall that before applying any factoring rule, we must always factor out what is common to each term.

Now we are ready to factor the trinomial, $8x^2 - 3x - 5$.

Step 1 $a \cdot c = 8(-5) = -40 = m \cdot n$
$b = -3 = m + n$
m and n are -8 and 5.

Step 2 $3(8x^2 \overbrace{- 8x + 5x}^{-3x} - 5)$

Step 3 $3[(8x^2 - 8x) + (5x - 5)]$
$3[8x(x - 1) + 5(x - 1)]$

Step 4 $3(x - 1)(8x + 5)$

Factoring by Inspection—An Alternate Approach

In the first part of this section we studied a systematic procedure for determining if a quadratic trinomial will factor and how to factor it. In many instances we are able to determine how the trinomial will factor by inspecting the problem rather than applying this procedure.

Factoring by inspection is accomplished as follows. Factor $7x + 2x^2 + 3$:

Step 1 Write the trinomial in standard form:

$2x^2 + 7x + 3$

Step 2 Determine the possible combinations of first-degree factors of the first term.

$(2x \quad)(x \quad)$

Step 3 Combine with the factors of step 2 all the possible factors of the last term.

$$(2x \quad 3)(x \quad 1)$$
$$(2x \quad 1)(x \quad 3)$$

Step 4 Determine the possible symbol, ($+$ or $-$), between the terms in each binomial.

$$(2x + 3)(x + 1)$$
$$(2x + 1)(x + 3)$$

The rules of signed numbers given in chapter 1 provide the answer to step 4.

1. If the third term is preceded by a $+$ sign and the middle term is preceded by a $+$ sign, then the symbols will be

$$(\quad + \quad)(\quad + \quad)$$

2. If the third term is preceded by a $+$ sign and the middle term is preceded by a $-$ sign, then the symbols will be

$$(\quad - \quad)(\quad - \quad)$$

3. If the third term is preceded by a $-$ sign, then the symbols will be

$$(\quad + \quad)(\quad - \quad)$$
or
$$(\quad - \quad)(\quad + \quad)$$

note

It is assumed that the first term is preceded by a $+$ sign or no sign. If it is preceded by a $-$ sign, these rules could still be used if a (-1) is first factored out of all the terms.

Step 5 Determine which factors, if any, yield the correct middle term.

$$(2x + 3)(x + 1)$$

$+3x$

$+2x$

$$(+3x) + (+2x) = +5x$$

$$(2x + 1)(x + 3)$$

$+x$

$+6x$

$$(+x) + (+6x) = \boxed{+7x}$$

The second set of factors gives us the correct middle term. Therefore, $(2x + 1)(x + 3)$ is the factorization of $2x^2 + 7x + 3$.

Examples Factor the following quadratic trinomials by inspection.

1. **Step 1** $6x^2 + 17x + 5$

   Step 2 $(6x \quad)(x \quad)$
   $(3x \quad)(2x \quad)$

   Step 3 $(6x \quad 5)(x \quad 1)$
   $(6x \quad 1)(x \quad 5)$
   $(3x \quad 5)(2x \quad 1)$
   $(3x \quad 1)(2x \quad 5)$

   Step 4 $(6x + 5)(x + 1)$
   $(6x + 1)(x + 5)$
   $(3x + 5)(2x + 1)$
   $(3x + 1)(2x + 5)$

   Step 5 $(6x + 5)(x + 1)$
   $+5x$
   $+6x$ $\qquad (+5x) + (+6x) = +11x$

   $(6x + 1)(x + 5)$
   $+x$
   $+30x$ $\qquad (+x) + (+30x) = +31x$

   $(3x + 5)(2x + 1)$
   $+10x$
   $+3x$ $\qquad (+10x) + (+3x) = +13x$

   $(3x + 1)(2x + 5)$
   $+2x$
   $+15x$ $\qquad (+2x) + (+15x) = \boxed{+17x}$

The last set of factors gives us the correct middle term. Hence,

$$(3x + 1)(2x + 5)$$

is the factorization of $6x^2 + 17x + 5$.

2. **Step 1** $4x^2 - 5x + 1$

Step 2 $(4x \quad)(x \quad)$
$(2x \quad)(2x \quad)$

Step 3 $(4x \quad 1)(x \quad 1)$
$(2x \quad 1)(2x \quad 1)$

Step 4 $(4x - 1)(x - 1)$
$(2x - 1)(2x - 1)$

Step 5 $(4x - 1)(x - 1)$

inner: $-x$; outer: $-4x$

$(-x) + (-4x) = \boxed{-5x}$

$(2x - 1)(2x - 1)$

inner: $-2x$; outer: $-2x$

$(-2x) + (-2x) = -4x$

The first set of factors gives us the factorization: $4x^2 - 5x + 1 = (4x - 1)(x - 1)$.

3. $13x - 5 + 6x^2$

Step 1 $6x^2 + 13x - 5$

Step 2 $(6x \quad)(x \quad)$
$(2x \quad)(3x \quad)$

Step 3 $(6x \quad 5)(x \quad 1)$
$(6x \quad 1)(x \quad 5)$
$(2x \quad 5)(3x \quad 1)$
$(2x \quad 1)(3x \quad 5)$

Step 4 $(6x + 5)(x - 1)$ or $(6x - 5)(x + 1)$
$(6x + 1)(x - 5)$ or $(6x - 1)(x + 5)$
$(2x + 5)(3x - 1)$ or $(2x - 5)(3x + 1)$
$(2x + 1)(3x - 5)$ or $(2x - 1)(3x + 5)$

Step 5 $(6x + 5)(x - 1)$ or $(6x - 5)(x + 1)$

inner: $+5x$; outer: $-6x$ or inner: $-5x$; outer: $+6x$

$(+5x) + (-6x) = -x$ or $(-5x) + (+6x) = +x$

$(6x + 1)(x - 5)$ or $(6x - 1)(x + 5)$

inner: $+x$; outer: $-30x$ or inner: $-x$; outer: $+30x$

$(+x) + (-30x) = -29x$ or $(-x) + (+30x) = +29x$

$(2x + 5)(3x - 1)$ or $(2x - 5)(3x + 1)$

Inner: $+15x$; outer: $-2x$ — or — inner: $-15x$; outer: $+2x$

$(+15x) + (-2x) = \boxed{+13x}$ or $(-15x) + (+2x) = -13x$

$(2x + 1)(3x - 5)$ or $(2x - 1)(3x + 5)$

Inner: $+3x$; outer: $-10x$ — or — inner: $-3x$; outer: $+10x$

$(+3x) + (-10x) = -7x$ or $(-3x) + (+10x) = +7x$

The factorization of $6x^2 + 13x - 5$ is $(2x + 5)(3x - 1)$.

4. **Step 1** $x^2 + 7x + 12$

Step 2 $(x \quad)(x \quad)$

Step 3 $(x \quad 12)(x \quad 1)$
$(x \quad 6)(x \quad 2)$
$(x \quad 3)(x \quad 4)$

Step 4 $(x + 12)(x + 1)$
$(x + 6)(x + 2)$
$(x + 3)(x + 4)$

Step 5 $(x + 12)(x + 1)$

Inner: $+12x$; outer: $+x$ — $(+12x) + (+x) = +13x$

$(x + 6)(x + 2)$

Inner: $+6x$; outer: $+2x$ — $(+6x) + (+2x) = +8x$

$(x + 3)(x + 4)$

Inner: $+3x$; outer: $+4x$ — $(+3x) + (+4x) = \boxed{+7x}$

The factorization of $x^2 + 7x + 12$ is $(x + 3)(x + 4)$.

Exercise Set 8.4

Factor the common quantity.

Examples 1. $x(2x + 1) - 4(2x + 1)$
$(2x + 1)(x - 4)$
2. $3x(2x - 5) + 2(2x - 5)$
$(2x - 5)(3x + 2)$

1. $x(2x + 1) + 3(2x + 1)$
2. $2(3x + 1) - x(3x + 1)$
•3. $4e(2e - 3) - 2(2e - 3)$
4. $R(3R - 4) + (3R - 4)$
5. $8p(p + 6) - (p + 6)$
6. $3V(4 - V) + 6(4 - V)$
7. $6E(3E + 2R) - R(3E + 2R)$
8. $7L(5L - 6W) - 3W(5L - 6W)$

Factor each trinomial. If a trinomial will not factor, state that as your answer.

Examples 1. $x^2 + 12x + 32$
$a \cdot c = 1 \cdot 32 = 32 = m \cdot n$
$b = 12 = m + n$
m and n are 4 and 8.
$$x^2 + \overbrace{4x + 8x}^{12x} + 32$$
$(x^2 + 4x) + (8x + 32)$
$x(x + 4) + 8(x + 4)$
$(x + 4)(x + 8)$
2. $6x^2 + 23x + 15$
$a \cdot c = 6 \cdot 15 = 90 = m \cdot n$
$b = 23 = m + n$
m and n are 5 and 18.
$$6x^2 + \overbrace{5x + 18x}^{23x} + 15$$
$(6x^2 + 5x) + (18x + 15)$
$x(6x + 5) + 3(6x + 5)$
$(6x + 5)(x + 3)$
3. $12x^2 + 12x - 9$
$3(4x^2 + 4x - 3)$ common factor of 3
$a \cdot c = 4(-3) = -12 = m \cdot n$
$b = 4 = m + n$
m and n are -2 and 6.
$$3[4x^2 \overbrace{- 2x + 6x}^{4x} - 3]$$
$3[(4x^2 - 2x) + (6x - 3)]$
$3[2x(2x - 1) + 3(2x - 1)]$
$3(2x - 1)(2x + 3)$

•9. $2x^2 + x - 6$
10. $3a^2 + 7a - 6$
11. $2x^2 + 3x + 1$
12. $4E^2 - 5E + 1$
13. $2R^2 - 7R + 6$
14. $R^2 - 4R + 6$

15. $5H^2 + 7H - 6$
16. $2x^2 - x - 1$
17. $9I^2 - 6I + 1$
18. $8x^2 - 17x + 2$
19. $5x^2 + 4x + 6$
20. $2L^2 - 11L + 12$
21. $6a^2 + 13a + 6$
22. $2r^2 + 13r + 18$
23. $4x^2 + 20x + 21$
24. $7R^2 + 20R - 3$
25. $4x^2 - 2x + 5$
26. $4E^2 - 4E - 3$
•27. $9y^2 - 21y - 8$
28. $6L^2 - 23L - 4$
29. $10V^2 + 7V - 6$
30. $10L^2 + 9L + 2$
31. $2H^2 - 9H + 10$
32. $7x^2 - 3x + 6$
33. $4e^2 + 14e + 12$
34. $5R^2 - 9R - 2$
35. $4I^2 + 10I + 6$
36. $6E^2 - 17E + 12$
37. $6a^2 + 7a - 3$
38. $3x^2 + 12x + 12$
39. $2i^2 + 6i - 20$
40. $3a^2 + 8a - 4$
41. $6x^2 + 5x - 6$
42. $3L^2 - 19L + 20$
43. $4x^2 + 12x + 9$
•44. $9z^2 - 30z + 25$
45. $7a^2 - 36a + 5$
46. $3x^2 + 2x + 4$
47. $15P^2 + 2P - 1$
48. $12x^2 + 13x - 4$
49. $2a^3 - 8a^2 - 20a$
50. $4x^2 + 10x + 4$
51. $2I^3 + 15I^2 + 7I$
52. $9x^2 - 21x - 8$
53. $8a^2 - 14a - 15$
54. $8E^2 - 18E + 9$
55. When a stone is thrown vertically into the air, the height s of the stone at any instant in time t is given by

$$s = -16t^2 + 32t - 16$$

Factor the right side.

8.5 Factoring Other Types of Trinomials

In exercise set 8.4 we saw, as an example, the factorization of the trinomial

$$x^2 + 12x + 32$$

The steps involved were as follows:

Step 1 $a \cdot c = 1 \cdot 32 = 32 = m \cdot n$ and $b = 12 = m + n$
Therefore, m and n are 4 and 8.

Step 2 We replaced the middle term value with 4 and 8.

$$x^2 + \overbrace{4x + 8x}^{12x} + 32$$

Step 3 We placed parentheses around the first and second terms and around the third and fourth terms, and then factored out what was common to each pair.

$$(x^2 + 4x) + (8x + 32)$$
$$x(x + 4) + 8(x + 4)$$

Step 4 We factored out the common quantity of each term and placed the remaining values from each term in a second pair of parentheses.

$$(x + 4)(x + 8)$$

This problem is a special case of the general trinomial because the coefficient a is equal to 1. When we are factoring a trinomial where $a = 1$, we can eliminate several of the steps in our factoring procedure. In fact, once we find the values for m and n, the problem is essentially finished.

Consider again the example $x^2 + 12x + 32$, which factored into $(x + 4)$ and $(x + 8)$. On closer inspection we see that the final factored form has the m and n values, 4 and 8, simply placed inside the parentheses.

Examples

1. $x^2 + 6x - 16$ $\quad m \cdot n = a \cdot c$, but since a is equal to 1, $a \cdot c = 1 \cdot c = c = m \cdot n$. Therefore, our conditions for m and n are: $m \cdot n = c$ and $m + n = b$. In our problem, $m \cdot n = -16$ and $m + n = 6$. Hence, m and n are 8 and -2, and we have

   $(x + 8)(x - 2)$
2. $x^2 - 7x - 30$ $\quad m \cdot n = -30$ and $m + n = 7$. Therefore, m and n are -10 and 3, and the polynomial factors into $(x - 10)(x + 3)$.

Many other types of trinomials will factor by our m and n procedure. Let us examine the problem $x^2y^2 - 7xy - 30$. If we consider (xy) as a quantity, we have $(xy)^2 - 7(xy) - 30$, and our problem looks like the previous example. Since m and n are -10 and 3, it factors into $(xy - 10)(xy + 3)$.

Examples

1. $a^2b^2 - 4ab - 21 = (ab)^2 - 4(ab) - 21$
   $m \cdot n = -21$ and $m + n = -4$. Therefore, m and n are -7 and 3, and the factored form is $(ab - 7)(ab + 3)$.
2. $r^2s^2 + 10rs - 24 = (rs)^2 + 10(rs) - 24$
   $m \cdot n = -24$ and $m + n = 10$. Therefore, m and n are 12 and -2, and the factored form is $(rs + 12)(rs - 2)$.

The following are some additional types of trinomials that will factor by this procedure.

Examples

1. $x^4 + 2x^2 - 3 = (x^2)^2 + 2(x^2) - 3$
   $m \cdot n = -3$ and $m + n = 2$. Therefore, m and n are 3 and -1, and the polynomial factors into $(x^2 + 3)(x^2 - 1)$.

   As previously noted, we must always inspect all factors of an answer to see if any one might factor further. In this example, we see that the second quantity, $x^2 - 1$, is the difference of two squares and factors into

   $(x + 1)(x - 1)$

Therefore, the completely factored form would be:

$(x^2 + 3)(x + 1)(x - 1)$

2. $x^4 - 13x^2 + 36 = (x^2)^2 - 13(x^2) + 36$

$m \cdot n = 36$ and $m + n = -13$. Therefore, m and n are -9 and -4, and the polynomial factors into $(x^2 - 9)(x^2 - 4)$.

Similar to the previous example, we see that in this example both quantities are the difference of two squares, and the completely factored form is:

$$(x^2 - 9)(x^2 - 4) = (x + 3)(x - 3)(x + 2)(x - 2)$$

---

note

All the examples in this section are special cases of the general trinomial where $a = 1$. If you are in doubt as to how to factor the problem, then use the four-step procedure from the previous section.

Exercise Set 8.5

Factor each of the following. If something will not factor, state that as your answer.

•1. $x^2 + 11x - 12$
2. $y^2 + 13y - 30$
3. $a^2 - 14a + 24$
4. $R^2 + 5R - 24$
5. $B^2 + 8B + 12$
6. $L^2 - 2L - 24$
7. $x^2 - x - 12$
8. $2I^2 + 6I - 20$
•9. $3e^2 - 18e - 48$
10. $a^2 + 5a + 7$
11. $x^2 - 4x + 6$
12. $y^2 + 17y + 30$
13. $4b^2 - 4b - 24$
14. $5s^2 - 15s - 50$
15. $x^2y^2 - 3xy - 18$
16. $a^2b^2 - 4ab - 21$
17. $R^2 - 9R + 4$
18. $s^2 - 7s + 12$
•19. $R^2S^2 - RS - 30$
20. $x^2y^2 + 13xy + 13$
21. $4L^2M^2 - 32LM + 28$
22. $3x^2y^2 + 21xy + 36$
•23. $x^4 - 7x^2 + 6$
24. $y^4 - 5y^2 + 4$
25. $a^4 - 7a^2 - 18$
26. $s^4 - 21s^2 + 80$
27. $x^4 - 23x^2 - 50$
28. $y^4 - 35x^2 - 36$
29. $3x^2 + 12x + 12$
30. $2i^2 + 6i - 20$
•31. $2a^3 - 8a^2 - 20a$
32. $4x^2 + 10x + 4$
33. $2I^3 + 15I^2 + 7I$
34. $16x^2y^2 + 56xy + 49$
35. $R^4 + 5R^2 - 36$
36. $s^4 - 6s^2 - 27$
37. $2x^4 - 16x^2 - 18$
38. $x^4 + 6x^2 - 7$
39. $2x^4 - 14x^2 - 16$
40. $3x^4 - 51x^2 + 48$

8.6 Other Types of Factoring

In section 8.2 we factored problems involving the difference of two squares. To be able to factor problems of this type, we identified the two terms as squares and applied the rule. In this section, we factor *the sum and difference of two cubes* in a similar fashion.

Consider the indicated product of $(a - b)(a^2 + ab + b^2)$. If we carry out the multiplication,

$$\begin{aligned}(a - b)(a^2 + ab + b^2) &= a^3 + a^2b + ab^2 - a^2b - ab^2 - b^3 \\ &= a^3 - b^3\end{aligned}$$

Therefore, $(a - b)(a^2 + ab + b^2) = a^3 - b^3$ and $(a - b)(a^2 + ab + b^2)$ is the factored form of $a^3 - b^3$.

The difference of two cubes factors as:

$$a^3 - b^3 = (a - b)(a^2 + ab + b^2)$$

If we can write a two-term polynomial as a first term cubed minus a second term cubed, then we can factor it as the difference of two cubes.

(1st term)^3 − (2nd term)^3
= (1st term − 2nd term) [(1st term)^2 + (1st term · 2nd term) + (2nd term)^2]

Examples

Factor completely.

1. $x^3 - 27$

   Rewrite the first term as a cube and the second term as a cube.

   $$x^3 - 27 = (x)^3 - (3)^3$$

   The first term is x and the second term is 3. Then, write down the rule for factoring the difference of two cubes:

   (−   ) [(   )^2 + (   )(   ) + (   )^2]
   ↑   ↑   ↑   ↑   ↑   ↑
   1st   2nd   1st   1st   2nd   2nd

   Now substitute the x where the first term appears according to the rule, and the 3 where the second term appears.

   (x − 3) [(x)^2 + (x)(3) + (3)^2]
   ↑   ↑   ↑   ↑   ↑   ↑
   1st   2nd   1st   1st   2nd   2nd

   Finally, simplify:

   $$(x - 3)(x^2 + 3x + 9)$$

   Therefore, $x^3 - 27 = (x - 3)(x^2 + 3x + 9)$.

2. $8x^3 - y^3 = (2x)^3 - (y)^3$

   First term $= 2x$ and   second term $= y$

   Then,

   $$\begin{aligned}&(\quad - \quad) [(\quad)^2 + (\quad)(\quad) + (\quad)^2] \\ &= (2x - y) [(2x)^2 + (2x)(y) + (y)^2] \\ &= (2x - y)(4x^2 + 2xy + y^2)\end{aligned}$$

3. $2a^3 - 54b^3 = 2(a^3 - 27b^3)$
Factor out the common quantity of 2:

$$2(a^3 - 27b^3) = 2\,[(a)^3 - (3b)^3]$$

Then,

$$\begin{aligned} &2(\quad - \quad)\,[(\quad)^2 + (\quad)(\quad) + (\quad)^2] \\ &= 2(a - 3b)\,[(a)^2 + (a)(3b) + (3b)^2] \\ &= 2(a - 3b)(a^2 + 3ab + 9b^2) \end{aligned}$$

4. $a^{15} - 64b^3 = (a^5)^3 - (4b)^3$
Then,

$$\begin{aligned} &(\quad - \quad)\,[(\quad)^2 + (\quad)(\quad) + (\quad)^2] \\ &= (a^5 - 4b)\,[(a^5)^2 + (a^5)(4b) + (4b)^2] \\ &= (a^5 - 4b)(a^{10} + 4a^5b + 16b^2) \end{aligned}$$

note

In example 4, we observe that a number raised to a power that is a multiple of three can be written as a cube by dividing the exponent by 3. The quotient is the exponent of the number inside the parentheses and the 3 is the exponent outside the parentheses. For example, $y^{12} = (y^4)^3$ or $z^{24} = (z^8)^3$.

If we carry out the indicated multiplication in $(a + b)(a^2 - ab + b^2)$, we have:

$$\begin{aligned} (a + b)(a^2 - ab + b^2) &= a^3 - a^2b + ab^2 + a^2b - ab^2 + b^3 \\ &= a^3 + b^3 \end{aligned}$$

Therefore, $(a + b)(a^2 - ab + b^2) = a^3 + b^3$ and $(a + b)(a^2 - ab + b^2)$ is the factored form of $a^3 + b^3$.

The sum of two cubes factors as:

$$a^3 + b^3 = (a + b)(a^2 - ab + b^2)$$

If we can write a two-term polynomial as a first term cubed plus a second term cubed, then it will factor as the sum of two cubes.

$$\begin{aligned} &(\text{1st term})^3 + (\text{2nd term})^3 \\ &\quad = (\text{1st term} + \text{2nd term})\,[(\text{1st term})^2 - (\text{1st term})(\text{2nd term}) + (\text{2nd term})^2] \end{aligned}$$

note

The sum of two cubes will factor, whereas the sum of two squares will not factor.

Examples Factor completely.

1. $a^3 + 8 = (a)^3 + (2)^3$

   Write the rule for the sum of two cubes:

   $$(\underset{\text{1st}}{\uparrow} + \underset{\text{2nd}}{\uparrow})[(\underset{\text{1st}}{\uparrow})^2 - (\underset{\text{1st}}{\uparrow})(\underset{\text{2nd}}{\uparrow}) + (\underset{\text{2nd}}{\uparrow})^2]$$

   Substitute:

   $$(\underset{\text{1st}}{\overset{a}{\uparrow}} + \underset{\text{2nd}}{\overset{2}{\uparrow}})[(\underset{\text{1st}}{\overset{a}{\uparrow}})^2 - (\underset{\text{1st}}{\overset{a}{\uparrow}})(\underset{\text{2nd}}{\overset{2}{\uparrow}}) + (\underset{\text{2nd}}{\overset{2}{\uparrow}})^2]$$

   Simplify:

   $$(a + 2)(a^2 - 2a + 4)$$

   Therefore, $a^3 + 8 = (a + 2)(a^2 - 2a + 4)$.

2. $x^3 + 125 = (x)^3 + (5)^3$

   1st term $= x$ and 2nd term $= 5$

   Then,

   $$\begin{aligned}&(\quad + \quad)[(\quad)^2 - (\quad)(\quad) + (\quad)^2]\\ &= (x + 5)[(x)^2 - (x)(5) + (5)^2]\\ &= (x + 5)(x^2 - 5x + 25)\end{aligned}$$

3. $8a^3 + b^{21} = (2a)^3 + (b^7)^3$

   Then,

   $$\begin{aligned}&(\quad + \quad)[(\quad)^2 - (\quad)(\quad) + (\quad)^2]\\ &= (2a + b^7)[(2a)^2 - (2a)(b^7) + (b^7)^2]\\ &= (2a + b^7)(4a^2 - 2ab^7 + b^{14})\end{aligned}$$

4. $x^3y^3 + z^3 = (xy)^3 + (z)^3$

   Then,

   $$\begin{aligned}&(\quad + \quad)[(\quad)^2 - (\quad)(\quad) + (\quad)^2]\\ &= (xy + z)[(xy)^2 - (xy)(z) + (z)^2]\\ &= (xy + z)(x^2y^2 - xyz + z^2)\end{aligned}$$

When we factor trinomials where the squared term has a coefficient other than 1, the three-term polynomial is changed to a four-term polynomial and the factorization is carried out by grouping the terms in pairs. We apply this technique of grouping when we factor four-term polynomials.

Consider the following example: $ax + ay + bx + by$. We observe that this is a four-term polynomial and we will try to factor it by grouping.

$$ax + ay + bx + by = (ax + ay) + (bx + by)$$

There is a common factor of a in the first two terms and a common factor of b in the second two terms.

$$(ax + ay) + (bx + by) = a(x + y) + b(x + y)$$

The quantity $(x + y)$ is common to both terms, and factoring it out, we have:

$$a(x + y) + b(x + y) = (x + y)(a + b).$$

Therefore, we have factored the polynomial by grouping.

note

In factoring trinomials, if we were able to find the integers m and n, then we were sure at the polynomial would factor. We have no such assurances when trying to factor a four-term polynomial.

Examples

Factor completely.

1. $ax + 2ay + bx + 2by = (ax + 2ay) + (bx + 2by)$
   $= a(x + 2y) + b(x + 2y)$
   $= (x + 2y)(a + b)$
2. $3ac + 6ad + 2bc + 4bd = (3ac + 6ad) + (2bc + 4bd)$
   $= 3a(c + 2d) + 2b(c + 2d)$
   $= (c + 2d)(3a + 2b)$
3. $2ax - 2ay + bx - by = (2ax - 2ay) + (bx - by)$
   $= 2a(x - y) + b(x - y)$
   $= (x - y)(2a + b)$
4. $6ax + by + 3ay + 2bx = 6ax + 3ay + 2bx + by$
   $= (6ax + 3ay) + (2bx + by)$
   $= 3a(2x + y) + b(2x + y)$
   $= (2x + y)(3a + b)$

note

Sometimes the terms must be rearranged so the pairs will have a common factor (see example 4).

5. $a^2x + a^2y - 9x - 9y = (a^2x + a^2y) + (-9x - 9y)$
   $= a^2(x + y) - 9(x + y)$
   $= (x + y)(a^2 - 9)$
   $= (x + y)(a + 3)(a - 3)$

note

In example 5, $(a^2 - 9)$ is the difference of two squares and will factor further.

Exercise Set 8.6

Factor completely. If a problem will not factor, state that as your answer.

Examples 1. $27x^3 + y^3$

$= (3x)^3 + (y)^3$

$(\quad + \quad)[(\quad)^2 - (\quad)(\quad) + (\quad)^2]$

$= (3x + y)[(3x)^2 - (3x)(y) + (y)^2]$

$= (3x + y)(9x^2 - 3xy + y^2)$

2. $16R^3 - 54$
$= 2[8R^3 - 27]$
$= 2[(2R)^3 - (3)^3]$
$2(\quad - \quad)[(\quad)^2 + (\quad)(\quad) + (\quad)^2]$
$= 2(2R - 3)[(2R)^2 + (2R)(3) + (3)^2]$
$= 2(2R - 3)(4R^2 + 6R + 9)$

3. $3ax + 6bx + 2ay + 4by$
$= (3ax + 6bx) + (2ay + 4by)$
$= 3x(a + 2b) + 2y(a + 2b)$
$= (a + 2b)(3x + 2y)$

1. $r^3 + s^3$
2. $L^3 + 8$
•3. $8x^3 + y^3$
4. $27r^3 + 8$
5. $h^3 - k^3$
6. $p^3 - q^3$
7. $64x^3 - y^3$
8. $r^3 - 27$
9. $ac + ad + bc + bd$
10. $rt + ru + st + su$
11. $27x^3 - 8y^3$
12. $64a^3 - 8$
13. $6ax - 2by + 3bx - 4ay$
•14. $5ax - 3by + 15bx - ay$
15. $8a^3 + 27b^3$
16. $64s^3 + 1$
17. $2a^3 + 16$
18. $3x^3 + 81$
19. $2x^3 - 16$
20. $81a^3 - 3b^3$
21. $x^5 + 27x^2y^3$
22. $16a^3 + 2b^3$
23. $20x^2 + 5xz - 12xy - 3yz$
24. $2ax^2 - bx^2 + 6a - 3b$
25. $8a^2b^3 - a^5$
26. $2x^3 - 54y^3$
27. $54r^3 + 2s^3$
28. $b^5 + 64b^2c^3$
29. $b^2x - b^2y - 16x + 16y$
30. $x^2y^2 - x^2 + 4y^2 - 4$
31. $x^3y^3 - z^3$
32. $x^3y^9 - 1$
•33. $a^{15}b^6 - 8c^9$
34. $x^{18}y^9 - 27z^3$
35. $a^3b^3 + 8$
36. $x^3y^6 + z^3$
37. $a^2x^2 - x^2 - 9a^2 + 9$
38. $4a^2x^2 - b^2x^2 - 4a^2y^2 + b^2y^2$
39. $x^9y^{12} + z^{15}$
40. $a^{12}b^{15} + c^{24}$

Chapter Summary

1. **Common factors** are factors that appear in all the original terms.
2. A polynomial is in **completely factored form** when
   a. The expression is written as a product of polynomials with integral coefficients.
   b. No polynomial, other than a monomial, in factored form contains another polynomial factor with integral coefficients.
3. The **difference of two squares** factors as follows:
   $$a^2 - b^2 = (a + b)(a - b)$$
4. The **sum of two squares** will not factor.
5. The **trinomial** $ax^2 + bx + c$ will factor if and only if we can find a pair of integers, m and n, whose product is $a \cdot c$ and whose sum is b.
6. The **difference of two cubes** factors as:
   $$a^3 - b^3 = (a - b)(a^2 + ab + b^2)$$

7. The **sum of two cubes** factors as:

$$a^3 + b^3 = (a + b)(a^2 - ab + b^2)$$

8. We try to factor **four-term polynomials** by grouping.

Chapter Review

[8.1] Supply the missing factors in the following.

1. $3x + 9 = 3(\quad)$
2. $9x^2 - 18x = 9x(\quad)$
3. $-4y^3 + 8y^2 = -4y^2(\quad)$
4. $7e + 14i - 28r = 7(\quad)$
5. $-5E^2 - 15E + 30E^3 = -5E(\quad)$

Write in completely factored form.

6. $3E^2 - 3EI + 3I$
7. $x^2y + xyz + xy^2z$
8. $-r^3e + e^3r$
9. $3R^3 - 2R^2 + 5R^4$
10. $4y^2 + 8y + 12y^3$
11. $x^4 + 3x^3 + 9x^2$
12. $16R^3S^2 - 12R^4S^3 + 24R^2S^2$
13. $10a^4b^3 + 15a^2b^2 - 20a^3b^2$
14. $2(a + b) + x(a + b)$
15. $y(x - 3z) + 4(x - 3z)$
16. $a(3R + 1) + b(3R + 1)$
17. $2a(x - 3y) - 3b(x - 3y)$

[8.2]

18. $R^2(a - 2b) - S^2(a - 2b)$
19. $x^2(R + S) - 9(R + S)$
20. $a^2(x - y) - b^2(x - y)$
21. $x^2(3a + b) - 4(3a + b)$
22. $R^2 - 16I^2$
23. $25 - E^2$
24. $9x^2 - y^4$
25. $x^4 - 16y^2$
26. $25A^2 - 9B^2$
27. $36S^2 - 49R^2$
28. $x^4 - 16$
29. $y^4 - 81$

[8.4]

30. $4x^2 + 4x + 1$
31. $9r^2 - 36r + 36$
32. $4x^2 - 5x + 1$
33. $9i^2 + 9i - 10$
34. $8a^2 - 2a - 3$
35. $24x^2 + 22x + 3$
36. $8a^2 - 18a + 9$
37. $2a^2 + 15a + 18$

[8.5]

38. $x^2 - 9x + 14$
39. $2a^3 - 8a^2 - 10a$
40. $a^2 + 14a + 24$
41. $x^2 - 4x - 32$
42. $a^2 - 16a - 36$
43. $3x^2 - 9x - 30$
44. $x^3 - x^2 - 6x$
45. $x^3 - 4x^2 - 21x$
46. $x^4 - 5x^2 + 4$
47. $x^4 - 17x^2 + 16$
48. $x^4 - 1$
49. $x^4 - 81$
50. $a^2b^2 + ab - 6$
51. $a^2b^2 + 10ab + 24$
52. $a^2b^2 - 9ab + 18$
53. $a^2b^2 - 8ab - 20$

[8.6]

54. $R^3 + 8S^3$
55. $16x^3 - 54$
56. $27a^3 + 125b^3$
57. $x^3y^3 - 1$
58. $2x^9 + 250$
59. $64x^{12} - y^{15}$
60. $a^3b^6 + c^9$
61. $6ax - 3ay - 2bx + by$
62. $4ax + 6by + 8ay + 3bx$
63. $ax^2 + 3bx^2 - 4a - 12b$
64. $ax^2 - 2bx^2 + 4a - 8b$

Chapter Test

Factor each of the following. If an exercise will not factor, state that as your answer.

1. $R^2 + 5R + 6$
2. $3a^2 + 6ab - 9a$
3. $3x^2 + 9x - 30$
4. $x(m - n) + y(m - n)$
5. $x^2 + 7x + 10$
6. $2xy - 4x^2y^2 - 8x^3y^3$
7. $y^2 - 2y - 8$
8. $3a^2 + 7a + 4$
9. $x^2 - 25y^2$
10. $a^2 + 3a + 7$
11. $a(x + 2y) - 4b(x + 2y)$
12. $x^2 - 7x + 5$
13. $16a^2 - b^2$
14. $25x^2y^2 - 16z^2$
15. $a^2 - 7a + 10$
16. $6R^2 + 13R + 6$
17. $2a^2 - 18b^2$
18. $a^2b^2 + 10ab + 16$
19. $b^3 + 27$
20. $6ax - 2ay + 3bx - by$
21. $x^2 + 3x + 9$
22. $12x^2 + 17x + 12$
23. $x^2 + 2x + 1$
24. $y^2 + 7y + 6$
25. $a^2 - 4a + 16$
26. $5y^2 + 20y + 15$
27. $4z^2 - 9$
28. $3x^2 - x - 4$
29. $s^2 - 7s - 78$
30. $16a^4 - 81b^4$
31. $s^2 - 12s - 45$
32. $80x^5 - 5x$
33. $4a^2 + 12a + 5$
34. $R^2 + R - 56$
35. $2x^4 - 32$
36. $a^2b^2 - 2ab - 8$
37. $x^4 - 12x^2 + 27$
38. $8x^3 - y^3z^3$
39. $2a^3 - 16b^6$
40. $5ax^2 - 20a + bx^2 - 4b$

9
Algebraic Fractions

A technician using a scanning electron microscope, which has magnification power of 750,000 to one. He is using the microscope to scan a wire bond on an integrated circuit at a magnification of 700 times.
(Photograph courtesy of Ford Motor Company)

9.1 Introduction

In chapter 2 we learned how to add, subtract, multiply, divide, and simplify numerical fractions. In this chapter we perform these same basic operations on algebraic fractions—fractions whose numerator, denominator, or both are algebraic expressions.

Examples

The following are algebraic fractions:

$$\frac{2x}{x+1}, \quad \frac{x^2-2}{x^2+x-1}, \quad \frac{3}{3x-1}, \quad \frac{5x^2+3xy-y^2}{2x+3y}$$

Algebraic fractions play an important role in mathematics since they frequently appear in formulas used in physics, chemistry, and all branches of engineering. Many formulas and other equations that we wish to solve also involve algebraic fractions.

Example

The formula for the joint conductance of resistors in parallel is:

$$\frac{1}{R_t} = \frac{1}{R_1} + \frac{1}{R_2}.$$

Solving the above formula for R_t, R_1, and R_2 involves the basic operations on algebraic fractions.

Since algebraic expressions are representations of numbers, the basic operations used for fractions involving whole numbers, or integers, are used in the same way on algebraic fractions.

9.2 Algebraic Fractions in Lowest Terms

In chapter 2 we used the fundamental principle of fractions to produce *equivalent fractions*—fractions such that if the numerator and denominator of one fraction are multiplied or divided by the same nonzero number, the other fraction is obtained. In symbols, this is:

$$\frac{a}{b} = \frac{ac}{bc} \qquad (b, c \neq 0)$$

note

Recall that division by zero is not allowed. Therefore, it is necessary that $b, c \neq 0$.

Examples

1. Given the fraction $\frac{7}{8}$. Multiply the numerator and the denominator by 2 to get $\frac{7}{8} = \frac{7 \cdot 2}{8 \cdot 2} = \frac{14}{16}$. So $\frac{7}{8}$ is equivalent to $\frac{14}{16}$.

   note

   Recall, by the property of proportions, $\frac{a}{b} = \frac{c}{d}$ if $a \cdot d = b \cdot c$. Then $\frac{7}{8} = \frac{14}{16}$ since $7 \cdot 16 = 8 \cdot 14$.

2. Given the fraction $\frac{12}{15}$. Divide the numerator and denominator by 3 to get $\frac{12}{15} = \frac{\frac{12}{3}}{\frac{15}{3}} = \frac{4}{5}$. So $\frac{12}{15}$ is equivalent to $\frac{4}{5}$.

When working with algebraic fractions we obtain equivalent fractions by again using the fundamental principle of fractions. The only difference is that we multiply or divide the numerator and denominator by the same algebraic expression.

Examples

1. Given the algebraic fraction $\frac{5x}{6x^2}$, if we divide the numerator and denominator by x we get $\frac{5x}{6x^2} = \frac{\frac{5x}{x}}{\frac{6x^2}{x}} = \frac{5}{6x}$. So $\frac{5x}{6x^2} = \frac{5}{6x}$, since

   $5x \cdot 6x = 6x^2 \cdot 5.$

2. Given the algebraic fraction $\frac{x - 1}{x + 2}$, if we multiply the numerator and denominator by x^2 we get

   $$\frac{x - 1}{x + 2} = \frac{x^2(x - 1)}{x^2(x + 2)} = \frac{x^3 - x^2}{x^3 + 2x^2}$$

When performing the operations of addition, subtraction, multiplication, and division on algebraic fractions it is desirable that, as with the operations on arithmetic fractions in chapter 2, the result be reduced to lowest terms. To reduce an algebraic fraction to lowest terms we use the fundamental principle of fractions by dividing the numerator and denominator by algebraic expressions that are common factors of the numerator and denominator.

To summarize, we reduce an algebraic fraction to lowest terms using the following procedure:

1. Write the numerator and denominator in the completely factored form.
2. Divide the numerator and denominator by the factor, or factors, that are *common* to both.

Examples

Simplify the following fractions by reducing them to lowest terms.

1. $\frac{36}{90}$

   Note that $\frac{36}{90} = \frac{(3)(3)(2)(2)}{(3)(3)(2)(5)}$ and the common factors are 3, 3, and 2. Divide the numerator and denominator by $(3)(3)(2) = 18$ to get:

   $$\frac{36}{90} = \frac{\frac{36}{18}}{\frac{90}{18}} = \frac{2}{5}$$

2. $\frac{6xy^2}{3x^2y}$

   Note that $\frac{6xy^2}{3x^2y} = \frac{(3)(2)(x)(y^2)}{(3)(x^2)(y)}$ and the common factors are 3, x, and y. Divide the numerator and denominator by $3xy$ to get

   $$\frac{6xy^2}{3x^2y} = \frac{\frac{6xy^2}{3xy}}{\frac{3x^2y}{3xy}} = \frac{2y}{x}$$

note

This reduction can also be done by the procedure sometimes called "cancellation":

$$\frac{6xy^2}{3x^2y} = \frac{\cancel{(3)}(2)\cancel{(x)}(y)\cancel{(y)}}{\cancel{(3)}(x)\cancel{(x)}\cancel{(y)}} = \frac{2y}{x}$$

3. $\dfrac{4x}{4x^2 + 16x}$

Note that $\dfrac{4x}{4x^2 + 16x} = \dfrac{4x}{4x(x + 4)}$ when we factor the denominator. Thus, the common factors are 4 and x. Dividing the numerator and denominator by $4x$ we get

$$\frac{4x}{4x^2 + 16x} = \frac{\dfrac{4x}{4x}}{\dfrac{4x(x + 4)}{4x}} = \frac{1}{x + 4}$$

4. $\dfrac{a^2 - a - 12}{a^2 - 3a - 18}$

Completely factor the numerator and denominator to get $\dfrac{a^2 - a - 12}{a^2 - 3a - 18} = \dfrac{(a - 4)(a + 3)}{(a - 6)(a + 3)}$. The factor $(a + 3)$ is common to the numerator and the denominator. Divide both by this common factor to get

$$\frac{a^2 - a - 12}{a^2 - 3a - 18} = \frac{\dfrac{(a - 4)(a + 3)}{(a + 3)}}{\dfrac{(a - 6)(a + 3)}{(a + 3)}} = \frac{a - 4}{a - 6}$$

So, $\dfrac{a^2 - a - 12}{a^2 - 3a - 18} = \dfrac{a - 4}{a - 6}$.

note

A common mistake that is made when reducing an algebraic fraction is to reduce (cancel) the individual *terms* of the numerator and denominator. For example consider the algebraic fraction $\dfrac{a - 4}{a - 6}$. Many students want to "cancel" the a's since they appear both in the numerator and denominator. This cancellation is incorrect since it would imply $\dfrac{a - 4}{a - 6} = \dfrac{-4}{-6} = \dfrac{2}{3}$, for *any* a. But if $a = 2$, then $\dfrac{a - 4}{a - 6} = \dfrac{2 - 4}{2 - 6} = \dfrac{-2}{-4} = \dfrac{1}{2}$ and $\dfrac{2}{3} \neq \dfrac{1}{2}$. Thus, $\dfrac{a - 4}{a - 6}$ and $\dfrac{2}{3}$ *are not equivalent fractions*. In fact, $\dfrac{a - 4}{a - 6}$ cannot be simplified further since a is not a factor of $a - 4$ and $a - 6$. Remember always that the fundamental principle of fractions states we can multiply or divide by *common factors* (*not terms*) to obtain equivalent fractions. Therefore, you *must* factor before reducing by the common factor.

5. $\dfrac{R^2 - 25}{R^2 - 2R - 15}$

Factor the numerator and denominator to get $\dfrac{R^2 - 25}{R^2 - 2R - 15} =$ $\dfrac{(R + 5)(R - 5)}{(R + 3)(R - 5)}$. Note that the common factor is $(R - 5)$ and divide the numerator and denominator by $(R - 5)$ to get

$$\frac{R^2 - 25}{R^2 - 2R - 15} = \frac{\dfrac{(R + 5)(R - 5)}{(R - 5)}}{\dfrac{(R + 3)(R - 5)}{(R - 5)}} = \frac{R + 5}{R + 3}.$$

Thus, $\dfrac{R^2 - 25}{R^2 - 2R^2 - 15} = \dfrac{R + 5}{R + 3}$.

Consider the algebraic expressions $(x - 5)$ and $(5 - x)$. Each of these expressions is the negative, or opposite, of the other. That is, $5 - x = -x + 5 = -1(x - 5) = -(x - 5)$. Then

$$\frac{x - 5}{5 - x} = \frac{x - 5}{-(x - 5)} = \frac{1}{-1} = -1$$

In general, when the numerator and denominator contain factors that have the same terms separated by a minus sign, but in reverse order, we use the following procedure:

1. Express the numerator and denominator in descending powers of x.
2. Factor -1 from any polynomial whose leading coefficient is negative $[a - b = -b + a = (-1)(b - a)]$.

Examples

Simplify the following fractions by reducing to lowest terms.

1. $\dfrac{4 - x}{x^2 - 16}$

Factor the denominator to get

$$\frac{4 - x}{x^2 - 16} = \frac{4 - x}{(x - 4)(x + 4)}.$$

Express the numerator in descending powers of x and factor -1 to get

$$\frac{4 - x}{(x - 4)(x + 4)} = \frac{-x + 4}{(x - 4)(x + 4)} = \frac{-1(x - 4)}{(x - 4)(x + 4)}$$

Divide the numerator and denominator by $x - 4$:

$$\frac{4 - x}{x^2 - 16} = \frac{\dfrac{-1(x - 4)}{(x - 4)}}{\dfrac{(x - 4)(x + 4)}{(x - 4)}} = \frac{-1}{x + 4}$$

So $\dfrac{4 - x}{x^2 - 16} = \dfrac{-1}{x + 4}$.

2. $\dfrac{1 - x^2}{2x^2 + x - 3}$

Factor the numerator and denominator:

$$\frac{1 - x^2}{2x^2 + x - 3} = \frac{(1 - x)(1 + x)}{(x - 1)(2x + 3)}.$$

Now $1 - x = -x + 1 = -1(x - 1)$, so

$$\frac{(1 - x)(1 + x)}{(x - 1)(2x + 3)} = \frac{-1(x - 1)(1 + x)}{(x - 1)(2x + 3)}.$$

Divide the numerator and denominator by $x - 1$ to get

$$\frac{1 - x^2}{2x^2 + x - 3} = \frac{\dfrac{-1(x - 1)(x + 1)}{(x - 1)}}{\dfrac{(x - 1)(2x + 3)}{(x - 1)}}$$

$$= \frac{-1(x + 1)}{(2x + 3)}$$

$$= \frac{-x - 1}{2x + 3}.$$

So, $\dfrac{1 - x^2}{2x^2 + x - 3} = \dfrac{-x - 1}{2x + 3}$.

Exercise Set 9.1

Obtain equivalent fractions in exercises 1 to 12 by multiplying the numerator and denominator by the given factor.

1. $\dfrac{5}{6}$ by 3

2. $\dfrac{7}{8}$ by -4

•3. $\dfrac{2a}{3b}$ by $3ab$

4. $\dfrac{5x}{4y}$ by $5xy$

•5. $\dfrac{5}{x - 3}$ by $2x$

6. $\dfrac{-4}{x + 4}$ by $-3x$

7. $\dfrac{4}{x + 3}$ by $x - 3$

8. $\dfrac{-5}{2x - 1}$ by $2x + 1$

•9. $\dfrac{x + 5}{x - 6}$ by $x + 2$

10. $\dfrac{2x + 5}{x + 7}$ by $x - 7$

11. $\dfrac{a + b}{a - b}$ by $(b - a)$

12. $\dfrac{3a + 4b}{2a - b}$ by $b - 2a$

Obtain equivalent fractions in exercises 13 to 24 by dividing the numerator and denominator by the given factor. (*Factor first,* where possible.)

13. $\dfrac{14}{16}$ by 2

14. $\dfrac{24}{32}$ by -8

•15. $\dfrac{5a^2}{15a^3}$ by $5a^2$

16. $\dfrac{16a^2b}{28ab^2}$ by $4ab$

17. $\dfrac{x - 2}{x^2 + x - 6}$ by $x - 2$

18. $\dfrac{x^2 - 4}{x^2 + 4x + 4}$ by $x + 2$

•19. $\dfrac{3x - 9}{3x^2 + 3x - 36}$ by $3(x - 3)$

20. $\dfrac{5x^2 - 20}{5x^2 - 5x - 10}$ by $5(x + 2)$

21. $\dfrac{x^2 + 14x + 49}{2x^2 + 11x - 21}$ by $x + 7$

22. $\dfrac{x^2 - 36}{3x^2 - 19x + 6}$ by $x - 6$

23. $\dfrac{2x^2 - 50}{x^2 - x - 20}$ by $5 - x$

24. $\dfrac{x^2 - 6x + 8}{2x^2 - 11x + 12}$ by $4 - x$

In exercises 25 to 30, change the given fraction to an equivalent fraction having the given numerator or denominator.

Examples 1. Change $\dfrac{3}{4x}$ to a fraction having denominator $24x$.

Set up the problem as

$$\frac{3}{4x} = \frac{?}{24x}.$$

Note that to obtain the denominator $24x$ it was necessary to multiply the denominator of the original fraction, $4x$, by 6. (*Note:* Divide $4x$ into $24x$.) To get an equivalent fraction, multiply the numerator of the given fraction by 6 also. Thus,

$$\frac{3}{4x} = \frac{3 \cdot 6}{4x \cdot 6} = \frac{18}{24x}.$$

2. Change $\frac{4a + b}{15x}$ to a fraction having numerator $12a + 3b$.

We want

$$\frac{4a + b}{15x} = \frac{12a + 3b}{?}.$$

Factoring

$$12a + 3b = 3(4a + b)$$

shows that $4a + b$ of the original fraction was multiplied by 3 to get $12a + 3b$. Therefore, multiply the denominator of the given fraction by 3 to get

$$\frac{4a + b}{15x} = \frac{(4a + b) \cdot 3}{15x \cdot 3} = \frac{12a + 3b}{45x}$$

•25. Change $\frac{5x}{3}$ to a fraction having denominator 21.

26. Change $\frac{4a}{5}$ to a fraction having numerator $16a$.

27. Change $\frac{4a - 3b}{10x}$ to a fraction having denominator $20ax$.

28. Change $\frac{6ab}{2 + 3b}$ to a fraction having numerator $18a^2b^2$.

•29. Change $\frac{x - 5}{x + 2}$ to a fraction having denominator $x^2 - 4$.

30. Change $\frac{a + 7}{2a - 1}$ to a fraction having numerator $a^2 - 49$.

31. Change $\frac{4 - 3y}{x + y}$ to a fraction having denominator $x^2 - y^2$.

32. Change $\frac{2x + 3}{x - 3}$ to a fraction having numerator $4x^2 + 12x + 9$.

Reduce the following fractions to their lowest terms.

33. $\frac{54}{72}$

34. $\frac{75}{125}$

35. $\frac{16a^2b}{20ab^2}$

36. $\frac{15a^2x^3}{35ax^2}$

•37. $\frac{20ab^2c^3}{4a^2bc}$

38. $\frac{72x^4y^3z^2}{9x^3yz^3}$

39. $\frac{x^2 - 9}{x^2 + 6x + 9}$

40. $\frac{x^2 - 10x + 25}{x^2 - 25}$

•41. $\frac{x^2 - 3x - 10}{x^2 - x - 6}$

42. $\frac{y^2 - y - 42}{y^2 + 12y + 36}$

43. $\dfrac{8a - 12b}{16a + 24b}$

44. $\dfrac{a^2 - 4b^2}{2a + 4b}$

45. $\dfrac{m^2 - n^2}{(m + n)^2}$

•46. $\dfrac{(x - y)^2}{y^2 - x^2}$

47. $\dfrac{2y^2 - 3y - 9}{4y^2 - 13y + 3}$

48. $\dfrac{12 + 9m + 3m^2}{4 + 7m + 3m^2}$

49. $\dfrac{(4y + 1)(y - 4)}{(4 - y)(2y - 5)}$

50. $\dfrac{(6p - 1)(2p + 3)(p + 6)}{(p - 7)(1 - 6p)(2p + 3)}$

51. $\dfrac{16xy - y^2}{y^3 - 16y}$

52. $\dfrac{3ab - 9b^2}{a^2 - 9b^2}$

53. $\dfrac{3a^2 - 10ab + 3b^2}{6a^2 - 18ab}$

54. $\dfrac{18m^2 - 68m + 64}{9m^2 - 16m}$

55. $\dfrac{(x - 2y)^3}{2y^2 + 3xy - 2x^2}$

56. $\dfrac{(3a - 2b)^2}{4b^2 - 12ab + 9a^2}$

57. $\dfrac{4a^2 + 12ab + 8b^2}{a^2 - 4b^2}$

58. $\dfrac{2 - 32x^2}{8 - 28x - 16x^2}$

59. $\dfrac{m^2n^2 - 16m^2}{mn^2 + mn - 12m}$

60. $\dfrac{16x^2 - 20xy - 6y^2}{96x^2 - 6y^2}$

Reduce the following fractions to lowest terms and evaluate each before and after the reduction by substituting the given values for the variables.

Example $\dfrac{6x - 12}{x^2 - 4}$ when $x = 3$

Substitute into the original fraction:

$$\frac{6(3) - 12}{(3)^2 - 4} = \frac{18 - 12}{9 - 4} = \frac{6}{5}$$

Reduce to lowest terms:

$$\frac{6x - 12}{x^2 - 4} = \frac{6(x - 2)}{(x + 2)(x - 2)} = \frac{6}{x + 2}$$

Then $\dfrac{6}{x + 2} = \dfrac{6}{3 + 2} = \dfrac{6}{5}$.

61. $\dfrac{5a - 10}{a^2 - 25}$ when $a = 2$

62. $\dfrac{x^2 - 16}{9x + 36}$ when $x = -2$

63. $\dfrac{x^2 + 2x - 8}{x^2 + x - 12}$ when $x = -3$

64. $\dfrac{y^2 - 5y - 14}{y^2 - 49}$ when $y = -1$

65. $\dfrac{2x^2 + xy - y^2}{3x^2 + xy - 2y^2}$ when $x = 3$, $y = -2$

66. $\dfrac{4a^2 + 10ab + 6b^2}{12a^2 + 8ab - 4b^2}$ when $a = 1$ and $b = -2$

67. Can $\dfrac{3a + 4}{3a + 2}$ be reduced? (If you are not sure, reduce as you think it can be and substitute 3 for a in your resulting fraction and in the original. You should get the same result in both if the fraction can be reduced.)

9.3 Multiplication and Division of Algebraic Fractions

Recall that when multiplying two arithmetic fractions the product is another fraction whose numerator is the product of the numerators of the given fractions and the denominator is the product of the denominators of the given fractions. That is, if a, b, c, and d are whole numbers or integers,

$$\frac{a}{b} \cdot \frac{c}{d} = \frac{a \cdot c}{b \cdot d} \qquad (b, d \neq 0)$$

note

Since division by zero is not allowed we say that b and d cannot be zero.

This same procedure is followed when multiplying two algebraic fractions.

Examples

Find the indicated products.

1. $\dfrac{3x}{4} \cdot \dfrac{5}{2y}$

Multiplying the numerators and denominators we get

$$\frac{3x}{4} \cdot \frac{5}{2y} = \frac{3x \cdot 5}{4 \cdot 2y} = \frac{15x}{8y} \qquad (y \neq 0)$$

note

$y \neq 0$ since then the denominators $2y$ and $8y$ would be zero and, again, division by zero is not allowed.

2. $\dfrac{x + 1}{x - 3} \cdot \dfrac{4}{x + 2}$

Multiplying the numerators and denominators we get

$$\frac{x + 1}{x - 3} \cdot \frac{4}{x + 2} = \frac{(x + 1) \cdot 4}{(x - 3)(x + 2)}$$

$$= \frac{4x + 4}{x^2 - x - 6} \qquad (x \neq 3, -2)$$

note

Since $x - 3 = 0$ when $x = 3$, and $x + 2 = 0$ when $x = -2$, then we must exclude 3 and -2 as possible values of x since the denominator would be zero when x is 3 or -2.

We should *always* be aware of the necessary restrictions placed on the variable to eliminate division by zero when the variable appears in the denominator. However, through the remainder of this chapter we shall assume the denominators do not represent zero.

Remember that we want the resulting product to be stated in lowest terms. This result can be accomplished before multiplying by dividing out any common factors that may be present in either the numerators or denominators.

Examples

Perform the indicated multiplication and state the answer in reduced form.

1. $\frac{4}{9x} \cdot \frac{3x^2}{2}$

   We can see that the numerator of the first fraction and denominator of the second fraction have a common factor, 2, and that the denominator of the first fraction and the numerator of the second fraction have a common factor, $3x$. Divide the numerator and denominator by 2 and $3x$ to get

   $$\frac{4}{9x} \cdot \frac{3x^2}{2} = \frac{4 \cdot 3x^2}{9x \cdot 2} = \frac{(2)(2)(3)(x)(x)}{(3)(3)(x)(2)} = \frac{2x}{3}$$

2. $\frac{(x + 1)}{(x - 3)} \cdot \frac{(x - 3)^2}{(x - 2)}$

   We can see that the denominator of the first fraction and the numerator of the second fraction have a common factor of $(x - 3)$. Divide by $(x - 3)$ to get

   $$\frac{(x + 1)}{(x - 3)} \cdot \frac{(x - 3)^2}{(x - 2)} = \frac{(x + 1)(x - 3)(x - 3)}{(x - 3)(x - 2)}$$

   $$= \frac{(x + 1)(x - 3)}{(x - 2)}$$

In general, when multiplying two algebraic fractions we use the following procedure:

1. Write the numerator and denominator of each fraction in completely factored form, as an indicated product of a single fraction.
2. Divide the numerator and the denominator by any common factors.
3. Multiply the remaining factors (or leave as indicated products).

note

At this point, if all common factors in step 2 have been divided out, the answer will be in the most reduced form.

Examples

Perform the indicated multiplications and write the answer reduced to lowest terms.

1. $\dfrac{x+1}{x-3}\cdot\dfrac{x^2-x-6}{x^2-1}$

   Factor the numerator and denominator of the second fraction:

   $$\frac{(x+1)\cdot(x-3)(x+2)}{(x-3)\cdot(x+1)(x-1)}$$

   Note that the common factors of $(x+1)$ and $(x-3)$ exist in the numerator and denominator of each fraction. Divide out each of these factors:

   $$\frac{(x+1)(x-3)(x+2)}{(x-3)(x+1)(x-1)}=\frac{x+2}{x-1}$$

   Thus,

   $$\frac{x+1}{x-3}\cdot\frac{x^2-x-6}{x^2-1}=\frac{x+2}{x-1}$$

2. $\dfrac{x^2-8x+16}{x^2+3x-10}\cdot\dfrac{x^2-4}{x^2-5x+4}$

   Factor both fractions to get

   $$\frac{(x-4)^2}{(x+5)(x-2)}\cdot\frac{(x+2)(x-2)}{(x-4)(x-1)}$$

   There are common factors of $(x-2)$ and $(x-4)$, so divide out these factors to get

   $$\frac{(x-4)^2(x+2)(x-2)}{(x+5)(x-2)(x-4)(x-1)}=\frac{(x-4)(x+2)}{(x+5)(x-1)}$$

   Therefore,

   $$\frac{x^2-8x+16}{x^2+3x-10}\cdot\frac{x^2-4}{x^2-5x+4}=\frac{(x-4)(x+2)}{(x+5)(x-1)}$$

note

A reminder—we cannot reduce this fraction further since there are no common factors and *it is incorrect to reduce by common terms* of the numerator and denominator.

3. $\dfrac{9-x^2}{2x-10}\cdot\dfrac{x^2-6x+5}{x^2+x-12}$

   Factor both algebraic fractions to get

   $$\frac{(3-x)(3+x)(x-5)(x-1)}{2(x-5)(x-3)(x+4)}$$

The only common factor *appears* to be $x - 5$. However, recall that $3 - x = -x + 3 = -1(x - 3)$. Therefore,

$$\frac{(3 - x)(3 + x)(x - 5)(x - 1)}{2(x - 5)(x - 3)(x + 4)} = \frac{-1(x - 3)(x + 3)(x - 5)(x - 1)}{2(x - 5)(x - 3)(x + 4)}$$

and we can now divide out $x - 5$ and $x - 3$ in both numerator and denominator. Thus,

$$\frac{-1(x - 3)(x + 3)(x - 5)(x - 1)}{2(x - 5)(x - 3)(x + 4)} = \frac{-1(x + 3)(x - 1)}{2(x + 4)}$$

$$= \frac{-1(x - 1)(x + 3)}{2(x + 4)}$$

$$= \frac{(1 - x)(x + 3)}{2(x + 4)}$$

Exercise Set 9.2

Perform the indicated multiplications and reduce your answer to lowest terms.

1. $\frac{5}{6} \cdot \frac{3x}{10y}$
2. $\frac{7a}{12b} \cdot \frac{9b}{28}$
•3. $\frac{21ab}{16c} \cdot \frac{8c^2}{3ab^2}$
4. $\frac{18x^2y^2}{5ab} \cdot \frac{25a^2b}{12xy}$
5. $\frac{24abc}{7xyz^2} \cdot \frac{14x^2yz}{9a^2b}$
6. $\frac{80x^2yz^3}{11mn^2} \cdot \frac{33mn^2}{25xyz}$
7. $\frac{x - y}{2x + y} \cdot \frac{2x - y}{x + y}$
8. $\frac{3a + b}{a - b} \cdot \frac{a + 3b}{3a + b}$
•9. $\frac{x^2 - y^2}{12} \cdot \frac{10}{x - y}$
10. $\frac{15}{a + b} \cdot \frac{a^2 - b^2}{27}$
•11. $\frac{x^2 - 4}{25y} \cdot \frac{24y^2}{y + 2}$
12. $\frac{16a^2}{b^2 - 9} \cdot \frac{b - 3}{12a^3}$
13. $\frac{2y^2 - 50}{13y} \cdot \frac{39y}{4y + 20}$
14. $\frac{3x^2}{5x - 15} \cdot \frac{2x^2 - 18}{18x}$
•15. $\frac{9 - x^2}{x + y} \cdot \frac{4x + 4y}{x - 3}$
16. $\frac{b^2 - a^2}{2a + 4b} \cdot \frac{a + 2b}{a - b}$
•17. $\frac{x^2 + 2xy + y^2}{16y^2} \cdot \frac{12y}{x^2 - y^2}$
18. $\frac{25a^3}{x^2 - 10x + 25} \cdot \frac{x^2 - 25}{5a}$
19. $\frac{6m + 12n}{10x + 5} \cdot \frac{100x^2 - 25}{9m^2 - 81n^2}$
20. $\frac{25a + 10b}{16x - 12y} \cdot \frac{16x^2 - 9y^2}{25a^2 - 4b^2}$
•21. $x^2 - 4x + 4 \cdot \frac{18}{x^2 - 4}$
22. $\frac{21}{a^2 - 9} \cdot a^2 + a - 12$
23. $\frac{2x^2 - 15x + 7}{x^2 - 9x + 8} \cdot \frac{x^2 - 2x + 1}{x^2 - 49}$
24. $\frac{a^2 - 5a - 14}{a^2 - 9a - 36} \cdot \frac{a^2 + 10a + 21}{a^2 + 4a - 77}$
25. $\frac{4y^2 - 4}{3y^2 - 13y - 10} \cdot \frac{y^2 - 6y + 5}{4y + 4}$

26. $\dfrac{a^2 - 5a + 6}{a^2 - 9a + 20} \cdot \dfrac{a^2 - 2a - 15}{a^2 + a - 12} \cdot \dfrac{a^2 - 16}{a^2 - 9}$

27. $\dfrac{x^2 + 8x + 16}{x^2 - 16} \cdot \dfrac{x^2 - 25}{x^2 + 9x + 20} \cdot \dfrac{x^2 + 5x}{x^2 - 5x}$

•28. $\dfrac{6m^2 - 7m + 2}{6m^2 + 5m + 1} \cdot \dfrac{2m^2 + m}{4m^2 - 1} \cdot \dfrac{12m^2 - 5m - 3}{12m^2 - 17m + 6}$

29. $\dfrac{4x^2 - y^2}{x^2 + 2x} \cdot \dfrac{x^2 - xy}{10x - 5y} \cdot \dfrac{x^2 + 4x + 4}{2x^2 - xy - y^2}$

30. $\dfrac{a^2 - b^2}{a^2 + 14ab + 49b^2} \cdot \dfrac{2a + 14b}{b - a} \cdot \dfrac{a^2 - 49b^2}{10a + 10b}$

Recall, when dividing two arithmetic fractions, that we "invert the second fraction and multiply." That is,

$$\frac{a/b}{c/d} = \frac{a}{b} \div \frac{c}{d} = \frac{a}{b} \cdot \frac{d}{c} = \frac{ad}{bc}$$

The same procedure is used to divide one algebraic fraction by another. Keep in mind that the reciprocal of $\dfrac{c}{d}$ is $\dfrac{d}{c}$ (simply invert the fraction, or turn it upside down). Then the reciprocal of $\dfrac{x + 1}{x + 2}$ is $\dfrac{x + 2}{x + 1}$ and the reciprocal of $3x^2$ is $\dfrac{1}{3x^2}$, because $3x^2 = \dfrac{3x^2}{1}$.

In general, to divide one algebraic fraction by another we follow these steps:

1. Change the operation to multiplication and write the reciprocal of the divisor.
2. Proceed as in multiplication.

Examples

Perform the indicated divisions and write the answer in reduced form.

1. $\dfrac{3ab}{5} \div \dfrac{9abc}{10}$

   The reciprocal of $\dfrac{9abc}{10}$ is $\dfrac{10}{9abc}$.

   So $\dfrac{3ab}{5} \div \dfrac{9abc}{10} = \dfrac{3ab}{5} \cdot \dfrac{10}{9abc} = \dfrac{(3)(10)ab}{(5)(9)abc}$.

   Note that the common factors are 3, 5, a, and b. Divide the numerator and denominator by $15ab$ to get

   $$\frac{(3)(10)ab}{(5)(9)abc} = \frac{2}{3c}.$$

   Therefore, $\dfrac{3ab}{5} \div \dfrac{9abc}{10} = \dfrac{2}{3c}$.

2. $\dfrac{x^2 - 4}{5} \div \dfrac{x - 2}{15}$

The reciprocal of $\dfrac{x - 2}{15}$ is $\dfrac{15}{x - 2}$.

Multiply $\dfrac{x^2 - 4}{5} \cdot \dfrac{15}{x - 2} = \dfrac{(x - 2)(x + 2) \cdot 15}{5(x - 2)} = 3(x + 2)$.

So $\dfrac{x^2 - 4}{5} \div \dfrac{x - 2}{15} = 3(x + 2)$.

3. $\dfrac{3R^2 - R - 2}{R^4 - R^3} \div \dfrac{3R^2 + 5R + 2}{R^3 - 2R^2}$

The reciprocal of $\dfrac{3R^2 + 5R + 2}{R^3 - 2R^2}$ is $\dfrac{R^3 - 2R^2}{3R^2 + 5R + 2}$, so multiply

$\dfrac{3R^2 - R - 2}{R^4 - R^3} \cdot \dfrac{R^3 - 2R^2}{3R^2 + 5R + 2}$.

Factor wherever possible to get

$\dfrac{(3R + 2)(R - 1)}{R^3(R - 1)} \cdot \dfrac{R^2(R - 2)}{(3R + 2)(R + 1)}$.

Reduce to lowest terms:

$$\frac{(3R + 2)(R - 1)R^2(R - 2)}{R^3(R - 1)(3R + 2)(R + 1)} = \frac{R - 2}{R(R + 1)}$$

Therefore,

$$\frac{3R^2 - R - 2}{R^4 - R^3} \div \frac{3R^2 + 5R + 2}{R^3 - 2R^2} = \frac{R - 2}{R(R + 1)}$$

Exercise Set 9.3

Write the reciprocal of the following expressions.

1. $\dfrac{2a}{3b}$
2. $\dfrac{x - y}{x + y}$
•3. $2ab$
4. $4a - b$
5. $\dfrac{5x - 3y}{4}$
6. $\dfrac{7}{x^2 - y^2}$
7. $\dfrac{x^2 - 2x + 1}{x - 3}$
8. $\dfrac{a^2 - b^2}{a^2 - a - 6}$
9. $\dfrac{1}{4abc}$
10. $\dfrac{7}{x^2 - y^2}$

Express the following indicated quotients as a product.

Example $\frac{2x + y}{4} \div \frac{x - y}{8} = \frac{2x + y}{4} \cdot \frac{8}{x - y}$

11. $\frac{15a}{7b} \div \frac{a^2}{14b^2}$

12. $\frac{x^2 - 1}{3} \div \frac{x + 1}{15}$

13. $\frac{24x - 8y}{x^2 - y^2} \div \frac{3x - y}{x + y}$

14. $\frac{x^2 - 16}{18x} \div \frac{x^2 + 4x + 4}{10x^2}$

15. $\frac{a^2 - a - 42}{a^2 + a - 12} \div \frac{2a^2 + a - 16}{5a - 10}$

16. $\frac{2x^2 + x - 3}{4x + 1} \div \frac{3x^2 + 2x + 6}{16x^2 - 1} \div \frac{1 - x^2}{25x}$

17. $\frac{15ab}{7a - 4b} \div \frac{25a^2b}{49a^2 - 16b^2} \cdot \frac{ab}{7a + 4b}$

18. $\frac{4x - y}{3x + 1} \div 4x^2 - 4y^2$

19. $\frac{x^2 + 5x - 24}{x^2 + 10x + 25} \div x - 3$

Find the indicated quotients and express the answers reduced to lowest terms.

20. $\frac{4a}{5} \div \frac{5}{3}$

21. $\frac{16b}{7a} \div \frac{5}{4}$

22. $\frac{14}{3a} \div \frac{6}{15a}$

23. $\frac{5x}{9y} \div \frac{20x^2}{6y^2}$

24. $\frac{3ab}{8a^2} \div \frac{15b^3}{16a^2}$

•25. $\frac{28m}{15m} \div \frac{7m^2}{3m}$

26. $\frac{12a^2b^2}{27ab} \div \frac{24a^3b^2}{36ab}$

27. $\frac{20mn}{9n^3} \div \frac{4m}{3m^2n^2}$

28. $\frac{x - 5}{2x - 1} \div \frac{x + 3}{x + 7}$

29. $\frac{3x}{x - 2} \div \frac{5x + 2}{4x - 3}$

30. $\frac{p - 9}{p^2 - 9} \div \frac{p + 1}{p - 1}$

31. $\frac{r^2 - 16}{r + 4} \div \frac{r - 1}{r^2 + 8r + 16}$

32. $\frac{a + 2}{3a} \div \frac{a^2 + 2a}{6a - 3}$

•33. $\frac{6}{3m + 4} \div \frac{4m + 4}{9m^2 + 12m}$

34. $\frac{p^2 + 2p + 1}{4p - p} \div \frac{p^2 - 1}{16p^2 - 1}$

35. $\frac{a^2 - 5a + 6}{a^2 - 9a + 20} \div \frac{a^2 - 3a + 2}{a^2 - 5a + 4}$

36. $\frac{6r^2 - r - 7}{12r^2 - 16r - 35} \div \frac{r^2 - r - 2}{2r^2 + r - 10}$

37. $\dfrac{3a^2 + 10a - 8}{36a^2 - 16} \div \dfrac{a^2 + 3a - 4}{9a^2 + 15a + 6}$

38. $\dfrac{25 - x^2}{x^2 + 7x + 10} \div \dfrac{2x^2 - 20x + 50}{x^2 + 4x + 4}$

39. $\dfrac{9 - x}{3x + 2} \div \dfrac{x^2 - 18x + 81}{4 - 9x^2}$

40. $3x^2 - 2x - 8 \div \dfrac{x^2 - 4}{x + 2}$

41. $8a^2 - 16a \div \dfrac{a^3 - 16a}{a - 4}$

42. $\dfrac{3x - 4}{2x + 1} \div (6x^2 - 5x - 4)$

•43. $\dfrac{m^2 - 3m - 10}{m^2 - 4} \div (2m^2 - 9m - 5)$

44. $\dfrac{x^2 - 25}{4x - 8} \cdot \dfrac{8 - 2x^2}{5x^2 + 5x - 10} \div \dfrac{5 + x}{15x^2 - 15}$

45. $\dfrac{6a^2 - 7a + 2}{6a^2 + 5a + 1} \div \dfrac{12a^2 - 17a + 6}{12a^2 - 5a - 3} \cdot \dfrac{2a^2 + a}{4a^2 - 1}$

9.4 Least Common Denominator (LCD)

From chapter 2, recall that to add or subtract fractions having the same denominator, you add or subtract the numerators and place this sum, or difference, over the common denominator. That is,

$$\frac{a}{b} + \frac{c}{b} = \frac{a + c}{b} \qquad (b \neq 0)$$

Example The sum of $\dfrac{3}{11} + \dfrac{4}{11} = \dfrac{3 + 4}{11} = \dfrac{7}{11}$

Recall that if fractions do not have the same denominator, it is necessary to change each of the fractions to an equivalent fraction by applying the fundamental principle of fractions so they do have the same denominator. In most cases, there are many numbers we can use as a common denominator. However, the most convenient denominator to use is the smallest (least) number that is divisible by each of the denominators—this is called the *least common denominator,* LCD. Recall also, that the LCD contains the greatest number of *prime factors* of any one of the given denominators.

Example

Find the LCD of the fractions whose denominators are 6, 8, and 15.

$6 = 2 \cdot 3, \quad 8 = 2^3, \quad 15 = 3 \cdot 5$

The LCD is given by $2^3 \cdot 3 \cdot 5 = 120$.

The procedure for finding the LCD of algebraic fractions is similar to that used for arithmetic fractions. To find the LCD:

1. Write each denominator in completely factored form.
2. Take each *different* factor that appears in any one of the factorizations in step 1.
3. Obtain the LCD by raising each factor of step 2 to the *greatest* exponent that factor has in step 1 and forming the product of these factors.

Examples

Find the LCD of the following algebraic fractions.

1. $\dfrac{5}{16a}; \dfrac{3}{8a^2}$

   Write each denominator as a product of prime factors.

   $16a = 2 \cdot 2 \cdot 2 \cdot 2 \cdot a = 2^4 \cdot a$

   $8a^2 = 2 \cdot 2 \cdot 2 \cdot a \cdot a = 2^3 \cdot a^2$

   The factors to be considered are 2 and a. Since the greatest power of 2 is 4 and of a is 2, then the LCD is

   $2^4 \cdot a^2 = 16a^2$

2. $\dfrac{17}{50x^3y^2}; \dfrac{19}{20x^2y^5}$

   Factor each denominator to get

   $50x^3y^2 = 2 \cdot 5 \cdot 5 \cdot x^3 \cdot y^2 = 2 \cdot 5^2 \cdot x^3 \cdot y^2$

   $20x^2y^5 = 2 \cdot 2 \cdot 5 \cdot x^2 \cdot y^5 = 2^2 \cdot 5 \cdot x^2 \cdot y^5$

   The factors to be considered are 2, 5, x, and y. Since the greatest power of 2 is 2, of 5 is 2, of x is 3, and of y is 5, the LCD is

   $2^2 \cdot 5^2 \cdot x^3 \cdot y^5 = 4 \cdot 25x^3y^5 = 100x^3y^5$

3. $\dfrac{x + 2}{x^2 + x - 12}; \dfrac{x - 3}{x^2 + 2x - 8}$

   Factor the denominators to get

   $x^2 + x - 12 = (x + 4)(x - 3)$

   $x^2 + 2x - 8 = (x + 4)(x - 2)$

   The factors are $(x + 4)$, $(x - 3)$, and $(x - 2)$, and since the highest power of each is 1, the LCD is given by $(x + 4)(x - 3)(x - 2)$.

note

Multiplying these factors out would cause unnecessary work and difficulty when finding equivalent fractions later, so we *always* leave this common denominator in factored form.

4. $\dfrac{5x}{x^2 - 2x + 1}$; $\dfrac{3}{x^2 - x}$; $\dfrac{x - 2}{x^2 + 11x - 12}$

Factor the denominators:

$x^2 - 2x + 1 = (x - 1)^2$
$x^2 - x = x(x - 1)$
$x^2 + 11x - 12 = (x + 12)(x - 1)$

The factors to be considered are x, $(x - 1)$, and $(x + 12)$. Since the greatest power of x is 1, of $(x - 1)$ is 2, and of $(x + 12)$ is 1, the LCD is given by $x(x - 1)^2(x + 12)$.

Now that we can find the least common denominator of a group of fractions, let us review the process by which we write each fraction as an equivalent fraction with a common denominator. To do this, we again use the fundamental principle of fractions. An equivalent fraction is obtained when we multiply the numerator and denominator by the same algebraic expression.

We use the following steps to form equivalent fractions with a common denominator LCD:

1. Find the LCD of the given fractions.
2. Divide the given denominators *into* the LCD.
3. Multiply numerator and denominator of the given fraction by the result of Step 2 to form an equivalent fraction.

Examples

Find the LCD and write the equivalent fraction with this common denominator.

1. $\dfrac{7}{15x^2}$; $\dfrac{5}{24x^3}$; $\dfrac{1}{6x}$

$15x^2 = 3 \cdot 5 \cdot x^2$
$24x^3 = 2 \cdot 2 \cdot 2 \cdot 3 \cdot x^3 = 2^3 \cdot 3 \cdot x^3$
$6x = 2 \cdot 3 \cdot x$

The LCD is $2^3 \cdot 3 \cdot 5 \cdot x^3 = 120x^3$. Divide each denominator into $120x^3$ to get

$$\frac{120x^3}{15x^2} = 8x; \quad \frac{120x^3}{24x^3} = 5; \quad \frac{120x^3}{6x} = 20x^2$$

Then

$$\frac{7 \cdot 8x}{15x^2 \cdot 8x} = \frac{56x}{120x^3}$$

$$\frac{5 \cdot 5}{24x^3 \cdot 5} = \frac{25}{120x^3}$$

$$\frac{1 \cdot 20x^2}{6x \cdot 20x^2} = \frac{20x^2}{120x^3}$$

So $\dfrac{7}{15x^2} = \dfrac{56x}{120x^3}$; $\dfrac{5}{24x^3} = \dfrac{25}{120x^3}$; $\dfrac{1}{6x} = \dfrac{20x^2}{120x^3}$

2. $\dfrac{x + 3}{x^2 + 6x - 7}$; $\dfrac{x - 6}{x^2 - 3x + 2}$

Factoring the denominators we get

$$x^2 + 6x - 7 = (x + 7)(x - 1)$$
$$x^2 - 3x + 2 = (x - 2)(x - 1)$$

So, the LCD is $(x + 7)(x - 2)(x - 1)$.
Then

$$\frac{x + 3}{(x + 7)(x - 1)} = \frac{(x + 3) \cdot (x - 2)}{(x + 7)(x - 1) \cdot (x - 2)}$$
$$= \frac{x^2 + x - 6}{(x + 7)(x - 1)(x - 2)};$$
$$\frac{x - 6}{(x - 2)(x - 1)} = \frac{(x - 6) \cdot (x + 7)}{(x - 2)(x - 1) \cdot (x + 7)}$$
$$= \frac{(x^2 + x - 42)}{(x + 7)(x - 1)(x - 2)}$$

3. $\dfrac{3}{a^2 - 4a + 4}$; $\dfrac{2}{a^2 - 1}$; $\dfrac{5}{a^2 - 3a + 2}$

Factor the denominators:

$$a^2 - 4a + 4 = (a - 2)^2$$
$$a^2 - 1 = (a + 1)(a - 1)$$
$$a^2 - 3a - 2 = (a - 2)(a - 1)$$

So, the LCD is $(a - 2)^2(a + 1)(a - 1)$.
Then

$$\frac{3}{(a - 2)^2} = \frac{3 \cdot (a + 1)(a - 1)}{(a - 2)^2 \cdot (a + 1)(a - 1)}$$
$$= \frac{3(a^2 - 1)}{(a - 2)^2(a + 1)(a - 1)}$$
$$= \frac{3a^2 - 3}{(a - 2)^2(a + 1)(a - 1)};$$
$$\frac{2}{(a + 1)(a - 1)} = \frac{2 \cdot (a - 2)^2}{(a + 1)(a - 1)(a - 2)^2}$$
$$= \frac{2(a - 2)^2}{(a + 1)(a - 1)(a - 2)^2}$$
$$= \frac{2a^2 - 8a + 8}{(a + 1)(a - 1)(a - 2)^2};$$
$$\frac{5}{(a - 2)(a - 1)} = \frac{5 \cdot (a - 2)(a + 1)}{(a - 2)(a - 1) \cdot (a - 2)(a + 1)}$$
$$= \frac{5(a^2 - a - 2)}{(a - 2)^2(a + 1)(a - 1)}$$
$$= \frac{5a^2 - 5a - 10}{(a - 2)^2(a - 1)(a + 1)}$$

Exercise Set 9.4

Find the least common denominator (LCD) of the given algebraic fractions and convert to equivalent fractions with the common denominators.

Example $\frac{3}{15x}; \frac{8}{25x^2}$

Factor the denominators:

$15x = 3 \cdot 5 \cdot x;\ 25x^2 = 5^2 \cdot x^2$

The LCD has factors 3, 5, and x, so the LCD is $3 \cdot 5^2 \cdot x^2 = 75x^2$.

Now, $\frac{3}{15x} = \frac{3 \cdot 5x}{15x \cdot 5x} = \frac{15x}{75x^2}$

and $\frac{8}{25x^2} = \frac{8 \cdot 3}{25x^2 \cdot 3} = \frac{24}{75x^2}$.

1. $\frac{1}{6x}; \frac{1}{9x}$
2. $\frac{3}{8a}; \frac{5}{12a^2}$
•3. $\frac{5}{16x^2}; \frac{8}{24x}$
4. $\frac{14}{33ab}; \frac{7}{22a}$
5. $\frac{5a}{6b^2}; \frac{7a}{14b}$
6. $\frac{3x}{28y^2}; \frac{10x}{35y^3}$
7. $\frac{13}{9z}; \frac{15}{7z^4}$
8. $\frac{17}{32a^2}; \frac{15}{64a^4}$
9. $\frac{16}{27abc}; \frac{1}{18a^2bc^2}$
10. $\frac{15}{5x^3yz^2}; \frac{17}{25xy^2z^3}$
11. $\frac{3}{4x^2}; \frac{5}{3x}; \frac{9}{8x^3}$
12. $\frac{7}{10a^2}; \frac{5}{12a^3}; \frac{3}{9a}$
•13. $\frac{6}{35ab}; \frac{7}{14a^2b}; \frac{1}{4ab^3}$
14. $\frac{8}{21x^3y}; \frac{11}{15xy^2}; \frac{7}{14xy}$
15. $\frac{5}{4x - 2}; \frac{3}{4}$
16. $\frac{8}{x - 4}; \frac{7}{3x - 12}$
17. $\frac{1}{a^2 + a}; \frac{1}{a^2 - 1}$
18. $\frac{1}{(z - 1)^2}; \frac{1}{z^2 - 1}$
•19. $\frac{a}{a^2 - 5a + 6}; \frac{2a - 1}{a^2 - 4}$
20. $\frac{7y}{y^2 - y - 12}; \frac{3y}{y^2 - 6y + 9}$
21. $\frac{4R}{R^2 - 9}; \frac{2R}{R^2 - R - 6}$
22. $\frac{4a^2}{a^2 - 3a + 2}; \frac{2a}{a^2 - 1}$
23. $\frac{E}{E^2 - 9}; \frac{2}{E^2 + E - 6}; \frac{E}{E^2 - 4E + 4}$
24. $\frac{3}{m - n}; \frac{5}{m + n}; \frac{5}{m^2 - 2mn + n^2}$
25. $\frac{2a}{a^2 - a - 2}; \frac{3a}{a^2 - 4}; \frac{a + 1}{a^2 - 4a - 5}$
26. $\frac{p - 1}{p^2 - 25}; \frac{p - 2}{10p + 50}; \frac{p}{5p - 25}$

27. $\frac{x-3}{x^2+6x+9}; \frac{x}{2x+6}; \frac{x-1}{x^2-x-12}$

28. $\frac{z+9}{4z-12}; \frac{z-1}{2z^2-18}; \frac{z}{z^2+6z+9}$

29. $\frac{5x}{49-x^2}; \frac{3x}{x^2-5x-14}; \frac{7}{x+7}$

30. $\frac{y}{2y+10}; \frac{3y+1}{25-y^2}; \frac{y^2}{y^2-10y+25}$

9.5 Addition and Subtraction of Algebraic Fractions

Keeping in mind the basic method for finding the least common denominator, which we discussed in section 9.4, let us now consider the operations of addition and subtraction of algebraic fractions. Recall that when adding or subtracting arithmetic fractions with common denominators, we merely add or subtract the numerators and place this sum or difference over the common denominator. The same procedure is followed with algebraic fractions.

Examples

Perform the indicated operations.

1. $$\frac{3}{x-2} + \frac{5}{x-2} = \frac{3+5}{x-2} = \frac{8}{x-2} \qquad (x \neq 2)$$

2. $$\frac{5a}{4x} + \frac{3a}{4x} - \frac{6a}{4x} = \frac{5a+3a-6a}{4x} = \frac{2a}{4x} = \frac{a}{2x} \qquad (x \neq 0)$$

note

Always reduce the answer to lowest terms, as was done here.

3. $$\frac{x-1}{x+6} + \frac{2x+3}{x+6} = \frac{(x-1)+(2x+3)}{x+6} = \frac{3x+2}{x+6} \qquad (x \neq -6)$$

4. $$\frac{7y - 3}{2y + 1} - \frac{y - 6}{2y + 1} = \frac{(7y - 3) - (y - 6)}{2y + 1}$$
$$= \frac{7y - 3 - y + 6}{2y + 1}$$
$$= \frac{6y + 3}{2y + 1}$$
$$= \frac{3(2y + 1)}{2y + 1}$$
$$= 3 \qquad \left(y \neq -\frac{1}{2}\right)$$

note

In example 4, it is important to group in the numerator the expressions to be subtracted. This procedure reminds us that when we subtract a group, we remove the parentheses and change the signs. Failure to do this results in a common error—*failure to change the signs.*

5. $$\frac{2t - 7}{3t - 1} + \frac{3t + 8}{3t - 1} - \frac{4t + 5}{3t - 1} = \frac{(2t - 7) + (3t + 8) - (4t + 5)}{3t - 1}$$
$$= \frac{2t - 7 + 3t + 8 - 4t - 5}{3t - 1}$$
$$= \frac{t - 4}{3t - 1} \qquad \left(t \neq \frac{1}{3}\right)$$

note

We placed parentheses about each numerator to remind us to change the signs of *all* terms when the expression is preceded by a subtraction sign.

When the denominators of the algebraic fractions to be added or subtracted are not the same, we use the following procedure to perform the indicated operations.

1. Find the LCD of the denominators of the fractions involved.
2. Change each fraction to an equivalent fraction with the common denominator—that is, the least common denominator, LCD.
3. Add or subtract the algebraic expressions and place this sum or difference over the common denominator.

Examples Perform the indicated operations and write your answer reduced to lowest terms.

1. $\dfrac{7}{4a^2} + \dfrac{5}{8a}$

Since $4a^2 = 2 \cdot 2 \cdot a \cdot a = 2^2 \cdot a^2$ and $8a = 2 \cdot 2 \cdot 2 \cdot a = 2^3a$, then the least common denominator is $2^3 \cdot a^2 = 8a^2$. To get equivalent fractions with this common denominator we must

a. multiply the numerator and denominator of the first fraction by 2.

b. multiply the numerator and denominator of the second fraction by a.

Thus,

$$\frac{7}{4a^2} + \frac{5}{8a} = \frac{7 \cdot 2}{4a^2 \cdot 2} + \frac{5 \cdot a}{8a \cdot a}$$
$$= \frac{14}{8a^2} + \frac{5a}{8a^2}$$
$$= \frac{14 + 5a}{8a^2}$$

2. $\dfrac{2x}{x - 1} + \dfrac{3x}{x^2 - 1}$

Factor the denominator of the second fraction:

$$x^2 - 1 = (x - 1)(x + 1)$$

and since the factor $(x - 1)$ is the denominator of the first fraction, the LCD is $(x + 1)(x - 1)$. Now we must multiply the numerator and denominator of the first fraction by $(x + 1)$ to get

$$\frac{2x}{x - 1} = \frac{2x}{x - 1} \cdot \frac{x + 1}{x + 1}$$
$$= \frac{2x(x + 1)}{(x - 1)(x + 1)}$$
$$= \frac{2x^2 + 2x}{(x - 1)(x + 1)}$$

The second fraction remains the same since its denominator *is the LCD*. Therefore,

$$\frac{2x}{x - 1} + \frac{3x}{x^2 - 1} = \frac{2x^2 + 2x}{(x - 1)(x + 1)} + \frac{3x}{(x - 1)(x + 1)}$$
$$= \frac{2x^2 + 2x + 3x}{(x - 1)(x + 1)}$$
$$= \frac{2x^2 + 5x}{(x - 1)(x + 1)}$$
$$= \frac{x(2x + 5)}{(x - 1)(x + 1)}$$

note

The expression in the numerator of the sum is factorable. You should always factor in case one or more is a factor of the denominator, which means the fraction could be reduced. Remember, we always want to state the answer in the most reduced form.

3. $\dfrac{4}{3x - 6} - \dfrac{x + 2}{x^2 - x - 2}$

Factor denominators:

$$\frac{4}{3x - 6} = \frac{4}{3(x - 2)} \quad \text{and} \quad \frac{x + 2}{x^2 - x - 2} = \frac{x + 2}{(x - 2)(x + 1)}$$

So, the LCD is $3(x - 2)(x + 1)$, and

$$\frac{4}{3(x - 2)} = \frac{4 \cdot (x + 1)}{3(x - 2)(x + 1)} = \frac{4x + 4}{3(x - 2)(x + 1)},$$

$$\frac{x + 2}{(x - 2)(x + 1)} = \frac{(x + 2) \cdot 3}{(x - 2)(x + 1) \cdot 3} = \frac{3x + 6}{3(x - 2)(x + 1)}$$

Therefore,

$$\begin{aligned}
\frac{4}{3x - 6} - \frac{x + 2}{x^2 - x - 2} &= \frac{4x + 4}{3(x - 2)(x + 1)} - \frac{3x + 6}{3(x - 2)(x + 1)} \\
&= \frac{(4x + 4) - (3x + 6)}{3(x - 2)(x + 1)} \\
&= \frac{4x + 4 - 3x - 6}{3(x - 2)(x + 1)} \\
&= \frac{x - 2}{3(x - 2)(x + 1)} \\
&= \frac{1}{3(x + 1)}
\end{aligned}$$

4. $\dfrac{x}{x^2 - 5x + 6} + \dfrac{3}{x^2 - 7x + 12}$

Factor the denominators:

$$\frac{x}{x^2 - 5x + 6} = \frac{x}{(x - 3)(x - 2)} \quad \text{and}$$

$$\frac{3}{x^2 - 7x + 12} = \frac{3}{(x - 4)(x - 3)}$$

The LCD is $(x - 3)(x - 2)(x - 4)$. Therefore,

$$\begin{aligned}
\frac{x}{(x - 3)(x - 2)} &= \frac{x \cdot (x - 4)}{(x - 3)(x - 2)(x - 4)} \\
&= \frac{x^2 - 4x}{(x - 3)(x - 2)(x - 4)}
\end{aligned}$$

and

$$\begin{aligned}
\frac{3}{(x - 4)(x - 3)} &= \frac{3 \cdot (x - 2)}{(x - 4)(x - 3)(x - 2)} \\
&= \frac{3x - 6}{(x - 4)(x - 3)(x - 2)}
\end{aligned}$$

Then

$$\frac{x}{x^2 - 5x + 6} + \frac{3}{x^2 - 7x + 12}$$
$$= \frac{x^2 - 4x}{(x - 3)(x - 2)(x - 4)} + \frac{3x - 6}{(x - 3)(x - 2)(x - 4)}$$
$$= \frac{(x^2 - 4x) + (3x - 6)}{(x - 3)(x - 2)(x - 4)}$$
$$= \frac{x^2 - x - 6}{(x - 3)(x - 2)(x - 4)}$$
$$= \frac{(x - 3)(x + 2)}{(x - 3)(x - 2)(x - 4)}$$
$$= \frac{x + 2}{(x - 2)(x - 4)}$$

note

In examples 3 and 4 we were able to reduce by a common factor in the numerator and denominator.

5. $\dfrac{5}{3x + 9} + \dfrac{x - 1}{x^2 + 6x + 9} - \dfrac{4}{x^2 - 9}$

Factor the denominators:

$$\frac{5}{3x + 9} = \frac{5}{3(x + 3)};$$
$$\frac{x - 1}{x^2 + 6x + 9} = \frac{x - 1}{(x + 3)^2};$$
$$\frac{4}{x^2 - 9} = \frac{4}{(x + 3)(x - 3)}$$

Then the LCD is $3(x - 3)(x + 3)^2$ and we have

$$\frac{5}{3(x + 3)} = \frac{5 \cdot (x + 3)(x - 3)}{3(x + 3)(x + 3)(x - 3)}$$
$$= \frac{5(x^2 - 9)}{3(x + 3)^2(x - 3)}$$
$$= \frac{5x^2 - 45}{3(x + 3)^2(x - 3)};$$
$$\frac{x - 1}{(x + 3)^2} = \frac{(x - 1) \cdot 3(x - 3)}{(x + 3)^2 \cdot 3(x - 3)}$$
$$= \frac{3(x - 1)(x - 3)}{3(x + 3)^2(x - 3)}$$
$$= \frac{3x^2 - 12x + 9}{3(x + 3)^2(x - 3)};$$

$$\frac{4}{(x+3)(x-3)} = \frac{4 \cdot 3(x+3)}{(x+3)(x-3) \cdot 3(x+3)}$$

$$= \frac{12(x+3)}{3(x+3)^2(x-3)}$$

$$= \frac{12x+36}{3(x+3)^2(x-3)}$$

Therefore,

$$\frac{5}{3x+9} + \frac{x-1}{x^2+6x+9} - \frac{4}{x^2-9}$$

$$= \frac{5x^2-45}{3(x+3)^2(x-3)} + \frac{3x^2-12x+9}{3(x+3)^2(x-3)} - \frac{12x+36}{3(x+3)^2(x-3)}$$

$$= \frac{(5x^2-45)+(3x^2-12x+9)-(12x+36)}{3(x+3)^2(x-3)}$$

$$= \frac{5x^2-45+3x^2-12x+9-12x-36}{3(x+3)^2(x-3)}$$

$$= \frac{8x^2-24x-72}{3(x+3)^2(x-3)}$$

$$= \frac{8(x^2-3x-9)}{3(x+3)^2(x-3)} \quad \text{(not reducible)}$$

6. $\dfrac{3x}{x-1} + \dfrac{5x}{x+1}$

Notice that neither of the denominators will factor. Therefore, the LCD is the product of the denominators. The LCD is $(x-1)(x+1)$. To get equivalent fractions we must multiply the numerator and denominator of (a) the first fraction by $(x+1)$ to get

$$\frac{3x}{x-1} \cdot \frac{x+1}{x+1} = \frac{3x(x+1)}{(x-1)(x+1)} = \frac{3x^2+3x}{(x-1)(x+1)}$$

(b) the second fraction by $(x-1)$ to get

$$\frac{5x}{x+1} \cdot \frac{x-1}{x-1} = \frac{5x(x-1)}{(x+1)(x-1)} = \frac{5x^2-5x}{(x+1)(x-1)}$$

Then,

$$\frac{3x}{x-1} + \frac{5x}{x+1} = \frac{3x(x+1)}{(x-1)(x+1)} + \frac{5x(x-1)}{(x-1)(x+1)}$$

$$= \frac{(3x^2+3x)+(5x^2-5x)}{(x-1)(x+1)}$$

$$= \frac{8x^2-2x}{(x-1)(x+1)}$$

$$= \frac{2x(4x-1)}{(x-1)(x+1)} \quad \text{(not reducible)}$$

When we add or subtract *two fractions,* we often use an important property of fractions that may be useful to technicians. This property states

$$\frac{a}{b} + \frac{c}{d} = \frac{ad + bc}{bd} \quad \text{and} \quad \frac{a}{b} - \frac{c}{d} = \frac{ad - bc}{bd}$$

The property is useful in particular when the given denominators have no common factors, as was the case in the previous example 5.

Examples

Perform the indicated operations.

1. $\frac{3}{4} + \frac{7}{9} = \frac{3 \cdot 9 + 4 \cdot 7}{4 \cdot 9} = \frac{27 + 28}{36} = \frac{55}{36}$

2. $\frac{5}{2a + 1} - \frac{2}{a - 3} = \frac{5(a - 3) - 2(2a + 1)}{(2a + 1)(a - 3)}$

$$= \frac{5a - 15 - 4a - 2}{(2a + 1)(a - 3)}$$

$$= \frac{a - 17}{(2a + 1)(a - 3)}$$

---

note

This property should be used *only* when adding or subtracting *two fractions*.

Exercise Set 9.5

Combine the given algebraic fractions and reduce your answer to lowest terms.

1. $\frac{4}{3x} + \frac{5}{2x}$

2. $\frac{6}{5a} + \frac{7}{4a}$

3. $\frac{7}{10a} - \frac{8}{15a}$

4. $\frac{9}{14y} - \frac{1}{21y}$

5. $\frac{5}{6z} + \frac{4}{8z} - \frac{1}{4z}$

•6. $\frac{13}{12b} - \frac{2}{9b} + \frac{5}{4b}$

7. $\frac{4a}{5} + \frac{7a}{15} - \frac{a}{9}$

8. $\frac{3x}{8} - \frac{2x}{5} + \frac{7x}{10}$

9. $\frac{2x - 1}{16} + \frac{x + 2}{24}$

10. $\frac{5a + 3}{12} - \frac{a - 4}{10}$

11. $\frac{5b + 1}{6} + \frac{3b - 2}{9} - \frac{b + 1}{12}$

12. $\frac{a - 1}{5} - \frac{2a + 3}{15} + \frac{3a - 1}{25}$

13. $\frac{3a + 1}{a} + \frac{2a - 3}{3a}$

14. $\frac{2x + 5}{6x} - \frac{x - 5}{9x}$

•15. $\frac{5m + 1}{2m} + \frac{m - 1}{m} - \frac{3m}{4}$

16. $\frac{p - 3}{5p^2} + \frac{3p - 5}{10p} - \frac{2p}{15}$

17. $\frac{4}{4ab} + \frac{5}{a^2b}$

18. $\frac{9}{5x^2y^2} - \frac{3}{10xy}$

19. $\frac{6}{5ab} + \frac{7}{15a^2b} - \frac{1}{10ab}$

20. $\dfrac{11}{6ab^2} - \dfrac{3}{8a^2} + \dfrac{7}{12b}$

21. $\dfrac{10}{x} + \dfrac{7}{y} - \dfrac{1}{xy}$

22. $\dfrac{4}{5y^2} - \dfrac{3}{2x} + \dfrac{5}{4xy}$

23. $\dfrac{3}{4x} + \dfrac{5}{6y} - \dfrac{2}{8x^2y^2}$

24. $\dfrac{9}{2m^2} + \dfrac{3}{4mn^2} + \dfrac{11}{12n}$

25. $\dfrac{4}{x - 1} + \dfrac{5}{x + 3}$

26. $\dfrac{7}{2x - 3} - \dfrac{6}{x - 5}$

27. $\dfrac{7}{4x - 6} - \dfrac{12}{3x + 9}$

28. $\dfrac{15}{5y - 10} + \dfrac{14}{2y + 4}$

29. $\dfrac{1}{a - 1} + \dfrac{1}{2a + 1} - \dfrac{1}{2a - 2}$

30. $\dfrac{4}{2z - 1} - \dfrac{5}{4z + 4} + \dfrac{6}{z + 1}$

31. $\dfrac{5}{a^2 - b^2} + \dfrac{6}{a^2 - ab}$

32. $\dfrac{12}{x^2 - 4} - \dfrac{7}{4x - 8}$

33. $\dfrac{1}{2a - b} + \dfrac{1}{2a + b} - \dfrac{3}{4a^2 - b^2}$

34. $\dfrac{2}{x - 2y} + \dfrac{13}{4x + 8y} + \dfrac{1}{x^2 - 4y^2}$

35. $\dfrac{x}{x - 1} + \dfrac{3x}{x^2 - 1}$

36. $\dfrac{2y}{y^2 - 16} - \dfrac{5y}{2y - 8}$

37. $\dfrac{2y}{x^2 - xy} + \dfrac{4}{x^3 - x^2y}$

38. $\dfrac{1}{6b^2 - 4bc} - \dfrac{2}{6c^2 - 9bc}$

•39. $\dfrac{a}{a + b} - \dfrac{2ab}{b^2 - a^2} + \dfrac{b}{a - b}$

40. $\dfrac{z - 1}{3z + 9} + \dfrac{8}{3z - 9} + \dfrac{5}{3 - z}$

41. $\dfrac{y - 1}{y^2 - 1} - \dfrac{y + 2}{y^2 - 4}$

42. $\dfrac{a^2 - ab}{(a - b)^2} + \dfrac{2a}{a + b}$

43. $\dfrac{4}{x^2 - x - 6} + \dfrac{5}{x^2 - 9}$

44. $\dfrac{8}{x^2 - 25} - \dfrac{7}{x^2 + 3x - 10}$

45. $\dfrac{2a - 3}{a^2 - 5a + 6} - \dfrac{3}{a - 2}$

46. $\dfrac{2p}{p^2 - 9p + 20} - \dfrac{5p - 2}{p - 5}$

47. $\dfrac{5}{z^2 - 4} - \dfrac{z}{z^2 - 1} + \dfrac{4}{z^2 + z - 2}$

48. $\dfrac{4a}{a^2 + 2a - 15} + \dfrac{3a}{2a^2 + 11a + 5} - \dfrac{5a}{2a^2 - 5a - 3}$

49. $\dfrac{5R}{R^2 + 8R + 15} + \dfrac{3R}{R^2 - 9} + \dfrac{R - 1}{R^2 - 2R - 15}$

50. $\dfrac{y - 1}{y^2 - 25} + \dfrac{y}{5y - 25} - \dfrac{y - 2}{10y + 50}$

51. In order for a lens maker to manufacture lenses that will refract light by exactly the right amounts, he uses the following formula: $\dfrac{1}{f} = (u - 1)\left(\dfrac{1}{R_1} + \dfrac{1}{R_2}\right)$. Combine and simplify the right side.

52. In electricity, the total resistance of any parallel circuit may be given by $\frac{1}{R_t} = \frac{I_1}{E_1} + \frac{I_2}{E_2} + \frac{I_3}{E_3}$. Combine the expression on the right side.

53. Given men A, B, and C can complete a given job in a, b, and c hours, respectively. Working together they can complete in 1 hour $1/a + 1/b + 1/c$ of the job. Obtain a single expression for what they can do together in 1 hour by combining.

54. For any number of capacitors in series, $\frac{1}{C} = \frac{1}{C_1} + \frac{1}{C_2} + \frac{1}{C_3} + \ldots$, find the sum of the three terms on the right side.

55. In electricity, the reciprocal of the equivalent resistance of a set of resistors connected in parallel is equal to the sum of the reciprocals of the individual resistances: $\frac{1}{R} = \frac{1}{R_1} + \frac{1}{R_2} + \frac{1}{R_3} + \frac{1}{R_4}$.
Combine the terms on the right side.

56. In thermodynamics, the efficiency of a Carnot heat engine is given by Eff $= 1 - \frac{T_2}{T_1}$, where T_1 and T_2 are temperatures of the reservoirs in the engine. Make a single fraction of the right side by combining terms.

9.6 Fractional Equations

An algebraic equation that contains at least one algebraic fraction is called a **fractional equation.** The basic operations for solving equations, which you learned in chapter 7, apply to fractional equations once the fractions in the equation are eliminated. We eliminate the fractions by using a procedure that employs the least common denominator, LCD: To eliminate the denominators, multiply each term of both members of the equation by the least common denominator, LCD.

Examples

Solve the following fractional equations for the variable involved.

1. $\frac{x - 3}{4} = \frac{x}{8}$

The LCD of the denominators 4 and 8 is 8. Multiply both sides by 8 to get

$$\frac{8}{1} \cdot \frac{x - 3}{4} = \frac{8}{1} \cdot \frac{x}{8}$$

Reduce each side to get

$$2(x - 3) = x$$
$$2x - 6 = x$$
$$x = 6$$

Check Substitute 6 for x in the original equation:

$$\frac{6 - 3}{4} = \frac{6}{8}$$
$$\frac{3}{4} = \frac{3}{4}$$

2. $\frac{t}{4} - \frac{t-4}{5} = \frac{7}{10}$

The LCD of 4, 5, and 10 is 20. Multiply each term of both sides by 20:

$$\frac{20}{1} \cdot \frac{t}{4} - \frac{20}{1} \cdot \frac{t-4}{5} = \frac{20}{1} \cdot \frac{7}{10}$$

Reduce to get

$$\begin{aligned} 5t - 4(t - 4) &= 2 \cdot 7 \\ 5t - 4t + 16 &= 14 \\ t + 16 &= 14 \\ t &= -2 \end{aligned}$$

note

A common error that is made when multiplying elements like $-4(t - 4)$ is to get an answer like $-4t - 16$. Don't forget you are multiplying -4 times each term in the group $(t - 4)$ and the correct result is $-4t + 16$.

3. $\frac{5}{3a} + \frac{4}{9} = \frac{4}{12a}$

The LCD is $36a$, so multiply each term by $36a$ to get

$$36a \cdot \frac{5}{3a} + 36a \cdot \frac{4}{9} = 36a \cdot \frac{4}{12a}$$

Reduce each term:

$$\begin{aligned} 12 \cdot 5 + 4a \cdot 4 &= 3 \cdot 4 \\ 60 + 16a &= 12 \\ 16a &= -48 \\ a &= -3 \end{aligned}$$

4. $\frac{4}{x} = \frac{3}{x-1}$

The LCD of x and $(x - 1)$ is $x(x - 1)$. Multiply each term of the equation by $x(x - 1)$ to get

$$x(x-1) \cdot \frac{4}{x} = x(x-1) \cdot \frac{3}{x-1}$$

Reduce each term:

$$\begin{aligned} (x - 1) \cdot 4 &= x \cdot 3 \\ 4x - 4 &= 3x \\ x &= 4 \end{aligned}$$

note

The following example demonstrates the need to check every solution of the fractional equations where the LCD contains the variable involved. Checking is necessary for validity because of *extraneous solutions*—solutions that satisfy the equation obtained by multiplying both sides by the LCD but that do not satisfy the original equation. This result occurs for solutions that would make the LCD equal to zero.

5. $\frac{3}{y} = \frac{4}{y^2 - 2y} - \frac{2}{y - 2}$

Factor the denominator $y^2 - 2y$ to get $y(y - 2)$. Determine that the LCD of the denominators is $y(y - 2)$, and then multiply each term by the LCD:

$$y(y - 2) \cdot \frac{3}{y} = y(y - 2) \cdot \frac{4}{y(y - 2)} - y(y - 2) \cdot \frac{2}{y - 2}$$

Finally, reduce each term:

$$\begin{aligned} (y - 2) \cdot 3 &= 4 - y \cdot 2 \\ 3y - 6 &= 4 - 2y \\ 5y &= 10 \\ y &= 2 \end{aligned}$$

Check Substitute 2 for y in the original equation:

$$\frac{3}{2} = \frac{4}{2^2 - 2(2)} - \frac{2}{2 - 2} = \frac{4}{0} - \frac{2}{0}$$

Since division by zero is not defined, then $y \neq 2$. It does not satisfy the equation. Thus, there is no solution to the equation.

6. $\frac{x}{3} - 4 = \frac{7}{2}$

Multiply all terms of the equation by the LCD of 2 and 3, 6:

$$\begin{aligned} 6 \cdot \frac{x}{3} - 6 \cdot 4 &= 6 \cdot \frac{7}{2} \\ 2x - 24 &= 3 \cdot 7 \\ 2x - 24 &= 21 \\ 2x &= 45 \\ x &= \frac{45}{2} \end{aligned}$$

It is important to recognize that multiplying each term by the LCD of the denominators eliminates *all* the fractions in the problem. This is the reason for multiplying by the LCD.

In any technical field, equations and formulas involving the fractional expressions and *more than one* variable are common. In such equations it is often desirable to solve for one variable in terms of the other variables given. The procedures for finding such solutions are identical to those used in solving fractional equations.

Examples

1. Solve $\frac{a}{3} + \frac{3x}{2} = c$ for x.

   The LCD of the denominators is $2 \cdot 3 = 6$. Multiply each term by 6:

   $$6 \cdot \frac{a}{3} + 6 \cdot \frac{3x}{2} = 6 \cdot c$$

   Reduce where possible:

   $$2 \cdot a + 3 \cdot 3x = 6c$$
   $$2a + 9x = 6c$$

   Add $-2a$ to both sides of the equation, $9x = 6c - 2a$, then divide both sides by 9:

   $$x = \frac{6c - 2a}{9}$$

   (*Note:* we cannot reduce this expression. *Always check that this is so.*)

2. Solve $\frac{a}{4} - \frac{b}{x} = 3c$ for x.

   Multiply each term by the LCD, $4x$:

   $$4x \cdot \frac{a}{4} - 4x \cdot \frac{b}{x} = 4x \cdot 3c$$

   Reducing where possible, the equation becomes

   $x \cdot a - 4 \cdot b = 4 \cdot 3 \cdot x \cdot c$ and $ax - 4b = 12cx$.

   To get the terms containing x together, add $4b$ to and subtract $12cx$ from both sides. Then

   $$ax - 12cx = 4b.$$

   Factor on the left side,

   $$(a - 12c)x = 4b,$$

   and divide both sides by the coefficient of x, $a - 12c$, to get

   $$x = \frac{4b}{a - 12c}$$

3. Solve $\frac{1}{a} = \frac{1}{b} + \frac{1}{c}$ for c.

The LCD of the denominators is abc, so multiply each term by abc to get:

$$abc \cdot \frac{1}{a} = abc \cdot \frac{1}{b} + abc \cdot \frac{1}{c}$$

Then reduce to get

$$bc = ac + ab.$$

Since we are solving for c, we must get the terms containing c together on one side of the equation. Therefore, subtract ac from both sides to get

$$bc - ac = ab.$$

Factor the left side to get

$$(b - a)c = ab.$$

Finally, divide each side by $b - a$:

$$c = \frac{ab}{b - a}$$

4. Solve $P = \frac{HD + 20}{100 + E}$ for H.

Multiply both sides of the equation by $100 + E$ to get

$$(100 + E) \cdot P = (100 + E) \cdot \frac{HD + 20}{100 + E}$$

Reduce on the right side to get

$$(100 + E) \cdot P = HD + 20.$$

Subtract 20 from both sides,

$$100P + PE - 20 = HD,$$

and divide each side by D:

$$H = \frac{100P + PE - 20}{D}$$

Exercise Set 9.6

Solve the following fractional equations for the variable.

1. $\frac{y}{4} = \frac{2}{3}$

2. $\frac{p}{6} = \frac{7}{9}$

•3. $\frac{a}{3} + \frac{5}{2} = 6$

4. $\frac{4x}{5} - \frac{2}{3} = 4$

5. $\frac{z}{8} + 3 = \frac{1}{4}$

6. $\frac{3R}{4} - 5 = \frac{5}{6}$

7. $\dfrac{3a}{6} + \dfrac{2a}{5} = a$

8. $\dfrac{5x}{8} - \dfrac{x}{12} = 3x$

9. $\dfrac{x + 1}{7} - \dfrac{2x - 3}{21} = 1$

10. $\dfrac{3a + 1}{9} + \dfrac{1}{12} = \dfrac{a - 1}{6}$

11. $\dfrac{3}{2x} = \dfrac{4}{5} + \dfrac{2}{x}$

12. $\dfrac{4}{b} - \dfrac{7}{3b} = \dfrac{2}{5}$

13. $\dfrac{2}{3R} + \dfrac{3}{2R} + \dfrac{1}{R} = 4$

14. $\dfrac{3}{w} - \dfrac{6}{5w} + \dfrac{1}{2w} = 5$

15. $\dfrac{4}{6y} + 5 = \dfrac{1}{9y} + 2$

16. $\dfrac{16}{5a} - 1 = 5 + \dfrac{3}{4a}$

17. $\dfrac{5}{3b} - \dfrac{1}{2} = \dfrac{7}{6b}$

18. $3 - \dfrac{5}{9x} = \dfrac{4}{6x}$

19. $\dfrac{a - 4}{3a} = \dfrac{2a - 1}{4a}$

20. $\dfrac{5 - x}{8x} = \dfrac{2x + 5}{6x}$

•21. $\dfrac{3p + 2}{7p} - 3 = \dfrac{p}{14p}$

22. $\dfrac{R + 2}{10R} + \dfrac{4R - 1}{4R} = 2$

23. $\dfrac{4}{x - 4} = \dfrac{5}{x + 4}$

24. $\dfrac{9}{3 - x} = \dfrac{8}{2x + 1}$

25. $\dfrac{10}{3z + 1} = \dfrac{3}{5}$

26. $\dfrac{3}{2a + 3} = \dfrac{1}{7}$

27. $\dfrac{5}{4x + 2} - \dfrac{3}{2x + 1} = 5$

28. $1 + \dfrac{5}{3a - 9} = \dfrac{10}{a - 3}$

29. $\dfrac{b - 1}{b^2 - 4} = \dfrac{6}{b + 2}$

30. $\dfrac{6}{R + 3} = \dfrac{4R + 3}{R^2 - 9}$

31. $\dfrac{5}{a^2 - 25} + \dfrac{3}{a - 5} = \dfrac{4}{a + 5}$

32. $\dfrac{6}{2p - 4} + \dfrac{4}{p - 2} = \dfrac{1}{p^2 - 4}$

33. $\dfrac{5}{x^2 + x - 6} = \dfrac{2}{x^2 + 3x - 10}$

34. $\dfrac{8}{a^2 - 6a + 8} = \dfrac{1}{a^2 - 16}$

•35. $\dfrac{1}{a - 3} + \dfrac{2}{a + 4} = \dfrac{6}{a^2 + a - 12}$

36. $\dfrac{5}{2x - 1} - \dfrac{7}{3x + 2} = \dfrac{9}{6x^2 + x - 2}$

Solve the following equations and formulas for the indicated letter.

37. $\dfrac{2}{x} + \dfrac{1}{y} = 3$ for x

38. $\dfrac{5}{I} - 6 = \dfrac{8}{E}$ for I

39. $\dfrac{1}{c} = \dfrac{1}{c_1} + \dfrac{1}{c_2}$ for c_1

40. $\dfrac{1}{R} = \dfrac{1}{R_1} + \dfrac{1}{R_2} + \dfrac{1}{R_3}$ for R

41. $\dfrac{1}{8} = \dfrac{1}{a} + \dfrac{1}{b}$ for a

42. $\dfrac{1}{x} - \dfrac{1}{y} - \dfrac{1}{z} = 6$ for y

43. $\dfrac{3}{a} - \dfrac{4}{b} = \dfrac{5}{ab}$ for a

•44. $\dfrac{5}{m} + 4 = \dfrac{6a}{2m} + 3b$ for m

45. $\dfrac{x + 4}{2} + \dfrac{y - 3}{5} = \dfrac{2}{10}$ for x

46. The principal amount of money, P, in a savings account paying interest rate r, over a given period of time, t, which pays interest I, is given by $P = \dfrac{I}{rt}$. Solve the equation for r.

47. The pressure per square inch of steam or water in a pipe, p, is given by $p = \dfrac{P}{LD}$, where P = the total pressure on a diametral plane, L = the length of the pipe in inches, and D = the diameter of the pipe in inches. Solve for D.

48. The safe internal unit pressure, p, in a pipe of a certain thickness is given by $p = \dfrac{2st}{D}$, where s = unit tensile stress, t = thickness of the pipe, and D = diameter of the pipe. Solve for s.

49. The pitch diameter of a gear, D, is given by $D = \dfrac{OD \times N}{N + 2}$, where OD = outside diameter of the gear and N = the number of teeth in the gear. Solve for N.

50. The rule governing the speeds of two gears, one the driver gear and the other the driven gear, is given by $\dfrac{T_A}{T_B} = \dfrac{R_B}{R_A}$, where T_A = number of teeth in the driven gear, T_B = number of teeth in the driver gear, R_B = revolutions per minute of the driven gear, and R_A = revolutions per minute of the driver gear. Solve for T_B.

51. Given F is the force on the large piston and f is the force on the small piston of a hydraulic press, then $\dfrac{F}{f} = \dfrac{A}{a}$, where A is the area of the large piston and a is the area of the small piston. Solve for f.

52. The kinetic energy of a body, KE, is computed by $KE = \frac{Wv^2}{2g}$, where v is the velocity expressed in feet per second, W is the weight in pounds, and g is acceleration due to gravity. Solve for W.

53. Charles's law regarding the relationship between pressure (P), volume (V), and the absolute temperature (T) of a gas as it expands when heated is given by $\frac{P_1V_1}{T_1} = \frac{P_2V_2}{T_2}$, where P_1, V_1, and T_1 are the initial conditions of the gas n and P_2, V_2, and T_2 are the final conditions. Solve for T_2.

54. The coefficient of linear expansion of a solid when heated (k) is given by $k = \frac{L_t - L_o}{L_o t}$, where L_o is the length at 0°C, L_t is the length at t°C, and t is any given temperature in Celsius. Solve for L_t. Solve for L_o.

55. The coefficient of linear expansion when metal is heated α (alpha) is given by $\alpha = \frac{L_2 - L_1}{L_1 (t_2 - t_1)}$. Solve for L_1.

56. Carnot's ideal efficiency of any heat engine operating between the temperature limits T_1 and T_2 is given by $CE = \frac{T_1 - T_2}{T_1}$. Solve for T_1.

•57. A formula for resistors in parallel is given by $R = \frac{R_1R_2}{R_1 + R_2}$. Solve for R_1.

58. In electronics, a formula for the unknown resistance in a battery (R_x) is given by $R_x = R_m \left(\frac{E_1}{E_2} - 1\right)$. Solve for E_2.

9.7 Complex Fractions

In the previous chapters and sections we have been working with simple arithmetic and simple algebraic fractions—that is, fractions with a single number or single expression in the denominator. Now we consider a complex fraction, that is, a fraction whose numerator or denominator, or both, contain fractions. The following are all complex fractions.

Examples

$$\frac{\frac{3}{4}}{\frac{5}{6}}, \quad \frac{\frac{1}{3} + 2}{1 - \frac{1}{2}}, \quad \frac{x}{\frac{1}{x} + 3}, \quad \frac{\frac{2}{x} + \frac{3}{x}}{x - 6}, \quad \frac{\frac{x - 1}{x - 2}}{\frac{x + 3}{x}}$$

It is desirable to reduce a complex fraction to a simple fraction by eliminating the fractions within the fraction.

Consider the complex fraction

$$\frac{\frac{3}{a} - 1}{1 + \frac{4}{b}}$$

For our discussion, let us call $\frac{3}{a} - 1$ and $1 + \frac{4}{b}$ the **primary numerator** and **primary denominator** of the fraction, respectively. Then, in the fractions $\frac{3}{a}$, $\frac{4}{b}$, and $\frac{1}{1}$, we call 3, 4, and 1 the **secondary numerators** and a, b, and 1 the **secondary denominators.**

Before we simplify this fraction, let us consider the complex fraction $\frac{\frac{3}{4}}{\frac{5}{6}}$. Now, $\frac{3}{4}$ is the primary numerator and $\frac{5}{6}$ is the primary denominator, while 4 and 6 are the secondary denominators. The complex fraction represents an indicated division, so we divide the primary numerator by the primary denominator. Thus,

$$\frac{\frac{3}{4}}{\frac{5}{6}} = \frac{3}{4} \div \frac{5}{6} = \frac{3}{4} \cdot \frac{6}{5} = \frac{18}{20} = \frac{9}{10}$$

Using this method, let us simplify the complex fraction

$$\frac{\frac{3}{a} - 1}{1 + \frac{4}{b}}$$

We first change the primary numerator, $\frac{3}{a} - 1$, and the primary denominator, $1 + \frac{4}{b}$, to single fractions. Thus,

$$\frac{3}{a} - 1 = \frac{3}{a} - \frac{a}{a} = \frac{3 - a}{a}$$

$$1 + \frac{4}{b} = \frac{b}{b} + \frac{4}{b} = \frac{b + 4}{b}$$

and we have

$$\frac{\frac{3}{a} - 1}{1 + \frac{4}{b}} = \frac{\frac{3 - a}{a}}{\frac{b + 4}{b}} = \frac{3 - a}{a} \div \frac{b + 4}{b}$$

$$= \frac{3 - a}{a} \cdot \frac{b}{b + 4}$$

$$= \frac{b(3 - a)}{a(b + 4)}$$

note

It is wrong to invert *before* combining the primary numerator and primary denominator into single fractions.

An alternate method for changing a complex fraction to a simple fraction—one that is often easier and more quickly performed—involves finding the least common multiple (LCM) of the secondary denominators. After we find it, we multiply the primary numerator and denominator by this LCM and reduce the resulting simple fraction to lowest terms.

note

The LCM is the same number as the LCD used in our previous work.

Examples

Reduce the complex fractions to simple fractions.

1. $\dfrac{\frac{4}{5}}{\frac{7}{5}}$

   The LCM of the secondary denominators is 5. Multiply the primary numerator and primary denominator by 5 to get

   $$\frac{\frac{4}{5}}{\frac{7}{5}} = \frac{\frac{4}{5} \cdot 5}{\frac{7}{5} \cdot 5} = \frac{4}{7}$$

2. $\dfrac{\frac{3}{4}}{\frac{5}{6}}$

   The LCM of the secondary denominators, 4 and 6, is 12. Multiply the primary numerator and primary denominator by 12:

   $$\frac{\frac{3}{4}}{\frac{5}{6}} = \frac{\frac{3}{4} \cdot 12}{\frac{5}{6} \cdot 12} = \frac{3 \cdot 3}{5 \cdot 2} = \frac{9}{10}$$

   (as we determined previously)

3. $\dfrac{\dfrac{5a}{6b}}{\dfrac{10a}{27b}}$

The LCM of the secondary denominators, $6b$ and $27b$, is $54b$. Multiply the primary numerator and primary denominator by $54b$:

$$\frac{\dfrac{5a}{6b} \cdot 54b}{\dfrac{10a}{27b} \cdot 54b} = \frac{5a \cdot 9}{10a \cdot 2} = \frac{45a}{20a} = \frac{9}{4}$$

4. $\dfrac{\dfrac{1}{2} + \dfrac{1}{4}}{1 - \dfrac{1}{8}}$

The LCM of the secondary denominators, 2, 4, and 8, is 8. Multiply the primary numerator and primary denominator by 8 to get

$$\frac{\dfrac{1}{2} + \dfrac{1}{4}}{1 - \dfrac{1}{8}} = \frac{\left(\dfrac{1}{2} + \dfrac{1}{4}\right) \cdot 8}{\left(1 - \dfrac{1}{8}\right) \cdot 8}$$

$$= \frac{\dfrac{1}{2} \cdot 8 + \dfrac{1}{4} \cdot 8}{1 \cdot 8 - \dfrac{1}{8} \cdot 8}$$

$$= \frac{4 + 2}{8 - 1} = \frac{6}{7}$$

5. $\dfrac{\dfrac{3}{a} - 1}{1 + \dfrac{4}{b}}$

The LCM of the secondary denominators, a and b, is ab. Multiply the primary numerator and primary denominator by ab to get

$$\frac{\dfrac{3}{a} - 1}{1 + \dfrac{4}{b}} = \frac{\left(\dfrac{3}{a} - 1\right) \cdot ab}{\left(1 + \dfrac{4}{b}\right) \cdot ab}$$

$$= \frac{\dfrac{3}{a} \cdot ab - 1 \cdot ab}{1 \cdot ab + \dfrac{4}{b} \cdot ab}$$

$$= \frac{3b - ab}{ab + 4a}$$

$$= \frac{b(3 - a)}{a(b + 4)}$$

(as we previously determined)

6. $\dfrac{\dfrac{4}{x} - \dfrac{3}{y}}{\dfrac{5}{x} + \dfrac{7}{y}}$

The LCM of the secondary denominators, x and y, is xy. Multiply the primary numerator, $\frac{4}{x} - \frac{3}{y}$, and the primary denominator, $\frac{5}{x} + \frac{7}{y}$, by xy to get

$$\frac{\dfrac{4}{x} - \dfrac{3}{y}}{\dfrac{5}{x} + \dfrac{7}{y}} = \frac{\left(\dfrac{4}{x} - \dfrac{3}{y}\right) \cdot xy}{\left(\dfrac{5}{x} + \dfrac{7}{y}\right) \cdot xy} = \frac{\dfrac{4}{x} \cdot xy - \dfrac{3}{y} \cdot xy}{\dfrac{5}{x} \cdot xy + \dfrac{7}{y} \cdot xy} = \frac{4y - 3x}{5y + 7x}$$

7. $\dfrac{\dfrac{a^2 - x^2}{x}}{\dfrac{a + x}{x^2}}$

The LCM of the secondary denominators, x and x^2, is x^2. Multiply the primary numerator, $\frac{a^2 - x^2}{x}$, and the primary denominator, $\frac{a + x}{x^2}$, by x^2. So,

$$\frac{\dfrac{a^2 - x^2}{x}}{\dfrac{a + x}{x^2}} = \frac{\left(\dfrac{a^2 - x^2}{x}\right) \cdot x^2}{\left(\dfrac{a + x}{x^2}\right) \cdot x^2}$$

$$= \frac{(a^2 - x^2) \cdot x}{a + x}$$

$$= \frac{(a + x)(a - x) \cdot x}{a + x}$$

$$= (a - x) \cdot x$$

note

In example 7 note that we reduced *before* performing the multiplication in the numerator. *Always look to do this.*

Exercise Set 9.7

Simplify the given complex fractions.

1. $\dfrac{\frac{2}{3}}{\frac{4}{5}}$

2. $\dfrac{\frac{7}{8}}{\frac{5}{6}}$

3. $\dfrac{\frac{4}{3}}{\frac{8}{9}}$

4. $\dfrac{\frac{9}{10}}{\frac{7}{6}}$

•5. $\dfrac{1 + \frac{3}{5}}{2 - \frac{1}{5}}$

6. $\dfrac{5 - \frac{3}{4}}{1 + \frac{5}{8}}$

7. $\dfrac{7}{2 + \frac{4}{5}}$

8. $\dfrac{10}{4 - \frac{11}{12}}$

•9. $\dfrac{4 + \frac{3}{5}}{6}$

10. $\dfrac{10 - \frac{7}{8}}{3}$

11. $\dfrac{\frac{3}{4} + \frac{5}{8}}{\frac{1}{2} - \frac{1}{4}}$

12. $\dfrac{\frac{6}{7} - \frac{5}{14}}{\frac{3}{14} - \frac{5}{7}}$

13. $\dfrac{x + \frac{1}{4}}{x - \frac{3}{4}}$

14. $\dfrac{y - \frac{5}{6}}{y + \frac{1}{2}}$

15. $\dfrac{\frac{1}{a} + 3}{\frac{2}{a} - 4}$

16. $\dfrac{5 - \frac{3}{b}}{6 + \frac{5}{b}}$

17. $\dfrac{\frac{3}{a^2} + 4}{5 - \frac{3}{a}}$

18. $\dfrac{\frac{5}{x} - 5}{6 + \frac{4}{x^3}}$

19. $\dfrac{a - \frac{3}{b}}{a + \frac{4}{b}}$

20. $\dfrac{x + \frac{4}{y}}{x - \frac{5}{y}}$

•21. $\dfrac{\frac{1}{x} + \frac{1}{y}}{\frac{1}{x} - \frac{1}{y}}$

22. $\dfrac{\frac{3}{x^2} - \frac{4}{y}}{\frac{5}{x} + \frac{2}{y^2}}$

23. $\dfrac{x + y}{\frac{1}{x} + \frac{1}{y}}$

24. $\dfrac{a - b}{\frac{2}{a} + \frac{3}{b}}$

25. $\dfrac{\frac{4}{a^2} + \frac{5}{b}}{a - b}$

26. $\dfrac{\frac{6}{x^2} - \frac{5}{y^2}}{x + y}$

27. $\dfrac{\frac{1}{x + y} - \frac{1}{x - y}}{\frac{1}{x + y} + \frac{1}{x - y}}$

28. $\dfrac{\frac{a}{a + b} - \frac{b}{a - b}}{\frac{b}{a + b} + \frac{a}{a - b}}$

•29. $\dfrac{\frac{x + y}{x - y} - \frac{x - y}{x + y}}{x^2 - y^2}$

30. $\dfrac{\dfrac{b}{a+b} - \dfrac{a}{a-b}}{a^2 - b^2}$

31. $\dfrac{\dfrac{2}{ab} + \dfrac{3}{ab^2}}{a^2b^2}$

32. $\dfrac{\dfrac{5}{x^2y^2} - \dfrac{4}{xy}}{xy}$

33. $\dfrac{\dfrac{2}{a+3} + \dfrac{1}{a-2}}{\dfrac{3}{a-2} - \dfrac{4}{a+3}}$

34. $\dfrac{\dfrac{7}{x-4} - \dfrac{5}{x+3}}{\dfrac{5}{x+3} + \dfrac{9}{x-4}}$

35. $\dfrac{\dfrac{5}{x^2 - x - 12}}{\dfrac{4}{x+3} - \dfrac{5}{x-4}}$

36. $\dfrac{\dfrac{7}{b-7} + \dfrac{8}{b-5}}{\dfrac{6}{b^2 - 12b + 35}}$

37. $\dfrac{3 - \dfrac{4}{a+4}}{\dfrac{5}{a+4} - \dfrac{6}{a-1}}$

38. $\dfrac{\dfrac{6}{x-5} + 7}{\dfrac{8}{x-5} - \dfrac{9}{x+3}}$

•39. $\dfrac{\dfrac{y^2 - y - 6}{y^2 + 2y + 1}}{\dfrac{y^2 - 2y - 8}{y^2 + 7y + 6}}$

40. $\dfrac{\dfrac{a^2 + 3a - 10}{a^2 - 5a - 14}}{\dfrac{a^2 + 6a + 5}{a^2 - 8a + 7}}$

41. A refrigeration coefficient of performance formula for the ideal refrigerator is given by

$$c = \frac{1}{\dfrac{T_2}{T_1} - 1}$$

Simplify the right side.

42. In electronics, a formula for coupled inductance in parallel with fields-aiding is given by

$$L_t = \frac{1}{\dfrac{1}{L_1 + M} + \dfrac{1}{L_2 + M}}$$

Simplify the right side.

43. In electronics, a formula for self-inductance of circuits in parallel is given by

$$L_t = \frac{1}{\dfrac{1}{L_1} + \dfrac{1}{L_2} + \dfrac{1}{L_3}}$$

Simplify the right side.

44. Coupled inductance with circuits connected in parallel with opposing fields is given by

$$L_t = \frac{1}{\dfrac{1}{L_1 - M} + \dfrac{1}{L_2 - M}}$$

Simplify the right side.

45. The change in frequency of a sound caused by relative motion between source and listener is called the Doppler effect. The formula for the Doppler effect in sound is given by

$$f = f_s \sqrt{\frac{1 + \frac{v}{c}}{1 - \frac{v}{c}}}$$

where f is the observed frequency, f_s is the frequency of the source, and v is the relative velocity between source and observer. Simplify the expression under the radical sign.

Chapter Summary

1. The **fundamental principle of fractions** is used to *reduce* algebraic fractions *to lowest terms* and to build to higher terms for addition and subtraction.
2. To reduce an algebraic fraction to lowest terms we divide the numerator and denominator by any common factors.
3. To multiply two algebraic fractions we multiply the numerators and place the product over the product of the denominators.
4. To divide two algebraic fractions we *invert the divisor* and proceed as in multiplication of fractions.
5. To find the **least common denominator** (LCD) of two or more algebraic fractions we do the following:
   a. Write each denominator in completely factored form.
   b. Take each different factor that appears in the factorizations and write them as a product.
   c. Obtain the LCD by raising each factor to the greatest power it has in step a.
6. To obtain an **equivalent fraction** we multiply or divide the numerator and denominator of the given fraction by a *common factor*.
7. To add or subtract algebraic fractions we must:
   a. State each fraction as an equivalent fraction with the same denominator (LCD).
   b. Add or subtract numerators and place this sum or difference over the common denominator.
8. We can add or subtract *two* algebraic fractions by using the property:

$$\frac{a}{b} \pm \frac{c}{d} = \frac{ad \pm bc}{bd}$$

9. A **fractional algebraic equation** is an equation in which one or more of the terms is a fraction.
10. To *solve a fractional equation* we do the following:
    a. Multiply each term of the equation by the least common denominator (LCD) to eliminate the fractions.
    b. Use the basic procedures to solve the resulting equation.
11. A **complex fraction** is a fraction whose numerator, denominator, or both, contains a fraction.

12. In the *complex fraction*

$$\frac{\frac{a}{b}}{\frac{c}{d}}$$

$\frac{a}{b}$ is the *primary numerator*, $\frac{c}{d}$ is the *primary denominator*, and b and d are the *secondary denominators*.

13. To *simplify a complex fraction* we do one of the following:
    a. Divide the primary numerator by the primary denominator after making a single fraction of each.
    b. Multiply the primary numerator and the primary denominator by the LCD of the secondary denominators.

Chapter Review

[9.2] Reduce the following algebraic fractions to lowest terms.

1. $\frac{18ab^2}{6a^2b}$
2. $\frac{45x^2yz^3}{30xy^3z^2}$
3. $\frac{x^2 - 49}{x^2 + 14x + 49}$
4. $\frac{x^2 - 3x - 18}{x^2 + x - 42}$
5. $\frac{18a - 6b}{15a - 5b}$
6. $\frac{x^2 - y^2}{y - x}$
7. $\frac{3p^2 - 8p + 4}{5p^2 - 9p - 2}$
8. $\frac{2R^2 - 32}{6R^2 + 22R - 8}$
9. $\frac{20 - 9n + n^2}{8 + 2n - n^2}$

[9.3] Perform the indicated multiplication and reduce the product to lowest terms.

10. $\frac{15}{8a} \cdot \frac{12a}{5}$
11. $\frac{24b}{7a} \cdot \frac{21a^2}{8b^2}$
12. $\frac{x - 3y}{2x + y} \cdot \frac{2x - y}{x - 3y}$
13. $\frac{5x - 10}{x + 3} \cdot \frac{3x + 9}{15}$
14. $\frac{m^2 - n^2}{14x} \cdot \frac{35x^2}{5m + 5n}$
15. $\frac{y}{y^2 - 1} \cdot \frac{y + 1}{y^2 - y}$
16. $\frac{b^2 - 36}{b^2 + 12b + 36} \cdot \frac{2b + 12}{b - 6}$
17. $\frac{5 - x}{6a - 3} \cdot \frac{a - 1}{x^2 - 25}$
18. $\frac{x^2 + x - 2}{x^2 - 4x - 12} \cdot \frac{x^2 - 7x + 6}{x^2 - 1}$

[9.3] Find the indicated quotients and state the answer reduced to lowest terms.

19. $\frac{14a}{9} \div \frac{7}{3}$
20. $\frac{10}{9b} \div \frac{35}{12b^2}$
21. $\frac{24ab}{7} \div \frac{16a^2b^2}{21}$
22. $\frac{2x - 1}{3x + 4} \div \frac{3x}{7x}$

23. $\dfrac{x + 6}{x^2 - 4} \div \dfrac{(x + 6)^2}{x + 2}$

24. $\dfrac{4a - 8}{2a + 1} \div \dfrac{6a - 12}{10a + 5}$

25. $\dfrac{x^2 + 16x + 64}{x^2 + 9x + 8} \div \dfrac{x^2 - 64}{x + 1}$

26. $\dfrac{1 - b}{(b + 3)^2} \div \dfrac{b^2 - 2b + 1}{b^2 - 9}$

27. $\dfrac{9a^2 + 15a + 6}{a^2 + 3a - 4} \div \dfrac{36a^2 - 16}{3a^2 + 10a - 8}$

[9.4] Find the least common denominator of the given algebraic fractions and convert them to equivalent fractions having the common denominators.

28. $\dfrac{3}{9x^2}; \dfrac{5}{12x}$

29. $\dfrac{5}{14ab}; \dfrac{8}{21a^2b^2}$

30. $\dfrac{x}{x^2 - 9}; \dfrac{2x}{2x + 6}$

31. $\dfrac{4y}{y^2 - 2y - 15}; \dfrac{9y}{y^2 - 25}$

[9.5] Add or subtract as indicated and reduce your answer to lowest terms.

32. $\dfrac{4x}{15} + \dfrac{8x}{21}$

33. $\dfrac{25}{16a} - \dfrac{13}{12a}$

34. $\dfrac{5}{3x + 1} + \dfrac{9}{4x - 3}$

35. $\dfrac{4x}{x + 4} - \dfrac{7x}{x^2 - 16}$

36. $\dfrac{4}{ab^2} + \dfrac{12}{5a^2b} - \dfrac{3}{4ab}$

37. $4a + \dfrac{6}{a + 1}$

38. $\dfrac{7}{x^2 + 1} - 10$

39. $\dfrac{x + 1}{4x} - \dfrac{3x - 5}{8x^2} + \dfrac{5x + 4}{12x}$

40. $\dfrac{4y}{y^2 - 7y - 18} + \dfrac{9y}{y^2 - 4}$

41. $\dfrac{5x}{x + 3} + \dfrac{2}{x^2 - 9} - \dfrac{7x}{x - 3}$

42. $\dfrac{9}{x} - \dfrac{4}{x - 5} + \dfrac{3}{x + 4}$

43. $\dfrac{5}{x^2 + 2xy + y^2} + \dfrac{5}{x^2 - xy - 2y^2}$

[9.6] Solve the following equations for the variable involved.

44. $\dfrac{x}{8} - 3 = \dfrac{2x}{12} + 1$

45. $\dfrac{4a + 1}{4} + \dfrac{5}{6} = 3a - \dfrac{5}{8}$

46. $\dfrac{12}{4a} - \dfrac{5}{6a} = 4$

47. $\dfrac{7}{a + 1} - \dfrac{1}{a^2 - 1} = \dfrac{2}{a + 1}$

48. $\dfrac{1}{y + 3} - \dfrac{6}{y} = \dfrac{7}{y} + \dfrac{2}{y + 3}$

49. $\dfrac{7}{2c} + \dfrac{5}{6c^2} = \dfrac{7}{12c^2}$

[9.6] Solve for the indicated variable.

50. $\dfrac{a}{x} + \dfrac{b}{x} = 3$ for x

51. $\dfrac{y}{x + 1} = \dfrac{3y}{x - 3}$ for x

52. $\dfrac{3}{4 - y} = \dfrac{a}{b}$ for y

53. $\dfrac{4}{y - c} - \dfrac{5}{y + b} = 0$ for y

54. An equation for tensile and compressional stresses is given by $y = \frac{Fl}{Ae}$. Solve for l.

55. The efficiency of a screw jack is calculated by the formula $E = \frac{Wp}{2+LF}$. Solve for L.

56. A formula for the theoretical mechanical advantage (TMA) of a chain fall is given by $TMA = \frac{2R}{R - r}$. Solve for R.

57. The total reaction force F of air against the plane at the bottom of a vertical loop in centripetal force is given by $F = \frac{mv^2}{r} + mg$. Solve for m.

[9.7] Simplify the given complex fractions and reduce to lowest terms.

58. $\dfrac{\frac{4}{7}}{\frac{9}{4}}$

59. $\dfrac{\frac{3}{4} - \frac{1}{2}}{\frac{1}{4} + \frac{3}{2}}$

60. $\dfrac{\frac{4}{5} + 1}{2 - \frac{1}{5}}$

61. $\dfrac{\frac{1}{x} - \frac{1}{y}}{\frac{1}{x} + \frac{1}{y}}$

62. $\dfrac{\frac{4}{x^2} + \frac{3}{x}}{\frac{2}{x^2} - \frac{5}{x}}$

63. $\dfrac{\frac{a}{b} - \frac{b}{a}}{\frac{a}{b} + \frac{b}{a}}$

64. $\dfrac{\frac{1}{x} - \frac{1}{y}}{\frac{1}{xy}}$

65. $\dfrac{\frac{1}{x} - \frac{1}{y}}{\frac{1}{x^2} - \frac{1}{y^2}}$

66. $\dfrac{\frac{a^2}{b^2} - 2a - 3}{\frac{a - 3b}{ab}}$

Chapter Test

Reduce the following to lowest terms.

1. $\frac{36}{42}$

2. $\frac{35ab^2}{28a^2b}$

3. $\frac{x^2 - 25}{x^2 + 3x - 10}$

4. $\frac{6a - 3b}{3a^2 + 3b^2}$

5. $\frac{x^2 - x - 20}{x^2 + x - 12}$

6. $\frac{2a^2 - 5ab - 3b^2}{6a^2 + 7ab + 2b^2}$

Perform the indicated operations and reduce your answer to lowest terms.

7. $\frac{3a}{4b} \cdot \frac{a}{b}$

8. $\frac{x + 1}{x - 3} \cdot \frac{x - 2}{x + 3}$

9. $\frac{a^2 - 1}{a + 2} \cdot \frac{a - 1}{a + 1}$

10. $\frac{x^2 - 7x + 6}{x^2 - 1} \cdot \frac{x + 1}{x + 6}$

11. $\frac{2a}{3b} \div \frac{2a^2}{21b^2}$

12. $\frac{x^2 - y^2}{4} \div \frac{x^2 + xy}{4x - 4}$

13. $\frac{2y^2 - 5y - 3}{2y^2 + y} \div \frac{y - 3}{y^4}$

14. $\dfrac{x+3}{y} - \dfrac{3}{y}$

15. $\dfrac{3}{x} + \dfrac{2}{3x}$

16. $\dfrac{3}{a} + \dfrac{4}{2b}$

17. $\dfrac{2}{xy^2} - \dfrac{3}{x^2y}$

18. $\dfrac{3}{a-b} + \dfrac{1}{a+b}$

19. $\dfrac{a}{a^2-1} - \dfrac{1}{a^2+a}$

20. $\dfrac{3}{y^2-25} + \dfrac{4}{y^2-4y-5}$

21. $\dfrac{3x}{x^2+x-12} - \dfrac{2x}{x^2-x-20}$

Simplify the complex fraction.

22. $\dfrac{\frac{5}{6}}{\frac{2}{9}}$

23. $\dfrac{\frac{2}{3}+\frac{5}{6}}{\frac{1}{3}-\frac{1}{6}}$

24. $\dfrac{1+\frac{1}{2}}{3-\frac{3}{4}}$

25. $\dfrac{2-\frac{a}{b}}{1+\frac{3}{b}}$

26. $\dfrac{x-\frac{x}{y}}{y-\frac{y}{x}}$

27. $\dfrac{\frac{4}{y}+3}{2-\frac{1}{y}}$

Solve the equation for the unknown.

28. $\dfrac{x}{2} = 3 + \dfrac{2x}{3}$

29. $\dfrac{y+1}{2} = \dfrac{2y-9}{5} + 3$

30. $\dfrac{a-2}{2a} = \dfrac{3}{2}$

31. $\dfrac{x}{3} + \dfrac{5}{9} = \dfrac{1}{3}$

32. $\dfrac{3}{x} = \dfrac{4}{x+3}$

33. $\dfrac{5}{x+10} - \dfrac{5}{x} = \dfrac{-3}{x}$

Solve for the given variable in terms of the other variables.

34. $\dfrac{1}{x} + 3 = \dfrac{2}{y} - 4$ for x

35. $\dfrac{1}{f} = \dfrac{1}{p} + \dfrac{1}{q}$ for p

36. $\dfrac{1}{R} = \dfrac{1}{R_1} + \dfrac{1}{R_2} + \dfrac{1}{R_3}$ for R_2

37. Acceleration a is given by the change in velocity per unit of time. Thus, if the initial velocity is v_1 and the final velocity is v_2, then $a = \dfrac{v_2 - v_1}{t}$. Solve for the final velocity, v_2.

38. The pressure p in pounds per square inch under a liquid surface is given by $p = \dfrac{hD}{144}$, where h = the density of the liquid in pounds per cubic foot and D = depth of the liquid in feet. Solve for D.

39. In refrigeration, the ideal coefficient of performance, n, is given by $n = \dfrac{Q_2}{Q_1 - Q_2}$, where Q_1 is the exhausted heat and Q_2 is the extracted heat. Solve for Q_2.

40. The relation between the distance of the object, D_o, the distance of the image, D_i, and the focal length, f, for any spherical mirror, is given by $\dfrac{1}{D_o} + \dfrac{1}{D_i} = \dfrac{1}{f}$. Solve for D_o.

10
Exponents, Roots, and Radicals

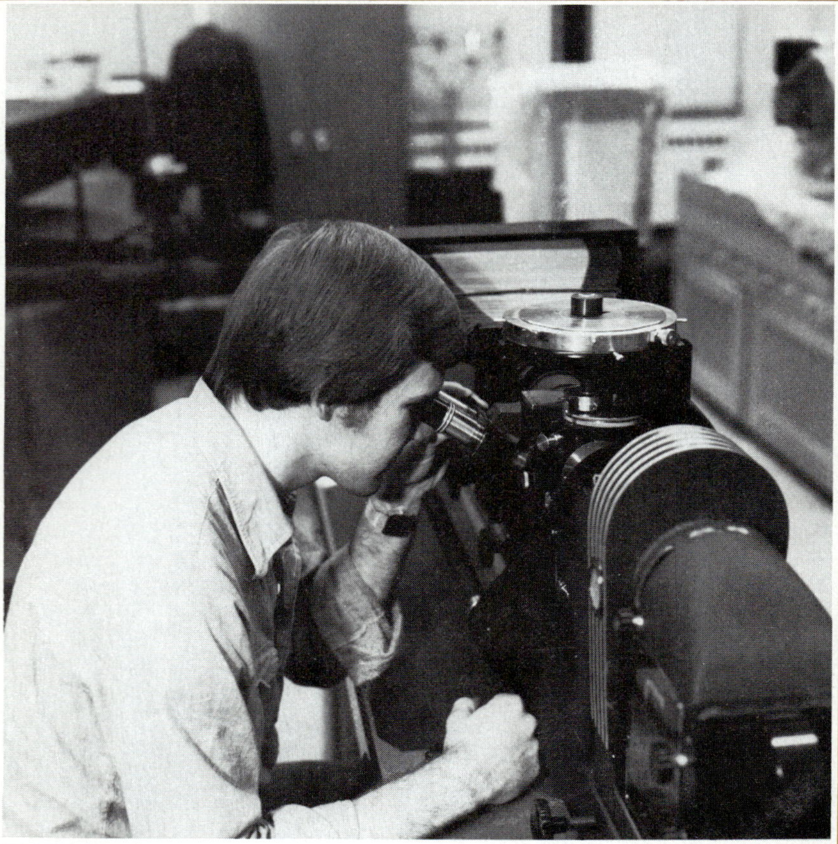

A technician using a metallograph to study and photograph microstructures from 50x to 2000x magnification.
(Photograph by Kenneth Walters, courtesy of Henry Ford Community College)

10.1 Integer Exponents

In section 6.4 we discussed some of the basic laws of exponents where the exponents were all positive integers. In this section, we are going to extend those same ideas to exponents that are negative integers and zero. First, we shall review the laws from chapter 6.

1. $a^m \cdot a^n = a^{m+n}$ To multiply factors having *like bases,* add the exponents to get the exponent of the given base in the product.
2. $a^m \div a^n = a^{m-n}$ To divide quantities having *like bases,* subtract the exponent of the denominator from the exponent of the numerator to get the exponent of the given base in the quotient.
3. $(ab)^n = a^n b^n$ When a group of factors is raised to a power, raise each of the factors in the group to that power.
4. $(a^m)^n = a^{m \cdot n}$ When raising a number to a power and then raising that to another power, we multiply the exponents to get the power of the given base.
5. $\left(\dfrac{a}{b}\right)^n = \dfrac{a^n}{b^n}$ When a fraction is raised to a power, raise the numerator and denominator to that power.

Examples

1. $x^4 \cdot x^3 = x^{4+3} = x^7$
2. $\dfrac{x^5}{x^3} = x^{5-3} = x^2$
3. $(xy)^3 = x^3y^3$
4. $(x^4)^3 = x^{4 \cdot 3} = x^{12}$
5. $x^a \cdot x^{3a} = x^{a+3a} = x^{4a}$
6. $(2a^2b^3)^3 = 2^3(a^2)^3(b^3)^3 = 8a^6b^9$
7. $\dfrac{R^{3n}}{R^n} = R^{3n-n} = R^{2n}$
8. $\left(\dfrac{R}{S}\right)^4 = \dfrac{R^4}{S^4}$
9. $\left(\dfrac{y^2}{z}\right)^5 = \dfrac{(y^2)^5}{z^5} = \dfrac{y^{10}}{z^5}$

Consider the following example: $a^0 \cdot a^m = a^{0+m} = a^m$. We observe that when a^0 is a factor, our answer is the other factor. For this result to occur, a^0 must be equal to 1: $a^0 \cdot a^m = 1 \cdot a^m = a^m$.

The fact that $a^0 = 1$ can also be seen when a number is divided by itself. By the definition of division, $\dfrac{a^m}{a^m} = 1$ $(a \neq 0)$, and by our second law of exponents, $\dfrac{a^m}{a^m} = a^{m-m} = a^0$. Both must yield the same result, hence $a^0 = 1$ $(a \neq 0)$.

note

We shall leave 0^0 undefined in this text.

Examples

1. $x^0 = 1$
2. $(x^2 + y^2)^0 = 1$
3. $10^0 = 1$
4. $(-5)^0 = 1$
5. $\left(\frac{3}{4}\right)^0 = 1$
6. $2x^0 = 2 \cdot 1 = 2$
7. $3a^0b = 3 \cdot 1 \cdot b = 3b$

note

The exponent 0 is associated only with the x in example 6 and only with the a in example 7. That is, only with the base directly to the left of it. In example 2 the quantity $(x^2 + y^2)$ is the base raised to the 0 power.

If we examine the first law of exponents with respect to negative exponents, we observe that $a^m \cdot a^{-m} = a^{m+(-m)} = a^0 = 1$. For this result to be consistent with our rule of reciprocals, $a^m \cdot \frac{1}{a^m} = 1$, then a^{-m} must be equal to $\frac{1}{a^m}$. That is,

$$a^{-m} = \frac{1}{a^m} \; (a \neq 0).$$

In words: *A number with a negative exponent is equivalent to the reciprocal of the same number with the exponent made positive.*

Examples

1. $x^{-4} = \frac{1}{x^4}$
2. $(ab)^{-7} = \frac{1}{(ab)^7} = \frac{1}{a^7b^7}$
3. $2x^{-3} = 2 \cdot \frac{1}{x^3} = \frac{2}{x^3}$

note

The exponent -3 is associated only with the x, that is, only with the base directly to the left of it.

4. $-a^{-3} = \frac{1}{-a^3}$

note

Only the sign of the exponent changes, not the sign of the number.

5. $\frac{1}{y^{-4}} = \frac{1}{\frac{1}{y^4}} = 1 \cdot \frac{y^4}{1} = y^4$

6. $\dfrac{a^{-3}}{b^{-2}} = \dfrac{\dfrac{1}{a^3}}{\dfrac{1}{b^2}} = \dfrac{1}{a^3} \cdot \dfrac{b^2}{1} = \dfrac{b^2}{a^3}$

7. $\dfrac{a^{-4}}{b^2} = \dfrac{\dfrac{1}{a^4}}{\dfrac{b^2}{1}} = \dfrac{1}{a^4} \cdot \dfrac{1}{b^2} = \dfrac{1}{a^4b^2}$

8. $\dfrac{2x^{-2}}{R^{-2}} = \dfrac{2 \cdot \dfrac{1}{x^2}}{\dfrac{1}{R^2}} = \dfrac{\dfrac{2}{x^2}}{\dfrac{1}{R^2}} = \dfrac{2}{x^2} \cdot \dfrac{R^2}{1} = \dfrac{2R^2}{x^2}$

9. $\dfrac{R^3}{R^0} = R^{3-0} = R^3$

10. $S^5 \cdot S^0 = S^{5+0} = S^5$

11. $(a^3)^0 = a^{3\cdot 0} = a^0 = 1$

12. $(B^{-2})^{-3} = B^{(-2)(-3)} = B^6$

13. $(2a^3b^{-2})^{-3} = 2^{-3}(a^3)^{-3}(b^{-2})^{-3}$

$= 2^{-3}a^{(3)(-3)}b^{(-2)(-3)}$

$= 2^{-3}a^{-9}b^6$

$= \dfrac{1}{2^3} \cdot \dfrac{1}{a^9} \cdot b^6$

$= \dfrac{b^6}{8a^9}$

14. $\dfrac{c^{-4}}{c^{-7}} = c^{(-4)-(-7)} = c^{(-4)+(7)} = c^3$

15. $\dfrac{a^2b^{-4}c^0}{a^{-5}b^2c^{-3}} = a^{2-(-5)}b^{(-4)-2}c^{0-(-3)}$

$= a^7b^{-6}c^3$

$= a^7 \cdot \dfrac{1}{b^6} \cdot c^3$

$= \dfrac{a^7c^3}{b^6}$

16. $\left(\dfrac{3a^{-2}}{9a^{-5}}\right)^{-2} = \left(\dfrac{3}{3^2} \cdot a^{(-2)-(-5)}\right)^{-2}$

$= (3^{1-2}a^{(-2)+(5)})^{-2}$

$= (3^{-1}a^3)^{-2} = (3^{-1})^{-2}(a^3)^{-2}$

$= 3^{(-1)\cdot(-2)} \cdot a^{3(-2)} = 3^2 \cdot a^{-6}$

$= 3^2 \cdot \dfrac{1}{a^6} = \dfrac{9}{a^6}$

note

Except in certain notations such as scientific notation, engineering notation, and so on, we do not leave zero or negative exponents in our answers.

Exercise Set 10.1

Write each of the given expressions in simplest form with only positive exponents.

1. x^0	2. $(2y)^0$	•3. $5a^0$
4. $7x^0$	•5. $(3B)^0$	6. S^{-2}
7. R^{-5}	8. $(2x)^{-3}$	•9. $(3P)^{-2}$
10. $4z^{-2}$	11. $9C^{-4}$	12. $\dfrac{1}{x^{-5}}$
•13. $\dfrac{5}{x^{-4}}$	14. $\dfrac{1}{2y^{-3}}$	15. $\dfrac{1}{3x^{-2}}$
16. 5^0	17. $(-12)^0$	18. $(2x^2 + y^3)^0$
19. $(x^2 - y^2)^0$	20. $(a^3)^0$	21. $(B^0)^5$
22. $\dfrac{1}{(2x - y)^0}$	23. $x^{-4}x^7$	24. $y^{-2}y^{10}$
25. a^5a^{-11}	26. $R^{-2}R^{-5}$	27. $(-5)^{-3}$
28. $\dfrac{3^{-2}}{3^{-5}}$	•29. $\dfrac{2^{-6}}{2^{-3}}$	30. $x^{-2}y^4$
31. a^2b^{-7}	32. $R^{-5}S^{-4}$	33. $P^0R^{-2}S^5$
34. $x^{-3}y^2z^{-4}$	35. $R^{-2}S^{-5}$	36. $2x^{-4}y^2$
37. $3a^{-5}b^{-4}$	38. $(2a^2)^{-2}$	39. $(3x^{-2})^3$
40. $(2a^2b^{-3})^{-2}$	•41. $(3xy^{-4})^{-3}$	42. $\dfrac{a^{-2}b^3}{a^3b^{-5}}$
43. $\dfrac{R^2S^{-4}}{R^{-3}S^5}$	44. $\dfrac{4y^{-3}}{4^{-1}y^2}$	45. $\dfrac{5^0R^2}{5^{-2}R^{-3}}$
46. $\dfrac{4^{-1}a^{-2}b^3c^0}{2a^{-3}b^{-1}c^{-2}}$	47. $\dfrac{9x^{-3}y^{-2}z}{3x^{-2}y^3z^{-3}}$	48. $\dfrac{3R^{-1}S^{-2}}{9R^{-3}S^2}$
•49. $\dfrac{8a^{-2}b^{-5}}{2a^{-1}b^4}$	50. $\dfrac{2x^{-1}y^{-2}}{3x^{-2}y^2}$	51. $\dfrac{6R^{-2}S^{-3}}{2R^2S^{-3}}$
52. $\dfrac{2a^{-1}b^0c^2}{5a^3b^{-1}c^{-3}}$	•53. $\left(\dfrac{6x^{-3}}{2x^{-1}}\right)^{-2}$	54. $\left(\dfrac{3R^{-2}}{2^{-1}R^2}\right)^{-3}$
55. $\dfrac{(RS^{-1})^{-3}}{RS^3}$	56. $\dfrac{(2a^2b^{-1})^{-2}}{(4a^{-3}b)^{-1}}$	
57. $\dfrac{(6a^3b^{-4})^{-2}}{(3a^2b^{-5})^{-1}}$	58. $\dfrac{3R^2(S^{-2}T)^{-2}}{6^{-1}R^{-4}(S^{-1}T)^0}$	

When we studied scientific notation, we observed that 10 raised to a negative exponent represented a fraction with a denominator of 10 to a power. Convert the following into fractional form.

Example $10^{-5} = \dfrac{1}{10^5} = \dfrac{1}{100{,}000}$

59. 10^{-1} 60. 10^{-3} 61. 10^{-4} 62. 10^{-2} 63. 10^{-6}

Convert the following fractions into 10 to a power.

Example $\dfrac{1}{100} = \dfrac{1}{10^2} = 10^{-2}$

64. $\dfrac{1}{1{,}000}$ 65. $\dfrac{1}{10}$ 66. $\dfrac{1}{1{,}000{,}000}$ •67. $\dfrac{1}{1{,}000{,}000{,}000}$ 68. $\dfrac{1}{10{,}000}$

10.2 Roots of a Number

In chapter 1 we discussed a square root as one of two equal factors of a number. That is: $\sqrt{a} = b$ *if* $b \cdot b = b^2 = a$, *and* b *is said to be the square root of* a. This concept can be extended to find cube roots (third roots of a number), fourth roots, fifth roots, and so on.

A cube root is one of three equal factors of a number. The symbol that is used to express the cube root is $\sqrt[3]{}$. The 3 is called the **index** of the radical expression, and it denotes what root we are looking for. The fourth root is indicated by $\sqrt[4]{}$, the fifth root by $\sqrt[5]{}$, and so on. General notation for the nth root is $\sqrt[n]{}$. Therefore, the nth root of a is written as $\sqrt[n]{a}$, where n is the index, $\sqrt{}$ is the **radical symbol,** and a is the **radicand.** The nth root of a number is one of n equal factors of the number such that $\sqrt[n]{a} = b$ if

$$\overbrace{b \cdot b \cdot b \cdots b}^{n} = b^n = a$$

Examples

1. $\sqrt[4]{x}$ is read "the fourth root of x."
2. $\sqrt[7]{a}$ is read "the seventh root of a."
3. $\sqrt{y}$ is read "the square root of y."

note

If there is no index associated with a radical symbol, it is understood to be 2.

4. $\sqrt[5]{32} = 2$ since $2 \cdot 2 \cdot 2 \cdot 2 \cdot 2 = 2^5 = 32$
5. $\sqrt[3]{27} = 3$ since $3 \cdot 3 \cdot 3 = 3^3 = 27$

If we recall our work with signed numbers, we observe that there are two different numbers that could be the square root of a given positive number. Consider the following example: $2^2 = 4$ and $(-2)^2 = 4$. Therefore, the square root of 4 could be either 2 or -2. To avoid the ambiguity of two different answers for the problem, we define *the principal square root of a positive number to be positive*. Hence, the principal square root of 4 is 2, that is, $\sqrt{4} = 2$.

Examples

1. $\sqrt{16} = 4$ since $4 \cdot 4 = 4^2 = 16$
2. $\sqrt{49} = 7$ since $7 \cdot 7 = 7^2 = 49$
3. $\sqrt{0.81} = 0.9$ since $(0.9)(0.9) = (0.9)^2 = 0.81$
4. $\sqrt{36} = 6$ since $6 \cdot 6 = 6^2 = 36$

note

In these examples, 16, 49, 0.81, and 36 are called **perfect squares** because their square root is a rational number.

When we want to express the negative value of a square root, we use the $-\sqrt{}$ symbol.

Examples

1. $-\sqrt{16} = -4$
2. $-\sqrt{49} = -7$
3. $-\sqrt{0.81} = -0.9$
4. $-\sqrt{36} = -6$

If we exclude even roots of negative numbers, which we discuss later in this chapter, we can extend the idea of principal root to all other roots: *The nth principal root of a number, denoted by $\sqrt[n]{}$ has the same sign as the number itself.*

Examples

1. $\sqrt[4]{16} = 2$ since $2 \cdot 2 \cdot 2 \cdot 2 = 2^4 = 16$
2. $\sqrt[3]{-27} = -3$ since $(-3)(-3)(-3) = (-3)^3 = -27$
3. $\sqrt[4]{81} = 3$ since $3 \cdot 3 \cdot 3 \cdot 3 = 3^4 = 81$
4. $\sqrt[5]{-32} = -2$ since $(-2)(-2)(-2)(-2)(-2) = (-2)^5 = -32$
5. $\sqrt[3]{-125} = -5$ since $(-5)(-5))(-5) = (-5)^3 = -125$

Now let us try to find an answer for the $\sqrt{2}$. We could use 1.414, but when we square 1.414 we do not get 2 as an answer. No matter how many places we carry our answer to, when we square that result, it will not equal 2. In other words, the $\sqrt{2}$ can never be exactly expressed as a decimal number. Such a number is called an **irrational number.** An irrational number is a nonzero number that cannot be expressed as the ratio of two integers. Another number that is irrational is π (pronounced pi), which is used in geometric formulas involving circles. A property of irrational numbers that we are interested in is the fact that we can never express them as a terminating decimal number.

When you are working with irrational numbers in a problem, the answer will have to be approximated to as many decimal places as are required in the problem.

Examples

Find an approximation for the following roots to the indicated number of decimal places.

1. $\sqrt{3}$ to 3 decimal places is 1.732.
2. $\sqrt{17}$ to 4 decimal places is 4.1231.
3. $\sqrt[3]{11}$ to 3 decimal places is 2.224.
4. $\sqrt[3]{-0.026}$ to 3 decimal places is -0.296.
5. $\sqrt[5]{-14}$ to 3 decimal places is -1.695.

Application

A technician wishing to determine the diameter of a cylinder bore (D) required to produce H horsepower from N cylinders of an engine turning 1,000 rpm uses the formula $D = \sqrt{\dfrac{H}{(0.4)N}}$. What will be the bore diameter required to produce 40 horsepower from a 4-cylinder engine?

Solution Substitute 40 for H and 4 for N:

$$D = \sqrt{\frac{(40)}{(0.4)(4)}} = \sqrt{\frac{40}{1.6}} = \sqrt{25} = 5.0$$

Therefore, a bore diameter of 5.0 in is required.

Exercise Set 10.2

Find the indicated root.

1. $\sqrt{100}$	2. $\sqrt{36}$	3. $\sqrt{25}$	•4. $\sqrt{64}$
5. $-\sqrt{144}$	6. $-\sqrt{81}$	•7. $-\sqrt{121}$	8. $-\sqrt{16}$
9. $\sqrt[3]{8}$	10. $\sqrt[3]{27}$	11. $\sqrt[3]{125}$	12. $\sqrt[3]{64}$
•13. $\sqrt[3]{-27}$	14. $\sqrt[3]{-125}$	15. $\sqrt[4]{16}$	16. $\sqrt[4]{81}$
17. $-\sqrt[4]{81}$	18. $-\sqrt[4]{625}$	19. $\sqrt[5]{32}$	20. $\sqrt[6]{64}$
21. $-\sqrt[5]{243}$	22. $\sqrt[10]{1}$	23. $\sqrt[14]{1}$	24. $\sqrt{0.04}$
•25. $\sqrt{1.44}$	26. $-\sqrt{1.69}$	27. $-\sqrt{0.64}$	28. $\sqrt[3]{0.008}$
29. $\sqrt[3]{0.027}$	30. $\sqrt[3]{-0.064}$	31. $\sqrt[3]{-0.125}$	32. $-\sqrt[3]{0.027}$
•33. $-\sqrt[3]{0.064}$	34. $-\sqrt[3]{-0.125}$	35. $-\sqrt[3]{-0.064}$	36. $\sqrt[9]{-1}$

Solve the following problems. Round to the accuracy of the given data.

37. The formula for finding the length of an edge (e) of a cube when the volume (v) is known is $e = \sqrt[3]{v}$. What is the length of the edge of a cube whose volume is 729 cm^3?
38. What is the length of the edge of a cube whose volume is 216 in^3? (Refer to problem 37.)
39. The current I (amperes) in a circuit is found by the formula:

$$\text{amps} = \sqrt{\frac{\text{watts}}{\text{ohms}}}.$$

What is the current of a circuit that has 3.00 ohms resistance and uses 1,728 watts?
40. What is the current of a circuit that has 2.00 ohms resistance and uses 512 watts? (Refer to problem 39.)
41. The slant height (S) of a right circular cone (shown in the figure on the left) is found by the formula:

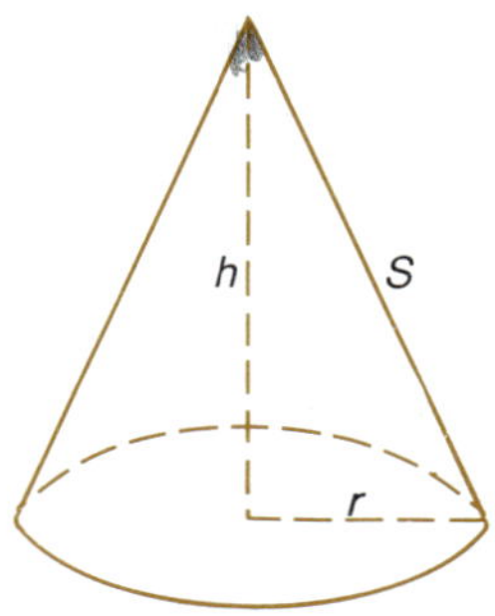

$$S = \sqrt{r^2 + h^2},$$

where r is the radius of the base and h is the height of the cone.
What is the slant height of a right circular cone whose base radius is 5.0 cm and height is 12.0 cm?
42. What is the slant height of a right circular cone whose base radius is 3.0 ft and height is 4.0 ft? (Refer to problem 41.)
•43. The impedance z is a measure in an AC circuit of how much the flow of current is retarded by all the circuit elements. The impedance is related to the total resistance (R) and reactance (x) by the following formula:

$$z = \sqrt{R^2 + x^2}.$$

Find the impedance (z ohms) of a circuit if the total resistance is $1{,}42\tilde{0}$ ohms and the reactance is 615 ohms.

Calculator Problems

Find the following roots using the rules of approximate numbers.

44. $\sqrt{6}$	45. $\sqrt{17}$	46. $\sqrt{101}$	47. $-\sqrt{1.4}$
48. $-\sqrt{2.5}$	•49. $\sqrt{0.00036}$	50. $\sqrt[3]{2}$	51. $\sqrt[3]{3}$
•52. $\sqrt[3]{-26}$	53. $\sqrt[3]{-16}$	54. $\sqrt[5]{31}$	

10.3 Multiplication Rule for Radicals

In this section we develop rules for simplifying radicals. Consider the following example:

$$\sqrt{4} \cdot \sqrt{25} = 2 \cdot 5 = 10$$

We also observe that

$$\sqrt{4 \cdot 25} = \sqrt{100} = 10$$

From this example, we conclude that $\sqrt{4} \cdot \sqrt{25} = \sqrt{4 \cdot 25}$. To generalize: *When we multiply two square roots, we multiply the radicands and put the product under a common square root symbol* ($\sqrt{a}\sqrt{b} = \sqrt{ab}$).

Examples

1. $\sqrt{3}\,\sqrt{5} = \sqrt{3 \cdot 5} = \sqrt{15}$
2. $\sqrt{6}\,\sqrt{7} = \sqrt{6 \cdot 7} = \sqrt{42}$
3. $\sqrt{3}\,\sqrt{a} = \sqrt{3a}$

In the previous section we excluded even roots of negative numbers from our discussion. We will continue to consider all variables to be representing positive real numbers whenever the index of the radical is even. The idea of a negative number under a radical symbol with an even index will be taken up later in this chapter.

Our multiplication rule for radicals can be extended to radicals with any index as follows: *When we multiply two radicals having a common index, we multiply the radicands and put the product under a common radical symbol with the original index.* $\sqrt[n]{a}\,\sqrt[n]{b} = \sqrt[n]{ab}$.

Examples

1. $\sqrt[3]{3}\,\sqrt[3]{2} = \sqrt[3]{3 \cdot 2} = \sqrt[3]{6}$
2. $\sqrt[5]{7} \cdot \sqrt[5]{9} = \sqrt[5]{7 \cdot 9} = \sqrt[5]{63}$
3. $\sqrt[4]{2}\,\sqrt[4]{a} = \sqrt[4]{2a}$
4. $\sqrt[3]{2}\,\sqrt[3]{x}\,\sqrt[3]{y} = \sqrt[3]{2xy}$
5. $\sqrt[3]{3}\,\sqrt[4]{5}$ cannot be multiplied together in this form since they do not have the same index.

An important use of the multiplication rule is in simplifying radicals. Consider the following example: $\sqrt{12}$ can be factored into $\sqrt{4 \cdot 3}$, and by our multiplication rule, we can write $\sqrt{12} = \sqrt{4 \cdot 3} = \sqrt{4}\,\sqrt{3} = 2\sqrt{3}$. We can simplify the radical because the radicand contains a perfect-square integer factor, 4. In the example, $2\sqrt{3}$ is called the *simplified form* of $\sqrt{12}$.

Examples

Simplify the following expressions.

1. $\sqrt{50} = \sqrt{25 \cdot 2} = \sqrt{25}\,\sqrt{2} = 5\sqrt{2}$
2. $\sqrt{28} = \sqrt{4 \cdot 7} = \sqrt{4}\,\sqrt{7} = 2\sqrt{7}$
3. $\sqrt{9a} = \sqrt{9 \cdot a} = \sqrt{9}\,\sqrt{a} = 3\sqrt{a}$
4. $\sqrt{x^2y} = \sqrt{x^2 \cdot y} = \sqrt{x^2}\,\sqrt{y} = x\sqrt{y}$
5. $\sqrt{a^3} = \sqrt{a^2 \cdot a} = \sqrt{a^2}\,\sqrt{a} = a\sqrt{a}$
6. $\sqrt{y^5} = \sqrt{y^2y^2y} = \sqrt{y^2}\,\sqrt{y^2}\,\sqrt{y} = y \cdot y \cdot \sqrt{y} = y^2\sqrt{y}$
7. $\sqrt{a^3b^4} = \sqrt{a^2 \cdot a \cdot b^2 \cdot b^2} = \sqrt{a^2}\,\sqrt{a}\,\sqrt{b^2}\,\sqrt{b^2} = a\sqrt{a}\,bb =$ $ab^2\sqrt{a}$

We can extend the procedure of simplifying square roots to radicals of indices other than 2.

Examples

Simplify the following radical expressions.

1. $\sqrt[3]{81} = \sqrt[3]{27 \cdot 3} = \sqrt[3]{27}\,\sqrt[3]{3} = 3\,\sqrt[3]{3}$

note

Since the index is 3, we can simplify the radical because the radicand has a factor that is a perfect-cube integer, $27 = 3^3$.

2. $\sqrt[4]{32} = \sqrt[4]{16 \cdot 2} = \sqrt[4]{16}\,\sqrt[4]{2} = 2\,\sqrt[4]{2}$

note

Since the index is 4, we can simplify the radical because the radicand has a factor that is a perfect fourth power integer, $16 = 2^4$.

3. $\sqrt[3]{x^5} = \sqrt[3]{x^3 \cdot x^2} = \sqrt[3]{x^3}\,\sqrt[3]{x^2} = x\,\sqrt[3]{x^2}$
4. $\sqrt[5]{y^{10}} = \sqrt[5]{y^5 \cdot y^5} = \sqrt[5]{y^5}\,\sqrt[5]{y^5} = y \cdot y = y^2$

note

In example 4, the power (10) was evenly divisible by the index (5), and the radical was eliminated. When the power of a factor is evenly divisible by the index, that factor will no longer remain under the radical symbol.

5. $\sqrt[4]{x^7y^4} = \sqrt[4]{x^4 \cdot x^3 \cdot y^4} = \sqrt[4]{x^4}\,\sqrt[4]{x^3}\,\sqrt[4]{y^4} = x\,\sqrt[4]{x^3}\,y = xy\,\sqrt[4]{x^3}$
6. $\sqrt[3]{a^7b^2} = \sqrt[3]{a^3a^3ab^2} = \sqrt[3]{a^3}\,\sqrt[3]{a^3}\,\sqrt[3]{ab^2} = a \cdot a\,\sqrt[3]{ab^2} = a^2\,\sqrt[3]{ab^2}$

note

No simplification relative to b is possible because the power of b is less than the value of the index.

7. $\sqrt[3]{54x^3y^5} = \sqrt[3]{27 \cdot 2 \cdot x^3 \cdot y^3 \cdot y^2} = \sqrt[3]{27}\,\sqrt[3]{x^3}\,\sqrt[3]{y^3}\,\sqrt[3]{2y^2}$
   $= 3xy\,\sqrt[3]{2y^2}$
8. $\sqrt{a^2 + b^2}$ Will not simplify because the radicand cannot be factored. *The radicand must always be in a factored form before we can simplify.*

note

$\sqrt{a^2 + b^2} \neq \sqrt{a^2} + \sqrt{b^2}$

Example 1. $\sqrt{9 + 16} \neq \sqrt{9} + \sqrt{16}$. The $\sqrt{9 + 16} = \sqrt{25}$ $= 5$, whereas $\sqrt{9} + \sqrt{16} = 3 + 4 = 7$.

2. $\sqrt{a^2 + 2ab + b^2} = \sqrt{(a + b)(a + b)}$
$= \sqrt{(a + b)^2}$
$= a + b$

A radical can be simplified if the radicand contains a perfect-nth-power factor, where n is the index of the radical. We can simplify a radical if the radicand has a factor (or factors) whose power is equal to or greater than the index.

Examples Perform the indicated multiplication and simplify.

1. $\sqrt{2}\sqrt{6} = \sqrt{2 \cdot 6} = \sqrt{12} = \sqrt{4 \cdot 3} = \sqrt{4}\sqrt{3} = 2\sqrt{3}$
2. $\sqrt[3]{a^2}\sqrt[3]{a} = \sqrt[3]{a^2 \cdot a} = \sqrt[3]{a^3} = a$
3. $\sqrt[4]{8a^2b^3}\sqrt[4]{4a^3b}$
$= \sqrt[4]{8a^2b^3 \cdot 4a^3b} = \sqrt[4]{32a^5b^4} = \sqrt[4]{16 \cdot 2 \cdot a^4 \cdot a \cdot b^4}$
$= \sqrt[4]{16}\sqrt[4]{a^4}\sqrt[4]{b^4}\sqrt[4]{2a} = 2ab\sqrt[4]{2a}$

note

A common error in problems involving radicals is forgetting to carry along the correct index for the radical symbol.

Exercise Set 10.3

Perform any indicated operations and express the given radicals in simplest form.

1. $\sqrt[3]{16}$
2. $\sqrt{125}$
•3. $\sqrt[3]{48}$
4. $\sqrt[4]{32}$
5. $\sqrt[5]{64}$
6. $\sqrt[3]{24}$
7. $\sqrt{63}$
8. $\sqrt{28}$
9. $\sqrt{75}$
10. $\sqrt[5]{a^7}$
11. $\sqrt[3]{b^8}$
12. $\sqrt[3]{x^9}$
13. $\sqrt[5]{y^{15}}$
14. $\sqrt{c^7}$
15. $\sqrt{D^5}$
16. $\sqrt[3]{a^{12}}$
17. $\sqrt[3]{b^{15}}$
18. $\sqrt{4a^2b^3}$
19. $\sqrt{9ab^4c^3}$
20. $\sqrt{27a^3b^5}$
•21. $\sqrt{24x^5yz^3}$
22. $\sqrt[3]{8r^2s^8}$
23. $\sqrt[3]{16L^4M^5}$
24. $\sqrt{18R^4}$
25. $\sqrt[4]{S^5}$
•26. $\sqrt[5]{64x^{10}y^{14}}$
27. $\sqrt[3]{81R^5S^{11}}$
28. $\sqrt{6}\sqrt{3}$
29. $\sqrt{27}\sqrt{6}$
30. $\sqrt{20}\sqrt{40}$
31. $\sqrt{15}\sqrt{15}$
32. $\sqrt{11}\sqrt{11}$
33. $\sqrt{6}\sqrt{10}$
34. $\sqrt{14}\sqrt{21}$
35. $\sqrt{18}\sqrt{24}$
36. $\sqrt{4}\sqrt{6}$
37. $\sqrt{9}\sqrt{6}$
38. $\sqrt{25}\sqrt{15}$
39. $\sqrt{20}\sqrt{20}$
40. $\sqrt{2}\sqrt{10}$
41. $\sqrt{5}\sqrt{15}$
42. $\sqrt{2a}\sqrt{3a}$
43. $\sqrt{5x}\sqrt{15x}$
44. $\sqrt[3]{a^2}\sqrt[3]{a}$
45. $\sqrt[3]{b^2}\sqrt[3]{b^2}$
46. $\sqrt[5]{b^4}\sqrt[5]{b^3}$
47. $\sqrt[5]{a}\sqrt[5]{a^4}$
48. $\sqrt{6x}\sqrt{14xy}$
•49. $\sqrt{2a}\sqrt{24b^2}$
50. $\sqrt[3]{5R^2S}\sqrt[3]{75R^2S^2}$
51. $\sqrt[3]{3ab^2}\sqrt[3]{18a^2b^2}$
52. $\sqrt[4]{8a^3b}\sqrt[4]{4a^2b^2}$
53. $\sqrt[4]{27R^2S^3}\sqrt[4]{9RS}$
54. $\sqrt[3]{25x^5y^7}\sqrt[3]{15xy^3}$
•55. $\sqrt[3]{16a^{11}b^4}\sqrt[3]{12a^4b^6}$
56. $\sqrt[4]{8xy}\sqrt[4]{4x^3y^3}$

In problems 57–60, round to the accuracy of the given data.

57. The moment of inertia for a rectangle is given by the formula $I = \dfrac{bh^3}{12}$. If we know the values of I and b, we can solve for h as follows: $h = \sqrt[3]{\dfrac{12I}{b}}$. Find h if $I = 2.0\ \text{cm}^4$ and $b = 3.0$ cm.

58. Use problem 57 to find h if $I = 9.00 \text{ in}^4$ and $b = 4.00$ in.

59. The moment of inertia for a circle is given by the formula $I = \dfrac{\pi r^4}{4}$. If we know the value of I, we can solve for r as follows: $r = \sqrt[4]{\dfrac{4I}{\pi}}$. Find r if $I = 12.56 \text{ in}^4$ and we use 3.14 for π.

60. Use problem 59 to find r if $I = 63.585 \text{ mm}^4$.

10.4 Quotient Rule for Radicals

The following example helps us develop a rule for division involving radicals:

$$\sqrt{\frac{4}{9}} = \sqrt{\left(\frac{2}{3}\right)^2} = \frac{2}{3}$$

We also observe that $\dfrac{\sqrt{4}}{\sqrt{9}} = \dfrac{2}{3}$. From our example, we conclude that $\sqrt{\dfrac{4}{9}} = \dfrac{\sqrt{4}}{\sqrt{9}}$. To generalize: *The square root of a fraction can be written as the square root of the numerator over the square root of the denominator:* $\sqrt{\dfrac{a}{b}} = \dfrac{\sqrt{a}}{\sqrt{b}}$.

Examples

1. $\sqrt{\dfrac{16}{25}} = \dfrac{\sqrt{16}}{\sqrt{25}} = \dfrac{4}{5}$

2. $\sqrt{\dfrac{36}{49}} = \dfrac{\sqrt{36}}{\sqrt{49}} = \dfrac{6}{7}$

3. $\sqrt{\dfrac{81}{100}} = \dfrac{\sqrt{81}}{\sqrt{100}} = \dfrac{9}{10}$

The quotient rule for radicals can be extended as follows to radicals with any index: *The nth root of a quotient can be written as the nth root of the numerator over the nth root of the denominator:*

$$\sqrt[n]{\frac{a}{b}} = \frac{\sqrt[n]{a}}{\sqrt[n]{b}}$$

Examples

1. $\sqrt[3]{\dfrac{8}{27}} = \dfrac{\sqrt[3]{8}}{\sqrt[3]{27}} = \dfrac{2}{3}$

2. $\sqrt[5]{\dfrac{32}{a^5}} = \dfrac{\sqrt[5]{32}}{\sqrt[5]{a^5}} = \dfrac{2}{a}$

3. $\sqrt[7]{\dfrac{b^7}{c^{14}}} = \dfrac{\sqrt[7]{b^7}}{\sqrt[7]{c^{14}}} = \dfrac{b}{c^2}$

note
As a general rule, fractions are never left under a radical symbol.

When a problem has been simplified and a radical still remains in the denominator, evaluating the resulting problem is usually complex. For this reason we adopt as standard form, that we will not leave radicals in the denominators of our answers. We eliminate such radicals through a procedure called **rationalizing the denominator.** This procedure changes the denominator from a radical (irrational number) to a rational number.

Consider the following example:

$$\sqrt{\frac{4}{5}} = \frac{\sqrt{4}}{\sqrt{5}} = \frac{2}{\sqrt{5}}$$

We want to eliminate the radical symbol in the denominator. To get rid of this radical, we multiply the numerator and denominator by a radical that will give a perfect square radicand in the denominator. In this example,

$$\frac{2}{\sqrt{5}} \cdot \frac{\sqrt{5}}{\sqrt{5}} = \frac{2\sqrt{5}}{\sqrt{25}} = \frac{2\sqrt{5}}{5}$$

Using the fundamental principle of fractions we multiplied numerator and denominator by the same number, $\dfrac{\sqrt{5}}{\sqrt{5}} = 1$, and our answer is equivalent to what we started with.

Examples

1. $\dfrac{5}{\sqrt{7}} = \dfrac{5}{\sqrt{7}} \cdot \dfrac{\sqrt{7}}{\sqrt{7}} = \dfrac{5\sqrt{7}}{\sqrt{49}} = \dfrac{5\sqrt{7}}{7}$

2. $\dfrac{3}{\sqrt{5}} = \dfrac{3}{\sqrt{5}} \cdot \dfrac{\sqrt{5}}{\sqrt{5}} = \dfrac{3\sqrt{5}}{\sqrt{25}} = \dfrac{3\sqrt{5}}{5}$

3. $\dfrac{4}{\sqrt{6}} = \dfrac{4}{\sqrt{6}} \cdot \dfrac{\sqrt{6}}{\sqrt{6}} = \dfrac{4\sqrt{6}}{\sqrt{36}} = \dfrac{4\sqrt{6}}{6} = \dfrac{2\sqrt{6}}{3}$

note
When dealing with a fraction, always check to see that the answer is in a most reduced form.

4. $\dfrac{\sqrt{2}}{\sqrt{3}} = \dfrac{\sqrt{2}}{\sqrt{3}} \cdot \dfrac{\sqrt{3}}{\sqrt{3}} = \dfrac{\sqrt{6}}{\sqrt{9}} = \dfrac{\sqrt{6}}{3}$

5. $\dfrac{2}{3\sqrt{5}} = \dfrac{2}{3\sqrt{5}} \cdot \dfrac{\sqrt{5}}{\sqrt{5}} = \dfrac{2\sqrt{5}}{3\sqrt{25}} = \dfrac{2\sqrt{5}}{3 \cdot 5} = \dfrac{2\sqrt{5}}{15}$

The following example helps us develop a general rule for rationalizing a denominator which has a single term.

$$\sqrt[3]{\frac{1}{a}} = \frac{\sqrt[3]{1}}{\sqrt[3]{a}} = \frac{1}{\sqrt[3]{a}}$$

At this point, we have simplified the problem, but a radical still remains in the denominator. We must now determine what we can do to the fraction to remove the radical from the denominator, yet keep our answer equivalent to what we now have. In making our determination we must keep the following observations in mind:

1. We can multiply numerator and denominator by the same number and form an equivalent fraction.
2. If we multiply by a radical, the indices must be the same in order to carry out the multiplication.
3. To bring a factor out from under the radical symbol and not leave any of the factor behind, the index must divide evenly into the power.

Now we can rationalize the fraction as follows:

$= \dfrac{1}{\sqrt[3]{a}} \cdot \dfrac{\sqrt[3]{}}{\sqrt[3]{}}$ Indices are the same.

$= \dfrac{1}{\sqrt[3]{a}} \cdot \dfrac{\sqrt[3]{a^2}}{\sqrt[3]{a^2}}$ Multiplying numerator and denominator by the same number.

$= \dfrac{\sqrt[3]{a^2}}{\sqrt[3]{a^3}}$ The sum of the powers of a add to the index in the denominator.

$= \dfrac{\sqrt[3]{a^2}}{a}$ The index divides evenly into that power and the radical is eliminated.

We can summarize the rationalization procedure for a denominator of one term as follows:

1. We multiply numerator and denominator by a radical with the same index as the radical we wish to eliminate from the denominator.
2. The exponent of the factor under the radical must be such that when we add it to the original exponent of the factor under the radical in the denominator, the sum will equal the index of the radical.
3. Carry out the multiplication and reduce the fraction if possible.

Examples

1. $\dfrac{a}{\sqrt[5]{b^2}} = \dfrac{a}{\sqrt[5]{b^2}} \cdot \dfrac{\sqrt[5]{b^3}}{\sqrt[5]{b^3}} = \dfrac{a\sqrt[5]{b^3}}{\sqrt[5]{b^5}} = \dfrac{a\sqrt[5]{b^3}}{b}$

2. $\dfrac{x^2}{\sqrt[7]{y^4}} = \dfrac{x^2}{\sqrt[7]{y^4}} \cdot \dfrac{\sqrt[7]{y^3}}{\sqrt[7]{y^3}} = \dfrac{x^2\sqrt[7]{y^3}}{\sqrt[7]{y^7}} = \dfrac{x^2\sqrt[7]{y^3}}{y}$

3. $\dfrac{x}{\sqrt[4]{x}} = \dfrac{x}{\sqrt[4]{x}} \cdot \dfrac{\sqrt[4]{x^3}}{\sqrt[4]{x^3}} = \dfrac{x\sqrt[4]{x^3}}{x} = \sqrt[4]{x^3}$

4. $\dfrac{\sqrt[5]{a^3}}{\sqrt[5]{b^2}} = \dfrac{\sqrt[5]{a^3}}{\sqrt[5]{b^2}} \cdot \dfrac{\sqrt[5]{b^3}}{\sqrt[5]{b^3}} = \dfrac{\sqrt[5]{a^3b^3}}{\sqrt[5]{b^5}} = \dfrac{\sqrt[5]{a^3b^3}}{b}$

5. $\dfrac{R}{S\sqrt[6]{T}} = \dfrac{R}{S\sqrt[6]{T}} \cdot \dfrac{\sqrt[6]{T^5}}{\sqrt[6]{T^5}} = \dfrac{R\sqrt[6]{T^5}}{S\sqrt[6]{T^6}} = \dfrac{R\sqrt[6]{T^5}}{ST}$

6. $\dfrac{1}{\sqrt[5]{a^2b}} = \dfrac{1}{\sqrt[5]{a^2b}} \cdot \dfrac{\sqrt[5]{a^3b^4}}{\sqrt[5]{a^3b^4}} = \dfrac{\sqrt[5]{a^3b^4}}{\sqrt[5]{a^5b^5}} = \dfrac{\sqrt[5]{a^3b^4}}{ab}$

7. $\dfrac{x}{\sqrt[7]{x^4y^6}} = \dfrac{x}{\sqrt[7]{x^4y^6}} \cdot \dfrac{\sqrt[7]{x^3y}}{\sqrt[7]{x^3y}} = \dfrac{x\sqrt[7]{x^3y}}{\sqrt[7]{x^7y^7}} = \dfrac{x\sqrt[7]{x^3y}}{xy} = \dfrac{\sqrt[7]{x^3y}}{y}$

Exercise Set 10.4

Express the given radicals in simplest form.

1. $\sqrt{\dfrac{9}{25}}$
2. $\sqrt{\dfrac{25}{36}}$
3. $\sqrt{\dfrac{25}{49}}$
4. $\sqrt{\dfrac{81}{100}}$
5. $\sqrt{\dfrac{3}{4}}$
6. $\sqrt{\dfrac{5}{9}}$
•7. $\sqrt[3]{\dfrac{8}{27}}$
8. $\sqrt[3]{\dfrac{1}{8}}$
9. $\sqrt[4]{\dfrac{16}{81}}$
10. $\sqrt[3]{\dfrac{27}{125}}$
11. $\sqrt{\dfrac{a^2}{b^2}}$
12. $\sqrt{\dfrac{64}{R^2}}$
13. $\sqrt{\dfrac{y^4}{16}}$
14. $\sqrt[3]{\dfrac{3a^6}{b^3}}$
15. $\sqrt[3]{\dfrac{x}{y^{12}}}$
16. $\sqrt[5]{\dfrac{R^4}{S^{10}}}$
17. $\sqrt[7]{\dfrac{a^{21}}{b^7}}$
18. $\sqrt{\dfrac{R^2}{S^2}}$
19. $\sqrt[3]{\dfrac{8x^3}{y^6}}$
20. $\sqrt[3]{\dfrac{27x^9}{y^{12}}}$
21. $\sqrt[5]{\dfrac{32x^4}{y^5}}$
22. $\sqrt[4]{\dfrac{R^4S^9}{T^{12}}}$
•23. $\sqrt[4]{\dfrac{a^9b^{13}}{c^8}}$
24. $\sqrt[5]{\dfrac{x^3y^2}{z^{15}}}$

Express the given radicals in simplest form with all denominators rationalized.

25. $\sqrt{\frac{1}{2}}$ 26. $\sqrt{\frac{1}{3}}$ 27. $\sqrt{\frac{4}{7}}$ 28. $\sqrt{\frac{9}{11}}$

29. $\sqrt{\frac{1}{15}}$ 30. $\sqrt{\frac{1}{14}}$ 31. $\sqrt{\frac{4}{75}}$ •32. $\sqrt{\frac{5}{12}}$

33. $\frac{2}{\sqrt{2}}$ 34. $\frac{6}{\sqrt{3}}$ 35. $\frac{10}{\sqrt{8}}$ 36. $\frac{15}{\sqrt{27}}$

37. $\sqrt[3]{\frac{8}{9}}$ 38. $\sqrt[3]{\frac{4}{25}}$ 39. $\sqrt[5]{\frac{32}{81}}$ 40. $\sqrt[3]{\frac{27}{16}}$

•41. $\sqrt[4]{\frac{16}{125}}$ 42. $\sqrt[4]{\frac{3}{4}}$ 43. $\sqrt{\frac{x^2}{y}}$ 44. $\sqrt{\frac{1}{a}}$

45. $\sqrt{\frac{1}{R}}$ 46. $\sqrt{\frac{a^2}{b^3}}$ 47. $\sqrt[3]{\frac{R^3}{S^2}}$ 48. $\sqrt[3]{\frac{x^6}{y}}$

49. $\sqrt[3]{\frac{a}{b^2}}$ 50. $\sqrt[3]{\frac{L}{M}}$ •51. $\sqrt[5]{\frac{x^5y^{10}}{z^3}}$ 52. $\frac{ab}{\sqrt[3]{a^2}}$

53. $\frac{xy}{\sqrt[5]{y^3}}$ 54. $\frac{\sqrt{a^5}}{\sqrt{a}}$ 55. $\frac{\sqrt[3]{b^5}}{\sqrt[3]{b^2}}$ 56. $\sqrt[3]{\frac{a^3}{b^2c}}$

57. $\sqrt[3]{\frac{8}{xy^2}}$ 58. $\sqrt[3]{\frac{R^2}{S^2T}}$ 59. $\sqrt[5]{\frac{1}{xy^3}}$ 60. $\sqrt[4]{\frac{16}{x^3y}}$

61. $\frac{a}{\sqrt[5]{a^2b^4}}$ 62. $\frac{R}{\sqrt[5]{R^2S}}$ 63. $\frac{ab}{\sqrt[3]{ab^2}}$ 64. $\frac{xy^2}{\sqrt[5]{x^4y}}$

•65. $\frac{x^2y}{\sqrt[7]{x^5y^2}}$ 66. $\frac{R^3S^2}{\sqrt[5]{R^7S}}$

In problems 67–68, round to the accuracy of the given data.

67. If we wish to construct a sphere of specific volume, we can find the length of radius necessary by the following formula: $r = \sqrt[3]{\frac{3V}{4\pi}}$. Find the radius necessary for a sphere to have a volume of 113.04 m³. (Use 3.14 for π.)

68. Use problem 67 to find r if $V = 904.32$ ft³. (Use 3.14 for π.)

10.5 Adding and Subtracting Radicals

We have learned that whenever we perform the operations of addition or subtraction, we must have similar terms. This condition applies also when we deal with terms that involve radicals. *We can add or subtract only similar radicals.*

For radicals to be similar, the following must be true:

1. The radicals must have the same index.
2. The radicands must be the same.

The addition and subtraction of radicals follows the exact same procedure as the addition and subtraction of algebraic expressions; that is, after determining that the radicals are similar, we perform the operations of addition and subtraction only on the numerical coefficients.

Examples

1. $5\sqrt{2} + 3\sqrt{2} = (5 + 3)\sqrt{2} = 8\sqrt{2}$
2. $4\sqrt{7} - 2\sqrt{7} = 2\sqrt{7}$
3. $\sqrt[3]{5} + 6\sqrt[3]{5} = 7\sqrt[3]{5}$
4. $3\sqrt{11} - 8\sqrt{11} = -5\sqrt{11}$
5. $2\sqrt{a} + 3\sqrt{a} = 5\sqrt{a}$
6. $4\sqrt[5]{x^2} + 3\sqrt[5]{x^2} = 7\sqrt[5]{x^2}$
7. $7\sqrt[3]{a^2b} - 5\sqrt[3]{a^2b} = 2\sqrt[3]{a^2b}$

Consider the example: $\sqrt{27} + 4\sqrt{3}$. It appears at first that we cannot carry out this indicated addition since the radicals are not similar; however, the $\sqrt{27}$ can be simplified: $\sqrt{27} = \sqrt{9 \cdot 3} = \sqrt{3^2}\,\sqrt{3} = 3\sqrt{3}$. Therefore, the problem becomes

$$\sqrt{27} + 4\sqrt{3} = 3\sqrt{3} + 4\sqrt{3} = 7\sqrt{3}$$

and we can now add the radicals together. Keep this fact in mind: All radicals that are to be added or subtracted must be in simplest form.

Examples

1. $\sqrt{45} + \sqrt{20} = \sqrt{3^2 \cdot 5} + \sqrt{2^2 \cdot 5}$
   $= 3\sqrt{5} + 2\sqrt{5}$
   $= 5\sqrt{5}$

2. $\sqrt{32} + 5\sqrt{8} = \sqrt{2^5} + 5\sqrt{2^3}$
   $= 2^2\sqrt{2} + 5 \cdot 2\sqrt{2}$
   $= 4\sqrt{2} + 10\sqrt{2}$
   $= 14\sqrt{2}$

3. $4\sqrt[3]{81} - \sqrt[3]{24} = 4\sqrt[3]{3^4} - \sqrt[3]{2^3 \cdot 3}$
   $= 4 \cdot 3\sqrt[3]{3} - 2\sqrt[3]{3}$
   $= 12\sqrt[3]{3} - 2\sqrt[3]{3}$
   $= 10\sqrt[3]{3}$

4. $\sqrt{24} - \sqrt{27} = \sqrt{2^3 \cdot 3} - \sqrt{3^3}$
$= 2\sqrt{2 \cdot 3} - 3\sqrt{3}$
$= 2\sqrt{6} - 3\sqrt{3}$

note

After simplifying the radicals, we still cannot carry out the indicated subtraction since the radicals are not similar.

5. $4\sqrt{12} + 3\sqrt{24} - \sqrt{27} = 4\sqrt{2^2 \cdot 3} + 3\sqrt{2^3 \cdot 3} - \sqrt{3^3}$
$= 4 \cdot 2\sqrt{3} + 3 \cdot 2\sqrt{2 \cdot 3} - 3\sqrt{3}$
$= 8\sqrt{3} + 6\sqrt{6} - 3\sqrt{3}$

We observe that the first and third radicals are similar; and we have:

$$5\sqrt{3} + 6\sqrt{6}.$$

6. $\sqrt[3]{16x^2y} + \sqrt[3]{54x^2y} = \sqrt[3]{2^4x^2y} + \sqrt[3]{2 \cdot 3^3 \cdot x^2y}$
$= 2\sqrt[3]{2x^2y} + 3\sqrt[3]{2x^2y}$
$= 5\sqrt[3]{2x^2y}$

7. $3\sqrt{3R} - \sqrt{12R} + 5\sqrt{48R} = 3\sqrt{3R} - \sqrt{2^2 \cdot 3R} + 5\sqrt{2^4 \cdot 3R}$
$= 3\sqrt{3R} - 2\sqrt{3R} + 5 \cdot 2^2\sqrt{3R}$
$= 3\sqrt{3R} - 2\sqrt{3R} + 5 \cdot 4\sqrt{3R}$
$= 3\sqrt{3R} - 2\sqrt{3R} + 20\sqrt{3R}$
$= 21\sqrt{3R}$

Exercise Set 10.5

Perform the indicated operations and simplify.

1. $5\sqrt{3} + 4\sqrt{3}$
2. $8\sqrt{7} - 2\sqrt{7}$
3. $6\sqrt{5} + 4\sqrt{5}$
4. $9\sqrt{6} - 6\sqrt{6}$
5. $2\sqrt{3} + 7\sqrt{3} - 3\sqrt{3}$
6. $5\sqrt{5} - 4\sqrt{5} + 6\sqrt{5}$
7. $\sqrt{7} + 5\sqrt{7} - 3\sqrt{7}$
8. $2\sqrt{10} + 11\sqrt{10} - 9\sqrt{10}$
9. $\sqrt{a} + 2\sqrt{a}$
10. $3\sqrt{x} + 4\sqrt{x}$
11. $5\sqrt{R} - 4\sqrt{R} + 7\sqrt{R}$
12. $6\sqrt{y} - \sqrt{y} + 4\sqrt{y}$
13. $5\sqrt{xy} + 2\sqrt{xy}$
14. $3\sqrt{x} + 2\sqrt{y} - \sqrt{x}$
15. $5\sqrt{R} + 2\sqrt{RS} + 3\sqrt{R}$
16. $\sqrt{ab} + 2\sqrt{ab} + 3\sqrt{a}$
17. $5\sqrt{xy} - \sqrt{xy} + 3\sqrt{y}$
18. $\sqrt{20} + 3\sqrt{5}$
•19. $\sqrt{8} + 5\sqrt{2}$
20. $\sqrt{12} + \sqrt{75}$
21. $\sqrt{48} - \sqrt{27}$
22. $2\sqrt{3} + 3\sqrt{12}$
23. $5\sqrt{7} + 4\sqrt{63}$
24. $5\sqrt{3} + \sqrt{27} - \sqrt{12}$
25. $4\sqrt{2} - \sqrt{8} + \sqrt{50}$
26. $\sqrt{75} - 4\sqrt{3} + 2\sqrt{27}$
•27. $\sqrt{12} + \sqrt{18} + \sqrt{50}$
28. $\sqrt{63} - \sqrt{28} + \sqrt{24}$
29. $\sqrt{50a} + \sqrt{8a}$
30. $\sqrt{32R} - \sqrt{18R}$
31. $3\sqrt{9x} - 5\sqrt{4x}$
32. $2\sqrt{4x^2y} + 3\sqrt{25x^2y}$
33. $2\sqrt{8R} + 4\sqrt{50R} - 7\sqrt{2R}$
34. $3\sqrt{48S} - 2\sqrt{12S} + \sqrt{3S}$

35. $\sqrt{50R} + 3\sqrt{12R} - \sqrt{18R}$

36. $4\sqrt{25x^2y} + 3\sqrt{81x^2y} - 2\sqrt{2y}$

37. $3\sqrt[3]{4} + 5\sqrt[3]{4}$

38. $7\sqrt[5]{2} - 4\sqrt[5]{2} + 3\sqrt[5]{2}$

39. $9\sqrt[4]{3} + 6\sqrt[4]{3} + 2\sqrt[4]{3}$

40. $7\sqrt[3]{11} - 11\sqrt[3]{7}$

41. $\sqrt[3]{16} + \sqrt[3]{54}$

42. $\sqrt[3]{24} - \sqrt[3]{81}$

•43. $\sqrt[3]{81} + 2\sqrt[3]{250}$

44. $\sqrt[3]{8a^2} + \sqrt[3]{27a^2}$

45. $\sqrt[4]{16x^3} + \sqrt[4]{81x^3}$

46. $\sqrt[4]{625R} - \sqrt[4]{81R}$

47. $\sqrt[3]{64x^2y} - \sqrt[3]{27x^2y}$

48. $\sqrt[3]{x^6y} + 2x^2\sqrt[3]{y}$

•49. $3a\sqrt[3]{b^2} - \sqrt[3]{a^3b^2}$

50. $\sqrt[3]{16R^2S} + \sqrt[3]{54R^2S}$

In problems 51–53, round to the accuracy of the given data.

51. The height (h) of a given figure can be found by first finding b from the formula: $b = \sqrt{c^2 - s^2}$. Use the accompanying figure to find h if $c = 10.0$ m and $s = 6.0$ m.

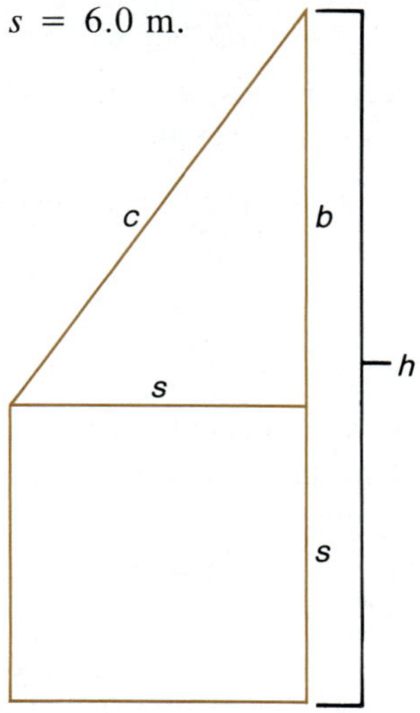

52. Find the height of the figure in problem 51 if $c = 13.0$ ft and $s = 5.0$ ft.

53. The following figure is made up of 9 equal squares, each square having an area of 7.29 cm². What are the dimensions of the figure?

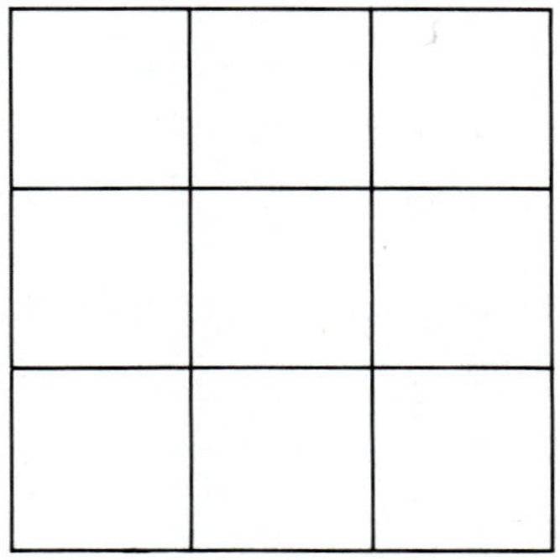

10.6 Additional Operations with Radicals

In section 10.3 we learned the procedure for multiplying two radicals. We shall now combine that procedure with the *distributive law,* $a(b + c) = ab + ac$, to perform multiplication of radical expressions of more than one term.

Examples

1. $\sqrt{2}(\sqrt{3} + \sqrt{5}) = \sqrt{2}\,\sqrt{3} + \sqrt{2}\,\sqrt{5}$
   $= \sqrt{6} + \sqrt{10}$

   No further simplification can be performed since the radicals are not similar and therefore cannot be added.

2. $\sqrt{3}(\sqrt{6} + \sqrt{21}) = \sqrt{3}\,\sqrt{6} + \sqrt{3}\,\sqrt{21}$
   $= \sqrt{18} + \sqrt{63}$
   $= \sqrt{3^2 \cdot 2} + \sqrt{3^2 \cdot 7}$
   $= 3\sqrt{2} + 3\sqrt{7}$

   After we multiply radicals, we should always check to see if the resulting radicals can be simplified.

3. $\sqrt[5]{a^2}\,(\sqrt[5]{a^3} + \sqrt[5]{a^4}) = \sqrt[5]{a^2}\,\sqrt[5]{a^3} + \sqrt[5]{a^2}\,\sqrt[5]{a^4}$
   $= \sqrt[5]{a^5} + \sqrt[5]{a^6}$
   $= a + a\sqrt[5]{a}$

4. $(\sqrt{2} + \sqrt{3})(\sqrt{2} + 5\sqrt{3})$

   In this example, since quantities are being multiplied together, we will multiply each term in the first pair of parentheses by each term in the second pair of parentheses.

   $(\sqrt{2} + \sqrt{3})(\sqrt{2} + 5\sqrt{3}) = \sqrt{2}\,\sqrt{2} + \sqrt{2} \cdot 5\sqrt{3} + \sqrt{3}\,\sqrt{2}$
   $+ \sqrt{3} \cdot 5\sqrt{3}$
   $= \sqrt{4} + 5\sqrt{6} + \sqrt{6} + 5\sqrt{9}$
   $= \sqrt{2^2} + 5\sqrt{6} + \sqrt{6} + 5\sqrt{3^2}$
   $= 2 + 5\sqrt{6} + \sqrt{6} + 5 \cdot 3$
   $= 2 + 5\sqrt{6} + \sqrt{6} + 15$
   $= 17 + 6\sqrt{6}$

5. $(3 - \sqrt{2})(3 + \sqrt{2}) = 9 + 3\sqrt{2} - 3\sqrt{2} - \sqrt{4}$
   $= 9 + 3\sqrt{2} - 3\sqrt{2} - 2$
   $= 9 - 2$
   $= 7$

   We observe that when we simplified, there were no longer any radicals in the problem.

6. $(\sqrt{5} + \sqrt{2})(\sqrt{5} - \sqrt{2}) = \sqrt{5}\sqrt{5} - \sqrt{5}\sqrt{2} + \sqrt{2}\sqrt{5} - \sqrt{2}\sqrt{2}$

$$
\begin{aligned}
&= \sqrt{25} - \sqrt{10} + \sqrt{10} - \sqrt{4} \\
&= 5 - \sqrt{10} + \sqrt{10} - 2 \\
&= 5 - 2 \\
&= 3
\end{aligned}
$$

Again we observe that the radicals are eliminated from the problem.

7. $(\sqrt{a} - \sqrt{b})(\sqrt{a} + \sqrt{b}) = \sqrt{a}\sqrt{a} + \sqrt{a}\sqrt{b} - \sqrt{b}\sqrt{a} - \sqrt{b}\sqrt{b}$

$$
\begin{aligned}
&= \sqrt{a^2} + \sqrt{ab} - \sqrt{ab} - \sqrt{b^2} \\
&= a + \sqrt{ab} - \sqrt{ab} - b \\
&= a - b
\end{aligned}
$$

In preceding examples 5, 6, and 7, the factors involved are called **conjugate factors.** A conjugate factor is used to rationalize the denominator of a fraction when the denominator contains two terms, where one or both terms contain a square root.

In example 7,

$$
\begin{aligned}
(\sqrt{a} - \sqrt{b})(\sqrt{a} + \sqrt{b}) &= (\sqrt{a})^2 - (\sqrt{b})^2 \\
&= a - b
\end{aligned}
$$

note

Therefore, instead of multiplying the two conjugates together, we can simply write our answer as the square of the second term subtracted from the square of the first term. The idea of conjugate factors is derived from the factorization of the difference between two squares.

To determine what the conjugate of a given factor is, we write the two terms of the factor and change the sign between them—that is, addition to subtraction, or subtraction to addition.

Examples

Form the conjugates of the given expressions.

1. $\sqrt{7} + 2$ The conjugate is $\sqrt{7} - 2$.
2. $\sqrt{11} - \sqrt{6}$ The conjugate is $\sqrt{11} + \sqrt{6}$.
3. $5 - 2\sqrt{3}$ The conjugate is $5 + 2\sqrt{3}$.
4. $\sqrt{a} + \sqrt{b}$ The conjugate is $\sqrt{a} - \sqrt{b}$.

If we want to rationalize the fraction

$$\frac{1}{3 - \sqrt{2}}$$

We recall from example 5 (page 352) that when we multiplied $3 - \sqrt{2}$ by $3 + \sqrt{2}$, no radicals were left in the answer. This result is precisely what we want to occur in our denominator here. Therefore, to rationalize this fraction, we apply the fundamental principle of fractions and multiply numerator and denominator by $3 + \sqrt{2}$, which is the conjugate of the denominator:

$$\frac{1}{3 - \sqrt{2}} = \frac{1}{3 - \sqrt{2}} \cdot \frac{3 + \sqrt{2}}{3 + \sqrt{2}}$$

$$= \frac{1(3 + \sqrt{2})}{(3)^2 - (\sqrt{2})^2}$$

$$= \frac{3 + \sqrt{2}}{9 - 2}$$

$$= \frac{3 + \sqrt{2}}{7}$$

Examples

Rationalize the denominators.

1. $$\frac{2}{\sqrt{7} + 2} = \frac{2}{\sqrt{7} + 2} \cdot \frac{\sqrt{7} - 2}{\sqrt{7} - 2}$$

$$= \frac{2(\sqrt{7} - 2)}{(\sqrt{7})^2 - (2)^2}$$

$$= \frac{2\sqrt{7} - 4}{7 - 4}$$

$$= \frac{2\sqrt{7} - 4}{3}$$

2. $$\frac{5}{\sqrt{11} - \sqrt{6}} = \frac{5}{\sqrt{11} - \sqrt{6}} \cdot \frac{\sqrt{11} + \sqrt{6}}{\sqrt{11} + \sqrt{6}}$$

$$= \frac{5(\sqrt{11} + \sqrt{6})}{(\sqrt{11})^2 - (\sqrt{6})^2}$$

$$= \frac{5(\sqrt{11} + \sqrt{6})}{11 - 6}$$

$$= \frac{5(\sqrt{11} + \sqrt{6})}{5}$$

$$= \sqrt{11} + \sqrt{6}$$

note

As a last step in dealing with fractions, reduce if possible.

3. $$\frac{\sqrt{3}}{5 - 2\sqrt{3}} = \frac{\sqrt{3}}{5 - 2\sqrt{3}} \cdot \frac{5 + 2\sqrt{3}}{5 + 2\sqrt{3}}$$

$$= \frac{\sqrt{3}(5 + 2\sqrt{3})}{(5)^2 - (2\sqrt{3})^2}$$

$$= \frac{5\sqrt{3} + 2\sqrt{9}}{5^2 - 2^2(\sqrt{3})^2}$$

$$= \frac{5\sqrt{3} + 2 \cdot 3}{25 - 4 \cdot 3}$$

$$= \frac{5\sqrt{3} + 6}{25 - 12}$$

$$= \frac{5\sqrt{3} + 6}{13}$$

4. $$\frac{\sqrt{ab}}{\sqrt{a} + \sqrt{b}} = \frac{\sqrt{ab}}{\sqrt{a} + \sqrt{b}} \cdot \frac{\sqrt{a} - \sqrt{b}}{\sqrt{a} - \sqrt{b}}$$

$$= \frac{\sqrt{ab}(\sqrt{a} - \sqrt{b})}{(\sqrt{a})^2 - (\sqrt{b})^2}$$

$$= \frac{\sqrt{a^2b} - \sqrt{ab^2}}{a - b}$$

$$= \frac{a\sqrt{b} - b\sqrt{a}}{a - b}$$

note

We cannot reduce the fraction in example 4 because there are no common factors. The a's and b's in the numerator and denominator are parts of terms, not factors.

Exercise Set 10.6

Perform the indicated operations and simplify.

1. $3(\sqrt{2} + \sqrt{3})$
2. $4(\sqrt{5} - \sqrt{7})$
3. $5(2\sqrt{6} + \sqrt{2})$
4. $3(3\sqrt{7} - \sqrt{11})$
5. $\sqrt{2}(\sqrt{3} + \sqrt{7})$
6. $\sqrt{5}(\sqrt{7} - \sqrt{3})$
7. $\sqrt{3}(2\sqrt{2} + \sqrt{5})$
8. $3\sqrt{2}(2\sqrt{3} - \sqrt{11})$
•9. $\sqrt{6}(\sqrt{2} + \sqrt{3})$
10. $\sqrt{5}(\sqrt{15} - \sqrt{10})$

11. $\sqrt{14}(\sqrt{21} + \sqrt{10})$

12. $2\sqrt{7}(\sqrt{35} - 3\sqrt{14})$

•13. $3\sqrt{3}(\sqrt{3} + 2\sqrt{15})$

14. $\sqrt{a}(\sqrt{ab} + \sqrt{b})$

15. $\sqrt{R}(3\sqrt{R} + \sqrt{S})$

16. $\sqrt[3]{4}(4\sqrt[3]{2} - \sqrt[3]{6})$

•17. $\sqrt[3]{9}(5\sqrt[3]{12} - \sqrt[3]{3})$

18. $3\sqrt[3]{6}(\sqrt[3]{4} + \sqrt[3]{9})$

19. $2\sqrt[3]{12}(\sqrt[3]{2} + \sqrt[3]{9})$

20. $\sqrt[3]{a^2}(\sqrt[3]{a^2} + \sqrt[3]{a})$

21. $\sqrt[5]{R^3}(\sqrt[5]{R^2} + \sqrt[5]{R^4})$

22. $\sqrt[5]{x^3}(\sqrt[5]{x} + \sqrt[5]{x^3})$

23. $\sqrt[5]{y^2}(3\sqrt[5]{y^2} + \sqrt[5]{y^4})$

24. $(5 + \sqrt{3})(4 - \sqrt{3})$

25. $(3 + \sqrt{2})(4 + \sqrt{2})$

26. $(5 - \sqrt{5})(5 - \sqrt{5})$

27. $(3 + \sqrt{a})(5 + \sqrt{a})$

28. $(3 - 4\sqrt{R})(4 - 3\sqrt{R})$

•29. $(7 + 2\sqrt{y})(6 + 5\sqrt{y})$

30. $(\sqrt{3} + \sqrt{2})(\sqrt{3} - \sqrt{2})$

31. $(\sqrt{7} + \sqrt{5})(\sqrt{7} - \sqrt{5})$

32. $(2 + \sqrt{6})(2 - \sqrt{6})$

33. $(5 - \sqrt{3})(5 + \sqrt{3})$

34. $(2 + \sqrt{5})^2$

35. $(3 - \sqrt{7})^2$

36. $(\sqrt{a} + \sqrt{b})(2\sqrt{a} - \sqrt{b})$

37. $(\sqrt{R} + 2\sqrt{S})(\sqrt{R} + \sqrt{S})$

38. $(\sqrt{x} + \sqrt{y})^2$

39. $(\sqrt{a} - \sqrt{b})^2$

40. $(\sqrt{L} - \sqrt{M})(\sqrt{L} + \sqrt{M})$

41. $(2\sqrt{a} - \sqrt{b})(2\sqrt{a} + \sqrt{b})$

42. $(x\sqrt{y} + \sqrt{z})(x\sqrt{y} - \sqrt{z})$

43. $(a\sqrt{b} + c)(a\sqrt{b} - c)$

44. $(2\sqrt{x} + y)^2$

45. $(3\sqrt{R} + \sqrt{S})^2$

Form the conjugates of the given expressions.

46. $7 + \sqrt{5}$

•47. $11 - \sqrt{3}$

48. $2\sqrt{3} - \sqrt{5}$

49. $5\sqrt{7} - \sqrt{2}$

50. $\sqrt{a} - 2\sqrt{b}$

•51. $\sqrt{R} + 3\sqrt{S}$

52. $x\sqrt{y} + \sqrt{z}$

53. $R\sqrt{S} - \sqrt{T}$

Express the given radicals in simplest form, with all denominators rationalized.

54. $\dfrac{1}{\sqrt{2} + 3}$

•55. $\dfrac{1}{\sqrt{3} - 2}$

56. $\dfrac{7}{2 + \sqrt{7}}$

57. $\dfrac{6}{3 - \sqrt{6}}$

58. $\dfrac{2}{\sqrt{2} + \sqrt{5}}$

59. $\dfrac{3}{\sqrt{6} - \sqrt{3}}$

60. $\dfrac{1}{\sqrt{a} + b}$

61. $\dfrac{1}{\sqrt{R} - \sqrt{S}}$

62. $\dfrac{x}{\sqrt{x} - y}$

63. $\dfrac{a}{a - \sqrt{b}}$

64. $\dfrac{3}{2\sqrt{3} - \sqrt{5}}$

65. $\dfrac{4}{2\sqrt{3} - \sqrt{6}}$

66. $\dfrac{1 + \sqrt{5}}{1 - \sqrt{5}}$

•67. $\dfrac{\sqrt{3} - \sqrt{7}}{\sqrt{3} + \sqrt{7}}$

68. $\dfrac{\sqrt{R} + S}{\sqrt{R} - S}$

In problems 69–70, round to the accuracy of the given data.

69. To find the design constant B for the most efficient helical compression spring with static load, the following formula is used: $B = \dfrac{C^3(5C + 1.23)}{2.46\sqrt{C} + 0.615}$. Find B if $C = 5.20$.

70. Refer to problem 69 to find B if $C = 4.00$.

10.7 Fractional Exponents

In this section, we develop the idea of a fraction used as an exponent. Consider the following example: $\left(a^{\frac{1}{2}}\right)^2 = a^{\left(\frac{1}{2}\right)\cdot 2} = a^1$. When we raise a number to a power and then raise that to another power, we multiply the exponents. In the previous section we observed that $(\sqrt{a})^2 = a$. For the rules of exponents and the procedures for radicals to be consistent, the following must be true: $a^{\frac{1}{2}} = \sqrt{a}$.

We can generalize our observation by saying that $a^{\frac{1}{n}}$ *represents the nth principal root of a; that is,* $a^{\frac{1}{n}} = \sqrt[n]{a}$.

Examples

1. $5^{\frac{1}{2}} = \sqrt{5}$
2. $a^{\frac{1}{3}} = \sqrt[3]{a}$
3. $(49)^{\frac{1}{2}} = \sqrt{49} = 7$
4. $(-8)^{\frac{1}{3}} = \sqrt[3]{-8} = -2$
5. $(a^9)^{\frac{1}{3}} = \sqrt[3]{a^9} = a^3$

Once you become acquainted with fractional exponents, the process of changing the fractional exponent to radical form for simplification will become unnecessary.

When we examine the problem, $(\sqrt[3]{a})^2 = \left(a^{\frac{1}{3}}\right)^2 = a^{\frac{2}{3}}$, we observe that *when a number is raised to a fractional exponent, the numerator of the fractional exponent indicates the power to which the base is to be raised, and the denominator indicates the root to be taken*. We can state this idea as follows: If $a > 0$ or if $a \neq 0$ and n is odd, $a^{\frac{m}{n}} = (\sqrt[n]{a})^m = \sqrt[n]{a^m}$.

note

Our restriction of "$a > 0$ or n is odd" is the same restriction we placed on radicals. That is, we are not allowing negative numbers under a radical with an even index.

We calculate $a^{\frac{m}{n}}$ by first finding the nth principal root of a and then raising the resulting number to the mth power.

Examples

1. $(16)^{\frac{3}{4}} = (\sqrt[4]{16})^3 = 2^3 = 8$
2. $(27)^{\frac{2}{3}} = (\sqrt[3]{27})^2 = 3^2 = 9$
3. $9^{\frac{3}{2}} = (\sqrt{9})^3 = 3^3 = 27$
4. $(-8)^{\frac{1}{3}} = (\sqrt[3]{-8})^1 = (-2)^1 = -2$
5. $(x^6)^{\frac{1}{3}} = \sqrt[3]{x^6} = x^2$

Since the expression $a^{\frac{1}{n}}$ represents the nth principal root of a, we shall restrict the use of fractional exponents to the following cases:

1. If the base is greater than zero ($a > 0$), we can use any fractional exponent.
2. If the base is less than zero ($a < 0$), then the reduced form of $\frac{m}{n}$ must have a denominator that is an odd integer.
3. If the base is zero ($a = 0$), our fractional exponent must be positive.

Examples

1. $\left(3^{\frac{1}{2}}\right)\left(3^{\frac{1}{3}}\right) = 3^{\frac{1}{2}+\frac{1}{3}} = 3^{\frac{3}{6}+\frac{2}{6}} = 3^{\frac{5}{6}}$
2. $\dfrac{2^{\frac{1}{2}}}{2^{\frac{1}{4}}} = 2^{\frac{1}{2}-\frac{1}{4}} = 2^{\frac{2}{4}-\frac{1}{4}} = 2^{\frac{1}{4}}$
3. $x^{\frac{2}{3}} \cdot x^{\frac{3}{4}} = x^{\frac{2}{3}+\frac{3}{4}} = x^{\frac{8}{12}+\frac{9}{12}} = x^{\frac{7}{12}}$
4. $\left(y^{\frac{1}{2}}\right)^{\frac{4}{3}} = y^{\frac{1}{2}\cdot\frac{4}{3}} = y^{\frac{2}{3}}$
5. $(a^4b^2)^{\frac{1}{3}} = (a^4)^{\frac{1}{3}}(b^2)^{\frac{1}{3}} = a^{\frac{4}{3}}b^{\frac{2}{3}}$
6. $\dfrac{R^{\frac{2}{3}}}{R^{\frac{1}{2}}} = R^{\frac{2}{3}-\frac{1}{2}} = R^{\frac{4}{6}-\frac{3}{6}} = R^{\frac{1}{6}}$
7. $\dfrac{a^{\frac{1}{5}}}{a^{-\frac{2}{5}}} = a^{\frac{1}{5}-\left(-\frac{2}{5}\right)} = a^{\frac{3}{5}}$
8. $b^{-\frac{1}{4}} \cdot b^{\frac{3}{4}} = b^{\left(-\frac{1}{4}\right)+\left(\frac{3}{4}\right)} = b^{\frac{2}{4}} = b^{\frac{1}{2}}$

9. $\left(x^{-\frac{2}{3}}\right)^{-\frac{1}{2}} = x^{\left(-\frac{2}{3}\right)\cdot\left(-\frac{1}{2}\right)} = x^{\frac{1}{3}}$

10. $\dfrac{S^{-\frac{1}{4}}T^{\frac{2}{5}}}{S^{\frac{3}{4}}T^{-\frac{4}{5}}} = S^{\left(-\frac{1}{4}\right)-\left(\frac{3}{4}\right)} \cdot T^{\left(\frac{2}{5}\right)-\left(-\frac{4}{5}\right)} = S^{-\frac{4}{4}}T^{\frac{6}{5}}$

$$= S^{-1}T^{\frac{6}{5}} = \frac{1}{S^1} \cdot T^{\frac{6}{5}} = \frac{T^{\frac{6}{5}}}{S}$$

Exercise Set 10.7

Change the given expression to radical form.

1. $2^{\frac{1}{2}}$ 2. $5^{\frac{1}{3}}$ •3. $6^{\frac{2}{3}}$ 4. $7^{\frac{3}{4}}$ 5. $a^{\frac{1}{4}}$

6. $R^{\frac{2}{5}}$ 7. $X^{\frac{3}{5}}$ 8. $S^{\frac{4}{7}}$ 9. $T^{\frac{9}{5}}$ 10. $B^{\frac{5}{3}}$

Change the given expression to fractional exponent form.

11. $\sqrt{7}$ 12. $\sqrt{5}$ 13. $\sqrt{R}$ 14. $\sqrt[3]{a}$ •15. $\sqrt[3]{x^2}$

16. $\sqrt[5]{R^3}$ 17. $\sqrt[4]{S^3}$ 18. $\sqrt[7]{a^2}$ 19. $\sqrt[5]{x^4}$ 20. $\sqrt[3]{y}$

Evaluate the given expressions.

21. $(36)^{\frac{1}{2}}$ 22. $(81)^{\frac{1}{2}}$ 23. $(25)^{\frac{1}{2}}$

24. $(16)^{\frac{1}{2}}$ 25. $(a^6)^{\frac{1}{3}}$ 26. $(R^{10})^{\frac{1}{2}}$

•27. $(S^{12})^{\frac{1}{6}}$ 28. $(x^4)^{\frac{1}{4}}$ 29. $(8)^{\frac{1}{3}}$

30. $(32)^{\frac{1}{5}}$ 31. $(-27)^{\frac{1}{3}}$ 32. $(-8)^{\frac{1}{3}}$

33. $(64)^{\frac{1}{3}}$ 34. $(27)^{\frac{2}{3}}$ 35. $(16)^{\frac{3}{4}}$

36. $(125)^{\frac{2}{3}}$ •37. $(64)^{\frac{5}{6}}$ 38. $(9)^{\frac{3}{2}}$

39. $(16)^{\frac{3}{2}}$ 40. $(16)^{\frac{5}{4}}$ 41. $(81)^{\frac{3}{4}}$

42. $(25)^{-\frac{1}{2}}$ 43. $(9)^{-\frac{1}{2}}$ 44. $(16)^{-\frac{3}{4}}$

45. $(27)^{-\frac{2}{3}}$ 46. $(-8)^{-\frac{1}{3}}$ •47. $(-27)^{-\frac{1}{3}}$

•48. $(-32)^{-\frac{3}{5}}$

Perform the indicated operations and simplify.

49. $2^{\frac{1}{2}} \cdot 2^{\frac{3}{2}}$ 50. $a^{\frac{3}{4}} \cdot a^{\frac{1}{4}}$ 51. $R^{\frac{1}{3}} \cdot R^{\frac{2}{3}}$

52. $S^{\frac{1}{2}} \cdot S^{\frac{1}{4}}$ •53. $X^{\frac{1}{4}} \cdot X^{\frac{1}{3}}$ 54. $y^{\frac{2}{3}} \cdot y^{\frac{3}{4}}$

55. $B^{\frac{1}{2}} \cdot B^{\frac{2}{5}}$

56. $\dfrac{2^{\frac{3}{2}}}{2^{\frac{1}{2}}}$

57. $\dfrac{x^{\frac{3}{4}}}{x^{\frac{1}{2}}}$

58. $\dfrac{R^{\frac{2}{3}}}{R^{\frac{1}{6}}}$

•59. $\dfrac{S^{\frac{3}{4}}T^{\frac{1}{2}}}{S^{\frac{1}{4}}T^{\frac{1}{4}}}$

60. $\dfrac{xy^{\frac{3}{4}}}{x^{\frac{1}{2}}y^{\frac{1}{4}}}$

•61. $\dfrac{RS}{R^{\frac{1}{2}}S^{\frac{1}{2}}}$

62. $\left(a^{\frac{2}{3}}\right)^{\frac{1}{2}}$

63. $\left(R^{\frac{3}{4}}\right)^{4}$

64. $\left(T^{\frac{1}{2}}\right)^{\frac{1}{2}}$

65. $\left(T^{-\frac{1}{4}}\right)^{-\frac{2}{3}}$

66. $\left(y^{-\frac{1}{2}}\right)^{-\frac{2}{5}}$

•67. $\left(16a^{4}\right)^{\frac{3}{4}}$

68. $(a^2b^4)^{\frac{1}{2}}$

69. $(x^3y^{12})^{\frac{1}{3}}$

70. $(R^2S)^{\frac{2}{3}}$

71. $(8a^6b^3)^{\frac{2}{3}}$

72. $\dfrac{x^{-\frac{1}{2}}x}{x^{\frac{1}{3}}}$

73. $\dfrac{R^2R^{\frac{1}{2}}}{R^{\frac{1}{3}}}$

74. $\dfrac{S^{\frac{2}{3}} \cdot S^{\frac{5}{3}}}{S^{-\frac{2}{3}}}$

•75. $\dfrac{y^{\frac{3}{4}} \cdot y^{\frac{1}{4}}}{y^{-\frac{1}{4}}}$

10.8 Imaginary and Complex Numbers

In this section, we examine what happens when we try to take an even root of a negative number. The expression $\sqrt{-4}$ has no meaning in the system of real numbers because there is no real number that, when multiplied by itself, equals a negative number. However, there are situations in which an answer for such a problem is required. Therefore, we define a new kind of number to provide the required result. We define j as a number such that $j^2 = -1$, where j is called an **imaginary number.**

If $j^2 = -1$, then $\sqrt{-1} = j$ and we are able to use this fact to find the square root of any negative number.

Examples

1. $\sqrt{-4} = \sqrt{(4)(-1)} = \sqrt{2^2j^2} = 2j$
2. $\sqrt{-25} = \sqrt{(25)(-1)} = \sqrt{5^2j^2} = 5j$
3. $\sqrt{-2} = \sqrt{(2)(-1)} = \sqrt{2j^2} = \sqrt{2}j$

To check our results, we square the answer to get the original radicand back.

Examples

1. $(2j)^2 = 2^2 \cdot j^2 = 4 \cdot (-1) = -4$
2. $(5j)^2 = 5^2 \cdot j^2 = 25(-1) = -25$
3. $(\sqrt{2}j)^2 = (\sqrt{2})^2j^2 = 2(-1) = -2$

Consider the following examples:

$$\begin{aligned}(\sqrt{-4})^2 &= (\sqrt{(4)(-1)})^2 \\ &= (\sqrt{4}\sqrt{-1})^2 \\ &= (2j)^2 \\ &= 2^2j^2 \\ &= (4)(-1) = -4\end{aligned} \qquad \begin{aligned}(\sqrt{-4})^2 &= \sqrt{-4}\sqrt{-4} \\ &= \sqrt{(-4)(-4)} \\ &= \sqrt{16} \\ &= 4\end{aligned}$$

We observe that depending on the order in which we apply the rules of exponents, we get different results. Therefore, *whenever we are dealing with the square root of a negative number, we will express the answer in terms of* j *before proceeding*. Hence, $(\sqrt{-4})^2 = (2j)^2 = 2^2j^2 = 4 \cdot (-1) = -4$.

Examples

1. $\sqrt{-2}\sqrt{-3} = j\sqrt{2} \cdot j\sqrt{3} = j^2\sqrt{6} = -\sqrt{6}$
2. $\sqrt{-5}\sqrt{-5} = j\sqrt{5} \cdot j\sqrt{5} = j^2 \cdot 5 = (-1) \cdot 5 = -5$
3. $\sqrt{-2}\sqrt{-8} = j\sqrt{2} \cdot j\sqrt{8} = j^2\sqrt{16} = (-1)(4) = -4$
4. $\sqrt{-6}\sqrt{-3} = j\sqrt{6} \cdot j\sqrt{3} = j^2\sqrt{18} = (-1) \cdot 3\sqrt{2} = -3\sqrt{2}$

If we apply the rules of exponents to j, we have:

$$\text{a cycle}\left\{\begin{aligned} j &= j \\ j^2 &= -1 \\ j^3 &= j^2j = (-1)j = -j \\ j^4 &= j^2j^2 = (-1)(-1) = 1\end{aligned}\right. \qquad \text{a cycle}\left\{\begin{aligned} j^5 &= j^4j = 1 \cdot j = j \\ j^6 &= j^4j^2 = 1(-1) = -1 \\ j^7 &= j^4j^3 = 1(-j) = -j \\ j^8 &= j^4j^4 = 1 \cdot 1 = 1\end{aligned}\right.$$

The powers of j go through the cycle of: j, -1, $-j$, and 1. Using this fact, it is possible to simplify j raised to any positive integer power.

Examples

1. $j^{10} = j^4 \cdot j^4 \cdot j^2 = 1 \cdot 1 \cdot (-1) = -1$
2. $j^{15} = j^4 \cdot j^4 \cdot j^4 \cdot j^3 = 1 \cdot 1 \cdot 1 \cdot (-j) = -j$
3. $j^{20} = j^4j^4j^4j^4j^4 = 1 \cdot 1 \cdot 1 \cdot 1 \cdot 1 = 1$

From these examples we can see that *when we simplify j to a positive integer power, the resulting power of j is the remainder when we divide the original power by 4.*

Examples

1. $j^{50} = j^2 = -1$ because $50 \div 4 = 12$ *remainder 2*
2. $j^{79} = j^3 = -j$ because $79 \div 4 = 19$ *remainder 3*
3. $j^{21} = j$ because $21 \div 4 = 5$ *remainder 1*

Now let us define a new type of number, one that combines the system of real numbers and the system of imaginary numbers. These new numbers are called **complex numbers** and are composed of a real part denoted by a and an imaginary part denoted by bj. *A complex number takes the form of* $a + bj$, *where a and b are real numbers and j represents the* $\sqrt{-1}$.

Examples

1. $2 + 5j$ is a complex number whose real part is 2 and whose imaginary part is 5.
2. $3 - 4j$ is a complex number whose real part is 3 and whose imaginary part is -4.
3. $0 + 2j$ is called a **pure imaginary number** because the real part is zero; that is, whenever the real part of a complex number is zero, the complex number is a pure imaginary number.
4. $9 + 0j$ is a **real number** because the imaginary part is zero; that is, whenever the imaginary part of a complex number is 0, the complex number is a real number.
5. $6 - \sqrt{-4} = 6 - 2j$ is a complex number whose real part is 6 and whose imaginary part is -2.

Exercise Set 10.8

Simplify the following radicals.

•1. $\sqrt{-9}$ 2. $\sqrt{-16}$ 3. $\sqrt{-36}$ 4. $\sqrt{-81}$
5. $\sqrt{-12}$ 6. $\sqrt{-18}$ •7. $\sqrt{-24}$ 8. $\sqrt{-30}$
9. $\sqrt{-20}$ 10. $\sqrt{-15}$ 11. $\sqrt{-6}$ 12. $\sqrt{-32}$

Perform the indicated operations.

•13. $(3j)^2$ 14. $(4j)^2$ 15. $(6j)^2$
16. $(9j)^2$ 17. $(\sqrt{3}j)^2$ 18. $(\sqrt{2}j)^2$
•19. $(2\sqrt{2}j)^2$ 20. $(3\sqrt{2}j)^2$ 21. $\sqrt{-3}\sqrt{-5}$
22. $\sqrt{-7}\sqrt{-11}$ 23. $\sqrt{-2}\sqrt{-2}$ 24. $\sqrt{-6}\sqrt{-6}$
25. $(\sqrt{-5})^2$ 26. $(\sqrt{-4})^2$ •27. $\sqrt{-3}\sqrt{-6}$
28. $\sqrt{-4}\sqrt{-9}$ 29. $\sqrt{-16}\sqrt{-1}$ 30. $\sqrt{-3}\sqrt{-12}$
31. $\sqrt{-2}\sqrt{-18}$ 32. $\sqrt{-3}\sqrt{-15}$ 33. $\sqrt{-2}\sqrt{-14}$
34. $\sqrt{-5}\sqrt{-6}$ 35. $\sqrt{-3}\sqrt{-7}$ 36. $\sqrt{-1}\sqrt{-1}$

Simplify the following expressions.

37. j^2 38. j^3 •39. j^4 40. j^{15} •41. j^{19}
42. j^{49} •43. j^{63} 44. j^{44} •45. j^{60} 46. j^{100}

Simplify the following expressions and choose which statement best describes the expression: (a) a real number; (b) an imaginary number; and (c) a complex number.

•47. $6 + \sqrt{-4}$ 48. $7 - \sqrt{-9}$ 49. $0 + \sqrt{-1}$
50. $5 + (\sqrt{-1})^2$ •51. $6 - 2j$ 52. $0 - \sqrt{-4}$
•53. $11 + (5j)^2$ 54. $5 - \sqrt{-25}$ 55. $4 - \sqrt{-16}$

Chapter Summary

1. The basic laws of exponents apply to exponents that are positive, zero, and negative integers and also to all other rational numbers.
2. Any number, other than zero, raised to the **zero power** is equal to 1, $a^0 = 1$, $a \neq 0$.
3. A number with a **negative exponent** is equivalent to the reciprocal of the same number with the exponent made positive, $a^{-m} = \dfrac{1}{a^m}$, $a \neq 0$.
4. $\sqrt[n]{a} = b$ if

$$\overbrace{b \cdot b \cdot b \cdots b}^{n} = b^n = a$$

5. The **principal root** of a number has the same sign as the number itself.
6. When we multiply two radicals having a *common index,* we multiply the radicands and put the product under a common radical symbol with the original index: $\sqrt[n]{a}\,\sqrt[n]{b} = \sqrt[n]{ab}$.
7. We can bring a *factor* out from under the radical symbol if it is raised to a power equal to or greater than the index.
8. The nth root of a quotient can be written as the nth root of the numerator over the nth root of the denominator:

$$\sqrt[n]{\frac{a}{b}} = \frac{\sqrt[n]{a}}{\sqrt[n]{b}}$$

9. We eliminate radicals from the denominator of a fraction by **rationalizing** the denominator.
10. We can add or subtract only **similar radicals.**
11. **Conjugate factors** are used to rationalize a denominator when the denominator has two terms, one or both of which contain a square root.
12. When a number is raised to a **fractional exponent,** the numerator of the fractional exponent indicates the power to which the base is to be raised and the denominator indicates the root to be taken. That is, if $a > 0$ or if $a \neq 0$ and n is odd, $a^{\frac{m}{n}} = (\sqrt[n]{a})^m = \sqrt[n]{a^m}$
13. j is called an **imaginary number** such that $j^2 = -1$ and $j = \sqrt{-1}$.
14. **Complex numbers** are composed of a real part, denoted by a, and an imaginary part, denoted by bj: $(a + bj)$.

Chapter Review

[10.1] Write each of the given expressions in simplest form, with only positive exponents.

1. $(3R)^0$
2. $(a + b)^0$
3. $7x^0$
4. $4C^{-3}$
5. $\dfrac{1}{3S^{-2}}$
6. $\dfrac{5^{-2}}{5^{-3}}$
7. $X^{-3}T^{-2}$
8. $(2x^2y)^{-3}$
9. $\dfrac{4a^{-2}b^3}{6a^{-4}b}$
10. $\left(\dfrac{3x^{-2}}{12x^{-4}}\right)^{-1}$

[10.2] Find the principal root.

11. $\sqrt{81}$ 12. $\sqrt{0.25}$ 13. $-\sqrt{9}$ 14. $-\sqrt{49}$ 15. $\sqrt[3]{-27}$
16. $\sqrt[3]{-1}$ 17. $\sqrt[4]{81}$ 18. $\sqrt[3]{-125}$ 19. $\sqrt[5]{-32}$ 20. $\sqrt[7]{128}$

[10.3] Perform the indicated operations and express the given radicals in simplest form.

21. $\sqrt[3]{32}$ 22. $\sqrt{27}$ 23. $\sqrt{40}$ 24. $\sqrt[4]{a^7}$
25. $\sqrt{18R^2S^3}$ 26. $\sqrt[3]{81x^4y^6z}$ 27. $\sqrt{2}\sqrt{14}$
28. $\sqrt[3]{4}\sqrt[3]{6}$ 29. $\sqrt{18}\sqrt{10}$ 30. $\sqrt[4]{2a^3b}\sqrt[4]{8a^2b^5}$

[10.4] Express the given radicals in simplest form with all denominators rationalized.

31. $\sqrt{\frac{16}{17}}$ 32. $\sqrt{\frac{7}{18}}$ 33. $\sqrt[3]{\frac{3}{16}}$ 34. $\sqrt{\frac{R}{S}}$ 35. $\sqrt{\frac{x}{y^3}}$

36. $\frac{a}{\sqrt{ab}}$ 37. $\frac{2R}{\sqrt{RS}}$ 38. $\sqrt[3]{\frac{x}{y^2}}$ 39. $\sqrt[3]{\frac{ab}{c^2}}$ 40. $\sqrt[7]{\frac{a^5}{b^3c^4}}$

[10.5] Perform the indicated operations and simplify.

41. $3\sqrt{7} + 4\sqrt{7}$ 42. $\sqrt{18} + 5\sqrt{2}$
43. $3\sqrt{20} - \sqrt{45}$ 44. $2\sqrt{75} - \sqrt{3} + 5\sqrt{27}$
45. $\sqrt{50R} - 2\sqrt{8R}$ 46. $7\sqrt{9x} - 2\sqrt{4x}$
47. $11\sqrt[3]{4} + 5\sqrt[3]{4}$ 48. $9\sqrt[3]{16} - 2\sqrt[3]{54}$
49. $\sqrt[4]{81x^4y} - \sqrt[4]{16x^4y}$ 50. $3R\sqrt[3]{S} + 2\sqrt[3]{R^3S}$

[10.6]

51. $\sqrt{3}(\sqrt{5} - \sqrt{7})$ 52. $\sqrt{10}(\sqrt{14} + \sqrt{6})$
53. $\sqrt[3]{R^2}(\sqrt[3]{R} + \sqrt[3]{R^2})$ 54. $(5 + \sqrt{7})(3 - \sqrt{7})$
55. $(6 - \sqrt{3})^2$ 56. $(\sqrt{5} + \sqrt{6})(\sqrt{5} - \sqrt{6})$
57. $(\sqrt{3} + \sqrt{5})^2$ 58. $(2\sqrt{a} - \sqrt{b})(2\sqrt{a} + \sqrt{b})$
59. $(\sqrt{s} + 2\sqrt{t})(\sqrt{s} - 3\sqrt{t})$ 60. $(2\sqrt{R} - \sqrt{S})^2$

Express the given radicals in simplest form with all denominators rationalized.

61. $\frac{1}{\sqrt{3} - 2}$ 62. $\frac{2}{\sqrt{6} + 4}$ 63. $\frac{1}{\sqrt{R} + S}$

64. $\frac{x}{\sqrt{xy} + x}$ 65. $\frac{R}{R + \sqrt{S}}$ 66. $\frac{\sqrt{2} + 3}{\sqrt{2} - 3}$

[10.7] Evaluate the given expressions.

67. $(36)^{\frac{1}{2}}$ 68. $(16)^{\frac{1}{4}}$ 69. $8^{\frac{2}{3}}$

70. $(25)^{\frac{3}{2}}$ 71. $(-8)^{\frac{1}{3}}$ 72. $(32)^{\frac{-2}{5}}$

Perform the indicated operations and simplify.

73. $R^{\frac{2}{5}} \cdot R^{\frac{3}{5}}$

74. $S^{\frac{1}{3}} \cdot S^{\frac{3}{4}}$

75. $\dfrac{R^{\frac{1}{2}}}{R^{\frac{1}{4}}}$

76. $\left(a^{\frac{3}{2}}\right)^{\frac{3}{4}}$

77. $(x^{-3})^{-\frac{2}{3}}$

78. $(16a^4b^8)^{\frac{3}{4}}$

79. $\dfrac{R^2S^2}{R^{\frac{1}{2}}S^{\frac{3}{2}}}$

80. $\dfrac{x^{-\frac{2}{3}}x}{x^{-\frac{1}{2}}}$

81. $\dfrac{a^{\frac{3}{4}} \cdot a^{-\frac{2}{3}}}{a^{-\frac{1}{3}}}$

82. $\left(B^{-\frac{2}{3}}\right)^{-\frac{9}{10}}$

[10.8] Perform the indicated operations and simplify.

83. $\sqrt{-49}$ 84. $\sqrt{-28}$ 85. $(5j)^2$ 86. $(\sqrt{5}j)^2$

87. $\sqrt{-4}\sqrt{-25}$ 88. $\sqrt{-9}\sqrt{-16}$ 89. $\sqrt{-3}\sqrt{-7}$

90. $\sqrt{-2}\sqrt{-6}$ 91. $\sqrt{-3}\sqrt{-12}$ 92. $\sqrt{-2}\sqrt{-18}$

Chapter Test

Perform any indicated operations, simplify, and leave all denominators rationalized.

1. $\sqrt{12} + \sqrt{75}$
2. $\sqrt{24}\sqrt{6}$
3. $-\sqrt{25}$
4. $\sqrt{3}(\sqrt{12} + \sqrt{6})$
5. $\sqrt{\dfrac{x^2}{y}}$
6. $\dfrac{4^{-2}}{4^{-3}}$
7. $(16)^{\frac{3}{4}}$
8. $B^{\frac{3}{4}} \cdot B^{\frac{1}{2}}$
9. $\sqrt[5]{64R^7S^{10}T^3}$
10. $\left(y^{\frac{3}{4}}\right)^{\frac{2}{3}}$
11. $(6 - \sqrt{7})(6 + \sqrt{7})$
12. $\dfrac{1}{\sqrt{5} - 3}$
13. $\dfrac{z^{\frac{5}{2}}}{z^{\frac{3}{2}}}$
14. $(3 - \sqrt{5})^2$
15. $\sqrt{-64}$
16. $\sqrt[3]{-8}$
17. $(-27)^{-\frac{2}{3}}$
18. $3\sqrt{9R} - 2\sqrt{4R}$
19. $(S^{-4})^{-\frac{1}{2}}$
20. $\sqrt[5]{\dfrac{R}{S^2T^3}}$
21. $(\sqrt{x} - 2\sqrt{y})(\sqrt{x} + 2\sqrt{y})$
22. $\dfrac{a}{\sqrt{a} + \sqrt{b}}$
23. $\sqrt{-27}$
24. $\dfrac{5a^{-3}b^{-2}c^0}{10a^{-4}b^{-5}}$

25. $(7j)^2$

26. $\dfrac{2}{\sqrt{6}}$

27. $\dfrac{3}{9 + \sqrt{6}}$

28. $(3\sqrt{R} - \sqrt{S})^2$

29. $(\sqrt{-2})^2$

30. $\dfrac{x^3y^2}{x^{\frac{1}{3}}y^{\frac{1}{2}}}$

31. $\sqrt{-8}\,\sqrt{-2}$

32. $\dfrac{R^{\frac{3}{4}}R^{-\frac{2}{3}}}{R^{-\frac{1}{3}}}$

33. $\sqrt{-3}\,\sqrt{-5}$

In problems 34–35, round to the accuracy of the given data.

34. The following figure is made up of 12 equal squares, each of which has an area of 3.24 in^2. What are the dimensions of the figure?

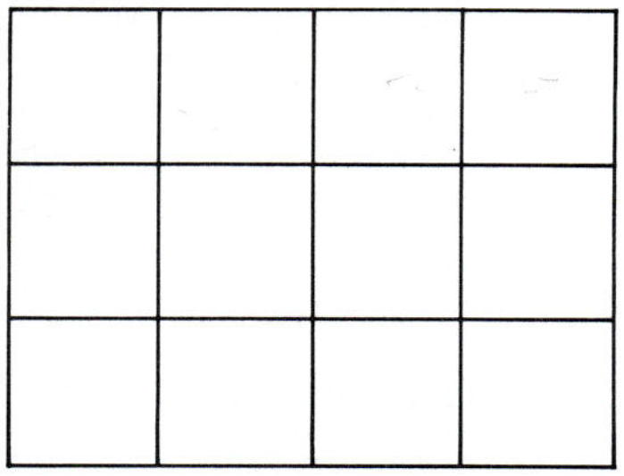

35. Find the height (h) of the given figure by finding b from the formula: $b = \sqrt{c^2 - s^2}$

If $c = 5.0$ yd and $s = 3.0$ yd find h.

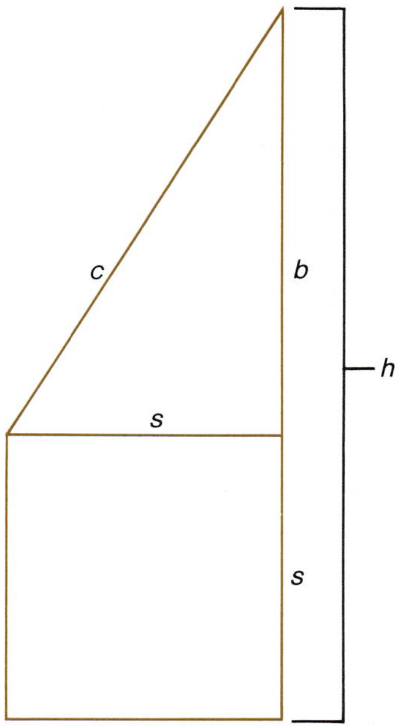

11
Quadratic Equations

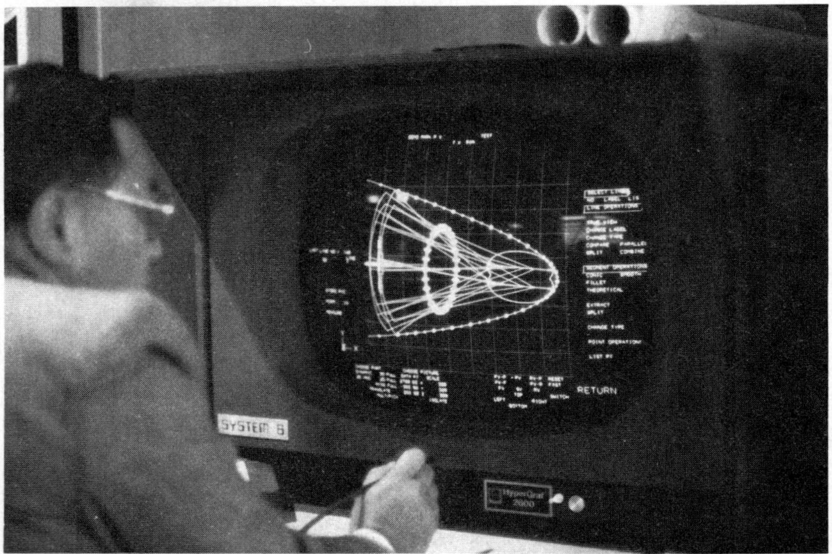

A technician plots the parabolic area of driver sight lines to help position the car instrument cluster in plain view.
(Photograph by Terry Wesner, courtesy of Ford Motor Company)

11.1 General Quadratic Form

In chapter 7 we studied solutions of linear equations, where the exponent of the unknown was 1. In this chapter we learn how to solve equations where the greatest power of the unknown is 2; these are called **quadratic equations.** Quadratic equations are used by technicians in many situations, including those dealing with area, acceleration, electronics, physics, and so on.

A quadratic equation is also called a **second-degree equation.** The *degree* of an equation refers to the highest sum of powers of the variables in any term. In an equation with one variable, the degree is the greatest power of the variable. Therefore, the greatest power of the variable in a quadratic equation is two.

When a quadratic equation is written with descending powers of the variable on one side of the equation and only zero on the other side, then the equation is said to be in *standard form*. The standard form of a quadratic equation is given by:

$$ax^2 + bx + c = 0 \qquad a \neq 0$$

note

It is necessary that $a \neq 0$ since the equation would no longer contain x^2 and therefore would not be a quadratic equation. The constants b and c can be zero, however, and the equation remains a quadratic. When b or c is zero, we have an **incomplete quadratic equation.**

The following are examples of quadratic equations in standard form.

Examples

1. $x^2 - 2x - 8 = 0$
2. $3x^2 + 5x + 6 = 0$
3. $x^2 + 5x = 0$
4. $x^2 - 4x = 0$
5. $x^2 - 9 = 0$
6. $x^2 - 1 = 0$
7. $x^2 = 0$

note

Preceding examples 1 and 2 are complete quadratic equations, whereas examples 3, 4, 5, 6, and 7 are incomplete quadratic equations since b or c, or both b and c, are zero.

The procedure for finding the roots of a quadratic equation make it necessary that quadratic equations *always* be stated in standard form. Thus, when an equation is not in standard form, as

$$3x^2 = 5x + 2$$

we must subtract $(5x + 2)$ from both sides of the equation to form the equivalent equation

$$3x^2 - 5x - 2 = 0$$

which is now in standard form.

Examples Arrange in standard form.

1. $x^2 - 3 = 2x$ By subtracting $2x$ from both sides we have
$$x^2 - 2x - 3 = 0$$
2. $3x^2 + 3 = -6x$ By adding $6x$ to both sides we have
$$3x^2 + 6x + 3 = 0$$
3. $x^2 = 16$ By subtracting 16 from both sides we have
$$x^2 - 16 = 0$$

Some of the equations we deal with require simplification before they can be put into standard form.

Examples

1. $$2x^2 + 3x = (x + 1)(x - 1)$$
First, we simplify the right side.
$$2x^2 + 3x = x^2 - x + x - 1$$
$$2x^2 + 3x = x^2 - 1$$
Then we move all the terms to one side.
$$2x^2 + 3x - x^2 + 1 = 0$$
Combine similar terms, and arrange in order.
$$x^2 + 3x + 1 = 0$$
2. $$2x(3x - 1) = (x + 2)(x + 1)$$
Simplify both sides.
$$6x^2 - 2x = x^2 + x + 2x + 2$$
$$6x^2 - 2x = x^2 + 3x + 2$$
All terms on one side.
$$6x^2 - 2x - x^2 - 3x - 2 = 0$$
Arrange in standard form.
$$5x^2 - 5x - 2 = 0$$
3. $$2 = \frac{1}{x} - \frac{1}{x^2}$$
Simplify (clear the fractions by multiplying both sides by the LCD).
$$x^2 \cdot 2 = \left(\frac{1}{x} - \frac{1}{x^2}\right)x^2$$
$$2x^2 = x - 1$$
All terms on one side and arrange in order.
$$2x^2 - x + 1 = 0$$

Sometimes, after we have simplified the equation, we find that we are not dealing with a quadratic equation. Consider the following.

Examples

1. $$2x(3x^2 - 1) = x - 3$$

   Simplify the left side.

   $$6x^3 - 2x = x - 3$$

   All terms on one side.

   $$6x^3 - 2x - x + 3 = 0$$
   $$6x^3 - 3x + 3 = 0$$

   We observe that the degree of the equation is 3, not 2; thus, we do not have a quadratic equation.

2. $$(x + 2)^2 = x(x - 4)$$

   Simplify both sides.

   $$(x + 2)(x + 2) = x(x - 4)$$
   $$x^2 + 2x + 2x + 4 = x^2 - 4x$$
   $$x^2 + 4x + 4 = x^2 - 4x$$

   All terms on one side.

   $$x^2 + 4x + 4 - x^2 + 4x = 0$$
   $$8x + 4 = 0$$

   Note that the degree of the equation is 1; thus, we have a linear equation.

In previous examples we saw how to put a quadratic equation in standard form, that is, into the form

$$ax^2 + bx + c = 0$$

Once the equation is in standard form, we can identify the constants a, b, and c.

Examples

Determine the values of a, b, and c.

1. $3x^2 + 2x + 5 = 0$ $\quad a = 3, b = 2, c = 5$
2. $2x^2 - 3x - 7 = 0$ $\quad a = 2, b = -3, c = -7$
3. $x^2 - x + 11 = 0$ $\quad a = 1, b = -1, c = 11$
4. $x^2 + 2x = 0$ $\quad a = 1, b = 2, c = 0$
5. $x^2 - 25 = 0$ $\quad a = 1, b = 0, c = -25$

Exercise Set 11.1

In the following problems:

1. Determine if the equation is a quadratic equation.
2. If it is a quadratic equation, state it in standard form.
3. When the equation is in standard form, determine the values of a, b, and c.

Examples 1.

$$2x(x - 4) = x^2 + 2$$
$$2x^2 - 8x = x^2 + 2$$
$$2x^2 - 8x - x^2 - 2 = 0$$
$$x^2 - 8x - 2 = 0$$
$$a = 1, b = -8, c = -2$$

2.

$$(x + 1)^2 = x(x - 3)$$
$$(x + 1)(x + 1) = x(x - 3)$$
$$x^2 + x + x + 1 = x^2 - 3x$$
$$x^2 + 2x + 1 = x^2 - 3x$$
$$x^2 + 2x + 1 - x^2 + 3x = 0$$
$$5x + 1 = 0$$ not a quadratic equation

1. $x^2 = 9x - 18$
2. $2e^2 = 5e$
3. $x^2 - x = 5$
4. $y^2 = 4$
•5. $x^2 = 6x$
6. $3x^2 - 6x = -3$
7. $12r^2 = 8r + 15$
8. $x^2 + 8 = -6x$
9. $12 - 25a + 12a^2 = 0$
10. $R^2 = 3R + 10$
11. $9 - 6x = -x^2$
12. $3t^2 = 9t$
13. $x = 12 - 6x^2$
14. $x^2 = 9x - 18$
15. $12x - 28 = -x^2$
16. $4t^2 = 16t$
•17. $50N + 24 = 25N^2$
18. $E^2 = 49$
19. $-12x^2 + 17x - 6 = 0$
20. $-18 + v^2 - 3v = 0$
21. $2x^2 - 4x + 3 = 2x^2 + 7$
22. $x^2 + 1 = 4x + 7 + x^2$
23. $(x + 3)(x - 4) = 11$
24. $2x(3x - 4) = 7$
•25. $(x + 1)^2 = x(x + 3)$
26. $2x^2(3x - 1) = x(x - 5)$
27. $x^2(x^2 + 1) = (x - 2)^2$
28. $(x + 3)(x - 1) = (x + 4)(x + 2)$
29. $(2x - 1)(x + 1) = 2x(3x - 4)$
30. $(x + 5)^2 - (x + 3) = 0$
31. $\frac{3}{x^2} = 6$
32. $\frac{10}{N} = 5N$
33. $\frac{3}{x} = \frac{x}{9}$
34. $\frac{a}{2} + \frac{4}{a} = 3$
35. $\frac{x}{3} + \frac{9}{x} = -3$
36. $\frac{B}{2} + \frac{1}{2B} = 2$
•37. $x - 4 = \frac{8}{x}$
38. $\frac{3}{2x} - \frac{1}{x^2} = 4$
•39. $\frac{2}{x + 1} = \frac{1}{x - 2}$
40. $\frac{1}{2x + 1} = \frac{2}{x - 3}$

11.2 Solutions by Factoring and Extracting the Root

Recall from chapter 7 that a root or solution of an equation is a replacement value of the variable that forms a true statement. Consider the quadratic equation $x^2 - x - 6 = 0$. For a value of x to be a solution of the equation, the value of $x^2 - x - 6$ must be zero when the number is substituted for x.

Examples

1. If we replace x with 2 in the equation we have

$$\begin{aligned}(2)^2 - (2) - 6 &= 4 - 2 - 6\\ &= 2 - 6\\ &= -4 \quad \text{which does not satisfy the equation.}\end{aligned}$$

2. If x is replaced with (-2) we have

$$\begin{aligned}(-2)^2 - (-2) - 6 &= 4 - (-2) - 6\\ &= 6 - 6\\ &= 0 \quad \text{and we observe that } (-2) \text{ is a solution of the equation.}\end{aligned}$$

3. If $x = 3$ we have

$$\begin{aligned}(3)^2 - (3) - 6 &= 9 - 3 - 6\\ &= 6 - 6\\ &= 0 \quad \text{and we see that 3 is a solution of the equation.}\end{aligned}$$

We conclude that a solution to a quadratic equation in standard form is a value of the variable that makes the left side of the equation equal to zero.

The first way to solve a quadratic equation that we will consider is by factoring. To solve a quadratic equation by factoring, we use the following mathematical idea: if $r \cdot s = 0$, then $r = 0$ and/or $s = 0$. That is: *If the product of two or more factors is zero, then at least one of the factors must be equal to zero.*

Consider the quadratic equation $x^2 - x - 6 = 0$. We can factor the left side into $(x + 2)(x - 3) = 0$. We know that for this product to be zero, either $(x + 2)$ or $(x - 3)$ must be equal to zero. Setting each of the factors equal to zero and solving, we have

$$\begin{aligned}x + 2 &= 0 \quad \text{or} \quad & x - 3 &= 0\\ x &= -2 & x &= 3\end{aligned}$$

Thus, the solutions are -2 or 3. These are the same values that previously satisfied the equation.

To solve a quadratic equation by factoring:

1. Put the equation in standard form.
2. Factor the left side of the equation.
3. Set each of the variable factors equal to zero and solve the resulting equations.

Examples

Solve the quadratic equations.

1. $x^2 + 5x + 6 = 0$ — The equation is in standard form.

   $(x + 2)(x + 3) = 0$ — Factor.

   $x + 2 = 0 \qquad x + 3 = 0$ — Set the factors equal to zero.

   $x = -2 \text{ or } \qquad x = -3$ — Solve.

2.
$$x^2 - 3 = 2x$$
$x^2 - 2x - 3 = 0$ — Standard form.
$(x - 3)(x + 1) = 0$ — Factor.
$x - 3 = 0 \qquad x + 1 = 0$ — Set the factors equal to zero.
$x = 3$ or $x = -1$ — Solve.

3.
$$3x^2 + 3 = -6x$$
$3x^2 + 6x + 3 = 0$ — Standard form.
$3(x^2 + 2x + 1) = 0$
$3(x + 1)(x + 1) = 0$ — Factor.
$x + 1 = 0 \qquad x + 1 = 0$ — Set the factors equal to zero.
$x = -1 \qquad x = -1$ — Solve.

note

In example 3, the factors involving x are the same. Therefore, there is only one solution.

4.
$$x^2 = 2x$$
$x^2 - 2x = 0$ — Standard form
$x(x - 2) = 0$ — Factor.
$x = 0 \qquad x - 2 = 0$ — Set the factors equal to zero.
$x = 0$ or $x = 2$ — Solve.

5. $2x^2 - 3x - 2 = 0$ — Standard form
$(2x + 1)(x - 2) = 0$ — Factor.
$2x + 1 = 0 \qquad x - 2 = 0$ — Set the factors equal to zero.
$2x = -1 \qquad x = 2$
$x = -\frac{1}{2}$ or $x = 2$ — Solve.

6.
$$x^2 = 16$$
$x^2 - 16 = 0$ — Standard form
$(x - 4)(x + 4) = 0$ — Factor
$x - 4 = 0 \qquad x + 4 = 0$ — Set the factors equal to zero.
$x =$ or $x = -4$ — Solve.

An alternate method of finding the solution in example 6 is also a useful procedure for dealing with incomplete quadratic equations where the constant b is zero. In the equation $x^2 = 16$, if we take the square root of both sides, we have:

$$x^2 = 16$$
$$x = \pm\sqrt{16}$$
$$x = \pm 4$$

and we have the same two answers, 4 and -4, that we had by factoring.

To solve an incomplete quadratic where $b = 0$:

1. Solve the equation for the square variable.
2. Take the square root of both sides. Remember that, when we take the square root of both sides, we have a positive and a negative result.

Examples Solve the quadratic equation.

1. $3x^2 - 12 = 0$ solving for x^2
$$3x^2 = 12$$
$$x^2 = 4$$
$$x = \pm\sqrt{4}$$ taking the square root
$$x = \pm 2$$

2. $9x^2 - 16 = 0$ solving for x^2
$$9x^2 = 16$$
$$x^2 = \frac{16}{9}$$
$$x = \pm\sqrt{\frac{16}{9}}$$ taking the square root
$$x = \pm\frac{4}{3}$$

3. $x^2 - 7 = 0$ solving for x^2
$$x^2 = 7$$
$$x = \pm\sqrt{7}$$ taking the square root

note

In examples 1 and 2 the original problem can be factored. In example 3, the problem will not factor and we must use extracting the root to solve for x.

Exercise Set 11.2

Solve the following equations.

1. $x^2 - 7x + 12 = 0$
2. $e^2 + 6e - 16 = 0$
3. $x^2 - 3x - 4 = 0$
4. $r^2 + 3r + 2 = 0$
•5. $P^2 + 5P - 14 = 0$
6. $x^2 + x - 42 = 0$
7. $x^2 + 2x = 0$
8. $2a^2 = 5a$
9. $2y^2 = 16$
10. $3y^2 - 27 = 0$
•11. $9a^2 - 25 = 0$
12. $7a^2 = 28$
13. $3y^2 = 6$
14. $5R^2 - 10 = 0$
15. $R^2 - 9 = 0$
16. $e^2 = 25$
17. $R^2 - 16 = 0$
18. $L^2 = 49$
•19. $2A^2 + 6A = 0$
20. $3N^2 = 9N$
21. $2y^2 - 18 = 0$
22. $6V^2 - 24V = 0$
23. $x^2 - 16x + 64 = 0$
24. $x^2 + 14x + 49 = 0$
25. $6y^2 - 11y - 10 = 0$
26. $3R^2 + 6R = -3$
27. $3V^2 = 7V + 6$
28. $-6x = -3x^2 - 3$
29. $12e^2 - 15 = 8e$
30. $6x^2 + 1 = 5x$
31. $2w^2 = w + 3$
32. $r^2 = 9r - 18$

33. $e = 12 - 6e^2$

34. $R^2 = 54 + 15R$

35. $18x^2 = 20$

36. $12 = 25x - 12x^2$

37. $2r^2 = 9 + 7r$

38. $9y^2 + 25 + 30y = 0$

39. $9v^2 + 20 = -27v$

40. $6P^2 - 7P = 20$

41. $3N^2 = 8 - 10N$

In problems 42–48, consider all given values to be exact.

42. An object with initial velocity v undergoes an acceleration a for time t. The displacement s of the object for this time is given by the equation

$$s = vt + \frac{1}{2}at^2.$$

a. Find t when $s = 6$, $v = 1$, $a = 2$.

b. Find t when $s = 6$, $v = 3$, $a = 6$.

43. The current in a circuit flows according to the equation $i = 12 - 12t^2$, where i is the current and t is the time in seconds. Find t when i is (a) 0 amp, (b) 9 amp.

44. The power output of a generator armature is given by $P_o = E_g I - r_g I^2$.

a. Find I when $P_o = 120$, $E_g = 22$, $r_g = 1$.

b. Find I when $P_o = 120$, $E_g = 16$, $r_g = 0.5$

45. When a square metallic plate is heated, the following relationship holds:

$$\frac{\Delta A}{\Delta S} = 2s + \Delta S,$$

where ΔA is the change in area, ΔS is the increase in length of each side, and s is the length of the side. Find ΔS when $\Delta A = 27$ and $s = 3$. (Δ is read "delta.")

46. The output power P of a 100-volt electric generator is defined by

$$P = 100I - 5I^2,$$

where I is amperes. Find I when (a) $P = 500$, (b) $P = 375$.

•47. A ball rolls down a slope and travels a distance $d = 6t + \frac{t^2}{2}$ feet in t seconds. Find t when (a) $d = 14$ ft, (b) $d = 32$ ft.

48. Kinetic energy (KE) in foot-pounds is determined by $KE = \frac{1}{2}mv^2$, where m is the weight in pounds and v the velocity in feet per second. Find v when $KE = 2.1 \times 10^6$ ft-lb and $m = 4.2$ lb.

11.3 The Quadratic Formula

In the previous section we considered quadratic equations that could be solved by factoring or extracting the root. Many quadratic equations cannot be solved by these means, however, and we need another procedure to determine these solutions.

Given the quadratic equation $ax^2 + bx + c = 0$, we can solve the equation in general, thereby deriving a formula by which *any* quadratic equation may be solved.

This formula is given by

$$x = \frac{-b \pm \sqrt{b^2 - 4ac}}{2a}$$

where a is the coefficient of the x^2 term, b is the coefficient of the x term, and c is the constant term. This equation is called the **quadratic formula.**

note

The $\pm$ symbol in the quadratic formula is a convenient way of combining the two solutions into one formula. That is, $\dfrac{-b \pm \sqrt{b^2 - 4ac}}{2a}$ is the same as $\dfrac{-b - \sqrt{b^2 - 4ac}}{2a}$ and $\dfrac{-b + \sqrt{b^2 - 4ac}}{2a}$.

To solve a problem by using the quadratic formula, we need only substitute the numerical values for a, b, and c into the formula and simplify. You must make sure that the sign of the number is also used in the substitution. Consider the equation $x^2 - 2x - 8 = 0$, where $a = 1$, $b = -2$, and $c = -8$. Then,

$$x = \frac{-(-2) \pm \sqrt{(-2)^2 - 4(1)(-8)}}{2(1)} = \frac{2 \pm \sqrt{4 - (-32)}}{2}$$

$$= \frac{2 \pm \sqrt{36}}{2} = \frac{2 \pm 6}{2}$$

Therefore,

$$x = \frac{2 + 6}{2} \quad \text{or} \quad x = \frac{2 - 6}{2}$$

$$x = \frac{8}{2} \qquad\qquad x = \frac{-4}{2}$$

$$x = 4 \qquad\qquad x = -2$$

note

Since $x^2 - 2x - 8 = 0$ factors into $(x - 4)(x + 2) = 0$, we can find the same answers, 4 and -2, by factoring. Since factoring is an operation that we are already familiar with, the quadratic formula should be used only when the problem will not factor.

To solve a quadratic equation by the quadratic formula:

1. Put the equation in standard form.
2. Identify a, b, and c.
3. Substitute the values of a, b, and c into the formula.
4. Simplify.

Examples

Solve by using the quadratic formula.

1. $x^2 = 4 - x$

$x^2 + x - 4 = 0$ Standard form

$a = 1, b = 1, c = -4$

$$x = \frac{-(1) \pm \sqrt{(1)^2 - 4(1)(-4)}}{2(1)} = \frac{-1 \pm \sqrt{1 - (-16)}}{2}$$

$$= \frac{-1 \pm \sqrt{17}}{2}$$

$$x = \frac{-1 + \sqrt{17}}{2} \quad \text{or} \quad \frac{-1 - \sqrt{17}}{2}$$

note

If a decimal approximation is required for this example, then, using $\sqrt{17} \approx 4.1$, we have $\frac{-1 + 4.1}{2} = 1.55$ and $\frac{-1 - 4.1}{2} = -2.55$.

2. $x^2 + 2x - 2 = 0 \quad a = 1, b = 2, c = -2$

$$x = \frac{-(2) \pm \sqrt{(2)^2 - 4(1)(-2)}}{2(1)} = \frac{-2 \pm \sqrt{4 - (-8)}}{2}$$

$$= \frac{-2 \pm \sqrt{12}}{2}$$

$$= \frac{-2 \pm 2\sqrt{3}}{2} = \frac{2(-1 \pm \sqrt{3})}{2} = -1 \pm \sqrt{3}$$

3. $3x^2 - x - 5 = 0 \quad a = 3, b = -1, c = -5$

$$x = \frac{-(-1) \pm \sqrt{(-1)^2 - 4(3)(-5)}}{2(3)} = \frac{1 \pm \sqrt{1 - (-60)}}{6}$$

$$= \frac{1 \pm \sqrt{61}}{6}$$

4. $2x^2 - 3x - 4 = 0 \quad a = 2, b = -3, c = -4$

$$x = \frac{-(-3) \pm \sqrt{(-3)^2 - 4(2)(-4)}}{2(2)} = \frac{3 \pm \sqrt{9 - (-32)}}{4}$$

$$= \frac{3 \pm \sqrt{41}}{4}$$

5. $x^2 - 7 = 0 \qquad a = 1, b = 0, c = -7$

$$x = \frac{-(0) \pm \sqrt{(0)^2 - 4(1)(-7)}}{2(1)} = \frac{\pm\sqrt{0 - (-28)}}{2} = \frac{\pm\sqrt{28}}{2}$$

$$= \frac{\pm 2\sqrt{7}}{2} = \pm\sqrt{7}$$

note

In example 5, we could have solved for x^2:

$$x^2 - 7 = 0$$
$$x^2 = 7$$

and extracted the root to get $x = \pm\sqrt{7}$.

6. $4x^2 - 3x = 0 \qquad a = 4, b = -3, c = 0$

$$x = \frac{-(-3) \pm \sqrt{(-3)^2 - 4(4)(0)}}{2(4)} = \frac{3 \pm \sqrt{9 - 0}}{8} = \frac{3 \pm 3}{8}$$

$$\frac{3 + 3}{8} = \frac{6}{8} = \frac{3}{4} \quad \text{or} \quad \frac{3 - 3}{8} = \frac{0}{8} = 0$$

note

In example 6, we could have solved by factoring:

$$4x^2 - 3x = 0$$
$$x(4x - 3) = 0$$
$$x = 0 \quad \text{or} \quad 4x - 3 = 0$$
$$4x = 3$$

So, $x = 0$ or $x = \frac{3}{4}$

7. $x^2 + x + 6 = 0 \qquad a = 1, b = 1, c = 6$

$$x = \frac{-(1) \pm \sqrt{(1)^2 - 4(1)(6)}}{2(1)} = \frac{-1 \pm \sqrt{1 - 24}}{2}$$

$$= \frac{-1 \pm \sqrt{-23}}{2} = \frac{-1 \pm j\sqrt{23}}{2}$$

note

Since $\sqrt{-23} = j\sqrt{23}$, our answers are complex numbers. This result occurs whenever the number under the radical symbol is negative. Written as complex numbers, the answers would be

$$-\frac{1}{2} \pm \frac{\sqrt{23}}{2}j.$$

Exercise Set 11.3

Solve the following quadratic equations by using the quadratic formula.

•1. $x^2 - 3x + 2 = 0$
2. $r^2 - 2r + 1 = 0$
3. $L^2 + 6L + 9 = 0$
4. $x^2 + 10x + 16 = 0$
5. $x^2 - 25 = 0$
6. $2x^2 - 8 = 0$
•7. $5x^2 - 10 = 0$
8. $3x^2 - 9 = 0$
9. $x^2 - 9x + 4 = 0$
10. $e^2 - 5e = 6$
11. $x^2 + 2x - 6 = 0$
12. $y^2 + y + 1 = 0$
13. $R^2 + 1 = 8R$
14. $2S^2 = 7S - 6$
15. $3E^2 = 5E + 6$
16. $x^2 = 4x$
17. $-x^2 = -3x$
18. $2x^2 = 7x$
19. $5x^2 - 9x = 0$
20. $4t^2 = 8t - 3$
21. $3a^2 = -9a - 2$
22. $x^2 - 9x = 6$
23. $3r^2 = r + 10$
24. $3I^2 + 5I = 4$
25. $x^2 + 8x + 16 = 0$
26. $2v^2 + 5v = -2$
27. $4x^2 + 25 = 20x$
28. $4x^2 - 7 = 12x$
29. $4x^2 + 12x + 9 = 0$
30. $4t^2 = 9t - 6$
31. $3a^2 - 2a - 7 = 0$
32. $4I^2 = 8 - 2I$
33. $3x^2 = 18 - 6x$
34. $9x^2 + 4 = 12x$
•35. $3r^2 - 3r = -8$

Solve the following problems using the quadratic formula. Consider all given values to be exact.

36. Using the formula $s = vt + \frac{1}{2}at^2$, find t when (a) $s = 8$, $v = 3$, $a = 4$; (b) $s = 5$, $v = 4$, $a = 2$.
37. Using the formula $P = 100I - 5I^2$, find the amperes I when the power of the 100-volt electric generator is (a) $P = 80$, (b) $P = 64$.
38. The distance s through which an object will fall in t seconds is $s = \frac{1}{2}gt^2$ ft, where $g = 32$ ft/s^2. Find t (correct to one-tenth of a second) when (a) $s = 64$, (b) $s = 96$, (c) $s = 120$.
39. If a projectile is fired vertically into the air, the distance in feet above the ground that it attains in t seconds is given by $s = 160t - 16t^2$. Find t when (a) $s = 0$, (b) $s = 40$, (c) $s = 160$.
40. In a certain electric circuit the relationship between i (in amperes), E (in volts), and R (in ohms) is given by $i^2R + iE = 8{,}000$. Find i ($i > 0$) when (a) $R = 2$ and $E = 80$, (b) $R = 4$ and $E = 60$.

Calculator Problems

Use your calculator; approximate the solutions to two decimal places.

41. $x^2 - 6x + 3 = 0$
42. $x^2 + 5x - 1 = 0$
43. $2x^2 - 3x - 4 = 0$
44. $5x^2 - x - 1 = 0$
45. $3x^2 + 2x - 2 = 0$
46. $3x^2 - 11 = 0$
47. $5x^2 - 2 = 0$
48. $3x^2 + 7x - 2 = 0$
•49. $4x^2 - 5x - 19 = 0$
50. $6x^2 + 8x - 5 = 0$

11.4 Verbal Problems

Many problems encountered by technicians are stated verbally and need to be translated into algebraic equations. Some of the mathematical models we use to determine technical situations are quadratic equations and, therefore, have two solutions. We may find that one, but not both, of the solutions of the equation fits the technical situation.

Consider the following example: A piece of lumber is to be divided into two pieces so that one piece is 5 in longer than the other. If the product of the two lengths is 104 in^2, what are the two lengths? We approach the problem with the same procedure we used in section 7.5. We let x equal the length of the shorter piece, and then $x + 5$ is the length of the longer piece. We can next write the equation and solve for the unknown:

$$\begin{aligned} x(x + 5) &= 104 \\ x^2 + 5x &= 104 \\ x^2 + 5x - 104 &= 0 \\ (x + 13)(x - 8) &= 0 \end{aligned}$$

Thus, $x = -13$ or 8. But, since we cannot have a negative length, 8 is our only solution. Relating this back to the original problem, we have that the lengths are 8 in and $8 + 5$ or 13 in.

Examples

1. Find two numbers whose sum is 13 and whose product is 40.

First Number	*Second Number*	*Equation*
x	$13 - x$	$x(13 - x) = 40$

Solution

$$\begin{aligned} x(13 - x) &= 40 \\ 13x - x^2 &= 40 \\ 0 &= x^2 - 13x + 40 \\ 0 &= (x - 5)(x - 8) \end{aligned}$$

answer: $x = 5,\quad 13 - 5 = 8,$
or if $x = 8,\quad 13 - 8 = 5$
Therefore, the two numbers are 5 and 8.

2. Four times an integer subtracted from the square of the integer is 21. Find the integer.

The Number	*Equation*
x	$x^2 - 4x = 21$

Solution

$$\begin{aligned} x^2 - 4x &= 21 \\ x^2 - 4x - 21 &= 0 \\ (x - 7)(x + 3) &= 0 \\ 7 \quad \text{or} \quad &-3 \end{aligned}$$

Since there were no stipulations that the answers were to be positive or negative, both results are the solutions.

3. The perimeter of a rectangle is 52 in and its area is 153 in^2. What are its dimensions? The formula for the perimeter of a rectangle is $P = 2L + 2W$. If we substitute 52 in place of P we have $52 = 2L + 2W$. Dividing both sides of the equation by 2 we have $26 = L + W$. We can use this fact to establish the unknowns.

Width of Rectangle	*Length of Rectangle*	*Equation*
W	$26 - W$	$153 = W(26 - W)$

Solution

$$\begin{aligned} W(26 - W) &= 153 \\ 26W - W^2 &= 153 \\ 0 &= W^2 - 26W + 153 \\ 0 &= (W - 9)(W - 17) \end{aligned}$$

The dimensions are 9 in and 17 in.

4. The length of a rectangle is 3 m more than three times its width. The area of the rectangle is 168 m^2. Find the dimensions.

Width of Rectangle	*Length of Rectangle*	*Equation*
W	$3W + 3$	$W(3W + 3) = 168$

Solution

$$\begin{aligned} W(3W + 3) &= 168 \\ 3W^2 + 3W &= 168 \\ 3W^2 + 3W - 168 &= 0 \\ 3(W^2 + W - 56) &= 0 \\ 3(W + 8)(W - 7) &= 0 \end{aligned}$$

Width is -8 or 7. Since we cannot have a negative width, 7 is our only answer. The dimensions are 7 m = width and $3 \cdot 7 + 3 = 21 + 3$ or 24 m = length.

Exercise Set 11.4

Write an equation for each problem and solve for the unknown quantities. Consider all given values to be exact.

1. Find two numbers whose sum is 11 and whose product is 24.
2. Find two numbers whose difference is 6 and whose product is 72.
•3. The sum of a number and its square is 56. Find the number.
4. The difference of a number and its square is 56. Find the number.
5. Three times a number added to the square of the number is 28. Find the number.
6. Seven times a number is 12 more than the number squared. Find the number.
•7. The difference of eight times a number and the square of the number is 16. Find the number.
8. Four times a number is 12 less than the number squared. Find the number.
9. The product of two consecutive numbers is 8 more than 3 times the smaller number. Find the numbers.
10. The product of two consecutive numbers is 8 less than 8 times the next consecutive number. Find the numbers.
11. The product of two consecutive odd integers is 143. Find the integers.
12. The product of two consecutive integers is 132. Find the integers.

•13. The product of two consecutive even integers is 168. Find the integers.
14. The product of two consecutive integers is 306. Find the integers.
15. One integer is 6 more than a second integer. The product of the two integers is 91. Find the integers.
16. One integer is 8 less than a second integer. The product of the two integers is 153. Find the integers.
•17. The product of two consecutive even integers is 4 more than 2 times their sum. Find the integers.
18. The product of two consecutive odd integers is 5 more than 6 times the lesser integer. Find the integers.
•19. The sum of two integers is -13 and their product is 36. Find the integers.
20. The sum of two integers is -3 and their product is -70. Find the integers.
21. One number is 1 more than 3 times the other. Their product is 14. Find the numbers.
22. One number is 2 more than the other and their product is -1. Find the numbers.
•23. The difference of two positive numbers is 5 and the sum of their squares is 97. Find the numbers.
24. The sum of two numbers is 11. The difference of their squares is 5 more than their product. Find the numbers.
•25. The area of a rectangular piece of sheet metal is 117 in². If the sum of the length and the width is 22 in, what are the dimensions of the metal plate? [*Hint:* Let L be the length of the piece. Then what is the width of the piece in terms of L? $(A = LW)$]
26. The length of a rectangular-shaped die is 7 in longer than its width. What are the length and the width of the rectangular face whose area is 78 in²?
27. The length of a rectangle is 2 m less than twice the width. If the area is 24 m², what are the dimensions of the rectangle?
28. The area of a rectangle is 21 ft². What are its dimensions if the length is 5 ft less than 4 times the width?
•29. The area of a rectangle is numerically equal to twice the length. If the length is 3 ft more than the width, what are the dimensions of the rectangle?
30. The length of a rectangle is 3 less than twice the width. If the area is numerically 5 times the length, find the dimensions of the rectangle.
31. The height of a book page is 3 in more than the width. If the area of the page is 6 less than 10 times the width, find the width of the page.
32. The height of a book page is 4 in more than the width. If there is a margin of 1 in all around the printed matter of the page, and the area of the printed matter is 32 in², what is the height of the page?
33. A rectangular part has a width that is 8 mm shorter than its length. If the part has a cross-section area of 105 mm², what are its dimensions?
34. A triangular-shaped plate has an altitude that is 5 in longer than its base. If the area of the plate is 52 in², what is the length of the base, b, and the altitude, h, if the area of a triangle, A, is given by $A = \frac{1}{2}bh$?
35. A metal bar is to be divided into two pieces so that one piece is 4 in shorter than the other. If the sum of the squares of the two lengths is 208 in², find the two lengths.

36. A metal strip is shaped into a right triangle, where $c^2 = a^2 + b^2$. Find x when $a = x$, $b = x + 14$, and $c = x + 16$. (*Hint:* Substitute for a, b, and c in the relationship and solve for x.)
37. The area of a triangle is 54 in². If the altitude in one-third the base, what are the lengths of the altitude and base? (area = 1/2 × base × altitude)
•38. The hypotenuse of a right triangle is 10 mm long. One leg is 2 mm longer than the other. What are the lengths of the two legs? (In a right triangle, the square of the hypotenuse is equal to the sum of the squares of the legs.)
39. The perimeter of a rectangle is 38 cm, and its area is 88 cm². What are its dimensions?
40. 360 m of fencing is used to enclose a rectangular field. Find the dimensions of the field if the length is 2 times the width.
41. The length of a rectangle is 2 m more than 3 times the width. Its area is 16 m². What are the dimensions?
42. If two metal bars are the same length, and the length of one is increased by 3 cm and if the second is decreased by 3 cm, the product of these two lengths is 27. Find the original length.

The formula $h = vt - 16t^2$ gives the height (h) that an object will travel in t seconds if it is propelled directly upward at a velocity of v feet per second. Use this formula in exercises 43 through 46.

43. If a ball is thrown upward at 96 ft/s, how long will it take the ball to reach a height of 80 ft?
44. How long before the ball of exercise 43 hits the ground? (*Hint:* $h = 0$ when this happens.)
45. If a projectile is fired upward with an initial velocity of 144 ft/s, how long will it take before the projectile strikes an object 288 ft directly overhead? (Refer to exercise 43).
46. How long before the projectile in exercise 46 falls back to the ground?
•47. The formula for the area A of a trapezoid is $A = \frac{1}{2}h(b_1 + b_2)$, where h is the altitude (height) of the trapezoid and b_1 and b_2 are the parallel bases.

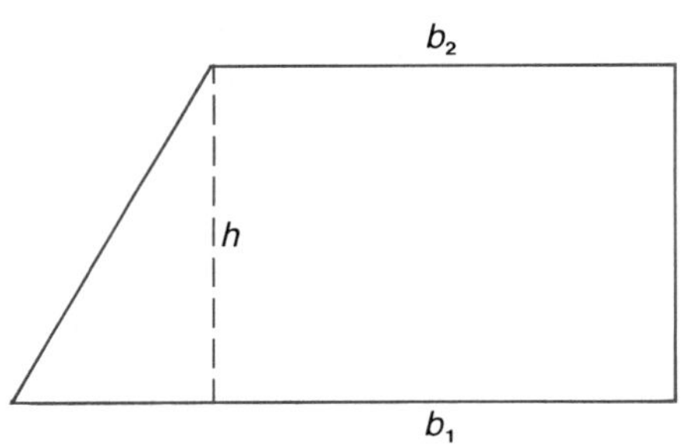

If the area of the trapezoid is 63 in², base b_1 is 10 in long and the altitude h is 1 in less than the length of b_2, find the length of h and b_2.

48. If the area of a trapezoid is 21 ft², base b_2 is 5 ft long, and base b_1 is 6 ft longer than the altitude h, find the altitude of the trapezoid.
49. The altitude and base b_2 have the same length. If the area of the trapezoid is 24 m² and base b_1 is twice as long as base b_2, find the dimensions of the trapezoid.
50. One base of a trapezoid is 3 times the length of the other base. If the altitude is twice as long as the shorter base and the area is 36 cm², find the dimensions of the trapezoid.
•51. The volume of a box (rectangular solid) is given by $V = lwh$, where l is the length, w is the width, and h is the height of the box.

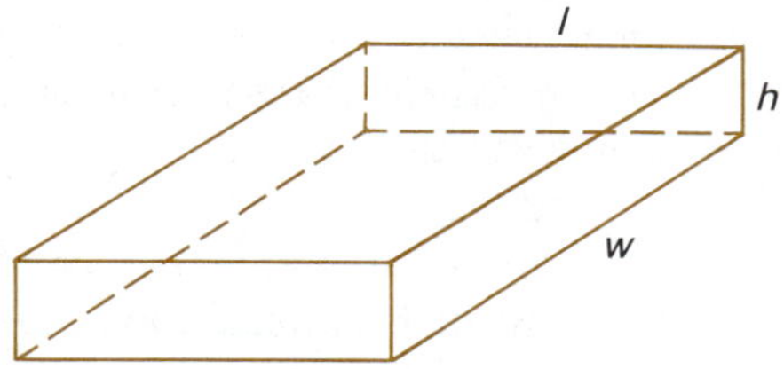

If a box is 4 ft tall, the length is 1 ft longer than the width, and the volume is 224 ft³, find the length and width of the box.
52. A box is 9 in long and has a volume of 162 in³. If the width of the box is twice the height, find the width and the height of the box.
53. A storage room is 3 times as long as it is wide. If the room contains 336 ft³ of space and has a ceiling 7 ft high, find the length and width of the room.
54. A cardboard box has a volume of 108 in³. If the length of the box is 1 in more than 2 times its width and the box has a height of 3 in, find the length and width of the box.
55. When two capacitors are connected in parallel, the total capacitance of the circuit, C_t, is given by $C_t = C_1 + C_2$. If the number of microfarads in C_1 is the square of the number in C_2, and the total capacitance is 42 microfarads, find the capacitance of each capacitor.
56. When a square piece of metal is heated, its area changes 4 mm² and the length of its side increases by 0.2 mm. Find the original length of each side. (*Hint:* The area A of a square is given by $A = s^2$, where s is the length of the side. Let s be the original length of the side and s^2 be the original area.)

Chapter Summary

1. The **standard form** of a **quadratic equation** is

   $$ax^2 + bx + c = 0 \qquad a \neq 0$$

2. If $b = 0$ or $c = 0$, then we have an **incomplete quadratic** equation.
3. We learned three ways of finding the solutions of a quadratic equation:
   a. Factoring
   b. Extracting the root
   c. Using the quadratic formula
4. The **quadratic formula** is given by:

   $$x = \frac{-b \pm \sqrt{b^2 - 4ac}}{2a}$$

   where a is the coefficient of the x^2 term, b is the coefficient of the x term, and c is the constant term.

Chapter Review

[11.1] State the following in standard form and determine the values for a, b, and c.

1. $3x^2 + 6 = 2x$
2. $5e^2 = 4e$
3. $6 = 7x^2$
4. $5x^2 = 8 - 4x$
5. $7 = 3x - x^2$
6. $4y + 3 = -8y^2$

[11.2] Find the solution by factoring.

7. $x^2 + 8x + 15 = 0$
8. $r^2 - 5r = 0$
9. $2e^2 = 18$
10. $i^2 - 9 = 0$
11. $x^2 = 2x - 1$
12. $R^2 - R = 42$
13. $x^2 + 7x - 8 = 0$
14. $2x^2 - 3x + 1 = 0$
15. $6x^2 = -x + 1$

[11.3] Find the solution by factoring or by use of the quadratic formula (factor, if possible).

16. $E^2 - 2E - 3 = 0$
17. $2r^2 - 3r = 5$
18. $x^2 - 5x + 6 = 0$
19. $4e^2 - 8e = 21$
20. $R^2 - 7R + 6 = 0$
21. $2x^2 = 9 - 7x$

[11.4]

22. A metal bar is to be divided into two pieces so that one piece is 3 in longer than the other. If the sum of the squares of the two lengths is 117 in^2, find the two lengths.
23. The difference of a number and 2 times its reciprocal is $\frac{31}{4}$. Find the number.
24. Find two consecutive positive integers whose product is 72.

Chapter Test

Solve the following equations by any means.

1. $x^2 - 3x - 10 = 0$
2. $x^2 = 81$
3. $x^2 + 3x - 5 = 0$
4. $x^2 - 10 = 0$
5. $2x^2 + x - 3 = 0$
6. $x^2 = 3x$
7. $3x^2 - x - 5 = 0$
8. $4x^2 - 36 = 0$
9. $x^2 - 2x - 4 = 0$
10. $x^2 + 4x + 1 = 0$
11. $x^2 - 3 = 0$
12. $x^2 = 9x$

13. $2x^2 - 3x - 4 = 0$

14. $2x^2 - 5x = 0$

15. $5x^2 + x - 4 = 0$

16. $2x^2 - 12 = 0$

17. $5x^2 - 16 = 0$

18. $5x^2 = 4x$

19. $x^2 + 9x + 14 = 0$

20. $4x^2 - 4x + 3 = 0$

21. Find two consecutive even numbers the sum of whose squares is 100.

22. The sum of a number and its reciprocal is $\frac{25}{12}$. Find the number.

23. The width of a rectangular-shaped die is 4 in shorter than its length. What are the length and the width of the rectangular face whose area is 45 in^2?

24. A rectangular part has a length that is 10 mm longer than its width. If the part has a rectangular face area of 96 mm^2, what are the length and width of this face?

12
Elements of Geometry

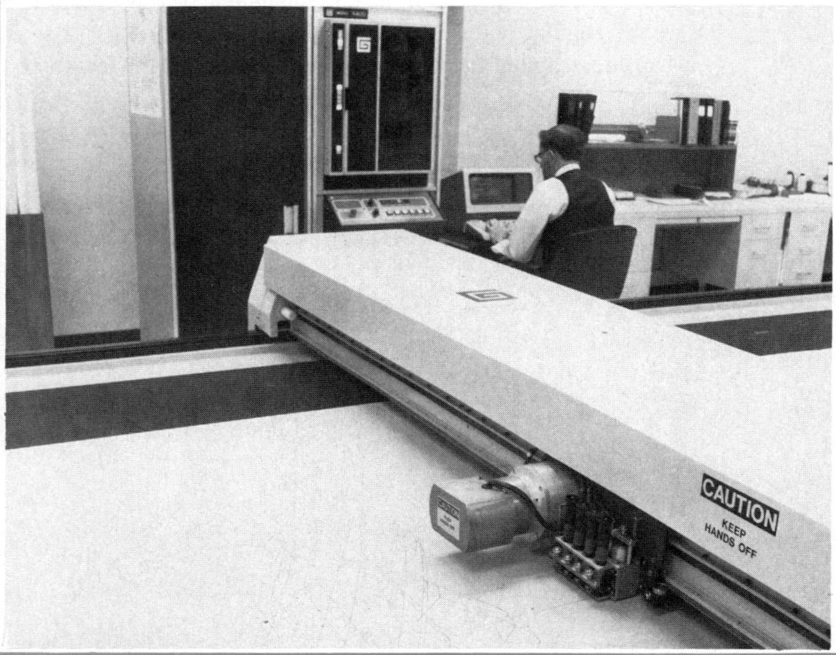

A technician using an automated drafting machine is now able to complete a drawing in three minutes that used to take more than a full week.
(Photograph courtesy of Ford Motor Company)

12.1 Introduction

Geometry had its beginning in either Egypt or Babylonia with the measurement of simple geometric figures. The Greek historian Herodotus wrote that the Egyptians were often forced to resurvey (remeasure) their lands because landmarks were destroyed when the Nile river flooded its banks. This explanation by Herodotus may explain why the name "geometry" has been applied to this field: the Greek word *geo* means "earth" and *metron* means "measure." The word *geometry* then takes on the meaning "earth measurement."

About 300 B.C. a Greek teacher of mathematics at the University of Alexandria, named Euclid, gathered the elementary facts of geometry into 13 books, called Euclid's *Elements*. These geometric concepts formed the basis for a rigorous, formal study of geometry. Another mathematician, Archimedes, then applied the geometry to mechanics. The plane and solid geometry of the texts we still use were almost completed when he died.

Geometry is concerned with the properties of points, lines, planes, triangles, squares, rectangles, circles, and other geometric figures. Engineers, designers, and draftspeople use geometric figures in layout plans for construction. These drawings, called *mechanical drawings,* are used in science, engineering, navigation, and industry.

Only those concepts and properties of geometry that are most important and useful in technical areas will be discussed in this chapter. It is not our intent to present a rigorous study of the field, so very few geometric proofs will be presented. Rather, the study and exercises will be developed so that the student who has had no complete, formal course in geometry will be able to solve problems that might be encountered in any future study of technical areas.

12.2 Points, Lines, and Planes

The three basic geometric elements in the study of geometry are the *point,* the *line,* and the *plane* (or *flat surface*). We will use these words and concepts without defining them and assume that the student has an intuitive understanding of them. We then use these undefined terms to define other terms.

A **point** is represented by a dot (·) and has only position. It has no length, width, or thickness. In other words, a point has no dimension. Points are designated in drawings by a capital letter next to the dot. See figure 12.1. Notice that a point can be formed by the intersection of two straight lines, as with point C in figure 12.1**b.**

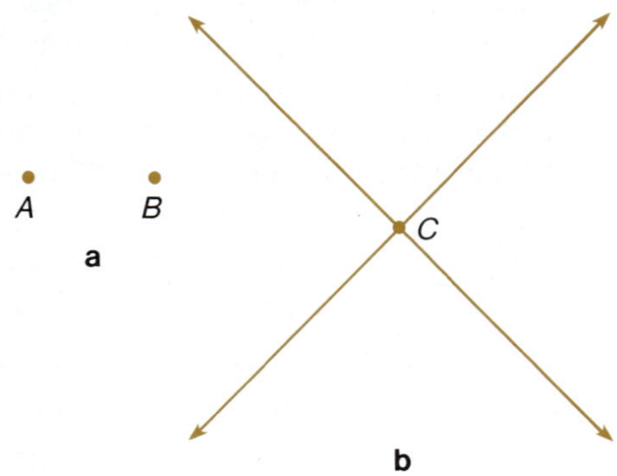

Figure 12.1

A **line** may be represented by the path of a pencil on a piece of paper, the path of a piece of chalk on a blackboard, or a string lying on the desk. Lines consist of a set of infinitely many points that together make up the geometric figure called the line. There are three kinds of lines, as illustrated in figure 12.2: straight, curved, and broken. Notice that arrows have been placed at the ends of the given lines. This is to indicate that these lines extend indefinitely far in each direction.

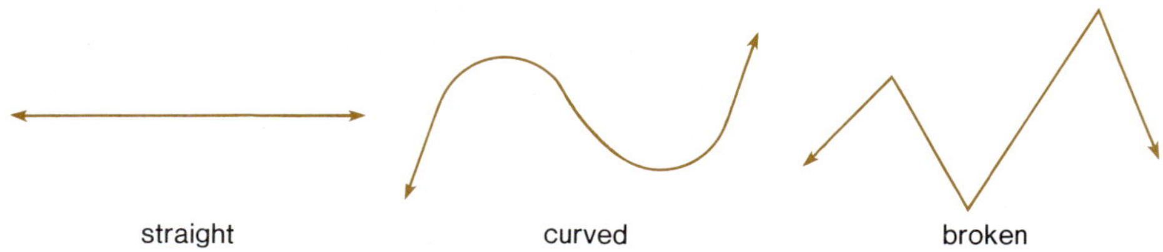

Figure 12.2

note

In this chapter we will work mostly with straight lines, so the word *line* will mean a *straight line* unless otherwise stated.

A line is designated by a lowercase letter or by two of its points. Consider figure 12.3. The line in figure 12.3 may be referred to as (1) line AB, or (2) line l. When two letters are used to denote a line, the symbol of a two-headed arrow is commonly placed above the letters—that is, $\overleftrightarrow{AB}$ means line AB.

Figure 12.3

The part of a line bounded by two specific points on the line is called a **line segment.** In figure 12.3 the portion of line l that lies between points A and B, including A and B, is the line segment AB, denoted by $\overline{AB}$.

note

A line segment *always* has a specific length whereas a line does not.

Given a specific point on a line, it is obvious that, with respect to that point, there are two "sides" of the line. Each of these sides is called a **half-line.** A **ray** is one of the half-lines plus the dividing point. To illustrate, let point A be a point of a horizontal line. Suppose point B is another point of the line to the left of A, and point C is another point of the line to the right of point A. See figure 12.4.

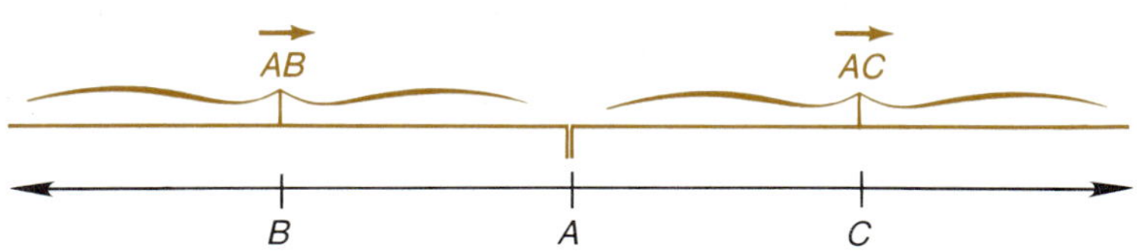

Figure 12.4

Then point A combined with the half-line to the right of point A that includes point C is called *ray AC*, denoted by $\overrightarrow{AC}$. Similarly, point A combined with the half-line to the left of point A that includes point B is called *ray AB*, denoted by $\overrightarrow{AB}$. We call point A the *endpoint* of each ray.

An important property of lines that is intuitively obvious is: *Through any two points, one and only one straight line can be drawn.*

A **plane,** or **flat surface,** is suggested by a flat sheet of paper, the top of a desk, the floor, walls, or ceiling of a room, or small portions of the earth's surface. A plane consists of infinitely many points that together make up the surface that is a plane. A plane is always perfectly flat. A straight line always fits perfectly in, or on, a plane.

12.3 Angles

The geometric figure that is formed by two rays having a common endpoint is called an **angle.** The common endpoint is called the **vertex** of the angle. Each ray is called a **side** of the angle. See figure 12.5. An angle in a plane obviously has an *interior* and an *exterior*, as indicated in figure 12.5. The symbol we use for an angle is $\angle$.

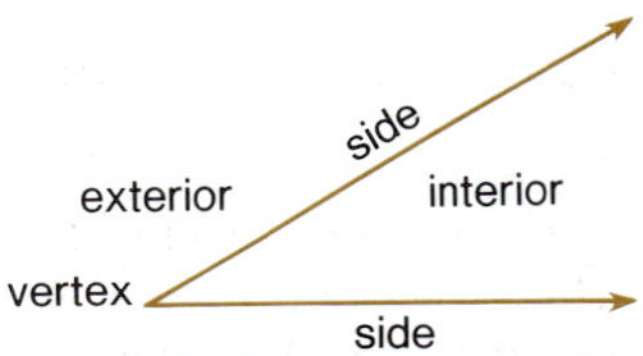

Figure 12.5

An angle can be referred to by the letter at its vertex, by a point on each of its sides together with its vertex, or by any designated letter in the interior of the angle. See figure 12.6, where the sides of the angle are $\overrightarrow{BA}$ and $\overrightarrow{BC}$ with the vertex at point B. The angle in figure 12.6 can be named:

1. Angle B, denoted by $\angle B$
2. Angle ABC, denoted by $\angle ABC$, or angle CBA, denoted by $\angle CBA$
3. Angle α, denoted by $\angle\alpha$

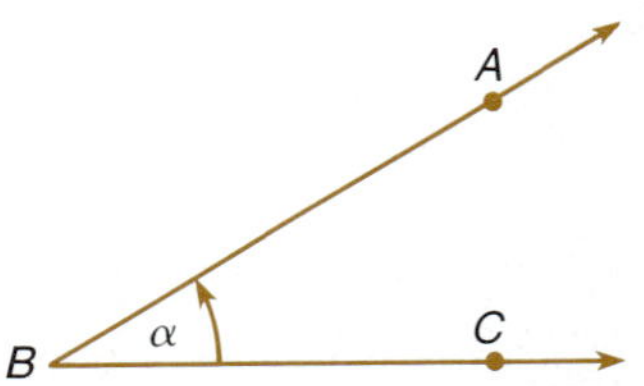

Figure 12.6

note

When a letter is used and placed in the interior of the angle (between the two sides) it is common practice to use Greek letters such as α (alpha), β (beta), γ (gamma), θ (theta), and so on. Sometimes a number is used in this way.

When naming the angle by using a point on each side and its vertex, *always* place the letter at the vertex between the other two letters, as we did with letter B, which is at the vertex in the angle ABC in figure 12.6.

If we think of one side (ray) of an angle rotating away from the other side, as indicated by the arrow in figure 12.6, the amount of this rotation determines the measure of the size of the angle, given in either *degrees, radians,* or *grads*. We shall limit our discussion in this chapter to measures of angles in degrees. Radian measures of angles will be discussed in a later chapter.

When one side of the angle makes a complete rotation so that it again coincides with the other side, the angle is said to have a measure of 360 degrees, denoted by 360°. An angle of measure 1° is 1/360th of a complete rotation. See figure 12.7. The measure of the angle depends on the size of the opening between the two sides measured at the same distance from the vertex.

Figure 12.7

The instrument used to measure angles in degrees is the **protractor.** The protractor can also be used to draw an angle of a given measure. To draw an angle of a given size, such as 30°, proceed as follows:

1. Draw line *DE*.

2. Locate a point on *DE* where you desire the vertex of the angle to be; call it *F*.

3. Place the protractor so that its base *AB* lies along the line *DE*, with its center, *C*, coinciding with point *F*.

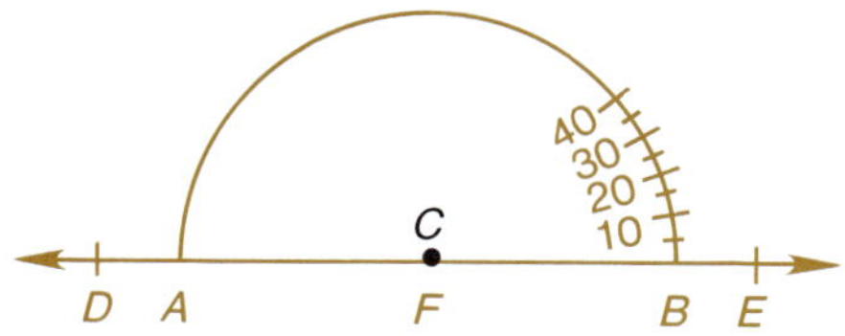

4. Starting at 0° on the right side of the protractor, find the desired number of degrees, 30. Place a dot on the paper at this point, called *G*.

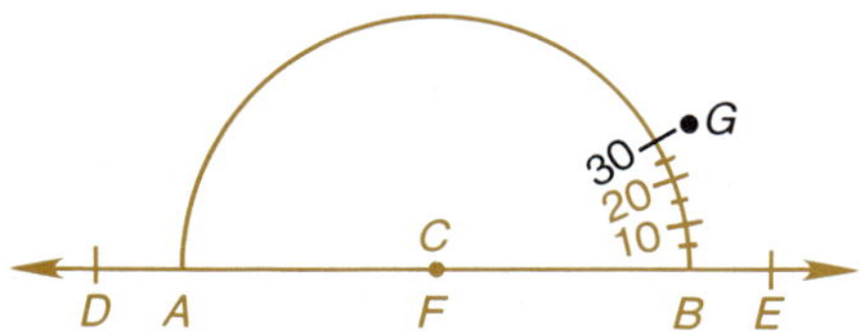

5. Draw a line from the vertex (at *F*) through the dot just located, *G*.

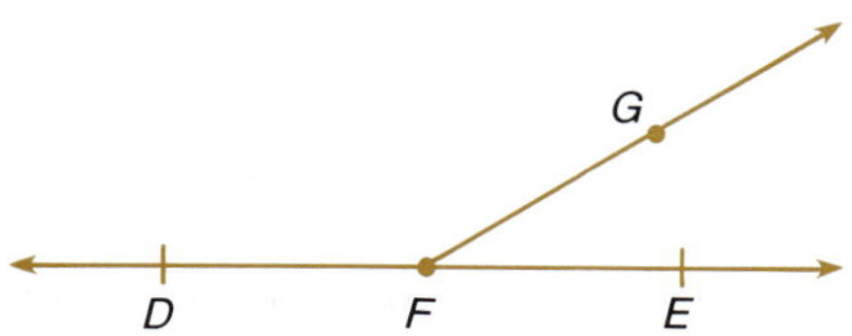

To measure the size of a given angle with a protractor we follow much the same procedure we used to draw the angle.

1. Place the base of the protractor along one side of the angle, with point *C* at the vertex.
2. The measure of the angle is determined by the number on the protractor where the other side crosses the protractor's curved number scale.

note

If the side of the angle is too short to cross the curve of the protractor, extend the side with a pencil line.

The basic unit of measure of an angle is the degree. A degree can be divided into 60 equal parts called **minutes,** denoted by $'$. The minute can also be divided into 60 equal parts called **seconds,** denoted by $''$. For example, 25 degrees, 15 minutes, 21 seconds is written

$25°15'21''$

When it is necessary to measure angles of less than $1°$, decimal parts of a degree are often used instead of minutes and seconds: $1° = 60'$ and $1' = 60''$; $1' = (1/60)°$ and $1'' = (1/60)'$.

Examples

1. Convert $13°24'$ to degrees.

$$24' = \left(\frac{24}{60}\right)^{\circ} = \left(\frac{2}{5}\right)^{\circ} = 0.4°. \text{ Then } 13°24' = 13.4°.$$

2. Convert $9'45''$ to minutes.

$$45'' = \left(\frac{45}{60}\right)' = \left(\frac{3}{4}\right)' = 0.75', \text{ and so } 9'45'' = 9.75'.$$

3. Convert $34.7°$ to degrees and minutes.
   $0.7° = (0.7)(60') = 42.0'$, and so $34.7° = 34°42'$.
4. Convert $4.15'$ to minutes and seconds.
   $0.15' = (0.15)(60'') = 9.00''$. Then $4.15' = 4'9''$.

To make the conversions used in these examples, we follow these rules:

1. To convert to a decimal part of a degree, divide by 60 (as in examples 1 and 2).
2. To convert a decimal part of a degree to minutes, multiply by 60 (as in examples 3 and 4).

Examples

1. $12°15' = 12.25°$ since $15' = \left(\frac{15}{60}\right)^{\circ} = \left(\frac{1}{4}\right)^{\circ} = 0.25°$
2. $56°12' = 56.2°$ since $12' = \left(\frac{12}{60}\right)^{\circ} = \left(\frac{1}{5}\right)^{\circ} = 0.2°$
3. $42.6° = 42°36'$ since $0.6° = (0.6)(60') = 36.0'$
4. $75.35° = 75°21'$ since $0.35° = (0.35)(60') = 21.00'$

Angles are named according to their size.

1. An **acute** angle is an angle whose measure is between $0°$ and $90°$.
2. A **right** angle is an angle whose measure is $90°$.
3. An **obtuse** angle is an angle whose measure is between $90°$ and $180°$.
4. A **straight** angle is an angle whose measure is $180°$.

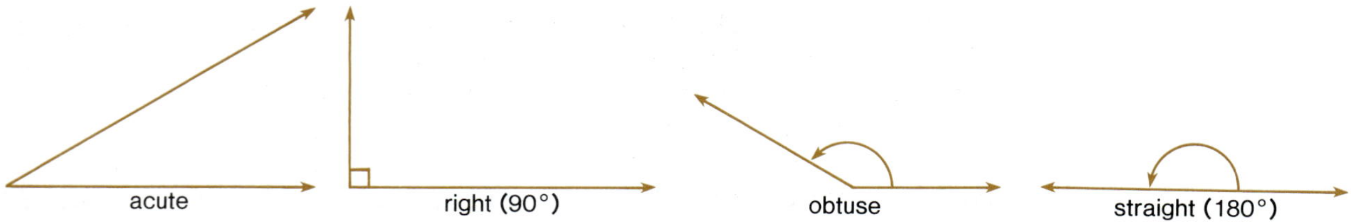

Figure 12.8

Examples of each are given in figure 12.8.

note
We place a small "box", □, in the interior of the right angle. This is the symbol that is always used to indicate that an angle is a right angle.

Two angles are given special names as they are related to each other by (1) their position in the plane and (2) the sum of the measure of their sizes. By definition, given angles A and B:

1. $\angle A$ and $\angle B$ are **adjacent angles** if they share a common vertex and a common side, but their interiors have no points in common.
2. $\angle A$ and $\angle B$ are **vertical angles** if they are nonadjacent angles formed by two intersecting lines.
3. $\angle A$ and $\angle B$ are **complementary angles** if the sum of their measures is 90°.
4. $\angle A$ and $\angle B$ are **supplementary angles** if the sum of their measure is 180°.

See figure 12.9 for examples of each of these angles.

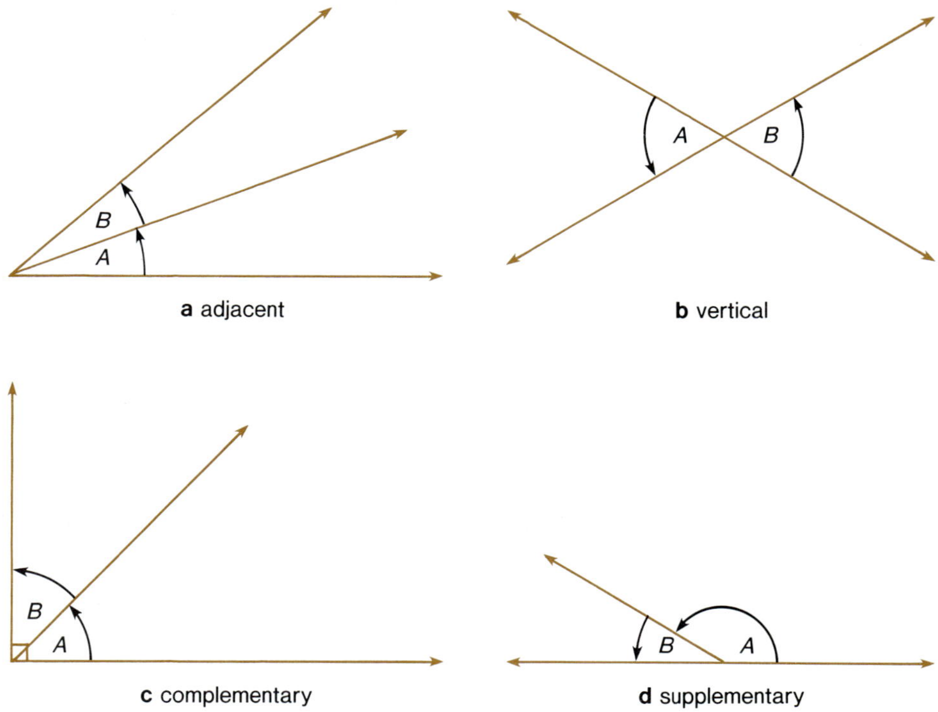

Figure 12.9

In figure 12.9**c** and **d** the angles have been drawn as adjacent angles so the ideas of complementary and supplementary can be better visualized. This is not a necessary condition, however, for these relationships to exist. That is, angles having measure 35° and 55° are complementary since the sum of their measure is 90°, whether or not they are adjacent, and angles having measure 76° and 104° are supplementary since the sum of their measure is 180°, whether they are adjacent or nonadjacent. We shall henceforth denote the measure of an angle A by $m \angle A$.

An important fact about **vertical** angles is demonstrated in figure 12.10.

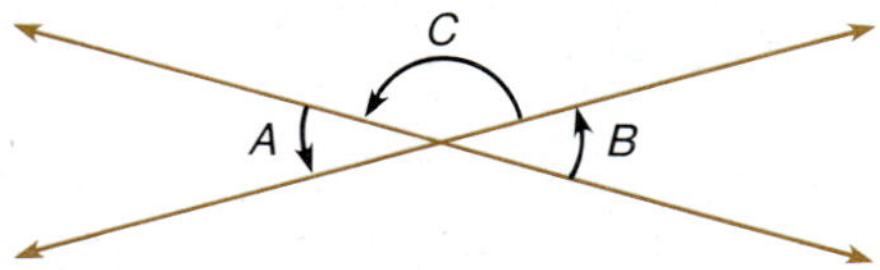

Figure 12.10

We can see that $m \angle A + m \angle C = 180°$ and that $m \angle B + m \angle C = 180°$. Then $m \angle A + m \angle C = m \angle B + m \angle C$, and subtracting $m \angle C$ from each side we find that $m \angle A = m \angle B$. Thus, vertical angles have the same measure.

note

When geometric figures have the same size and shape they are often called **congruent.** Thus, by the property of vertical angles just discussed, vertical angles are congruent angles.

Exercise Set 12.1

Identify each of the following geometric figures.

1. 

2. 

3.

4. 

5. 

6. 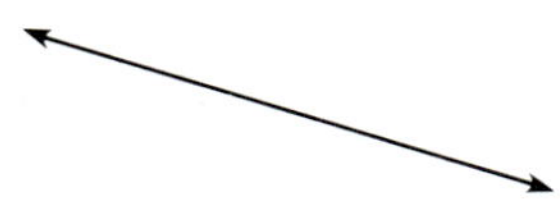

7. 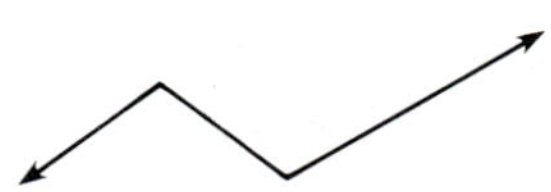

8. 

•9. 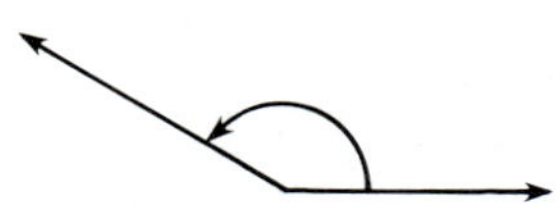

10. 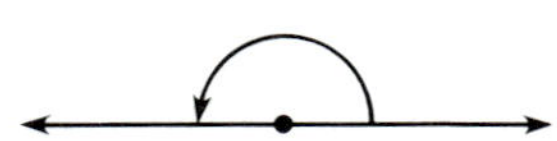

11.

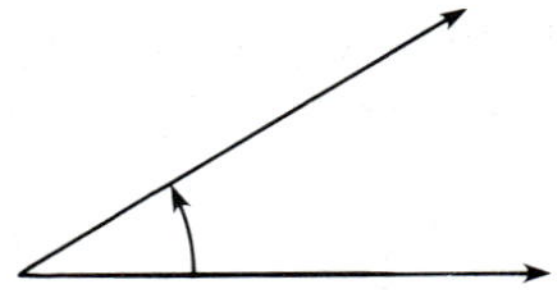

12.

13.

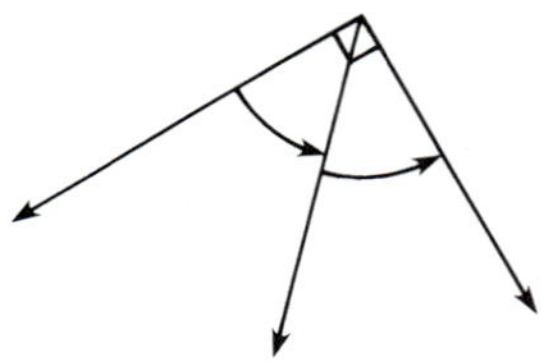

14.

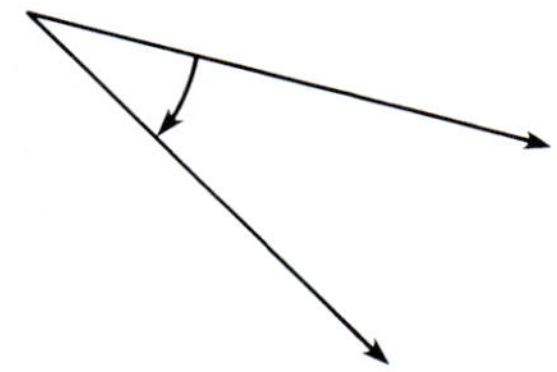

15.

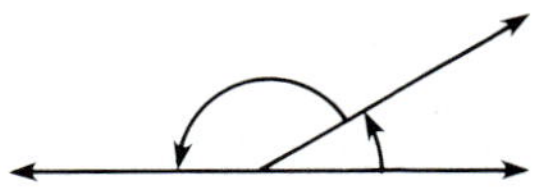

•16.

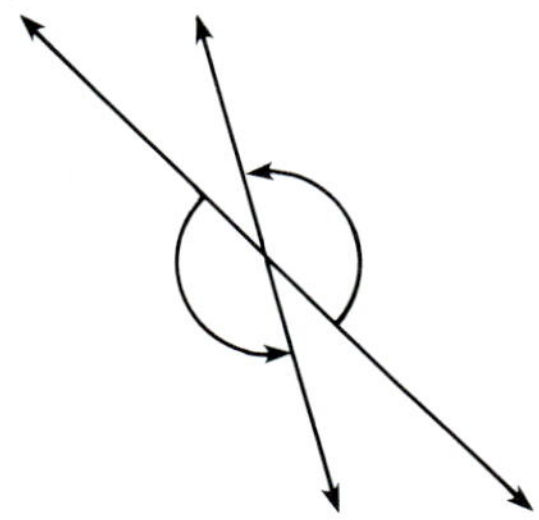

•17. Identify the following in the accompanying art:

a. Two pairs of adjacent angles
b. One pair of complementary angles
c. Two pairs of supplementary angles
d. One obtuse angle
e. One acute angle
f. Two right angles
g. One straight angle

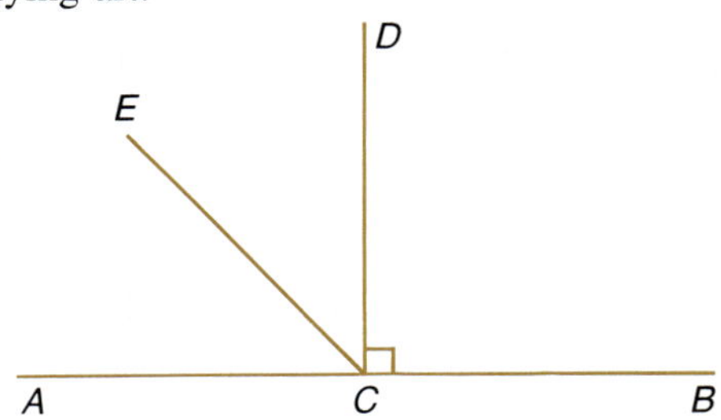

Name each angle in *four* different ways.

18.

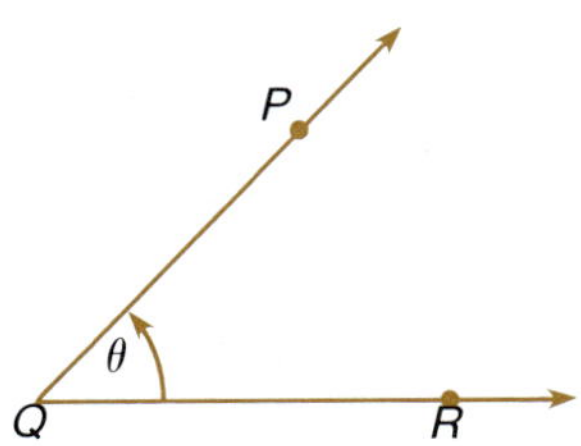

•19.

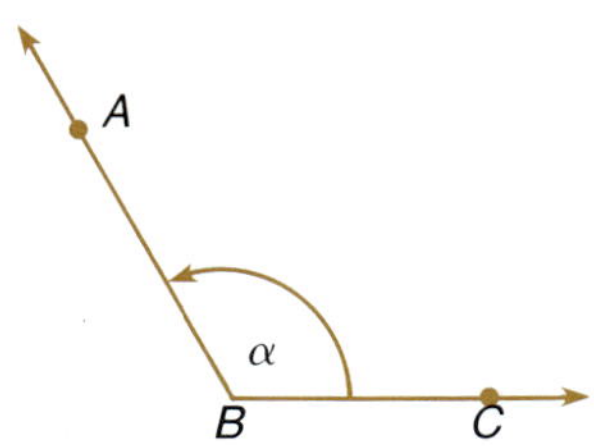

20.

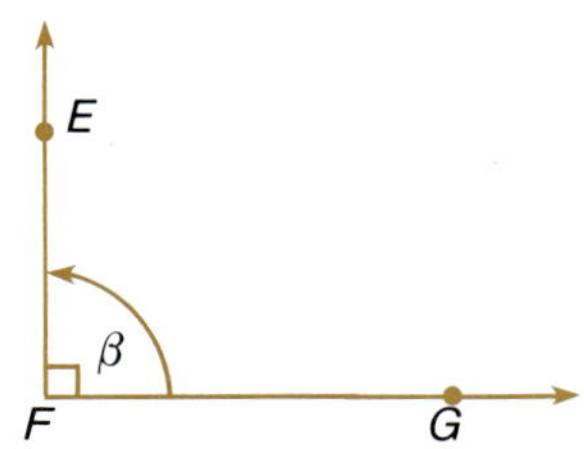

Write the measure of the following angles using decimal parts of a degree. Round off to the nearest hundredth of a degree where necessary.

21. 9°6′ 22. 24°30′ •23. 17°20′
24. 43°25′ 25. 75°50′ •26. 32°18′

Write the measure of the following angles in degrees and minutes (to the nearest second where necessary).

Example Consider 25.28°.

$$0.28° = (0.28)(60) = 16.8'$$

but $0.8' = (0.8)(60) = 48.0''$. Thus, $25.28° = 25°16'48''$.

27. 32.4° 28. 8.9° 29. 10.05°
•30. 71.55° 31. 52.75° 32. 14.35°
33. 1.12° 34. 17.34° •35. 99.27°

Using a protractor, find the measure of each of the following angles. Describe the angle according to its measure.

36.

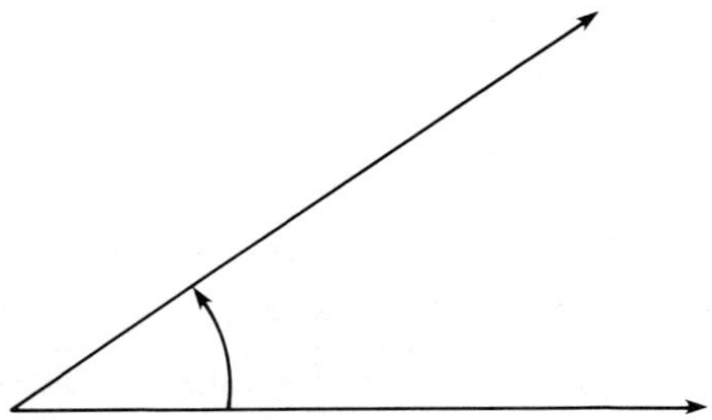

•37.

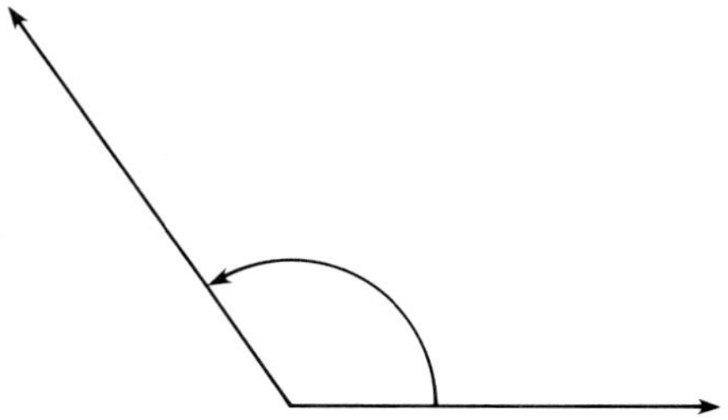

38.

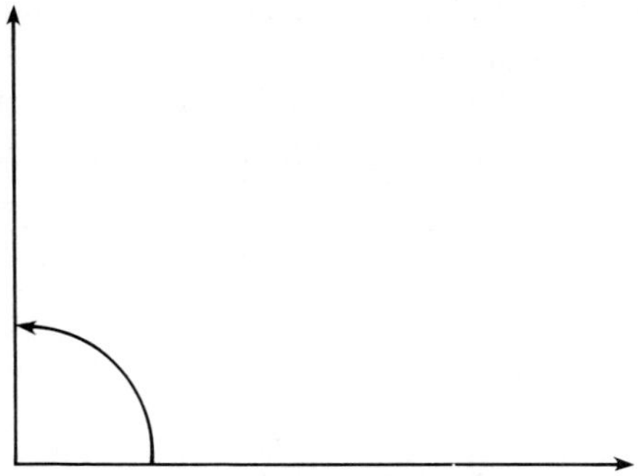

39.

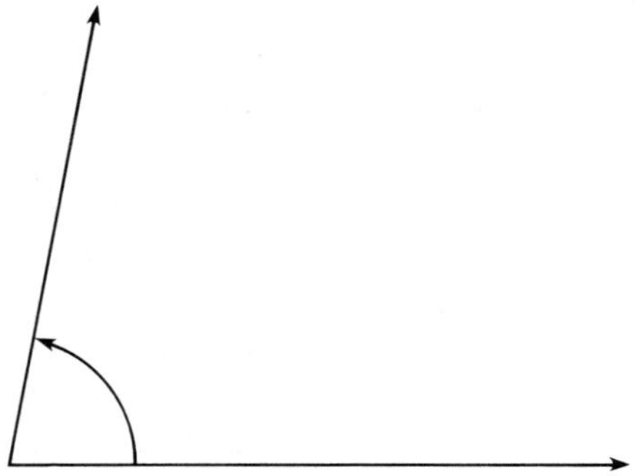

40.

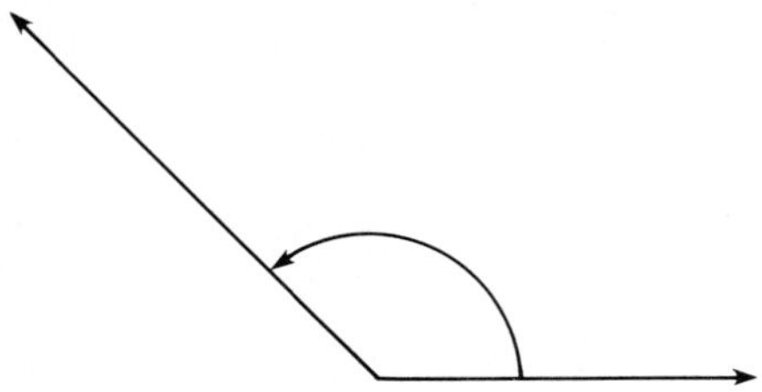

Draw angles of the given measure and describe the angle by its measure.

•41. 35°	42. 87°	43. 115°
•44. 150°	45. 123°	46. 16°
47. 175°	48. 63°	49. 47°

12.4 Lines and Angles

When lines lie in the same plane they are said to be **coplanar.** Coplanar lines can be related by their position in two basic ways:

1. *Intersecting* lines are lines that cross (have a single point in common).
2. *Parallel* lines are lines that are the same distance (d) apart at all points. That is, they will not intersect no matter how far they are extended.

See figure 12.11.

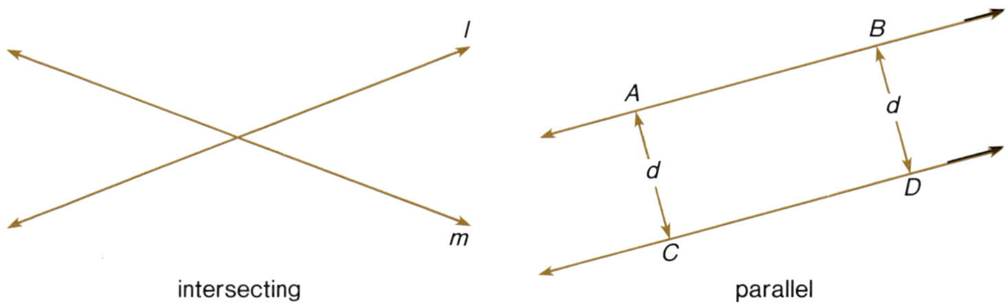

Figure 12.11

When two lines are parallel, the relationship is indicated by the symbol ‖. Therefore, in figure 12.11, since $\overleftrightarrow{AB}$ is parallel to $\overleftrightarrow{CD}$, the relationship is indicated by $\overleftrightarrow{AB} \parallel \overleftrightarrow{CD}$.

Lines can also be classified according to their position in a drawing:

1. A *vertical* line is a line actually, or apparently, located at a right angle with the plane of the horizon.
2. A *horizontal* line is a line that is parallel to the horizon.
3. An *oblique* line is a line that is neither vertical nor horizontal.

See figure 12.12.

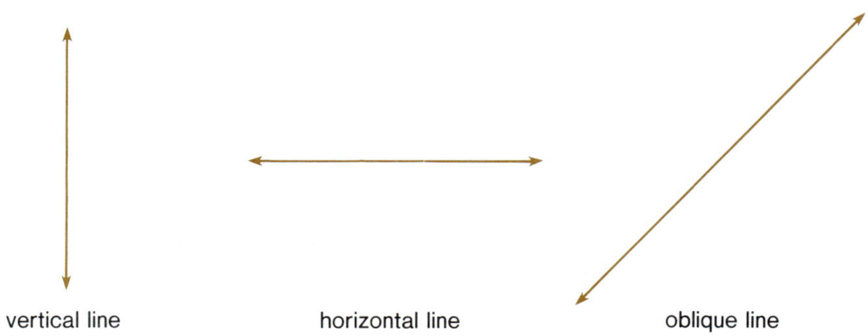

Figure 12.12

A line that is formed by attaching a weight to a piece of string and allowing it to hang free is called a **plumb** line. Vertical and horizontal coplanar lines form an angle of 90° (right angle) with each other.

When intersecting lines form an angle of 90° (right angle), as with vertical and horizontal lines, they are called **perpendicular** lines. We indicate that two lines are perpendicular by the symbol $\perp$. See figure 12.13.

note

Perpendicular lines do not necessarily have to be vertical and horizontal.

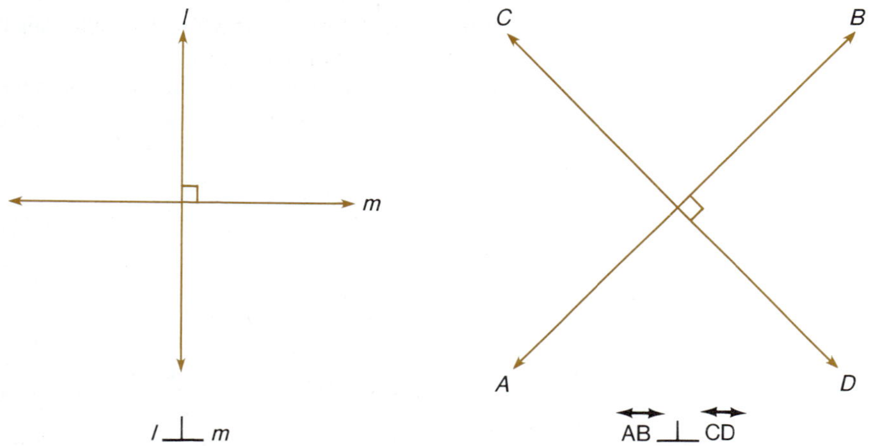

Figure 12.13

When two lines in a plane are cut by a third line in the plane, called a **transversal,** eight angles are formed. In figure 12.14 lines $\overleftrightarrow{AB}$ and $\overleftrightarrow{CD}$ are cut by transversal $\overleftrightarrow{EF}$. We call the four angles between the two lines—that is, angles 3, 4, 5, and 6—**interior** angles, and the four angles outside the two lines—that is, angles 1, 2, 7, and 8—**exterior** angles. The two nonadjacent interior angles on opposite sides of the transversal—$\angle 3$ and $\angle 6$, and $\angle 5$ and $\angle 4$—are called pairs of **alternate-interior** angles.

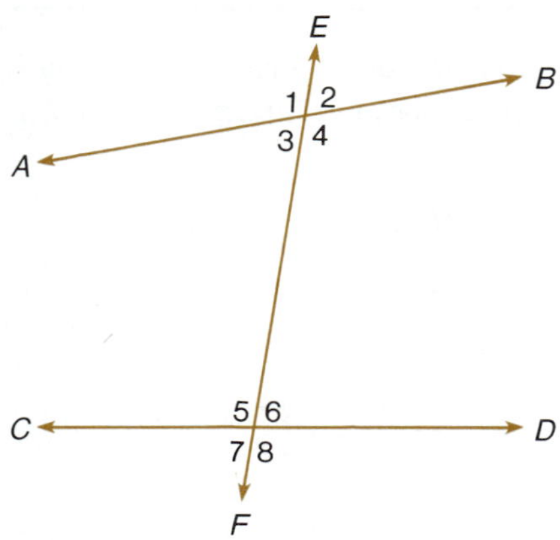

Figure 12.14

The two nonadjacent exterior angles on opposite sides of the transversal—$\angle 1$ and $\angle 8$, and $\angle 2$ and $\angle 7$— are called pairs of **alternate-exterior** angles. We call $\angle 1$ and $\angle 5$, $\angle 2$ and $\angle 6$, $\angle 3$ and $\angle 7$, and $\angle 4$ and $\angle 8$ **corresponding** angles since they occupy the same positions relative to the transversal and the other two lines.

If the two lines cut by the transversal are parallel lines, some very special relationships exist between the sets of angles just mentioned. Refer to figure 12.15. Given $\overleftrightarrow{AB} \parallel \overleftrightarrow{CD}$:

1. *Alternate-interior* angles have the same measure. Thus, $m\ \angle 3 = m\ \angle 6$, and $m\ \angle 4 = m\ \angle 5$.
2. *Alternate-exterior* angles have the same measure. Thus, $m\ \angle 1 = m\ \angle 8$, and $m\ \angle 2 = m\ \angle 7$.
3. *Corresponding* angles have the same measure. Thus, $m\ \angle 1 = m\ \angle 5$, $m\ \angle 2 = m\ \angle 6$, $m\ \angle 3 = m\ \angle 7$, and $m\ \angle 4 = m\ \angle 8$.

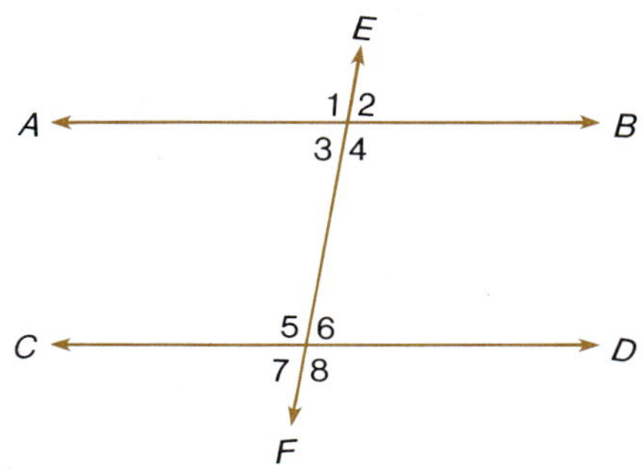

Figure 12.15

note

When two lines are cut by a transversal and any of the special relationships just described are true, the *lines are parallel*. That is, if (1) the alternate-interior angles, (2) the alternate-exterior angles, or (3) the corresponding angles have the same measure, then the lines are parallel.

Example Given lines l_1 and l_2 cut by transversal l_3 (figure 12.16), if $m\ \angle 1 = m\ \angle 2$, or $m\ \angle 3 = m\ \angle 4$, or $m\ \angle 1 = m\ \angle 4$, then l_1 is parallel to l_2.

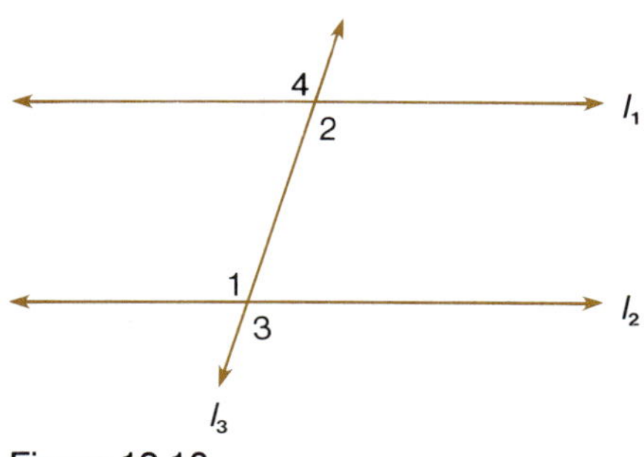

Figure 12.16

Exercise Set 12.2

1. Using your protractor, draw two oblique perpendicular lines.
2. Using a ruler, draw two oblique parallel lines. (*Hint:* The distance between two parallel lines is measured by the perpendicular to the two lines.)

Using the accompanying art, identify the following pairs of angles.

•3. $\angle A$ and $\angle D$
4. $\angle D$ and $\angle H$
•5. $\angle E$ and $\angle G$
6. $\angle C$ and $\angle G$
7. $\angle G$ and $\angle F$
8. $\angle B$ and $\angle C$
•9. $\angle B$ and $\angle F$
10. $\angle B$ and $\angle D$

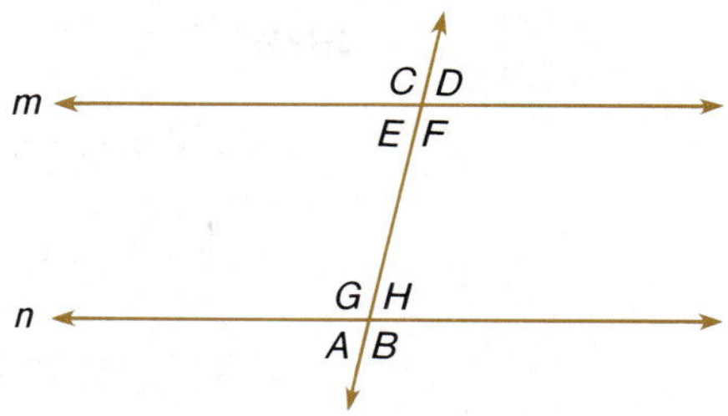

11. Which of the given pairs of angles in the preceding art are equal in size? ($m \parallel n$)
12. A man driving north turns east. Through what angles does he turn?
13. A woman driving east turns to drive northeast. Through what angles does she turn?
14. In the accompanying art, if $l \parallel m$, find the measure of the given angles.

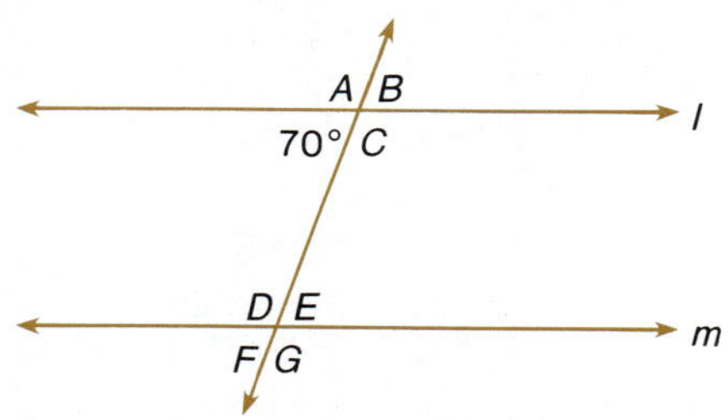

$m\,\angle A =$
$m\,\angle B =$
$m\,\angle C =$
$m\,\angle D =$
$m\,\angle E =$
$m\,\angle F =$
$m\,\angle G =$

15. In the accompanying art, must $\overleftrightarrow{AB}$ be parallel to $\overleftrightarrow{CD}$? Explain your answer.

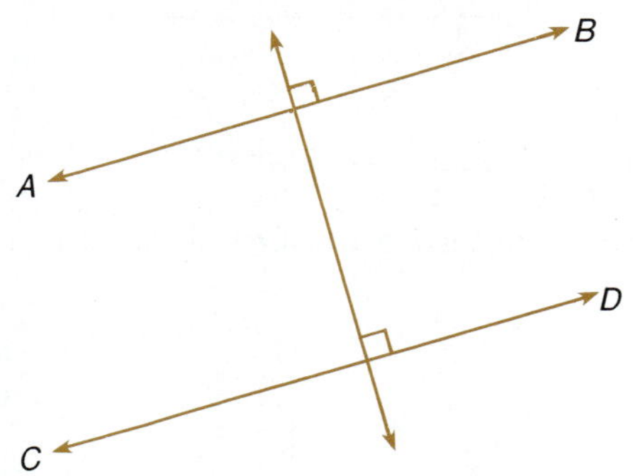

•16. Given that lines l_1, l_2, and l_3 are parallel lines cut by transversal m in the accompanying art, find the measure of each angle.

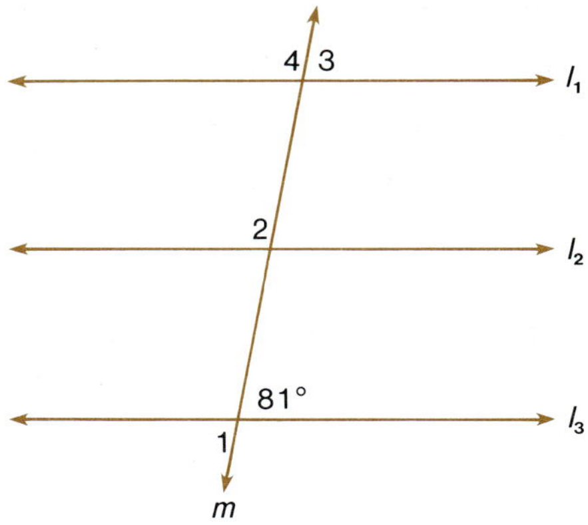

$m\angle 1 =$
$m\angle 2 =$
$m\angle 3 =$
$m\angle 4 =$

17. In the accompanying art, if $\overleftrightarrow{AB} \parallel \overleftrightarrow{DE} \parallel \overleftrightarrow{CE}$,

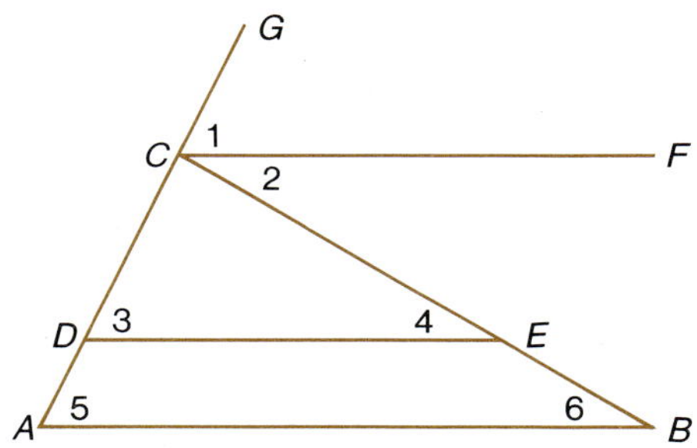

$m\angle 1 = m\angle \underline{\ ?\ } = m\angle \underline{\ ?\ }$
$m\angle 4 = m\angle \underline{\ ?\ } = m\angle \underline{\ ?\ }$
If $\overline{BC} \perp \overline{AG}$ and $m\angle 3 = 63°$, then $m\angle 2 =$ ___ .

18. In the accompanying art, if $\overline{AB} \parallel \overline{CD}$ and $\overline{AD} \parallel \overline{BC}$, name two pairs of congruent angles and explain why they are congruent.

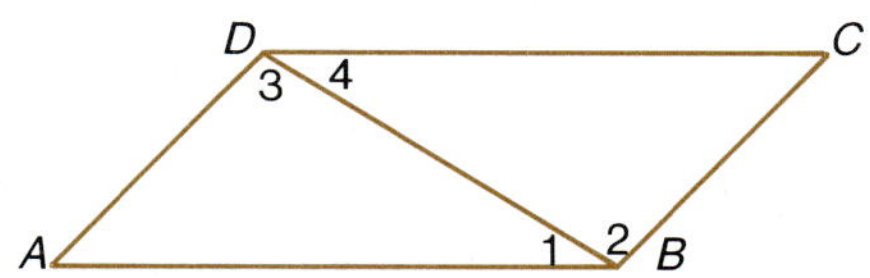

•19. In the accompanying art, if $\overline{BF} \perp \overline{AD}$, $\overline{GC} \perp \overline{AD}$, $\overline{EB} \parallel \overline{FC}$ and $\angle EBC = 138°$, find:

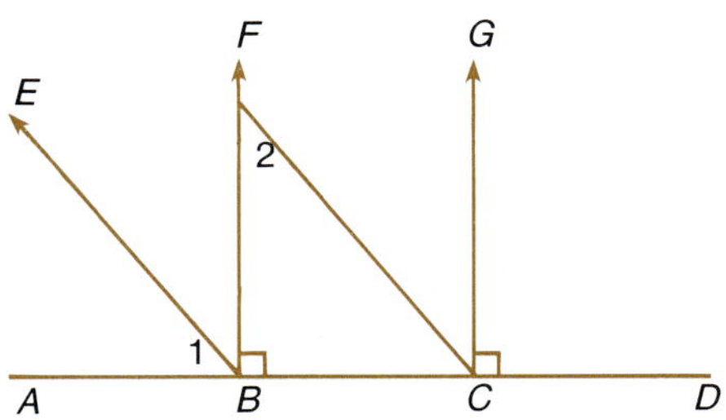

a. $m\angle 1 = \underline{\ ?\ }$, $m\angle 2 = \underline{\ ?\ }$

b. What condition exists between $\overline{BF}$ and $\overline{CG}$? Tell why.

12.5 Closed Plane Figures —Polygons

When straight, broken, or curved lines lying in a plane (coplanar lines) enclose a portion of the plane, the figure thus formed is called a *closed plane figure*. In this section we discuss some of the more important figures that fall into this group of geometric figures, together with the special properties each possesses.

When three or more *straight* line segments enclose a portion of a plane, the figure thus formed is called a **polygon.** Each line segment forming the polygon is called a **side** of the polygon. Each point where two sides meet is called a **vertex** of the polygon. The interior angles of a polygon are the angles formed by the sides whose interiors are in the interior of the polygon. Polygons are named according to their number of sides. See figure 12.17.

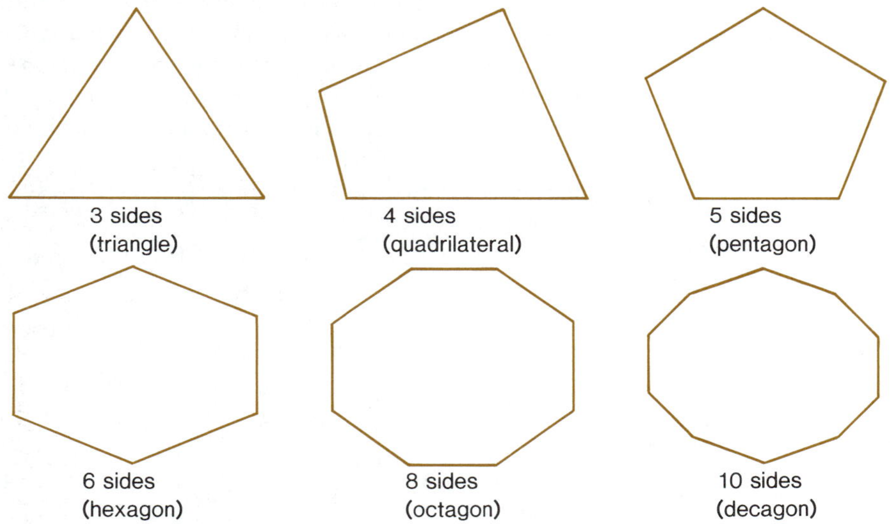

Figure 12.17

note

We shall consider here only **convex** polygons—polygons whose interior angles are less than 180°. A polygon that is not convex is called **concave.**

A polygon with all sides equal in length and all interior angles equal in measure is a **regular** polygon. Regular polygons are said to be **equilateral** and **equiangular.** Such polygons are used often in industry since nuts and bolt heads are usually regular polygons. A polygon whose sides are not all equal in length is called an **irregular** polygon. Our discussion will be limited in this book to the triangle and some of the special quadrilaterals.

By definition, a **triangle** is a polygon having three sides, three vertices, and three interior angles.

note

It should be noted that three line segments can form a triangle *only if* the longest line segment is shorter than the sum of the lengths of the other two sides.

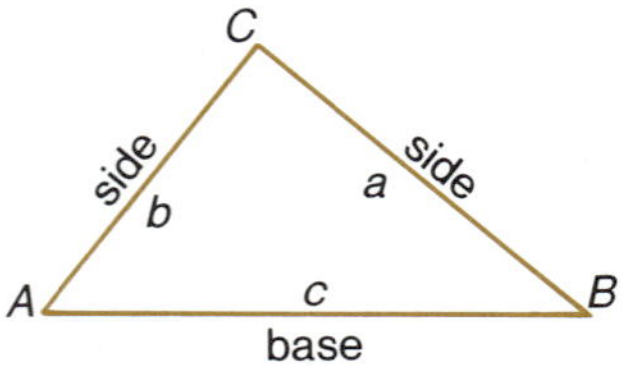

Figure 12.18

The sum of the measures of the three interior angles of a triangle is 180°. By convention, triangles are usually named by the letters associated with the three vertices. The triangle in figure 12.18, whose vertices are the points A, B, and C, is called triangle ABC, denoted by $\triangle ABC$. A side of a triangle can be named by its two endpoints or by a lowercase letter that is the same as the vertex opposite the side. For example, in $\triangle ABC$ in figure 12.18, the sides are $\overline{AB}$, $\overline{BC}$, and $\overline{AC}$, or c, a, and b, respectively.

The side on which the triangle "rests" is called its **base.** Thus, side $\overline{AB}$ (or c) is the base of $\triangle ABC$. Any of the sides of a triangle can be the base. The three angles of the triangle are referred to as $\angle A$, $\angle B$, and $\angle C$. From the previously mentioned property of triangles, $m\angle A + m\angle B + m\angle C = 180°$.

An important measurement in a triangle is its *altitude* (or *height*), which is denoted by h. By definition, the *altitude* of a triangle is the line segment drawn from any vertex of the triangle perpendicular to the side opposite that vertex. Thus, a given triangle can have three altitudes, each of possibly different length. The altitude we normally work with is the one drawn perpendicular to the base, as h in figure 12.19**a,** or an extension of the base, as in figure 12.19**b.** The measurement of the base and the altitude drawn to the base are important to the area of a triangle, which is discussed in chapter 13.

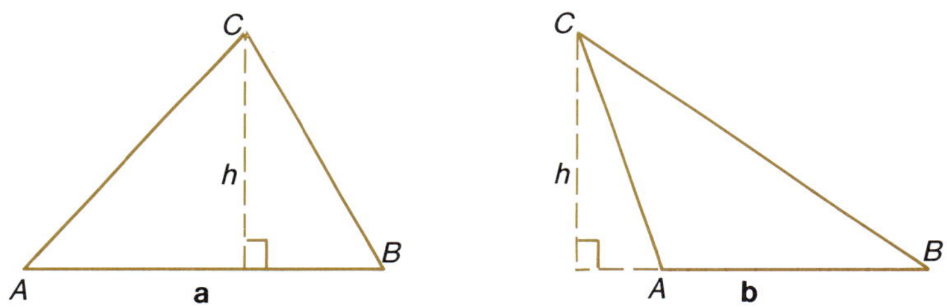

Figure 12.19

Triangles are classified according to either their sides or their angles. The following are triangles classified according to their sides:

1. A **scalene** triangle is a triangle with no sides equal in length (and thus no angles with equal measurements).
2. An **isosceles** triangle is a triangle with two sides equal in length (and two angles having the same measurement—called the **base angles**).
3. An **equilateral** triangle is a triangle with all its sides equal in length (and all three angles with the same measurement).

note

The angles opposite sides of a triangle with equal length have the same measure, and vice-versa. Refer to figure 12.20**b** and **c.**

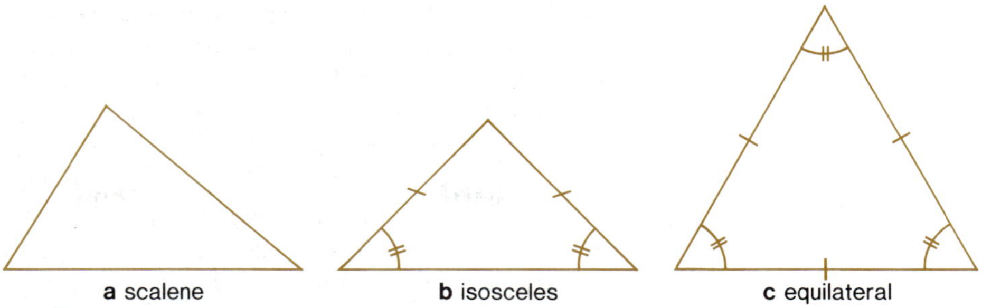

Figure 12.20

note

In figure 12.20, the convention is introduced of putting slash marks through sides that are of equal length and through angles with the same measure.

The following are triangles classified according to their angles:

1. An **acute** triangle is a triangle with three acute angles.
2. An **obtuse** triangle is a triangle with one obtuse angle.
3. An **equiangular** triangle is a triangle that has all three angles equal in measurement (thus, the three sides have the same length).
4. A **right** triangle is a triangle with one right angle.

See figure 12.21.

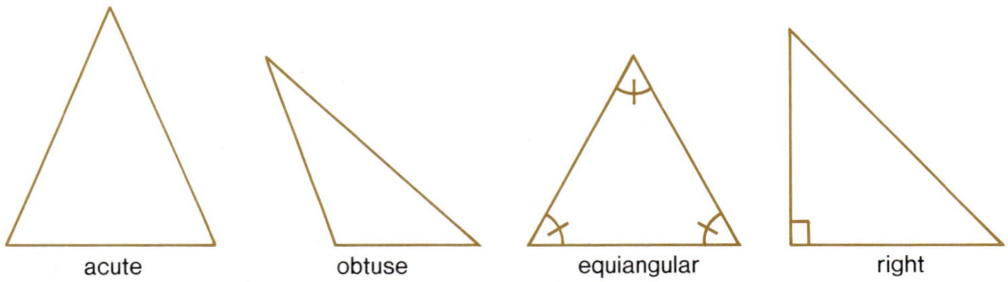

Figure 12.21

Recall that any triangle has three different altitudes—lines drawn from a vertex perpendicular to the side opposite the vertex. In any triangle these altitudes intersect one another in a single point. This point is called the **orthocenter** of the triangle.

A triangle has another set of lines that intersect in a single point. A **median** of a triangle is the line segment drawn from a vertex to the midpoint of the opposite side. The point of intersection of the three medians of a given triangle is called the **centroid** (center of gravity of the triangle). See figure 12.22.

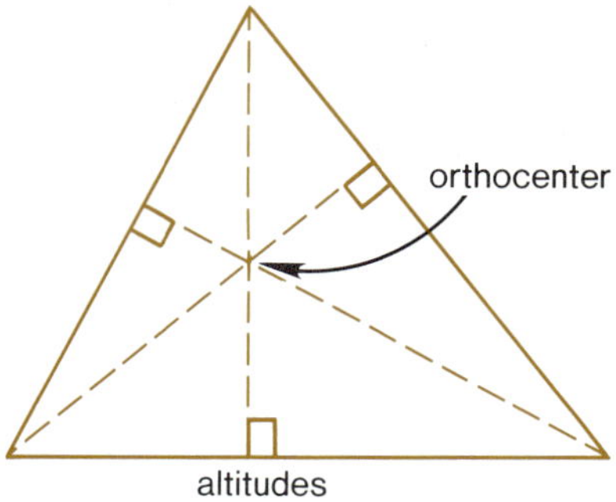

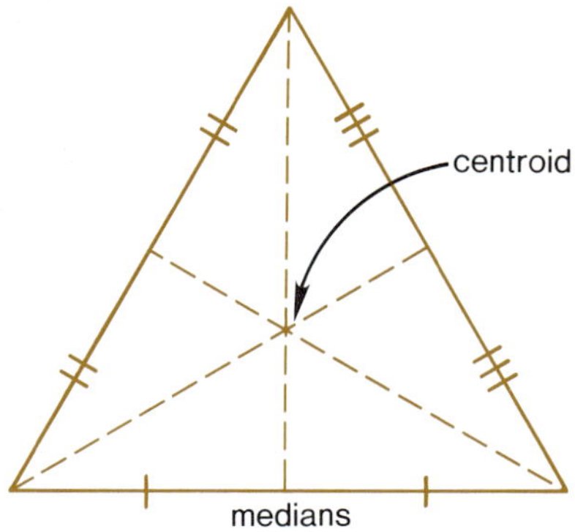

Figure 12.22

A polygon with four sides, four angles, and four vertices is called a **quadrilateral.** A **diagonal** of a polygon is a line segment drawn from one vertex to a nonadjacent vertex.

note

A quadrilateral is the polygon with the least number of sides that has a diagonal. Triangles do not have diagonals.

We can see in figure 12.23 that all quadrilaterals have two diagonals.

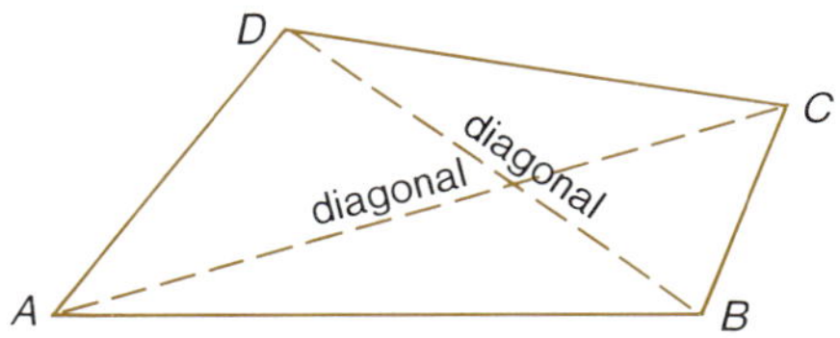

Figure 12.23

Quadrilaterals are named by the letters at their vertices, and, like triangles, they are identified by their angles and sides. Thus, figure 12.23 is called quadrilateral *ABCD*.

If we draw just one of the diagonals of quadrilateral *ABCD*, two triangles are formed: See figure 12.24. Triangles *ABC* and *ADC* are formed by diagonal $\overline{AC}$. Since the sum of the measure of the interior angles of a triangle is 180°, then 180° + 180° = 360° leads to the following property of all quadrilaterals: *The sum of the measure of the interior angles of a quadrilateral is 360°.*

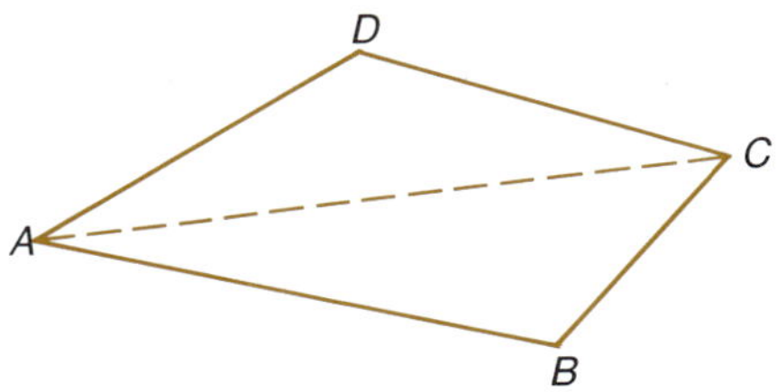

Figure 12.24

Two special classifications of quadrilaterals—parallelograms and trapezoids—have special properties. A **trapezoid** is a quadrilateral with one pair of opposite sides parallel. If the nonparallel sides of the trapezoid are equal in length, the trapezoid is said to be **isosceles.** In figure 12.25, trapezoid $ABCD$ has $\overline{AB} \parallel \overline{DC}$, while in isosceles trapezoid $MNOP$, $\overline{MN} \parallel \overline{OP}$ and $\overline{MP} = \overline{ON}$.

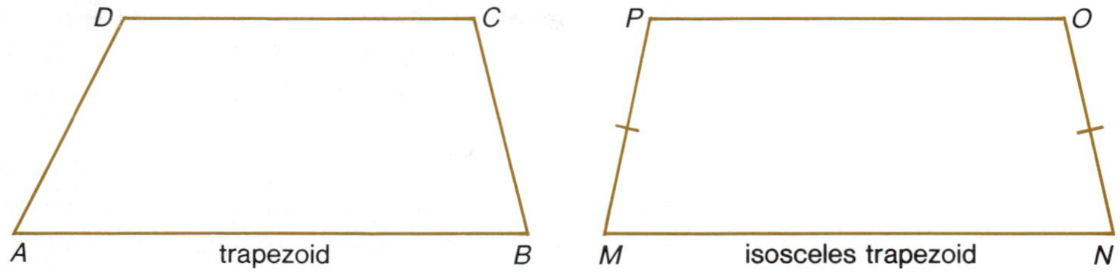

Figure 12.25

note

In isosceles trapezoid $MNOP$ the interior angles M and N are the *base angles.* They have the same measure.

The two parallel sides of a trapezoid are called its **bases.** Thus, in trapezoid $ABCD$ (figure 12.25), $\overline{AB}$ and $\overline{DC}$ are its bases.

A **parallelogram** is a quadrilateral in which both pairs of opposite sides are parallel and equal in length. In figure 12.26, quadrilateral $ABCD$ is a parallelogram since: (1) $\overline{AB} \parallel \overline{DC}$ and $\overline{AD} \parallel \overline{BC}$, and (2) $\overline{AB} = \overline{DC}$ and $\overline{AD} = \overline{BC}$.

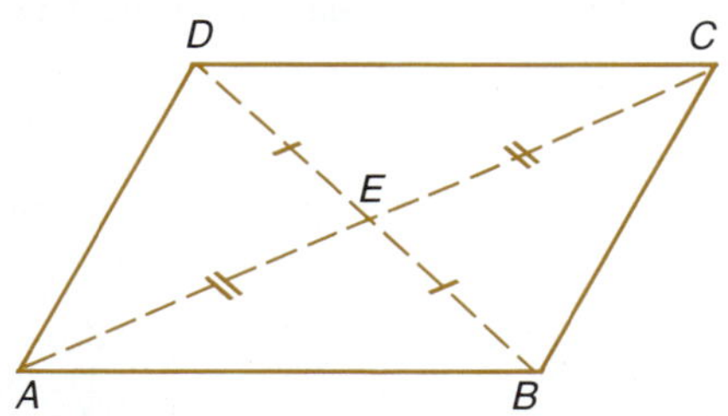

Figure 12.26

The altitude (height, h) of a parallelogram and a trapezoid is the perpendicular distance between a pair of parallel sides. As with the triangle, the altitude is denoted by h. See figure 12.27.

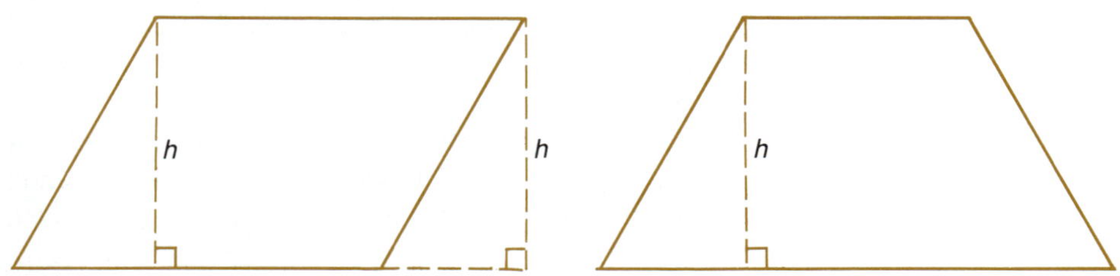

Figure 12.27

The parallelogram can "rest" on any one of its sides, so any one of the sides can be its base. Additional properties that a parallelogram possesses exist as a result of the properties of parallel lines cut by a transversal:

1. The opposite angles of a parallelogram have the same measure.
2. Consecutive angles of a parallelogram are supplementary.

Thus, in figure 12.26, $m\angle A = m\angle C$ and $m\angle B = m\angle D$, while $m\angle A + m\angle B = m\angle B + m\angle C = m\angle C + m\angle D = m\angle D + m\angle A = 180°$. It can be shown also that the diagonals of a parallelogram bisect one another. So, in figure 12.26, $\overline{AE} = \overline{EC}$ and $\overline{BE} = \overline{ED}$.

Some parallelograms have special properties and are given special names. A **rectangle** is a parallelogram having four right angles. Being a parallelogram, a rectangle has all of the properties of a parallelogram in addition to the following special properties:

1. All angles of a rectangle are right angles.
2. The diagonals of a rectangle are equal in length.

A special rectangle is the square. A **square** is a rectangle having all sides equal in length. A square, then, possesses all the properties of a rectangle with one additional property: *The diagonals of a square are perpendicular bisectors of each other.*

Another special parallelogram is the rhombus. A **rhombus** is a parallelogram having all sides equal in length. The rhombus has the same special property of a square that the diagonals are perpendicular bisectors of each other.

note

The angles do not have the same measure and they are not right angles.

See figure 12.28.

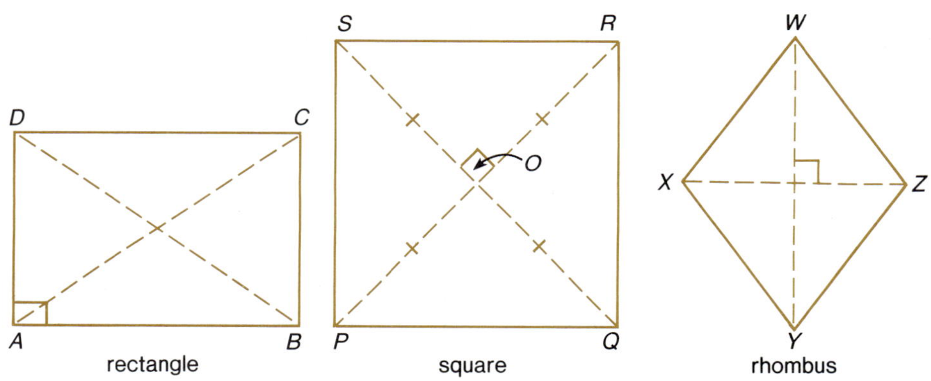

Figure 12.28

Of the properties described, rectangle *ABCD* in figure 12.28 has the following: (1) $m\angle A = m\angle B = m\angle C = m\angle D = 90°$; and (2) $\overline{AC} = \overline{BD}$. The square *PQRS* has the following: (1) $\overline{PR} \perp \overline{QS}$; and (2) $\overline{OS} = \overline{OQ}$, and $\overline{PO} = \overline{OR}$.

Using all possible diagonals drawn from one of the vertices of a polygon to form triangles and using the fact that the sum of the interior angles of a triangle is 180°, it is possible to determine the number of degrees in the sum of the interior angles of any polygon. Consider the pentagon (five sides) in figure 12.29. Three triangles are formed by drawing the two diagonals possible from one vertex, each containing 180° in the interior angles. The sum of the interior angles of a pentagon, therefore, must be $3 \times 180° = 540°$. Thus, in figure 12.29, $m\angle A + m\angle B + m\angle C + m\angle D + m\angle E = 540°$.

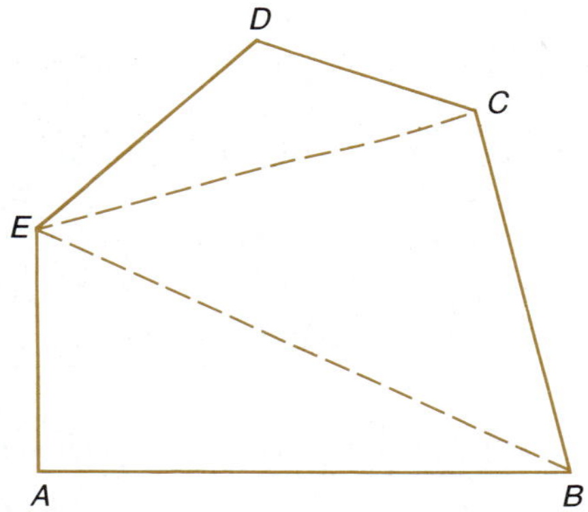

Figure 12.29

In general, for any polygon having n sides, the sum of the interior angles will always be given by $(n - 2)180°$.

Exercise Set 12.3

Identify the following figures.

•1.

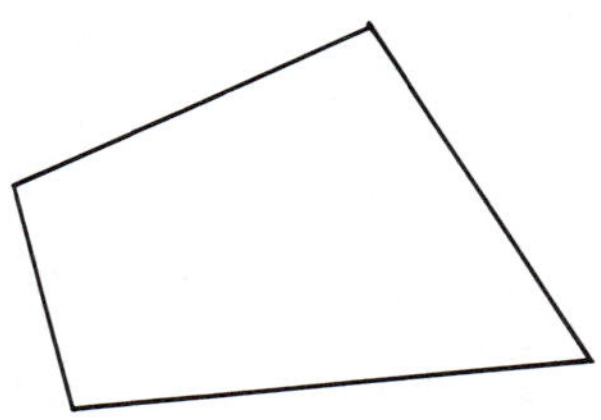

2.

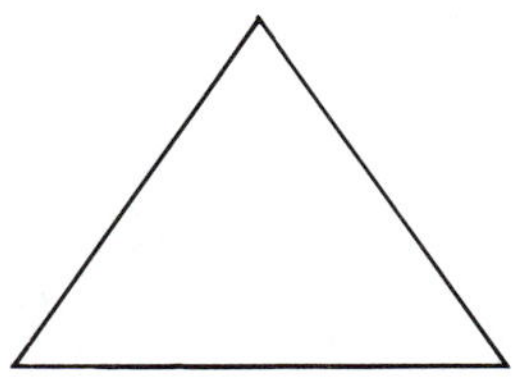

•3.

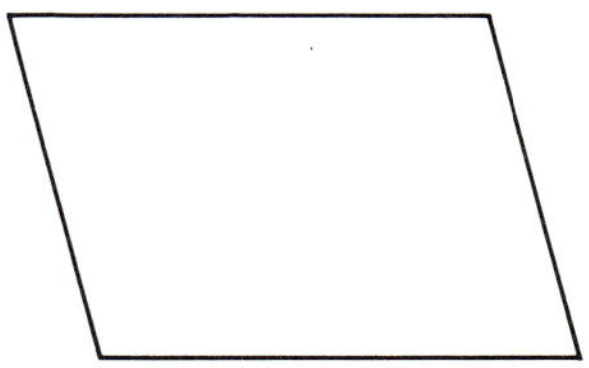

4.

•5.

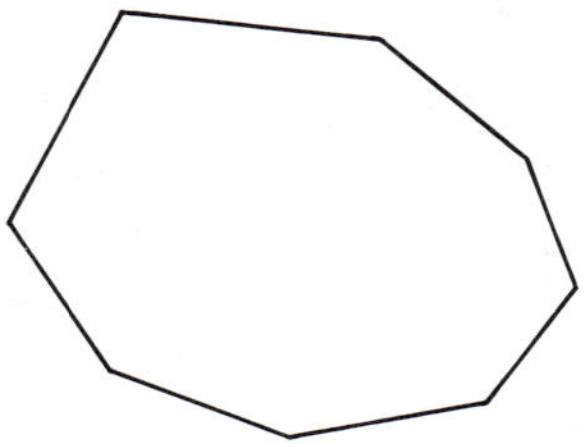

6.

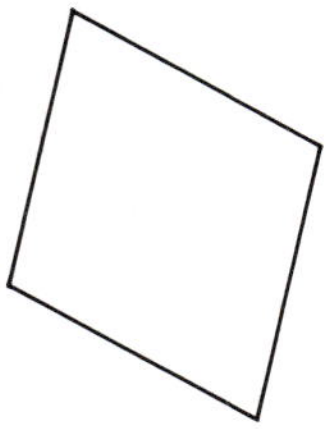

7.

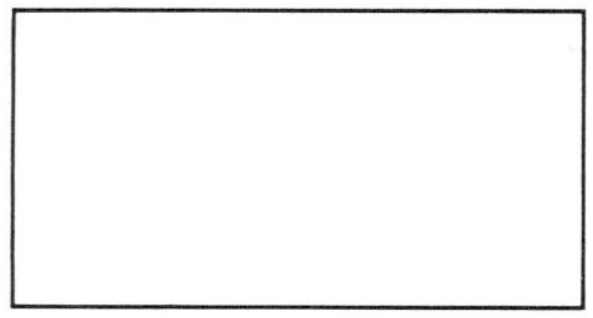

8.

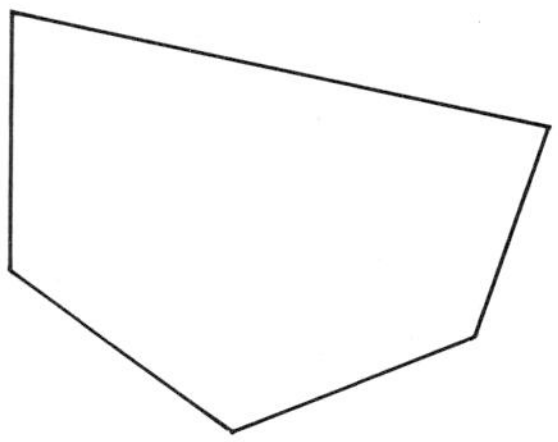

•9.

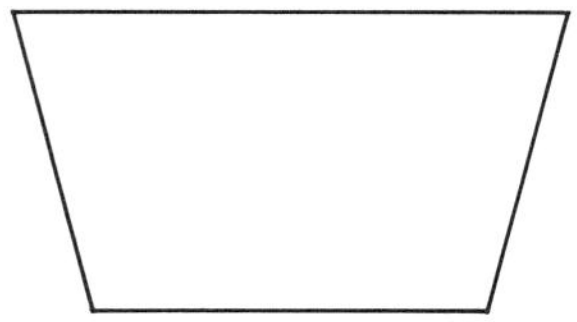

10.

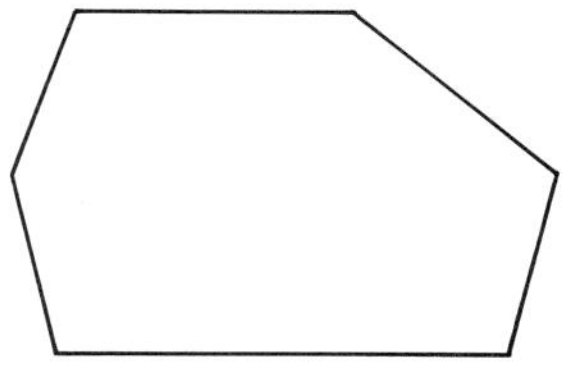

Identify the following triangles according to their sides.

11.

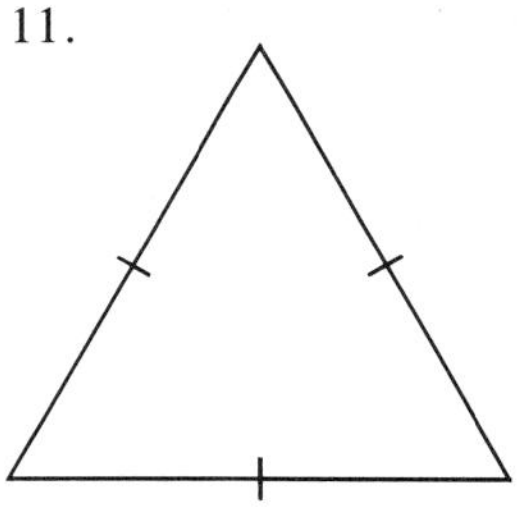

•12.

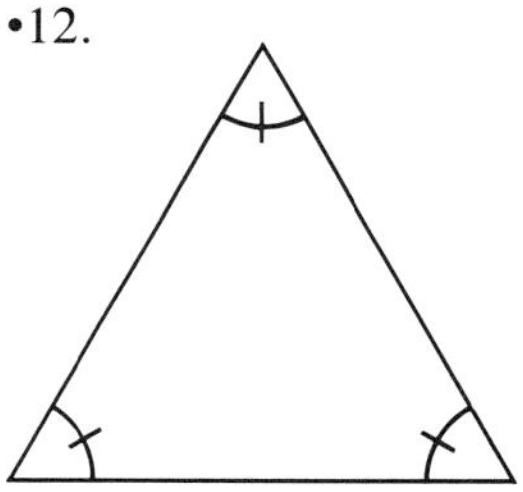

13.

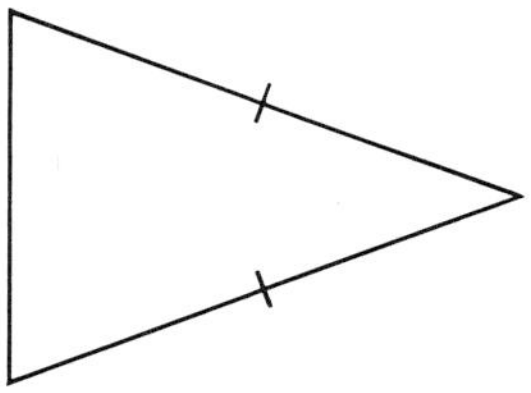

•14.

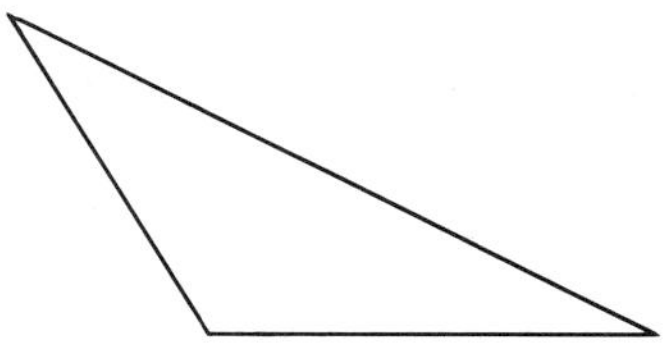

15.

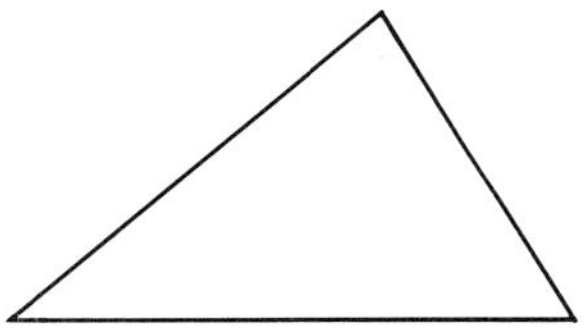

Identify the following triangles according to their angles.

•16.

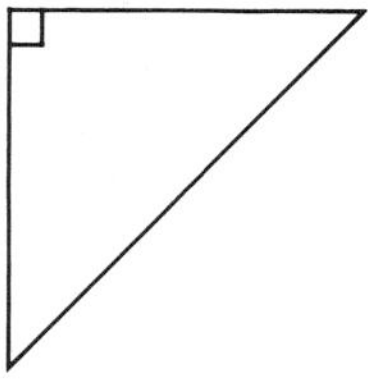

17.

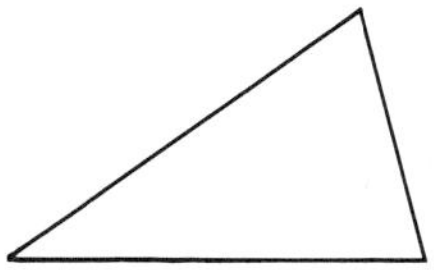

18.

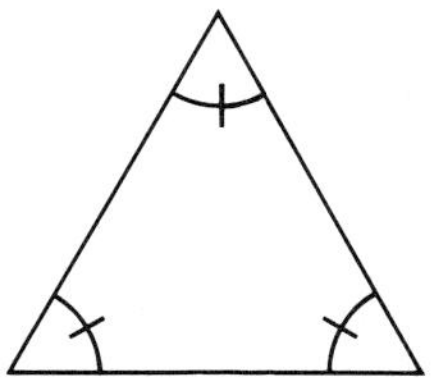

19.

20.

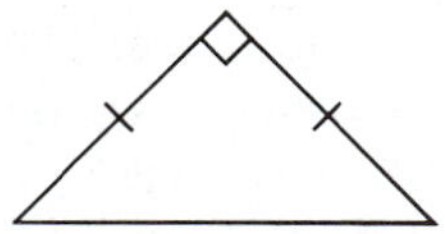

21.

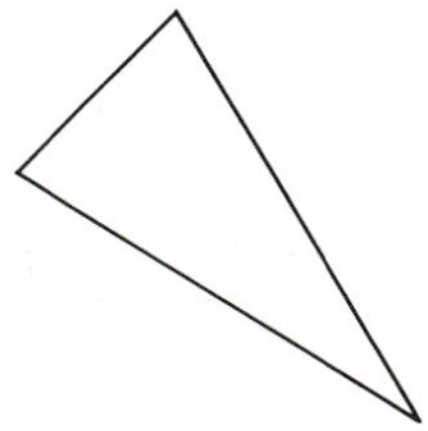

•22.

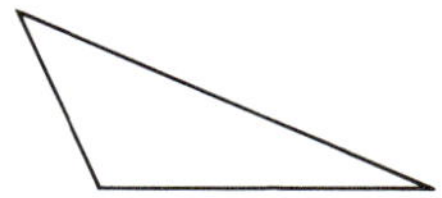

23. Identify each side of the given triangle in two ways.

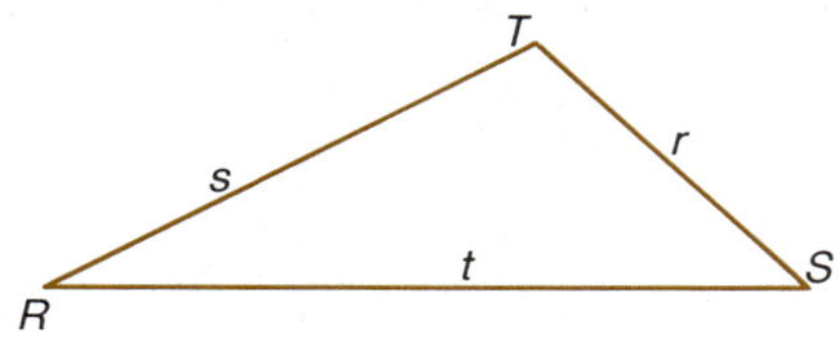

24. In the triangle (problem 23), name the triangle in *four* different ways.
25. Find the sum of the interior angles of (a) a hexagon and (b) an octagon.
•26. In the given triangle ABC, (a) $\overline{CD}$ is a(n) ____________; (b) $\overline{AE}$ is a(n) ____________; and (c) $\overline{BF}$ is a(n) ____________ of the triangle.

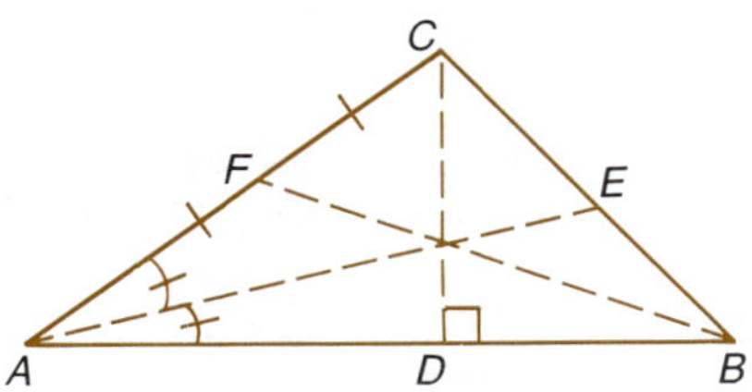

27. If the two diagonals of a quadrilateral are perpendicular bisectors of each other, what kind of a quadrilateral might the figure be?
•28. Three interior angles of a quadrilateral are 52°, 63°, and 79°. How many degrees are there in the fourth angle?
29. The interior angles of a quadrilateral are 46°, 46°, 134°, and 134°. What is the name of the quadrilateral?
30. Given a quadrilateral whose diagonals bisect the angles. Identify the possible quadrilaterals this describes.
31. Draw a trapezoid with bases of 2 in and 3 in, having one right angle.
32. Draw a square having sides 2 in long.
•33. Draw a parallelogram with sides of 1 in and 2 in. If the figure is *not* a rectangle, find the measure of the length of the altitude drawn between the 2 in sides. Draw a diagonal of the parallelogram and name the angles having equal measure.
34. Two angles of a triangle measure 23° and 69°. What is the measure of the third angle?

35. What is the measure of each interior angle of a *regular* (a) pentagon, (b) hexagon, (c) octagon?

36. What is the measure of each interior angle of an equilateral triangle?

37. One of the acute angles of a right triangle is 39°. What is the measure of the size of the other acute angle?

38. In an isosceles triangle, one of the equal base angles has measure 49°. What is the measure of the other angles of the triangle?

•39. One of the base angles of an isosceles trapezoid has measure 41°. What is the measure of the other three interior angles of the trapezoid?

40. In the accompanying figure, draw all possible diagonals from the vertex A. Count the number of triangles thus formed. What can you say about the sum of the interior angles of the polygon?

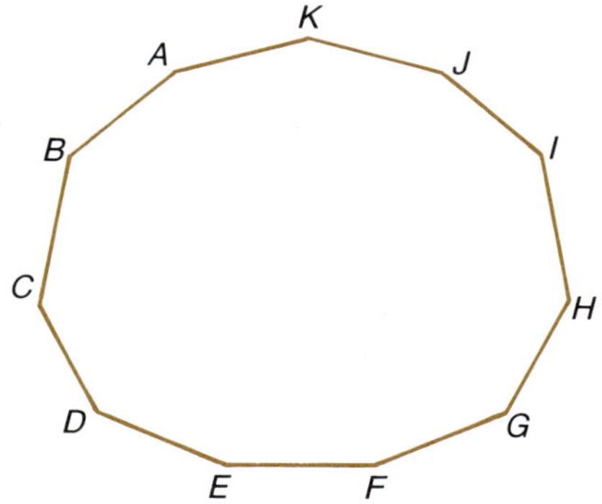

•41. In the accompanying figure, if $\overline{AB} \parallel \overline{DE}$, $\overline{AC} = \overline{BC}$, and $m\ \angle A = 52°$, find the measure of $\angle B$, $\angle 1$, $\angle 2$, and $\angle 3$.

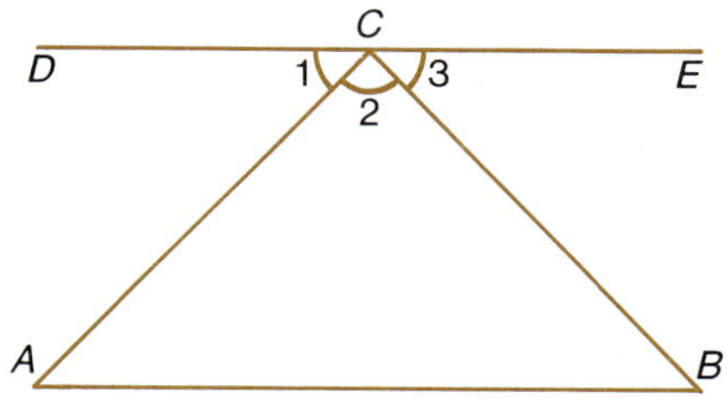

42. In the accompanying figure, if $m\ \angle\beta = m\ \angle r$, determine the measure of: (a) $\angle\beta$, (b) $\angle r$, (c) $\angle\alpha$, (d) $\angle\theta$.

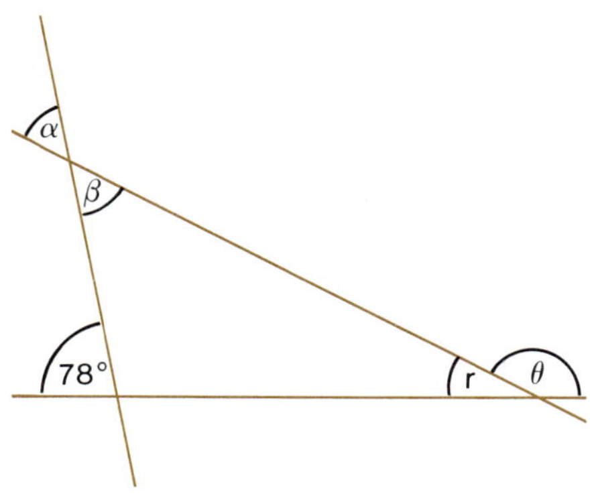

43. In the accompanying figure, find the measure of $\angle\alpha$.

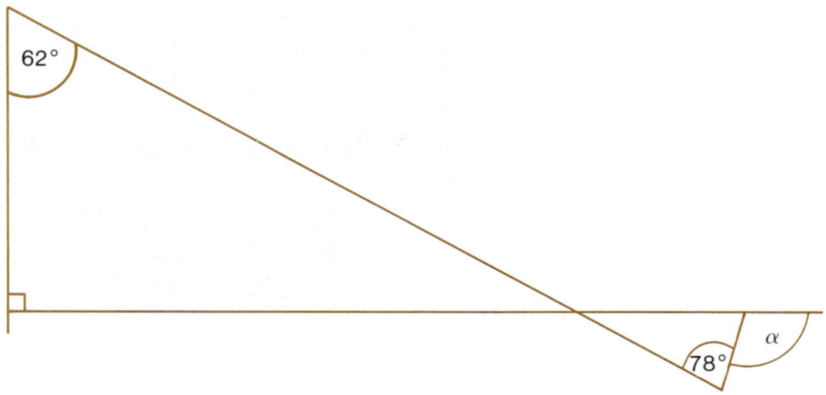

•44. Draw each triangle from the data given using a protractor and rule.

a. $m\angle A = 30°$, $m\angle B = 110°$, $c = 5.00$ cm
b. $m\angle B = 25°$, $c = 5.00$ cm, $b = 7.00$ cm
c. $m\angle C = 70°$, $a = 5.50$ cm, $b = 8.00$ cm
d. $m\angle A = 40°$, $m\angle B = 120°$, $a = 9.80$ cm
e. $a = 10$ cm, $b = 7.00$ cm, $c = 5.00$ cm
f. $a = 6.00$ cm, $b = 8.00$ cm, $c = 10.00$ cm
g. $a = 6.5$ cm, $b = 4.00$ cm, $c = 1.50$ cm

12.6 The Pythagorean Theorem

A right triangle has one property that is of great importance in the technical and physical world. Consider the right triangle *ABC* in figure 12.30, with the right angle at vertex *C* and sides *a*, *b*, and *c*. The longest side is the **hypotenuse**—the side opposite the right angle. The remaining two sides are the **legs** of the right triangle. The relationship among the three sides of a right triangle is associated with the Greek scholar Pythagoras since he was said to have given the first proof of the theorem. Because of this association, the theorem is called the **Pythagorean theorem.** It states that in a right triangle the square of the length of the hypotenuse is equal to the sum of the squares of the lengths of the legs. Using right $\triangle ABC$ in figure 12.30 to illustrate, the theorem states

$$c^2 = a^2 + b^2$$

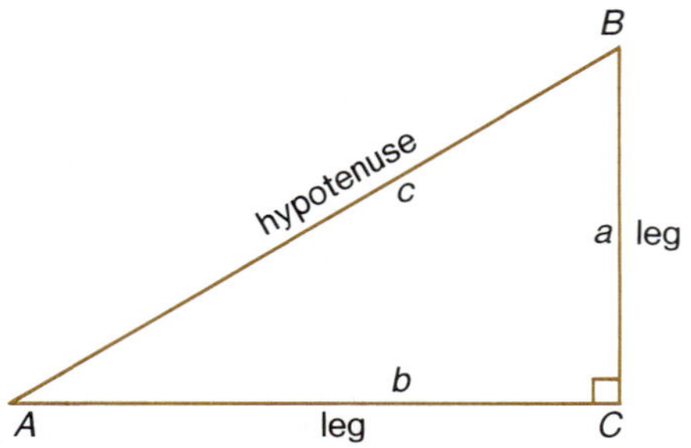

Figure 12.30

This statement can also take on the following forms:

$$a^2 = c^2 - b^2 \quad \text{and} \quad b^2 = c^2 - a^2$$

Taking the square root of both sides, $c = \sqrt{a^2 + b^2}$; $a = \sqrt{c^2 - b^2}$; and $b = \sqrt{c^2 - a^2}$.

note

It is common practice in mathematics to let c be the hypotenuse. The legs then are always denoted by a and b.

The formula is so important and will be used in so many areas of your technical and mathematical studies that it should be committed to memory. Students should be familiar with all the different forms.

Examples

1. Find the length of the hypotenuse of a right triangle whose legs are 6 cm and 8 cm.
   In the right $\triangle ABC$ in figure 12.31, we want c when given $a = 6$ cm and $b = 8$ cm.

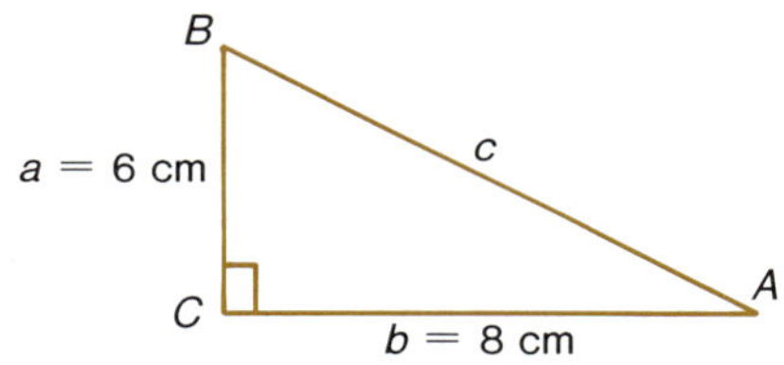

Figure 12.31

By the Pythagorean theorem,

$$\begin{aligned} c &= \sqrt{a^2 + b^2} \\ &= \sqrt{6^2 + 8^2} \\ &= \sqrt{36 + 64} \\ &= \sqrt{100} = 10 \end{aligned}$$

Therefore, $c = 10$ cm

2. Find the second leg of a right triangle whose hypotenuse is 13 in long and first leg is 5 in long.

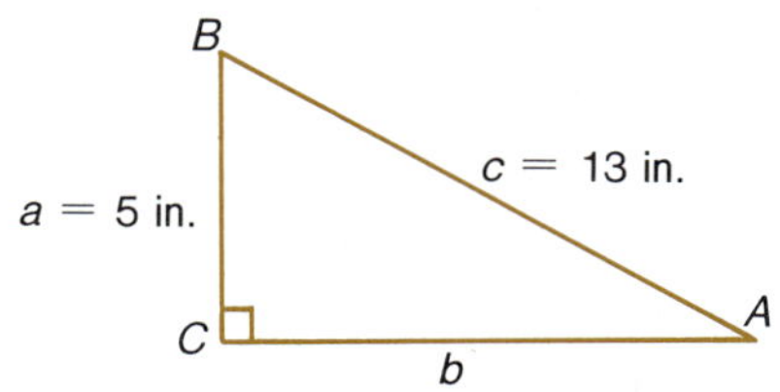

Figure 12.32

In $\triangle ABC$ in figure 12.32, we want to find b, given $c = 13$ in and $a = 5$ in. Using one form of the theorem,

$$\begin{aligned} b &= \sqrt{c^2 - a^2} \\ &= \sqrt{(13)^2 - 5^2} \\ &= \sqrt{169 - 25} \\ &= \sqrt{144} = 12 \text{ in} \end{aligned}$$

Therefore, $b = 12$ in

3. A 20.0 ft flag pole casts a shadow that is 15.0 ft long. What is the distance from the tip of the shadow to the top of the flag pole?

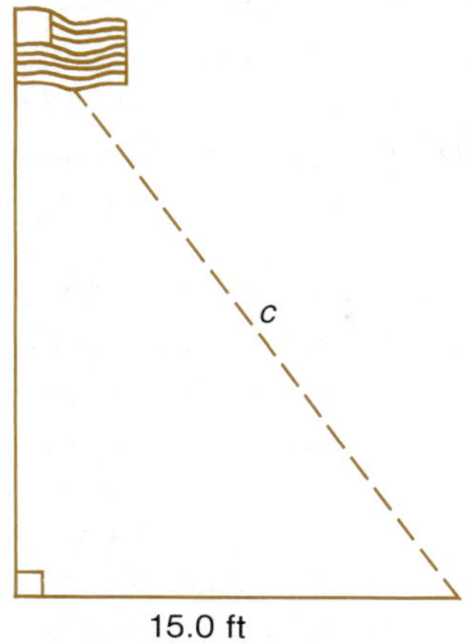

Figure 12.33

We want to find the hypotenuse c given legs of 20.0 ft and 15.0 ft (see figure 12.33).

$$\begin{aligned} c &= \sqrt{(20.0)^2 + (15.0)^2} \\ &= \sqrt{400 + 225} \\ &= \sqrt{625} \\ &= 25.0 \text{ ft} \end{aligned}$$

Therefore, it is 25.0 ft from the top of the shadow to the top of the flag pole.

4. An automobile travels east from an intersection a distance of 13.0 mi. The auto then turns north and travels to a point that is 21.5 mi on a direct line from the intersection. How far north did the auto travel?

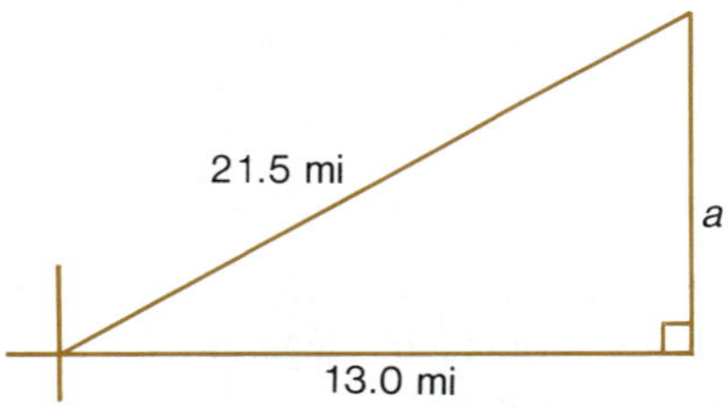

Figure 12.34

We want to find leg a in figure 12.34, given the hypotenuse $c = 21.5$ mi and leg $b = 13.0$ mi.

$$\begin{aligned} a &= \sqrt{(21.5)^2 - (13.0)^2} \\ &= \sqrt{462.25 - 169.0} \\ &= \sqrt{293.25} \\ &\approx 17.1 \text{ mi} \end{aligned}$$

Thus, the auto traveled approximately 17.1 mi north.

Exercise Set 12.4

Given right $\triangle ABC$ with hypotenuse c and legs a and b, find the missing side of the triangle. Round off to three significant digits.

	a	b	c		a	b	c
1.	3	4	?	2.	7	?	16
•3.	12	5	?	4.	15	8	?
5.	6	?	10	6.	?	16	20
7.	17	6	?	8.	10	24	?
•9.	?	10	15	10.	45	?	50
11.	?	24	25	12.	3.2	5.4	?
13.	31.2	?	40.9	14.	?	1.25	2.3
15.	4.33	2.14	?	16.	1.36	?	24.7
17.	0.705	1.78	?	18.	?	0.976	3.07
•19.	.0873	?	0.314	20.	0.0123	0.0713	?

Find the missing measure. Round off according to the rules for approximate numbers.

21. Find the length of the diagonal of a square with side (a) 4.0 cm long, (b) 3.0 in long, (c) 2.0 ft long.
22. Given diagonal d and side s of a square, express d in terms of s.
23. Given a rectangle with length 15 ft and width 20 ft, find the length of the diagonal.
24. A man 6.0 ft tall casts a shadow 3.4 ft long at a particular time of day. Find the distance from the top of the man's head to the tip of his shadow.
25. The distance from the top of a telephone pole to the place where a guy wire is anchored in the ground is 25.4 ft. If the guy wire is anchored at a point 10.0 ft from the base of the pole, how tall is the pole?
26. A roadway rises 6.0 ft in a horizontal distance of 350 ft. What is the length of this stretch of road?
•27. A roadway falls 7.001 ft as an automobile travels a distance of 430.0 ft. This fall takes place over what horizontal distance?
28. The accompanying plate is to have two holes jig-bored in a toolroom for accurate location. Find the distance d.

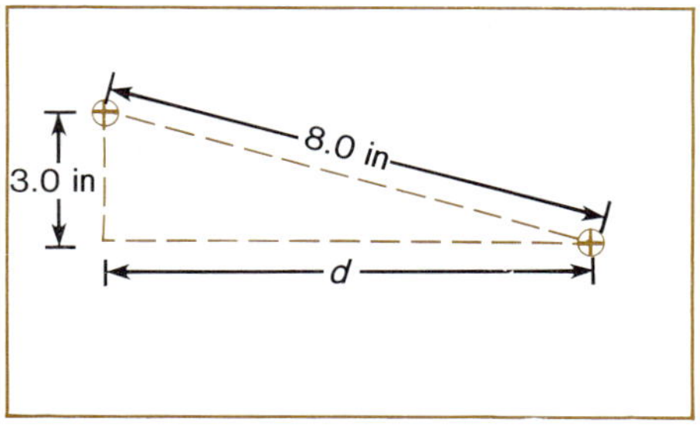

29. A surveyor runs a traverse having three courses that form a right triangle. If the two courses forming the legs are 732.0 m and 354.0 m, what is the distance along the hypotenuse?
30. Find the altitude of an isosceles triangle having a base 6.00 cm and equal sides 12.0 cm. (*Hint:* The altitude bisects the base.)
31. Construct an equilateral triangle whose sides are 3 in long. Calculate the height of the triangle. Measure the height using a ruler and compare results.
32. Calculate the length of the shortest stock to make the stair stringer in the accompanying figure. (*Hint:* State the lengths as decimals.)

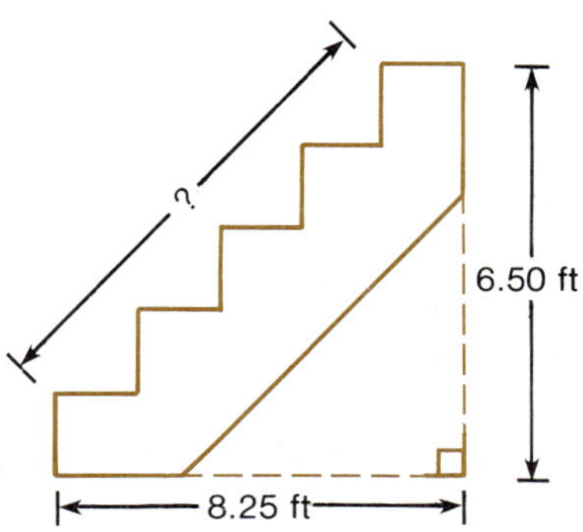

33. The accompanying figure is a roof truss. Find the height h of the truss.

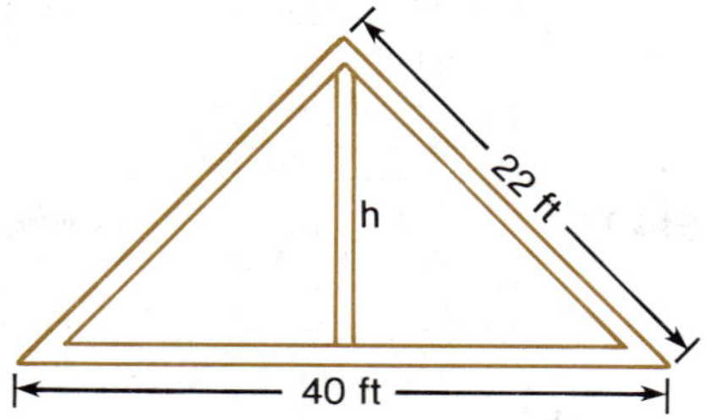

•34. A good proportion for a page being printed is for the diagonal to be twice as long as the width of the page. How long should the page be if the width is 3.50 in?

35. In a right triangle with angles 30°, 60°, and 90°, the hypotenuse is twice as long as the shortest side. Find the length of the other side of the triangle if the shortest side is 3.25 m long.

36. The accompanying figure illustrates the plumbing concepts: the travel, offset, and advance of pipe.

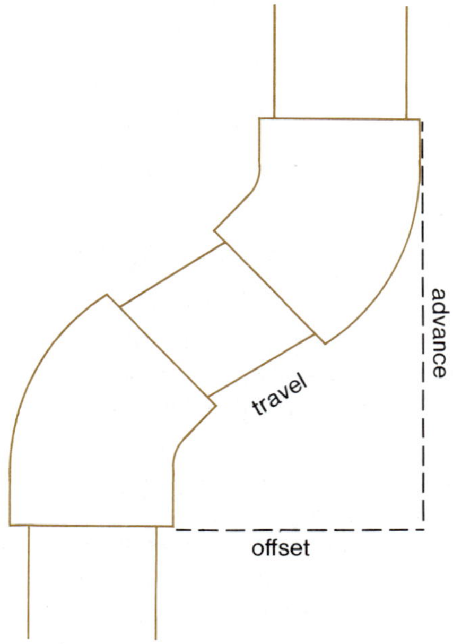

Find:

a. The travel if the advance is 12.0 m and the offset is 7.50 in.

b. The advance if the travel is 42 cm and the offset is 16 cm.

37. A woman rows a boat across a river that is 300.0 ft wide. The current carries the boat to a point 42.00 ft downstream from the point directly across from the starting point. How far did the woman actually row the boat?

38. Given isosceles right $\triangle ABC$, each of the equal angles have what measure? Show that the length of the hypotenuse is equal to $\sqrt{2}$ times a leg.

39. Given an alternating-current (AC) circuit, the impedance (apparent resistance) Z can be represented by the hypotenuse of a right triangle whose legs are the ohmic resistance R and the reactance X_L (the effective resistance to the circuit). Thus, $Z^2 = R^2 + X_L^2$, and so $Z = \sqrt{R^2 + X_L^2}$.

Find: a. The impedance when the coil has resistance 2.40 ohms and reactance 2.19 ohms.

b. The reactance of a coil when the impedance is 24 ohms and the resistance is 15 ohms.

12.7 The Circle

The last of the plane geometric figures we shall discuss is the circle. By definition, a **circle** is a closed curved line consisting of all points in the plane that are the same distance from a fixed point in the plane. Having no line segments, a circle *is not a polygon*.

note

A circle may be described by a point on the terminal side of an angle that is revolved through an angle of 360°. Thus, we say there are 360° in a circle.

The following are characteristics of the circle:

1. The fixed point in the plane is the **center** of the circle, denoted by O.
2. The line segment from the center to any point on the circle is the **radius** of the circle, denoted by r.
3. The line segment between two points on the circle passing through the center is the **diameter** of the circle, denoted by d.

   note

   The length of the diameter of a circle is two times the length of the radius. That is, $d = 2r$.

4. A line segment connecting any two points on the circle is a **chord** of the circle. The diameter of a circle is the longest chord of that circle.
5. A line passing through any two points on the circle is a **secant** of the circle.

   note

   A secant is a line, while a chord is a line segment.

6. A line in the same plane that touches the circle at only one point is a **tangent** line of the circle. The point that the line and circle have in common is called the **point of tangency.**
7. The length of the closed curved line that forms the circle (the distance around the circle) is the **circumference** of the circle, denoted by C. The circumference of a circle is discussed more extensively in the next chapter.

See figure 12.35.

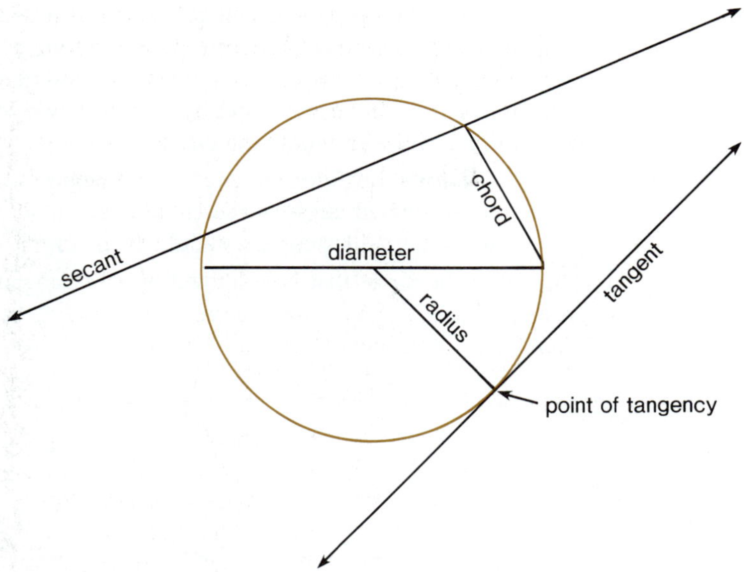

Figure 12.35

In relation to a circle it can be shown that:

1. The radius drawn to the point of tangency is perpendicular to the tangent line.
2. The two line segments drawn from the two points of tangency to a common point outside the circle are equal in length (as illustrated in figure 12.36).
3. A line drawn from the center to a point A outside the circle bisects the angle between the tangents drawn to the circle from the point.

In figure 12.36, $\overline{OB} \perp \overline{AB}$ and $\overline{OC} \perp \overline{AC}$, while $AB = AC$ and $m\,\angle\alpha = m\,\angle\beta$.

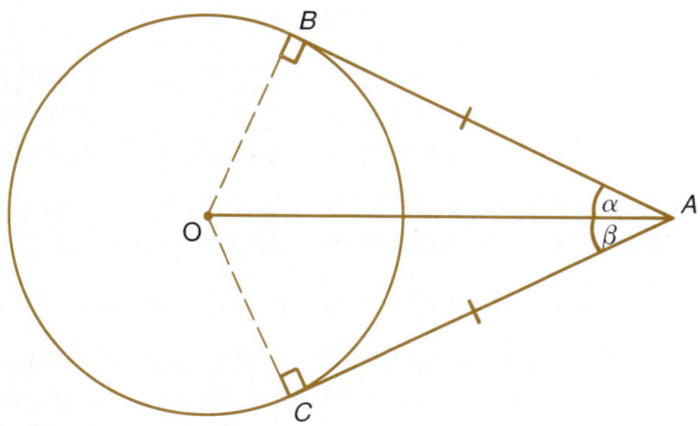

Figure 12.36

note

We denote the measure of $\overline{AB}$ and $\overline{AC}$ by AB and AC.

Now consider the arcs of a circle and angles within the circle. An **arc** of the circle is the curved line segment between any two points on the circle. The symbol used to represent an arc is $\frown$. Arcs of the circle are described by a chord of the circle, two radii drawn from the center, and by angles drawn within the circle. There are two kinds of angles drawn within the circle:

1. A **central angle** of the circle is the angle formed by two radii.
2. An **inscribed angle** of the circle is an angle whose vertex is on the circle and whose sides contain chords of the circle.

The arc of the circle that is described by these angles is called the **intercepted arc.** In figure 12.37, $\angle AOB$ is a *central angle* and $\angle ACB$ is an *inscribed angle* of the circle. They both intercept $\widehat{AB}$.

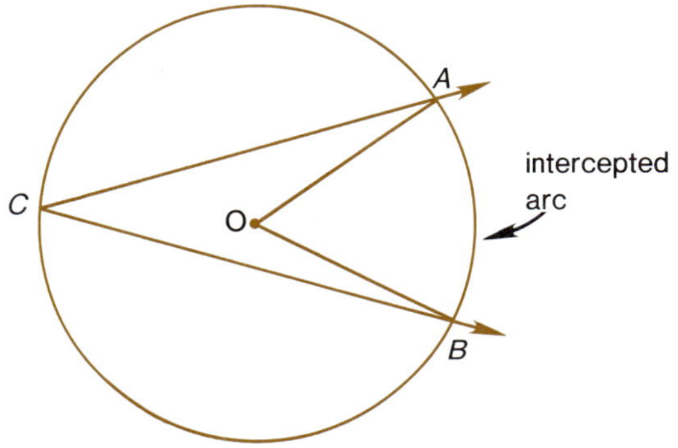

Figure 12.37

There are two ways to measure arcs: linear measure and degree measure. The length of an arc is a **linear measure** and is given in the same unit of measure as the radius of the circle. The measure of an arc is in degrees. *The number of degrees in an arc is the same as the central angle that intercepts the arc.*

note

The diameter of a circle, which can be treated as a central angle of the circle—a straight angle—then intercepts two equal arcs, each of measure 180°. Each such equal arc is called a *semicircle*.

An inscribed angle is measured by one-half the degree measure of the arc it intercepts. Thus, to find the measure of an inscribed angle of the circle intercepting the same arc as a central angle of the circle, divide by two. In figure 12.37, $m\ \angle ACB = \frac{1}{2} m\ \angle AOB$. From our discussion, we can determine that an inscribed angle that intercepts an arc of 180°, a semicircle or half-circle, is *always* a right angle. Therefore, an inscribed angle whose sides go to the endpoints of a diameter will be a right angle. See figure 12.38.

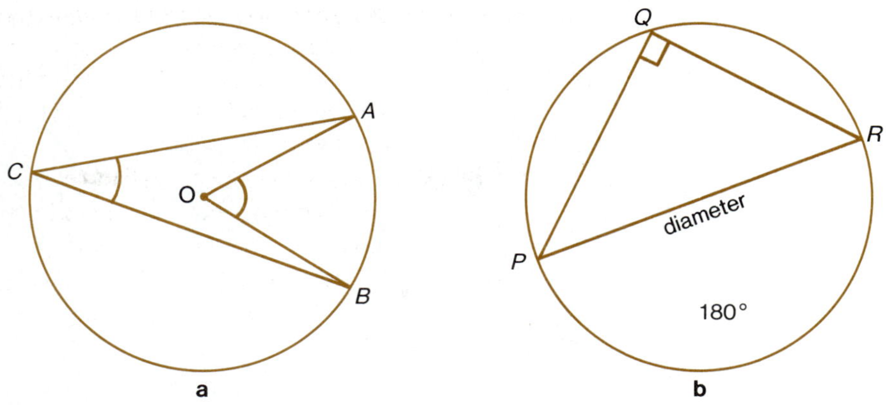

Figure 12.38

In figure 12.38**a,** given $\widehat{AB}$ is 76°, then central angle $\angle AOB = 76°$ and inscribed angle $\angle ACB = 38°$. In figure 12.38**b,** since inscribed $\angle PQR$ intercepts 180° $\widehat{PR}$, then $\angle PQR$ is a right angle and $\angle PQR = 90°$.

Exercise Set 12.5

Use the accompanying circle with center O to identify the following.

1. Two diameters
2. One right angle
3. Two tangent lines
4. Three chords
5. Two acute central angles
6. Two obtuse central angles
7. Two inscribed angles
8. Two points of tangency
9. One secant line

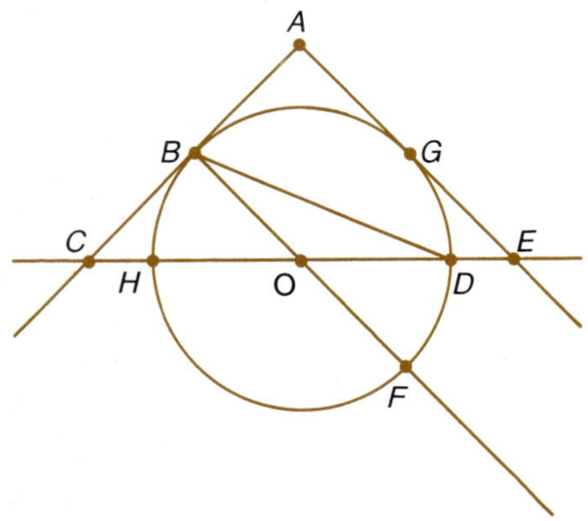

Using the preceding figure, and given $\angle HOB = 60°$, determine the measure of the following.

•10. $\widehat{HB}$ •11. $\widehat{BGD}$ 12. $\angle BOD$ •13. $\widehat{DF}$
•14. $\angle FBD$ 15. $\angle HDB$ 16. $\widehat{HF}$

Find the measure of angle θ in each of the following circles having center at O.

17.

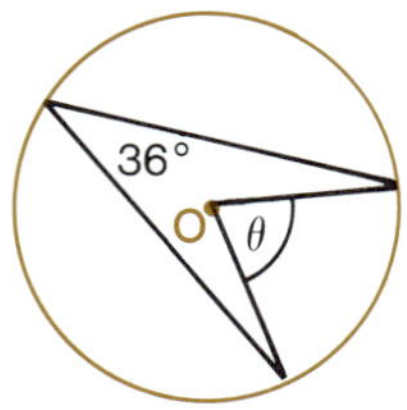

•18.

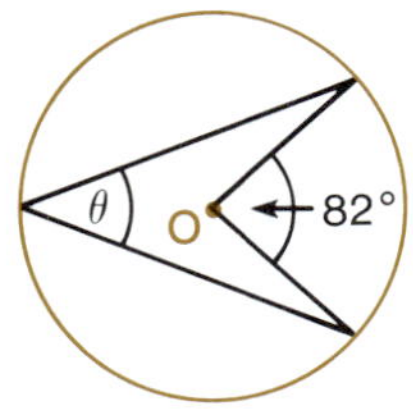

•19.

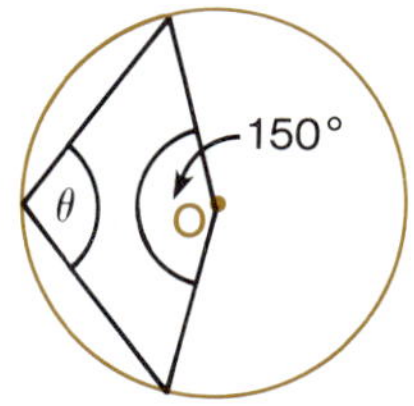

•20.

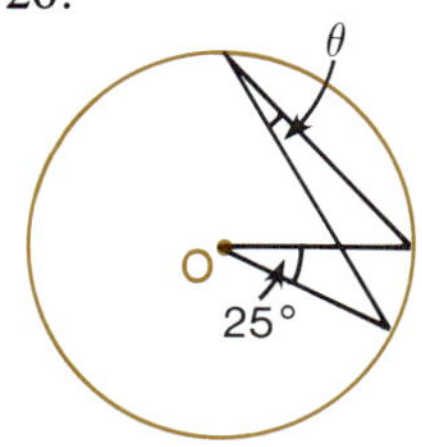

•21.

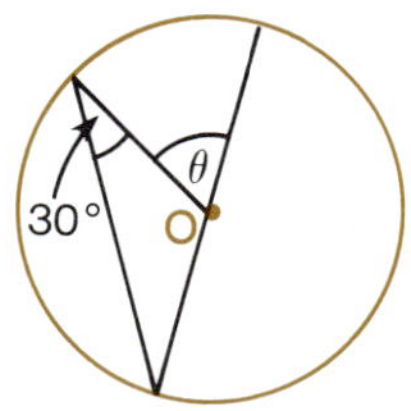

•22.

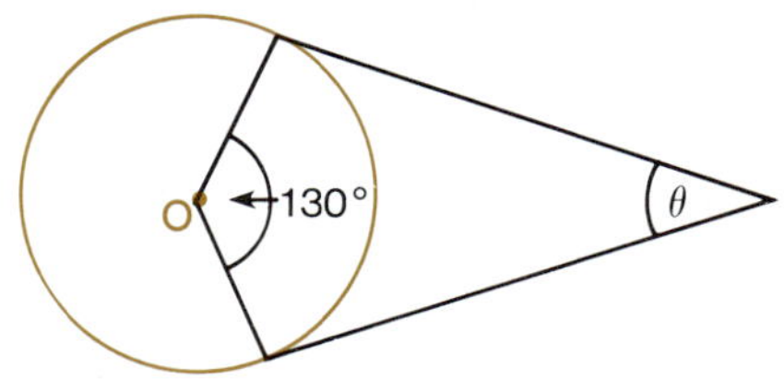

•23.

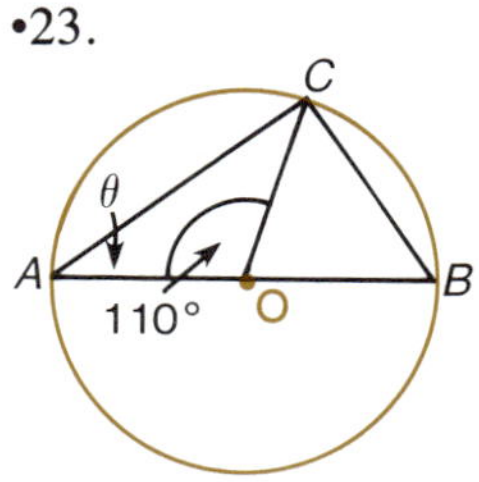

24.

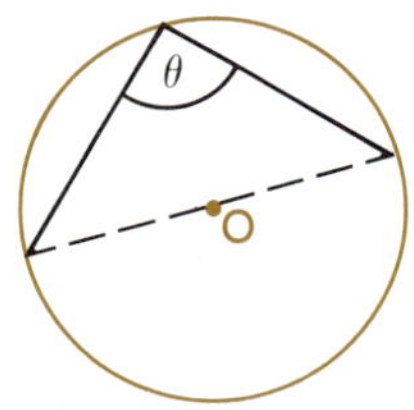

•25.

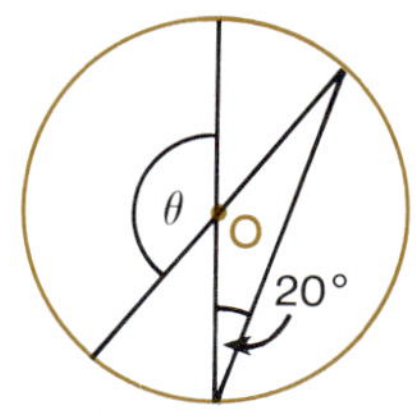

•26.

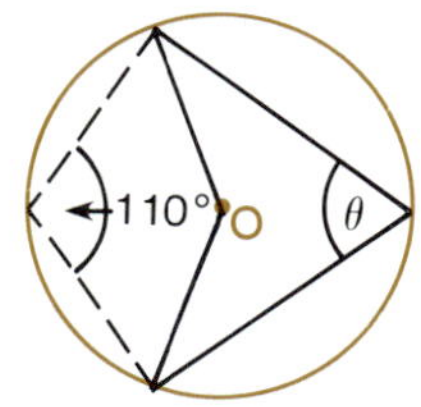

12.8 Congruent and Similar Polygons

Two important relationships that can exist between two or more polygons are congruence and similarity. By definition, two polygons are said to be **congruent** if they can be made to coincide. Such polygons must have the same size and the same shape—characteristics that can occur only if the corresponding parts of the polygons have the same measure. That is, if the corresponding angles are the same size and the corresponding sides have the same length, the polygons will be congruent. Corresponding parts of polygons have the same relative position in their respective figures.

Figure 12.39 shows congruent quadrilaterals $ABCD$ and $A'B'C'D'$ (read "A prime, B prime, C prime, D prime"). Given these congruent quadrilaterals, the corresponding angles that have the same measure are $\angle A$ and $\angle A'$, $\angle B$ and $\angle B'$, $\angle C$ and $\angle C'$, and $\angle D$ and $\angle D'$, and the corresponding sides that are equal in length are $\overline{AB}$ and $\overline{A'B'}$, $\overline{BC}$ and $\overline{B'C'}$, $\overline{CD}$ and $\overline{C'D'}$, and $\overline{DA}$ and $\overline{D'A'}$. Congruent plane figures are denoted by the symbol $\cong$. Thus, in figure 12.39 quadrilateral $ABCD \cong$ quadrilateral $A'B'C'D'$ which is read "quadrilateral $ABCD$ *is congruent to* quadrilateral $A'B'C'D'$." Congruence of polygons is used extensively in the proof of many geometric theorems. (This type of proof is beyond the scope of this text.) Congruence is used, too, in many branches of industry for making matching parts by patterns.

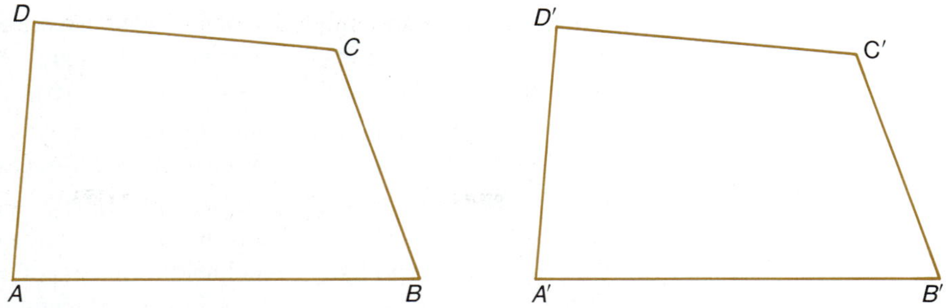

Figure 12.39

In chapter 3 we examined measurements that were proportional. Now we apply the concepts of proportionality to the sides of polygons. Comparing two polygons by congruence involves their having the same size and same shape. Comparing by *similarity* involves polygons having the same shape but disregards equality of size. In similar polygons corresponding angles have the same measure, but the measures of the sides are related proportionally. By definition, two polygons are **similar** if their *corresponding angles are equal in size and the measures of their corresponding sides are in proportion.*

note

To say the corresponding sides are proportional is to say the *ratios* of the measures of the corresponding sides are *equal.*

The symbol used to denote similarity of polygons is $\sim$. See figure 12.40. Given $\triangle ABC$ is similar to $A'B'C'$, denoted by $\triangle ABC \sim \triangle A'B'C'$, then $m\ \angle A = m\ \angle A'$, $m\ \angle B = m\ \angle B'$, and $m\ \angle C = m\ \angle C'$, and $\dfrac{AB}{A'B'} = \dfrac{BC}{B'C'} = \dfrac{AC}{A'C'}$.

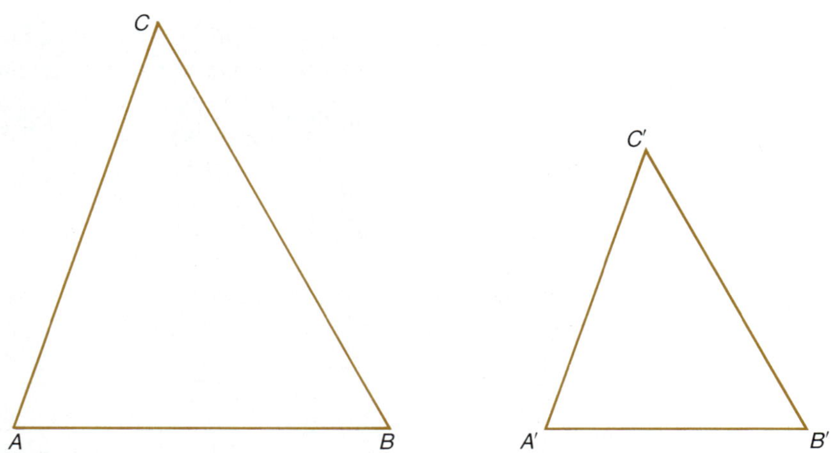

Figure 12.40

Congruence and similarity of polygons have extensive applications in triangles, and thus we limit our applications here to triangles. From previous definitions we can see that similar triangles have two basic properties:

1. Corresponding angles are equal in size.
2. Corresponding sides are proportional in measure.

To show that two triangles are similar, we need only show that *one* of these conditions is met. Conversely, if we know that two triangles are similar, we can use the properties of similarity to find unknown parts of one triangle by using known parts of the other. This property is used in solving many practical and technical problems.

Examples

1. Show that, given $\overline{BC} \parallel \overline{DE}$, then $\triangle ABC \sim \triangle ADE$ in figure 12.41. To solve this problem, show that the corresponding angles have the same measure:
   a. $\angle A$ is common to both triangles.
   b. Parallel lines $\overline{BC}$ and $\overline{DE}$ are cut by transversals $\overline{AC}$ and $\overline{AB}$. Since corresponding angles have the same measure, $m\ \angle ACB = m\ \angle ADE$ and $m\ \angle ABC = m\ \angle AED$.
   c. All three corresponding angles have the same measure; thus, $\triangle ABC \sim \triangle ADE$.

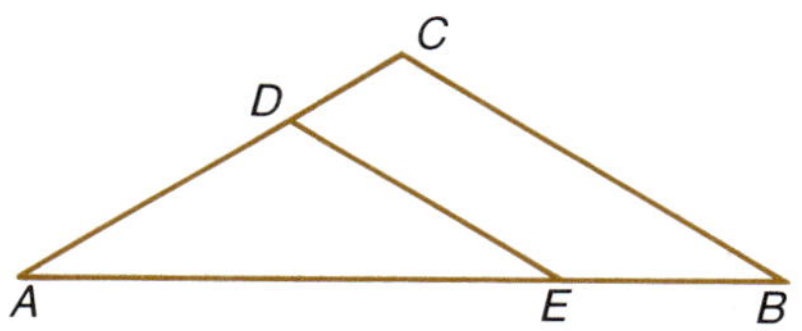

Figure 12.41

2. In figure 12.42, if $\triangle PQR \sim \triangle XYZ$, find the length of the missing sides. We want the length of $\overline{QR}$ and $\overline{XZ}$. Since corresponding sides are in proportion, then $\frac{PR}{XZ} = \frac{PQ}{XY} = \frac{QR}{ZY}$. Thus, $\frac{15}{XZ} = \frac{10}{8}$ and $\frac{10}{8} = \frac{QR}{6}$.

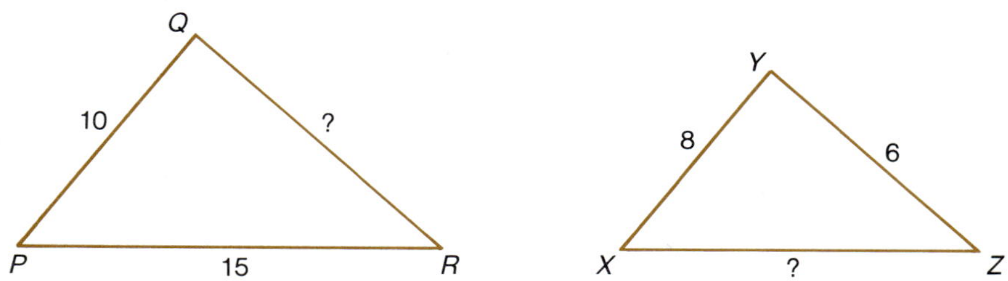

Figure 12.42

Solve the given proportions by cross-products:

a. $10 \cdot XZ = 8 \cdot 15$

$$XZ = \frac{8 \cdot 15}{10} = \frac{120}{10} = 12$$

b. $8 \cdot QR = 10 \cdot 6$

$$QR = \frac{10 \cdot 6}{8} = \frac{60}{8} = 7\frac{1}{2}$$

3. In figure 12.43, given two right triangles $\triangle ABC$ and $\triangle A'B'C'$, with $m\ \angle A = m\ \angle A'$, show that the triangles are similar. Given the right triangles with right angles $\angle C$ and $\angle C'$, then $m\ \angle A + m\ \angle B = 90°$ and $m\ \angle A' + m\ \angle B' = 90°$. Solving these equations for $\angle B$ and $\angle B'$, respectively, yields $m\ \angle B = 90° - m\ \angle A$, and $m\ \angle B' = 90° - m\ \angle A'$, and since $m\ \angle A = m\ \angle A'$, then $m\ \angle B = m\ \angle B'$. Hence, the two triangles have angles equal in measure and so are similar.

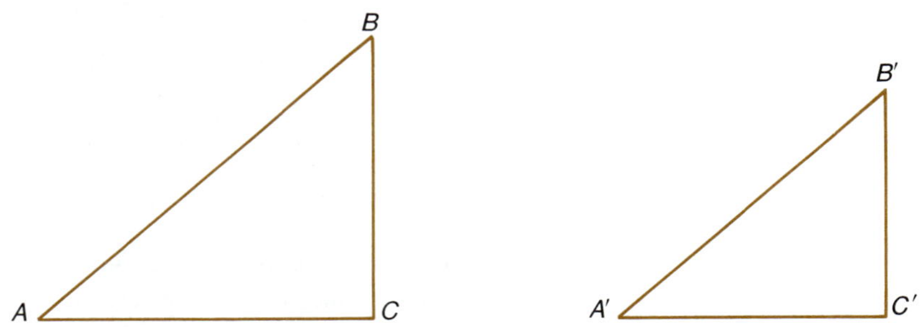

Figure 12.43

We have just proved an important theorem about right triangles that states: *Two right triangles having one acute angle equal in size are similar*.

4. A man 6 ft tall casts a shadow 4 ft long at the same time of day that a building casts a shadow 30 ft long. How tall is the building? Figure 12.44 illustrates the situation. Right angles are theoretically formed at the foot of the building and the man. The rays of the sun form the hypotenuse of the right triangles thus formed, and they are parallel. The shadows, then, are transversals cutting the parallel lines, and thus, the corresponding acute angles are equal in size at the tips of the shadows.

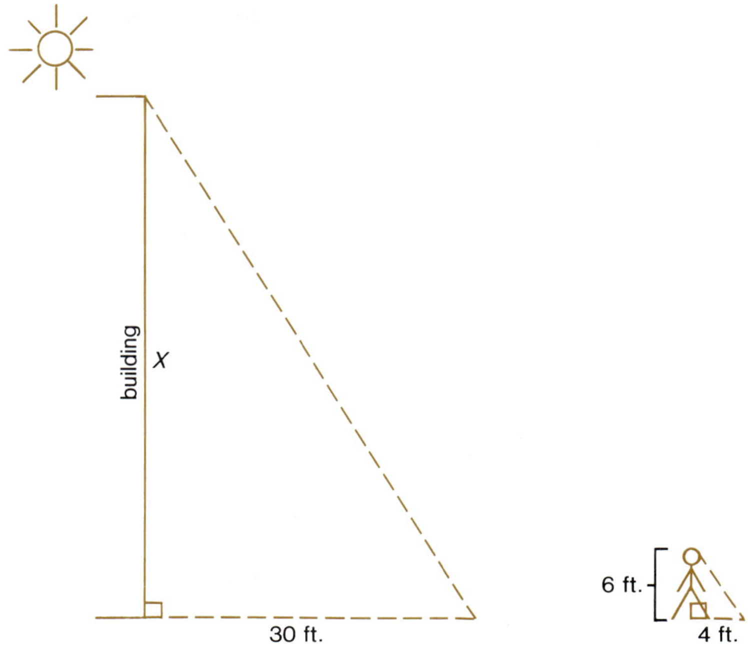

Figure 12.44

The result is right triangles with one acute angle having the same measure, and the triangles are similar. Let x = height of the building.

then, $\frac{x}{6} = \frac{30}{4}$ and so

$$4 \cdot x = 6 \cdot 30$$

$$x = \frac{6 \cdot 30}{4} = 45 \text{ ft}$$

Thus, the building is 45 ft tall.

Engineering and technical fields use scale drawings extensively. This application is one of the most practical ways of using similar figures to find missing measurements.

Example The straight-line distance from Detroit to Chicago is 220 mi and from Detroit to Sault Ste. Marie, Michigan, is 350 mi. If a map uses the scale 1 in = 50 mi, what will be the distance on the map from Detroit to Chicago and to Sault Ste. Marie? From figure 12.45,

$\frac{1}{50} = \frac{x}{350}$ and $\frac{1}{50} = \frac{y}{220}$

then $50 \cdot x = 350$ and $50 \cdot y = 220$.

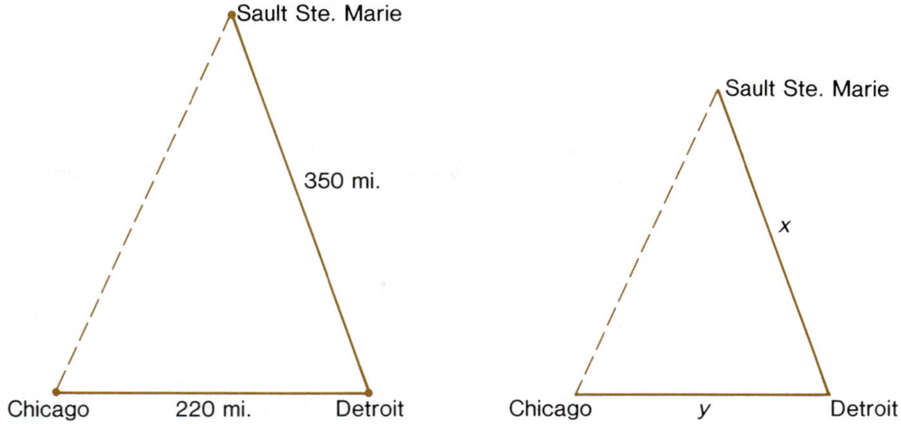

Figure 12.45

So $x = 7$ in and $y = 4.4$ in. Then, on the map, the distance from Detroit to Chicago is represented by 4.4 in and the distance from Detroit to Sault Ste. Marie is represented by 7 in.

Exercise Set 12.6

In exercises 1 through 4, given that the sets of triangles are similar, identify corresponding parts.

1.

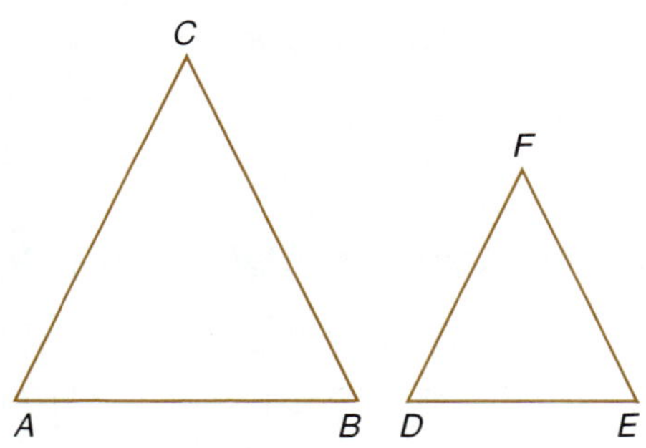

•2.

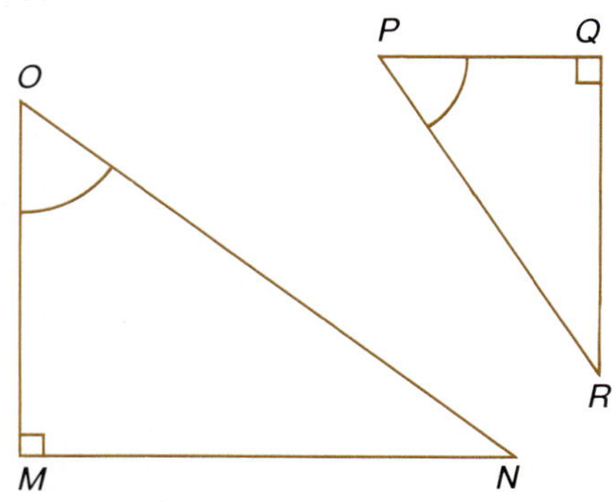

3.

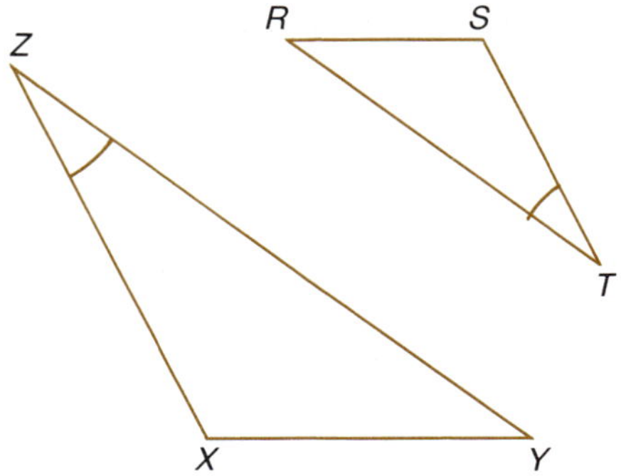

4.

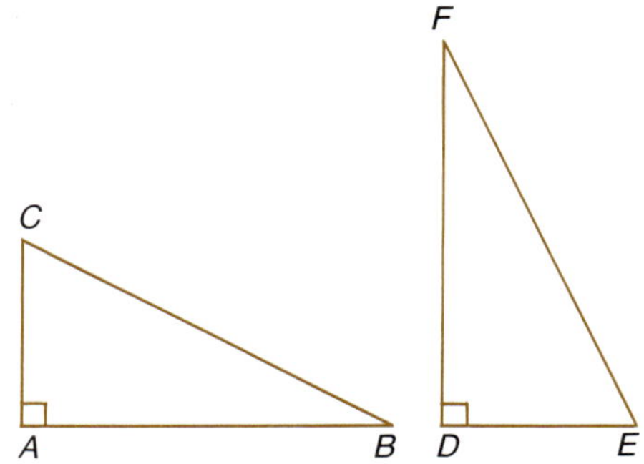

5. Given $\overline{AC} \parallel \overline{DE}$, show that $\triangle ABC \sim \triangle DBE$ and identify the corresponding parts.

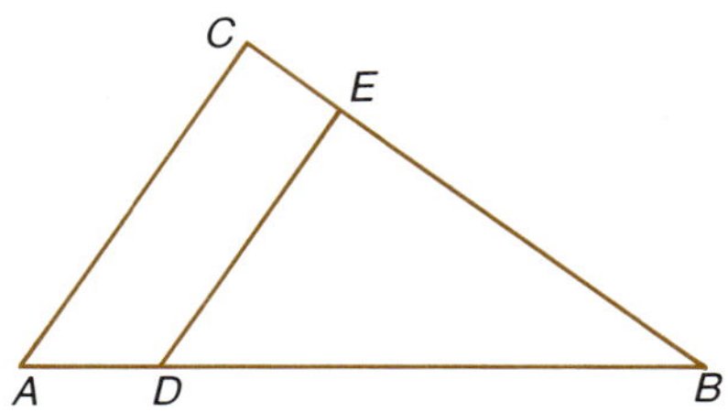

•6. Show $\triangle ABC \sim \triangle EDC$ and determine the corresponding parts. ($\overline{AE}$ and $\overline{BD}$ are straight.)

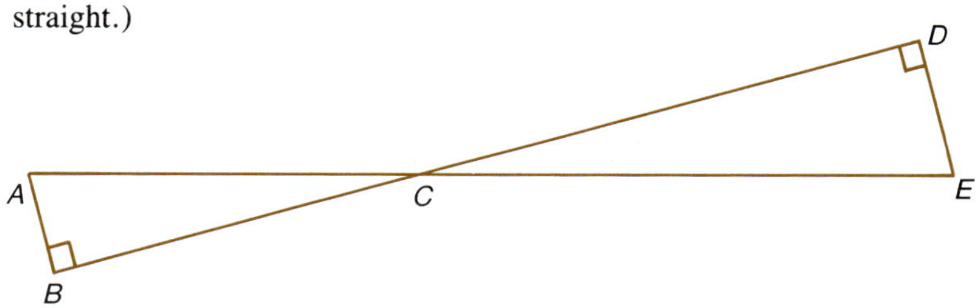

7. Show $\triangle PQR \sim \triangle QSR$ and $\triangle PQR \sim \triangle PSQ$. Identify the corresponding parts.

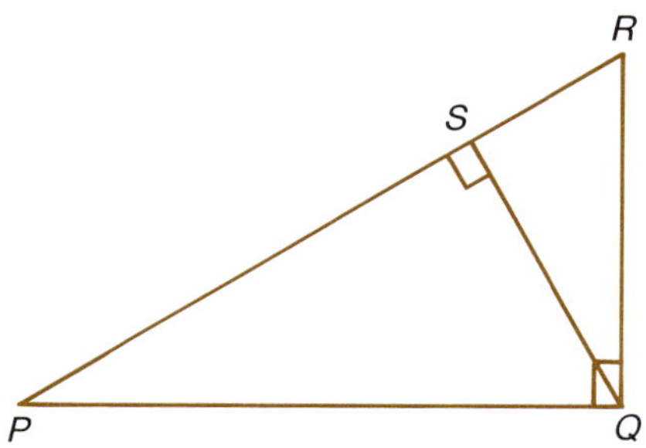

In exercises 8 through 19, use properties of similar triangles to tell why the sets of triangles are similar and find the unknown quantity in each problem.

8.

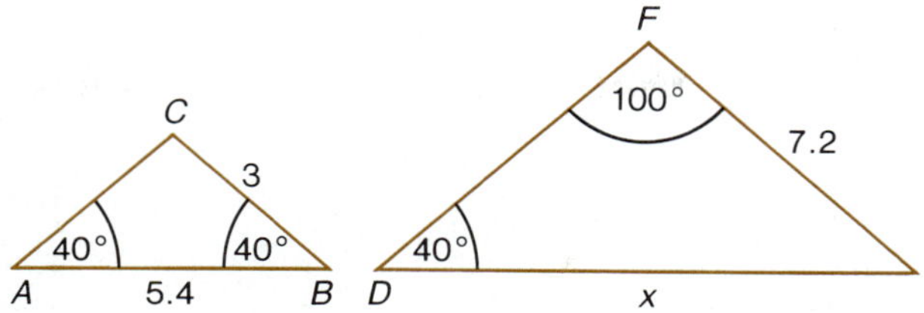

9.

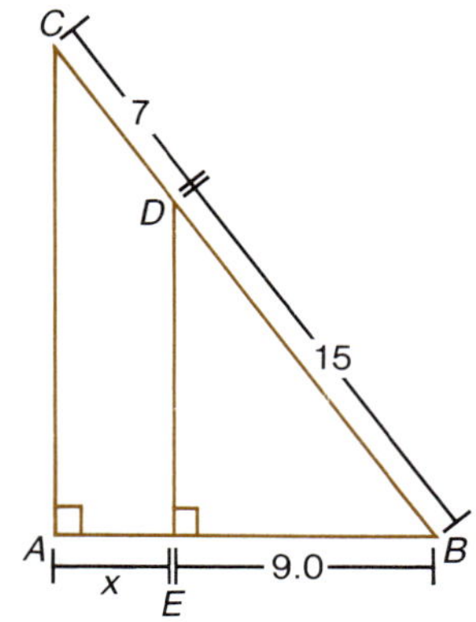

10.

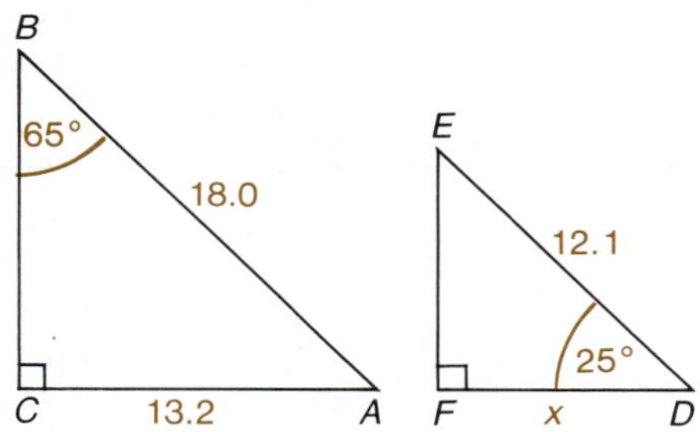

11.

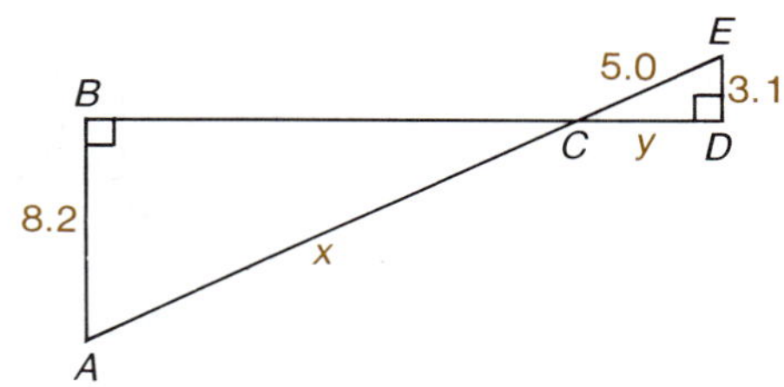

•12.

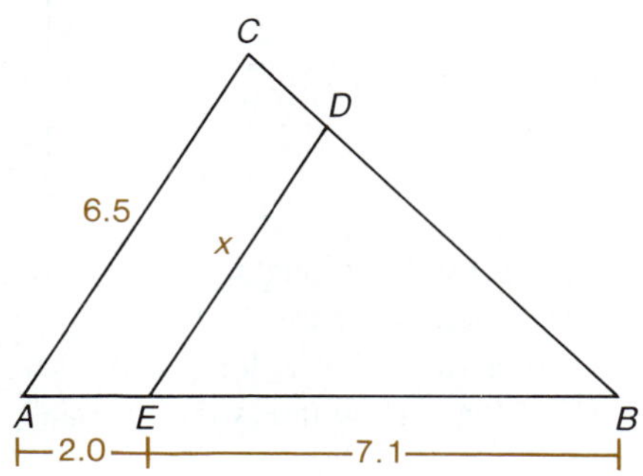

13.

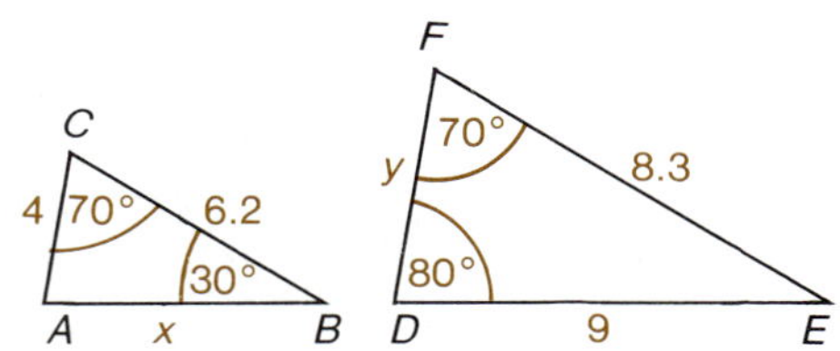

•14.

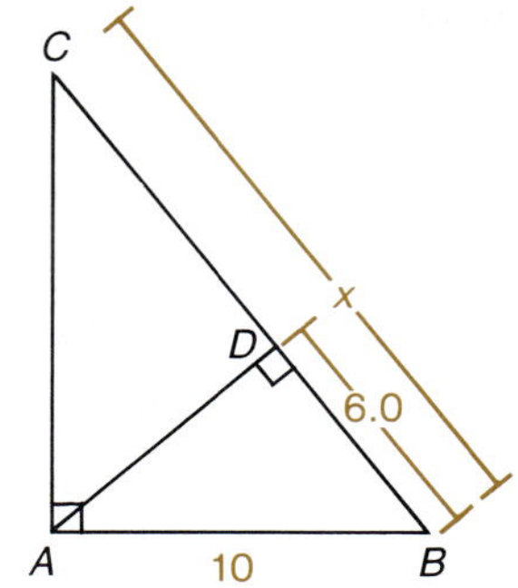

15.

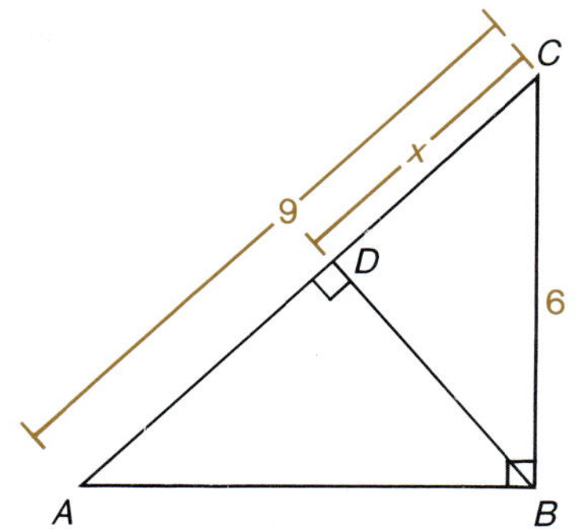

16.

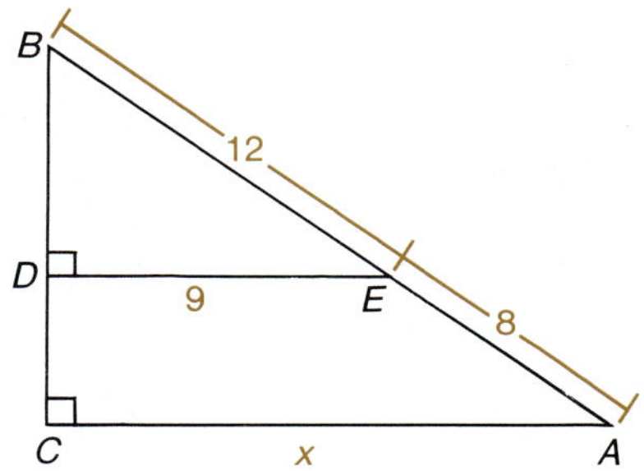

17.

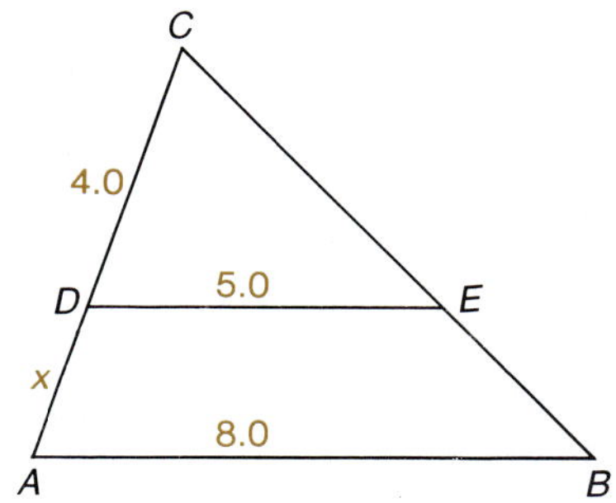

•18.

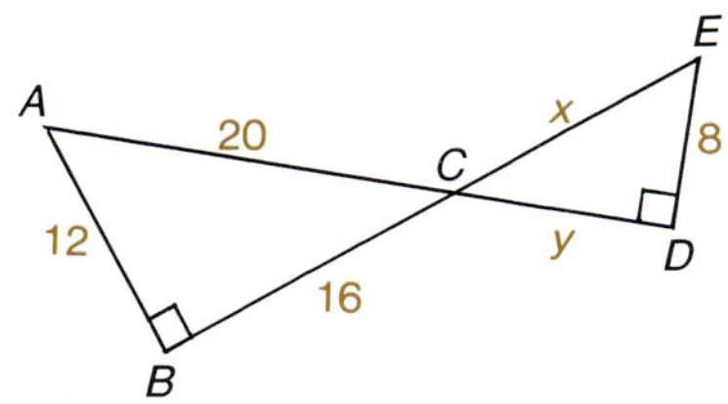

19.

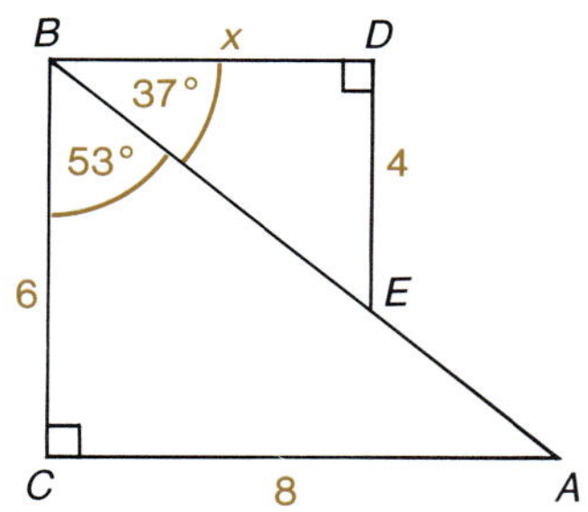

•20. A tree casts a shadow 24 ft long at the same time a boy 4.0 ft tall casts a shadow 5.0 ft long. How tall is the tree?

21. A woman casts a shadow 9 ft long at the same time that a yard stick casts a shadow 4 1/3 ft long. How tall is the woman?

22. Determine the width of the river in the accompanying figure, assuming the banks are parallel for the given distance.

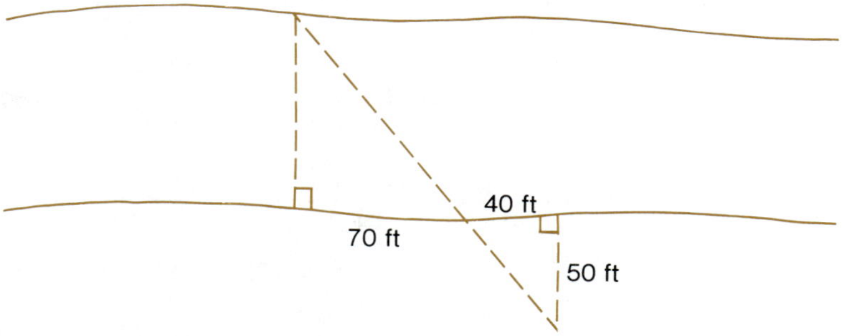

23. A farmer draws to scale a triangular piece of land. The sides of the triangle are 5, 10, and 12 in. If the longest side of the piece of land is 720 ft, how long are the other sides of the piece of land?
24. On a map, the scale is 1 cm = 40 mi. Find the distance on the map between Detroit and Cincinnati if the cities are 250 mi apart.
25. Using similar triangles, find the distance from Q to a rock at point P on the opposite banks of the stream in the given figure.

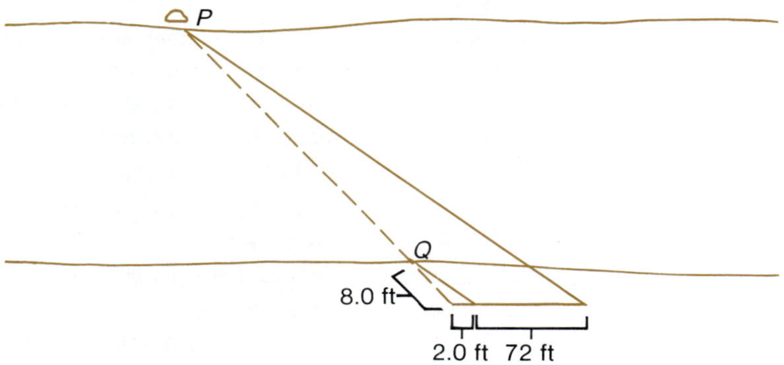

26. Find the distance AB across the lake if it is known that $m\overline{EB} = 250$ m, $m\overline{ED} = 90$ m, and $m\overline{CD} = 50$ m. We know $\overline{CD} \parallel \overline{AB}$.

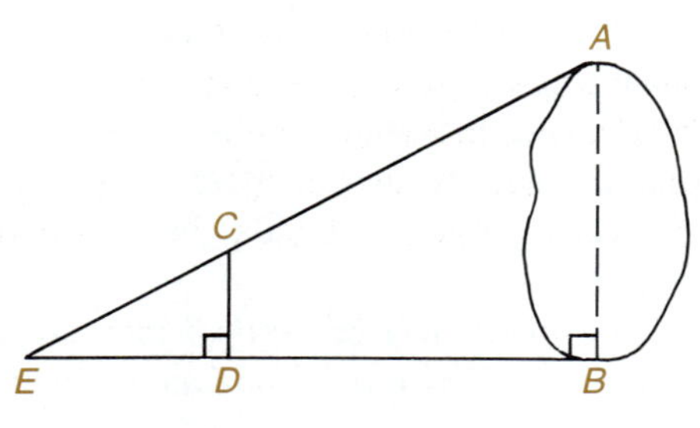

•27. In the following construction, studs $\overline{FG}$, $\overline{DE}$, and $\overline{BC}$ are perpendicular to $\overline{AC}$. Find the lengths of $\overline{DE}$, $\overline{FG}$, $\overline{DC}$, $\overline{EF}$, and $\overline{AB}$.

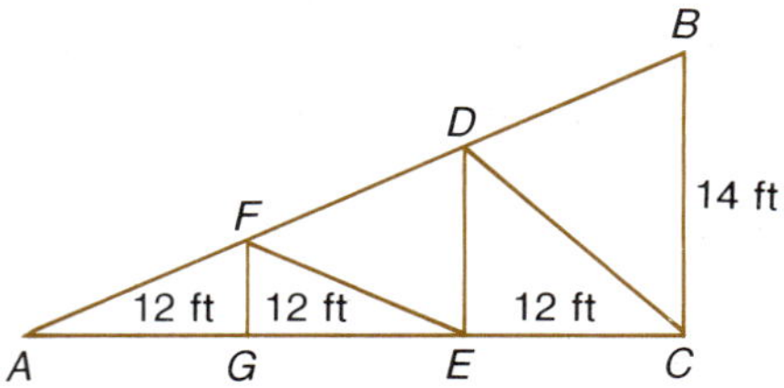

28. On a map, 4 in = 100 mi. Find the distance between two cities if they are 5.4 in apart on the map.
29. The blueprints for a factory have a scale of 1.5 in = 10 ft. Find the length and width of a room in the factory if the scale dimensions are 30 in long and 5.4 in wide.

Chapter Summary

1. The **point, line,** and **plane** are undefined geometric elements.
2. A **line segment** is a part of a line with specific length.
3. A **ray** is the part of a line extending infinitely far in one direction from a point on the line.
4. Two points determine a unique straight line.
5. An **angle** is a geometric figure formed by two rays with a common endpoint.
6. The size of an angle is measured in **degrees** or **radians.** When two angles have the same measure they are said to be congruent.
7. Angles are named by their measure:

   Acute—measure between 0° and 90°
   Right—measure 90°
   Obtuse—measure between 90° and 180°
   Straight—measure 180°
   Reflex—measure between 180° and 360°.

8. **Complementary** angles are two angles the sum of whose measure is 90°.
9. **Supplementary** angles are two angles the sum of whose measure is 180°.
10. **Adjacent** angles are angles that share a common vertex and a common side but whose interiors have no point in common.
11. **Vertical** angles are the nonadjacent angles formed by two intersecting lines. Vertical angles have the same measure.
12. **Coplanar** lines are two or more lines lying in the same plane. Coplanar lines either *intersect* (have a single point in common) or are parallel (have no point in common).
13. Lines in a plane may be *vertical* (at right angles with the horizon), *horizontal* (parallel to the horizon), or *oblique* (neither vertical nor horizontal).
14. **Perpendicular** lines intersect to form a right angle.
15. A **transversal** is a line that intersects two or more other lines.

16. When two parallel lines are cut by a transversal, the (a) *alternate-interior* angles have the same measure; (b) *alternate-exterior* angles have the same measure; and (c) *corresponding* angles have the same measure.
17. A **closed planar figure** (or *plane figure*) encloses a portion of the plane.
18. A **polygon** is a plane figure formed by intersecting straight lines, called *sides* of the polygon.
19. The *point of intersection* of any two sides of a polygon is called a **vertex.**
20. Polygons are named by the number of sides that form the polygon:

    Triangle—3 sides
    Quadrilateral—4 sides
    Pentagon—5 sides
    Hexagon—6 sides
    Octagon—8 sides
    Decagon—10 sides

21. A **regular polygon** is a polygon whose sides are all the same length and whose angles are all the same measure.
22. The sum of the measure of the interior angles of a triangle is 180°. The sum of the measure of the interior angles of a polygon is given by $(n - 2)180°$, where n is the number of sides.
23. The **altitude,** h, of a triangle is the length of the line segment from a vertex perpendicular to the opposite side.
24. Triangles are named by:
    a. the length of their sides:

       Scalene—all sides unequal in length
       Isosceles—two sides have the same length
       Equilateral—all three sides equal in length

    b. the measure of the angles:

       Acute—all angles are acute
       Right—one right angle
       Obtuse—one obtuse angle

25. A **parallelogram** is a quadrilateral whose opposite sides are parallel and equal in length.
26. A **rectangle** is a parallelogram with one right angle. (Note: Then *all* angles are right.)
27. A **square** is a rectangle whose sides are all equal in length.
28. A **rhombus** is a parallelogram having all sides equal in length.
29. A **trapezoid** is a quadrilateral with *one* pair of opposite sides parallel.
30. The **altitude,** h, of a parallelogram and a trapezoid is the perpendicular distance between the two parallel sides.
31. The *Pythagorean Theorem* states that in a right triangle, the sum of the squares of the lengths of the two legs equals the square of the length of the hypotenuse.
32. A **circle** is a closed, curved line consisting of all points in the plane that are the same distance from a fixed point in the plane, called the *center*.
33. The **radius,** r, of a circle is the line segment from its center to its circumference.

34. The **diameter,** *d*, of a circle is the line segment that passes through the center point and has its ends terminate on the circumference. Its length is two times the length of the radius.
35. A **chord** of the circle is a line segment joining two points on the circle. The *longest* chord of a circle is its diameter.
36. A **tangent** line to a circle is a line that touches the circle at *only one point*. The common point is called the *point of tangency*.
37. A **secant** line is a line passing through two points on a circle.
38. The *radius* drawn to the point of tangency is *perpendicular* to the *tangent line* at that point.
39. An **arc** of a circle is the curved line segment between any two points of the circle.
40. A **central angle** of a circle is an angle whose vertex is at the center.
41. An **inscribed angle** of a circle is an angle whose vertex is a point on the circle and whose sides are chords of the circle.
42. The *measure* of an *inscribed angle* of a circle is one-half the measure of the central angle intercepting the same arc.
43. The *measure* of a *central angle* of a circle is the same as the arc it intercepts.
44. An angle inscribed within a semicircle is always a right angle.
45. **Congruent polygons** have the *same size* and the *same shape*.
46. **Similar polygons** have the *same shape* but *different size*.
47. The symbol for *congruence* is $\cong$ and the symbol for *similarity* is $\sim$.
48. Triangles are similar if and only if (a) their corresponding angles have the same degree measure, or (b) the measures of their corresponding sides are in proportion.

Chapter Review

[12.2]

1. Ray *AB* is denoted by what symbol?
2. Line *AB* is denoted by what symbol?
3. Line segment *AB* is denoted by what symbol?
4. The figure formed by two rays with a common endpoint is called __________.
5. Draw an angle of measure (a) 47°, (b) 132°.
6. An angle is usually measured in what two units of measure?
7. Write the following angles in degrees correct to the nearest tenth of a degree: (a) 19°20′, (b) 73°42′, (c) 27°53′.
8. Write the following angles in minutes and seconds, correct to the nearest second where necessary: (a) 10.4°, (b) 37.25°, (c) 63.29°.
9. Give the names of four different angles according to their measure in degrees.
10. Two angles sharing a common vertex, with a common side between them, whose interiors have no points in common are called __________ angles.
11. What are the names of the two angles the sum of whose measure is (a) 90°, (b) 180°?

[12.4]

12. Lines in the plane are named by their position in the plane. Name three different kinds of lines in the plane.
13. Two lines that never meet, no matter how far they extend, are called ________ lines.
14. Two lines that intersect in a 90° angle are called ____________ lines.
15. In the accompanying figure, given lines l_1 and l_2 such that $l_1 \parallel l_2$, name the pairs of (a) alternate-exterior angles, (b) alternate-interior angles, (c) corresponding angles.

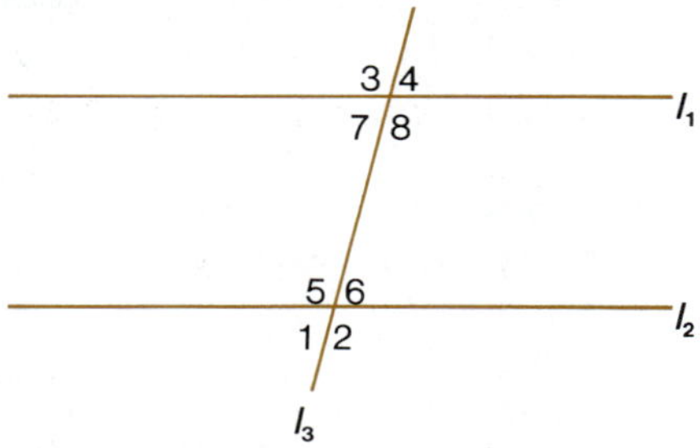

16. In the preceding figure, if $m\ \angle 5 = 105°$, determine the measure in degrees of (a) $\angle 8$, (b) $\angle 6$, (c) $\angle 2$, (d) $\angle 3$.
17. In the accompanying figure, given $\overline{AC} \parallel \overline{DE}$, $\overline{FC} \parallel \overline{AB}$, $\overline{AC} \perp \overline{BG}$, and $m\ \angle ABC = 36°$, find the measure of (a) $\angle 1$, (b) $\angle 2$, (c) $\angle 3$, (d) $\angle 4$.

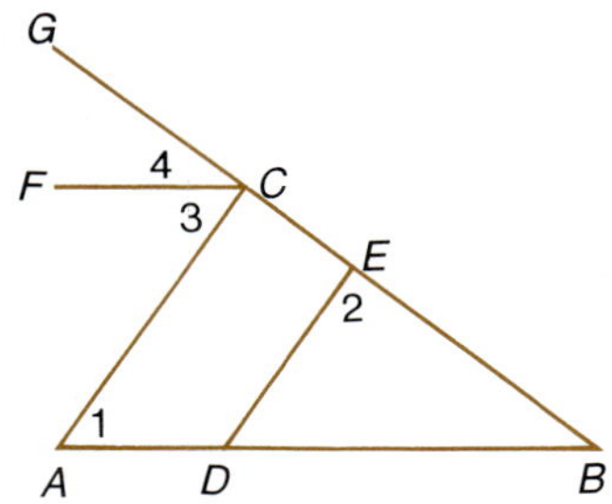

[12.5]

18. Give the names of the polygons with the following number of sides: (a) 5, (b) 4, (c) 3, (d) 6, (e) 8, (f) 10.
19. A polygon whose sides are all the same length and whose angles are all the same measure is called a ____________ polygon.
20. Triangles may be named according to their sides and their angles. Give the names of the triangles according to their (a) sides, (b) angles.
21. The name given to the line segment drawn from a vertex of a triangle perpendicular to the opposite side is ____________.
22. A ____________ is the line segment drawn from a vertex to the midpoint of the opposite side.
23. A ____________ is a line segment drawn in a polygon from one vertex to a nonadjacent vertex of the polygon.

24. Give the names of quadrilaterals having the following properties:
    a. Opposite sides are parallel and equal in length.
    b. One pair of opposite sides is parallel.
    c. All four sides are equal in length.
    d. Diagonals are equal in length.
    e. Diagonals are perpendicular bisectors of each other.
25. Find the sum of the interior angles of a polygon with seven sides.
26. What name can you give to a triangle that has two sides equal in length and one right angle?
27. What is the measure of the other two angles of an isosceles triangle if one of the equal angles is 35°?

[12.6]

28. Given right $\triangle ABC$ with legs a and b and hypotenuse c, find the missing side, correct to the nearest tenth, given:
    a. $a = 4$ and $b = 7$
    b. $a = 7$ and $c = 9$
    c. $b = 3.2$ and $c = 8.6$.
29. A girl 5 ft 3 in tall casts a shadow that is 6 ft 2 in. How far is it from the top of the girl's head to the tip of the shadow? (State your answer in feet to the nearest inch.)
30. Find the length of the equal sides of an isosceles triangle if the base is 8.00 cm long and the triangle has an altitude of 10.4 cm. (*Hint:* The altitude bisects the base.)
31. A 24-ft ladder is used to make an antenna connection on a building, and the bottom of the ladder cannot be farther than 6.0 ft away from the building. Where would the top of the ladder touch? Would this height be sufficient to make a connection 25 ft above the ground?

[12.7]

32. The distance from the center of a circle to any point on the circle is called the ____________.
33. A line that has just one point in common with a circle is called a ____________ line. The point is called ____________. What condition exists between the radius drawn to this point and the tangent line?
34. A line intersecting a circle in two points is called a ____________ line. If a line segment has its endpoints in the circle it is called a ____________.
35. The longest chord of a circle is the ____________.
36. Find the measure of the angle α in the following figures:

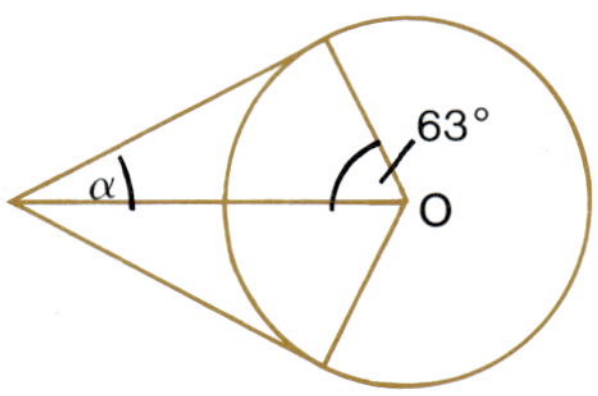

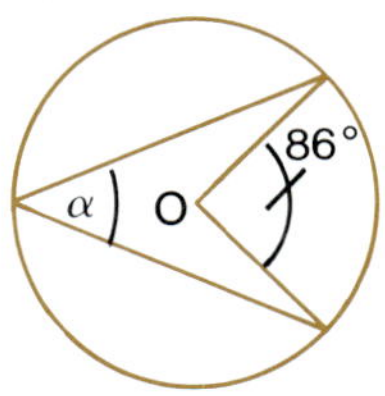

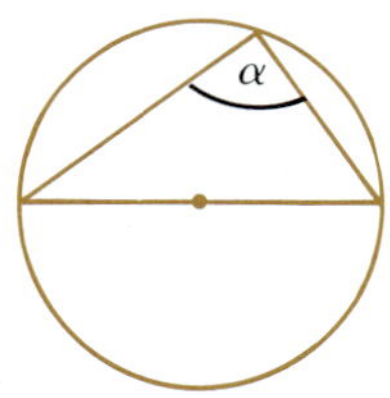

[12.8]

37. The symbol that indicates two plane figures are congruent is ______; that they are similar is ______.
38. Two figures that have the same shape but not the same size are called ______ figures.
39. In the accompanying figure $\triangle ABC$ is similar to $\triangle ADE$. Explain why.

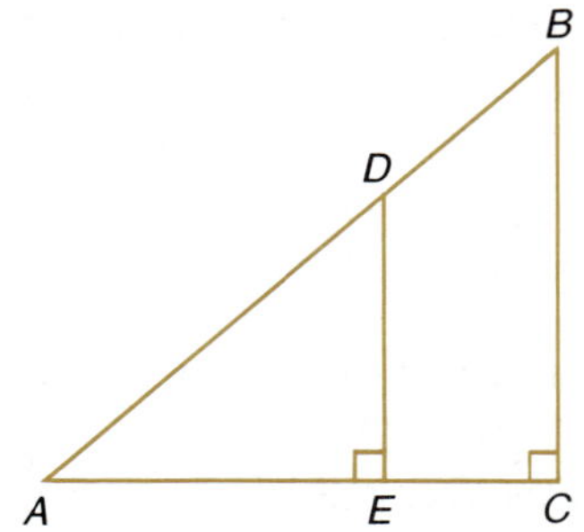

40. A woman 5 ft 1 in tall casts a shadow 3 ft 4 in long at the same time a tree casts a shadow 19.0 ft long. How tall is the tree? (State your answer in feet to the nearest inch.)
41. In similar triangles, the first triangle has sides 12 ft, 9 ft, and 15 ft. If the longest side of the second triangle is 25 ft, how long are its other two sides?
42. In a scale drawing of a machine part, 1 cm = 6 in. If one measurement on the drawing is 2.30 cm, what is that dimension on the part?
43. Given two right triangles that have one acute angle of 37° and a corresponding side that is equal in length, what can you say about the two triangles?

Chapter Test

1. Name the three basic geometric elements that are not defined.
2. Name three different kinds of lines.
3. What is the difference between a line and a line segment?
4. Two rays with a common endpoint form what geometric figure?
5. How many straight lines can be drawn through any two distinct points in the plane?
6. Using your protractor, draw an angle of (a) 70°, (b) 125°.
7. Convert 25°36′ to degrees.
8. Convert 122.8° to degrees and minutes.
9. In the given figure:

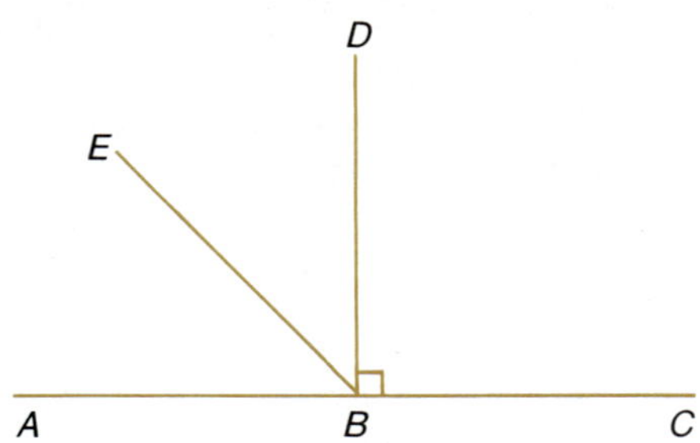

a. $\angle CBD$ has what measure?
b. Name two adjacent angles.
c. $\angle EBC$ is what kind of an angle?
d. $\angle ABE$ and $\angle EBD$ are called ______ angles.
e. Name $\angle ABC$.
f. What relationship exists between $\overline{BD}$ and $\overline{AC}$?

10. If $l_1 \parallel l_2$ in the given figure:

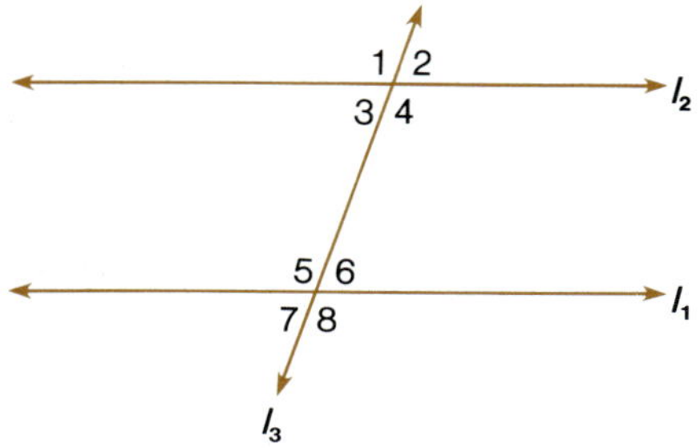

a. Find the measure of $\angle 3$ if $m\ \angle 2 = 79°$.
b. Find the measure of $\angle 4$ if $m\ \angle 6 = 82°$.
c. Find the measure of $\angle 1$ if $m\ \angle 8 = 85°$.
d. If $m\ \angle 5 = 113°$, name all other angles having the same measure and tell why.

11. What is the name given to a polygon that has all sides equal in length and whose interior angles have the same measure?
12. The line segment drawn from a vertex perpendicular to the opposite side of a triangle is called a(n) ____________.
13. Find the sum of the measures of the interior angles of a polygon with 7 sides.
14. In the accompanying parallelogram:

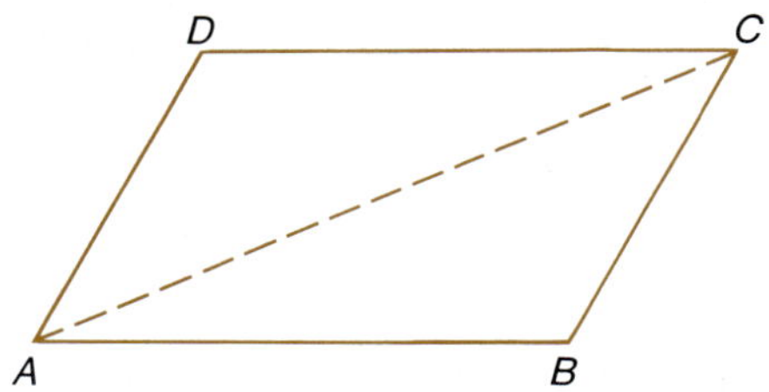

a. If $m\ \angle ABC = 120°$, then $m\ \angle BCD$ = ________ and $m\ \angle ADC$ = ________.
b. If $m\ \angle ACD = 28°$, then $m\ \angle BAC$ = ________.
c. What condition exists between $\triangle ADC$ and $\triangle ABC$?
d. $m\ \angle ADC + m\ \angle ACD + m\ \angle DAC$ = ________.

15. Name two quadrilaterals whose sides are all equal in length.
16. Name the given triangles by (a) their angles, (b) their sides.

a.

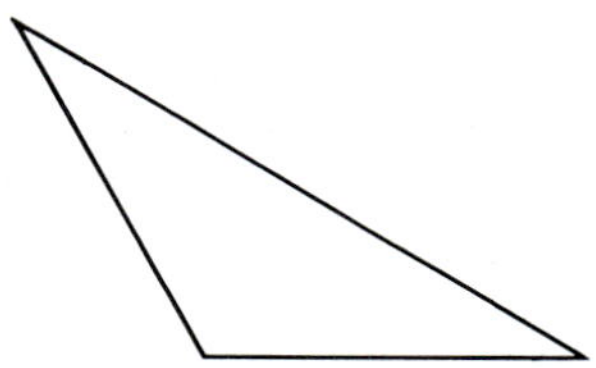

b.

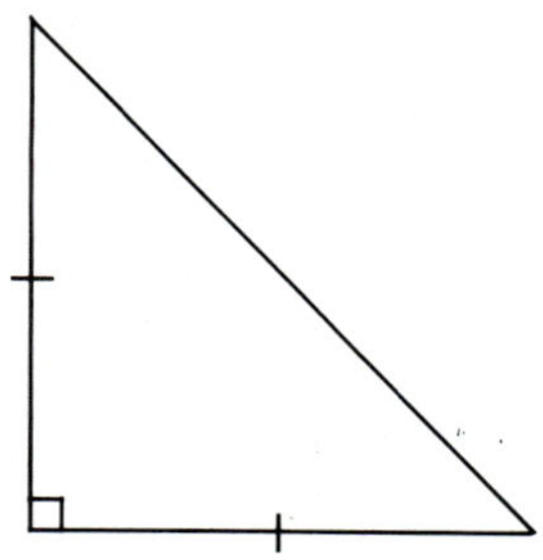

c.

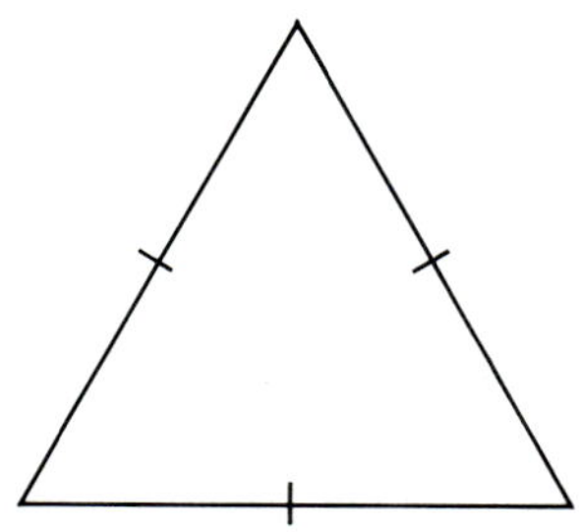

17. Given a rectangle with length 10.2 in and width 3.40 in, use your calculator to fing the length of a diagonal.
18. A guy wire attached to the top of a telephone pole is 24.0 ft long. If it is anchored in the ground a distance of 12.0 ft from the foot of the pole, how tall is the telephone pole?

19. Given the accompanying circle with center at O:

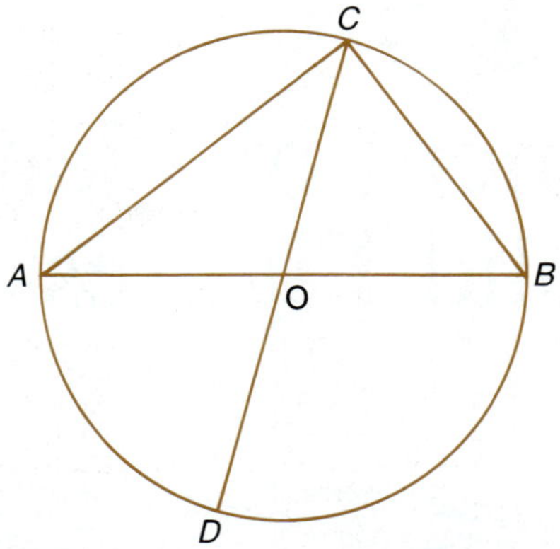

a. Name a diameter of the circle.
b. $\angle ACB$ has what measure?
c. Name $\angle CAB$.
d. Name $\angle AOD$.
e. If $m\ \angle ACD = 37°$, then $\widehat{AD} =$ _____°.
f. If $\widehat{DB} = 115°$, $m\ \angle BOD =$ _______°.
g. Name two chords of the circle.

20. Given the accompanying circle with center at O:

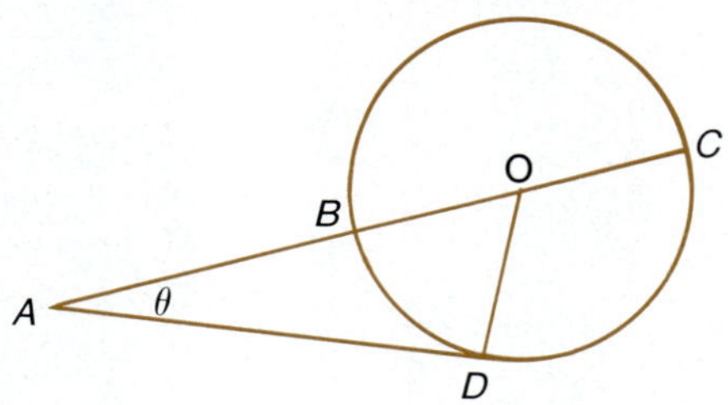

a. Name $\overline{AD}$.
b. Name point D.
c. Find the measure of θ and $m\ \angle AOD$.

21. In the accompanying figure:

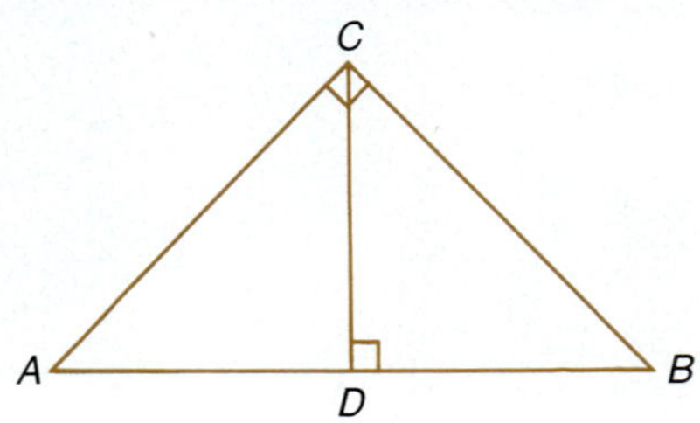

a. Explain why $\triangle ABC \sim \triangle CBD$.
b. If $AC = 6.0$ and $AB = 10.0$, find AD.

22. A woman draws a triangular flower garden to scale having sides 3, 7, and 9. If the shortest side of the garden is 15 ft, find the lengths of the other sides of the garden.

13
Measurement of Plane and Solid Figures

A technician using computer-aided design, or CAD, facilities as an aid in designing new components.
(Photograph courtesy of Ford Motor Company)

13.1 Perimeters of Plane Figures

In chapter 12 we defined some of the elementary geometric figures and their properties. In this chapter we examine terminology and formulas—involving the measurement of surfaces and volumes—related to these figures.

We consider first the measure of the perimeter of a closed plane figure. The word *perimeter* comes from two words—*peri,* meaning "around," and *meter,* meaning "measure." Thus, we define the **perimeter** of a closed plane figure, denoted by P, as the measure of the distance around the figure. The perimeter of any polygon can then be given by the sum of the lengths of its sides. (See figures 13.1 and 13.2.) The measure of the perimeter is given in linear units: inches, feet, centimeters, meters, miles, and so on.

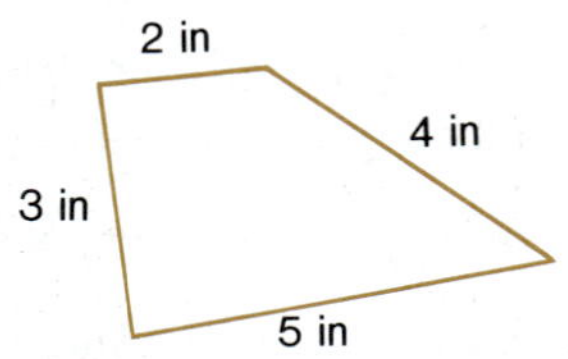

$$P = 2 \text{ in} + 3 \text{ in} + 4 \text{ in} + 5 \text{ in}$$
$$= 14 \text{ in}$$

Figure 13.1

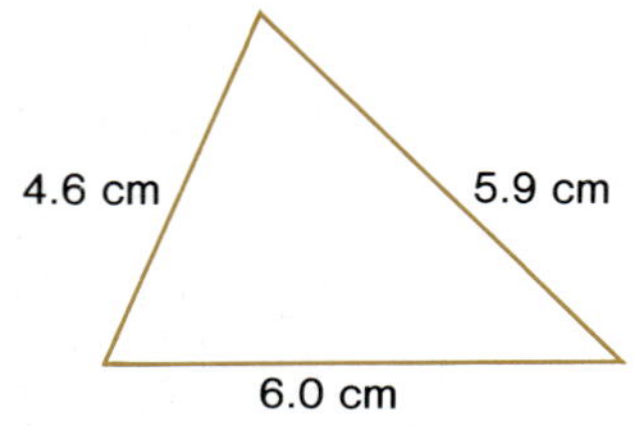

$$P = 4.6 \text{ cm} + 5.9 \text{ cm} + 6.0 \text{ cm}$$
$$= 16.5 \text{ cm}$$

Figure 13.2

Because of the special properties of some polygons, formulas can be used to find their perimeter:

1. The perimeter of any regular polygon with s length of sides and n number of sides is given by $P = ns$.
2. The perimeter of any parallelogram having sides of length a and b is given by $P = 2a + 2b$.

Examples

Find the perimeter of the following figures.

1. A square whose sides have length 3.2 cm. A square is a regular polygon with four sides, so

$$P = 4n = 4(3.2 \text{ cm}) = 12.8 \text{ cm} \approx 13 \text{ cm}$$

2. A rectangle with length 4 in and width $2\frac{1}{2}$ in. A rectangle is a parallelogram, so $P = 2l + 2w$ and

$$P = 2(4 \text{ in}) + 2(2\frac{1}{2} \text{ in})$$
$$= 8 \text{ in} + 5 \text{ in}$$
$$= 13 \text{ in}$$

3. An equilateral triangle with sides of length 13 ft. All sides have the same length and there are three sides, so

$$P = 3n = 3(13 \text{ ft}) = 39 \text{ ft}$$

4. A parallelogram whose sides have length 4 cm and 10 cm

$$\begin{aligned} P &= 2a + 2b \\ &= 2(4 \text{ cm}) + 2(10 \text{ cm}) \\ &= 8 \text{ cm} + 20 \text{ cm} \\ &= 28 \text{ cm} \end{aligned}$$

The perimeter of a circle is called the **circumference,** C. The formula for the circumference of a circle is derived from the fact that the ratio of the circumference to the diameter is the same for all circles. That is, given circumference C and diameter d, $\frac{C}{d} = k$, where k is a constant. This ratio, $\frac{C}{d}$, or constant, is represented by the symbol π (pi) and is approximately equal to 3.14159 to six significant digits. We usually use the value rounded to three significant digits, 3.14.

Thus, we can determine that $\frac{C}{d} = \pi$, and, multiplying both sides by d, we get the formula for the circumference of a circle. That is, $C = \pi d$. Since the diameter of a circle is twice the radius, given radius r, the circumference can also be given by $C = 2\pi r$.

Examples

1. The circumference of a circle with diameter 4.00 in is given by $C = \pi(4.00) = 3.14(4.00) = 12.56 \text{ in} \approx 12.6 \text{ in}$.
2. The circumference of a circle with radius 5.20 ft is given by $C = 2\pi(5.20) = 2(3.14)(5.20) = 32.656 \text{ ft} \approx 32.7 \text{ ft}$.
3. Given that the circumference of a circle is 21.98 cm, find the diameter.

   Since $21.98 = \pi d = 3.14d$, then $d = \frac{21.98}{3.14} = 7$ cm.

Now let us consider the perimeter of geometric figures that are combinations of the basic figures.

Examples

Find the perimeter of the following figures.

1. Figure 13.3 is a semicircle placed on a rectangle. (The dotted line where the figures are joined should not be counted in finding the perimeter.) The sum of the lengths of the sides of the rectangle is:

   $$3.0 \text{ cm} + 2(4.0 \text{ cm}) = 3.0 \text{ cm} + 8.0 \text{ cm} = 11.0 \text{ cm}$$

   The circumference of the semicircle, with $d = 3.0$ cm is:

   $$\frac{1}{2}C = \frac{1}{2}(3.14)(3 \text{ cm}) = 1.57 \text{ (3 cm)} = 4.71 \text{ cm} \approx 4.7 \text{ cm}$$

   Therefore, $P = 11 \text{ cm} + 4.7 \text{ cm} \approx 15.7 \text{ cm}$

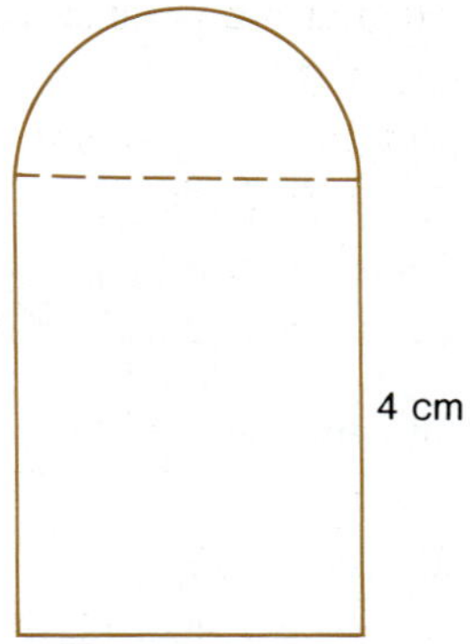

Figure 13.3

2. Given a machine part composed of a square, an equilateral triangle, and a semicircle (see figure 13.4). The formula for the perimeter of the figure is:

$$P = 2s + 2s + \frac{1}{2}\pi s = 4s + \frac{1}{2}\pi s$$

$$= 4(2.00) + \frac{1}{2}(3.14)(2.00) = 8.00 + 3.14 = 11.14 \text{ in}$$

$$\approx 11.1 \text{ in}$$

note

$C = \pi d$ and since $d = s$, then $C = \pi s$ and $\frac{1}{2}C = \frac{1}{2}\pi s$

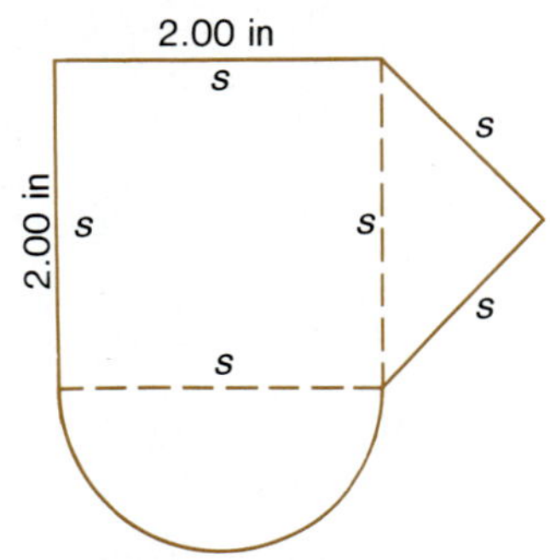

Figure 13.4

Exercise Set 13.1

In exercises 1 to 12, find the perimeters (circumferences for circles) of the given figures. (Round off each answer to three significant digits.)

1. A square with side 3.20 cm
2. An equilateral triangle with side 4.0 ft
3. A rectangle with length 3.0 in and width 1.25 in
4. An isosceles triangle with equal sides of 2.32 cm and a third side of 1.72 cm

•5. A parallelogram with sides $\frac{3}{8}$ in and $\frac{2}{3}$ in

6. A rhombus with side $2\frac{3}{4}$ cm
7. A quadrilateral with sides 1.53 ft, 2.76 ft, 4.70 ft, and 8.62 ft
8. A triangle with sides 4.72 m, 3.62 m, and 2.93 m
9. An isosceles trapezoid with equal sides 3.33 cm and bases 2.25 cm and 4.50 cm

•10. A circle with radius r = 14.2 mm

11. A circle with diameter d = 4.20 ft
12. A regular pentagon with side s = 14.34 m

Find the perimeters of the following geometric figures. (Round answers to the precision of the measured data.)

13.

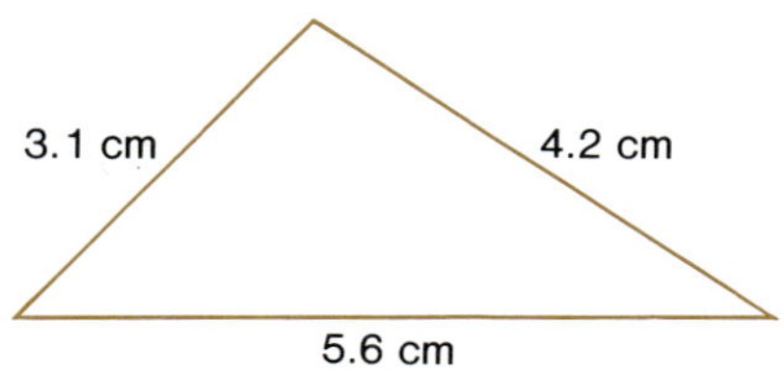

•14.

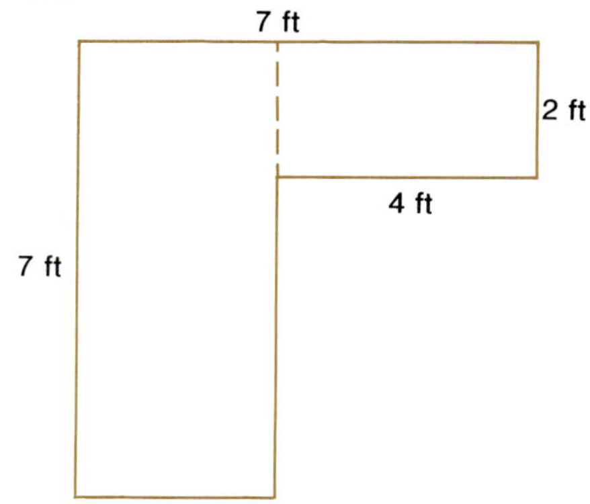

15.

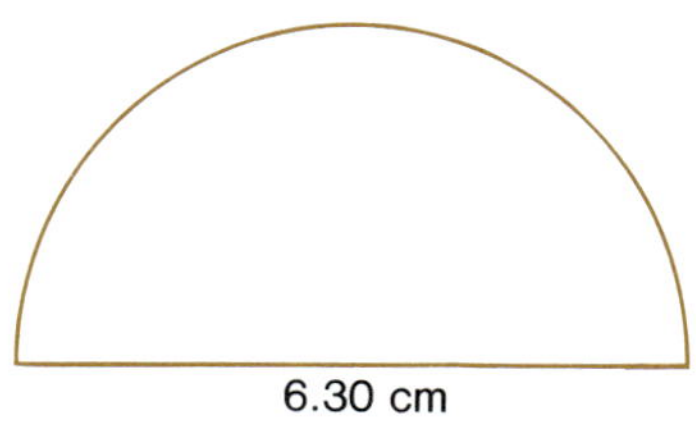

16.

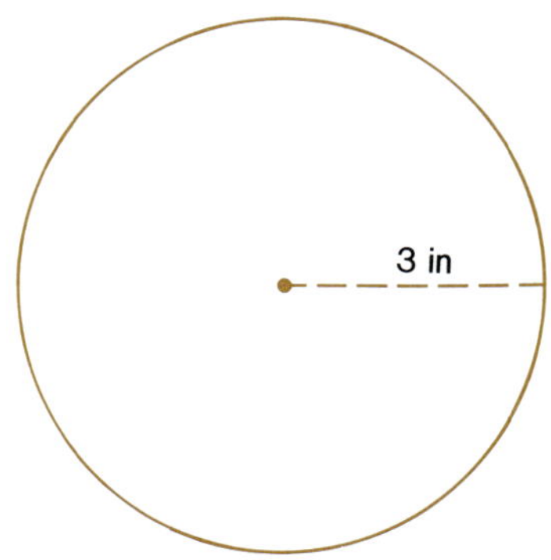

17.

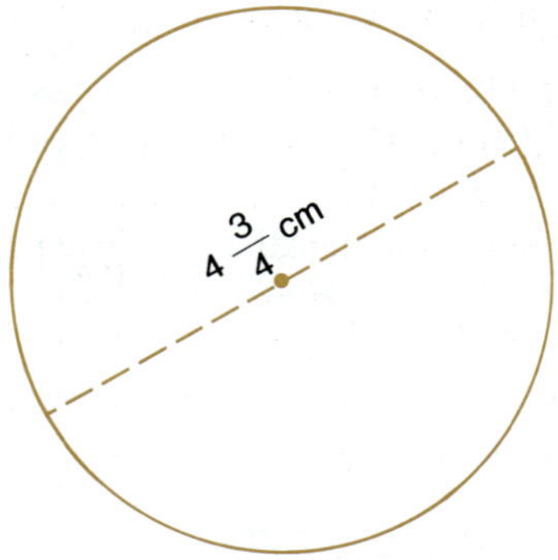

18.

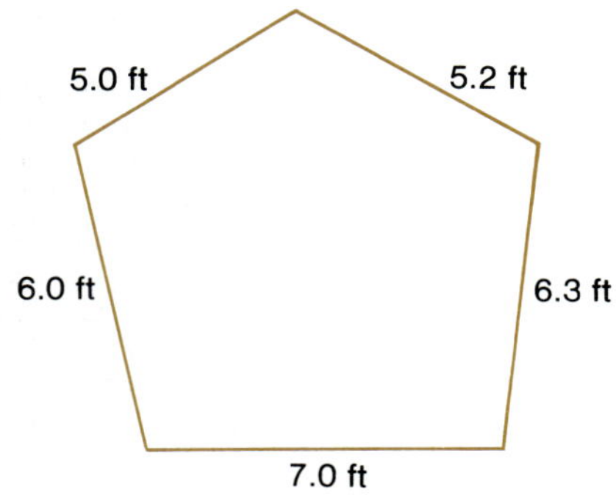

19.

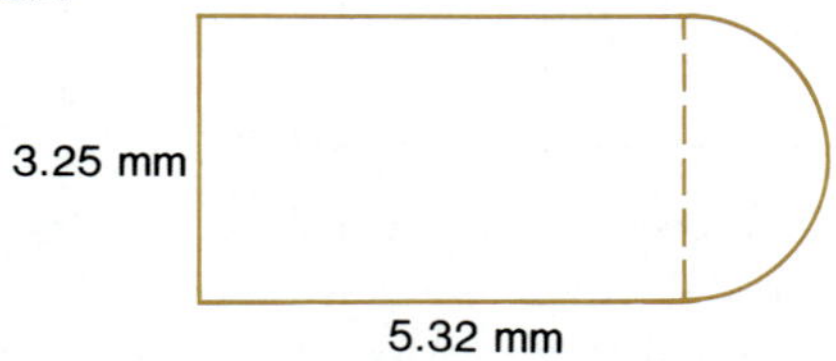

20.

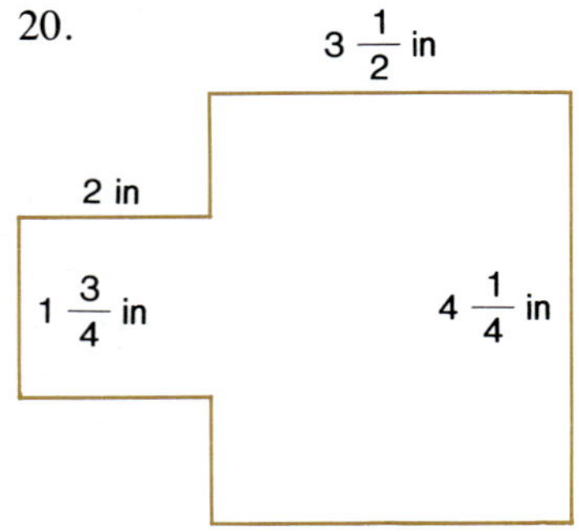

21.

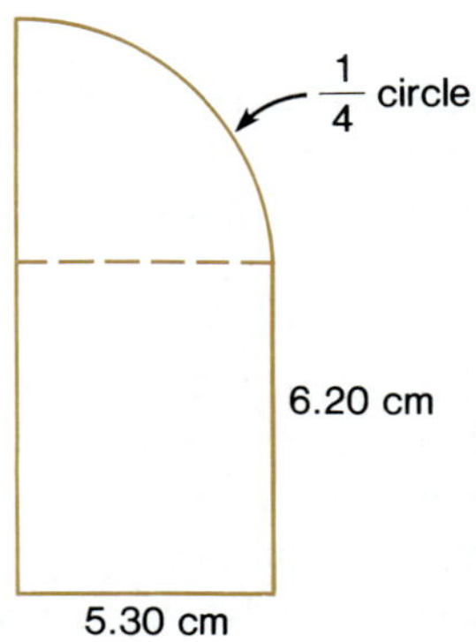

22.

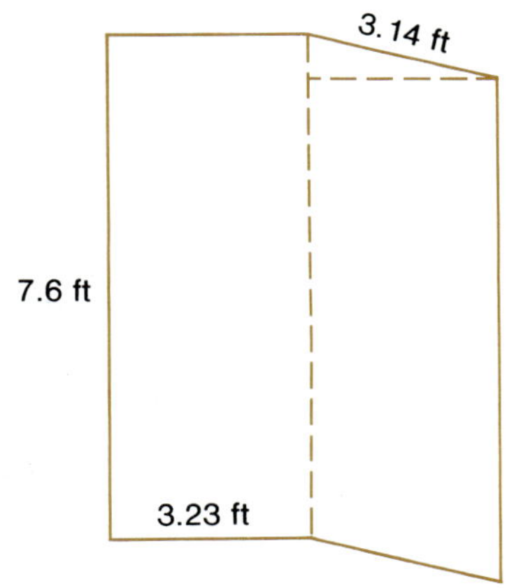

•23.

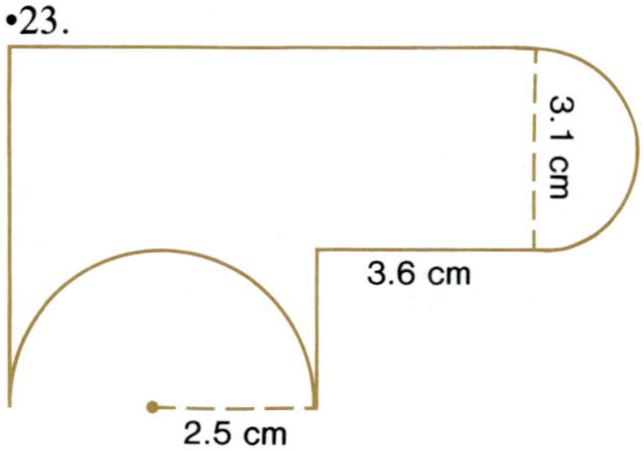

24.

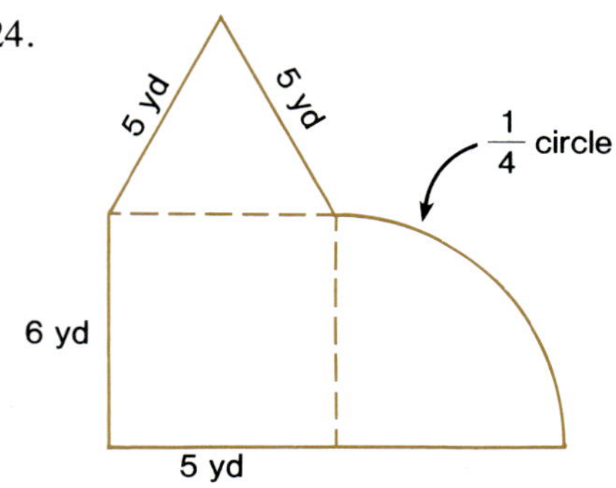

In exercises 25 to 40 round according to three significant digits.

25. The radius of a circle is 2.5 ft. What is the diameter of the circle?
26. The diameter of a circle is 5.36 cm. What is the radius?
•27. If the length of the diameter is tripled, does the circumference triple in length also?
28. The floor of a room is rectangular, with length 15.3 ft and width 12.6 ft. The room has a door 3 ft wide. How many feet of floor molding is necessary for the room?
29. A running track has the shape given in the accompanying figure. How long is two laps around the track?

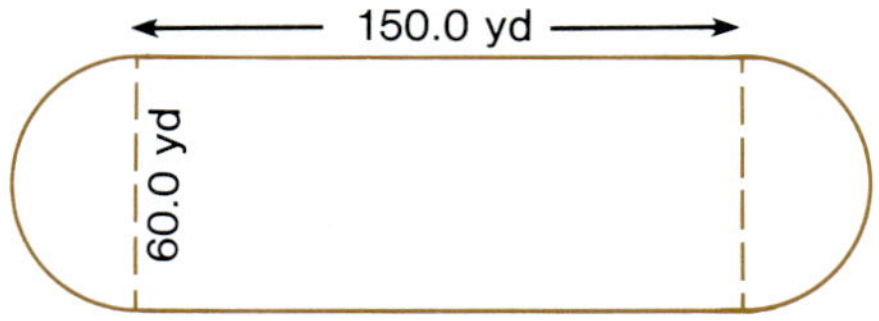

30. The circumference of a circular flower bed is 7.85 yd. What is the diameter of the flower bed?
31. A plot of ground in the shape of a parallelogram has sides 23 ft and 15 ft. If fencing costs $1.35 per ft, what will it cost to fence in the plot of ground?
32. A piece of wire 42.3 in long is to be bent into the shape of an equilateral triangle. How long is each side of the triangle? What does the 42.3 in represent?
33. Find the perimeter of a rectangular-shaped piece of metal that is 5.50 in long and 3.75 in wide.
34. A woman jogs around a circular track that has a diameter of 31.2 yd. How far does she jog in 15 laps around the track?
35. What is the pitch diameter of a gear if the circumference of the pitch circle is 20.42 in?
36. Find the radius of a cylindrical tank if the distance around the tank is 62.5 ft.
•37. An emery wheel has a radius of 4 in. How many inches will a point on a wheel travel in 1 min if the wheel revolves at a rate of 26 rpm?
38. What must the length of a label be if it is to fit around a can that is 3.00 in in diameter, allowing 0.250 in for pasting?
39. What must the length of a bar of steel be to make a rim for a carriage wheel 4.30 ft in diameter, allowing 0.004 ft for weld?
40. How long is the belt connecting the pulleys shown in the accompanying figure?

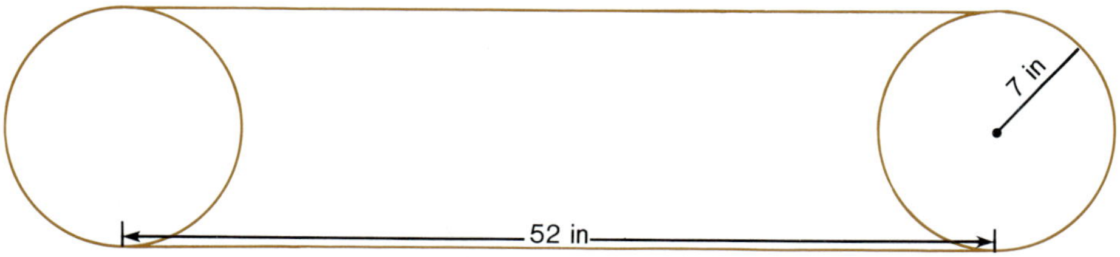

13.2 Areas of Plane Figures

The area of a closed plane figure is that portion of the plane enclosed by the figure. We define *the area of a geometric figure as the amount of the plane the figure encloses*. Whereas perimeter was measured in terms of length, area is measured in *square units* (or units square). Thus, the measure of an area is the number of square units in that area or surface.

Consider a square whose sides are each 1 inch in length (see figure 13.5). We say this square has an area of 1 square inch, which we denote by 1 in^2.

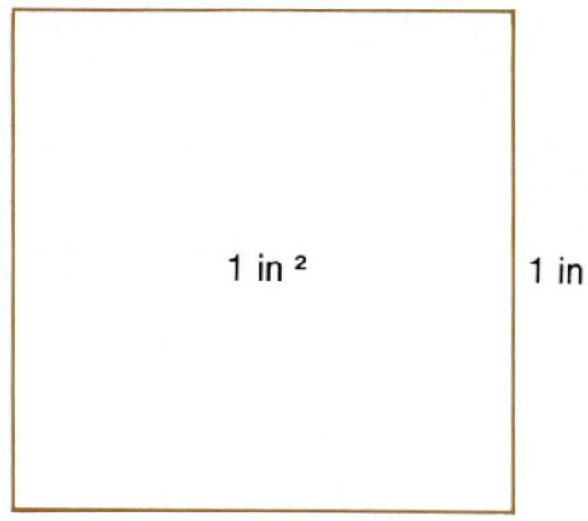

Figure 13.5

Now suppose we have a rectangle that is 3 in wide and 5 in long (see figure 13.6).

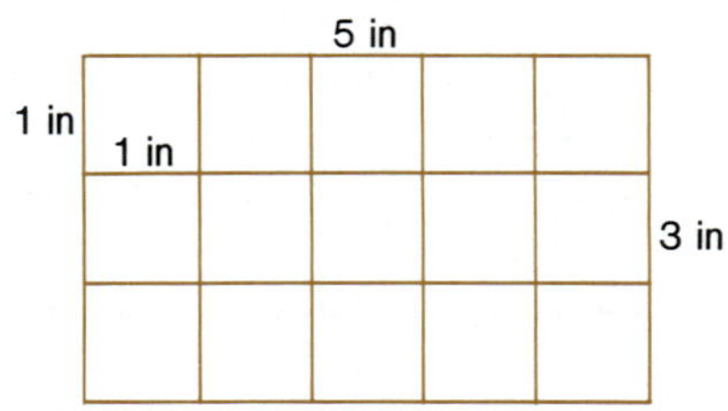

Figure 13.6

We can divide this rectangle into 15 squares, each 1 in on a side with an area of 1 in^2. Thus, the area of the rectangle is 15 in^2, which we read "15 square inches." The area A of any rectangle, then, is given by the product of its length times its width. If l is the length and w is the width of the rectangle, then

$$A = lw$$

note

Remember that the dimensions, l and w in this case, must be in the same unit of measure.

Examples

1. Find the area of a rectangle with length 3.0 ft and width 2.5 ft.

$$A = lw = (3.0 \text{ ft})(2.5 \text{ ft}) = 7.5 \text{ ft}^2$$

2. Find the area of figure 13.7. The dotted line separates the figure into two rectangles. The area of the figure will be the sum of the areas of the two rectangles thus formed, A_1 and A_2:

$$A_1 = (4.0 \text{ cm})(3.0 \text{ cm}) = 12 \text{ cm}^2$$
$$A_2 = (7.0 \text{ cm} - 3.0 \text{ cm})(2.0 \text{ cm}) = (4.0 \text{ cm})(2.0 \text{ cm}) = 8.0 \text{ cm}^2$$

Then

$$A = A_1 + A_2 = 12 \text{ cm}^2 + 8.0 \text{ cm}^2 = 20 \text{ cm}^2$$

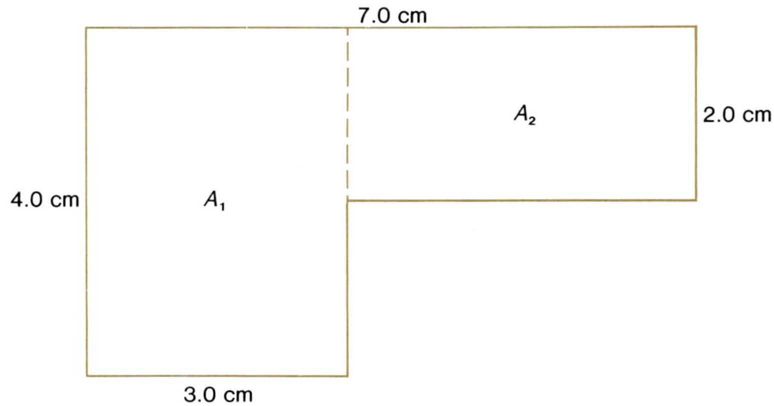

Figure 13.7

A square is a special rectangle whose length and width are the same. If we denote each of these sides by s, then

$$A = s^2$$

is the formula for the area of a square.

Example

What is the area of a square bolt head that is $\frac{1}{2}$ in on each side?

Since $s = \frac{1}{2}$ in

$$A = s^2 = \left(\frac{1}{2}\right)^2 = \left(\frac{1}{2}\right)\left(\frac{1}{2}\right) = \frac{1}{4} \text{ in}^2$$

The definition of the area of a rectangle can also be used to define the area of a parallelogram. Consider figure 13.8. We denote the perpendicular distance between parallel sides $\overline{AB}$ and $\overline{DC}$ by h, an *altitude* of the parallelogram, and the side $\overline{AB}$ by b, the *base* of the parallelogram. If $\triangle ADA'$ is moved right, to the position of $\triangle BCB'$, rectangle $A'DCB'$ is thus formed with area $A = m(\overline{A'B'})h$. But $m\overline{A'B'}$ is the same length as the base b. Therefore, the area of the parallelogram $ABCD$ is given by

$$A = (\text{base})(\text{altitude}) = bh$$

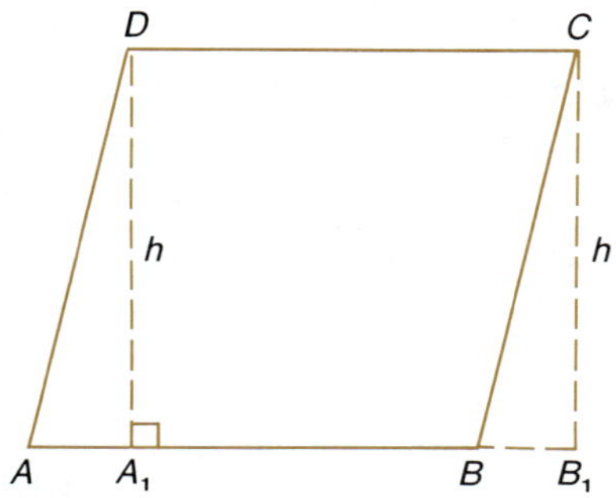

Figure 13.8

Examples

1. Find the area of a parallelogram with base $b = 1.5$ m and altitude $h = 3.6$ m.

   $A = bh = (1.5 \text{ m})(3.6 \text{ m}) = 5.4 \text{ m}^2$

2. Find the area of a plot of ground that has the shape of a parallelogram, with base $b = 36$ ft and altitude $h = 7.0$ yd.

   note
   The dimensions are not in the same unit of measure, so we must change one dimension to get the same unit measure for each.

   Since $36 \text{ ft} = \frac{36}{3} = 12$ yd, or $7.0 \text{ yd} = (7.0)(3) = 21$ ft, the area of the plot of ground is:

   $A = (12 \text{ yd})(7.0 \text{ yd}) = 84 \text{ yd}^2$ or $A = (36 \text{ ft})(21 \text{ ft}) = 756 \text{ ft}^2$

   note
   Since $756 \div 84 = 9$, we can conclude that $1 \text{ yd}^2 = 9 \text{ ft}^2$.

The formula for the area of a parallelogram can be used to derive a formula for the area of a triangle. One of the diagonals of the parallelogram $ABCD$, $\overline{DB}$, is drawn in figure 13.9. The resulting triangles, ABD and BCD, have the same size, so the area of one of the triangles, $\triangle ABD$, is one-half the area of the parallelogram. Since the area of the parallelogram is $A = bh$, then the area of a triangle is:

$$A = \frac{1}{2}(\text{base})(\text{altitude}) = \frac{1}{2}bh$$

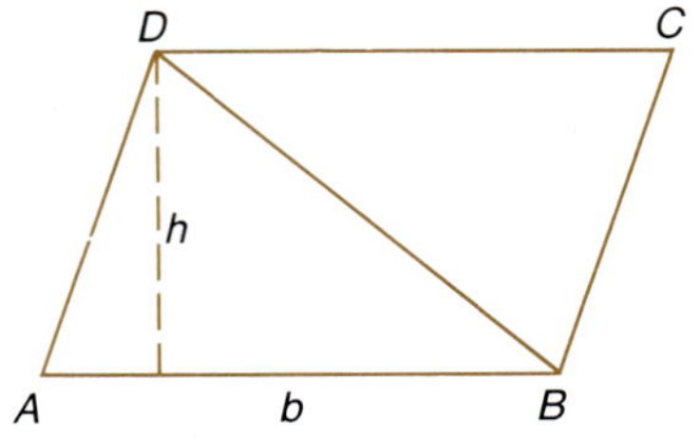

Figure 13.9

An alternate method for finding the area of a triangle involves using the lengths of the three sides. Named after a mathematician who lived in Alexandria, **Hero's formula** for finding the area of a triangle with sides a, b, and c is given by

$$A = \sqrt{s(s - a)(s - b)(s - c)}$$

where $s = \frac{1}{2}(a + b + c)$.

Examples

1. Find the area of a triangle whose sides are 8.0 in, 15 in, and 17 in in length.
   Now, using $A = \sqrt{s(s - a)(s - b)(s - c)}$,

   $$\begin{aligned} s = \frac{1}{2}(a + b + c) &= \frac{1}{2}(8.0 \text{ in} + 15 \text{ in} + 17 \text{ in}) \\ &= \frac{1}{2}(40 \text{ in}) \\ &= 20 \text{ in} \end{aligned}$$

   Then

   $$\begin{aligned} A &= \sqrt{20(20 - 8.0)(20 - 15)(20 - 17)} \\ &= \sqrt{20(12)(5)(3)} \\ &= \sqrt{3{,}600 \text{ in}^4} \\ &= 60 \text{ in}^2 \end{aligned}$$

2. Find the area of a triangle with base $b = 4.2$ cm and altitude $h = 2.1$ cm.

   $$A = \frac{1}{2}bh = \frac{1}{2}(4.2 \text{ cm})(2.1 \text{ cm}) = (2.1 \text{ cm})(2.1 \text{ cm}) \approx 4.4 \text{ cm}^2$$

3. Find the area of a tract of land with the shape of a right triangle (see figure 13.10), correct to three significant digits.
   In a right triangle, one leg is always the base and the other the altitude. In figure 13.10, let base $b = 432$ rd. Then the altitude h is the other leg. Using the Pythagorean theorem,

   $$\begin{aligned} h &= \sqrt{(763)^2 - (432)^2} = \sqrt{582{,}169 - 186{,}624} \\ &= \sqrt{395{,}545} \approx 628.924 \text{ rd} \end{aligned}$$

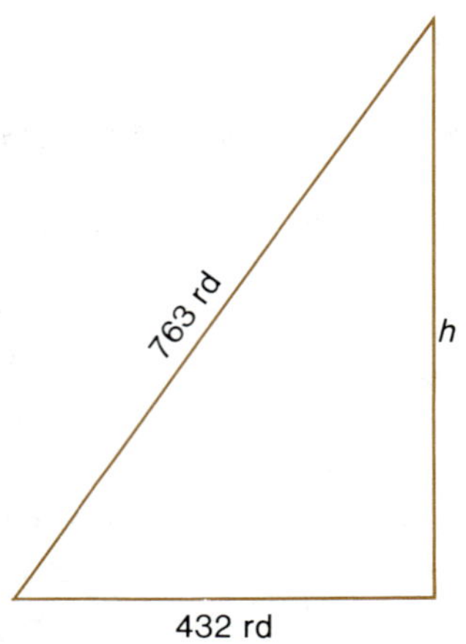

Figure 13.10

Then,

$$A = \frac{1}{2}(432 \text{ rd})(628.9 \text{ rd}) = (216 \text{ rd})(628.9 \text{ rd}) \approx 136{,}000 \text{ rd}^2$$

The formula for the area of a trapezoid can be determined using the area of a triangle. In figure 13.11, diagonal $\overline{BD}$ of the trapezoid $ABCD$ is drawn, thus forming two triangles ABD and BCD. We let b_1, side $\overline{AB}$, be the base of $\triangle ABD$ with altitude h, and b_2, side $\overline{DC}$, be the base of $\triangle BCD$ with the same altitude h. Then the areas of the two triangles are given by $A_1 = \frac{1}{2}b_1h$ and $A_2 = \frac{1}{2}b_2h$. Adding the two areas and factoring, we get the area of the trapezoid

$$A = \frac{1}{2}b_1h + \frac{1}{2}b_2h = \frac{1}{2}h(b_1 + b_2)$$

and so

$$A = \frac{1}{2}h(b_1 + b_2)$$

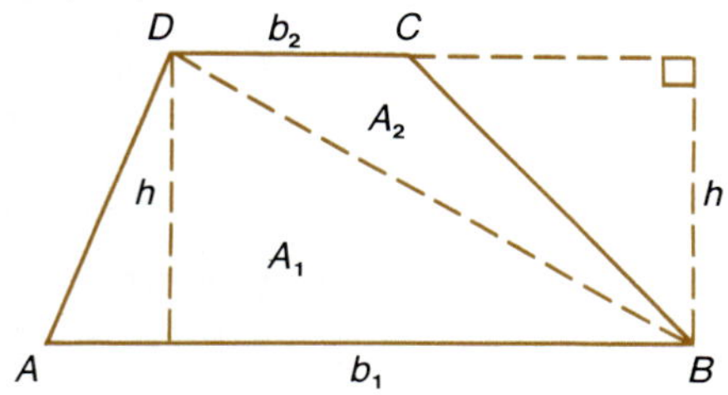

Figure 13.11

In words: The area of a trapezoid is given by one-half times the altitude times the sum of the bases (parallel sides).

Examples

1. Find the area of a trapezoid with altitude 6.0 cm and bases 5.0 cm and 4.0 cm.

$$A = \frac{1}{2}(6.0 \text{ cm})(5.0 \text{ cm} + 4.0 \text{ cm}) = (3.0 \text{ cm})(9.0 \text{ cm}) = 27 \text{ cm}^2$$

2. A flower bed is laid out in the form of an isosceles trapezoid, with equal sides 8.0 ft and bases of 6.0 ft and 10 ft. Find the area of the flower bed. See figure 13.12.

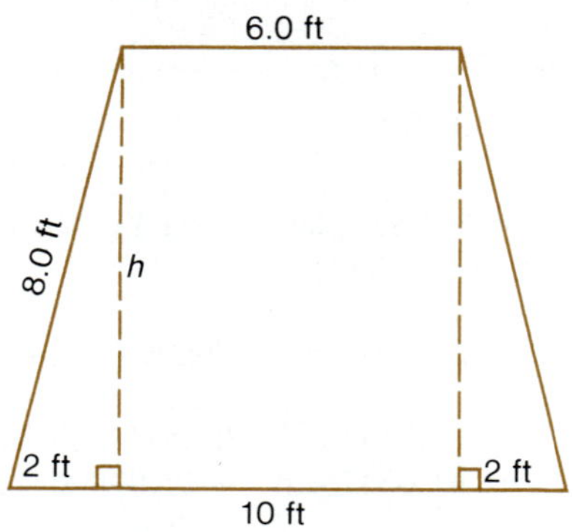

Figure 13.12

We must determine first the altitude h by dividing the trapezoid into a rectangle and two right triangles. The base of each triangle then is 2 ft, using the Pythagorean theorem:

$$h = \sqrt{8^2 - 2^2} = \sqrt{64 - 4} = \sqrt{60} \approx 7.7 \text{ ft}$$

Then,

$$A = \frac{1}{2}(7.7 \text{ ft})(10 \text{ ft} + 6.0 \text{ ft}) = \frac{1}{2}(7.7 \text{ ft})(16 \text{ ft})$$
$$= (8.0 \text{ ft})(7.7 \text{ ft}) \approx 62 \text{ ft}^2$$

The formula for the area of a circle is derived by methods beyond the scope of this book. Let us just state the formula, again using the constant π: The area of a circle equals π times the square of the radius. Then,

$$A = \pi r^2$$

where r is the radius of the circle.

Examples

1. Find the area of a circle with radius $r = 2.50$ in.

$$A = \pi r^2 = (3.14)(2.5 \text{ in})^2 = (3.14)(6.25 \text{ in}^2) \approx 19.7 \text{ in}^2$$

2. Find the area of a circular sheet of metal having diameter 4.60 ft.

Since the formula for the area of a circle requires that we know the radius, we must find the radius when the diameter is given. Recall that the radius is one-half of the diameter. Therefore, $r = \frac{1}{2}(4.6 \text{ ft}) = 2.30$ ft and

$$A = (3.14)(2.30 \text{ ft})^2 = (3.14)(5.29 \text{ ft}^2) \approx 16.6 \text{ ft}^2$$

3. The flat cast iron plate shown in figure 13.13 has a 3.75 in round hole in it. Find the area of the cast iron plate.

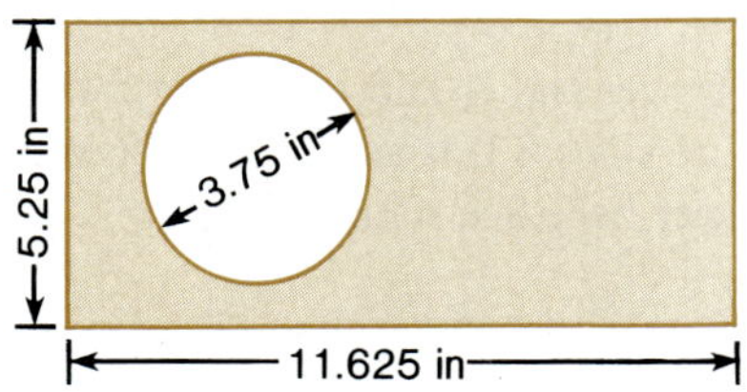

Figure 13.13

The area of the plate is equal to the area of the rectangle *minus* the area of the circle. Then

$$A = l \cdot w - \pi r^2 \qquad [\textit{Note: } r = \frac{1}{2}d = \frac{1}{2}(3.75) = 1.875]$$
$$= (5.25 \text{ in})(11.625 \text{ in}) - (3.14)(1.875)^2$$
$$= 61.03125 \text{ in}^2 - 11.039063 \text{ in}^2 = 49.992187 \text{ in}^2 \approx 50.0 \text{ in}^2$$

4. What is the area of the metal in the key section shown in figure 13.14?

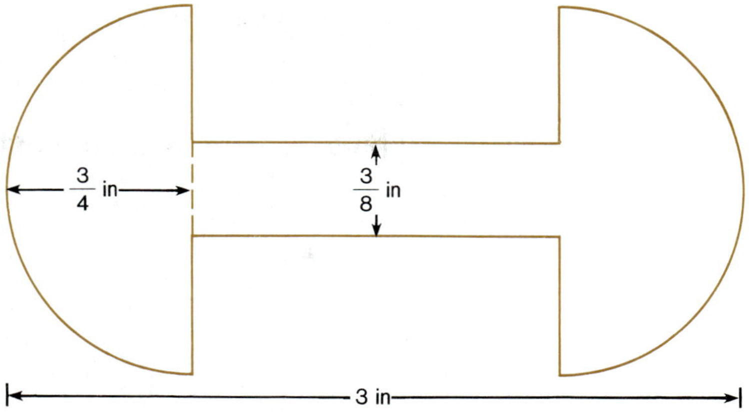

Figure 13.14

The area of the section is equal to the area of the rectangle *plus* the sum of the areas of the two semicircles (which make *one* complete circle).

$$A = l \cdot w + \pi r^2 = (1.50 \text{ in})\left(\frac{3}{8} \text{ in}\right) + (3.14)\left(\frac{3}{4} \text{ in}\right)^2$$
$$= (1.50 \text{ in})(0.375 \text{ in}) + (3.14)(0.75 \text{ in})^2$$
$$= 0.5625 \text{ in}^2 + 1.76625 \text{ in}^2 = 2.32875 \text{ in}^2 \approx 2.33 \text{ in}^2$$

(correct to nearest hundredth)

Exercise Set 13.2

In exercises 1 to 16, find the area of the given figures. Round off correct to three significant digits.

1. A rectangle with length 25 ft and width 16 ft
2. A rectangle with length 1.4 ft and width 7.3 in (*Note:* The units of measure are not the same.)
3. A square with side 16.0 cm in length
4. A square with side $4\frac{1}{3}$ in in length
5. A parallelogram with base 2.30 yd and altitude 4.50 yd
6. A parallelogram with base 4.30 cm and altitude 5.03 cm
7. A triangle with base 5.76 ft and altitude 2.57 ft
•8. An isosceles triangle with equal sides 13 in and base 10 in
9. A right triangle with legs 15.0 in and 2.50 ft
10. A trapezoid whose bases are 15 yd and 9.0 yd in length and whose altitude is 8.0 yd
11. A trapezoid with altitude 6 ft and bases 2 yd and 3 yd in length

12. A circle with radius 3.26 m

13. A circle with diameter 3.75 ft $\left(\text{Use } \pi = \frac{22}{7}.\right)$

•14. A triangle with sides of length 3.1 cm, 4.5 cm, and 6.8 cm

15. A triangle whose sides are 4.07 in, 3.80 in and 6.91 in in length

16. A triangle whose base is 6.39 ft and altitude is 9.50 ft in length

Find the area of the given figures. Round off according to the accuracy of the measured data.

17.

18.
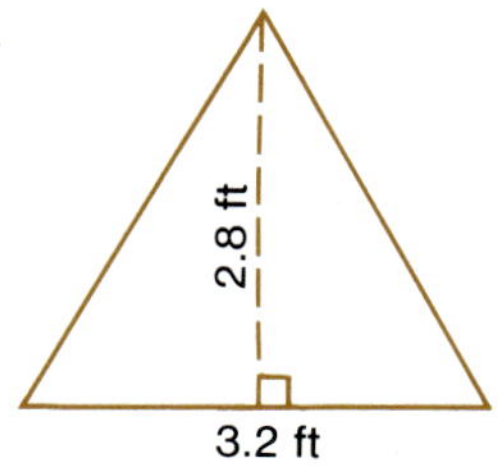

19.
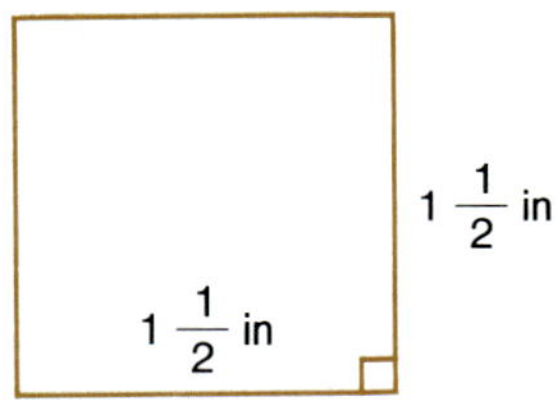

20.
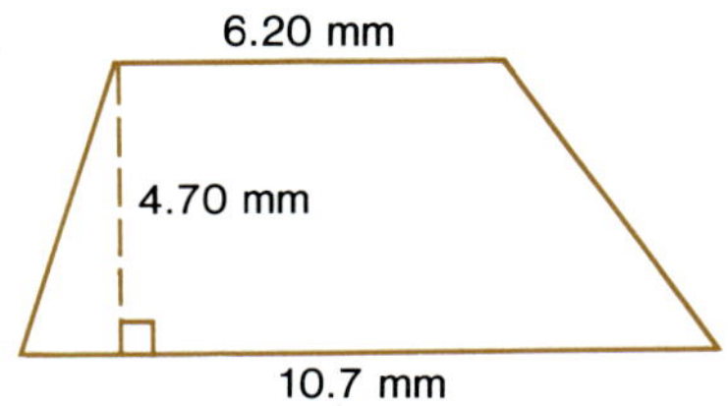

21.
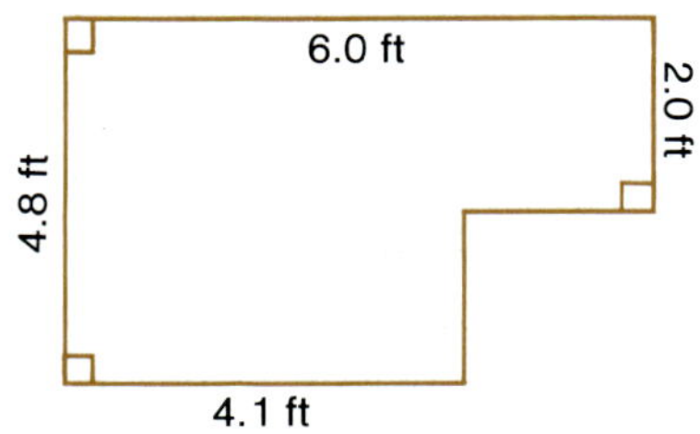

•22.
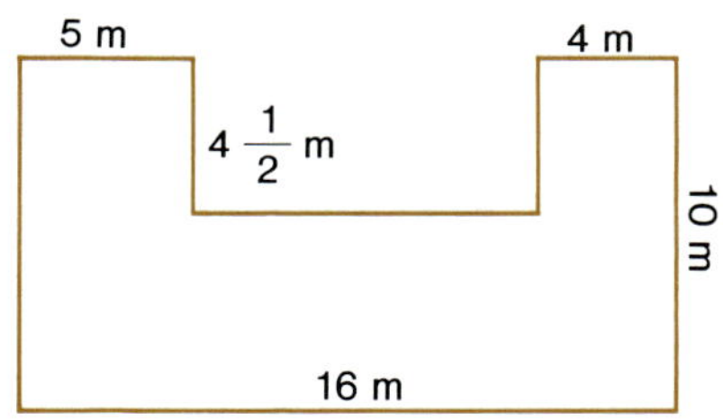

23.
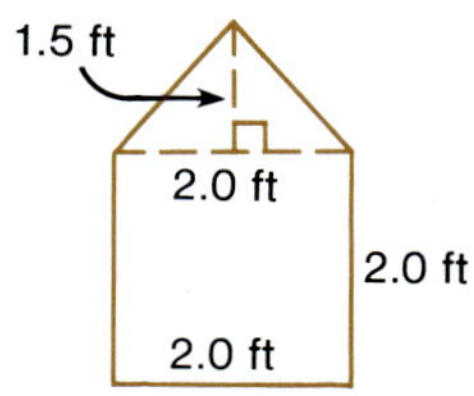

•24.
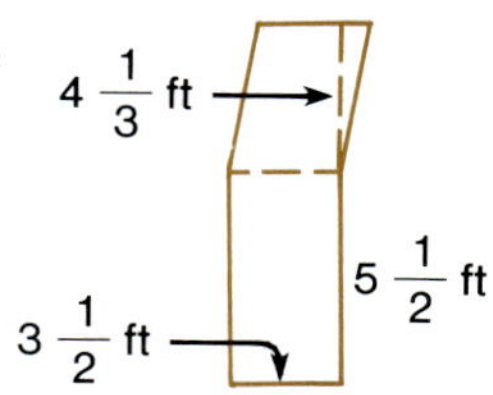

25.
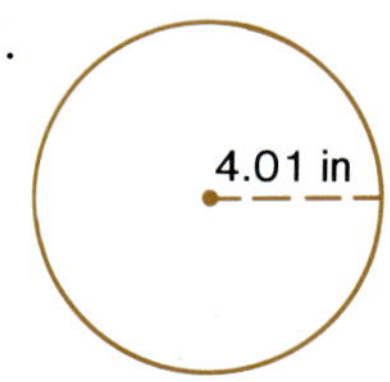

26.
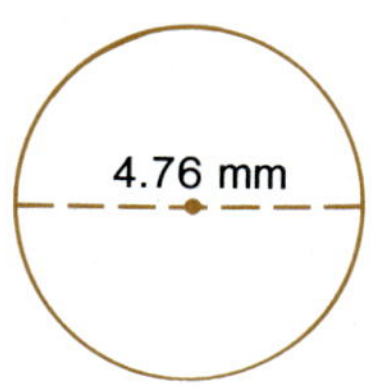

27.
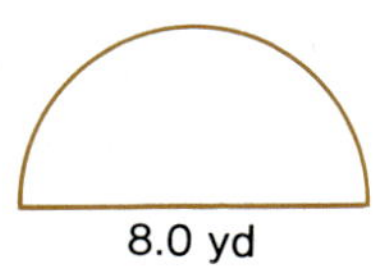

28.
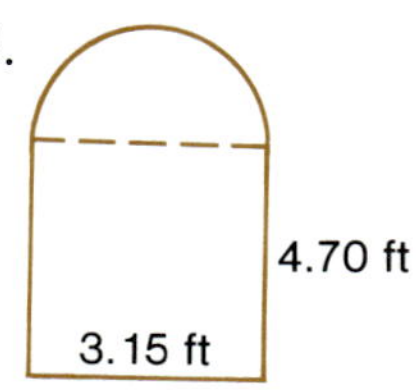

29.

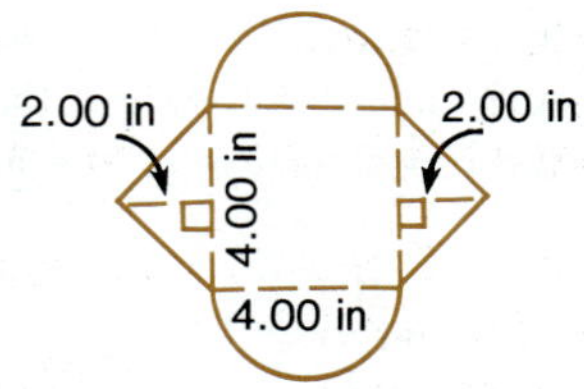

•30.

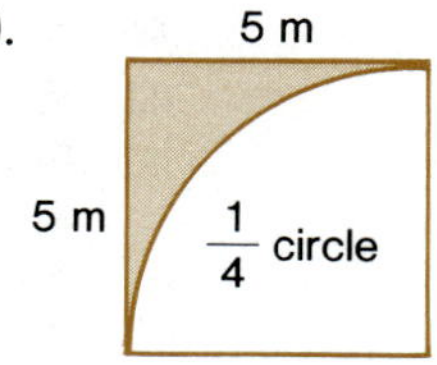

31.

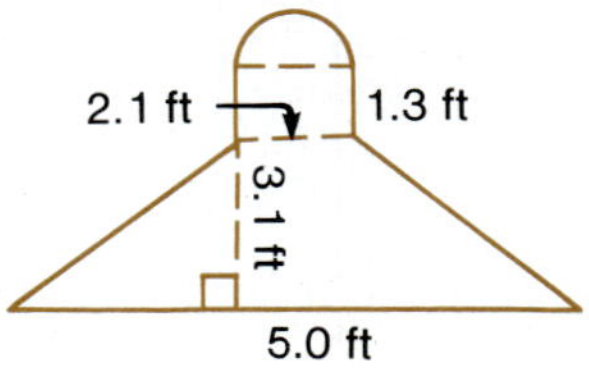

32.

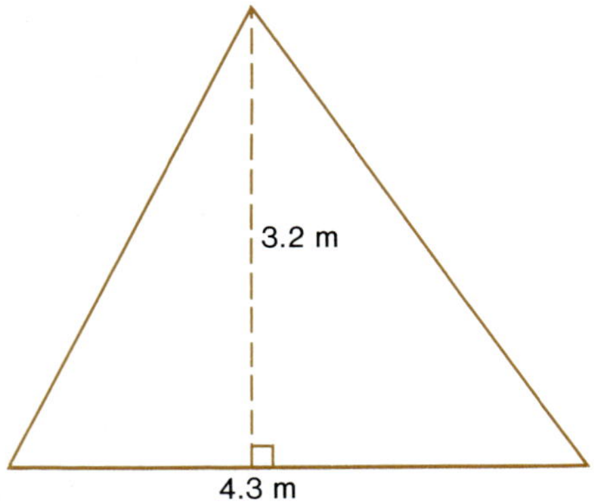

33.

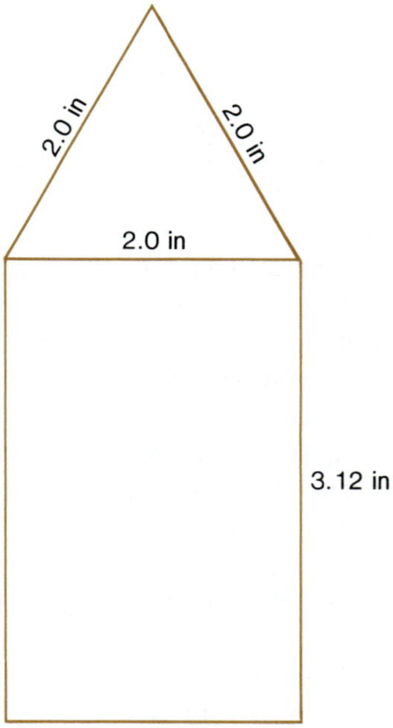

Round off each answer correct to three significant digits.

34. What is the area of a floor tile that measures 32.1 cm by 27.8 cm?
35. A city block measures 600 m long and 620 m wide. What is the area of the block?
36. If the length and width of a rectangle are doubled, what effect does this have on the area? (*Hint:* Use a specific example.)
37. A rectangular pasture is 350.0 ft long and has an area of 104,400 ft^2. How wide is the pasture?
38. What is the area of a square bolt head that is 3.2 cm on an edge?
39. A piece of wire 28 in long is bent into a square. What is the area enclosed by the wire?

40. A square flower bed has area 144 ft². How long is each side of the bed?
41. What happens to the area of a square if the length of its side is tripled?
42. Find the area of the cross section of the given steel I-beam if the steel is 0.50 in thick.

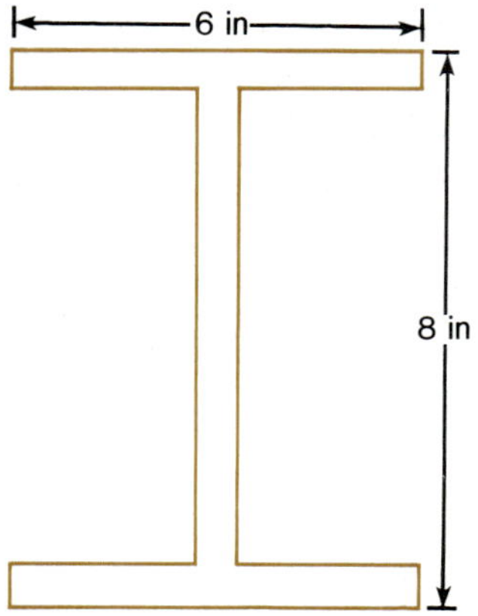

•43. What is the cost of placing clay over a tennis court 42 ft wide and 72 ft long at 30¢/sq yd?
44. Find the area of a rectangle in terms of x, if the length is $2x + 1$ and the width is $3x - 2$.
45. Find the cost of a rectangular-shaped rug that is 12 ft by 18 ft if the rug costs \$7.95/yd (Round your answer to the nearest cent².).
46. Find the cost of painting the walls and ceiling of a room that has a square floor 12 ft on each edge if the ceiling is 8 ft from the floor and paint costs 81¢/ft².
47. A metal cabinet is 50.0 in high, 20.0 in deep, and 24.0 in wide. Find the outside surface area of the cabinet.
48. The trapezoidal gable of a house is 12 m long at the bottom, 4.5 m long at the top, and 5 m high. What is the area of the gable?
49. The triangular gable of a house is 24.6 ft across the bottom and is 17.3 ft high. Find the area of the gable.
50. Find the area of a 30°-60°-90° right triangle if the hypotenuse is 10 in long. (*Hint:* Recall that the side opposite the 30° angle is one-half the length of the hypotenuse.)
51. A piece of wire 600.0 cm long is bent into an equilateral triangle. What is the length of each side? Find the area enclosed by the triangle.
52. Find the base of a triangle if its area is 576 ft² and its altitude is 9.00 ft.
•53. The area of a trapezoid is 4,256 in² and the bases are 68 in and 84 in, respectively. Find the altitude.
54. The bases of a trapezoid are 26 in and 84 in, respectively, and the altitude is 24 in. Find the side of a square with the same area.

55. If the base of a triangle is doubled and the altitude remains the same, how is the area changed?
56. The area of a square is 60 in^2. Find the length of its diagonal.
57. Find the area of a piston head that is 3.75 in in diameter.
58. The radius of a sodium (Na) atom is 1.86 Å. What is its area correct to four significant digits?
59. What is the total pressure on a circular piston head 6.00 cm in diameter if the pressure is 32.0 kg/cm^2?
60. Find the area of the shaded portion in the accompanying figure. (*Hint:* Find the difference in the areas of the rectangle and circle.)

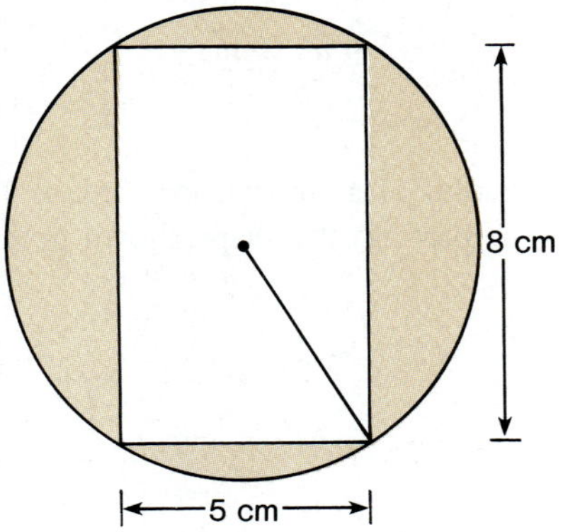

61. Find the area of a washer whose dimensions are given in the accompanying figure. (*Hint:* Find the difference in the areas of the two circles.)

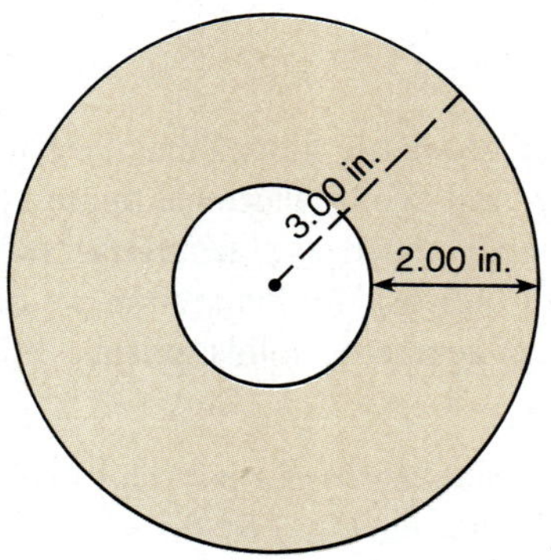

62. A wire 60 cm long is bent into the shape of a circle. Find the area of the region within the circle, correct to three significant digits.
63. If the diameter of a circle is doubled, what happens to the area?

64. Find the area of the disk in the accompanying figure after six holes, each 2 mm in diameter, have been drilled.

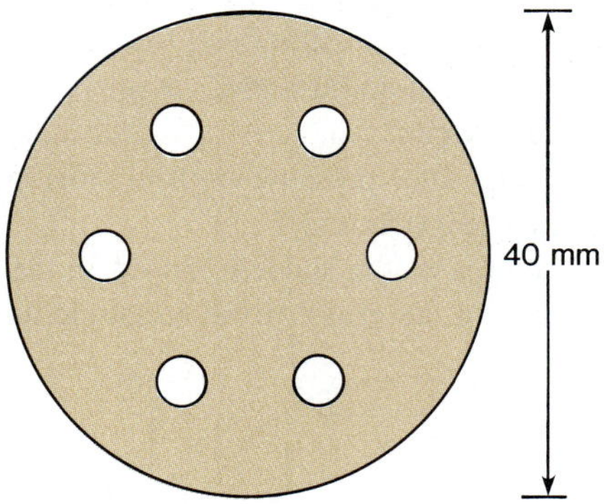

65. Find the radius of a circle whose area is 147 cm^2.

66. A machine part has the shape shown in the accompanying figure. Find the area of the part.

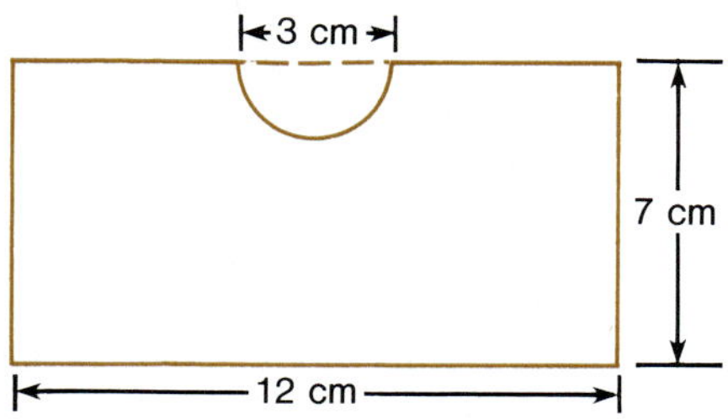

13.3 Solid Geometric Figures

Our discussion of geometric figures thus far has concerned one-dimensional figures—points and lines—and two-dimensional figures—plane figures. A geometric figure that is three-dimensional is called a **geometric solid.** The three dimensions are length, width, and depth (height). In this section we discuss the volume and surface area of the most important geometric solids: prisms, cylinders, pyramids, cones, and spheres.

note

Geometric solids enclose a portion of space. The **volume** of the solid is the amount of space enclosed by the solid.

When four or more planes intersect to enclose a portion of space, the solid thus formed is called a **polyhedron.** The plane surfaces are the **faces** of the polyhedron and the lines of intersection of the planes are the **edges** of the polyhedron. Figure 13.15 shows a polyhedron. We can determine that the polyhedron *ABCD* has four faces—triangles *ABC*, *BCD*, *ABD*, and *ADC*—and six edges—$\overline{AB}$, $\overline{BC}$, $\overline{BD}$, $\overline{CD}$, $\overline{AC}$, and $\overline{AD}$.

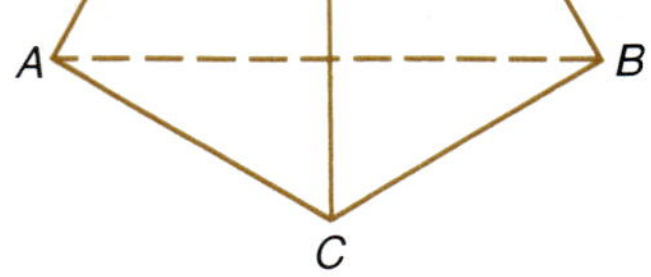

Figure 13.15

A prism is a special kind of polyhedron. By definition, a **prism** is a polyhedron whose ends, or bases, are parallel congruent polygons and whose sides, or faces, are parallelograms. A prism is named according to the shape of its bases: triangular, rectangular, hexagonal, and so on. One side of a base is called a **base edge.**

The prism in figure 13.16**a** is a triangular prism because its bases are the triangles ABC and DEF. Its faces are the parallelograms (in this case rectangles) $ABED$, $BCFE$, and $ADFC$. The base edges are $\overline{AB}$, $\overline{BC}$, $\overline{AC}$, $\overline{DE}$, $\overline{EF}$, and $\overline{DF}$. In figure 13.16**b** the prism is a rectangular prism (most often called a **rectangular solid**), whose bases are the rectangles $ABCF$ and $EDHG$. The faces are the rectangles $ABGH$, $BCDH$, $EFCD$, and $AGEF$; base edges are $\overline{AB}$, $\overline{BC}$, $\overline{CF}$, $\overline{AF}$, $\overline{GH}$, $\overline{HD}$, $\overline{DE}$, and $\overline{EF}$.

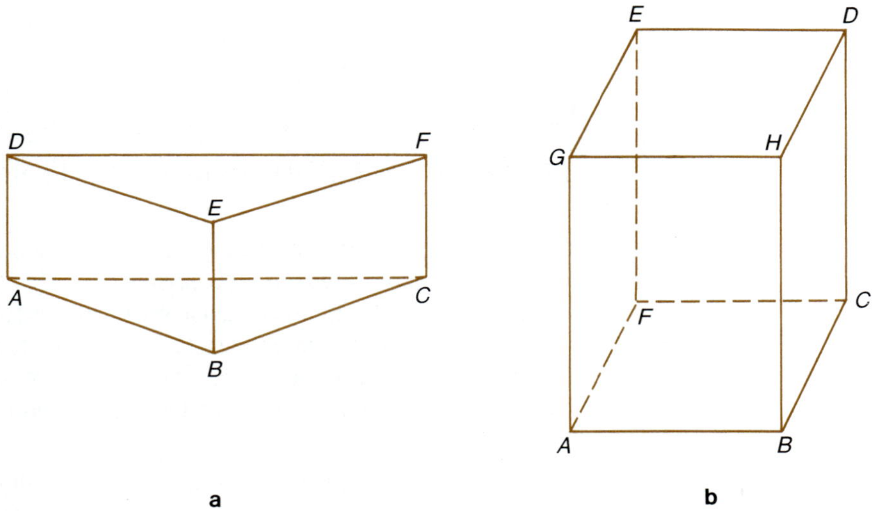

Figure 13.16

The sides of the parallelogram faces are called the **lateral edges.** When a lateral edge is perpendicular to the bases the prism is a right prism. We consider only right prisms in our discussion, and so the word *prism* will be assumed to mean a right prism.

note

Both prisms in figure 13.16 are right prisms.

The altitude (height) of a prism is the perpendicular distance between the two bases; again it is denoted by h. Thus, in our right prisms, a lateral edge is the altitude of the prism.

The lateral area of a prism, denoted by S, is the sum of the areas of the lateral faces. See figure 13.17. The lateral area is given by

$$\begin{aligned} S &= (m\overline{AB})h + (m\overline{BC})h + (m\overline{CF})h + (m\overline{AF})h \\ &= (m\overline{AB} + m\overline{BC} + m\overline{CF} + m\overline{AF})h \\ &= (\text{perimeter of the base})h \end{aligned}$$

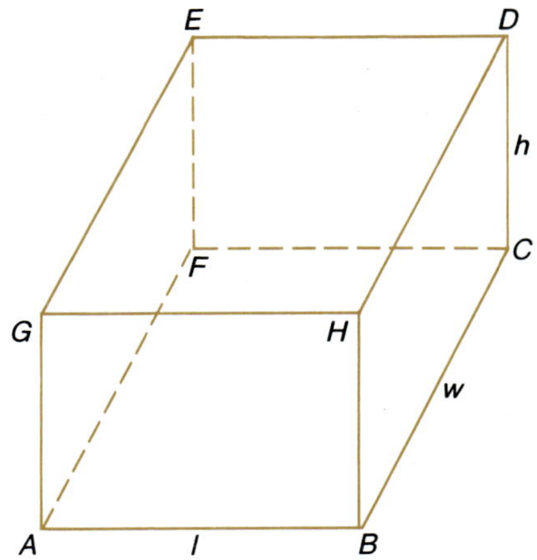

Figure 13.17

Therefore, the lateral area S of any prism can be given by

$$S = Ph$$

where P is the perimeter of the base and h is the altitude (height) of the prism. (*Note:* This formula applies *only to right prisms.*)

Again, consider the rectangular prism (rectangular solid) in figure 13.17. If we multiply the area of the base times the number of ''layers'' (linear units) in the altitude h we get the volume of the prism. Since the area of the base is given by length l times width w, then the volume V of the rectangular solid is given by

$$V = lwh$$

In general, given the area of the base B and altitude h of any prism, the volume is equal to the area of the base times the altitude. That is,

$$V = Bh$$

A special rectangular solid is the **cube,** which has all edges (e) the same length (see figure 13.18). From the preceding discussion and figure 13.18 we can determine that the volume V of a cube is given by $V = e \cdot e \cdot e$ and so $V = e^3$. The lateral surface area S is given by $S = 4e^2$ and the total surface area T (all six faces) is given by $T = 6e^2$.

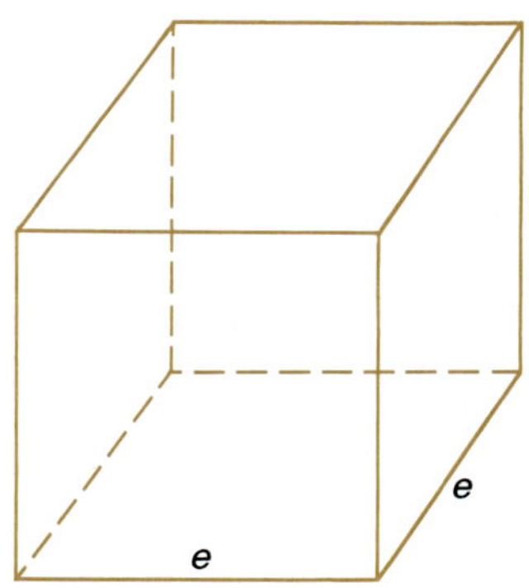

Figure 13.18

Examples

1. Find the lateral area and volume of a prism with an altitude of 9 cm and a rectangular base 6 cm long and 4 cm wide.

   a. $S = Ph$

   Now, $P = 2l + 2w$
   $= 2(6 \text{ cm}) + 2(4 \text{ cm})$
   $= 12 \text{ cm} + 8 \text{ cm}$
   $= 20 \text{ cm}$

   Then, $S = (20 \text{ cm})(9 \text{ cm}) = 180 \text{ cm}^2$

   b. $V = Bh$

   Now, $B = lw$
   $= (6 \text{ cm})(4 \text{ cm})$
   $= 24 \text{ cm}^2$

   Then, $V = (24 \text{ cm}^2)(9\text{cm}) = 216 \text{ cm}^3$

2. Find the lateral area and the volume of a prism with a right triangular base with legs 6.00 in and 4.20 in and altitude of 10.3 in.

   a. $S = Ph$

   Now, $P = 6.00 \text{ in} + 4.20 \text{ in} + \text{hypotenuse}$. The height of the hypotenuse is:

   $h = \sqrt{(6.00)^2 + (4.20)^2}$
   $= \sqrt{36 + 17.64}$
   $= \sqrt{53.64} \approx 7.32 \text{ in}$

   Then $P = 6.00 \text{ in} + 4.20 \text{ in} + 7.32 \text{ in} = 17.5 \text{ in}$ and

   $S = (17.5 \text{ in})(10.3 \text{ in}) \approx 180 \text{ in}^2$

   b. $V = Bh$

   Now $B = \frac{1}{2}(6 \text{ in})(4.2 \text{ in}) = (3 \text{ in})(4.2 \text{ in}) = 12.6 \text{ in}^2$, so

   $V = (12.6 \text{ in}^2)(10.3 \text{ in}) \approx 130 \text{ in}^3$

3. A metal box in the shape of a cube is to be coated with an enamel spray. If the box is 2 ft 6 in on each edge e and the cost of the spray is determined to be 14¢/ft², find the cost of coating the box.

   Now $T = 6e^2$, and since $e = 2 \text{ ft } 6 \text{ in} = 2.5 \text{ ft}$,
   then, $T = 6(2.5 \text{ ft})^2$
   $= 37.50 \text{ ft}^2$

   At 14¢/ft², the cost C is given by

   $C = (0.14)(36.50 \text{ ft}^2)$
   $= \$5.25$

In general, the total area T of any prism can be given by $T = S + 2B$, where S is the lateral area and B is the area of the base of the prism.

Exercise Set 13.3

In exercises 1 through 10 find the lateral area, total area, and volume of the given right prism (round off according to the accuracy of the measured data).

1. A cube with edge $e = 16.2$ ft
•2. A cube with edge $e = 2.87$ m
3. A rectangular solid with altitude $h = 4.2$ in and base of length $l = 7.1$ in and width $w = 5.7$ in
4. A rectangular solid with $h = 43.0$ mm, $l = 21.6$ mm, and $w = 15.6$ mm
5. A triangular prism with altitude $h = 9.3$ yd and a right triangular base with legs 7.0 yd and 6.7 yd
•6. A triangular prism with altitude $h = 5.25$ ft and right triangular base with legs 4.50 ft and 3.67 ft
7. A prism having altitude 4.0 cm and parallelogram base whose sides are 4.3 cm and 3.6 cm and whose altitude is 2.5 cm.
8. A prism with a parallelogram base whose sides are 5.23 m and 7.31 m and whose altitude is 8.79 m. The distance between the parallel sides 7.31 m of the base is 2.93 m
9. A right prism with a square base 15.0 mm on each edge and altitude 41.0 mm
10. A right prism with altitude 1.30 ft and square base of area 1.44 ft^2

In exercises 11 to 20 round off to the accuracy of the measured data.

11. A man wishes to lay a concrete apron in front of his garage. How many cubic meters of concrete is necessary if the apron is to be 3.25 m wide, 4.75 m long, and 0.120 m thick?
12. A boxcar that is 35.25 ft long and 8.500 ft wide is filled with grain to a depth of 5.667 ft. Find the volume of the wheat put into the boxcar. If 1 bushel occupies 1.250 ft^3 of space, how many bushels of grain are there in the boxcar?
13. One gal of paint will cover 350 ft^2 of wall space. How many gallons will be required to paint the walls of a room with an 8-ft high ceiling that is 26 ft $\times$ 18 ft? What is the cost if paint costs \$10.95/gal?
14. The lining of the bottom and sides of a cube-shaped container contains 245 ft^2 of metal. How many cubic feet of water will the container hold?
15. A box 2.0 in high is made from a piece of tin that is 20 in long and 14 in wide (see the accompanying figure). What is the volume of the box if the tin is bent to make the sides and ends?

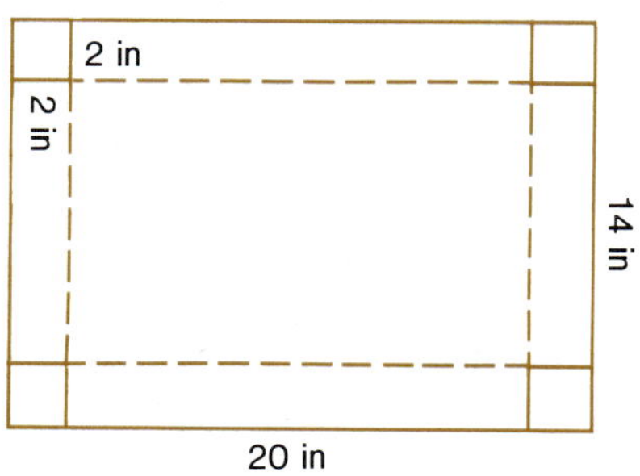

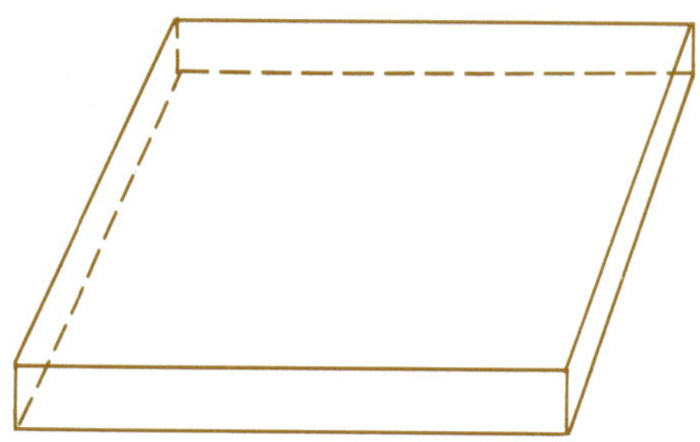

•16. Find the surface area of a steel bar that is 0.75 in × 1.50 in × 16 in long (exclude the ends).

17. If steel weighs 0.283 lb/in^3, find the weight of a piece of steel that is
   a. 0.5 in × 0.5 in × 9 in
   b. 0.57 in × 2.0 in × 6.0 ft
   c. 1.32 in × 2.15 in × 2.00 ft
   (*Hint:* Make all dimensions inches.)

18. Find the cost at 40¢/lb of the amount of sheet copper necessary to line the bottom and sides of a cubical box 7 ft on an edge if copper sheets weigh 12 oz per ft^2.

19. What length must be cut from a bar of steel 0.500 cm by 1.25 cm in cross section in order to make 1 cm^3? (*Note:* We want h when $V = 1$ cm^3).

20. Find the weight of a block of ice that is 3.50 ft long, 2.00 ft wide, and 1.75 ft high if ice has a density of 57 lb/ft^3.

13.4 Cylinders, Pyramids, Cones, and Spheres

Cylinders

Cylindrical objects are common in our daily life. Cylinders can be right or oblique and are named by their base. Our discussion is limited to a right cylinder (lateral surface perpendicular to the bases) with circular bases—called a *right circular cylinder*. By definition, a **right circular cylinder** (or a cylinder of revolution) is a solid formed by revolving a rectangle about one of its sides, called the **axis of revolution.**

From this definition, the two bases of a right circular cylinder are circles and the lateral surface is a curved surface. The axis of such a cylinder is perpendicular to the bases and, thus, is equal to the *altitude* h of the cylinder, which is the perpendicular distance between the two bases.

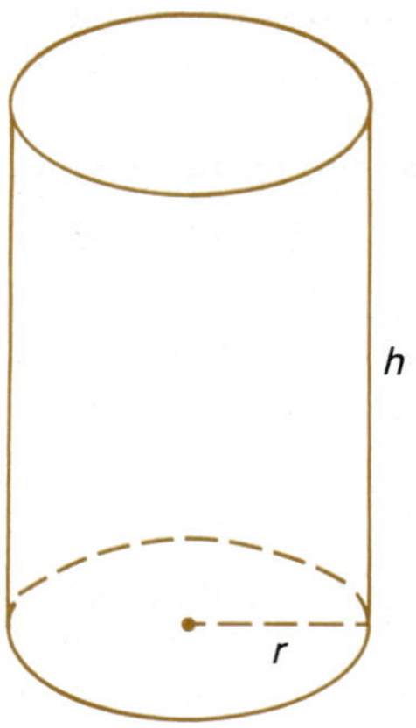

Figure 13.19

note

From now on in our discussion the word *cylinder* will mean right circular cylinder unless otherwise specified.

As with any prism, the lateral area S is determined by the perimeter of the base times the altitude h. That is, $S = Ph$. But, in this case, the bases are circles and the "perimeter" of a circle is its circumference C. Since we learned that $C = 2\pi r$, where r = the radius of the circle, then $S = 2\pi rh$, or $S = \pi dh$ since the diameter $d = 2r$. See figure 13.19.

In the same way, the volume of a prism is given by $V = Bh$, where B = area of the base. But the area of the base is πr^2 since the base is a circle; therefore,

$$V = \pi r^2 h$$

is the volume of a cylinder.

Examples

1. Find the lateral area and volume of a cylinder with base radius $r = 3.4$ mm and altitude $h = 5.3$ mm.

   $S = 2\pi rh = 2(3.14)(3.4 \text{ mm})(5.3 \text{ mm}) \approx 110 \text{ mm}^2$
   $V = \pi r^2 h = (3.14)(3.4 \text{ mm})^2(5.3 \text{ mm}) \approx 190 \text{ mm}^3$

2. A can in the shape of a right circular cylinder has a radius $r = 3.25$ in and a height of 6.50 in. Find the lateral area S and the total area T of the can.

   $S = 2\pi rh = 2(3.14)(3.25 \text{ in})(6.50 \text{ in}) \approx 133 \text{ in}^2$

   Since the cylinder has two circular bases and the area of a circle is πr^2, then

   $$\begin{aligned} T &= 2\pi r^2 + 2\pi rh \\ &= 2(3.14)(3.25 \text{ in})^2 + 2(3.14)(3.25 \text{ in})(6.5 \text{ in}) \\ &= 66.3 \text{ in}^2 + 130 \text{ in}^2 \approx 196.0 \text{ in}^2 \end{aligned}$$

3. Find the volume of a steel rod that is 2.100 cm in diameter and 124.3 cm long. (*Note:* Since $d = 2.1$ cm, then $r = \frac{d}{2} = \frac{2.100}{2} \approx 1.100$ cm)

   $V = \pi r^2 h = (3.14)(1.1 \text{ cm})^2(124.3 \text{ cm}) \approx 472.3 \text{ cm}^3$

Exercise Set 13.4

In exercises 1 through 6 find (a) the lateral area, (b) the total area, and (c) the volume of the given circular cylinder. (Round off according to the accuracy of the measured data.)

1. Base radius $r = 2.3$ in and altitude $h = 7.3$ in
2. Base radius $r = 7.25$ ft and altitude $h = 10.7$ ft
3. Base diameter $d = 1.63$ cm and altitude $h = 4.72$ cm
4. Base diameter $d = 6.17$ m and altitude $h = 5.25$ m
5. Base radius $r = 2.00$ cm and altitude $h = 0.200$ m (*Note:* Change dimensions to same unit of measure.)
•6. Base diameter $d = 1.6$ ft and altitude $h = 3.1$ yd

In exercises 7 to 21 round off to three significant digits.

7. Find the volume of a piece of 2.0-in diameter rod that is 14 in long.
•8. The density of copper is 8.90 g/cm³. Find the weight of a piece of copper wire that is 0.60 cm in diameter and 10 m long. (*Note:* 100 cm = 1 m)
9. Find the weight of a solid cylindrical piece of brass 1.2 in in diameter having length 8.0 in if brass weighs 0.3018 lb/in³.
10. If water weighs 62.5 lb/cu ft, find the weight of the water in a cylindrical tank that is 20.5 ft tall with a base radius of 3.00 ft.

11. Find the length of a 1.125-in diameter cylinder with a volume of 5.467 in³. (*Hint:* We want h, given $d = 1.125$ in and $V = 5.467$.)
12. Find the diameter of a cylindrical piece of steel 14.0 in in length if the volume is 8.42 in³.
13. In the tank shown in the accompanying art, how many square feet of sheet metal are used to form the side of the tank? (Allow nothing for overlap.)

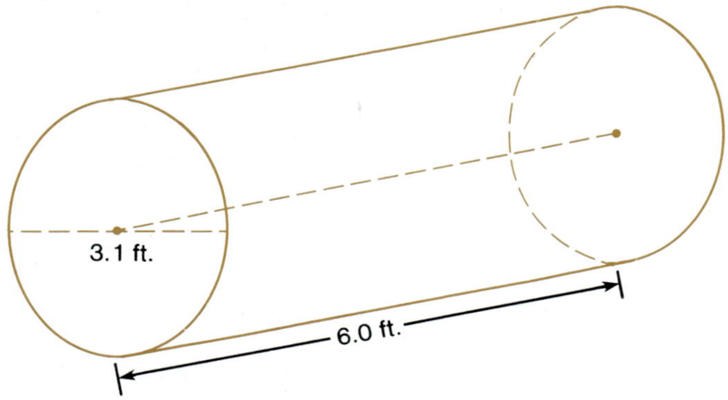

14. Find the volume of a piece of 1.25 in round stock that is 4.50 ft long. (*Note:* $d = 1.25$ in)
15. Find the weight of a cylindrical block of sealing wax that has radius $r = 3.0$ in and altitude $h = 2.5$ ft, if the wax has density 112 lb/cu ft.
•16. At a cost of 42¢/sq yd, how much would it cost (to the nearest cent) to paint the exterior of a storage tank, excluding the bottom, if the storage tank is 74.0 ft tall and has a base diameter of $d = 42.4$ ft?
17. A concrete sidewalk is to be laid around a circular flower bed as shown in the accompanying figure. If the concrete is to be 4 in thick, how many cubic feet of concrete is necessary? (*Hint:* Subtract the area of the inside circle from the area of the outside circle to get the area of the walk.)

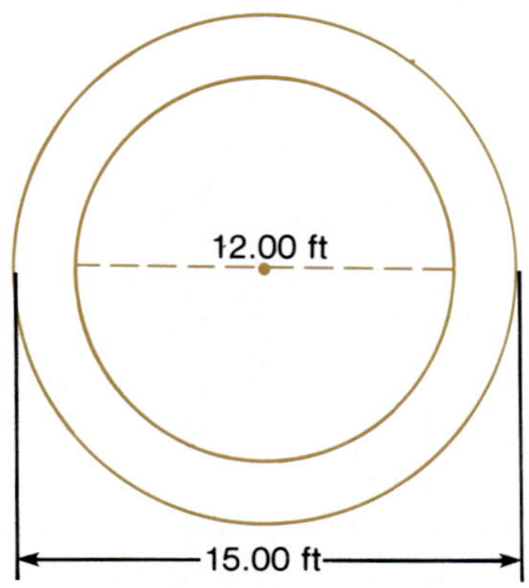

•18. A bearing bushing has an OD (outside diameter) of 3.9375 in and an ID (inside diameter) of 3.375 in (see the accompanying figure). How many cubic inches of steel make up the bushing if it is 7.500 in long?

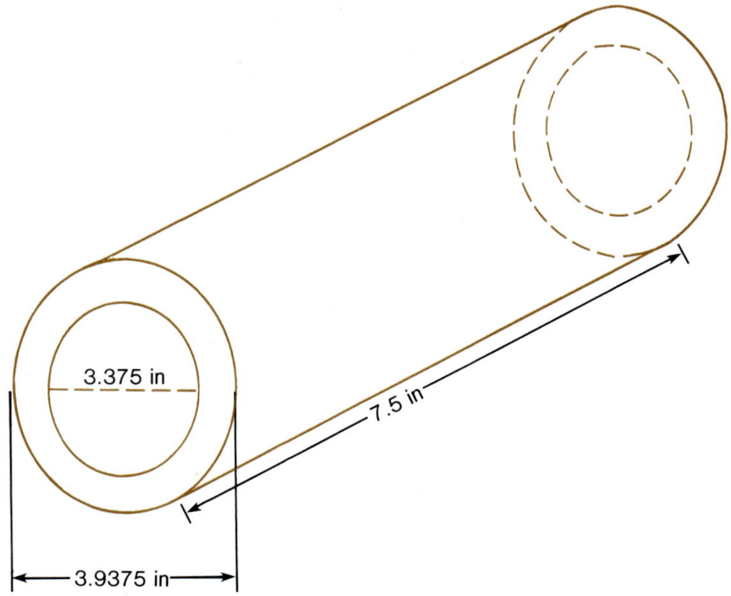

19. A special resistance wire has a diameter of 0.0370 in. The next smaller size has a diameter of 0.0345 in. Find the *difference* in the volume of the wire needed to make a length of wire 18.00 ft long.

•20. The bottom of a container is a cube with an edge of 4.0 in, and the top of the container is a cylinder with a diameter of 4.0 in and height of 4.0 in. Find the volume and total surface area of the container.

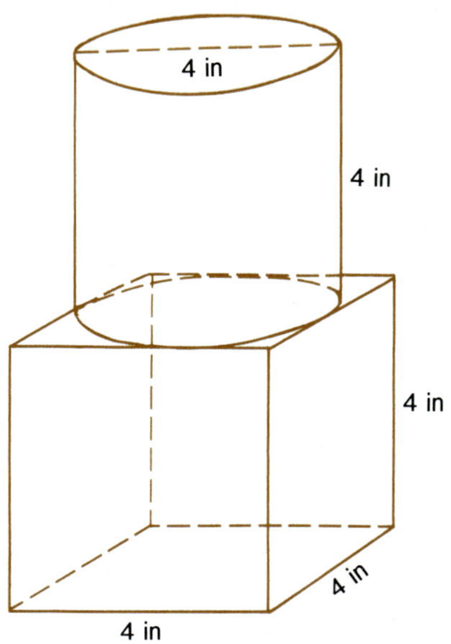

21. A steel washer 5.50 cm in diameter has a hole 1.75 cm in diameter and is 0.094 cm thick. Find the amount of steel necessary to make 1,000 such washers.

Pyramids

Pyramids and cones are similar figures in that the lateral surface of each extends upward to a point. The principal difference between the two is that the pyramid has a polygon for its base and a cone has a circle (see figure 13.20).

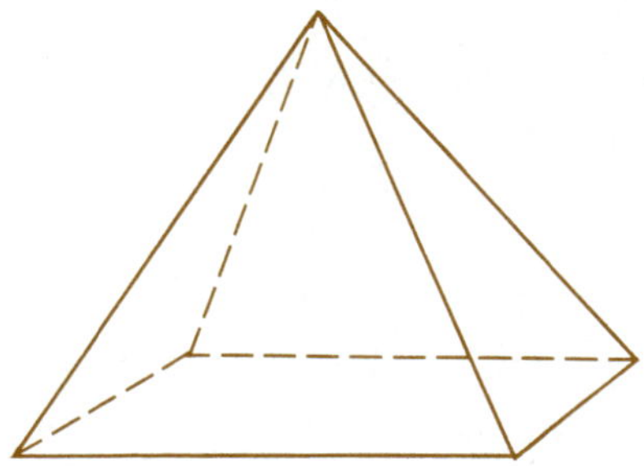

pyramid

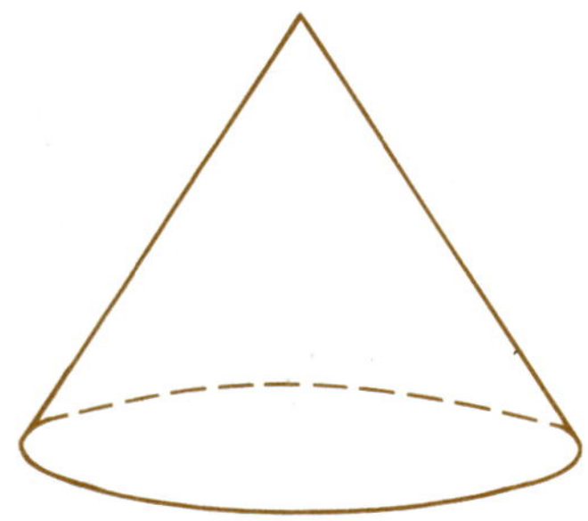

cone

Figure 13.20

By definition, a **pyramid** is a solid with a polygon base, whose sides (lateral faces) are triangles that meet at a common point. We call the common point the **vertex,** or **apex,** of the pyramid. The altitude (height) of the pyramid is the distance from the vertex perpendicular to the base. Pyramids, like prisms, are named according to the shape of the base. Thus, a triangular pyramid has a triangular base, a square pyramid has a square base, and so on. Figure 13.21 shows two different pyramids.

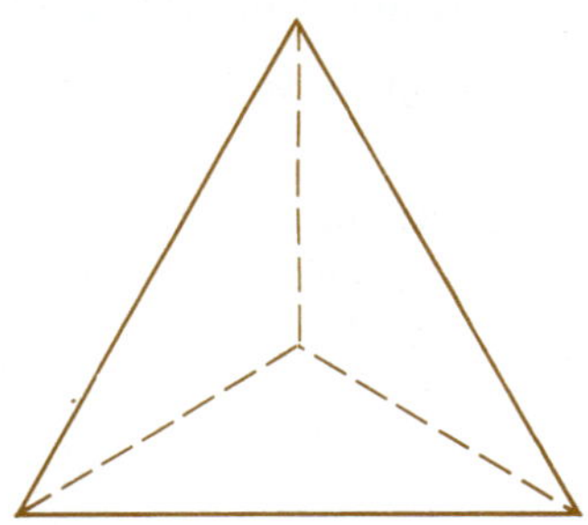

triangular pyramid

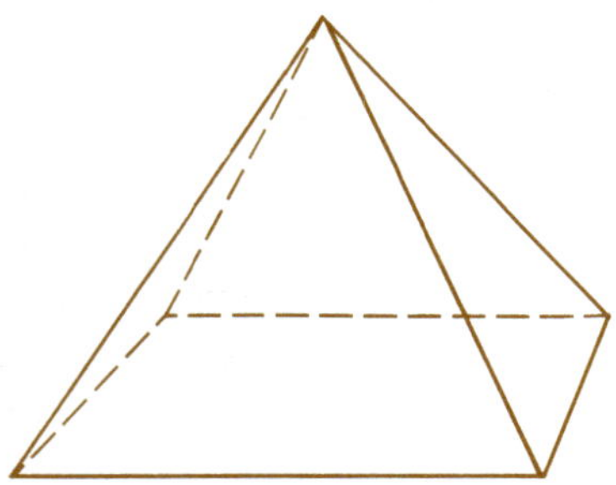

rectangular pyramid

Figure 13.21

The pyramid that we will discuss is the **right pyramid**—a pyramid whose base is a regular polygon and whose lateral faces are congruent isosceles triangles. The altitude, or height, of each of these triangular faces is called the **slant height** of the pyramid, denoted by l. Consider figure 13.22, a regular square pyramid. In this figure $\overline{EO}$ is the altitude of the pyramid and $\overline{EF}$ is a slant height of the pyramid. Point E is the vertex, or apex, of the pyramid.

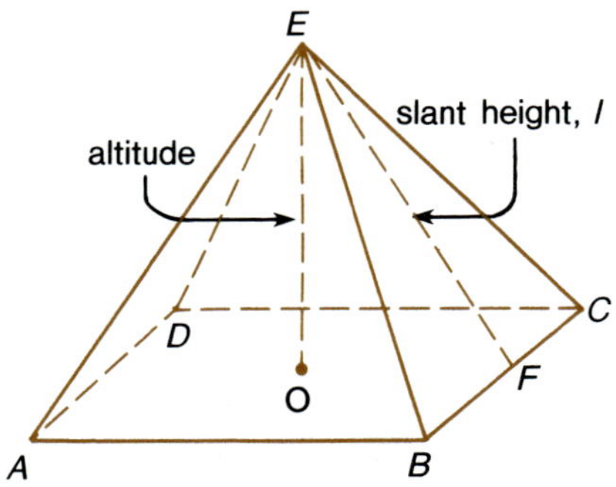

Figure 13.22

The lateral area of a pyramid is the sum of the areas of the triangles forming the faces of the pyramid. Since all faces are congruent triangles, the base is a regular polygon, and the area of each triangular face is one-half the product of the slant height l times the base of the triangle, then $S = \frac{1}{2}Pl$, where S is the lateral area, P is the perimeter of the base, and l is the slant height of the pyramid.

note

The total area T of a pyramid is given by $T =$ lateral area $+$ area of the base, so $T = S + B = \frac{1}{2}Pl + B$.

The volume of a pyramid is equal to one-third the volume of a prism with the same base and altitude. Thus, since the volume of a prism is given by $V = Bh$, the volume of a pyramid is given by

$$V = \frac{1}{3}Bh$$

where B is the area of the base and h is the altitude of the pyramid.

Examples

1. Find the (a) lateral area, (b) total area, and (c) volume of a square pyramid 11.0 cm on each side with a slant height of 7.37 cm and height of 4.90 (see figure 13.23).

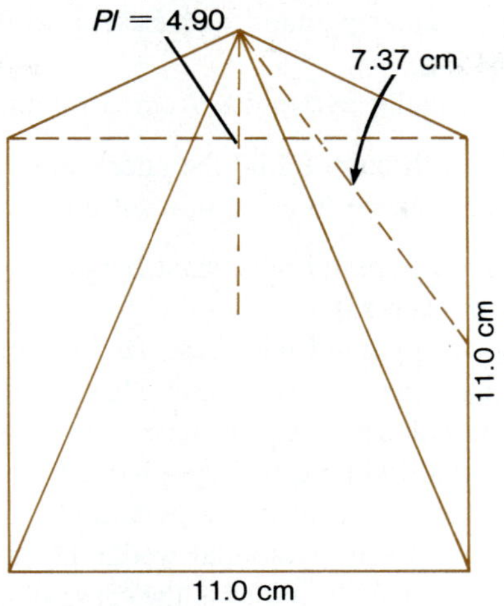

Figure 13.23

a. $S = \frac{1}{2}Pl$

$$P = 4e = 4(11.0\text{ cm}) = 44.0\text{ cm}$$

Then,

$$S = \frac{1}{2}(44.0\text{ cm})(7.37\text{ cm}) = (22.0\text{ cm})(7.37\text{ cm}) \approx 162\text{ cm}^2$$

b. $T = S + B$

$$= \frac{1}{2}Pl + e^2 = 162\text{ cm}^2 + (11.0\text{ cm})^2$$

$$= 162\text{ cm}^2 + 121.0\text{ cm}^2 \approx 283\text{ cm}^2$$

c. $V = \frac{1}{3}Bh$

$$= \frac{1}{3}(121.0\text{ cm}^2)(4.90\text{ cm}) \approx 198\text{ cm}^3$$

Exercise Set 13.5

In exercises 1 through 4 find the volume of the given pyramid. (Round all exercises according to the accuracy of the measured data.)

1. A square pyramid with base edge $e = 4.0$ in and altitude $h = 12$ in
2. A square pyramid with base edge $e = 2.4$ m and altitude $h = 8.3$ m
3. A rectangular pyramid with base 5.50 ft long and 4.75 ft wide and altitude $h = 9.67$ ft
•4. A rectangular pyramid 8.7 cm tall with base 2.1 cm long and 1.3 cm wide

In exercises 5 through 8 find the lateral area and the total area of the given pyramids. Round off correct to three significant digits.

5. A square pyramid with slant height $l = 520.0$ mm and a base of length 225.0 mm on each side.
•6. A square pyramid with base 16.3 cm on each side and slant height of 39.2 cm
7. A square pyramid with base edge 7.0 ft and slant height 9.2 ft
8. A pyramid with a square base with sides 5.2 m long and slant height of 8.0 m
9. The Great Pyramid of Egypt has a base of 762 ft^2 and is 484 ft tall. How many cubic feet are there in the pyramid?
10. A wrought iron pyramidal wedge has a square base 0.50 ft on each side and a height of 0.75 ft. What is the mass of the wedge if wrought iron has a density of 490 lb/cu ft?
11. A grain storage bin in the shape of an inverted pyramid has a rectangular base that is 8.0 ft $\times$ 10 ft. If the bin is 16 ft tall, how many cubic feet of grain can be stored in the bin?
•12. A church spire is pyramidal with a hexagonal base that is 6.0 ft on each side. If the slant height of each triangular face is 50 ft and a quart of paint will cover 225 ft^2, how many quarts of paint will be needed to paint the spire?
13. A tent is pyramidal in shape with a square base, each side of which is 9.500 ft. The tent has height of 10.25 ft. If the tent has slant height $l = 12.50$ ft, find (a) the volume of the space contained within the tent, (b) the amount of canvas needed to make the four sides and floor (excluding seams), and (c) the cost of waterproofing the sides if the waterproofing costs 56¢/sq ft (to the nearest cent).
14. Find the lateral area of a pyramid whose base is an equilateral triangle with 9.0 in sides and whose slant height is 15 in.

Cones

If a straight line passing through a fixed point is rotated so that the line follows the path of a circle, a conical surface is formed. If a plane cuts the surface of this conical surface, the solid geometric figure called a **cone** is formed. Like the pyramid a cone tapers to a point called the *vertex,* or *apex*. The straight line drawn from the vertex to the center of the base is called the **axis.** We limit our discussion here to the **right circular cone** and refer to it simply as a cone.

A cone has a circular base, and its *altitude* (height) is the perpendicular line from the vertex to the base. The *slant height l* is the length of a straight line drawn from the vertex to the circumference of the base (see figure 13.24).

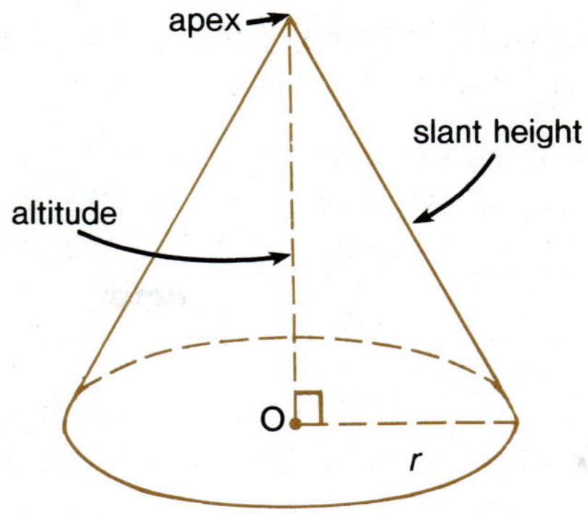

Figure 13.24

The lateral area and volume of a cone follow the same pattern as the pyramid. The lateral area is given by one-half times the circumference of the base times the slant height. Thus,

$$S = \frac{1}{2}Cl = \frac{1}{2}(2\pi r)l = \pi rl$$

where r is the radius of the base and l is the slant height of the cone.

note

The total area T of a cone is given by $T = S + B$ or $T = \pi rl + \pi r^2 = \pi r(l + r)$.

The volume of a cone is exactly one-third the volume of a right circular cylinder having the same base and altitude. Therefore, the volume V of a right circular cone is given by

$$V = \frac{1}{3}\pi r^2 h.$$

where h is the altitude and r is the radius of the base of the cone.

Examples

1. Find (a) the lateral area, (b) the total area, and (c) the volume of a cone having base radius 6.0 m, altitude 7.3 m, and slant height 9.4 m.
   a. $S = \pi rl = (3.14)(6.0 \text{ m})(9.4 \text{ m}) \approx 180 \text{ m}^2$
   b. $T = S + B$
   $= \pi rl + \pi r^2$
   $= \pi r(l + r)$
   $= (3.14)(6.0 \text{ m})\ (9.4 \text{ m} + 6.0 \text{ m})$
   $= (3.14)(6.0 \text{ m})(15.4 \text{ m}) \approx 290 \text{ m}^2$

c. $V = \frac{1}{3}\pi r^2 h$

$= \frac{1}{3}(3.14)(6.0 \text{ m})^2(7.3 \text{ m})$

$= \frac{1}{3}(3.14)(36 \text{ m}^2)(7.3 \text{ m})$

$= \frac{1}{3}(825.2)\text{m}^3 = 280 \text{ m}^3$

2. Find the surface area of a cone-shaped feed bin that is 15 ft high and has a base diameter of 10 ft.
Given $d = 10$ ft, then $r = 5$ ft and we need the slant height l. Use the right triangle, formed by the height of the cone and the radius of the base. The slant height is the hypotenuse of this triangle:
Then $l^2 = r^2 + h^2$,
so $l = \sqrt{5^2 + 15^2}$
$= \sqrt{25 + 225}$
$= \sqrt{250}$
≈ 15.8 ft.
And, $S = \pi r l$
$= \pi(5 \text{ ft})(15.8 \text{ ft})$
$\approx 248 \text{ ft}^2$ (correct to three significant digits)

Exercise Set 13.6

In exercises 1 through 4, find (a) the lateral area, and (b) the total area of the given cone. (Round off all answers to three significant digits.)

1. A cone with base radius 6.2 cm and slant height $l = 4.2$ cm
•2. A cone with base diameter 74.6 in and slant height $l = 22.7$ in
3. A cone with base radius 45.0 cm and altitude $h = 24.0$ cm
4. A cone with base diameter $d = 2.52$ ft and altitude $h = 1.83$ ft

In exercises 5 through 8 find the volume of the given cone.

5. A cone with a base radius $r = 6.3$ cm and altitude $h = 9.4$ cm
•6. A cone with base radius $r = 1.72$ m and altitude $h = 3.76$ m
7. A cone with base diameter $d = 8.40$ in and altitude $h = 7.50$ in
8. A cone with base diameter $d = 125.2$ mm and altitude $h = 432.6$ mm
9. Find the lateral area, total surface area, and volume of a conical pile of sand having a diameter of 14.0 in, an altitude of 20.0 in, and a slant height $l = 24.4$ in.
•10. Find the weight of a conical casting of iron 6.0 in in diameter with slant height 14 in, if cast iron weighs 0.26 lb/cu in.

11. A solid wood cone is made from a 1-cu ft block of wood. If the radius of the base is 1.00 in and the altitude of the cone is 10.0 in, how much of the wood from the block is not used in making the cone?
12. A water tank in the shape of an inverted cone has diameter at the top of 10.0 ft and height of 16.0 ft. How many pounds of water will the tank hold if 1 cu ft of water weighs 62.4 lb?
13. If the interior of the tank in problem 12 is to be coated with a rust prohibitive that costs 45¢/sq ft, what is the cost of rustproofing the tank?
14. A grain silo on a farm has the shape of a cone on top of a cylinder, as shown in the accompanying figure. Find the cost of painting the exterior of the silo if the paint costs 57¢/sq ft.

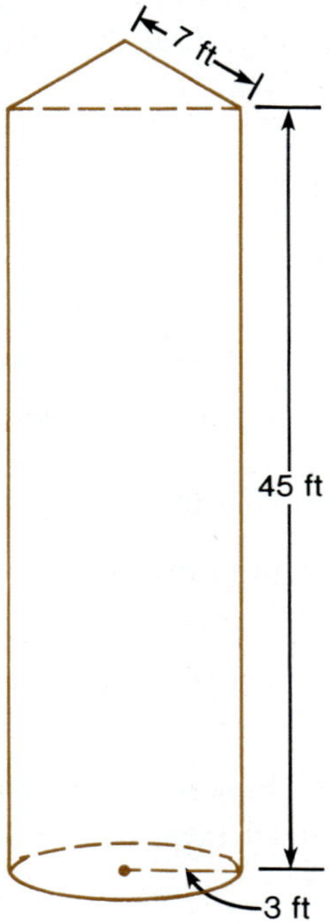

15. Find the surface area of a conical paper cup that has a diameter 4.2 in and height 4.5 in.
16. The top of a copper boiler is conical in shape, with a base diameter of 12 ft 4 in and altitude of 3 ft 6 in. How many square feet of copper are required to make the top? (Allow 2% for seams.)
17. What is the altitude of a cone whose base is 36 ft in diameter and whose slant height is 24 ft? Find the volume of the cone.

Sphere

The last geometric solid we consider is the sphere, whose curved surface contains every point that is equally distant from a point within the sphere, called the **center** of the sphere. A straight line segment passing through the center, with endpoints on the surface of the sphere, is called a **diameter** of the sphere. The line segment from the center to any point on the surface of the sphere is called the **radius** of the sphere. By definition, a **sphere** is the solid figure formed by rotating a circle about a diameter of the circle.

note

The radius and diameter of the circle have the same length as the radius and diameter of the sphere.

It is beyond the scope of this text to develop the formulas for the surface area and volume of a sphere. So, let us just say that, given radius r of the sphere, the surface area is given by

$$S = 4\pi r^2$$

and the volume is given by

$$V = \frac{4}{3}\pi r^3$$

Example

The radius of the moon is 1,080 mi. Find (a) the surface area and (b) the volume of the moon.

a. $S = 4\pi r^2 = 4(3.14)(1{,}080 \text{ mi})^2 = 14{,}600{,}000 \text{ mi}^2$

b. $V = \frac{4}{3}\pi r^3 = \frac{4}{3}(3.14)(1{,}080 \text{ mi})^3 = 5.27 \times 10^9 \text{ mi}^3$

Exercise Set 13.7

In exercises 1 through 4 find the surface areas and the volume of the spheres with the given radius or diameter. (In all exercises, round off to three significant digits.)

1. $r = 4.20$ in
2. $r = 7.32$ cm
3. $d = 3.50$ ft
•4. $d = 62.7$ m
5. Find the volume of a cast iron ball that has a diameter of 5 in.
6. If cast iron weighs 0.26 lb/cu in, find the weight of the ball in problem 5.
7. An iron ball 5 in in diameter has a 1-in thick coating of lead. Find the volume of the iron ball and of the lead in the coating.
8. The radius of the earth is 3,960 mi. Find the surface area of the earth. What is the volume of the earth? What is the ratio of the surface area of the moon to the surface area of the earth? What is the ratio of the volume of the earth to the volume of the moon? (Refer to example.)
9. What is the diameter of a sphere whose surface area is 154 in^2? (*Hint:* Find d when $S = 154$.)

•10. Find the total weight of twenty-five 1.25-in diameter steel ball bearings if the steel weighs 0.286 lb/cu in.

11. A hemisphere is half of a sphere. The top of a farm silo is a hemisphere whose diameter is 7.3 ft. Find the surface area of the top of the silo.

12. If the hemisphere in problem 11 is located on top of a cylindrical silo that is 35.0 ft tall (same diameter), find the volume of the interior of the silo.

13. A spherical weight 3 in in diameter is to be covered with tape 0.5 in wide. How many linear feet of tape will it take to cover the weight if there is no overlap?

•14. What is the radius of a sphere whose volume is 63 cm^3?

Chapter Summary

1. The **perimeter,** P, of a plane closed figure is the distance around the figure. P = the sum of the lengths of the sides.
2. The perimeter of a circle is called the **circumference,** denoted by C. $C = 2\pi r$ or $C = \pi d$, where r is the radius of the circle and d is the diameter of the circle.
3. The **area** of a plane closed figure is the measure of the portion of the surface enclosed by the figure, measured in square units or $units^2$. Area is denoted by A.
4. The **area** of a
   a. *Rectangle* is given by $A = lw$, where l is the length and w is the width of the rectangle.
   b. *Square* is given by $A = s^2$, where s is the length of each side.
   c. *Parallelogram* is given by $A = bh$, where b is the length of the base and h is the altitude of the parallelogram drawn between the base and its parallel side.
   d. *Triangle* is given by $A = \frac{1}{2}bh$, where b is the length of the base and h is the altitude of the triangle drawn to the base. Also,
   $A = \sqrt{s(s - a)(s - b)(s - c)}$,
   where a, b, and c are the lengths of the sides and $s = \frac{1}{2}(a + b + c)$.
   e. *Trapezoid* is given by $A = \frac{1}{2}h(b_1 + b_2)$, where h is the altitude of the trapezoid and b_1 and b_2 are the lengths of the parallel sides (bases) of the trapezoid.
   f. *Circle* is given by $A = \pi r^2$, where r is the length of the radius of the circle.
5. The **volume** of a
   a. *Right prism* is given by $V = Bh$, where B is the area of the base and h is the altitude of the prism.
   b. *Cube* is given by $V = e^3$, where e is the length of an edge of the cube.
   c. *Rectangular solid* is given by $V = lwh$, where l is the length and w the width of the base and h is the altitude (depth) of the solid.
   d. *Right circular cylinder* is given by $V = \pi r^2 h$, where r is the length of the radius of the base and h is the altitude of the cylinder.
   e. *Pyramid* is given by $V = \frac{1}{3}Bh$, where B is the area of the base and h is the altitude of the pyramid.

f. *Cone* is given by $V = \frac{1}{3}\pi r^2 h$, where r is the length of the radius of the base and h is the altitude of the cone.

g. *Sphere* is given by $V = \frac{4}{3}\pi r^3$, where r is the length of the radius of the sphere.

6. The **lateral area,** S, of a
   a. *Right prism* is given by $S = Ph$, where P is the perimeter of the base and h is the altitude of the prism.
   b. *Cube* is given by $S = 4e^2$, where e is the length of an edge of the cube.
   c. *Right circular cylinder* is given by $S = 2\pi rh$, where r is the length of the radius of the base and h is the altitude of the cylinder.
   d. *Pyramid* is given by $S = \frac{1}{2}Pl$, where P is the perimeter of the base and l is the slant height of the pyramid.
   e. *Cone* is given by $S = \pi rl$, where r is the length of the radius of the base and l is the slant height of the cone.
   f. *Sphere* (surface area) is given by $S = 4\pi r^2$, where r is the length of the radius of the sphere.

Chapter Review

[13.1]

1. The distance around a plane figure is called its ____________ , while the portion of the plane it encloses is called its ____________ .

In all exercises, round off to three significant digits.

Find (a) the perimeter and (b) the area of the following plane figures.

2.

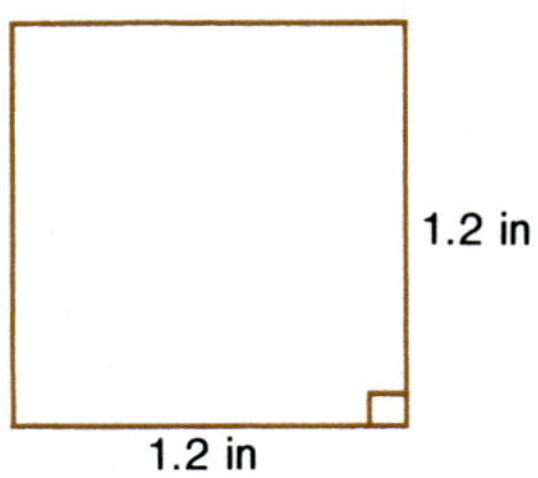

3.

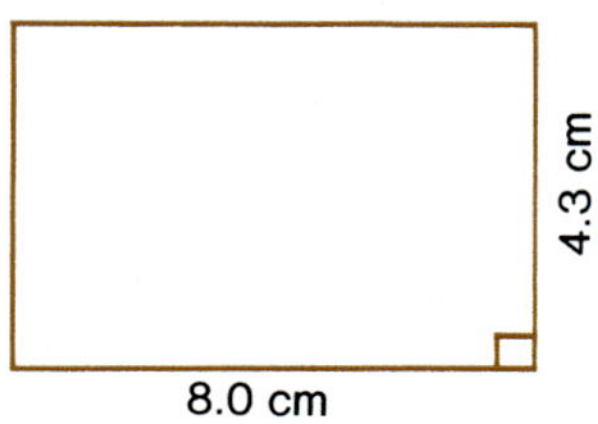

4.

5.

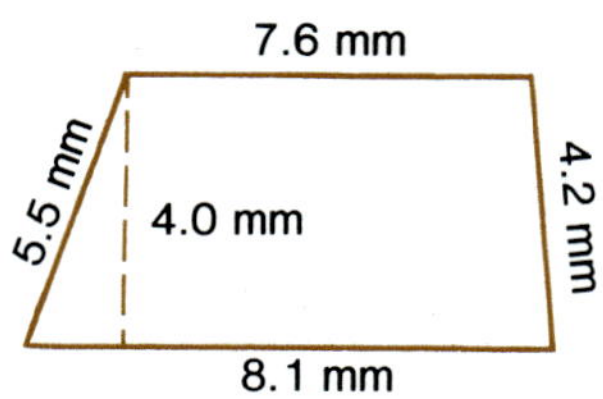

6.

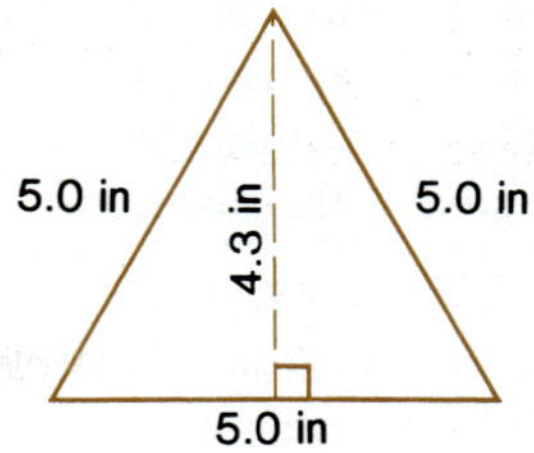

7.

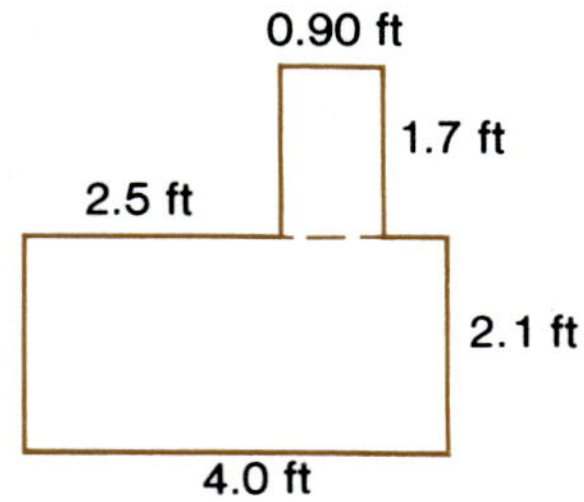

[13.2] Find the circumference and area of the following circles and parts of a circle.

8.

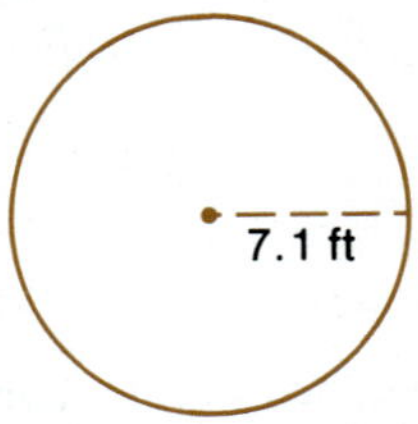

9.

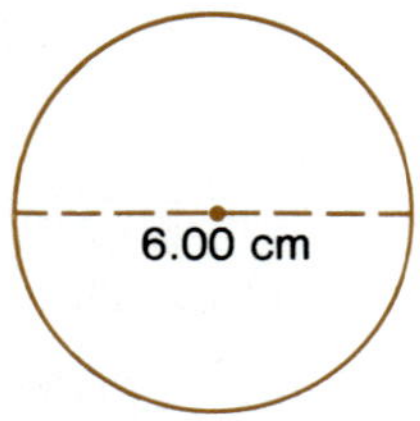

10.

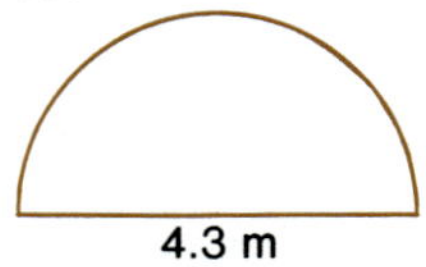

11.

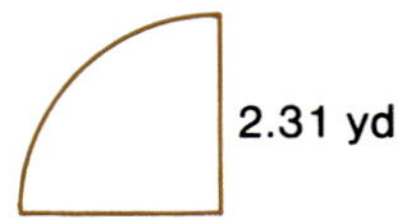

12. A triangular steel plate of the size and shape shown is to be carburized on both sides. Find the area to be carburized.

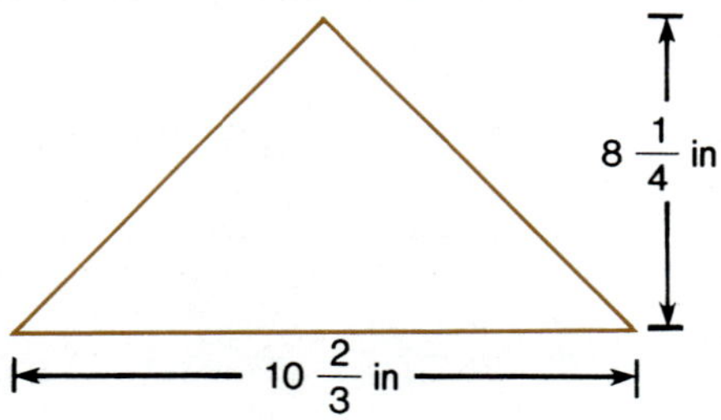

13. A room has length, width, and height dimensions of 20 ft × 15 ft × 8 ft. How many feet of floor molding is necessary if the room has one doorway 3 ft wide plus a heating vent and cold air return vent at the base of two walls that are 1 ft 8 in and 1 ft 6 in wide, respectively? Find the number of gallons of paint necessary to paint the four walls and ceiling of the room with two coats of paint if 1 gal of paint will cover 800 sq ft of surface (disregard the doorway and vents).

14. Find the number of feet of fencing necessary to enclose a plot of ground in the shape of the isosceles trapezoid given in the accompanying figure. Find the area of the ground the fencing encloses.

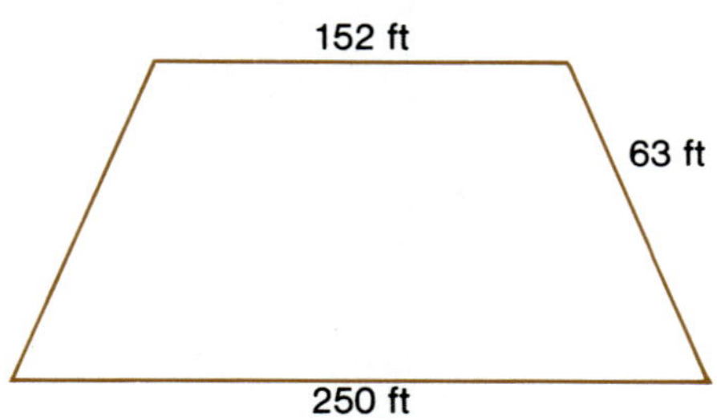

15. A pull-starter for a lawn mower has diameter 4.25 in. How much rope is necessary if you want the rope to go around the starter three times?
16. The flat cast iron plate shown in the accompanying figure has a 3.75 in round hole in it. Find the area of the cast iron.

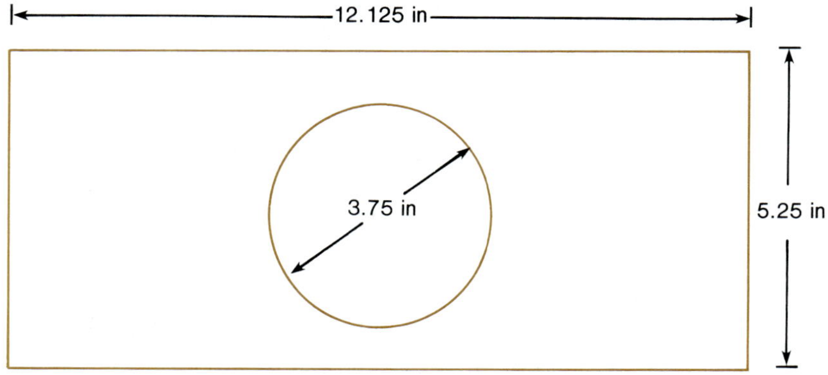

17. A 14-in diameter grinding wheel has a center hole 2 in in diameter. Find the area of a side of the wheel.
18. Find the cross-sectional area of the given steel I-beam.

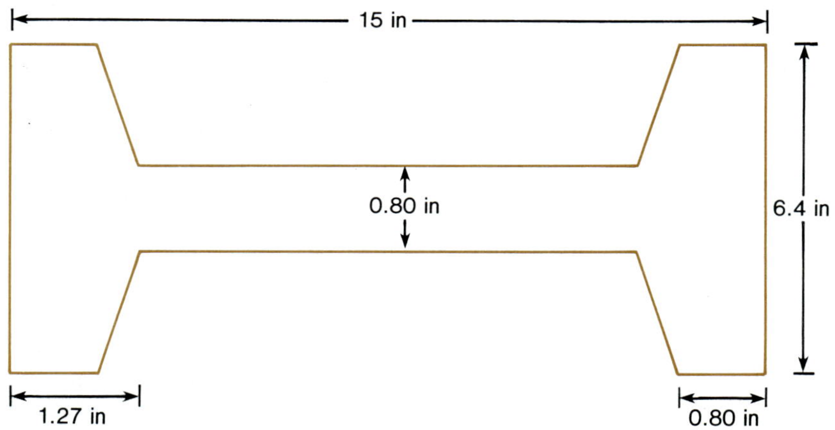

19. In the given figure, what is the area of the square? What is the area of the shaded portion of the figure?

Note: The unshaded portion is one-fourth of a circle.

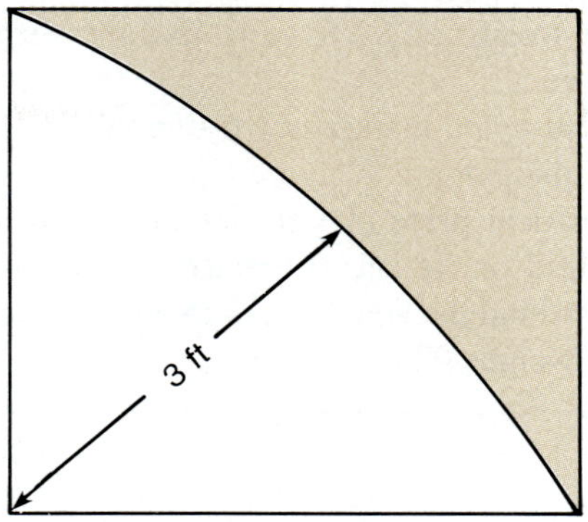

[13.3], [13.4]

20. The measure of the portion of space enclosed by a solid geometric figure is its ______________________ .

21. A solid geometric figure with trapezoid-shaped bases and faces that are parallelograms perpendicular to the bases is called a ______________________ .

22. The solid geometric figure that is the shape of a room in a house is called a ______________________ .

Find the lateral area and volume of the following geometric solids.

23.

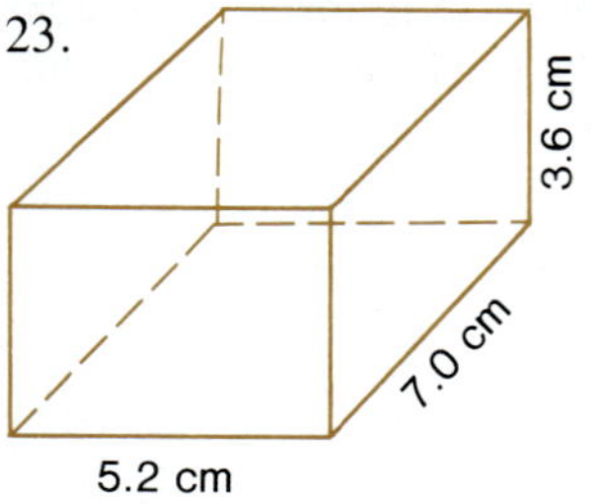

24.

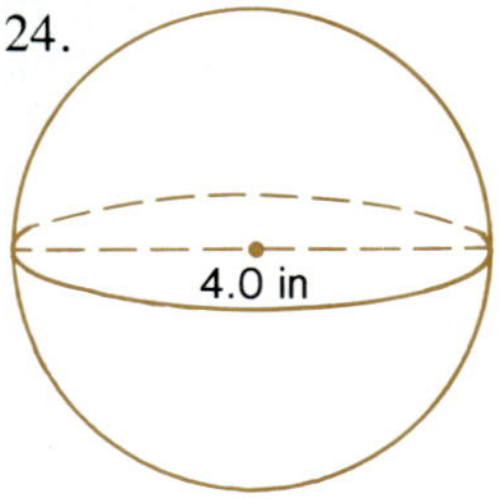

25.

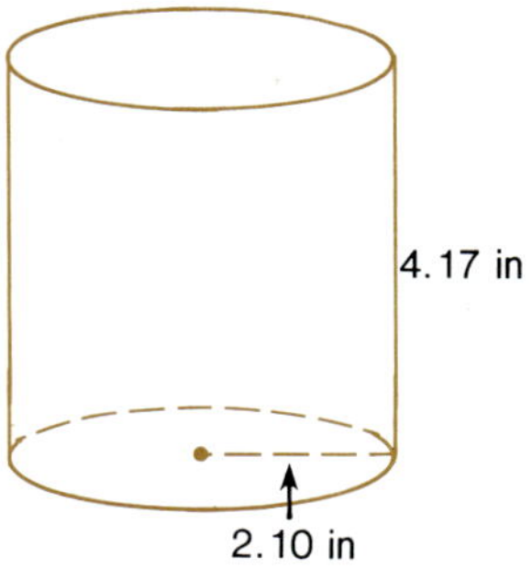

26.

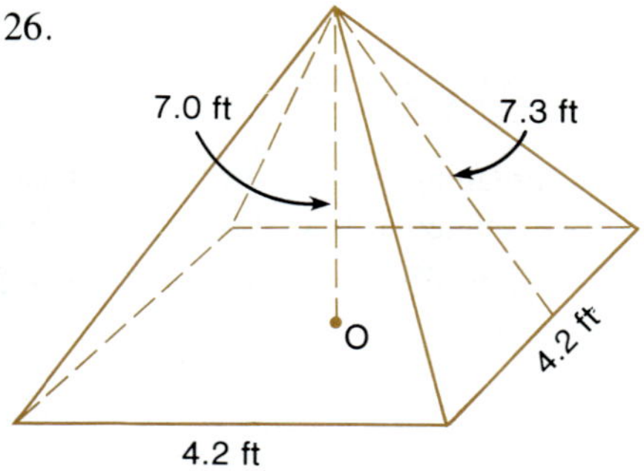

27.

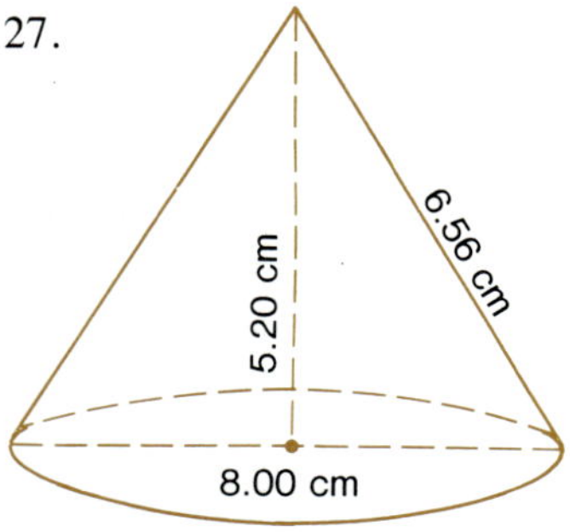

28. Find the total area of the solid in (a) problem 23, (b) problem 25, and (c) problem 27.
29. Find the volume of a piece of 0.625 in × 2.00 in steel bar 7.75 in long. How much surface area does the bar have?
30. If brass weighs 521.7 lb/cu ft, find the weight of a solid brass ball that is 1.2 ft in diameter.
31. A conical pylon marker is 1 ft 6 in tall and has a radius of 8 in. Find the surface area of the pylon.
32. A cylindrical piece of steel, as illustrated in the accompanying figure, is to be carbonized on its lateral surface as part of the heat treating process it is to go through. Find the area to be carbonized. How much does the piece of steel weigh if steel weighs 0.283 lb/in^3?

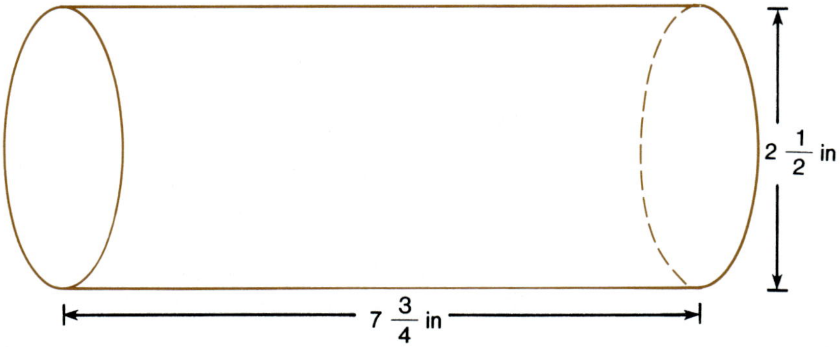

33. The wrought iron plate in the accompanying figure has a base of 8 in and a height of 6 in. If the plate is 0.375 in thick, what does it weigh if wrought iron weighs 0.2834 lb/in^3? Find the total surface area of the plate.

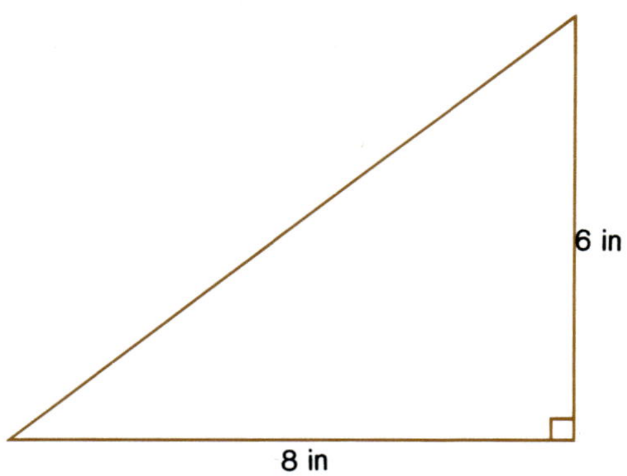

34. A tank for a hardening bath in the shape of an inverted pyramid has a depth of 2 ft 8 in and a rectangular base 3 ft 2 in × 2 ft 4 in. How many gallons of liquid will be necessary to fill the tank to the top if a gallon contains 231 in^3?

Chapter Test

Round off all answers to three significant digits.

1. Find the perimeter (or circumference) and area of the given figures.

a.

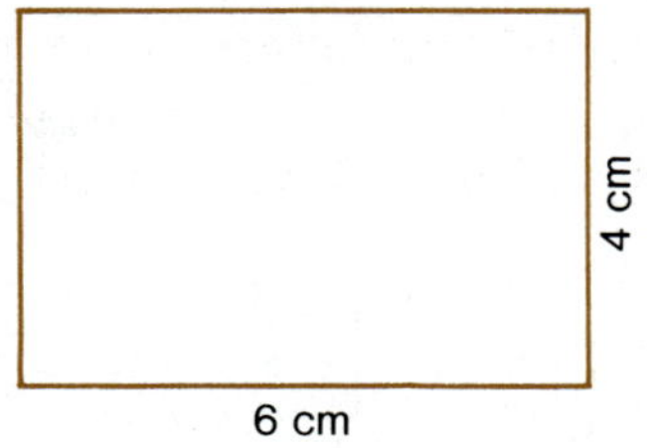

b.

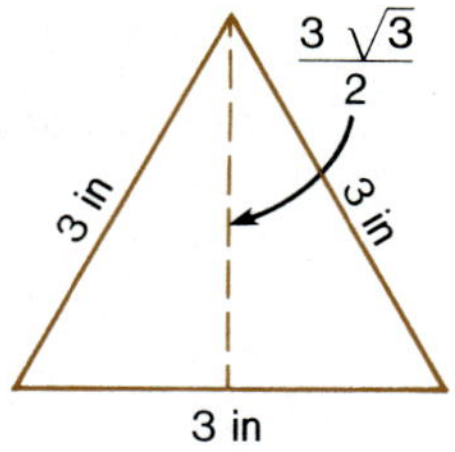

c.

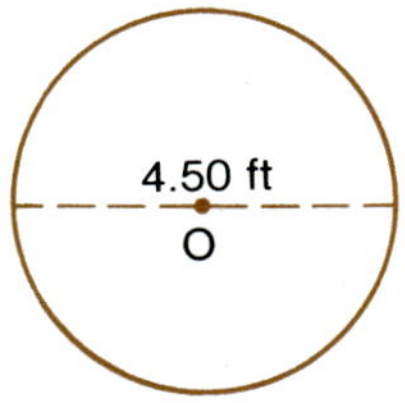

d.

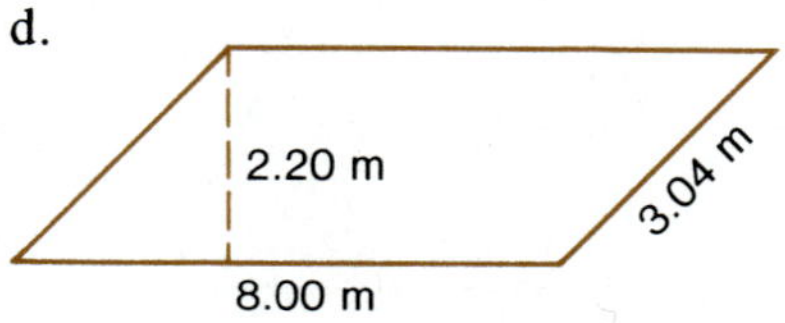

e.

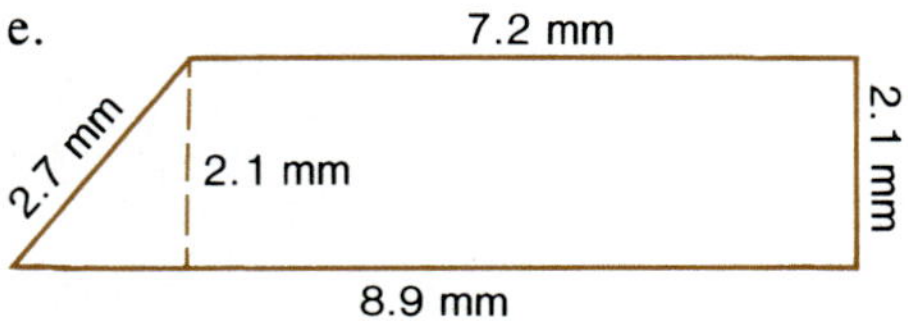

2. A man wishes to plant trees evenly spaced around the edge of his rectangular property, which is 420 ft by 246 ft. If the trees are to be spaced 6 ft apart, how many trees will he need?
3. If the area of a square is 5.125 in^2, find the perimeter of the square. (*Hint:* If $A = s^2$, then $s = \sqrt{A}$.)
4. Find the area of the end of a cylindrical bar 2.75 in in diameter.
5. A piece of wire 14.5 in long is to be bent into the shape of a circle. What is the radius of the circle, correct to three significant digits?
6. Calculate the area of a brass punching shaped as shown in the figure.

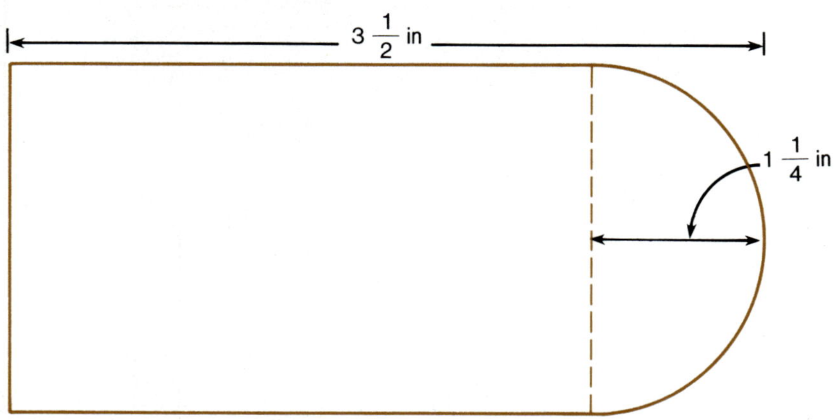

7. If the length of the bases of a trapezoid is doubled, how does this affect the area?
8. Sheet metal weighing 2.5 lb/ft^2 is cut into a triangle, as in the accompanying figure. Find the weight of the piece.

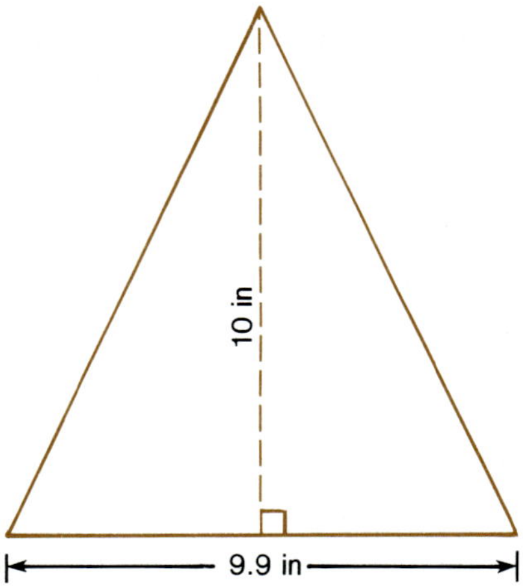

9. Find the (a) volume and (b) lateral area of the following geometric solids.

a.

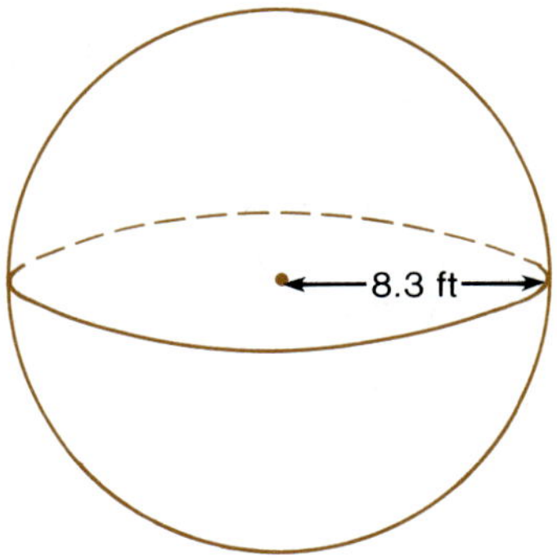

b.

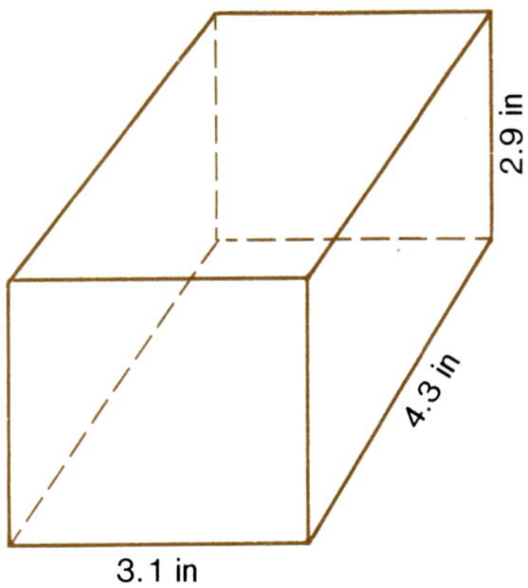

c.

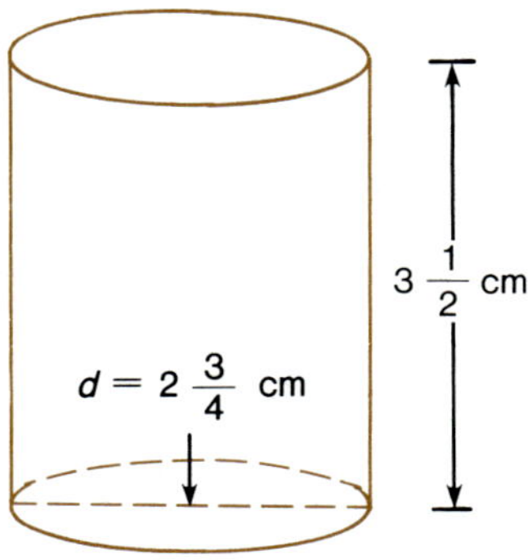

d.

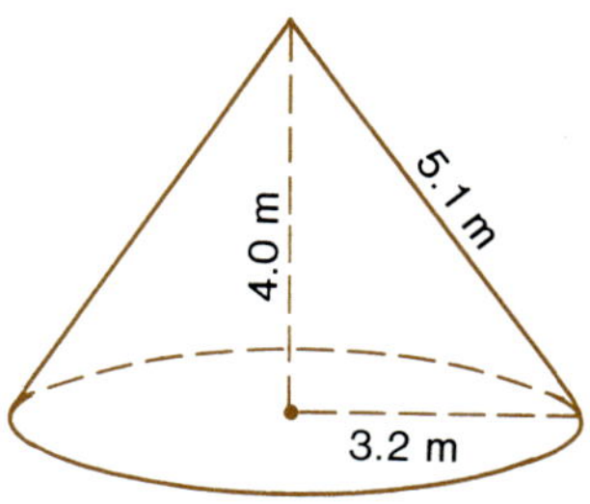

10. Find the weight of a solid piece of bronze in the shape of a rectangular solid 0.5625 in × 2 in × 1.750 ft if bronze weighs 0.3195 lb/in^3.

11. Find the cost of coating the bronze bar in problem 10 with a lacquer that costs \$0.125/sq in (round off to the nearest cent).
12. The pin in the accompanying figure is made of steel. Find the number of cubic inches in 400 such pins. (*Hint:* Find the volume of each cylinder and add the volumes.) What is the weight of each pin if steel weighs 0.283 lb/in^3?

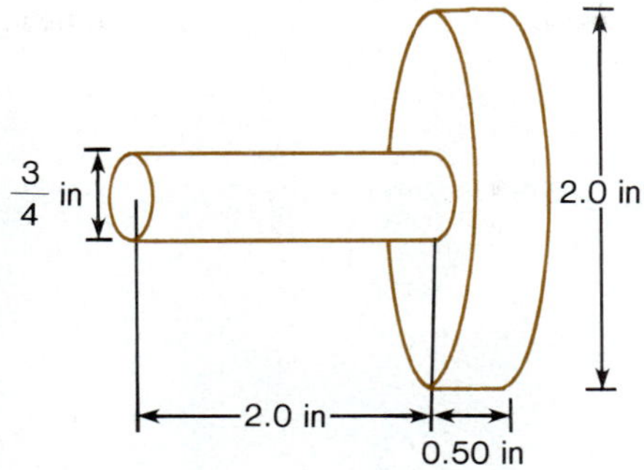

13. How many square inches of paper is necessary to make the labels for 1,000 soup cans with a base diameter of 2.5 in and a height of 3.625 in?
14. A canvas tent in the shape of a square pyramid has a slant height of 8 ft 6 in. If the base is 6 ft on each side and canvas costs \$0.75/sq ft, find the cost of making the tent, disregarding labor. The tent has a canvas floor (round off to the nearest cent).
15. A container in the shape of an inverted square pyramid has base sides of 6 in and an altitude of 8.5 in. How many ball bearings will it hold if each bearing occupies approximately 3.0 in^3 of space?
16. In the accompanying figure, find
    a. the volume of the cylindrical part
    b. the volume of the cone
    c. the total volume of the solid

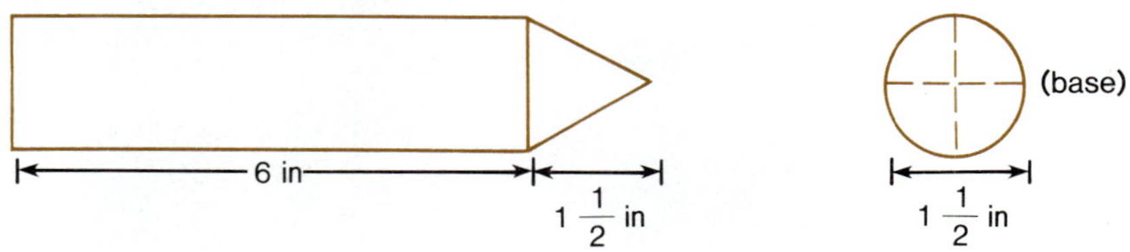

17. If the volume of a cone is 14.25 in^3 and the radius of the base has length 1.7 in, find the height of the cone.
18. A household funnel with a base radius 1.25 in and height 3.0 in is made of plastic. If the plastic used to make the funnel costs \$0.015/sq in, what is the cost of making 500 such funnels?
19. Find the weight of fifteen 1.25-in diameter ball bearings if the steel from which they are made weighs 0.286 lb/cu in.
20. If a gas chamber shaped like a sphere has a 10.0 ft radius, how many square feet are there on its surface?
21. If the radius of a sphere is tripled, what effect does this have on the volume?
22. Find the diameter of a sphere whose surface area is 146.3 cm^2.

14 Logarithms

A technician using a sine bar to drill holes on a bridgeport.
(Photograph by Kenneth Walters, courtesy of Henry Ford Community College)

14.1 Introduction

In engineering and technical fields the operations of multiplication and division, extracting a root, and raising to a power are used extensively. Today technicians can use the hand-held calculator to perform these operations. Without this instrument, the operations can be difficult and time-consuming. There is, however, an alternative to the hand-held calculator that can simplify the operations considerably, namely, **logarithms.**

A Scotsman by the name of John Napier is generally credited with conceiving the idea of the logarithm. He introduced the first logarithm tables in 1614. Though they were not the same logarithm tables used today, they established the use of this important tool in mathematics. Logarithms are used to:

1. Reduce the operation of multiplication to addition
2. Reduce division to subtraction
3. Reduce raising to a power to simple multiplication
4. Reduce extracting a root to simple division

Many long and difficult calculations can be made easier through the use of logarithms. They can provide a high degree of accuracy.

14.2 Definition of a Logarithm

Consider the expression $4^2 = 16$, which we call an **exponential equation.** A definite relationship exists between the *exponent,* 2, and the value, 16; that is, 2 is the power to which the *base*, 4, must be raised to get the *number* 16. In previous chapters we focused on the exponent. Now we direct our attention to the value 16.

In the exponential equation $4^2 = 16$, we call 2 a *logarithm*. In fact, the equation is read as "the logarithm of 16 to the base 4 is 2," which we indicate by the expression $\log_4 16 = 2$, where "log" is the abbreviation of the word *logarithm*. Thus, the **logarithm** of a number is the exponent, or power, to which another number, the base, must be raised to get that number.

In the logarithmic equation, $\log_4 16 = 2$, the number written below and to the right of "log" is the base, the number to be raised to the second power to get 16. The statements

$4^2 = 16$ and $\log_4 16 = 2$

are equivalent. An extremely important fact to understand and remember is that

A logarithm is an exponent.

Examples The following are equivalent equations.

1. $2^3 = 8$ and $\log_2 8 = 3$
2. $5^2 = 25$ and $\log_5 25 = 2$
3. $10^4 = 10{,}000$ and $\log_{10} 10{,}000 = 4$

To generalize: $b^x = N$, where x is the exponent to which base b must be raised to get N, and $\log_b N = x$, where x is the logarithm of N to the base b, mean the same thing. That is,

$\log_b N = x$ is equivalent to $b^x = N$.

Consider the exponential equation $b^x = N$. Suppose we let b be a negative number and x be a fraction. What kind of a number is N? To illustrate, let $b = -9$ and $x = \frac{1}{2}$. Then $(-9)^{\frac{1}{2}} = N$. But $(-9)^{\frac{1}{2}}$, written in radical form, is $\sqrt{-9}$, so $\sqrt{-9} = N$ makes N an *imaginary* number; therefore, if $b < 0$, it is possible for b^x to be an imaginary number.

Now suppose $b = 1$ and x is any real number. The resulting value for N is 1 for any x chosen, since 1 to any power is *always* 1. To illustrate,

$$1^2 = 1,\ 1^{\frac{1}{2}} = 1,\ 1^0 = 1$$

To avoid the possibility of N's being imaginary or always being equal to 1, we require that base b be some positive number and *not* equal to 1. That is, $b > 0$, $b \neq 1$.

Examples

1. $9^2 = 81$ expressed as a logarithmic equation is $\log_9 81 = 2$.
2. $\log_{\frac{1}{2}}\left(\frac{1}{8}\right) = 3$ expressed as an exponential equation is $\left(\frac{1}{2}\right)^3 = \frac{1}{8}$.

Logarithmic equations can be solved, using this relationship, when b, x, or N is unknown.

Examples

1. Find b if $\log_b 64 = 2$.
   Change $\log_b 64 = 2$ to exponential form: $b^2 = 64$. Since $64 = 8^2$, we have $b^2 = 8^2$ and $b = 8$.
2. Find N if $\log_5 N = 3$.
   Writing $\log_5 N = 3$ in exponential form, we get $5^3 = N$. Then $N = 125$.
3. Find x if $\log_3 81 = x$.
   Writing $\log_3 81 = x$ in exponential form, we get $3^x = 81$. Since $3^4 = 81$, then $3^x = 3^4$ and $x = 4$.

Exercise Set 14.1

In exercises 1 through 10 express each exponential equation as a logarithmic equation.

1. $10^2 = 100$
•2. $3^4 = 81$
3. $2^6 = 64$
4. $4^4 = 256$
5. $5^0 = 1$
6. $10^6 = 1{,}000{,}000$
7. $9^{\frac{1}{2}} = 3$
8. $(32)^{\frac{1}{5}} = 2$
9. $\left(\frac{1}{2}\right)^2 = \frac{1}{4}$
10. $x^y = u$

In exercises 11 through 20 express each logarithmic equation as an exponential equation.

11. $\log_{10}100{,}000 = 5$
12. $\log_4 16 = 2$
13. $\log_5 625 = 4$
•14. $\log_7 49 = 2$
15. $\log_9 1 = 0$
16. $\log_3 243 = 5$
17. $\log_8 512 = 3$
18. $\log_{64} 4 = \frac{1}{3}$
19. $\log_4 4 = 1$
20. $\log_x u = y$

In exercises 21 through 32 find the values of x.

21. $\log_x 25 = 2$
•22. $\log_x 32 = 5$
23. $\log_x \frac{1}{27} = -3$
24. $\log_x 512 = 3$
25. $\log_7 x = 4$
26. $\log_4 x = -2$
•27. $\log_{\frac{1}{4}} x = 4$
28. $\log_{10} x = 5$
29. $\log_5 125 = x$
30. $\log_6 216 = x$
31. $\log_{10}100{,}000 = x$
32. $\log_2 512 = x$

14.3 Properties of Logarithms

Since a logarithm is an exponent, let us review the laws of exponents learned in chapter 10 before we develop the basic properties for logarithms. These laws of exponents are as follows:

1. $x^m \cdot x^n = x^{m+n}$
2. $\dfrac{x^m}{x^n} = x^{m-n} \qquad (x \neq 0)$
3. $(x^m)^n = x^{mn}$
4. $(xy)^m = x^m \cdot y^m$
5. $x^0 = 1 \qquad (x \neq 0)$
6. $x^{-n} = \dfrac{1}{x^n} \qquad (x \neq 0)$

We use these properties to obtain the properties of logarithms. Let $\log_b x = m$ and $\log_b y = n$. Then

$$b^m = x \qquad \text{and} \qquad b^n = y$$

and

$$x \cdot y = b^m \cdot b^n = b^{m+n}$$

by the properties of exponents. Writing this as a logarithmic equation, we have $\log_b(xy) = m + n$, and by substitution,

$$\boxed{\log_b(xy) = \log_b x + \log_b y}$$

That is, the logarithm of the product of two numbers is equal to the sum of the logarithms of the two numbers.

In like fashion, we can show that

$$\log_b\left(\frac{x}{y}\right) = \log_b x - \log_b y$$

That is, the logarithm of the quotient of two numbers is equal to the logarithm of the numerator minus the logarithm of the denominator.

And, if r is a real number,

$$\log_b x^r = r \log_b x$$

That is, the logarithm of the rth power of a number is equal to r times the logarithm of the number.

Two special properties of logarithms which derive from these basic properties are:

1. $\log_b b = 1$ since $b^1 = b$
2. $\log_b 1 = 0$ since $b^0 = 1$, for any real number b.

Putting these derivations into words:

1. The logarithm of any number to a base equal to that number is always 1.
2. The logarithm of 1 to *any base* is always *zero*.

Examples

1. $\log_3 3 = 1$
2. $\log_{\frac{1}{3}}\frac{1}{3} = 1$
3. $\log_4 1 = 0$
4. $\log_{\frac{1}{2}} 1 = 0$

We use these properties of logarithms in the following examples:

Examples

1. $\log_3(2 \cdot 5) = \log_3 2 + \log_3 5$
2. $\log_4(3 \cdot 7) = \log_4 3 + \log_4 7$
3. $\log_{10}\left(\frac{3}{5}\right) = \log_{10} 3 - \log_{10} 5$
4. $\log_8\left(\frac{8}{7}\right) = \log_8 8 - \log_8 7$

   (*Note:* $\log_8 8 = 1$, so we have $1 - \log_8 7$.)
5. $\log_3 5^2 = 2 \cdot \log_3 5$
6. $\log_9(10)^3 = 3 \log_9(10)$
7. $\log_2\sqrt{7} = \log_2(7)^{\frac{1}{2}} = \frac{1}{2}\log_2 7$

note

When N is under a radical, it is changed to fractional exponential form, as in examples 7 and 8.

8. $\log_{10}\sqrt[3]{13} = \log_{10}(13)^{\frac{1}{3}} = \frac{1}{3}\log_{10}(13)$

9. $\log_{10}(12) = \log_{10}(2 \cdot 2 \cdot 3) = \log_{10}(2^2 \cdot 3)$
$= \log_{10}2^2 + \log_{10}3$
$= 2\log_{10}2 + \log_{10}3$

10. $\log_e\left(\frac{27}{8}\right) = \log_e 27 - \log_e 8$
$= \log_e 3^3 - \log_e 2^3$
$= 3\log_e 3 - 3\log_e 2$

note

In examples 9 and 10 each number is written in *prime factor* form before the properties of logarithms are applied.

Exercise Set 14.2

Use the properties of logarithms to write the following as a sum or difference of two logarithms or the product of a real number times a logarithm.

1. $\log_b(3 \cdot 5)$
•2. $\log_2(7 \cdot 5)$
3. $\log_3(2 \cdot 13)$
4. $\log_b\left(\frac{2}{3}\right)$
5. $\log_5\left(\frac{7}{11}\right)$
6. $\log_7\left(\frac{23}{17}\right)$
7. $\log_b 4^3$
•8. $\log_6 8^2$
9. $\log_{10}5^4$
10. $\log_b(24)$
11. $\log_{\frac{1}{2}}(36)$
12. $\log_4(45)$
•13. $\log_e\left(\frac{8}{9}\right)$
14. $\log_{10}\left(\frac{15}{16}\right)$
15. $\log_5\left(\frac{24}{25}\right)$
•16. $\log_b\frac{x^3}{y^2}$
17. $\log_b(xy)^3$
•18. $\log_b\sqrt[3]{x^2}$
19. $\log_b\sqrt{xy}$
•20. $\log_b\sqrt[4]{\frac{x}{y}}$

Write the following logarithmic expressions as the logarithm of a single number.

Example $2\log_3 7 + 3\log_3 5 = \log_3 7^2 + \log_3 5^3$
$= \log_3(7^2 \cdot 5^3)$
$= \log_3(49 \cdot 125)$
$= \log_3(6125)$

21. $\log_4 5 + \log_4 11$
22. $\log_3(17) + \log_3 3$
23. $\log_{10}(25) - \log_{10}12$
•24. $\log_e 11 - \log_e 16$
25. $2\log_5 11$
26. $3\log_6 4$
27. $2\log_2 7 + \log_2 5$
•28. $3\log_{10}3 + 2\log_{10}4$

29. $4 \log_e 2 - \log_e 10$

30. $5 \log_e 2 - 3 \log_e 4$

31. $\frac{1}{2}\log_3 4 - \frac{1}{3}\log_3 27$

32. $\frac{1}{5}\log_4 32 + \frac{1}{4}\log_4 81$

33. $\log_e 4 + 2 \log_e 3 - \log_e 2$

•34. $2 \log_{10} 5 - \frac{1}{2}\log_{10} 9 + 3 \log_{10} 2$

35. $4 \log_2 3 - \frac{1}{3}\log_2 8 + \log_2 7 - 2 \log_2 5$

36. $2\ \mathrm{lob}_b x + 3 \log_b y$

37. $4\ \mathrm{lob}_b x - 2 \log_b y$

38. $\frac{1}{3}\log_b x^2$

39. $\frac{1}{2}\log_b x + \frac{1}{2}\log_b y$

•40. $\frac{1}{4}\log_b x^3 - \frac{1}{4}\log_b y$

Given $\log_b 2 = 0.3010$, $\log_b 3 = 0.4771$, and $\log_b 10 = 1$, compute the following logarithms and write an equivalent exponential equation.

Example Find $\log_b 6$.

$$\begin{aligned} \log_b 6 &= \log_b(2 \cdot 3) \\ &= \log_b 2 + \log_b 3 \\ &= 0.3010 + 0.4771 \\ &= 0.7781 \end{aligned}$$

Then, $b^{0.7781} = 6$

41. $\log_b 20$

•42. $\log_b 30$

43. $\log_b 8$

44. $\log_b 9$

45. $\log_b 400$

•46. $\log_b 9000$

47. $\log_b \frac{2}{3}$

48. $\log_b \frac{3}{10}$

49. $\log_b \frac{27}{4}$

50. $\log_b \sqrt{2}$

51. $\log_b \sqrt[3]{3}$

52. $\log_b \sqrt[3]{16}$

53. $\log_b \sqrt{\frac{3}{4}}$

•54. $\log_b \sqrt[4]{\frac{10}{3}}$

•55. What is b?

14.4 Common Logarithms

When studying scientific notation, we learned that it is possible to write any positive number as a product of a number greater than or equal to 1 but less than 10 times a power of ten. Now we have also learned that the powers of ten are logarithms to the base 10. Logarithms to **base 10,** one of the most used kinds of logarithms, are called the **common logarithms.** The use of base 10 follows from our number system, which is base 10. Because the base will always be 10, we adopt the convention of eliminating it in a logarithmic statement. That is, we write $\log_{10} 5$ as "log 5" and assume that when the base is missing it is *always* 10.

Examples

1. log 4 means $\log_{10}4$
2. log 5.28 means $\log_{10}5.28$

A common logarithm of a number is the power to which 10 must be raised to get the number. That is, log 100 = 2 since $10^2 = 100$. Table 14.1 describes some other common logarithms.

Table 14.1 Common Logarithms

log 10,000 = 4 since $10^4 = 10{,}000$
log 1,000 = 3 since $10^3 = 1{,}000$
log 100 = 2 since $10^2 = 100$
log 10 = 1 since $10^1 = 10$
log 1 = 0 since $10^0 = 1$
log 0.1 = −1 since $10^{-1} = 0.1$
log 0.01 = −2 since $10^{-2} = 0.01$
log 0.001 = −3 since $10^{-3} = 0.001$
log 0.0001 = −4 since $10^{-4} = 0.0001$

From table 14.1 we can see that the logarithm of a number that lies between 100 and 1,000 must be a number between 2 and 3. That is, the logarithm of any number between 100 and 1,000 is 2 *plus* some decimal fraction. The logarithm of a number between 1,000 and 10,000 is a number between 3 and 4, that is, the number 3 *plus* some decimal fraction. The same relationship holds for the logarithms of numbers less than 1. The logarithm of a number that lies between 0.01 and 0.1 is a number that lies between −2 and −1; that is, the logarithm is −2 *plus* a decimal fraction.

These "decimal fractions" mentioned are *all positive*. Thus, if the logarithm of a number N that lies between 0.01 and 0.1 is −2 *plus* a decimal fraction, then log N, where $0.01 < N < 0.1$, is given by

$$\log N = -2 + (\text{a positive decimal})$$

Example

log 0.0345 = −2 + (0.5378), where 0.5378 is a positive number.

Thus, the logarithm of a number N consists of two parts: an integer and a decimal fraction. The integer is the **characteristic,** and the decimal fraction is the **mantissa** of the logarithm of the number N.

Example

In the preceding example, given log 0.0345 = −2 + (0.5378), −2 is called the characteristic and 0.5378 is called the mantissa of log 0.0345.

To determine the characteristic of the logarithm of a number, we write the number in scientific notation. The integral power of 10 is the characteristic.

Examples

Find the characteristic of the logarithms of the following numbers.

1. log 123
   Since $123 = 1.23 \times 10^2$, then the characteristic is 2.
2. log 1.23
   Since $1.23 = 1.23 \times 10^0$, then the characteristic is 0.
3. log 0.000123
   Since $0.000123 = 1.23 \times 10^{-4}$, then the characteristic is -4.

We find the mantissa, the logarithm of the number between 1 and 10, by using table C in the appendix. Table C gives the logarithms correct to four significant digits for significant digit numbers from 1.00 to 9.99. The following examples illustrate the method for finding the logarithm of such a number.

Examples

Find the logarithm of the given number using table C.

1. 3.41
   Read down the column headed N to find 34, the first two significant digits of the number. Move right on this row to the column headed 1, the third significant digit. Read the four-significant-digit number 5328. Thus, $\log 3.41 = 0.5328$.

note
The characteristic of 3.41 is 0, and we have found that $10^{0.5328} \approx 3.41$.

2. 7.86
   Read down the column headed N to 78. Read across this row to the column headed 6 and find the digits 8954. Therefore, $\log 7.86 = 0.8954$.

note
The characteristic of 7.86 is 0; thus, $10^{0.8954} \approx 7.86$.

To find the logarithm of a positive number N that does *not* lie between 1.00 and 9.99 we add the characteristic of the number to the mantissa (the logarithm of a number between 1.00 and 9.99). The following examples illustrate this method.

Examples Find the following logarithms using table C.

1. log 725

   Since $725 = 7.25 \times 10^2$, the characteristic is 2. The mantissa is given by log 7.25 = 0.8603. Then

   $$\log 725 = 2 + \log 7.25 = 2.8603$$

   Thus, $10^{2.8603} \approx 725$.

2. log 5,030,000

   Since $5{,}030{,}000 = 5.03 \times 10^6$, the characteristic is 6. The mantissa is given by log 5.03 = 0.7016. Therefore,

   $$\log 5{,}030{,}000 = 6 + \log 5.03 = 6.7016$$

   and $10^{6.7016} \approx 5{,}030{,}000$.

note

Note that any number having significant digits 503 has the same mantissa when written in scientific notation. The *only* change will be in the characteristic. Thus, log 5,030 = 3 + 0.7016 = 3.7016, but log 50,300 = 4 + 0.7016 = 4.7016.

3. log 0.0619

   Since $0.0619 = 6.19 \times 10^{-2}$, the characteristic is -2. The mantissa is given by log 6.19 = 0.7917. Therefore,

   $$\log 0.0619 = -2 + 0.7917$$

   (*Remember:* All mantissas are positive.) Since this is not a convenient way of writing the logarithm of a number, we express -2 by "8 − 10" and write

   $$\log 0.0619 = 8.7917 - 10$$

note

The form $-2 + 0.7917$ *cannot* be written -2.7917 since this makes the entire number negative, whereas only the 2 is actually negative; the 0.7917 is positive.

4. log 0.000582

   Since $0.000582 = 5.82 \times 10^{-4}$, the characteristic is -4. The mantissa is given by log 5.82 = 0.7649 so

   $$\log 0.000582 = -4 + \log 5.82 = -4 + 0.7649$$

   Since $-4 = 6 - 10$ we then write

   $$\log 0.000582 = 6.7649 - 10$$

note

In general, we express all negative characteristics as some positive number minus 10. If we then perform the indicated operations in the logarithms obtained in examples 3 and 4 we get

$$8.7917 - 10 = -1.2083$$
$$6.7649 - 10 = -3.2351$$

and the results are negative numbers. In computations requiring extensive use of the characteristic, this form of the given logarithm is often useful.

Linear interpolation is used to find the table value of the mantissa of the logarithm of a number having four (or more) significant digits. The procedure is based on an assumption that is not entirely correct: that the differences between the mantissas are proportional to the difference between the numbers with which they are associated. Nevertheless, the resulting number is an excellent approximation. Consider the following examples, which illustrate linear interpolation.

note

This method is used to interpolate on other tables.

Examples

Find the logarithm of the given number.

1. 52.36

The characteristic of 52.36 is 1 and the logarithm of 52.36 must lie between log 52.30 and log 52.40. Since 52.36 lies $\frac{6}{10}$ of the way from 52.30 to 52.40, we assume log 52.36 will lie $\frac{6}{10}$ of the way from log 52.30 to log 52.40. The following diagram helps us determine a proportion of the difference between log 52.30 and log 52.36.

$$10\left[\begin{array}{l} 6\left[\begin{array}{l}\log 52.30 = 1.7185 \\ \\ \log 52.36 = ?\end{array}\right]d \\ \\ \quad\ \log 52.40 = 1.7193\end{array}\right]8$$

note

The placement of the decimal point is unimportant in determining the differences we use.

Set up the proportion and solve for d:

$$\frac{6}{10} = \frac{d}{8}$$
$$10d = 48$$
$$d = 4.8$$

Since we require the mantissa to have only four significant digits, we round this off to $d = 5$. This is then *added* to the last digit of 1.7185 to obtain

$$\log 52.36 = 1.7190$$

2. 0.009782

The characteristic is -3 and log 0.009782 lies between log 0.009780 and log 0.009790. Set up the diagram:

log 0.009780 = 7.9903 − 10
log 0.009782 = ?
log 0.009790 = 7.9908 − 10

(Brackets: 2 and d span the first two lines; 10 and 5 span all three.)

The proportion $\frac{2}{10} = \frac{d}{5}$ yields $d = 1$. Now add 1 to the last digit of 7.9903 to obtain

$$\log 0.009782 = 7.9904 - 10$$

To find the common logarithm of a number on your calculator, use the following procedure:

1. Enter the number N.
2. Press the [log] key.
3. Read the log N on the display.

Examples

Using your calculator, find the given common logarithms.

1. log 35.6
   Press [3] [5] [•] [6] [log] and read **1.55145** on the display. Thus, $\log 35.6 \approx 1.5515$.
2. log 325,000
   Press [3] [2] [5] [0] [0] [0] [log] and read **5.5118834** on the display. Therefore, $\log 325{,}000 \approx 5.5119$.
3. log 0.00145
   Press [•] [0] [0] [1] [4] [5] [log] and read **−2.838632** on the display. Then $\log 0.00145 \approx -2.8386$.
4. log 0.0000378
   Press [•] [0] [0] [0] [0] [3] [7] [8] [log] and read **−4.4225082** on the display.

5. Given the formula $P = k \log T$, find P when $k = 3.12$ and $T = 1.36$.

Substitute:

$$\begin{aligned} P &= 3.12 \log 1.36 \\ &= 3.12(0.133539) \\ &= 0.4166414 \end{aligned}$$

Then $P \approx 0.417$.

note

To use the calculator, enter [1] [•] [3] [6] [log] [×] [3] [•] [1] [2] [=].

6. Given $B = K \log (W + H)$, find K when $B = 4.93$, $W = 15.8$, $H = 16.1$.

Substitute:

$$\begin{aligned} 4.93 &= K \log(15.8 + 16.1) \\ &= K \log(31.9) \\ &= K(1.5038) \end{aligned}$$

Then $K = \dfrac{4.93}{1.5038} = 3.2784$. Therefore, $K \approx 3.28$.

note

To use the calculator, enter [1] [5] [•] [8] [+] [1] [6] [•] [1] [log]. To find K, enter [4] [•] [9] [3] [÷] [1] [•] [5] [0] [3] [8] [=].

Exercise Set 14.3

In exercises 1 through 17 find the common logarithm of the number using table C. Interpolate where the number has four significant digits.

1. 23
2. 547
3. 7,360
•4. 80,200
5. 523,000
6. 79,400,000
7. 0.157
8. 0.00863
9. 0.0000941
•10. 0.000000107
11. 5,736
•12. 82,470
13. 179,400,000
14. 0.09163
15. 0.00007638
•16. 0.0008074
17. 0.0000002589

In exercises 18 through 42 find the common logarithm of the number using the calculator.

18. 5
19. 47
20. 620
21. 125,125
22. 6,730
23. 8,740
24. 24,300
25. 90,600
26. 143,000
27. 763,000
28. 4,060,000
29. 8,560,000
30. 59.3
31. 4.76
32. 104.0

33. 6.94
34. 89.1
35. 0.573
36. 0.763
37. 0.0831
38. 0.0123
39. 0.00240
40. 0.000992
41. 0.0000359
•42. 0.000000601

Determine the logarithm of the numbers involved in each of the problems in exercises 43 through 48.

43. An artificial earth satellite is orbiting above the earth's surface at a height of 405,000 mi.
44. The resistive force F exerted by water on a ship traveling at 1.5 statute mi/hr is 244,000 lb.
45. The power necessary to run a refrigerator compressor is 9,500,000 ft-lb/hr.
46. The angular velocity of the earth is 0.000073 rad/s and the earth's moment of inertia is 9.8×10^{37} kg/m^3.
47. The current rate of interest on money market certificates is 14.13%.
48. The mass density of air at 0°C is 0.001292 g/cm^3.
49. Given the formula $H = w \log T$, find (a) H when $w = 120$ and $T = 360$ and (b) w when $H = 16.4$ and $T = 32.1$.
50. Given $P_H = -\log A_H$, find P_H when $A_H = 0.00152$.
51. Given $\log R = \dfrac{D}{10}$, find D when $R = 1.43$.
52. The characteristic impedance Z_0 of an air-insulated parallel-conductor transmission line is given by $Z_0 = 276 \log\left(\dfrac{b}{a}\right)$, where b = the center-to-center distance between conductors, and a = the radius of the conductor (in the same units as b). See the accompanying figure. Find Z_0 if $b = 5.0$ cm when $a = 0.102$ cm.

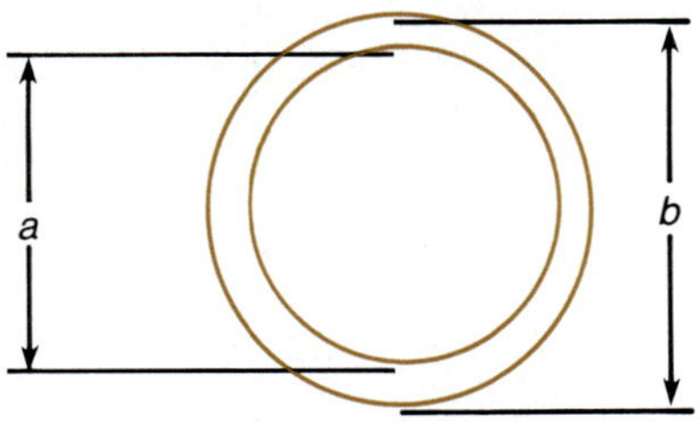

53. In comparing the "loudness" of two sounds, the number of decibels of sound, N_{dB}, is given by $N_{dB} = 10 \log\left(\dfrac{P_1}{P_2}\right)$, where P_1 and P_2 are the signal power levels to be compared. Find N_{dB} when $P_1 = 950$ and $P_2 = 25$.
•54. A decibel power gain G_{dB} of an electronic amplifier is given by $G_{dB} = 10 \log\left(\dfrac{P_o}{P_i}\right)$, where P_o = the power output and P_i = the power input. Find G_{dB} when $P_o = 20$ and $P_i = 0.005$.

14.5 Antilogarithms

Suppose we know the logarithm of a number N and wish to determine N. This procedure is called finding the **antilogarithm** (or **antilog**) of a number. To find N, we use table C working "backwards" through the table. The following examples illustrate.

Examples

Find the number N, given the following logs.

1. $\log N = 3.6395$

   The characteristic (the power of ten) is 3 and the mantissa is 0.6395. In table C the digits "6395" are under the column headed 6, the third digit of the number. Moving left to the column headed N we read "43," the first two digits of the number. Thus,

   $$\log 4.36 = 0.6395$$
   $$\text{and } N = 4.36 \times 10^3$$
   $$= 4360$$

2. $\log N = 7.8401 - 10$

   Now, $7.8401 - 10 = -3 + 0.8401$, so the characteristic (the power of ten) is -3, and the mantissa is 0.8401. In table C we find "8401" under the column headed 2, which is the third digit of the number. Reading left to the column headed N we read "69," which are the first two digits of the number. Thus,

   $$\log 6.92 = 0.8401$$
   $$\text{and } N = 6.92 \times 10^{-3}$$
   $$= 0.00692$$

note

Our number N is automatically stated in scientific notation.

3. $\log N = -5.5258$ (*Note:* .5258 is a negative number.)

   Add 10 to -5.5258 and subtract 10 from this result to get

   $$10 + (-5.5258) - 10 = 4.4742 - 10$$

   Thus,

   $$-5.5258 = 4.4742 - 10$$

   which we write $-6 + 0.4742$. The characteristic (the power of ten) is -6 and the mantissa is 0.4742. From table C we determine

   $$\log 2.98 = 0.4742$$

   Therefore,

   $$N = 2.98 \times 10^{-6}$$
   $$= 0.00000298$$

If the mantissa is not a number given in table C, we can interpolate to get the most accurate answer possible using the table, or use the nearest table value to get a more approximate answer. Which method we use depends on the degree of accuracy required.

The procedure for using linear interpolation to find the antilogarithm is the same as that used to find the logarithm of a number having four significant digits. Consider the following examples.

Examples

Find the number N whose logarithm is given.

1. Log $N = 0.2936$

   The characteristic is 0 and the mantissa lies between table mantissas 0.2923 and 0.2945. Set up the diagram:

$$10\left[\begin{array}{l} d\left[\begin{array}{ll}\log 1.96 & = 0.2923 \\ \\ \log N & = 0.2936\end{array}\right] 13 \\ \\ \qquad \log 1.97 \;\; = 0.2945\end{array}\right] 22$$

   The diagram yields the proportion

$$\frac{d}{10} = \frac{13}{22}$$

   Solving for d yields

$$22d = 130$$
$$d = 5.9$$

   which is rounded off to $d = 6$. Add the digit 6 to 1.96 to obtain $\log 1.966 = 0.2936$.

$$N = 1.966$$

2. Log $N = 4.7254$

   The characteristic is 4 and the mantissa 0.7254 lies between table mantissas 0.7251 and 0.7259. Set up the diagram:

$$10\left[\begin{array}{l} d\left[\begin{array}{ll}\log 5.31 \times 10^4 & = 4.7251 \\ \\ \log N & = 4.7254\end{array}\right] 3 \\ \\ \qquad \log 5.32 \times 10^4 \;\; = 4.7259\end{array}\right] 8$$

   The proportion is

$$\frac{d}{10} = \frac{3}{8}$$

   Solve for d:

$$8d = 30$$
$$d = 3.75$$

   which rounds off to $d = 4$. Add the digit 4 to 5.31 to obtain $\log (5.314 \times 10^4) = 4.7254$.

$$N = 5.314 \times 10^4 = 53{,}140$$

To find the antilog on the calculator we use either the [inv] [log] keys or the [y^x] key, depending on the keyboard. To use the [inv] [log] keys, use the following procedure:

1. Enter the logarithm.
2. Press the [inv] [log] keys.
3. Read the number on the display.

Examples

Find N, given the following.

1. $\log N = 4.3$
   Enter [4] [•] [3] [inv] [log] and read **19952.623** on the display. Then $N \approx 19952.62$ and the antilog of 4.3 is 19,952.62.
2. $\log N = 0.3143$
   Enter [•] [3] [1] [4] [3] [inv] [log] and read **2.0620538** on the display. Then $N \approx 2.06$ and the antilog of 0.3143 is 2.06.
3. $\log N = -3.9157$
   Enter [3] [•] [9] [1] [5] [7] [+/−] [inv] [log] and read **1.2142273 − 04** on the display. Then $N \approx 0.00012$ and the antilog of -3.9157 is 0.00012.

To use the [y^x] key we first write the logarithmic equation as an equivalent *exponential* equation.

Examples

Find N, given the following.

1. $\log N = 3.4$
   Since $\log N = 3.4$ is equivalent to $10^{3.4} = N$, then enter [1] [0] [y^x] [3] [•] [4] [=] and read **2511.8864** on the display. Thus, $N \approx 2511.89$ and the antilog of 3.4 is 2511.89.
2. $\log N = 0.9132$
   We want N such that $N = 10^{0.9132}$. Enter [1] [0] [y^x] [•] [9] [1] [3] [2] [=] and read **8.188417919** on the display. Therefore, $N \approx 8.19$ and the antilog of 0.9132 is 8.19.
3. $\log N = -1.37$
   We want N such that $N = 10^{-1.37}$. Enter [1] [0] [y^x] [1] [•] [3] [7] [+/−] [=] and read **0.042579518** on the display. Therefore, $N \approx 0.043$ and the antilog of -1.37 is 0.043.
4. $\log N = -4.5791$
   We want N such that $N = 10^{-4.5791}$. Enter [1] [0] [y^x] [4] [•] [5] [7] [9] [1] [+/−] [=] and read **.0000263572.** Then $N = 0.000026$ and the antilog of -4.5791 is 0.000026.

5. Given $D = 10 \log R$, find R when $D = 42.6$.
Substituting 42.6 for D, we obtain $42.6 = 10 \log R$ and then $\log R = \frac{42.6}{10} = 4.26$. Taking the antilog, we find $R = 18197.008 \approx 18{,}200$.

note

Using the calculator, enter [4] [•] [2] [6] [inv] [log] or [1] [0] [y^x] [4] [•] [2] [6] [=].

Exercise Set 14.4

Find the antilog N in each of the following using table C. If the mantissa is not on the table, interpolate.

1. 0.5416
2. 3.8445
3. 4.4997
•4. 7.9138
5. 8.7528 − 10
•6. 5.2175 − 10
7. 9.9987 − 10
8. 2.6243 − 10
9. −3.1445
10. −6.2013
11. −8.7620
•12. −5.6402
13. 8.8340
14. 2.5283
15. 3.9896 − 10
16. 7.7699 − 10
17. −4.2391
18. −7.4727

Find the antilog N in each of the following using your calculator.

19. $\log N = 2.7$
•20. $\log N = 4.6$
21. $\log N = 3.7050$
22. $\log N = 4.5145$
23. $\log N = 7.739$
24. $\log N = 12.93$
25. $\log N = 0.76$
26. $\log N = 0.1234$
27. $\log N = 0.9217$
28. $\log N = -3.0$
29. $\log N = -1.34$
30. $\log N = -2.263$
31. $\log N = -7.8134$
•32. $\log N = -4.9219$
33. $\log N = -5.2655$
•34. Given $P_H = -\log A_H$, find A_H when $P_H = 7.78$.
35. Given $H = w \log T$, find T when $w = 1.93$ and $H = 2.76$.
36. Given $\log R = \frac{D}{10}$, find R when $D = 43.4$.
•37. Given $G_{\text{dB}} = 10 \log \left(\frac{P_o}{P_i}\right)$, find P_o when $G_{\text{dB}} = 15$ and $P_i = 0.50$.
38. Given $Z_0 = 276 \log \left(\frac{b}{a}\right)$, find (a) b when $Z_0 = 492$ ohms and $a = 0.111$ cm and (b) a when $Z_0 = 542.6$ ohms and $b = 5.5$ cm.

14.6 Natural Logarithms

Since the base of our number system is 10 and we can write numbers in scientific notation using powers of ten, base 10 logarithms are convenient. Actually, however, almost any positive number can be used as the base of logarithms. Whatever base is used, the definition of the logarithm of a number remains the same: It is the exponent to which the base must be raised to obtain the number.

Another number base is often used in mathematics. This number is an irrational number that is approximately equal to 2.7182818284 . . . , denoted by e. That is,

$$e \approx 2.718$$

We call logarithms to **base e** the **natural logarithms** and denote the natural logarithm of a number by *ln*. Thus, the natural logarithm of a number N is given by $\ln N$. That is, $\log_e N$ is written by $\ln N$.

note

As with the previous logarithms, $\ln N = x$ is equivalent to $N = e^x$.

The properties of logarithms that apply to common logarithms also apply to the natural logarithms. Thus,

1. $\ln(MN) = \ln M + \ln N$
2. $\ln\left(\dfrac{M}{N}\right) = \ln M - \ln N$
3. $\ln M^n = n \ln M$

To find the natural logarithm of a number on the calculator we use the [ln x] key in the same way that we used the [log] key to find common logarithms.

Examples

1. Find ln 523.
   Enter [5] [2] [3] [ln x] and read **6.2595815** on the display. Then, $\ln 523 \approx 6.26$.

   note

   Remember, since $\ln N = x$ is equivalent to $N = e^x$, we have just determined that $e^{6.26} \approx 523$.

2. Find ln 0.0235.
   Enter [•] [0] [2] [3] [5] [ln x] and read **−3.7507549** on the display. Therefore, $\ln 0.0235 \approx -3.75$.

   note

   We have determined that $e^{-3.75} \approx 0.0235$.

3. Given the equation of decay $\ln\left(\dfrac{A_o}{A}\right) = kt$, where A_o is the original amount and A is the amount left after t years, find k when $A_o = 45$ g, $A = 15$ g, and $t = 1.75$ yr. Substituting, we have $\ln\left(\dfrac{45}{15}\right) = k(1.75)$. Then $\ln 3 = k(1.75)$, and

$$k = \frac{\ln 3}{1.75} = \frac{1.0986123}{1.75} = 0.63$$

note

Using your calculator, enter [3] [ln x] [÷] [1] [•] [7] [5] [=].

4. Given $v = c \ln\left(\frac{M}{m}\right)$, find v if $c = 3.24$, $M = 934$, and $m = 87$.

Substituting

$$v = 3.24 \ln\left(\frac{934}{87}\right) = 3.24 \ln(10.735632)$$

$$= 3.24(2.3735683) = 7.6903613.$$

So, $v \approx 7.69$.

note

On the calculator enter [9] [3] [4] [÷] [8] [7] [=] [ln x] [×] [3] [•] [2] [4] [=].

Suppose we know the natural logarithm of a number N and wish to find N. Again, since $\ln N = x$ is equivalent to $N = e^x$, we use the [e^x] key. If no [e^x] key is provided, use [inv] [ln x] after entering the logarithm.

Examples

1. Find N if $\ln N = 0.35$.
   Now $N = e^{0.35}$, so we enter [•] [3] [5] [e^x] and read **1.4190675** on the display. Then $N \approx 1.42$ and $\ln 1.42 \approx 0.35$.
2. Find N if $\ln N = -1.53$.
   $N = e^{-1.53}$ and we enter [1] [•] [5] [3] [+/−] [e^x] and read **2.1653567 −01** on the display. Therefore, $N \approx 0.217$ and $\ln 0.217 \approx -1.53$.
3. Using the decay equation $\ln\left(\frac{A_o}{A}\right) = kt$, find A_o if $A = 20.0$, $k = 0.0536$, and $t = 24.3$.

   Substituting, we have $\ln\left(\frac{A_o}{20.0}\right) = (0.0536)(24.3)$.

   Then

$$\frac{A_o}{20.0} = e^{1.30248}$$

$$A_o = 20.0\, e^{1.30248}$$

$$= 20.0(3.6784078)$$

$$= 73.568156$$

   Therefore, $A_o \approx 73.6$.

note

To use your calculator, enter [1] [•] [3] [0] [2] [4] [8] [e^x] [×] [2] [0] [•] [0] [=].

4. Given the equation $\ln(c + t) = -k$, find t when $c = 0.934$ and $k = 0.210$.
Substituting, we get $\ln(0.934 + t) = -0.210$.
Then

$$\begin{aligned} 0.934 + t &= e^{-0.210} \\ t &= e^{-0.210} - 0.934 \\ &= 0.81058 - 0.934 \\ &= -0.1234 \end{aligned}$$

So, $t \approx -0.123$.

note
Using a calculator, enter [.] [2] [1] [0] [+/−] [e^x] [−] [.] [9] [3] [4] [=].

5. Given the equation $L = e^{-t}$, find t when $L = 0.92$. Since $0.92 = e^{-t}$ is equivalent to $\ln 0.92 = -t$, then $-0.8338 = -t$. So, $t \approx 0.83$.

Exercise Set 14.5

In exercises 1 through 17, find the natural logarithm of the given number using the calculator. (Round all problems to three significant digits.)

1. ln 36
2. ln 456
3. ln 75.2
4. ln 84.9
5. ln 1.03
6. ln 4.19
7. ln 5230
8. ln 7190
9. ln 24,500
•10. ln 16,400
11. ln 0.149
12. ln 0.991
13. ln 0.0786
14. ln 0.0574
15. ln 0.000793
16. ln 0.0000455
17. ln 0.000813

Find N in the following problems.

18. $\ln N = 1.4$
19. $\ln N = 0.748$
•20. $\ln N = 5.6372$
21. $\ln N = 14.3$
22. $\ln N = -0.0173$
23. $\ln N = -2.93$
24. $\ln N = -8.0731$
25. $\ln N = -15.1$

•26. Given the formula of work w defined by $w = \dfrac{d}{\ln P}$, find
   a. w when $d = 1.2$ and $P = 11.3$
   b. d when $w = 0.93$ and $P = 3.12$
   c. P when $w = 0.81$ and $d = 5.32$

27. Given the formula for decay, $\ln\left(\dfrac{A}{A_o}\right) = -kt$, find
   a. k when $A_o = 24.2$, $A = 9.61$, $t = 51.3$
   b. A when $A_o = 51.3$, $k = 0.871$, $t = 30.1$

•28. Given the formula $P = e^h$, find h when
   a. $P = 31.2$
   b. $P = 0.473$

29. Given $P = 14.7\, e^{-0.3h}$, find h when
    a. $P = 3.76$
    b. $P = 23.9$
30. Given $A = ke^{-ct}$, find c if $A = 0.391$, $k = 1.69$, and $t = 115$.
31. The law of exponential growth is given by $I = I_o e^{kt}$. If $k = 0.3$, using exponential growth, what will the population of the United States be in the year 2000 if the population was 208 million in 1970?

note

I_o represents the original population, 208 million; I is the population we want to calculate, t is the time taken for the growth (the time from 1970 to the year 2000).

•32. Ten grams of uranium will decay to an amount of 2.5 g in 460,000 yr. What is the half-life of uranium? (*Hint:* Use $I = I_o e^{-kt}$. Half-life means the amount of time it takes for the substance to decay to one-half its original amount. Find constant k first and then use $I = 5$ and $I_o = 10$ to find t.)
33. If a current is given by the expression $i = \left(\frac{E}{R}\right)e^{-t/RC}$, find i when $E = 20$ volts, $R = 2$ ohms, $C = 2$ faradays, and $t = 3$ s.
34. An equation giving the amount A of radium that remains after t years of A_o mg of radium decaying through radioactivity is

$$t = \frac{\ln A_o - \ln A}{4.10 \times 10^{-4}}$$

Determine the number of years it will take for 800 mg of radium to decay so that 500 mg remain.
35. In problem 34, if it takes 7.32×10^2 years for an amount A_o to decay to 300 mg, how much radium was there originally?
36. In finding the insulation resistance by the leakage method, resistance R is found by the formula

$$R = 10^6 \cdot \frac{t}{C} \cdot \frac{1}{\ln \frac{V_o}{V}}$$

where t is the time in seconds, C is the capacity, and V is the voltage. Find R if $t = 100$ s, $V_o = 120$, $V = 112.5$, and $C = 0.090$.
37. The work W done by a volume of gas expanding at a constant temperature from volume V_0 to volume V_1 is given by $W = p_0V_0 \ln\frac{V_1}{V_0}$, where p_0 is the pressure when the volume is V_0. Find W when $p_0 = 20$ lb/in^2, $V_0 = 15$ in^3, and $V_1 = 24$ in^3.
38. A theory in learning states that the number of nonsense symbols a person can memorize is given by $S = 16(1 - e^{-t})$, where t is the time in minutes the person studies the symbols and S is the number memorized. Using the formula, how many can a person memorize in (a) 4 min? (b) 2 min? (c) 1 min? (d) 30 s?

39. When P dollars is invested at a rate of interest i per year compounded continuously, the number of dollars in the amount A after t years is given by $A = Pe^{it}$. Find:
   a. A when $P = \$5{,}000$, $i = 12\%$, $t = 1$ yr
   b. A when $P = \$500$, $i = 8\%$, $t = 3$ yr
   c. A when $P = \$2{,}000$, $i = 7\frac{1}{4}\%$, $t = 6$ months
   d. P if $A = \$450$, $i = 6\%$, $t = 1$ yr
   e. P if $A = \$5{,}120$, $i = 5\frac{1}{2}\%$, $t = 2$ yr
40. Using the formula for interest compounded continuously, $A = Pe^{it}$, find:
   a. t if $A = \$550$, $P = \$500$, $i = 5\%$
   b. i if $A = \$3{,}200$, $P = \$2{,}800$, $t = 2$ yr

14.7 Other Logarithmic Bases

Now consider the logarithms of numbers to other bases. Suppose we let

$$y = \log_b N$$

then

$$b^y = N$$

Taking the common logarithm of both sides

$$\log b^y = \log N$$

and using a property of logarithms,

$$y \log b = \log N$$

Solving for y we get

$$y = \frac{\log N}{\log b}, \qquad b > 0 \text{ and } N > 0$$

But, $y = \log_b N$, so we substitute to get

$$\boxed{\log_b N = \frac{\log N}{\log b}}$$

We now have a means for finding the logarithm of *any* number to *any* base using common logarithms.

Examples

1. Find $\log_3 8$.
   Since $b = 3$ and $N = 8$,
   $$\log_3 8 = \frac{\log 8}{\log 3} = \frac{0.9031}{0.4771} = 1.89289$$
   Then, $\log_3 8 \approx 1.89$.

   note

   Using the [y^x] key we can check our result since $3^{1.89289} \approx 8$. Press [3] [y^x] [1] [•] [8] [9] [2] [8] [9] and read . . . **8.0008854**

2. Find $\log_{15}7$.

Since $b = 15$ and $N = 7$,

$$\log_{15}7 = \frac{\log 7}{\log 15} = \frac{0.8451}{1.1761} = 0.71856$$

Therefore, $\log_{15}7 \approx 0.719$.

3. Find $\log_5 0.912$.

Since $b = 5$ and $N = 0.912$,

$$\log_5 0.912 = \frac{\log 0.912}{\log 5} = \frac{-0.040005}{0.69897} = -0.057$$

Therefore, $\log_5 0.912 \approx -0.057$.

Exercise Set 14.6

Find the value of the following logarithms.

1. $\log_3 5$	•2. $\log_4 7$	3. $\log_2 15$
4. $\log_8 12$	5. $\log_9 3$	6. $\log_{11} 8$
7. $\log_{16} 6$	8. $\log_{15} 1.3$	9. $\log_{13} 52.3$
10. $\log_{0.5} 5$	11. $\log_{\frac{1}{4}} 24$	12. $\log_{0.8} 0.12$
13. $\log_{4.2} 6.73$	•14. $\log_{9.3} 25.1$	15. $\log_{25.6} 124$

14.8 Solving Exponential Equations

We use the common logarithms or natural logarithms together with general properties of logarithms to solve an exponential equation where the unknown quantity is the exponent. The equations

$$2^x = 5, \quad 3^x = 20, \quad \left(\frac{1}{2}\right)^x = 5$$

are exponential equations where we must solve for the exponent.

Consider the equation $2^x = 5$. We take the common logarithm of both sides to get

$$\log 2^x = \log 5$$

Using the property of logarithms, $\log_a N^n = n \log_a N$, we have x

$$x \log 2 = \log 5$$

Solving for x we get

$$x = \frac{\log 5}{\log 2} = \frac{0.69897}{0.30103} = 2.321928$$

Therefore, $x \approx 2.32$.

We have determined that $2^{2.32} \approx 5$.

note

We could have used the natural logarithms in the same way to get

$$\ln 2^x = \ln 5$$
$$x \ln 2 = \ln 5$$
$$x = \frac{\ln 5}{\ln 2} \approx 2.32$$

Examples

Solve for x in the following exponential equations.

1. $3^x = 20$

$$\log 3^x = \log 20$$
$$x \log 3 = \log 20$$
$$x = \frac{\log 20}{\log 3} = \frac{1.30103}{0.47712} = 2.7268$$

Then, $x \approx 2.73$ and $3^{2.73} \approx 20$.

2. $\left(\frac{1}{2}\right)^x = 5$

$$\ln\left(\frac{1}{2}\right)^x = \ln 5$$
$$x \ln 0.5 = \ln 5$$
$$x = \frac{\ln 5}{\ln 0.5} = \frac{1.60943}{-0.69315} = -2.3219$$

Then, $x \approx -2.32$ and $\left(\frac{1}{2}\right)^{-2.32} \approx 5$.

Exercise Set 14.7

Solve for x.

1. $4^x = 3$
•2. $5^x = 9$
3. $3^x = 15$
4. $7^x = 5$
5. $8^x = 15$
6. $12^x = 31$
7. $\left(\frac{1}{2}\right)^x = 7$
8. $(0.6)^x = 0.14$
9. $(3.5)^x = 14.3$
10. $(8.7)^x = 20.4$
11. $\left(9\frac{2}{5}\right)^x = 6$
•12. $\left(15\frac{1}{4}\right)^x = 21$

Chapter Summary

1. A **logarithm** is an exponent.
2. The logarithmic equation $\log_b N = x$ is equivalent to the exponential equation $b^x = N$.
3. The properties of logarithms are:

$$\log_b(MN) = \log_b M + \log_b N$$
$$\log_b\left(\frac{M}{N}\right) = \log_b M - \log_b N$$
$$\log_b M^n = n \log_b M$$

4. A **common logarithm** is a logarithm of base 10.
5. To find the common logarithm of a number N, denoted by log N, on the calculator, use the [log] key.
6. The number N whose logarithm is given is the **antilogarithm** (called *antilog*) of that logarithm.
7. To find the antilog of a common logarithm on the calculator, use the [y^x], or [inv] [log] keys.
8. **Natural logarithms** are logarithms of base e ($e \approx 2.718$).
9. The natural logarithm of a number is denoted by ln N.
10. To find the natural logarithm of a number on the calculator use the [ln x] key.
11. To find the antilog of a natural logarithm on the calculator use the [e^x] key or use [inv] [ln x] keys if your calculator does not have the [e^x] key.
12. Given $b > 0$ and $N > 0$, $\log_b N = \dfrac{\log N}{\log b}$.

Chapter Review

[14.2] Express the following logarithmic equations as equivalent exponential equations.

1. $\log_3 9 = 2$
2. $\log_6 216 = 3$
3. $\log_{25} 5 = \frac{1}{2}$
4. $\log_p q = s$

Express the following exponential equations as equivalent logarithmic equations.

5. $5^3 = 125$
6. $(49)^{\frac{1}{2}} = 7$
7. $\left(\frac{1}{3}\right)^3 = \frac{1}{27}$
8. $p^q = s$

Solve the following equations for x.

9. $\log_x 81 = 4$
10. $\log_x \frac{1}{16} = -4$
11. $\log_6 x = 4$
12. $\log_{\frac{1}{2}} x = -3$
13. $\log_4 64 = x$
14. $\log_{\frac{1}{3}} \frac{1}{81} = x$

[14.3] Using properties of logarithms, write the following as the sum or difference of two logarithms or the product of a real number times a logarithm.

15. $\log_5 6$
16. $\log_4 21$
17. $\log_b\left(\frac{5}{7}\right)$
18. $\log_e(18)$
19. $\log_{10}(16)$
20. $\log_6\left(\frac{14}{25}\right)$
21. $\log_b\left(\frac{m^2}{n^3}\right)$
22. $\log_b \sqrt[4]{10}$
23. $\log_6 \sqrt[3]{x^5}$

Write the following logarithmic expressions as the logarithm of a single number.

24. $\log_3 7 + \log_3 13$

25. $\log_b 16 - \log_b 9$

26. $4 \log_7 3$

27. $2 \log_5 7 - 3 \log_5 2$

28. $\frac{1}{3}\log_x 8 + \frac{1}{2}\log_x 16$

29. $3 \log_2 5 + 2 \log_2 9 - \frac{1}{4}\log_2 16$

Given log 3 = 1.0986 and log 5 = 1.6094, compute the following.

30. $\log_b 15$

31. $\log_b\left(\frac{3}{5}\right)$

32. $\log_b 45$

33. $\log_b\sqrt{5}$

34. $\log_b \sqrt[3]{\frac{5}{3}}$

35. $\log_b 225$

[14.4] Using your calculator, find the following common logarithms.

36. log 5.73

37. log 92.6

38. log 357.2

39. log 8,720

40. log 0.713

41. log 0.00158

[14.5] Using your calculator, find N.

42. $\log N = 0.081$

43. $\log N = 0.993$

44. $\log N = 2.34$

45. $\log N = 30.9$

46. Given the formula $k \log Q = A - P$, find k when $Q = 34.9$, $A = 3.76$, and $P = 1.12$.

47. Using the formula of problem 46, find Q when $k = 0.516$, $A = 34.2$, and $P = 15.9$.

48. Given $\log G = \frac{H - T}{H}$, find

a. T when $H = 400$ and $G = 1.80$

b. G when $H = 320$ and $T = 250$

[14.6] Find the following natural logarithms using your calculator.

49. ln 271

50. ln 8.73

51. ln 0.315

52. ln 0.00925

53. ln 10,600

54. ln 521,000

Find N in the following problems using your calculator.

55. $\ln N = 0.819$

56. $\ln N = 5.73$

57. $\ln N = 24.9$

58. $\ln N = 67.3$

59. $\ln N = 0.023$

60. $\ln N = 0.000813$

61. Given: $A = A_o e^{it}$, find

a. A when $A_o = \$3,100$, $i = 0.02$, and $t = 5$

b. A_o when $A = \$60,000$, $i = 0.015$, and $t = 12$

c. t when $A = \$2,000$, $A_o = \$1,500$, $i = 0.01$

62. Given $w = \frac{d}{\ln p}$, find

a. w when $d = 14.2$ and $p = 2.14$

b. p when $w = 20.7$ and $d = 63.4$

63. Given the formula $R = c \ln\left(\frac{V}{T}\right)$, find:

a. R when $c = 0.437$, $V = 63.4$, and $T = 9.3$
b. V when $R = 620$, $c = 83.2$, and $T = 15.3$

[14.7] Evaluate the following logarithms using $\log_b N = \frac{\log N}{\log b}$.

64. $\log_2 15$ 65. $\log_{18} 4$ 66. $\log_{\frac{1}{4}} 3$
67. $\log_{0.9} 32$ 68. $\log_{1.8} 1.5$ 69. $\log_{21.5} 300$

[14.8] Solve for x in the following exponential equations.

70. $9^x = 15$ 71. $2^x = 1.83$ 72. $15^x = 23$
73. $\left(\frac{3}{5}\right)^x = 8$ 74. $(0.8)^x = 0.943$ 75. $(6.6)^x = 14.91$

Chapter Test

1. Logarithms to base 10 are called ____________ and logarithms to base e are called ____________.
2. The logarithmic equation $\log_5 625 = 4$ is equivalent to what exponential equation?
3. If $\log_4 \frac{1}{16} = x$, then $x =$ ____________.
4. If $\log N = 2.375$, $N =$ ____________.
5. The $\log_b b =$ ____________.
6. For any b, $\log_b 1 =$ ____________.
7. Write the logarithmic expression $3 \log_b 2 + \frac{1}{2} \log_b 4 - \log_b 5$ as the logarithm to the base b of a single number.
8. Given $\log_b 7 = 1.7713$, $\log_b 2 = 0.6309$, and $\log_b 3 = 1$, find

a. $\log_b 14$ b. $\log_b 12$ c. $\log_b 98$ d. $\log_b\left(\frac{3}{2}\right)$

Use your calculator to solve the following problems.

9. Find $\log_{14} 21$.
10. Solve the equation $8^x = 25.6$ for x.
11. Find A given $A = ke^{-ct}$ if $k = 0.17$, $c = 0.8$, and $t = 16$.
12. If $\ln A = \frac{h}{t}$, find t when $A = 55.6$ and $h = 143.1$.
13. If $P = 20 \log H$, find H when $P = 35$.
14. Given $\ln T = -kp$, find T if $k = 0.481$ and $p = 3.14$.
15. If $D = 10 \log\left(\frac{P_2}{P_1}\right)$, find D if $P_2 = 4.73$ and $P_1 = 2.09$.

15
Graphs

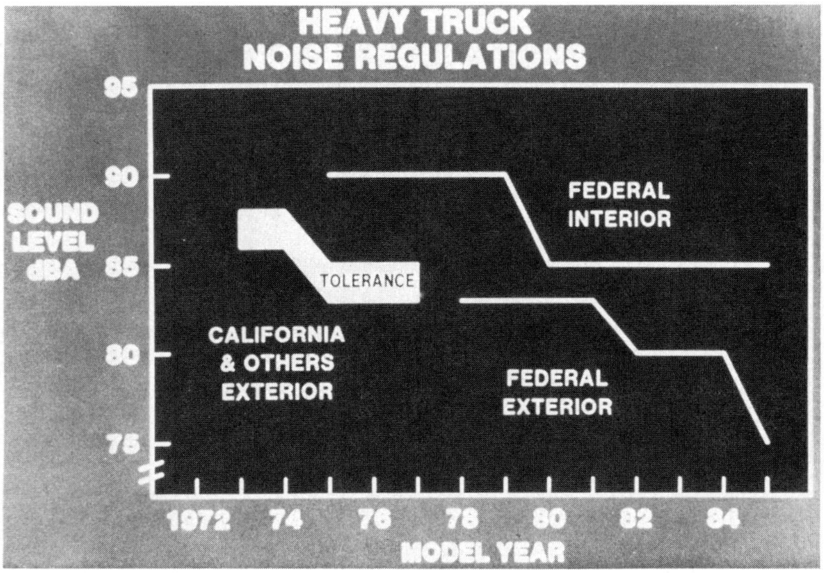

Representation of data showing the federal noise regulations for heavy trucks.
(Photograph courtesy of Ford Motor Company)

15.1 The Rectangular Coordinate System

In chapter 1 we saw how numbers can be represented on the number line. We used this technique to represent answers when we solved equations in one variable. In this chapter we examine equations in two variables whose solutions are ordered pairs of numbers in the form (x,y).

To be able to visualize our answers, we need two number lines. These two number lines are represented in figure 15.1. The horizontal number line is called the ***x*-axis,** and the vertical number line is called the ***y*-axis.** Together, the x and y axes form the **rectangular coordinate system.** The point at which the two axes intersect, (0,0), is called the **origin.** The two axes divide the coordinate system into four regions, called **quadrants,** which are numbered in a counterclockwise fashion.

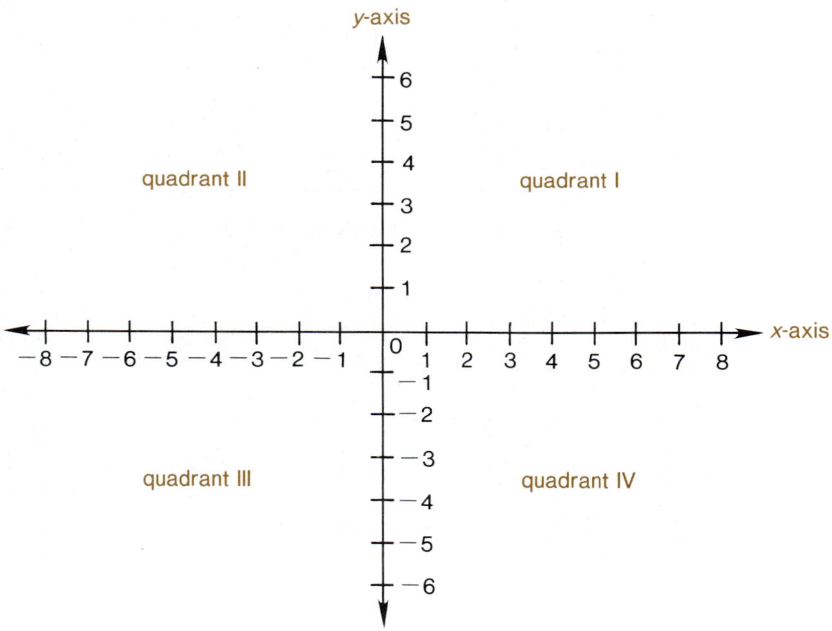

Figure 15.1

When we label the values along the x-axis, numbers to the right of the origin are positive and those to the left are negative. On the y-axis, positive numbers are above the origin and negative numbers are below.

We can graph any ordered pair, (x,y), as a point on the coordinate system if we consider the ordered pair as two instructions that will bring us to the proper location of the point. To graph the point (4,3), for example, we start at the origin and carry out the first instruction. Since the x-value, called the **abscissa,** is 4, we move four units to the right along the x-axis. The y-value, called the **ordinate,** now instructs us to move up from this position three units—and we have plotted the point (4,3) (see figure 15.2).

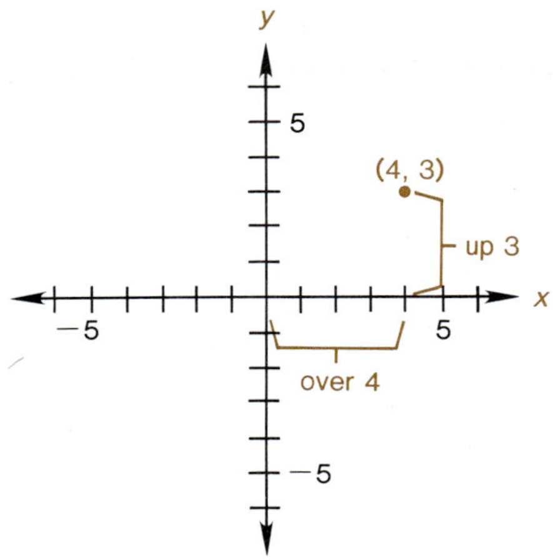

Figure 15.2

Examples Plot the following points. See figure 15.3.

1. (2,4)
2. (−1,−3)
3. (−4,3)
4. (5,0)
5. (2,−3)
6. (0,4)
7. (−3,0)

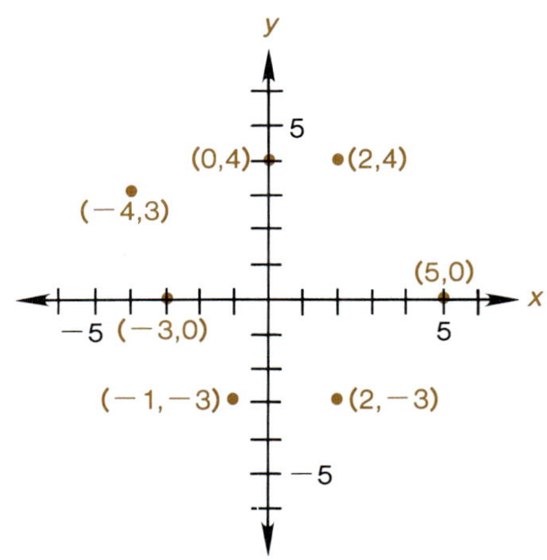

Figure 15.3

note

Whenever either one of the components of the ordered pair is zero, the point is located on one of the axes.

Consider the equation $y = x + 1$. This is an equation in two variables, and its solutions will be infinitely many ordered pairs of numbers, (x,y), that satisfy the equation. If we let $x = 1$, then $y = (1) + 1 = 2$, and we have the point (1,2) as a solution to the equation $y = x + 1$.

Examples Find the missing component in each ordered pair.

1. $y = 3x - 6$ a. (1,), b. (−2,), c. (0,), d. (2,)

a. $y = 3x - 6$
$y = 3(1) - 6$
$= 3 - 6$
$= -3$
Hence $(1, -3)$

b. $y = 3(-2) - 6$
$= -6 - 6$
$= -12$
$(-2, -12)$

c. $y = 3(0) - 6$
$= 0 - 6$
$= -6$
$(0, -6)$

d. $y = 3(2) - 6$
$= 6 - 6$
$= 0$
$(2,0)$

2. $y = x^2 - 1$ a. (3,), b. (−1,), c. (1,), d. (0,)

a. $y = (3)^2 - 1$
$= 9 - 1$
$= 8$
$(3,8)$

b. $y = (-1)^2 - 1$
$= 1 - 1$
$= 0$
$(-1,0)$

c. $y = (1)^2 - 1$
$= 1 - 1$
$= 0$
$(1,0)$

d. $y = (0)^2 - 1$
$= 0 - 1$
$= -1$
$(0, -1)$

In these examples, we observe that there is a definite relationship between the values of x and y. In the equations $y = 3x - 6$ and $y = x^2 - 1$, the value of y depends on the chosen value of x, and for this reason, we call x the **independent variable** and y the **dependent variable.** We arbitrarily pick a value for the independent variable and generate the corresponding value of the dependent variable.

We also observe that for each value of the independent variable, x, there corresponds one and only one value of the dependent variable, y. This is a very important mathematical relationship that we will consider later in this chapter.

Exercise Set 15.1

Plot the following ordered pairs on a coordinate system.

1. $(2,4)$ •2. $(5,1)$ 3. $(-1,3)$
4. $(-4,4)$ 5. $(-6,-1)$ •6. $(-2,-3)$
7. $(0,4)$ 8. $(0,2)$ 9. $(5,0)$
10. $(-4,0)$ 11. $(-7,0)$ 12. $(0,0)$
13. $\left(\frac{1}{2},3\right)$ •14. $\left(\frac{2}{3},-2\right)$ 15. $\left(\frac{3}{2},0\right)$

Use the coordinate system given to determine to the nearest integer the coordinates (x,y) of the given points.

16. A 17. B
18. C 19. D
20. E 21. F
22. G 23. H
24. I 25. J

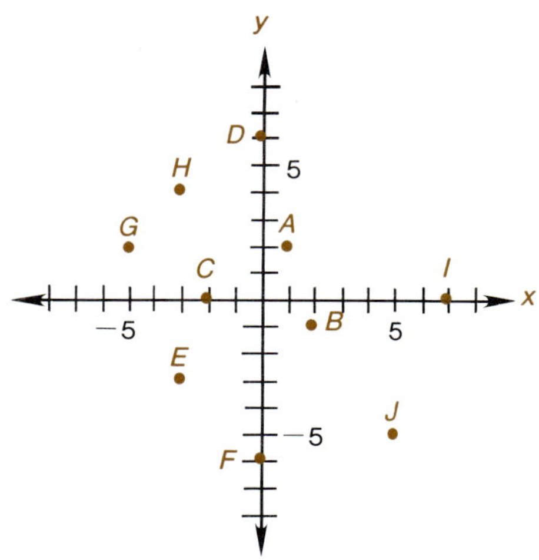

Find the missing component in each ordered pair using the given equation. Then plot the ordered pairs using a separate coordinate system for each problem.

26. $y = x + 4$ $(0,\ ), (-4,\ ), \left(\frac{5}{2},\ \right), (2,\ )$

27. $y = x - 3$ $(0,\ ), (3,\ ), (1,\ ), \left(-\frac{1}{2},\ \right)$

•28. $y = 2x + 1$ $(0,\ ), \left(-\frac{1}{2},\ \right), (2,\ ), (-1,\ )$

29. $y = 3x - 4$ $(0,\ ), (-1,\ ), \left(-\frac{1}{3},\ \right), (2,\ )$

30. $y = 2x$ $(0\ ), \left(\frac{1}{2},\ \right), (2,\ ), (-1,\ )$

31. $y = 3x$ $(0,\ ), (3,\ ), (-3,\ ), \left(\frac{1}{3},\ \right)$

32. $y = -x + 1$ $(0,\ ), (1,\ ), (-2,\ ), (-4,\ )$

33. $y = -x + 3$ $(0,\ ), (-3,\ ), (3,\ ), \left(\frac{5}{2},\ \right)$

34. $y = -2x + 3$ $(0,\ ), (3,\ ), (-2,\ ), \left(\frac{3}{2},\ \right)$

35. $y = x^2 + 1$ $(0,\ ), (1,\ ), (-1,\ ), \left(\frac{1}{2},\ \right)$

36. $y = x^2 - 4$ $(0,\ ), (2,\ ), (-3,\ ), (-2,\ )$

37. $y = x^2 - 9$ $(0,\ ), (-3,\ ), (2,\ ), (3,\ )$

38. $y = x^2 + 3$ $(0,\ ), (-2,\ ), (2,\ ), (-1,\ )$

State the quadrant in which each point lies.

Example $(-1,3)$ lies in the second quadrant because the x component is negative and the y component is positive.

39. $(-2,-5)$
40. $(4,-1)$
41. $(5,3)$
42. $(-7,-9)$
43. $\left(\frac{1}{2},-4\right)$
44. $\left(-\frac{2}{3},7\right)$
45. $\left(-\frac{5}{2},-\frac{3}{4}\right)$
46. $\left(\frac{7}{8},-\frac{7}{8}\right)$
47. $(5,-14)$
48. $(-3,-3)$

•49. In what quadrant does a point (x,y) lie in if:
 a. $x > 0$ and $y < 0$
 b. $x < 0$ and $y < 0$
 c. $x < 0$ and $y > 0$
 d. $x > 0$ and $y > 0$

•50. What is the value of x for any point on the y-axis?

51. What is the value of y for any point on the x-axis?

15.2 Functions

In many technical areas, mathematics is used to build a formula by which the technician is able to explain or predict what is going to happen in a particular situation. The given situation is represented by the independent variable and the response that we are determining is represented by the dependent variable.

In section 15.1 we observed that there was a definite relationship between the values of x and y in our equations. We also noted from the examples that for each x value there corresponded one and only one y value. This observation leads us to the following important mathematical concept: A **function** *is a rule or correspondence where each value of the independent variable gives exactly one value of the dependent variable.* In terms of the ordered pairs, (x,y), this means that each value of x is paired with a unique (only one) value of y. This concept is important because when we have an equation, and are determining the outcome to a given situation, we want the equation to be able to tell us the one exact response to expect. Consider the following example.

Example

Automotive engineers have developed a formula that, given the diameter of the bore of a 4-cylinder engine, we can determine the horsepower that the engine will produce at 1,000 rpm. This equation is given as $H = 1.6B^2$, where horsepower (H) depends on the cylinder bore (B). We say that horsepower is a function of the bore diameter. The following is a table of some of the values of the function.

Independent Variable Cylinder Bore (in inches)	Dependent Variable Horsepower	Ordered Pairs
2	$H = 1.6(2)^2 = 1.6(4) = 6.4$	(2, 6.4)
3	$H = 1.6(3)^2 = 1.6(9) = 14.4$	(3, 14.4)
3.5	$H = 1.6(3.5)^2 = 1.6(12.25) = 19.6$	(3.5, 19.6)
4	$H = 1.6(4)^2 = 1.6(16) = 25.6$	(4, 25.6)
4.5	$H = 1.6(4.5)^2 = 1.6(20.25) = 32.4$	(4.5, 32.4)

We see from our example that once a value of B (bore diameter) is chosen, the value of H (horsepower) is dependent on that choice and from our formula, only one value for horsepower is possible for a given bore diameter.

The type of relationship described in the example is called a function, and we say that the dependent variable is a function of the independent variable. The following are some other examples of a function.

Examples

1. $V = e^3$ The volume of a cube is a function of the length of its edge.
2. $S = 4\pi r^2$ The surface area of a sphere is a function of its radius.
3. $F = \frac{9}{5}C + 32$ The temperature in degrees Fahrenheit is a function of the temperature in degrees Celsius.
4. $P = 4s$ The perimeter of a square is a function of the length of its side.

Two words that are associated with the idea of a function are *domain* and *range*. *The* **domain** *of the function is the collection of all the numbers that can be used as values for the independent variable. The* **range** *of a function is the collection of all possible values that the dependent variable can take on.*

From our previous examples we can see that the domain of a function may be limited to only certain numbers. In our horsepower problem, for example, we cannot have a negative bore diameter. In examples on volume, surface area, and perimeter, we cannot choose negative values for the edge of a cube, the radius of a circle, or the side of a square. In our temperature example, we cannot choose values less than -273.16°C because at that value all molecular motion stops and values of temperature less than this have no physical meaning.

If we are dealing with a formula where it is not specified what the equation represents, the domain of the function will be all real numbers for which the function is defined.

Examples

Determine the domain of x.

1. $y = 2x + 1$	All real numbers
2. $y = \frac{1}{x}$	All real numbers except 0, $x \neq 0$, because at 0 the function is not defined
3. $y = \frac{1}{x + 2}$	All real numbers except -2, $x \neq -2$, because at -2 the function is not defined
4. $y = \sqrt{x}$	All real numbers greater than or equal to 0, $x \geq 0$, because numbers less than zero produce a negative number under the square root symbol
5. $y = \sqrt{x - 5}$	All real numbers greater than or equal to 5, $x \geq 5$, because numbers less than 5 produce a negative number under the square root symbol

---

We generally let f, g, h, or some other letter denote a function. This general notation allows us to express the dependent variable y as a function of the independent variable x. Our functional notation is $y = f(x)$, which is read "y is equal to f of x" and indicates that y is a function of x.

note

$f(x)$ is read "f of x" and does not mean f times x. Using a letter such as f indicates that a functional relationship exists between the variables.

A useful aspect of $f(x)$ is that it gives us a notation that will indicate values of the independent variable at which we wish to evaluate the function. The value of the function, $f(x)$, when we wish to evaluate it for $x = 3$, for example, is expressed as $f(3)$.

Examples

If $f(x) = x^2 + 3x$, find the following values and indicate the answer as an ordered pair.

1. $f(3) = (3)^2 + 3(3) = 9 + 9 = 18$ (3,18)
2. $f(-2) = (-2)^2 + 3(-2) = 4 + (-6) = -2$ (−2,−2)
3. $f(-3) = (-3)^2 + 3(-3) = 9 + (-9) = 0$ (−3,0)
4. $f(2) = (2)^2 + 3(2) = 4 + 6 = 10$ (2,10)
5. $f(0) = (0)^2 + 3(0) = 0 + 0 = 0$ (0,0)
6. In general $f(\ ) = (\ )^2 + 3(\ )$ and we substitute the given value of x in place of x throughout the equation.

note

y and $f(x)$ are two different ways of expressing the dependent value.

Exercise Set 15.2

Find the indicated value of the given function.

1. $f(x) = 2x; f(2), f(0), f(-1)$
2. $f(r) = 3r; f(3), f(0), f(-2)$
3. $f(t) = t + 5; f(5), f(-5), f(0)$
4. $f(a) = a - 4; f(4), f(2), f(0)$
5. $f(x) = 3x + 5; f(-3), f\left(-\frac{5}{3}\right), f(1)$
6. $f(s) = 5s + 2; f(-2), f(3), f\left(-\frac{2}{5}\right)$
7. $f(r) = r^2; f(-2), f(2), f\left(\frac{1}{2}\right)$
8. $f(a) = 2a^2; f(3), f(-2), f(0)$
9. $f(x) = x^2 - 1; f(1), f(-2), f(-1)$
•10. $f(t) = 2t^2 + 3; f(1), f(-1), f(0)$
11. $f(r) = r^2 + 5r + 6; f(-2), f(0), f(-3)$
12. $f(q) = \dfrac{1}{q + 1}; f(1), f(-1), f\left(\frac{1}{2}\right)$
13. $f(t) = \dfrac{5}{t - 2}; f(-2), f(4), f(2)$
14. $f(a) = \dfrac{a}{a^2 - 1}; f(0), f(1), f\left(\frac{1}{2}\right)$

Find the indicated values of the given function. Leave your answer as an ordered pair.

15. $f(x) = 8x - 4;\quad f(1), f\left(\frac{1}{2}\right), f(0), f(-1.4)$
16. $f(x) = 6 - 2x;\quad f(2), f(-2), f(3), f(2.7)$
17. $f(x) = x^2 - 4;\quad f(2), f(-2), f(-3), f(0.3)$
18. $f(x) = x^2 - 9;\quad f(4), f(3), f(-3), f(-3.4)$
19. $f(x) = x^2 + 3x;\quad f(1), f(-3), f(0), f(-0.5)$
20. $f(x) = x^2 - 5x;\quad f(5), f(4), f(0), f(-2.1)$
21. $f(x) = x^2 + 6x + 8;\quad f(-2), f(0), f(-4), f(-3.9)$

•22. $f(x) = x^2 - 7x + 10;\quad f(3), f(5), f(2), f(-1.4)$

23. $f(x) = \dfrac{x}{x + 1};\quad f(1), f(0), f(-1.1), f(-0.9)$

•24. $f(x) = \dfrac{x}{x - 6};\quad f(4), f(0), f(5.9), f(6.1)$

Determine the domain of the given function.

25. $f(x) = 5x + 1$

•26. $f(x) = 3 - 7x$

27. $f(x) = x^2 + 3x - 10$

28. $f(x) = x^2 + 4$

29. $f(x) = \sqrt{x + 3}$

30. $f(x) = \sqrt{x - 4}$

•31. $f(x) = \sqrt{5 - x}$

32. $f(x) = \dfrac{1}{x + 2}$

33. $f(x) = \dfrac{x}{x - 6}$

•34. $f(x) = \dfrac{2x}{x + 5}$

Evaluate the functions at the given values. Round the answer to the accuracy of the given data.

35. The formula for the horsepower (H) produced at 1,000 rpm by a 6-cylinder engine with bore diameter (B) in inches is given by $H = 2.4B^2$ or $f(B) = 2.4B^2$. Find $f(3.00)$, $f(3.50)$, and $f(4.00)$.
36. The volume of a cube can be expressed as a function of its edge by $f(e) = e^3$. Find $f(4.0)$, $f(3.30)$, and $f(2.50)$.
37. The surface area of a sphere can be expressed as a function of its radius by $f(r) = 4\pi r^2$. Find $f(3.70)$, $f(1.60)$, and $f(4.90)$. (Use 3.14 for π.)
38. The temperature in degrees Fahrenheit can be expressed as a function of degrees Celsius by $f(\text{C}) = \frac{9}{5}\text{C} + 32$. Find $f(10\tilde{0})$, $f(25)$, and $f(\tilde{0})$.
39. The perimeter of a square can be expressed as a function of the length of its side by $f(s) = 4s$. Find $f(2.90)$, $f(13.6)$, and $f(9\tilde{0})$.

15.3 The Line

In section 15.2 the idea of the function was examined. In this section, we start to visualize this concept in the form of a graph of the function. The first step in graphing a function is to make a table where we arbitrarily choose values for x, the independent variable, and then generate the corresponding values for y, the dependent variable. In making such a table, we are forming ordered pairs that are solutions to the equation and points on the graph.

The first graph that we examine is that of a linear function. A **linear function** is a function of the form

$$y = mx + b \qquad \text{or} \qquad f(x) = mx + b$$

where m and b are constants. This function gets its name from the fact that its graph is a straight line.

Consider the equation $y = 2x + 4$. We observe that this is a linear function whose graph will be a straight line. We then set up a table, arbitrarily choosing values for x and determining the corresponding values for y.

x	y	Ordered Pairs
-3	$y = 2(-3) + 4 = -6 + 4 = -2$	$(-3,-2)$
-2	$y = 2(-2) + 4 = -4 + 4 = 0$	$(-2,0)$
-1	$y = 2(-1) + 4 = -2 + 4 = 2$	$(-1,2)$
0	$y = 2(0) + 4 = 0 + 4 = 4$	$(0,4)$
1	$y = 2(1) + 4 = 2 + 4 = 6$	$(1,6)$
2	$y = 2(2) + 4 = 4 + 4 = 8$	$(2,8)$

We can generate as many ordered pairs as we wish, but with a linear function, only two or three are necessary. We now locate (plot) the ordered pairs on our coordinate system and pass a straight line through all the points, as in figure 15.4.

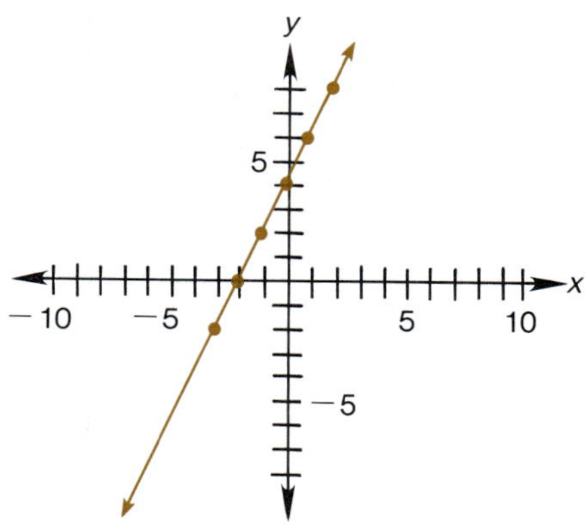

Figure 15.4

note

We have drawn only a portion of the graph. A line goes on forever in both directions.

We noted that we need only find two or three points to be able to graph a line. To be precise, we need only two points to graph a line, but we always choose a third point as a check on our work. In general, the most common points to determine are the points where the line crosses the axes. These points are called the ***x*- and *y*-intercepts** of the line. We observe from our example that when the line crosses the x-axis, the value of y is zero and when the line crosses the y-axis, the value of x is zero.

To find intercepts the following procedures are used:

1. To find the x-intercept let $y = 0$ and find the corresponding value of x. $(x,0)$
2. To find the y-intercept let $x = 0$ and find the corresponding value for y. $(0,y)$

note

A common error is to think that the x-intercept is where $x = 0$. This is incorrect. The x-intercept is the point $(x,0)$ where the line crosses the x-axis, which is where y is zero. The same sort of mistake is made concerning the y-intercept.

Examples

Graph the following linear functions.

1. $y = -2x + 3$

We find the y-intercept by letting $x = 0$:

$$y = -2(0) + 3$$
$$y = 0 + 3$$
$$y = 3 \qquad (0,3)$$

We find the x-intercept by letting $y = 0$:

$$0 = -2x + 3$$
$$-3 = -2x$$
$$\frac{3}{2} = x \qquad \left(\frac{3}{2},0\right)$$

We pick one more arbitrary point as a check, for example $x = 2$. Then

$$y = -2(2) + 3$$
$$= -4 + 3$$
$$= -1 \qquad (2,-1)$$

The graph of the three points is figure 15.5.

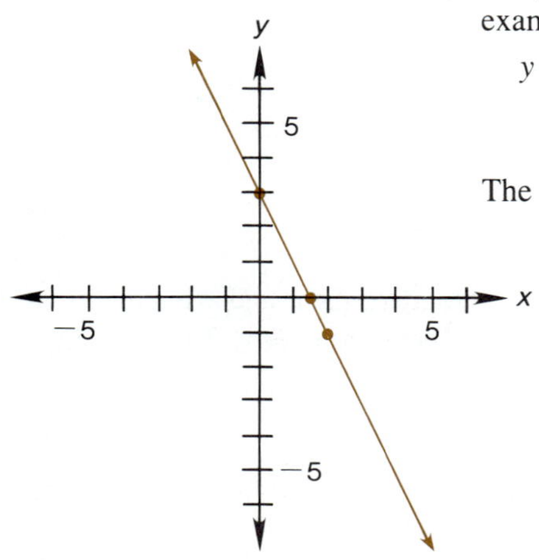

Figure 15.5

2. $y = 3x + 4$

x	y	
0	4	y-intercept
$-\frac{4}{3}$	0	x-intercept
-1	1	arbitrary check

The graph is figure 15.6.

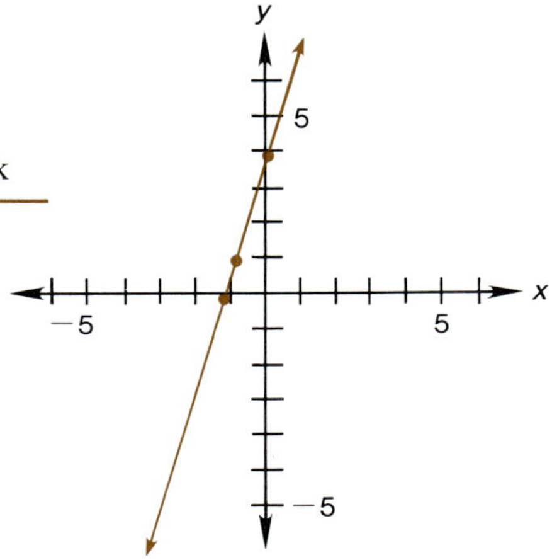

Figure 15.6

3. $y = 3x$

x	y	
0	0	x- and y-intercept
1	3	arbitrary choice
-1	-3	arbitrary check

The graph is figure 15.7.

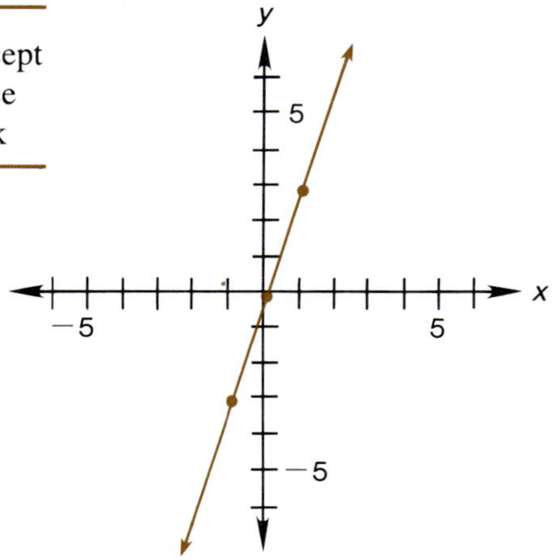

Figure 15.7

note

In this example, the x and y intercepts are the same point, the origin. We must therefore choose 2 more points.

4. $3y - 2x = 9$

x	y	
0	3	y-intercept
$-\frac{9}{2}$	0	x-intercept
3	5	arbitrary check

The graph is figure 15.8.

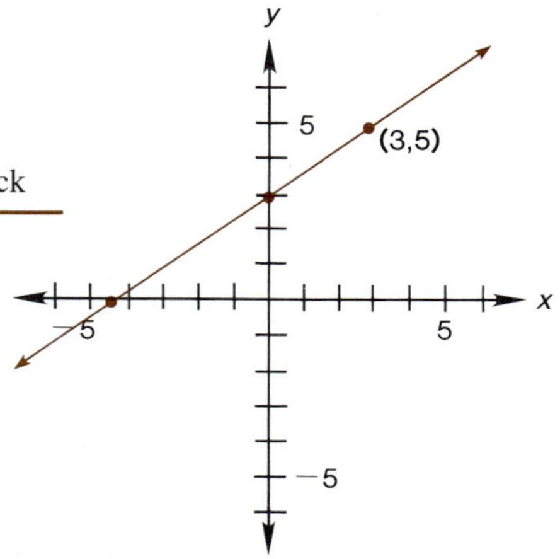

Figure 15.8

note

In example 4 we might wish to solve the equation for y:

$$3y - 2x = 9$$
$$3y = 2x + 9$$
$$y = \frac{2}{3}x + 3$$

We have expressed y as a function of x.

Exercise Set 15.3

Find the x- and y-intercepts. Leave the answer in the form of an ordered pair.

1. $y = 2x + 4$
•2. $y = 5x - 10$
3. $y = 3x + 1$
4. $y = 2x - 3$
5. $2x + 3y = 6$
6. $x + 4y = 12$
7. $(1.2)x + (2.4)y = 4.8$
8. $x - y = 4$
9. $y = 5x$
10. $y = -2x$
11. $3x - 2y = 0$
12. $y - 4x = 0$
13. $2x + 5y - 11 = 0$
•14. $(0.3)x - (0.4)y = 0.7$
15. $y = \frac{1}{2}x - \frac{3}{2}$
16. $y = \frac{2}{3}x - \frac{1}{3}$

Graph the given functions.

17. $y = 3x + 6$
•18. $y = x - 2$
19. $y = x + 5$
20. $y = 2x + 4$
21. $y = 2x - 8$
22. $y = 3x$
23. $y = x$
•24. $y = -2x$
25. $y = -3x$
26. $x + y = 0$
27. $y - 2x = 0$
•28. $(1.1)x + (2.2)y + 6.6 = 0$
29. $x - y - 1 = 0$
30. $3x - 2y = 7$
31. $(0.5)x + (0.2)y = 1.1$
32. $2x + 2y = 3$
33. $y = \frac{1}{2}x + \frac{1}{4}$
34. $y = \frac{1}{3}x - \frac{2}{3}$
35. $y = \frac{x}{5} + 2$
36. $y = 3 - \frac{x}{4}$

Solve the following equations for y in terms of x, thereby expressing y as a function of x. Leave the answers in the form $y = mx + b$.

Example $3y - 2x = 4$

Solution
$$3y - 2x = 4$$
$$3y = 2x + 4$$
$$y = \frac{2}{3}x + \frac{4}{3}$$

37. $y - 2x + 7 = 0$
•38. $y - 3x - 4 = 0$
39. $3y - 4x = 9$
40. $5y - 2x = 7$
41. $7x + 3y = 10$
42. $5x + 3y = 4$
43. $x - 5y + 7 = 2$
44. $x - 3y + 9 = 0$
45. $5x - 8y + 14 = 5$
•46. $7x + 5y - 11 = 7$

Graph the following functions by plotting the dependent variable (the variable for which the equation is solved) along the y-axis and the independent variable along the x-axis. Graph the functions only in the regions for which the equations would have meaning. (*Note:* lengths, areas, and volumes are never negative.)

47. Distance is equal to rate times time ($D = rt$). If we have a constant rate of 50 mi/h, plot the graph of D and t for the first 12 h.
•48. Use problem 47 to determine what the graph would look like if the rate were 37 mi/h.
49. Use problem 47 to determine what the graph would look like if the rate were 60 km/h.
50. The force in pounds needed to deform a spring is equal to the spring's constant times distance in inches ($F = Kx$). If K is 6 lb/in, plot the graph for F and x.
51. Use problem 50 to determine what the graph would look like if the spring constant, K, were equal to 9 lb/in.

•52. A drag racer knows that when he uses a 4.56 rear end, his speed, (S), can be found by multiplying 0.021 times the engine's rpm's, (R), that is, $S = (0.021)R$. If the maximum rpm's that the engine can safely turn is 7,200, plot the graph for S and R.

53. If the drag racer in problem 52 changes to a 3.70 rear end, the equation relating speed and rpm's will become $S = (0.026)R$. Plot this graph for S and R.

•54. The relationship between degrees Celsius (C) and degrees Fahrenheit (F) is given by the formula $F = \frac{9}{5}C + 32$. Plot the graph for F and C.

55. If in problem 54 we make degrees Celsius the dependent variable, the equation becomes $C = \frac{5}{9}(F - 32)$. Plot this graph for C and F.

15.4 The Parabola

In chapter 11 we studied solutions of quadratic equations. In this section we are going to graph **quadratic functions** in two variables of the form $y = ax^2 + bx + c$ or $f(x) = ax^2 + bx + c$, where a, b, and c are constants and $a \neq 0$.

A quadratic function involving the variables x and y can be identified as a second-degree polynomial, where only x is raised to the second power. The graph of such a function is called a **parabola.** As with any equation that we wish to graph and are uncertain about how it will look, we arbitrarily choose values for x and generate the corresponding y values to determine the appearance of the graph.

Consider the quadratic function given by $f(x) = x^2 + 2x - 8$. Since we do not know what the graph looks like, we need to choose a sufficient number of points.

x	$f(x)$ or y	(x, y)
-5	$y = (-5)^2 + 2(-5) - 8 = 25 + (-10) - 8 = 7$	$(-5, 7)$
-4	$y = (-4)^2 + 2(-4) - 8 = 16 + (-8) - 8 = 0$	$(-4, 0)$
-3	$y = (-3)^2 + 2(-3) - 8 = 9 + (-6) - 8 = -5$	$(-3, -5)$
-2	$y = (-2)^2 + 2(-2) - 8 = 4 + (-4) - 8 = -8$	$(-2, -8)$
-1	$y = (-1)^2 + 2(-1) - 8 = 1 + (-2) - 8 = -9$	$(-1, -9)$
0	$y = (0)^2 + 2(0) - 8 = 0 + 0 - 8 = -8$	$(0, -8)$
1	$y = (1)^2 + 2(1) - 8 = 1 + 2 - 8 = -5$	$(1, -5)$
2	$y = (2)^2 + 2(2) - 8 = 4 + 4 - 8 = 0$	$(2, 0)$
3	$y = (3)^2 + 2(3) - 8 = 9 + 6 - 8 = 7$	$(3, 7)$

Plotting all the points and passing a smooth curve through them, we have the graph of a quadratic function (see figure 15.9).

note

We do not connect the points with straight lines.

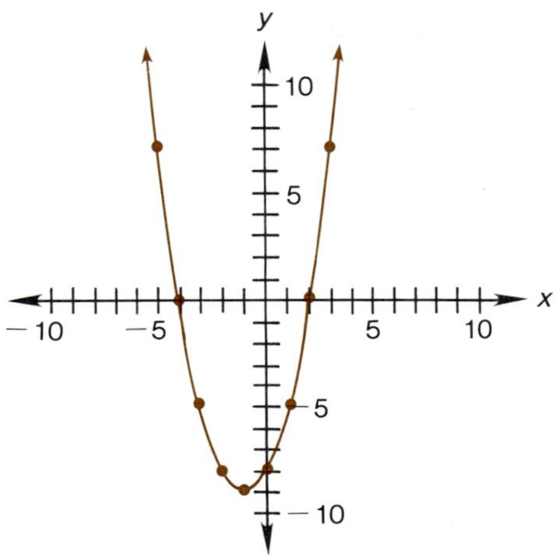

Figure 15.9

No matter how many points we plot, we cannot be sure that connecting all the points with a smooth curve will produce the correct graph and reveal all the important features of a curve. There are certain features of a parabola, therefore, that we always include in a graph, namely the x- and y-intercepts and the extreme point of the graph called the **vertex.**

We observe from the example that when the curve crosses the x-axis, the value of y is zero, and when the curve crosses the y-axis, the value of x is zero. This is the same observation that we made with the linear functions. Thus, we can generalize the idea of finding the x- and y-intercepts for any graph as follows:

1. To find the x-intercepts, if there are any, let $y = 0$ in the equation and solve for x. $(x,0)$
2. To find the y-intercepts, if there are any, let $x = 0$ in the equation and solve for y. $(0,y)$

Examples

Find the x- and y-intercepts.

1. $y = x^2 - 4$
   a. Let $y = 0$,
      $0 = x^2 - 4$
      $0 = (x - 2)(x + 2)$
      $x = 2$ and -2
      Therefore, the x-intercepts are $(2,0)$ and $(-2,0)$.
   b. Let $x = 0$,
      $y = (0)^2 - 4$
      $y = -4$
      Hence the y-intercept is $(0,-4)$.

2. $y = x^2 - 6x + 9$

a. Let $y = 0$,

$$0 = x^2 - 6x + 9$$
$$0 = (x - 3)(x - 3)$$
$$x = 3$$

Therefore, the x-intercept is (3,0).

b. Let $x = 0$,

$$y = (0)^2 - 6(0) + 9$$
$$y = 9$$

Hence, the y-intercept is (0,9).

3. $y = x^2 + 3x - 5$

a. Let $y = 0$ then $0 = x^2 + 3x - 5$. This will not factor. Therefore, we use the quadratic formula

$$x = \frac{-b \pm \sqrt{b^2 - 4ac}}{2a}$$

to find the x-intercepts:

$$x = \frac{-(3) \pm \sqrt{(3)^2 - 4(1)(-5)}}{2(1)}$$

$$= \frac{-3 \pm \sqrt{9 - (-20)}}{2} = \frac{-3 \pm \sqrt{29}}{2}$$

The x-intercepts are $\left(\frac{-3 + \sqrt{29}}{2},0\right)$ and $\left(\frac{-3 - \sqrt{29}}{2},0\right)$.

For graphing purposes we could use the approximations (1.2,0) and (−4.2,0).

b. Let $x = 0$,

$$y = (0)^2 + 3(0) - 5$$
$$y = -5$$

The y-intercept is (0, −5).

4. $y = -x^2 + 2x + 3$

a. Let $y = 0$,

$$0 = -x^2 + 2x + 3$$
$$0 = (-x + 3)(x + 1)$$
$$x = 3 \text{ and } -1$$

The x-intercepts are (3,0) and (−1,0).

b. Let $x = 0$,

$$y = -(0)^2 + 2(0) + 3$$
$$y = 3$$

The y-intercept is (0,3).

5. $y = x^2 + 1$
   a. Let $y = 0$,

$$0 = x^2 + 1$$
$$-1 = x^2$$
$$x = \pm\sqrt{-1} = \pm j$$

   Since there are no real solutions for x, the graph has no x-intercepts.

   b. Let $x = 0$,

$$y = (0)^2 + 1$$
$$y = 1$$

   The y-intercept is $(0,1)$.

There is one remaining point of interest on the graph that we want to find—the vertex. The **vertex** is the extreme point on the graph in that it is either the maximum or minimum value that the graph will attain. If the equation is in standard form, $f(x) = ax^2 + bx + c$, we can find the x value of the vertex by using the equation $x = \frac{-b}{2a}$. Once we have determined the x component, we can replace x with this value in the original function and generate the corresponding y value. Recall our example from the beginning of this section: $y = x^2 + 2x - 8$. The value of a is 1 and b is 2. Therefore,

$$x = \frac{-(2)}{2(1)} = \frac{-2}{2} = -1$$

We then substitute this value for x in the original equation and we have

$$y = (-1)^2 + 2(-1) - 8 = 1 + (-2) - 8 = -9$$

Hence, our vertex is the point $(-1,-9)$.

Examples

Find the vertex.

1. $y = x^2 - 4 \quad a = 1, b = 0$, then

$$x = \frac{-(0)}{2(1)} = \frac{-0}{2} = 0$$
$$y = (0)^2 - 4 = -4$$

   The vertex is $(0,-4)$.

2. $y = x^2 - 6x + 9 \quad a = 1, b = -6$, then

$$x = \frac{-(-6)}{2(1)} = \frac{6}{2} = 3$$
$$y = (3)^2 - 6(3) + 9 = 9 - 18 + 9 = 0$$

   The vertex is $(3,0)$.

3. $y = x^2 + 3x - 5 \quad a = 1, b = 3$, then

$$x = \frac{-(3)}{2(1)} = -\frac{3}{2}$$

$$y = \left(-\frac{3}{2}\right)^2 + 3\left(-\frac{3}{2}\right) - 5$$

$$= \frac{9}{4} + \left(-\frac{9}{2}\right) - 5$$

$$= \frac{9}{4} + \left(-\frac{18}{4}\right) - \frac{20}{4}$$

$$= \frac{9 - 18 - 20}{4}$$

$$= \frac{-29}{4}$$

The vertex is $\left(-\frac{3}{2}, -\frac{29}{4}\right)$.

4. $y = -x^2 + 2x + 3 \quad a = -1, b = 2$, then

$$x = -\frac{2}{2(-1)} = -\frac{2}{-2} = 1$$

$$y = -(1)^2 + 2(1) + 3 = -1 + 2 + 3 = 4$$

The vertex is $(1,4)$.

5. $y = x^2 + 1 \quad a = 1, b = 0$, then

$$x = \frac{0}{2(1)} = 0$$

$$y = (0)^2 + 1 = 1$$

The vertex is $(0,1)$.

Using function notation, we can generalize the idea of finding the vertex. The **vertex** of a quadratic function in standard form, $f(x) = ax^2 + bx + c$, is the point $\left(\frac{-b}{2a}, f\left(\frac{-b}{2a}\right)\right)$. If we identify the function as a quadratic, and find the x- and y-intercepts, the vertex, and two or three arbitrary points, we can be reasonably certain that the graph will be correct.

Examples Graph the following quadratic functions.

1. $y = x^2 - 4$ In previous examples we found the x- and y-intercepts and the vertex. We need only determine a few more points and we will be ready to graph the function.

x	y	
2	0	x-intercepts
-2	0	
0	-4	y-intercept and vertex
-1	-3	arbitrary points
1	-3	
-3	5	
3	5	

note
We pick arbitrary points on both sides of the vertex.

The graph of the function is figure 15.10.

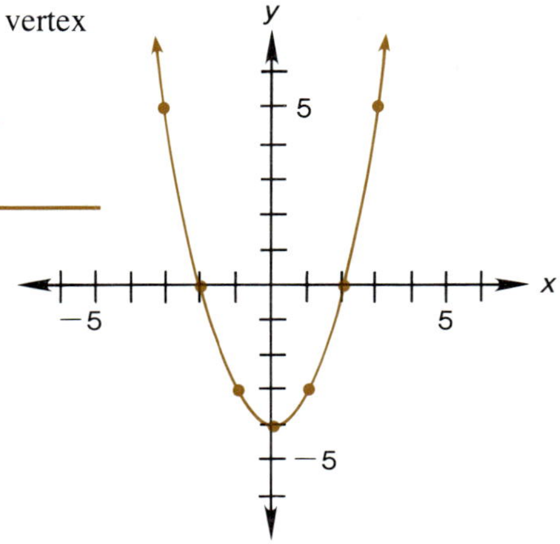

Figure 15.10

2. $y = x^2 - 6x + 9$

x	y	
3	0	x-intercept and vertex
0	9	y-intercept
1	4	arbitrary points
2	1	
4	1	
5	4	
6	9	

The graph of the function is figure 15.11.

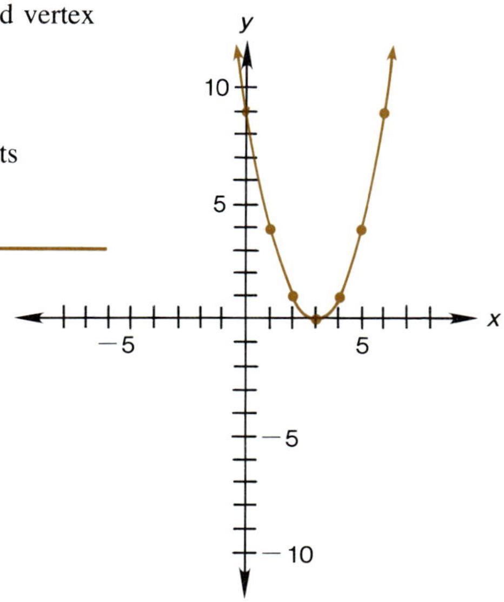

Figure 15.11

3. $y = x^2 + 3x - 5$

x	y	
$\frac{-3 + \sqrt{29}}{2} \approx 1.2$	0	x-intercepts
$\frac{-3 - \sqrt{29}}{2} \approx -4.2$	0	
0	-5	y-intercept
$-\frac{3}{2}$	$-\frac{29}{4}$	vertex
-5	5	arbitrary points
-2	-7	
-1	-7	
2	5	

The graph of the function is figure 15.12.

note

To plot the x-intercepts, we had to approximate the irrational x values.

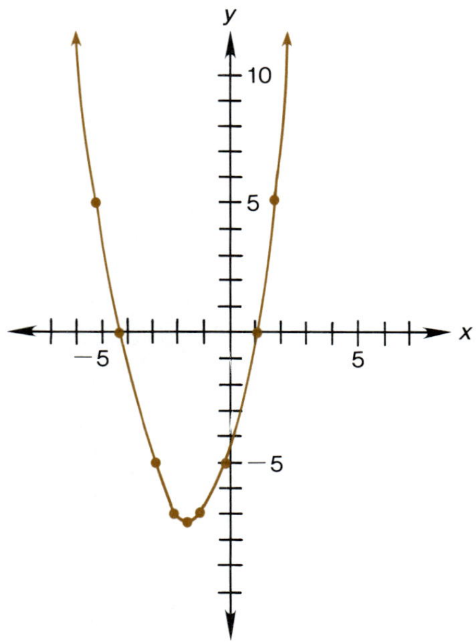

Figure 15.12

4. $y = -x^2 + 2x + 3$

x	y	
3	0	x-intercepts
-1	0	
0	3	y-intercepts
1	4	vertex
-2	-5	arbitrary points
2	3	
4	-5	

The graph of the function is figure 15.13.

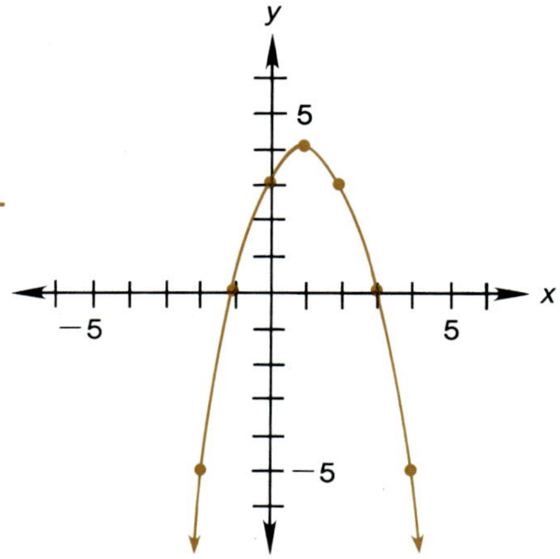

Figure 15.13

note

When the value of a, the coefficient of the x^2 term, is negative, the parabola opens downward. When a is positive, the parabola opens upward.

5. $y = x^2 + 1$

x	y	
0	1	y-intercept and vertex
-3	10	arbitrary points
-2	5	
-1	2	
1	2	
2	5	
3	10	

The graph of the function is figure 15.14.

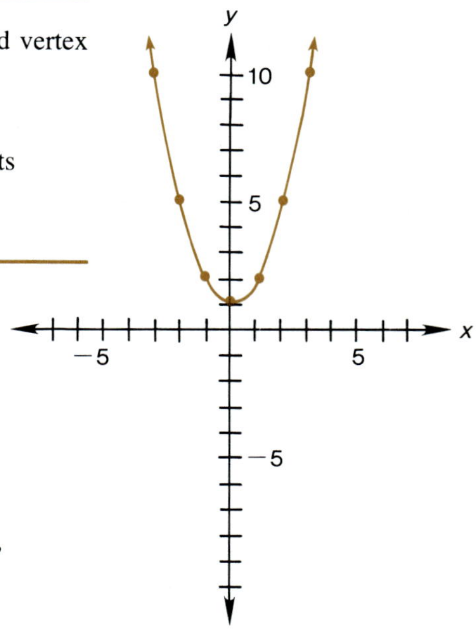

Figure 15.14

note

When we graph a parabola, we draw only a portion of the graph. The parabola continues to open up or down indefinitely, depending on the sign of a.

Quadratic functions are used in many physical situations. For example, if an object is thrown into the air, the path that the object follows is a parabola. Consider the following.

Example

A projectile is fired vertically into the air, and the distance in feet above the ground in t seconds is given by $S = 160t - 16t^2$. Find the highest point (vertex) of the projectile and the moment when the projectile will strike the ground. Graph the function.

a. For the vertex, $S = -16t^2 + 160t$ and $a = -16$, $b = 160$. Then the t value of the vertex is

$$t = \frac{-b}{2a} = \frac{-160}{2(-16)} = \frac{-160}{-32}$$

The height (S) is:

$$S = -16(5)^2 + 160(5) = -400 + 800 = 400 \text{ ft}$$

The maximum height is attained when $t = 5$ seconds and $S = 400$ ft.

b. The projectile will strike the ground when $S = 0$. Therefore, set $S = 0$ and solve for t:

$$\begin{aligned} 0 &= -16t^2 + 160t \\ &= -16t(t - 10) \\ t &= 0 \quad \text{and} \quad 10 \end{aligned}$$

The value $t = 0$ is the time the projectile was fired; hence, the value $t = 10$ is the time when the object will strike the ground.

c. To graph the function, as in figure 15.15, we plot time (t) along the x-axis since it is the independent variable, and distance (S) along the y-axis.

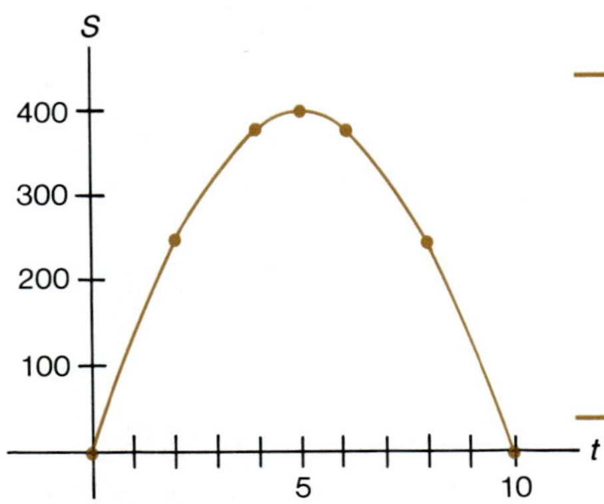

Figure 15.15

t	S	
5	400	vertex
0	0	x- and y-intercept
10	0	x-intercept
2	256	arbitrary points
4	384	
6	384	
8	256	

note

We used different scales on the x- and y-axes; also, values for negative time or distance are not represented on the graph since they do not have any meaning in this example.

Exercise Set 15.4

Find the x- and y-intercepts if they exist. If they do not exist, so state.

1. $y = x^2 - 16$
•2. $y = x^2 - 9$
3. $y = x^2 - 6x + 8$
4. $y = x^2 + 11x - 12$
5. $y = x^2 + 8x + 12$
6. $y = x^2 - 4x - 12$
7. $y = 5 - x^2$
8. $y = 7 - x^2$
9. $y = x^2 + 6x + 9$
•10. $y = x^2 - 4x + 4$
11. $y = x^2 + 5$
12. $y = x^2 + 6$
13. $y = x^2 + 4x + 6$
•14. $y = x^2 + 2x + 5$
15. $y = -x^2 + 8x - 16$
16. $y = 25 - x^2$
17. $y = 2x^2 + 3x + 1$
18. $y = 2x^2 - 7x + 6$
19. $y = -2x^2 - x + 6$
20. $y = -3x^2 + 7x + 6$

Find the vertex.

21. $y = x^2 - 16$
•22. $y = x^2 - 9$
23. $y = x^2 - 6x + 8$
24. $y = x^2 + 11x - 12$
25. $y = x^2 + 8x + 12$
26. $y = x^2 - 4x - 12$
27. $y = 5 - x^2$
28. $y = 7 - x^2$
29. $y = x^2 + 6x + 9$
30. $y = x^2 - 4x + 4$
31. $y = x^2 + 5$
32. $y = x^2 + 6$
33. $y = x^2 + 4x + 6$
34. $y = x^2 + 2x + 5$
35. $y = -x^2 + 8x - 16$
36. $y = 25 - x^2$
37. $y = 2x^2 + 3x + 1$
•38. $y = 2x^2 - 7x + 6$
39. $y = -2x^2 - x + 6$
40. $y = -3x^2 + 7x + 6$

Graph the following functions.

41. $y = x^2 - 16$
•42. $y = x^2 - 9$
43. $y = x^2 - 6x + 8$
44. $y = x^2 + 11x - 12$
45. $y = x^2 + 8x + 12$
46. $y = x^2 - 4x - 12$
47. $y = 5 - x^2$
•48. $y = 7 - x^2$
49. $y = x^2 + 6x + 9$
50. $y = x^2 - 4x + 4$
51. $y = x^2 + 5$
52. $y = x^2 + 6$
53. $y = x^2 + 4x + 6$
54. $y = x^2 + 2x + 5$
55. $y = -x^2 + 8x - 16$
56. $y = 25 - x^2$
57. $y = 2x^2 + 3x + 1$
58. $y = 2x^2 - 7x + 6$
59. $y = -2x^2 - x + 6$
•60. $y = -3x^2 + 7x + 6$

Graph the following functions by plotting the dependent variable (the variable for which the equation is solved) along the y-axis and the independent variable along the x-axis. Graph the function only in the regions for which the equations would have meaning. (*Note:* Distance and time are never negative.)

61. When a ball rolls down an inclined plane, it travels a distance $d = 6t + \frac{t^2}{2}$ feet in t seconds. Plot the graph of d and t. How long will it take the ball to travel 14 ft?
62. The output power P of a 100-volt electric generator is defined by $P = 100I - 5I^2$, where I is in amperes. Plot the graph of P and I.
63. The current in a circuit flows according to the equation $i = 12 - 12t^2$, where i is the current and t is the time in seconds. Plot the graph of i and t.
•64. If a projectile is fired vertically into the air with an initial velocity of 80 ft/s, the distance in feet above the ground in t seconds is given by $S = 80t - 16t^2$. Find the object's maximum height and when the object will strike the ground. Plot the graph of S and t.
65. Refer to problem 64. If the initial velocity is 96 ft/s, the equation will be $S = 96t - 16t^2$. Find the maximum height and when the projectile will strike the ground. Plot the graph of S and t.
66. The distance S through which an object will fall in t seconds is $S = 16t^2$. Plot the graph of S and t for the first 5 seconds.
67. An object is dropped from the top of the Empire State Building (1,250 ft tall). The distance that the object is from the ground is given by the equation $S = 1{,}250 - 16t^2$. Plot the graph of S and t and determine when the object will strike the ground.

15.5 Graphs of Other Functions

In the two preceding sections we examined the graphs and graphing techniques of two special functions. In this section we are going to sketch the curves of functions whose graphs are not lines or parabolas.

When we first approached the idea of graphing, we arbitrarily chose values for x and generated the corresponding y values. No matter how many points we plot, though, we cannot be sure that connecting all the points with a smooth curve will produce the correct graph and reveal all the important features of the curve. When we do not identify the function we are dealing with as a line or a parabola, our graph will consist of plotting a sufficient number of points to allow some pattern to emerge, and then sketching the graph accordingly.

We continue to find the x- and y-intercepts whenever they exist. As before, by letting $y = 0$ and solving for x, we find the x-intercepts and by letting $x = 0$ and solving for y, we find the y-intercept. The following are examples of plotting other functions.

Examples Graph the following functions.

1. $y = x^3 - 1$ The first points we will look for are the intercepts. If there are any intercepts, they may indicate further choices for x. Let $y = 0$; then

$$0 = x^3 - 1$$
$$1 = x^3$$
$$x = \sqrt[3]{1}$$
$$x = 1$$ The x-intercept is (1,0).

Let $x = 0$; then

$$y = (0)^3 - 1$$
$$y = -1$$ The y-intercept is $(0, -1)$.

x	y	
0	−1	y-intercept
1	0	x-intercept
−2	−9	arbitrary points
−1	−2	
2	7	

The graph of the function is figure 15.16.

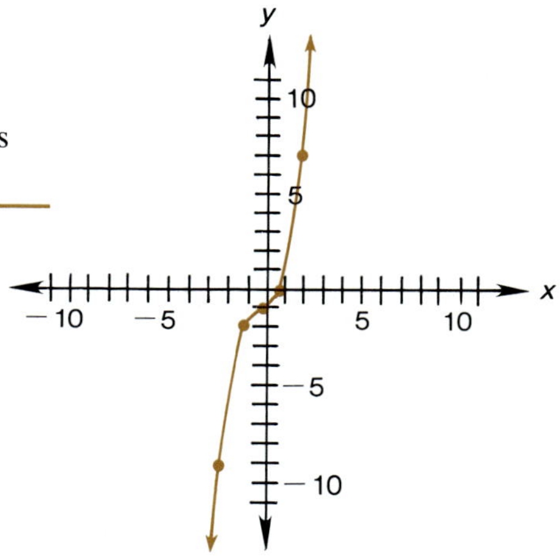

Figure 15.16

2. $y = x^3 - x$

Intercepts $0 = x^3 - x$
$= x(x^2 - 1)$
$= x(x - 1)(x + 1)$ $x = 0, 1, -1$

x-intercepts are $(0,0)(1,0)(-1,0)$.

$$y = (0)^3 - (0) = 0$$

y-intercept is (0,0).

x	y	
0	0	x- and y-intercept
1	0	x-intercepts
-1	0	
-2	-6	arbitrary points
$-\frac{1}{2}$	$\frac{3}{8}$	
$\frac{1}{2}$	$-\frac{3}{8}$	
2	6	

The graph of the function is figure 15.17.

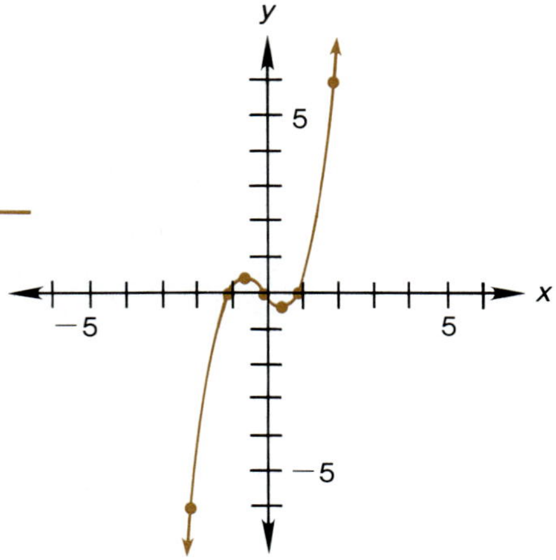

Figure 15.17

note

When the graph passes through the origin (0,0) it is both an x-intercept and y-intercept.

note

We chose values before, in between, and after each x-intercept.

3. $y = 2^x$ In this function, the independent variable x is an exponent. To determine points on the graph, it is necessary to be able to evaluate 2 to a power. For the y-intercept, we let $x = 0$ as follows:

$$\begin{aligned} y &= 2^x \\ &= 2^0 \\ &= 1 \end{aligned}$$ The y-intercept is (0,1).

For the x-intercept, we set y equal to zero,

$$\begin{aligned} y &= 2^x \\ 0 &= 2^x \end{aligned}$$

but there is no value of x that will satisfy this equation; hence, there is no x-intercept.

We can then select a few more arbitrary points as follows:

x	y
-3	$y = 2^{-3} = \frac{1}{2^3} = \frac{1}{8}$
-2	$y = 2^{-2} = \frac{1}{2^2} = \frac{1}{4}$
-1	$y = 2^{-1} = \frac{1}{2^1} = \frac{1}{2}$
0	$y = 2^0 = 1$ y-intercept
1	$2^1 = 2$
2	$2^2 = 4$
3	$2^3 = 8$

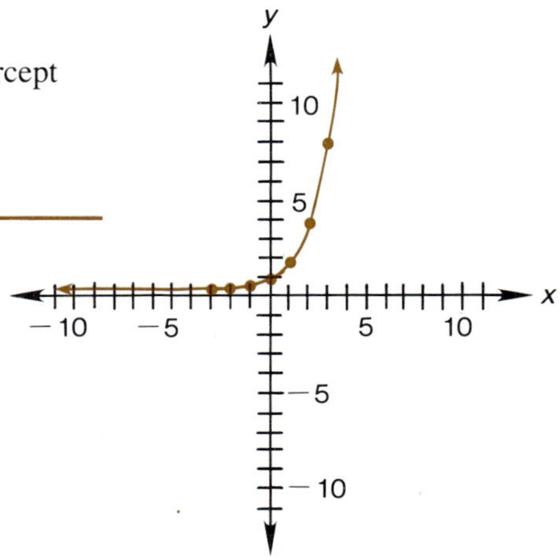

Figure 15.18

The function is called an **exponential function** and its graph is shown in figure 15.18.

We observed that the y-values could not be zero or negative, and we see from the graph that the curve gets closer and closer to the x-axis as the values of x decrease. The x-axis is a boundary condition in our graph in that the graph levels out to the x-axis. This boundary condition is called an **asymptote** of the graph.

4. $y = \dfrac{1}{x - 1}$ The most important feature of this graph will be what happens to the curve in the vicinity of $x = 1$, because at the point when $x = 1$, we have division by zero and our function will not exist. To get a good picture of what is happening to the graph around $x = 1$, we need to choose values very close to 1. In solving for the intercepts, we find the y-intercept

$$y = \frac{1}{(0) - 1} = \frac{1}{-1} = -1$$

to be $(0, -1)$, but when we set $y = 0$ and try to solve

$$0 = \frac{1}{x - 1}$$

we find there are no x-intercepts. For there to be an x-intercept, the quotient must be zero, which occurs when the numerator is equal to zero.

x	y
-10	$\frac{1}{-10-1} = -\frac{1}{11}$
-5	$\frac{1}{-5-1} = -\frac{1}{6}$
-1	$\frac{1}{-1-1} = -\frac{1}{2}$
0	$\frac{1}{0-1} = -1$ y-intercept
0.5	$\frac{1}{0.5-1} = \frac{1}{-0.5} = -2$
0.9	$\frac{1}{0.9-1} = \frac{1}{-0.1} = -10$
1	$\frac{1}{0}$ graph does not exist
1.1	$\frac{1}{1.1-1} = \frac{1}{0.1} = 10$
1.5	$\frac{1}{1.5-1} = \frac{1}{0.5} = 2$
2	$\frac{1}{2-1} = \frac{1}{1} = 1$
5	$\frac{1}{5-1} = \frac{1}{4}$
10	$\frac{1}{10-1} = \frac{1}{9}$

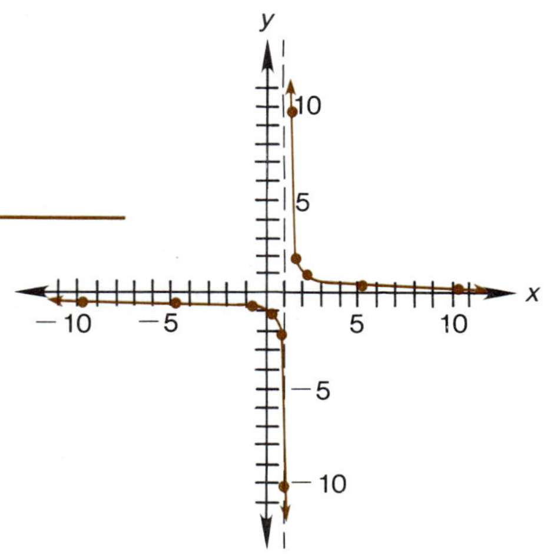

Figure 15.19

We observe that as we choose values closer to 1, the graph in figure 15.19 goes to plus or minus infinity, and the vertical line $x = 1$ is a boundary condition (asymptote) of the graph. As we increase the values of x in the positive direction or decrease them in the negative direction, the curve levels out to the x-axis (another asymptote), but since there are no x-intercepts, the graph only gets close to the x-axis—it never really intercepts it.

5. $y = \sqrt{x + 4}$ The most important consideration for this graph is that we choose values of x that will give us real answers for y. Therefore, we must choose values from the domain of x such that $x + 4 \geqslant 0$. It follows that if $x + 4 \geqslant 0$, then $x \geqslant -4$. Since the choice of the values that we pick is arbitrary, we pick values that will give us a perfect square under the radical.

x	y	
-4	$\sqrt{-4 + 4} = \sqrt{0} = 0$	x-intercept
-3	$\sqrt{-3 + 4} = \sqrt{1} = 1$	
0	$\sqrt{0 + 4} = \sqrt{4} = 2$	y-intercept
5	$\sqrt{5 + 4} = \sqrt{9} = 3$	
12	$\sqrt{12 + 4} = \sqrt{16} = 4$	

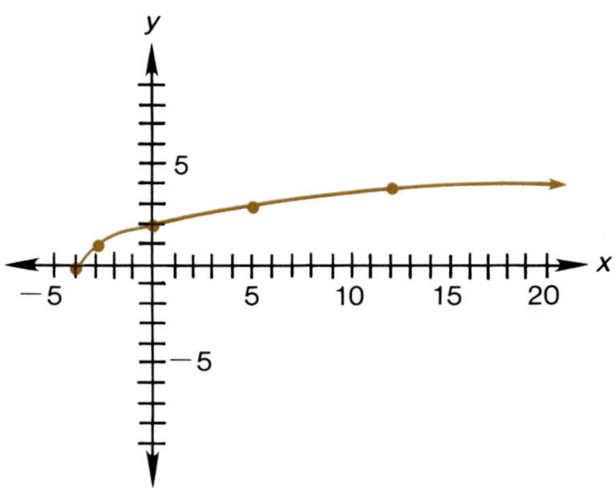

Figure 15.20

The graph of the function is figure 15.20.

note

Our graph does not exist for values of $x < -4$, and when we took the square root to find the value of y, we were interested only in the principal root.

Exercise Set 15.5

Graph the given functions.

1. $y = x^3 + 1$
•2. $y = x^3 - 8$
3. $y = x^3 - 4$
4. $y = x^3 + 5$
5. $y = x^4 - 1$
6. $y = x^4 - 16$
7. $y = x^3 - 4x$
8. $y = x^3 + 5x^2 + 6x$
9. $y = x^3 - 2x^2 - 8x$
•10. $y = x^3 + x^2 - 12x$
11. $y = x^3 + 4x^2 - 12x$
12. $y = 4^x$
13. $y = 3^x$
14. $y = \left(\frac{1}{3}\right)^x$
15. $y = \left(\frac{1}{2}\right)^x$
16. $y = 3^{\frac{x}{2}}$
17. $y = 2^{\frac{x}{2}}$
•18. $y = 2^{1 - x}$
19. $y = 2^{1 + x}$
20. $y = 4^{-x}$
21. $y = \frac{1}{x}$
22. $y = \frac{1}{x + 1}$
23. $y = \frac{1}{x - 2}$
24. $y = \frac{x}{x + 1}$
25. $y = \frac{x}{x - 3}$
•26. $y = \frac{x + 3}{x - 3}$
27. $y = \frac{2 - x}{x + 2}$
28. $y = \frac{2x + 1}{3x - 1}$
29. $y = \frac{x - 1}{2x + 2}$
30. $y = \sqrt{x + 1}$
31. $y = \sqrt{x}$
32. $y = \sqrt{x - 1}$
33. $y = \sqrt{x - 4}$
•34. $y = \sqrt{1 - x}$
35. $y = \sqrt{4 - x}$
36. $y = \sqrt{x^2}$
37. $y = \sqrt{x^2 + 1}$
38. $y = \sqrt{x^2 + 4}$
39. $y = \sqrt{x^2 - 1}$
40. $y = \sqrt{x^2 - 4}$

Graph the following functions by plotting the dependent variable (the variable for which the equation is solved) along the y-axis and the independent variable along the x-axis. Graph the function in the regions for which the equations would have meaning.

41. The profit P that a manufacturer makes by producing x items is given by $P = x^3 + 8x^2 + 12x$. Plot the graph of P and x.
42. If the population of a country increases 4% each year, then the population is given by the equation $P = I(1.04)^t$, where I is the initial population and t is time in years. If the initial population is 10,000, plot the graph of P and t.
43. A metal rod is heated to 100°C and is then cooled in air kept at 25°C. After t minutes, the temperature is given by

$$C = 25 + 80\left(\frac{1}{3}\right)^t.$$

Plot the graph of C and t.

44. The formula for the population of a certain country is given by $P = I \cdot 2^{\frac{t}{32}}$, where I is the initial population and t is time in years. Plot the graph of P and t if the initial population is 1,000,000.
45. Using problem 44, plot the graph of P and t if the initial population is 20,000,000.
46. The radius of a circle can be expressed as a function of its area by $R = \sqrt{\frac{A}{\pi}}$. Plot the graph of R and A using 3.14 for π.
47. The half-life of a radioactive material is the length of time it takes for half of the given quantity to decay. A certain radioactive substance has a half-life of 3 min. The amount of radioactive material left after t minutes is given by $M = A\left(\frac{1}{2}\right)^{\frac{t}{3}}$, where A is the initial amount of substance and t is time in minutes. If the initial amount is 8 g, plot the graph of M and t.
48. Using problem 47, plot the graph of M and t if A is 32 oz.
49. If P dollars are invested for n years at an interest rate i compounded yearly, then the amount A at the end of the n years is given by $A = P(1 + i)^n$. If P is 1,000 and i is 10%, plot the graph of A and n.
50. Using problem 49, plot the graph of A and n if P is 15,000 and i is 12%.
51. The side of a square can be expressed as a function of its area by $S = \sqrt{A}$. Plot the graph of S and A.
52. The perimeter P of a rectangle can be expressed as a function of its length x by $P = 2x + \frac{2}{x}$. Plot the graph of P and x.
53. A rowboat can go 7 mi/h in still water. If it goes 10 mi upstream, the time required for this trip can be expressed as a function of the rate r of the stream as follows: $t = \frac{10}{7 - r}$. Plot the graph of t and r.
54. The pressure p and the volume v of a certain container of gas according to Boyle's law satisfy the equation $P = \frac{2{,}000}{v}$. If the pressure is measured in pounds per square inch and the volume in cubic inches, plot the graph of P and v.
55. An educator estimates that the percent P of material remembered by a student in a class t months after the student learned the material is given by $P = \frac{100}{1 + t}$. Plot the graph of P and t.

15.6 Representation of Data

Graphing is an important tool for technicians. A graph allows a technician to be able to visualize data that has been accumulated or to see the correspondence between interrelated quantities. A graph is a means of summarizing data so a reader can quickly understand a relationship.

Data can be represented in many different ways. In most instances how we present data can be left up to our individual creativity. In this section we examine several common graphs, but you should remember that there are many variations possible for the representation of data.

The first graphing technique we examine is called the **bar graph.** The bar graph is made up of several parallel rectangles called **bars.** These bars are of lengths proportional to the quantities that they represent. The bars are equally spaced and their widths are all equal. They can be constructed either vertically or horizontally. To construct a bar graph, we choose a scale that will accommodate our data and then determine the length of each bar. Consider the following.

Example

The four tallest buildings in the United States are: Sears Tower (1,454 ft), the World Trade Center (1,350 ft), the Empire State Building (1,250 ft), and Standard Oil of Indiana (1,136 ft). Construct a bar graph of these data.

We want the graph to show clearly the differences in the heights of these buildings. If we want a 5-in graph, we must choose a scale that will allow us to draw the Sears Tower as approximately 5 in high, and then the other buildings will be of lesser heights. We therefore choose a scale of 300 ft = 1 in (figure 15.21).

Building	Height (ft)	Length of Bar (in.)
Sears Tower	1,454	$\frac{1{,}454}{300} \approx 4.8$
World Trade Center	1,350	$\frac{1{,}350}{300} = 4.5$
Empire State Building	1,250	$\frac{1{,}250}{300} \approx 4.2$
Standard Oil	1,136	$\frac{1{,}136}{300} \approx 3.8$

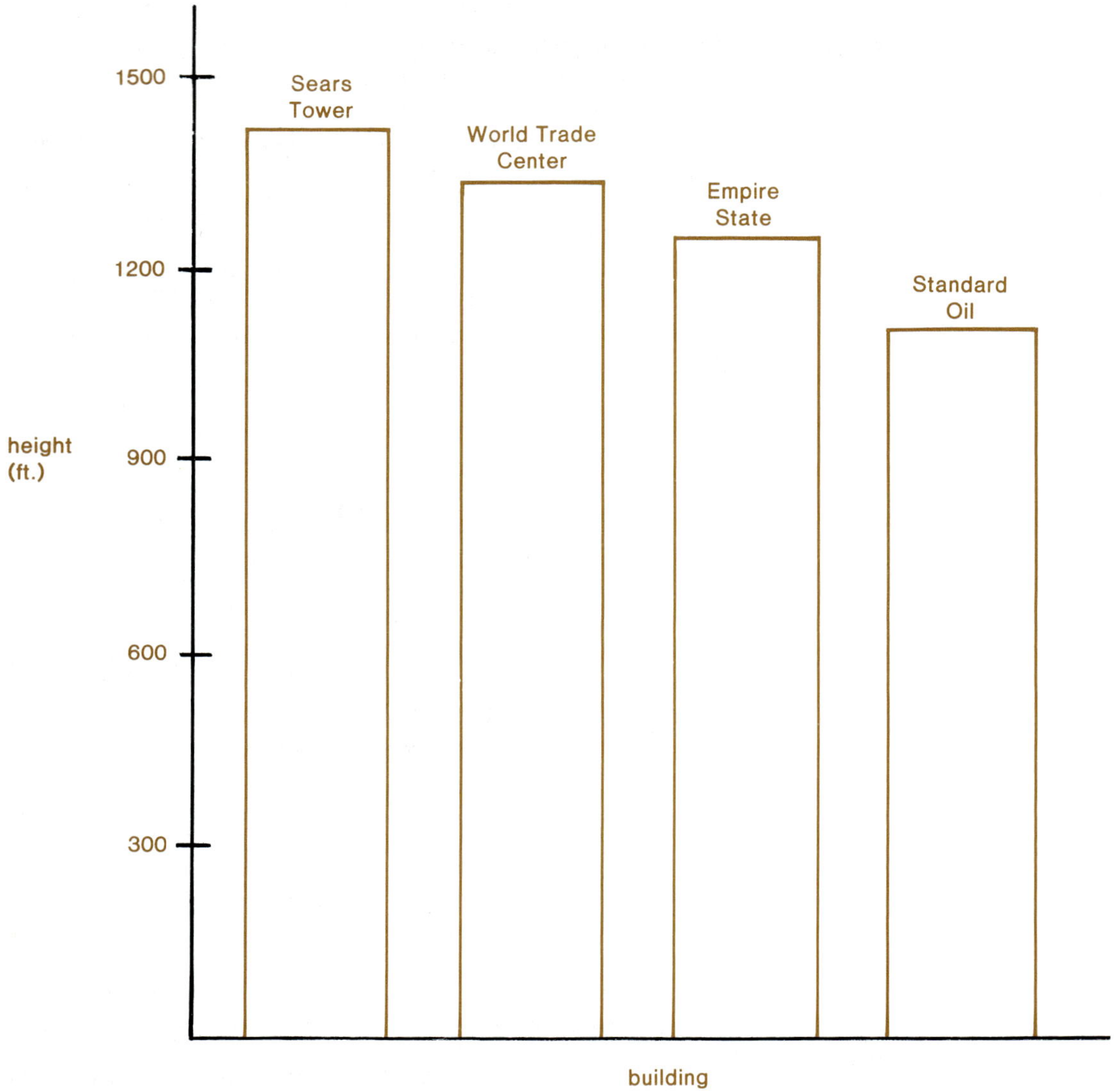

Figure 15.21

The **circle graph** is useful when we want to show the relationship between parts and the whole quantity they comprise. To construct a circle graph, we first express the data in terms of a percent of the whole quantity. The portion that each part will occupy in the graph will be its percent of 360°. (The whole circle has 360°.) Observe the following example.

Example The following table gives the results of the 1980 presidential election. The circle graph is shown in figure 15.22.

Candidate	Votes	% of Votes	Number of Degrees of the Circle
Carter	34,913,320	41.2	148
Reagan	43,201,220	50.9	183
Anderson	5,581,370	6.6	24
Others	1,102,690	1.3	5
TOTAL	84,798,600		

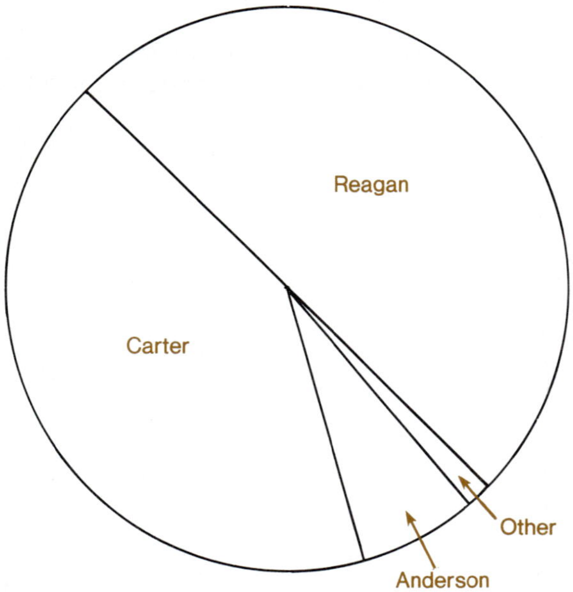

Figure 15.22

The last type of graph we examine is the **broken-line graph.** The broken-line graph is used to show changing conditions. We choose an appropriate scale for each quantity and then plot the different points. These points are then joined by straight lines. In many instances the changes will be over a certain interval of time, and if time is one of the quantities, it is generally placed horizontally; the other quantity is then placed vertically. In general, we place the independent variable along the horizontal axis and the dependent variable along the vertical axis. Consider the following example.

Example The following table gives the number of housing starts in the United States for the years 1973–1979.

Year	Number of Units
1973	2,056,000
1974	1,353,000
1975	1,170,000
1976	1,549,000
1977	1,990,000
1978	2,024,000
1979	1,750,000

To construct the broken-line graph, we place the number of houses on the vertical axis, and the year along the horizontal axis (see figure 15.23).

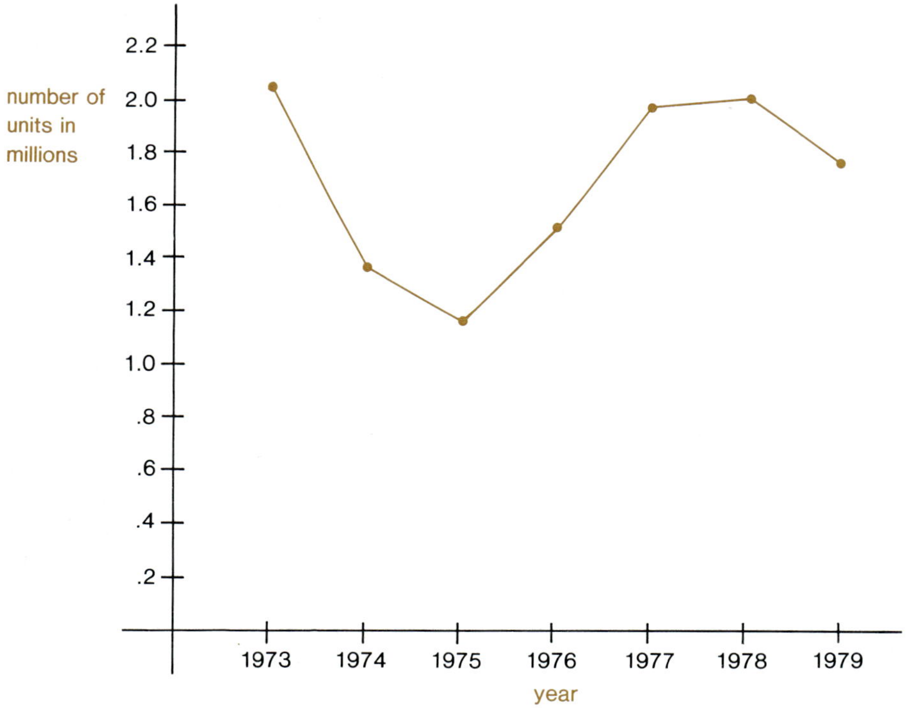

Figure 15.23

note

The scales that we use on the horizontal and vertical axes need not be the same.

Exercise Set 15.6

Construct a bar graph to represent the given data. Choose a scale you feel is most appropriate for the data. (*Note:* There are many different scales possible.)

1. The six largest U.S. businesses are:

Company	Assets
Exxon	\$49,490,000,000
General Motors	\$32,216,000,000
Mobil Oil	\$27,506,000,000
Ford Motor Company	\$23,525,000,000
Texaco	\$22,992,000,000
Standard Oil of California	\$18,103,000,000

•2. The consumer price indices for urban wage earners, with 1967 as a base of 100, are: 1960—88.7; 1965—94.5; 1970—116.3; 1975—161.2; 1980—248.0.

3. The estimated maximum speeds in miles per hour of the six fastest animals are: cheetah—70; pronghorn antelope—61, wildebeest—50; lion—50; Thomson's gazelle—50; quarterhorse—47.

4. The five tallest mountains in the United States are: Mt. McKinley—20,320 ft; Mt. St. Elias—18,008 ft; Mt. Foraker—17,400 ft; Mt. Bona—16,421 ft; and Mt. Blackburn—16,390 ft.

5. The estimated number of women in the working population over the last 50 years are: 1930—10,752,000; 1940—12,845,000; 1950—18,410,000; 1960—23,270,000; 1970—31,560,000; and 1980—44,010,000.

•6. The seven busiest airports in the world are as follows:

Airport	Passengers
O'Hare, Chicago	47,850,000
Hartsfield International, Atlanta	41,650,000
International, Los Angeles	34,925,000
Heathrow, London	27,975,000
Kennedy, New York	26,975,000
International, San Francisco	24,150,000
Dallas, Fort Worth	22,575,000

7. The following is a list of advertising expenditures in billions of dollars relative to the different media:

Media	Amount (in billions of dollars)
Radio	3.4
Television	10.4
Magazines	2.8
Newspapers	14.5
Direct mail	6.5
Billboards	0.6

8. The following is a projection of the number of people that will be in each age group in the year 2000:

Age	Numbers (in millions)
Under 19	100.7
20–34	55.4
35–44	41.4
45–54	35.9
55–64	23.0
65–74	17.1
75 & Over	13.6

Construct a circle graph to represent the given data.

9. An average family of 4 spends $10,642 per year. The following is a breakdown of their spending:

Item	Amount
Food	3,940
Housing	2,680
Transportation	1,212
Clothing	820
Personal and Medical Care	1,510
Other	480

10. The blue-collar work force has approximately 28,065,000 members. The following is a listing of the major categories:

Occupational Group	Number
Craft and kindred workers	11,850,000
Operators	9,240,000
Transport equipment operators	3,100,000
Nonfarm laborers	3,875,000

11. The approximate number of engineers in the United States is 1,073,000. This group is composed of the following areas:

Type of Engineer	Number
Aeronautical and Astronautical	64,000
Civil	164,000
Electrical and Electronic	366,000
Industrial	238,000
Mechanical	241,000

12. The average American pays $215.43 per year in direct health costs. This amount breaks down into the following categories.

Type of Expenditure	Payment
Hospital care	$36.74
Doctor bills	$58.21
Dentist bills	$43.92
Drugs	$54.64
Eyeglasses	$12.80
Others	$ 9.12

13. Of the 7,120,224 life insurance policy dividends paid in the United States last year, 1,823,472 were taken in cash, 2,691,792 were used to purchase additional paid-up life insurance, and 2,604,960 were left with the insurance companies to earn interest.

•14. The oil reserves in the Western hemisphere are estimated at 133 billions of barrels. The following is the approximate dispersement of the oil in billions of barrels: United States—38.7; South America—23.4; Canada—9.6; and Mexico—61.3.

15. In the 1980 Winter Olympic games, 38 gold medals were awarded in the following fashion: Soviet Union—10; East Germany—9; United States—6; Austria—3; Sweden—3; Liechtenstein—2; Others—5.

16. In the United States last year, there were 102,555 accidental deaths in the following categories:

Category	Number
Motor vehicle	49,235
Falls	13,760
Fires	6,235
Drowning	6,020
Firearms	1,935
Others	25,370

17. It is estimated that last year, the total vehicle fuel consumption was 126,087,094,000,000 gal. The breakdown of the use of fuel was as follows:

Vehicle	Fuel Consumed
Passenger vehicles	84,794,300,000,000
Buses	1,021,514,000,000
Single unit trucks	27,780,000,000,000
Tractor-trailer trucks	12,491,280,000,000

Construct a broken-line graph to represent the following data.

18. The temperatures recorded for Detroit on March 27 were as follows:

Temperature (°F)	48	49	51	44	39	35	34
Time (P.M.)	12	2	4	6	8	10	12

19. The following is a list of the number of mobile homes manufactured over a 10-year period.

Number made	401,000	498,000	574,000	563,000	328,000
Year	1970	1971	1972	1973	1974
Number made	214,000	251,000	264,000	275,000	276,000
Year	1975	1976	1977	1978	1979

•20. The Dow Jones Industrial Stock Averages for a 1-yr period were as follows:

Month	Average
October	915
November	822
December	838
January	875
February	863
March	785
April	805
May	850
June	867
July	935
August	932
September	932

21. The Annual Railroad Carloadings for the years 1970–1979 were as follows:

Year	Total
1970	27,200,000
1971	25,300,000
1972	26,100,000
1973	27,350,000
1974	26,175,000
1975	23,225,000
1976	23,450,000
1977	23,175,000
1978	23,375,000
1979	23,875,000

22. The following is a list of the number of home runs that Babe Ruth hit during the years that he was with the New York Yankees: 1920—54; 1921—59; 1922—35; 1923—41; 1924—46; 1925—25; 1926—47; 1927—60; 1928—54; 1929—46; 1930—49; 1931—46; 1932—41; 1933—34; 1934—22.

23. The following table gives the amount of material necessary to make a cubical container with a certain length of an edge.

Length of Edge (m)	Material (m^2)
1	6
2	24
3	54
4	96
5	150
6	216

Chapter Summary

1. Solutions of equations in the two variables x and y are **ordered pairs,** (x,y).
2. A **function** is a rule or correspondence where each value of the independent variable gives exactly one value of the dependent variable.
3. The **domain** of the function is the collection of all the numbers that can be used as values for the independent variable.
4. The **range** of a function is the collection of all possible values that the dependent variable can take on.
5. The graph of a **linear function** is a line.
6. The graph of a **quadratic function** is a parabola.
7. To find the ***x*-intercepts,** if there are any, we let $y = 0$ in the equation and solve for x, $(x,0)$.
8. To find the ***y*-intercepts,** if there are any, we let $x = 0$ in the equation and solve for y, $(0,y)$.
9. The **vertex** of a quadratic function in standard form, $f(x) = ax^2 + bx + c$, will be the point $\left(\frac{-b}{2a}, f\left(\frac{-b}{2a}\right)\right)$.
10. To graph an equation, we plot a sufficient number of points to allow some pattern to emerge, and then we sketch the graph accordingly.
11. The **bar graph** is used to compare similar things or events.
12. The **circle graph** shows the relationship between parts and the whole quantity they comprise.
13. The **broken-line graph** is used to show changing conditions.

Chapter Review

[15.1] Plot the following ordered pairs on a coordinate system.

1. $(1,5)$
2. $(4,-4)$
3. $(-1,-6)$
4. $(0,-4)$
5. $\left(2,\frac{1}{2}\right)$
6. $\left(5,-\frac{2}{3}\right)$

Determine to the nearest integer the coordinates (x,y) of the given points.

7. *A*
8. *B*
9. *C*
10. *D*
11. *E*
12. *F*

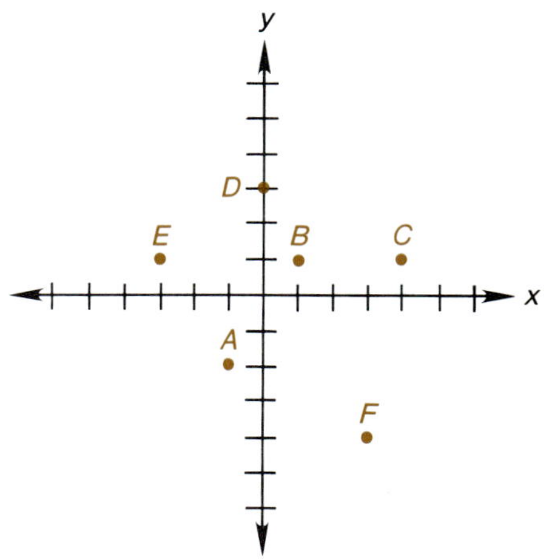

Find the missing component in each ordered pair using the given equation. Then plot the ordered pairs using a separate coordinate system for each problem.

13. $y = 3x + 4$ $\quad (2,\ ),\ (0,\ ),\ \left(\frac{4}{3},\ \right),\ \left(-\frac{3}{2},\ \right)$

14. $y = -2x + 5$ $\quad (-2,\ ),\ \left(\frac{5}{2},\ \right),\ (0,\ ),\ (3,\ )$

15. $y = x^2 + 2$ $\quad (1,\ ),\ (-1,\ ),\ (0,\ ),\ (2,\ )$

16. $y = x^2 - 16$ $\quad (4,\ ),\ \left(\frac{7}{2},\ \right),\ (0,\ ),\ (-4,\ )$

[15.2] Find the indicated value of the given function.

17. $f(x) = 2x - 3$ $\quad (-3,\ ),\ \left(\frac{3}{2},\ \right),\ (0,\ )$

18. $f(r) = r^2 + 3r$ $\quad (0,\ ),\ (-2,\ ),\ (-3,\ )$

19. $f(t) = t^2 + 2t - 1$ $\quad (2,\ ),\ (-1,\ ),\ (0,\ )$

20. $f(a) = \dfrac{a + 1}{2}$ $\quad \left(\frac{1}{2},\ \right),\ (0,\ ),\ (-1,\ )$

21. $f(s) = \dfrac{s}{s^2 + 1}$ $\quad (2,\ ),\ (-2,\ ),\ (0,\ )$

Determine the domain of the given function.

22. $f(x) = \sqrt{x + 4}$

23. $f(x) = \dfrac{1}{x + 1}$

[15.3] Graph the given functions and give the x- and y-intercepts.

24. $y = 2x - 3$

25. $y = \frac{1}{2}x + 4$

26. $2x - y = 7$

27. $y = 5x$

28. $5x - 2y + 4 = 0$

29. $6y - x - 2 = 0$

30. $\frac{1}{2}x - \frac{1}{4}y = \frac{3}{4}$

31. $\frac{x}{2} + y = 1$

[15.4] Graph the given functions and give the x- and y-intercepts and the vertex.

32. $y = x^2 - 1$

33. $y = x^2 + 6x + 8$

34. $y = x^2 - 8x + 12$

35. $y = 4 - x^2$

36. $y = x^2 + 3$

37. $y = x^2 + x + 3$

38. $y = 2x^2 + 7x + 3$

39. $y = -2x^2 - x + 1$

[15.5] Graph the given functions.

40. $y = x^3 - 8$

41. $y = 3$

42. $y = \dfrac{1}{x - 4}$

43. $y = x^4 - 1$

44. $y = \sqrt{3 - x}$

45. $y = \sqrt{x^2 - 9}$

[15.6]

46. Construct a broken-line graph to represent the following data: The temperatures recorded for Chicago on July 11 were as follows:

Temperature (°C)	22	24	29	28	23	20	17
Time (P.M.)	12	2	4	6	8	10	12

47. Construct a bar graph to represent the following data: The following is the percentage of unemployed workers in selected years since 1929:

Year	1929	1932	1941	1945	1950	1960	1970	1975	1980
Percent unemployed	3.2	23.6	9.9	1.9	5.3	5.5	4.9	8.5	7.8

48. Construct a circle graph to represent the following data: The following is the percentage of working women in various earnings groups:

Earnings group	Less than \$3,000	3,000–4,999	5,000–6,999	7,000–9,999	10,000–14,999	15,000 and over
Percent	3.0	5.7	17.5	29.8	30.8	13.4

Chapter Test

Graph the following functions and give their domains. If the function is linear, give the x- and y-intercepts. If the function is quadratic, give the x- and y-intercepts and the vertex.

1. $y = -2x + 1$
2. $3x - y = 4$
3. $5x = 2y$
4. $y = x^2 - 16$
5. $y = x^2 - 3x - 10$
6. $y = x^3 + 1$
7. $y = 2^{x+3}$
8. $y = 9 - x^2$
9. $5x - 2y + 7 = 0$
10. $y = \dfrac{1}{x + 3}$
11. $\dfrac{1}{2}x + \dfrac{3}{4}y - \dfrac{2}{3} = 0$
12. $y = x^2 + 4$
13. $y = x^4 - 9$
14. $\dfrac{x}{3} + \dfrac{y}{2} = 1$
15. $y = \sqrt{9 - x}$
16. $y = \dfrac{x}{2x - 4}$
17. $y = x^2 + 3x + 5$
18. $y = \sqrt{x^2 - 16}$
19. $y = \left(\dfrac{1}{2}\right)^{x-1}$
20. $y = 2x^2 + x - 3$

16
Simultaneous Linear Equations

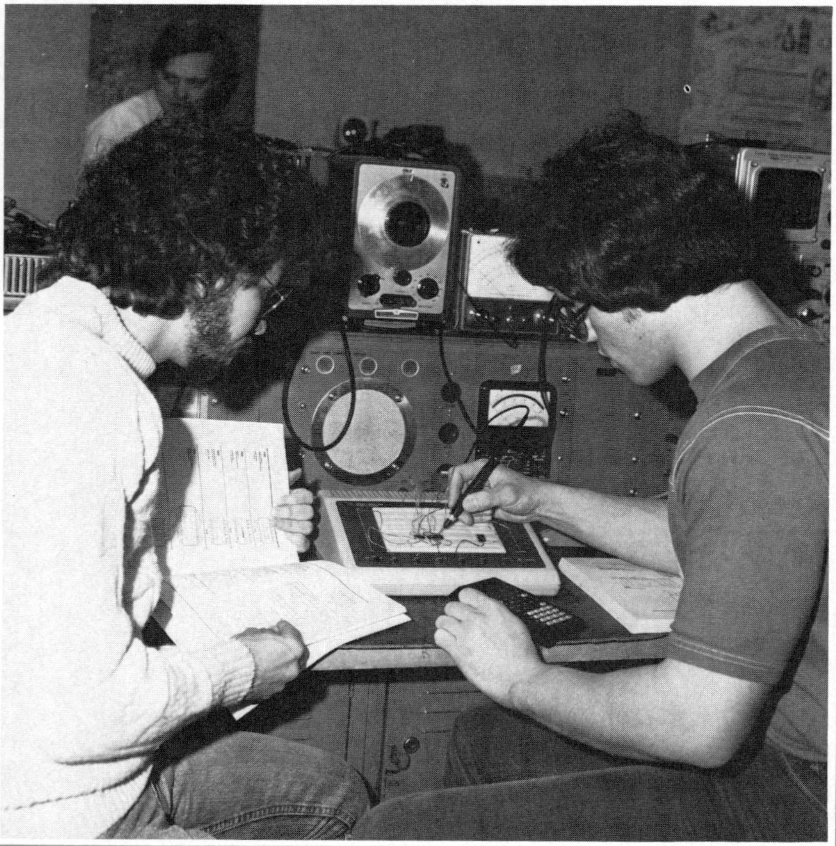

Technicians testing a digital logic circuit.
(Photograph by Kenneth Walters, courtesy of Henry Ford Community College)

16.1 Graphical Solution

There are many physical situations where it is necessary to deal with two equations involving the same two unknowns. Such a situation is called a **system of simultaneous equations.** The following are some examples.

Examples

$4x - y = 1$	$3R - 7S = 5$	$5a + 3b = -20$
$x + y = 4$	$5R - 2S = -11$	$7a + 2b = -17$

When solving such a system, we are seeking the ordered pairs (x,y) such that, when they are substituted into all of the original equations, the resulting statements will be true. These common points are points at which the graphs of the equations intersect.

Examples

Determine if the following are solutions to the given system of equations.

1. (1,3) is a solution of the system

$$\begin{aligned} 4x - y &= 1 \\ x + y &= 4 \end{aligned}$$

because if we substitute 1 for x and 3 for y in the system,

$$\begin{aligned} 4(1) - (3) &= 1 \\ 4 - 3 &= 1 \\ 1 &= 1 \end{aligned} \qquad \text{and} \qquad \begin{aligned} (1) + (3) &= 4 \\ 4 &= 4 \end{aligned}$$

Hence, the resulting statements are both true. Therefore, the point (1,3) satisfies both equations at the same time (simultaneously).

2. $R = -3$ and $S = -2$ is a solution of

$$\begin{aligned} 3R - 7S &= 5 \\ 5R - 2S &= -11 \end{aligned}$$

because when we substitute the values into the equations, both statements are true.

$$\begin{aligned} 3(-3) - 7(-2) &= 5 \\ -9 - (-14) &= 5 \\ -9 + 14 &= 5 \\ 5 &= 5 \end{aligned} \qquad \text{and} \qquad \begin{aligned} 5(-3) - 2(-2) &= -11 \\ -15 - (-4) &= -11 \\ -15 + 4 &= -11 \\ -11 &= -11 \end{aligned}$$

3. $a = -4$ and $b = 0$ is not a solution of

$$\begin{aligned} 5a + 3b &= -20 \\ 7a + 2b &= -17 \end{aligned}$$

because when substituted into the equations, both are not true:

$$\begin{aligned} 5(-4) + 3(0) &= -20 \\ -20 + 0 &= -20 \\ -20 &= -20 \\ &\text{TRUE} \end{aligned} \qquad \text{and} \qquad \begin{aligned} 7(-4) + 2(0) &= -17 \\ -28 + 0 &= -17 \\ -28 &= -17 \\ &\text{FALSE} \end{aligned}$$

The first method we examine of finding the solution of a system of equations is graphing. We are looking for the coordinates of points where the graphs intersect. This point of intersection will give the solution of the system. Consider the system $4x - y = 1$ and $x + y = 4$. We use the graphing techniques we learned in chapter 15 to graph both equations on the same coordinate plane.

note

Throughout this chapter, we deal only with systems of two linear equations, but a first step in dealing with any system of equations is to identify, if possible, the types of equations involved.

To graph the two equations, we determine the intercepts and one extra value as a check (see figure 16.1).

$4x - y = 1$ and $x + y = 4$

x	y		x	y
0	-1	y-intercept	0	4
$\frac{1}{4}$	0	x-intercept	4	0
1	3	arbitrary check	2	2

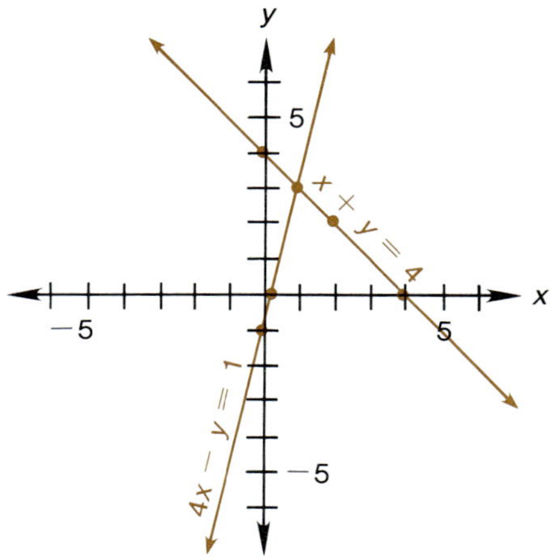

Figure 16.1

We now *approximate* the coordinates of the point of intersection. These appear to be (1,3), which is correct. It should be obvious that the graphical method of solution may not be exact, due to the fact that the drawing of the graph depends on measuring devices, and inaccuracies can thereby develop in the graphing process.

Examples

Graph the following system and approximate the coordinates of the point of intersection.

1. $5x - y = 5$ and $x + y = 2$

x	y		x	y
0	-5	y-intercept	0	2
1	0	x-intercept	2	0
$\frac{1}{2}$	$-\frac{5}{2}$	check	1	1

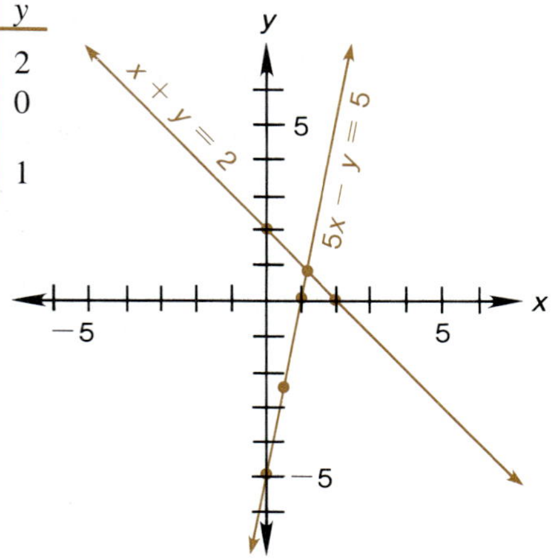

Figure 16.2

An estimate of the point of intersection (see figure 16.2) might be (1,1). We will see in the next section that the actual point of intersection is $\left(\frac{7}{6}, \frac{5}{6}\right)$.

2. $3x + y = 5$ and $x - 2y = 4$

x	y		x	y
0	5	y-intercept	0	-2
$\frac{5}{3}$	0	x-intercept	4	0
1	2	check	2	-1

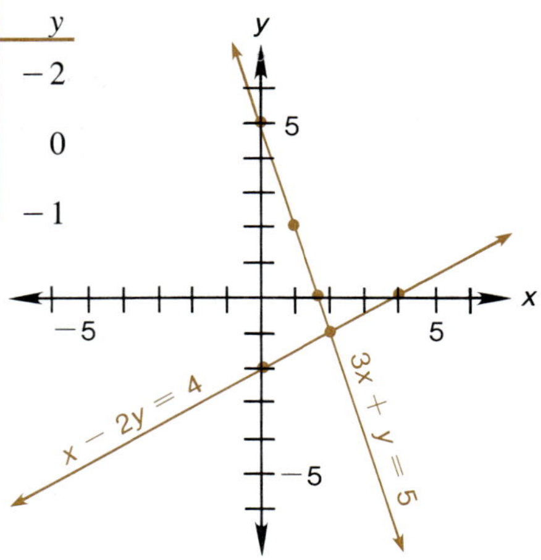

Figure 16.3

It appears that the point of intersection is $(2, -1)$ (see figure 16.3), and this is correct.

3. $\frac{1}{2}x + \frac{2}{3}y = \frac{1}{6}$ and $\frac{3}{4}x - \frac{1}{4}y = \frac{1}{2}$ The coefficients in this example are fractions, and as a first step we can simplify the problems by clearing all the fractions.

$\frac{1}{2}x + \frac{2}{3}y = \frac{1}{6}$ Multiply both sides by 6. $\rightarrow 6\left(\frac{1}{2}x + \frac{2}{3}y\right) = \frac{1}{6} \cdot 6$

$3x + 4y = 1$

$\frac{3}{4}x - \frac{1}{4}y = \frac{1}{2}$ Multiply both sides by 4. $\rightarrow 4\left(\frac{3}{4}x - \frac{1}{4}y\right) = \frac{1}{2} \cdot 4$

$3x - y = 2$

$3x + 4y = 1$

x	y	
0	$\frac{1}{4}$	y-intercept
$\frac{1}{3}$	0	x-intercept
-1	1	check

$3x - y = 2$

x	y
0	-2
$\frac{2}{3}$	0
-1	-5

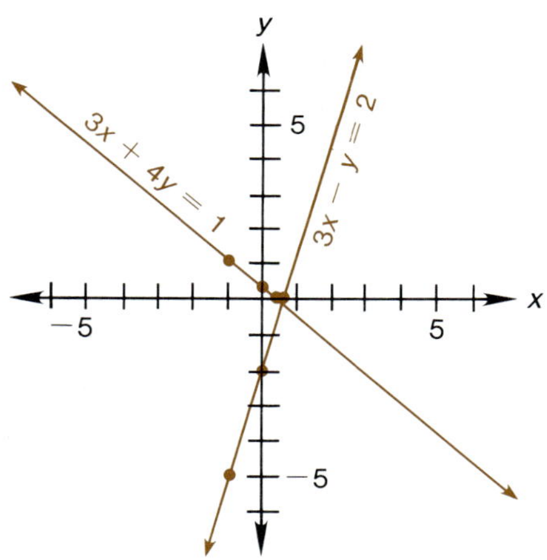

Figure 16.4

The point of intersection appears to be $(0.6, -0.2)$ (see figure 16.4), which is correct. A system of linear equations that intersect at one point is called a **consistent** and **independent** system of equations.

In chapter 12 on geometry, we learned that parallel lines never intersect. If this is the situation between the lines in our system of equations, we say that the system is **inconsistent** and there are no solutions (no points of intersection).

Example $3x - 4y = 12$ and $6x - 8y = -24$

x	y		x	y
0	-3	y-intercept	0	3
4	0	x-intercept	-4	0
2	$-\frac{3}{2}$	check	-2	$\frac{3}{2}$

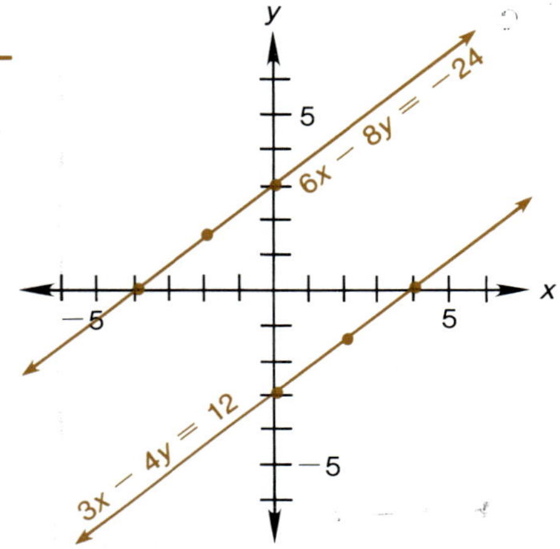

Figure 16.5

The lines appear to be parallel (they are), and they do not intersect (see figure 16.5).

In addition to getting one answer, or no answers, there is one more possible situation that can exist between two lines. If all the terms in the first equation are multiplied by the same constant to produce a second equation, then each equation produces the same line and the system is **dependent.** In a dependent system of equations, an unlimited number of solutions is possible because each ordered pair that satisfies the first equation also satisfies the second.

Example $2x - y = 3$ and $4x - 2y = 6$

x	y		x	y
0	-3	y-intercept	0	-3
$\frac{3}{2}$	0	x-intercept	$\frac{3}{2}$	0
1	-1	check	1	-1

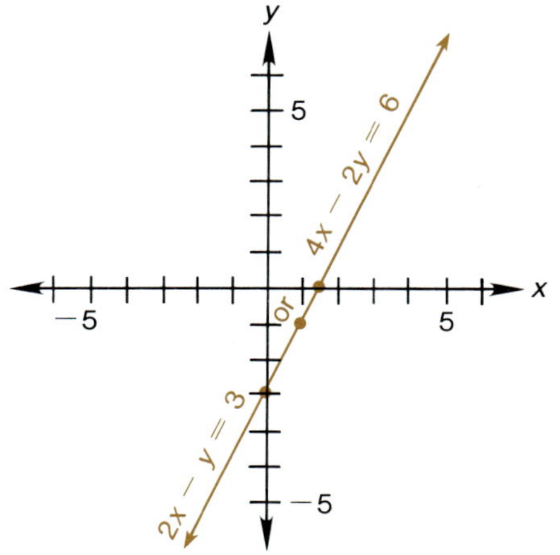

Figure 16.6

We observe that the intercepts and checkpoint are the same for both lines; therefore, the two equations are equivalent and the graph is a single line (see figure 16.6). Since both lines are the same, any point that satisfies one of the equations must also satisfy the other. There is no unique solution to a dependent system of equations.

It is helpful to know what type of system of equations we are dealing with before we graph the problem. We can find this information by observing the three ratios of the coefficients of x and y and the constants. The following are guidelines for a general system. If we are dealing with a system of two linear equations given by

$a_1x + b_1y = c_1$,

$a_2x + b_2y = c_2$ where a_2, b_2, and c_2 are not zero, then

1. It will be a consistent and independent system (one solution) if the ratios of the coefficients of x and y, $\frac{a_1}{a_2}$ and $\frac{b_1}{b_2}$, are not equal: $\frac{a_1}{a_2} \neq \frac{b_1}{b_2}$.
2. It will be an inconsistent system (no solutions) if the ratios of the coefficients of x and y are equal but their ratio is not equal to the ratio of the constants $\frac{c_1}{c_2}$: $\left(\frac{a_1}{a_2} = \frac{b_1}{b_2}\right) \neq \frac{c_1}{c_2}$.
3. It will be a dependent system (unlimited number of solutions) if all three ratios are equal: $\frac{a_1}{a_2} = \frac{b_1}{b_2} = \frac{c_1}{c_2}$.

Examples

Determine the number of solutions in each system.

1. $4x - y = 1$ and $x + y = 4$

   Note that $a_1 = 4$, $b_1 = -1$, $c_1 = 1$; $a_2 = 1$, $b_2 = 1$, $c_2 = 4$; and the ratios of the coefficients are $\frac{4}{1} \neq \frac{-1}{1}$, that is, $\frac{a_1}{a_2} \neq \frac{b_1}{b_2}$. Thus, as we saw in our previous examples, this system does have one solution, (1,3).

2. $5x - y = 5$ and $x + y = 2$

   Note that $a_1 = 5$, $b_1 = -1$, $c_1 = 5$; $a_2 = 1$, $b_2 = 1$, $c_2 = 2$. The ratios are $\frac{5}{1} \neq \frac{-1}{1}$, that is, $\frac{a_1}{a_2} \neq \frac{b_1}{b_2}$, and there is one solution, $\left(\frac{7}{6}, \frac{5}{6}\right)$.

3. $y + 3x = 4$ and $2y + 6x = 12$

   Note that $a_1 = 3$, $b_1 = 1$, $c_1 = 4$; $a_2 = 6$, $b_2 = 2$, $c_2 = 12$. The ratios are $\frac{3}{6} = \frac{1}{2} \neq \frac{4}{12}$, that is, $\frac{a_1}{a_2} = \frac{b_1}{b_2} \neq \frac{c_1}{c_2}$. The first two ratios are equal; therefore, we compare them to the third ratio and observe that this is an inconsistent system of equations, which has no solutions.

4. $2x - y = 3$ and $4x - 2y = 6$

   Note that $a_1 = 2$, $b_1 = -1$, $c_1 = 3$; $a_2 = 4$, $b_2 = -2$, $c_2 = 6$. The ratios are $\frac{2}{4} = \frac{-1}{-2} = \frac{3}{6}$, that is, $\frac{a_1}{a_2} = \frac{b_1}{b_2} = \frac{c_1}{c_2}$. This is a dependent system of equations with an unlimited number of solutions.

note

We can determine whether or not two lines are parallel by comparing the ratios of the coefficients to the ratio of the constants. This is a useful procedure for determining if lines are parallel since, in the graph, they may appear to be parallel but if the graph is extended, the two lines may intersect.

Exercise Set 16.1

Determine the number of solutions in each system. If the system has no solutions, state that it is inconsistent. If the system has an unlimited number of solutions, state that it is dependent.

1. $2x + y = 7$
   $x - 3y = 5$

•2. $x - 3y = 4$
   $x + 3y = 8$

3. $2x + 3y = -2$
   $2x - 3y = 4$

•4. $x - 2y = 1$
   $3x - 6y = 4$

5. $2x + y = -3$
$6x + 3y = 2$

6. $5x + 2y = 3$
$10x + 4y = 6$

7. $-2x + 3y = 4$
$4x - 6y = -8$

8. $5x - 7y = 9$
$x + 14y = 18$

9. $\frac{1}{2}x + \frac{2}{3}y = \frac{1}{4}$
$\frac{3}{2}x + 2y = \frac{3}{4}$

10. $-2x - 6y = 10$
$3x - 3y = 5$

Determine if the given ordered pair is a solution to the system of equations.

11. $x + y = 2$
$3x - y = 10$ $(3,-1)$

•12. $x + 2y = 4$
$x + 4y = 10$ $(-2,3)$

13. $x - y = 1$
$x + y = 5$ $(3,2)$

14. $3x - y = -10$
$3x + y = 4$ $(-1,7)$

15. $2x + y = 2$
$6x - y = 22$ $(3,-4)$

16. $x - 3y = -1$
$3x + y = -3$ $(-1,0)$

17. $3x + 4y = 14$
$2x + y = 6$ $(1,4)$

•18. $3x + y = 12$
$2x + 3y = 15$ $(2,6)$

Solve the following systems by graphing. State your answer to the precision that your instructor suggests. All answers in the back of the book are given exactly. If the system is inconsistent or dependent, so state.

19. $x - y = 1$
$x + y = 5$

20. $x + 2y = 4$
$x + 4y = 10$

21. $x - y = 3$
$2x + 3y = 11$

22. $3x - y = 10$
$x + y = 2$

23. $2x - y = 3$
$2x + y = -1$

24. $2x + y = 1$
$3x - 2y = -9$

25. $2x + 3y = 13$
$3x - 2y = 0$

26. $3x + y = 10$
$6x + 2y = 5$

27. $2x + y = 2$
$-6x - 3y = 4$

•28. $x - 2y = 4$
$3x + 2y = 4$

29. $2x + y = 6$
$3x + 4y = 4$

30. $3x + y = -3$
$x - 3y = -1$

31. $2x + y = 1$
$-3x + y = 1$

32. $2x - y = 3$
$6x - 3y = 9$

33. $2x + 3y = 4$
$-8x - 12y = -16$

34. $\frac{3}{2}x - \frac{1}{2}y = -5$
$x + \frac{1}{3}y = \frac{4}{3}$

35. $\frac{1}{3}x - \frac{1}{2}y = -\frac{5}{6}$
$\frac{1}{2}x + \frac{1}{3}y = 2$

36. $x - \frac{1}{3}y = -4$
$2x + \frac{3}{2}y = 5$

37. $5x - y = 7$
$2x + y = 4$

38. $-3x - y = 0$
$x + 3y = 5$

39. $5x + y = 10$
$9x + y = -20$

40. $3x - 2y = 7$
$x - y = 9$

41. $6x - 5y = 14$
$4x + y = 8$

42. $8x - 12y = 16$
$2x - 3y = 4$

43. $3x - 6y = 12$
$x - 2y = 6$

44. $3x + y = 10$
$2x - y = 5$

45. $\frac{1}{2}x - \frac{3}{4}y = 6$
$-\frac{2}{3}x + \frac{1}{4}y = 2$

46. $2x + 3y = 15$
$3x + y = 12$

47. $5x + y = 10$
$5x + 3y = 15$

48. $-3x + y = 4$
$2x - 4y = 5$

49. $x + 2y = 7$
$x - 2y = 14$

50. $3x + y = 5$
$6x - y = 10$

16.2 Solution by Elimination

In section 16.1, we observed that we could find a solution to a system of linear equations by graphing the lines and approximating the coordinates of the point of intersection. This technique unfortunately does not usually give the exact answer to the problem. For example, when we graphed the system of equations $5x - y = 5$ and $x + y = 2$, we approximated our answer to be $(1,1)$, but the true answer is actually $\left(\frac{7}{6}, \frac{5}{6}\right)$.

When the problem requires an *exact* solution, we must use algebraic methods to solve the system. We continue to graph the problems also even though the graph is only an approximation, because the graph allows us to visualize the system and verify the results that we find algebraically.

In chapter 7 we learned that we could add or subtract the same quantity from both sides of a linear equation and form an equivalent equation. Keeping this idea in mind, together with the fact that in an equation both sides represent the same number, we are ready to solve the system $5x - y = 5$ and $x + y = 2$ algebraically. In the equation $x + y = 2$, the $x + y$ and the 2 are equal; therefore, if we add the $x + y$ to the left-hand side of $5x - y = 5$ and add the 2 to the right-hand side, we are adding the same amount to both sides of the equation, and are forming an equivalent equation:

$$5x - y = 5$$
$$5x - y + (x + y) = 5 + (2)$$

Simplifying, we have

$$5x - y + x + y = 7$$
$$6x = 7$$

We have formed a linear equation with one unknown that can be easily solved.

$$6x = 7$$
$$x = \frac{7}{6}$$

At this point, the x-coordinate of our solution has been determined, and to complete our solution, we substitute this value of x into one of the original equations to generate the corresponding y value. If $x = 7/6$, then

$$5\left(\frac{7}{6}\right) - y = 5$$

$$\frac{35}{6} - y = 5$$

$$-y = \frac{30}{6} - \frac{35}{6}$$

$$-y = -\frac{5}{6}$$

$$y = \frac{5}{6}$$

Therefore, the exact (algebraic) solution to this system of equations is $\left(\frac{7}{6}, \frac{5}{6}\right)$.

In this problem, we were able to solve the system by combining the two original equations in such a way as to *eliminate* one of the variables and thereby form a single equation with only one unknown, which could easily be solved. This value is one of the components of the point of intersection; the other component is found by substituting this value into one of the original equations, thereby generating the other half of the ordered pair. You should realize that it makes no difference which variable is eliminated.

Examples

Solve the following systems of equations by elimination.

1. $2x + 3y = 5$ and $-2x + y = 3$

By adding the left sides together and the right sides together we can eliminate x:

$$\begin{array}{r} 2x + 3y = 5 \\ -2x + y = 3 \\ \hline 4y = 8 \end{array}$$

Solving the resulting equation we have $y = 2$. Substituting this into one of the original equations and solving for x:

$$2x + 3(2) = 5$$

$$2x + 6 = 5$$

$$2x = -1$$

$$x = -\frac{1}{2}$$

The solution is $\left(-\frac{1}{2}, 2\right)$.

2. $x + 3y = 4$ and $2x - 3y = 2$

The steps are as follows:

$$\begin{array}{rl} x + 3y = 4 & \\ \underline{2x - 3y = 2} & \\ 3x \quad = 6 & \text{Add the equations.} \\ x = 2 & \text{Solve for the single unknown.} \\ (2) + 3y = 4 & \text{Substitute this value into} \\ 3y = 2 & \text{an original equation.} \\ y = \frac{2}{3} & \end{array}$$

Therefore, the solution is $\left(2,\frac{2}{3}\right)$.

In the preceding examples we observe that a variable could be eliminated because when we added the original equations, one of the variables had coefficients that were opposites. Consider the system $2x - 5y = -1$ and $3x + y = 7$. When these equations are added,

$$\begin{array}{r} 2x - 5y = -1 \\ \underline{3x + \;\; y = 7} \\ 5x - 4y = 6 \end{array}$$

both variables still remain. *To eliminate one of the variables, we must multiply one or both equations by some constant so that equivalent equations are formed where one of the variables has coefficients that are opposites.* For example, the system could be solved as follows:

$$\begin{array}{lll} 2x - 5y = -1 & & 2x - 5y = -1 \\ 3x + y = 7 & \text{Multiply by 5.} & \underline{15x + 5y = 35} \\ & \text{Add the equations.} & 17x \quad = 34 \\ & \text{Solve for } x. & x = 2 \\ & \text{Substitute.} & 3(2) + y = 7 \\ & & 6 + y = 7 \\ & & y = 1 \end{array}$$

Solution (2,1)

Examples

Solve by elimination.

1. $3x + 4y = 14$
   $2x + y = 6$

Multiply by -4.

$$\begin{aligned} 3x + 4y &= 14 \\ -8x - 4y &= -24 \end{aligned}$$

Add the equations. $-5x = -10$

Solve for x. $x = 2$

Substitute.

$$\begin{aligned} 2(2) + y &= 6 \\ 4 + y &= 6 \\ y &= 2 \end{aligned}$$

Solution (2,2)

2. $7x + 5y = -9$ Multiply by 2. $14x + 10y = -18$
   $3x - 2y = -8$ Multiply by 5. $15x - 10y = -40$

Add the equation. $29x = -58$

Solve for x. $x = -2$

Substitute.

$$\begin{aligned} 7(-2) + 5y &= -9 \\ -14 + 5y &= -9 \\ 5y &= 5 \\ y &= 1 \end{aligned}$$

Solution $(-2,1)$

note

In this example, both equations have to be multiplied by a constant in order to form opposite coefficients of y. We could have eliminated x by multiplying the top equation by 3 and the bottom equation by (-7), forming opposite coefficients of 21 and (-21). The choice of which variable to eliminate is an arbitrary one, but for ease in solving the problem, it is best to eliminate the variable whose opposite coefficients are least in value.

3. $\frac{1}{2}x + \frac{2}{3}y = \frac{1}{6}$ and $\frac{3}{4}x - \frac{1}{4}y = \frac{1}{2}$. The first step in this problem is to clear the fractions.

$\frac{1}{2}x + \frac{2}{3}y = \frac{1}{6}$ Multiply both sides by 6.

$$\begin{aligned} 6\left(\frac{1}{2}x + \frac{2}{3}y\right) &= \frac{1}{6} \cdot 6 \\ 3x + 4y &= 1 \end{aligned}$$

$\frac{3}{4}x - \frac{1}{4}y = \frac{1}{2}$ Multiply both sides by 4.

$$\begin{aligned} 4\left(\frac{3}{4}x - \frac{1}{4}y\right) &= \frac{1}{2} \cdot 4 \\ 3x - y &= 2 \end{aligned}$$

We now solve the system:

$$\begin{aligned} 3x + 4y &= 1 \\ 3x - y &= 2 \end{aligned}$$

Multiply by (-1).

$$\begin{aligned} 3x + 4y &= 1 \\ -3x + y &= -2 \end{aligned}$$

Add the equations. $5y = -1$

Solve for y. $y = -\frac{1}{5}$

Substitute.

$$\begin{aligned} 3x - \left(-\frac{1}{5}\right) &= 2 \\ 3x + \frac{1}{5} &= 2 \\ 3x &= 2 - \frac{1}{5} \\ 3x &= \frac{9}{5} \\ x &= \frac{9}{5} \cdot \frac{1}{3} \\ x &= \frac{3}{5} \end{aligned}$$

Solution $\left(\frac{3}{5}, -\frac{1}{5}\right)$

4. $(0.2)x - (0.3)y = 1.1$
   $(0.1)x + (0.4)y = 0.8$

   We can clear the decimals by multiplying by 10.

$$\begin{aligned} 2x - 3y &= 11 \\ x + 4y &= 8 \end{aligned}$$

We now solve the system.

$$\begin{aligned} 2x - 3y &= 11 \\ x + 4y &= 8 \end{aligned}$$

Multiply by (-2).

$$\begin{aligned} 2x - 3y &= 11 \\ -2x - 8y &= -16 \end{aligned}$$

Add the equations. $-11y = -5$

Solve for y. $y = \frac{5}{11}$

Substitute.

$$\begin{aligned} x + 4\left(\frac{5}{11}\right) &= 8 \\ x + \frac{20}{11} &= 8 \\ x &= 8 - \frac{20}{11} \\ x &= \frac{88}{11} - \frac{20}{11} = \frac{68}{11} \end{aligned}$$

Solution $\left(\frac{68}{11}, \frac{5}{11}\right)$

5. $3R - 4S = 5$ $\qquad$ $3R - 4S = 5$

$4R - 2S = 6$ Multiply by (-2). $\quad -8R + 4S = -12$

Add the equations. $\quad -5R = -7$

Solve for R. $\quad R = \frac{7}{5}$

Substitute. $\quad 4\left(\frac{7}{5}\right) - 2S = 6$

$$\frac{28}{5} - 2S = 6$$

$$-2S = \frac{30}{5} - \frac{28}{5}$$

$$-2S = \frac{2}{5}$$

$$S = \left(\frac{2}{5}\right)\left(-\frac{1}{2}\right)$$

$$S = -\frac{1}{5}$$

Solution $R = \frac{7}{5}$ and $S = -\frac{1}{5}$.

note

We do not leave our answer as $\left(\frac{7}{5}, -\frac{1}{5}\right)$, since it is not clear which value in the parentheses represents which variable.

In section 16.1, the idea of a dependent or inconsistent system of equations was discussed. If the ratios of the coefficients have not been examined, which is usually the case, the following situation occurs.

Consider the following dependent system: $4x - 2y = 6$ and $2x - y = 3$. The solution is as follows:

$4x - 2y = 6$ $\qquad$ $4x - 2y = 6$

$2x - y = 3$ Multiply by (-2). $\quad -4x + 2y = -6$

Add the equations. $\quad 0x + 0y = 0$

$$0 = 0$$

At this point both variables have been eliminated, and we have formed the true statement that $0 = 0$. This result indicates that any ordered pair that satisfies either equation is a solution, and there are an unlimited number of solutions.

If we are dealing with an inconsistent system, the following situation occurs. Given the system $3x - 4y = 12$ and $6x - 8y = -24$, we solve as follows:

$3x - 4y = 12$	Multiply by (-2)	$-6x + 8y = -24$
$6x - 8y = -24$		$6x - 8y = -24$
	Add the equations.	$0 = -48$

The resulting statement, $0 = -48$, is false, thus indicating that there is no ordered pair that can satisfy this system.

In summary, to solve a system of linear equations whose form is

$$a_1x + b_1y = c_1$$
$$a_2x + b_2y = c_2$$

by elimination, we use the following procedure:

1. Multiply both sides of one or both equations, if necessary, by some nonzero constant such that the coefficients of one of the unknowns become opposites.
2. Add the two equations, thereby forming a single equation with one unknown.
3. Solve this newly formed equation for the unknown value.
4. Substitute the value found in step 3 into either of the original equations and determine the remaining unknown value.
5. If the results to step 2 are:
   a. $0 = 0$, the answer is a dependent system of equations.
   b. $0 =$ some number other than zero, the answer is an inconsistent system of equations.

Exercise Set 16.2

Solve the following systems by elimination. Graph the system to verify your solution. If a system is inconsistent or dependent, so state.

1. $x + 2y = 4$
   $x + 4y = 10$
2. $x - y = 1$
   $x + 5y = 5$
3. $x + y = 2$
   $3x - y = 10$
4. $x - y = 3$
   $2x + 3y = 11$
5. $3x - 2y = -9$
   $2x + y = 1$
•6. $5x - y = 4$
   $x + 3y = 2$
7. $4x + y = 11$
   $2x + 3y = 4$
•8. $3x - 2y = 6$
   $4x - 3y = 7$
9. $5x + 2y = 3$
   $3x - 3y = 4$
10. $4x + 7y = 11$
   $8x - 3y = -4$

Solve the following systems by elimination. If a system is inconsistent or dependent, so state. Leave your answer as an ordered pair, (x,y).

11. $3x + y = 10$
   $6x + 2y = 5$
12. $2x + y = 2$
   $-6x - 3y = 4$
13. $x - 2y = 4$
   $3x + 2y = 4$
14. $2x + y = 6$
   $3x + 4y = 4$
15. $3x + y = -3$
   $x - 3y = -1$
16. $2x + y = 1$
   $-3x + y = 1$

17. $8x - 4y = 12$
$2x - y = 3$

18. $x - 2y = 1$
$3x - 6y = 3$

19. $5x - 2y = 4$
$3x + 3y = 5$

20. $-2x + 3y = 6$
$4x + y = 1$

21. $\frac{1}{2}x + \frac{1}{3}y = 1$
$\frac{2}{3}x - \frac{1}{4}y = \frac{1}{12}$

•22. $\frac{2}{3}x - \frac{1}{4}y = 4$
$\frac{1}{3}x + y = 2$

23. $(0.4)x - (0.7)y = 0.7$
$(0.6)x - (0.5)y = 2.7$

•24. $x + (0.4)y = 3.4$
$(0.6)x - (1.4)y = 0.4$

25. $\frac{1}{2}x + \frac{2}{5}y = \frac{7}{10}$
$\frac{3}{2}x + \frac{6}{5}y = \frac{3}{10}$

•26. $\frac{6}{7}x - \frac{3}{5}y = \frac{9}{10}$
$\frac{2}{7}x - \frac{1}{5}y = \frac{5}{10}$

27. $10x + y = 5$
$2x + 5y = -23$

28. $\frac{7}{2}x - y = -\frac{17}{2}$
$\frac{2}{3}x + \frac{1}{3}y = 1$

29. $\frac{9}{2}x + \frac{1}{2}y = -10$
$\frac{1}{3}x + \frac{1}{4}y = \frac{3}{4}$

30. $10x - 5y = 7$
$2x - y = 4$

31. $-6x + 3y = 9$
$2x - y = -3$

32. $5x - 3y = 34$
$4x + 5y = 5$

33. $3x - 2y = 19$
$5x + 3y = 19$

34. $4x + 5y = -2$
$3x - 2y = -36$

Solve the following systems by elimination. If a system is inconsistent or dependent, so state.

35. $2R - S = 5$
$R + S = 7$

36. $T - 4U = 11$
$2T - 2U = 5$

37. $7a + 2b = -3$
$a - 4b = 5$

38. $6L - M = -4$
$3L + 2M = -1$

39. $2R + S = 13$
$3R + S = 17$

40. $2P + Q = 2$
$6P - Q = 22$

41. $2a + 3b = 13$
$3a - 2b = 0$

42. $5D + 2E = 17$
$3D - 7E = 2$

43. $8R - 4S = 12$
$2R - S = 3$

•44. $15a - 9b = 12$
$5a - 3b = 4$

45. $3N - 2M = 19$
$5N + 3M = 19$

46. $2L + 3M = -2$
$8L + 12M = -3$

47. $2D - 3E = 6$
$14D - 21E = -3$

16.3 Solution by Substitution

In chapter 7 we learned that if two quantities were equal, we could replace the one quantity with the other and thereby form an equivalent equation. We use this principle to solve systems of equations by *substitution*.

Consider the system $5x - y = 5$ and $x + y = 2$. We can solve the second equation for y as follows:

$$x + y = 2$$
$$y = 2 - x$$

Using the principle of substitution, we replace y in the other equation by its equal, $2 - x$, as follows:

$$5x - y = 5$$
$$5x - (2 - x) = 5$$

The resulting equation has one unknown and can be solved for x.

$$5x - 2 + x = 5$$
$$6x - 2 = 5$$
$$6x = 7$$
$$x = \frac{7}{6}$$

At this point the procedure is the same as before and the other unknown is found by substitution:

$$x + y = 2$$
$$\left(\frac{7}{6}\right) + y = 2$$
$$y = 2 - \frac{7}{6}$$
$$y = \frac{12}{6} - \frac{7}{6}$$
$$y = \frac{5}{6}$$

The solution is $\left(\frac{7}{6}, \frac{5}{6}\right)$.

Examples Solve by substitution.

1. $3x + 2y = 14$
   $2x - y = 7$

   Solve the bottom equation for y:

   $$2x - y = 7$$
   $$2x = 7 + y$$
   $$2x - 7 = y$$

   Substitute into the top equation:

   $$3x + 2y = 14$$
   $$3x + 2(2x - 7) = 14$$

Solve:

$$\begin{aligned} 3x + 4x - 14 &= 14 \\ 7x - 14 &= 14 \\ 7x &= 28 \\ x &= 4 \end{aligned}$$

Now solve for the remaining unknown:

$$\begin{aligned} y &= 2x - 7 \\ y &= 2(4) - 7 \\ y &= 8 - 7 \\ y &= 1 \end{aligned}$$

The solution is (4,1).

note

When we determine the second unknown value, we use the equation that we used for substitution.

2. $3x + 2y = 11$
   $x - 2y = 4$
   Solve the second equation for x,

$$\begin{aligned} x - 2y &= 4 \\ x &= 2y + 4 \end{aligned}$$

Substitute:

$$\begin{aligned} 3x + 2y &= 11 \\ 3(2y + 4) + 2y &= 11 \end{aligned}$$

Solve:

$$\begin{aligned} 6y + 12 + 2y &= 11 \\ 8y + 12 &= 11 \\ 8y &= -1 \\ y &= -\frac{1}{8} \end{aligned}$$

Determine the remaining unknown:

$$\begin{aligned} x &= 2y + 4 \\ x &= 2\left(-\frac{1}{8}\right) + 4 \\ x &= -\frac{1}{4} + 4 \\ x &= \frac{15}{4} \end{aligned}$$

The solution is $\left(\frac{15}{4}, -\frac{1}{8}\right)$.

3. $2R - 3S = 4$

$3R + 7S = 9$

Solve the first equation for R,

$$2R - 3S = 4$$
$$2R = 3S + 4$$
$$R = \frac{3S + 4}{2}$$

Substitute:

$$3\left(\frac{3S + 4}{2}\right) + 7S = 9$$

Solve:

$$\frac{9S + 12}{2} + 7S = 9$$
$$2\left[\frac{9S + 12}{2} + 7S\right] = 9 \cdot 2$$
$$2 \cdot \frac{9S + 12}{2} + 2 \cdot 7S = 18$$
$$9S + 12 + 14S = 18$$
$$23S + 12 = 18$$
$$23S = 6$$
$$S = \frac{6}{23}$$

Determine the remaining value:

$$R = \frac{3S + 4}{2}$$
$$R = \frac{3\left(\frac{6}{23}\right) + 4}{2}$$
$$R = \frac{\frac{18}{23} + \frac{92}{23}}{2}$$
$$R = \frac{\frac{110}{23}}{2}$$
$$R = \frac{55}{23}$$

The solution is $R = \frac{55}{23}$, $S = \frac{6}{23}$.

You should observe that the choice of which equation or which variable we use for substitution is arbitrary. Whenever possible, we choose the variable that has a coefficient of 1 or -1. If there are no variables with these coefficients, the choice will be the variable with the simplest coefficient.

note

After substitution, if the resulting equation simplifies to a statement that is always true, the system is dependent. If the equation simplifies to a statement that is never true, the system is inconsistent.

To summarize the procedure of solution by substitution:

1. Use one of the equations to solve for one of the unknowns in terms of the other.
2. Substitute the expression found in step 1 into the other equation.
3. Solve the newly formed equation for the unknown value.
4. Substitute the value found in step 3 into the expression obtained in step 1.

In many of the systems of equations, the equation may be in the form where it is solved for y in terms of x. This is usually the situation if the problem is stated in terms of functional relationships. If this is the case, the system of equations can be solved by a process of substitution called *comparison*. Consider the following system: $y = 2x + 1$ and $y = x - 4$. Since both equations are solved for y, we can set these two expressions equal to each other, $2x + 1$ and $x - 4$, and the solution is as follows:

$$\begin{aligned} y &= 2x + 1 \\ (x - 4) &= 2x + 1 && \text{comparison (substitution)} \\ x - 4 &= 2x + 1 \\ -4 &= x + 1 \\ -5 &= x \\ y &= x - 4 && \text{determine remaining unknown} \\ y &= (-5) - 4 \\ y &= -9 \end{aligned}$$

The solution is $(-5, -9)$.

Example Solve by comparison: $y = 3 - 2x$
$y = 2x + 4$

Substitute:

$$\begin{aligned} y &= 3 - 2x \\ (2x + 4) &= 3 - 2x \end{aligned}$$

Solve:

$$\begin{aligned} 2x + 4 &= 3 - 2x \\ 4x + 4 &= 3 \\ 4x &= -1 \\ x &= -\frac{1}{4} \end{aligned}$$

Determine remaining unknown:

$$y = 2x + 4$$

$$y = 2\left(-\frac{1}{4}\right) + 4$$

$$= -\frac{2}{4} + 4$$

$$= -\frac{1}{2} + \frac{8}{2}$$

$$= \frac{7}{2}$$

The solution is $\left(-\frac{1}{4}, \frac{7}{2}\right)$.

To summarize the procedure of solution by comparison

1. Solve both equations for the same unknown.
2. Form an equation by setting equal the two solutions from step 1.
3. Solve the newly formed equation for the unknown value.
4. Substitute the value found in step 3 into either of the original equations and determine the remaining unknown value.

When the method of solution for a system of equations is not specified, the method is usually selected by following these criteria. The process of elimination is generally the easiest. Substitution is most useful when the coefficient of one of the variables is 1 or -1, and comparison is convenient when both equations have been solved for the same variable. Graphical solutions are approximations and can be used when exact answers are not necessary or to verify the results that we find algebraically.

Exercise Set 16.3

Solve the following systems by substitution. Graph the system to verify your solution. If a system is inconsistent or dependent, so state.

1. $3x - y = 10$
   $x + y = 2$
2. $2x + 3y = 15$
   $3x + y = 12$
3. $-2x + 5y = 22$
   $2x + y = -10$
4. $5x + y = 10$
   $5x + 3y = 15$
5. $2x - y = 3$
   $2x + y = -1$
•6. $3x + y = 10$
   $2x - y = 5$
7. $2x + y = 13$
   $3x + y = 17$
8. $3x - y = -10$
   $3x + y = 4$
9. $2x + y = 2$
   $6x - y = 22$
10. $x - y = 3$
   $2x + 3y = 11$

Solve the following systems by substitution. If a system is inconsistent or dependent, so state.

11. $2x + y = 3$
$3x - y = 4$

12. $x + 5y = 7$
$2x + 3y = 5$

13. $5x - 3y = 4$
$x + 2y = -2$

14. $x + y = 4$
$3x - 5y = 7$

15. $R - S = 3$
$2R + 3S = 11$

16. $P - Q = 1$
$P + Q = 5$

17. $3x - y = 7$
$2x + 3y = -4$

•18. $-x + 3y = -3$
$4x + 5y = 11$

19. $3M + N = -3$
$M - 3N = -1$

20. $2T + U = 1$
$-3T + U = 1$

21. $2x + 3y = 5$
$4x + 6y = 7$

22. $-3x - 2y = 6$
$6x + 4y = 5$

23. $2x + 3y = 3$
$5x - 4y = 1$

24. $5x + 2y = 11$
$7x - 3y = 4$

25. $2x + 3y = 2$
$6x - 2y = 5$

26. $R - 2S = 4$
$3R + 2S = 4$

27. $2a + b = 6$
$3a + 4b = 4$

•28. $(0.6)x + (0.3)y = 1.2$
$(0.3)x - (0.5)y = 0.1$

29. $(0.8)x + (0.4)y = 0.4$
$(0.6)x - (0.4)y = -1.8$

30. $-x - 4y = 3$
$2x + 8y = -6$

31. $x + 4y = -3$
$-2x - 8y = 6$

32. $(0.5)R + (0.2)S = -0.3$
$(0.6)R - (1.4)S = 0.8$

33. $3x - 2y = 7$
$3x + 3y = 4$

34. $5x - 3y = 11$
$4x + 5y = 5$

Solve the following systems by comparison. If a system is inconsistent or dependent, so state.

35. $y = 2x + 1$
$y = -x + 3$

•36. $y = 5x + 3$
$y = 2x - 4$

37. $y = 3x - 4$
$y = 3x + 5$

38. $y = 5x - 7$
$y = 5x + 4$

39. $y = -3x + 2$
$y = 2x - 1$

•40. $y = \frac{1}{2}x + \frac{3}{2}$
$y = 2x + 5$

41. $y = \frac{3}{4}x - 7$
$y = -\frac{1}{2}x + \frac{2}{3}$

42. $y = -x - \frac{3}{2}$
$y = -3x + \frac{4}{7}$

43. $y = 5x - 3$
$y = 6x + 2$

44. $y = \frac{5}{7}x - \frac{1}{14}$
$y = \frac{3}{14}x - 1$

45. $y = -3x + 5$
$y = 3x - 5$

16.4 Applications

As we stated before, many problems that technicians encounter will be stated verbally and will need to be translated into an algebraic equation or system of equations. In chapter 7, we learned how to take a verbal statement and write an algebraic equation in one unknown, which could then be solved. While many problems involving linear equations can be solved using one unknown, the use of two unknowns sometimes makes the solution easier.

The use of two unknowns requires that we set up two equations that express the relationships between the unknowns. There is no standard procedure for solving a verbal problem, but the following guidelines should be useful.

1. Read the problem carefully.
2. Note what information is given and what information you are asked to find.
3. Determine any prior knowledge that may be useful in setting up the equations.
4. Let two variables represent the two unknown quantities.
5. Using the given conditions in the problem and the unknowns from Step 4, write the two equations which express the relationship between the two unknowns.
6. Solve this system of equations.
7. Check your results in the original statement of the problem.

Examples

1. The length of a rectangle is three times as long as its width, and its perimeter is 40 in. Find the dimensions.

note

In this example, we need the prior knowledge that the perimeter of a rectangle is given by the formula $P = 2w + 2l$.

Width	*Length*	*Equations*
x	y	$y = 3x$
		$2x + 2y = 40$

Solution

$3x - y = 0$	Multiply by 2.	$6x - 2y = 0$
$2x + 2y = 40$		$2x + 2y = 40$
	Add the equations.	$8x = 40$
	Solve for x.	$x = 5$

Determine the remaining value by substituting the unknown value into either of the original equations.

$$y = 3(5)$$
$$y = 15$$

Therefore, the width is 5 in and the length is 15 in.

2. A woman has \$10,000, part of which she invests at 11% and the rest at 8%. If her total income from the two investments is \$980, how much does she have invested at each rate?

note

The prior knowledge that is needed for this problem is that Interest = Principle × Rate × Time. Time in our problems will always be one year.

11% Investment	*8% Investment*	*Equations*
x	y	$x + y = 10{,}000$
		$(0.11)x + (0.08)y = 980$

Solution $x + y = 10{,}000$
$(0.11)x + (0.08)y = 980$ Multiply by 100 to clear the fractions. $x + y = 10{,}000$, $11x + 8y = 98{,}000$

Multiply the first equation by (-8).

$$\begin{aligned} -8x - 8y &= -80{,}000 \\ 11x + 8y &= 98{,}000 \end{aligned}$$

Add the equations. $3x = 18{,}000$

Solve for x. $x = 6{,}000$

Determine the remaining value by substituting the known value into one of the original equations:

$$\begin{aligned} (6{,}000) + y &= 10{,}000 \\ y &= 4{,}000 \end{aligned}$$

Hence, she has \$6,000 at 11% and \$4,000 at 8%.

note

Examples 1 and 2 were solved using one unknown in section 7.5.

3. The sum of the lengths of two pieces of wire is 11 m. If three times the length of the first wire is added to the length of the second wire, the result is 25 m. Find the lengths of the two pieces of wire.

First piece of wire	*Second piece of wire*	*Equations*
x	y	$x + y = 11$
		$3x + y = 25$

Solution $x + y = 11$ Multiply by (-1). $-x - y = -11$
$3x + y = 25$ $3x + y = 25$

Add the equations. $2x = 14$

Solve for x. $x = 7$

Determine the remaining unknown by substituting the known value into either of the original equations:

$$\begin{aligned} (7) + y &= 11 \\ y &= 4 \end{aligned}$$

Therefore, the two pieces of wire are 7 m and 4 m.

4. In a particular piece of machinery, three times the number of spot welds is 12 more than two times the number of bolts, while twice the number of bolts is 4 more than the number of spot welds. How many bolts and spot welds are there?

Bolts	*Spot welds*	*Equations*
x	y	$3y = 2x + 12$
		$2x = y + 4$

Solution $2x - 3y = -12$
$2x - y = 4$ Multiply by (-3).

$$\begin{aligned} 2x - 3y &= -12 \\ -6x + 3y &= -12 \\ \hline -4x &= -24 \\ x &= 6 \end{aligned}$$

Add the equations. Solve for x.

Determine the remaining unknown by substituting the known value into either of the original equations:

$$\begin{aligned} 2(6) &= y + 4 \\ 12 &= y + 4 \\ 8 &= y \end{aligned}$$

The number of bolts is 6 and the number of spot welds is 8.

note

In example 4, when we check our results in the original statement of the problem, the values for the number of bolts or spot welds must be whole numbers because we cannot have a fractional part or a negative number of bolts or spot welds.

5. The line with equation $y = mx + b$ passes through a point (x,y) if and only if the coordinates of the point satisfy the equation. Find the values of m and b such that the line will contain the points $(-1,-7)$ and $(2,5)$.

note

Ordered pairs (x,y) are values of x and y that can be substituted for x and y in the original equation, $y = mx + b$.

First unknown	*Second unknown*	*Equations*
m	b	$-7 = m(-1) + b$
		$5 = m(2) + b$

Solution $-m + b = -7$ Multiply by 2.
$2m + b = 5$

$$\begin{aligned} -2m + 2b &= -14 \\ 2m + b &= 5 \\ \hline 3b &= -9 \\ b &= -3 \end{aligned}$$

Add the equations. Solve for b.

Determine the remaining unknown by substituting the known value into either of the original equations:

$$\begin{aligned} 2m + (-3) &= 5 \\ 2m &= 8 \\ m &= 4 \end{aligned}$$

The unknown values are $m = 4$ and $b = -3$. Hence, the equation of the line passing through the given points is $y = 4x - 3$. You should observe from this example that, given any two points that lie on a line, we can determine the equation of that line.

6. What quantities of silver 65% pure and 45% pure must be mixed together to give 100 g of silver 50% pure?

Number of grams of 65%	*Number of grams of 45%*	*Equations*
x	y	$x + y = 100$ $(0.65)x + (0.45)y = (0.50)100$

Solution $x + y = 100$ $\qquad\qquad$ $x + y = 100$

$(0.65)x + (0.45)y = 50$ $\quad$ Multiply by 100 to clear the fractions. $\quad$ $65x + 45y = 5{,}000$

Multiply the first equation by (-45).

$$\begin{aligned} -45x - 45y &= -4{,}500 \\ 65x + 45y &= 5{,}000 \end{aligned}$$

Add the equations. $\quad 20x = 500$

Solve for x. $\quad x = 25$

Determine the remaining unknown by substituting the known value into either of the original equations:

$$\begin{aligned} (25) + y &= 100 \\ y &= 75 \end{aligned}$$

Therefore, 25 g of the 65% pure silver mixed with 75 g of the 45% pure silver will produce 100 g of 50% pure silver.

Exercise Set 16.4

Solve the following problems by setting up two equations with two unknowns.

1. The perimeter of a rectangle is 36 m and the length is 2 m more than three times the width. Find the dimensions.
•2. The perimeter of a room is 40 ft, and twice the length increased by three times the width is 48 ft. What are the dimensions?
3. The perimeter of a rectangle is 100 m. The rectangle is 22 m longer than it is wide. Find the dimensions.
4. The length of a rectangle is 2 ft more than twice its width, and its perimeter is 46 ft. Find the dimensions.
5. The length of a room is 2 ft less than twice its width. If its perimeter is 62 ft, what are the dimensions?
•6. Phil has $20,000, part of which he invests at 8% interest and the rest at 6%. If his total income from the two investments is $1,460, how much did he invest at each rate?
7. Nancy has $18,000. She invests part of her money at $7\frac{1}{2}\%$ interest and the rest at 9%. If her income from the two investments is $1,560, how much did she invest at each rate?
8. Greta has $18,000, part of which she invests at 10% interest and the rest at 8%. If her income from each investment is the same, what does she have invested at each rate?

9. Dick has \$30,000, part of which he invests at 9% interest and the rest at 7%. If his income from the 7% investment is \$820 more than that from the 9% investment, how much was invested at each rate?

•10. Lynne made two investments totaling \$25,000. On one investment she made an 18% profit, but on the other investment she took an 11% loss. If her net gain is \$2,180, how much was each investment?

11. Larry made two investments totaling \$21,000. One investment made him a 13% profit, but on the other investment he took a 9% loss. If his net loss is \$196, how much was in each investment?

•12. A piece of pipe is 19 ft long. The pipe must be cut such that one piece is 5 ft longer than the other piece. What are the final lengths of the two pieces of pipe?

13. A 12-ft board is cut into two pieces so that one piece is 4 ft longer than the other. How long is each piece?

14. A 24-ft rope is cut into two pieces so that one piece is twice as long as the other. How long is each piece?

15. The sum of two currents is 80 amps. If the greater current is 24 amps more than the lesser current, find their values.

16. A 50-ft extension cord is cut into two pieces such that one piece is 12 ft longer than the other piece. How long is each piece?

17. The sum of the number of teeth on two gears is 64, and their difference is 12. How many teeth are on each gear?

18. Two gears have a total of 83 teeth, and one gear has 15 fewer teeth than the other. How many teeth are on each gear?

19. Two electrical voltages have a total of 126 V. If one voltage is 32 volts more than the other, find the voltages.

20. The sum of two voltages is 85 and their difference is 32. Find the voltages.

21. The sum of two resistors in a series is 24 ohms and their difference is 14 ohms. How many ohms are in each resistor?

22. One resistor exceeds another resistor by 25 ohms, and their sum is 67 ohms. How many ohms are in each resistor?

23. A clothing store sells suits at \$125 and \$185. They observe that they sold 40 suits at \$5,720. How many suits of each type did they sell?

•24. Three times the number of bolts in a piece of machinery is 3 less than twice the number of spot welds, while seven times the number of bolts is 5 more than four times the number of spot welds. How many bolts and spot welds are there?

25. When two batteries are connected in series, we add the voltages together to get a total voltage of 27 V. When the batteries are connected in opposition, the total voltage 5 V is their difference. Find the voltages of the two batteries.

26. When two batteries are connected in series, the total voltage is 53 V. When the batteries are connected in opposition, the total voltage is 7 V. Find the voltage of the two batteries. (Refer to problem 25.)

27. Three times the number of teeth on a first gear is one more than the number of teeth on a second gear. Also, five times the number of teeth on the first gear is 4 less than twice the number of teeth on the second gear. Find the number of teeth on each gear.

•28. The specific gravity of an object is defined as its weight in air divided by the difference between its weight in air and its weight when submerged in water. If an object has a specific gravity of 3, and the sum of its weight in air and that in water is 30 lb, find its weights in air and in water.

29. The line with equation $y = mx + b$ passes through a point (x,y) if and only if the coordinates of the point satisfy the equation. Find the values of m and b such that the line will contain the points $(1, -1)$ and $(2,2)$.

•30. Find the values of m and b for the line that passes through the points $(2,1)$ and $(-1,7)$. (Refer to problem 29.)

31. Find the values of m and b for the line that passes through the points $(-2,3)$ and $(3, -7)$. (Refer to problem 29.)

32. Find the values of m and b for the line that passes through the points $(-1,7)$ and $(2, -2)$. (Refer to problem 29.)

33. An auto mechanic has two bottles of battery acid. One contains a 10% solution and the other a 4% solution. How many milliliters of each solution must he use to make 120 ml of a solution that is 6% acid?

•34. A metallurgist wishes to form 2,000 kg of an alloy that is 80% copper. This alloy is to be obtained by fusing some alloy that is 68% copper and some that is 83% copper. How many kilograms of each alloy must be used?

35. If a jeweler wishes to form 12 oz of 75% pure gold from sources that are 60% and 80% pure gold, how much of each source must he mix together?

36. A chemist wishes to make 1,000 liters of a 3.5% acid solution by mixing a 2.5% solution with a 25% solution. How many liters of each solution is necessary?

37. A pharmacist is able to fill 200 3-grain and 2-grain capsules using 500 grains of a certain drug. How many capsules of each kind does he fill?

38. A solution that is 38% silver nitrate is to be mixed with a solution that is 3% silver nitrate to obtain 100 cl of solution that is 5% silver nitrate. How many cl of each solution should be used in the mixture?

39. A druggist has two solutions, one 60% hydrogen peroxide and the other 30% hydrogen peroxide. How many liters of each should she mix to obtain 30 liters of a solution that is 40% hydrogen peroxide?

40. Two cars are 100 mi apart. If they drive towards each other, they will meet in 1 h, and if they drive in the same direction they will meet in 2 h. Find their speeds.

•41. A cyclist and a pedestrian are 20 mi apart. If they travel towards each other they will meet in 75 min, but if they travel in the same direction, the cyclist will overtake the pedestrian in 150 min. What are their speeds?

Chapter Summary

1. A **graphical solution** to a system of equations is found by estimating the coordinates of the point of intersection.
2. A **consistent** and **independent** system of equations has one solution.
3. A **dependent** system of equations has an unlimited number of solutions.
4. An **inconsistent** system of equations has no solution.
5. A system of equations can be solved by **elimination, substitution,** or **comparison** when an exact answer is required.

Chapter Review

[16.1] Solve the following systems by graphing. State your answer to the tenth's position. All answers in the back of the book are given exactly. If the system is inconsistent or dependent, so state.

1. $x + y = 4$
   $x - y = 2$
2. $3x - y = 2$
   $x + y = 6$
3. $x - 2y = 0$
   $2x - y = 6$
4. $x + y = 4$
   $-x + y = 8$
5. $\frac{1}{2}x + y = 5$
   $\frac{1}{3}x - y = 4$
6. $-3x + 5y = 16$
   $3x + 5y = 14$
7. $(0.3)x + (1.2)y = 5.6$
   $(2.1)x + (0.1)y = -3.2$
8. $(2.7)x - (3.1)y = 5.3$
   $(-3.5)x - (1.8)y = -2.3$

[16.2] Solve the following systems by elimination. If a system is inconsistent or dependent, so state.

9. $x - 2y = 3$
   $x + y = 3$
10. $x - y = 5$
    $3x + 2y = 25$
11. $2x + 3y = 1$
    $-x + y = 2$
12. $5a + b = 4$
    $5a - 3b = 8$
13. $R + 3S = 6$
    $2R - 3S = -6$
14. $\frac{1}{2}x + \frac{1}{4}y = \frac{7}{4}$
    $\frac{2}{3}x - \frac{1}{3}y = \frac{13}{3}$
15. $(0.3)x - (0.2)y = 1.1$
    $(0.3)x + (0.2)y = 1.9$
16. $x - 2y = 5$
    $-3x + 6y = 4$
17. $2L - M = 3$
    $3L + M = 7$
18. $3x - y = 2$
    $9x - 3y = 6$

[16.3] Solve the following systems by substitution. If a system is inconsistent or dependent, so state.

19. $x - y = 2$
    $2x - 3y = 1$
20. $2x + y = 7$
    $2x - y = 13$
21. $3x - 2y = 4$
    $6x - 4y = 5$
22. $2s - 3t = 3$
    $3s + t = 7$
23. $R + S = 4$
    $R - S = 2$
24. $a - b = 5$
    $\frac{1}{2}a + \frac{1}{3}b = \frac{25}{6}$
25. $12x - 9y = 15$
    $4x - 3y = 5$
26. $5x + 3y = 4$
    $2x - y = 5$

Solve the following systems by comparison. If a system is inconsistent or dependent, so state.

27. $y = 3x + 1$
$y = -2x - 5$

28. $y = 5x$
$y = -x + 3$

29. $y = 2x + 7$
$y = -4x - 1$

30. $y = \frac{1}{2}x - 3$
$y = \frac{2}{3}x + 4$

31. $y = (0.7)x - 1.6$
$y = (4.2)x + 5$

32. $y = (5.1)x + 3.7$
$y = (-3.9)x - 2.9$

[16.4] Solve the following problems by setting up two equations with two unknowns.

33. The perimeter of a rectangle is 64 ft. Find the dimensions if three times the width is 4 less than the length.
34. Bruce made two investments totaling $18,000. On one investment he made a 14% profit, but on the other investment he took a 23% loss. If his net loss is $70, how much was in each investment?
35. Four times the number of teeth on a first gear is 6 less than the number of teeth on a second gear. If the total number of teeth on both gears is 66, how many teeth are on each gear?
36. A chemist wishes to make 120 ml of a 75% acid solution by mixing a 60% solution with an 80% solution. How many milliliters of each solution is necessary?

Chapter Test

Solve the following systems of equations. If a system is inconsistent or dependent, so state.

1. $2x + y = 5$
$4x - y = 7$

2. $5a + 2b = 1$
$3a - 2b = 7$

3. $3R + 5S = 8$
$7R - S = 6$

4. $4y - 3z = -2$
$2y + z = 4$

5. $6L + 3M = 30$
$5L - 4M = 12$

6. $2x + y = 4$
$6x + 3y = 12$

7. $5R + 3S = -1$
$7R + 2S = 3$

8. $5x - 7y = 11$
$2x + 3y = 4$

9. $3x - 9y = 11$
$x - 3y = 5$

10. $x + \frac{3}{2}y = 4$
$x - \frac{7}{5}y = 4$

11. $4R - 3S = 11$
$2R + 5S = 9$

12. $5x - 2y = 9$
$2x - 3y = 14$

13. $(0.2)x + (0.3)y = 1.3$
$(0.3)x + (0.5)y = 2.3$

14. $-3x + 7y = 5$
$4x - 2y = 3$

15. $3R + 5S = 1$
$4R + 3S = 5$

16. $(1.2)x + (5.4)y = 1.8$
$(-2.4)x - (10.8)y = 3$

17. Norma invested part of her money at 6% and the rest at 8%, and the total income from both investments is \$2,740. If she had interchanged her investments, her income would have been \$2,580. How much did she invest at each rate?

18. A mechanic has two solutions of antifreeze, one is 60% antifreeze and the other is 30% antifreeze. How many liters of each should be mixed to obtain 30 liters of a solution that is 40% antifreeze?

17
Right Triangle Trigonometry

A technician using a chart-recorder to plot a graph of voltage as a function of time.
(Photograph by Kenneth Walters, courtesy of Henry Ford Community College)

17.1 Introduction

The study of trigonometry is the study of triangle measurement and is over 2,000 years old. Trigonometry was developed by the ancient Greeks to find the lengths of the sides and the measure of the angles of triangles. The annual flood of the Nile River made it necessary for the Egyptians to use trigonometry to reestablish land boundaries. The principles of trigonometry were used by the Babylonians in their study of astronomy. The word *trigonometry* comes from the Greek words *trigon* (meaning "triangle") and *metry* (meaning "measurement"). Today, trigonometry is used in the study of electronics, engineering, and other technical fields.

We can measure *directly* the length of a piece of pipe, the height of a man, the distance between towns, and so on. However, to measure

the height of a tree
the distance across a body of water
the force applied to objects
the strength of a construction

and many other such problems, we measure *indirectly* using triangles. Some of the basic properties of triangles, in particular right triangles, lend themselves to such measurement.

In chapter 12 we studied the measure of angles and the various kinds of triangles. To develop trigonometry, however, we place an angle in a very special position in the rectangular coordinate plane.

17.2 Angles in Standard Position

Recall that a *ray* is that portion of a straight line extending from a given point on the line, the *endpoint*, indefinitely far in one direction. An angle is then generated by the rotation of the ray about its endpoint from its initial position, called the *initial side* of the angle, to a terminated (final) position, called the *terminal side* of the angle. We call the fixed point, the endpoint of the ray, the *vertex* of the angle. The measurement of the size of the angle is the amount of rotation of the terminal side away from the initial side. See figure 17.1.

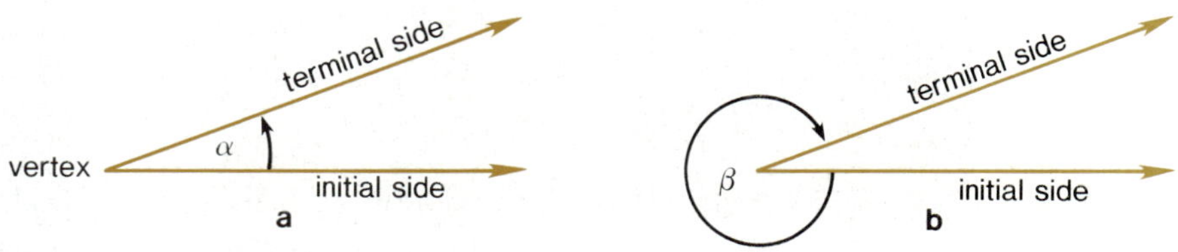

Figure 17.1

In trigonometry the measurement of an angle can be positive or negative. When the rotation of the terminal side away from the initial side is: *counterclockwise,* as with angle α in figure 17.1**a,** the angle is said to be *positive;* when it is *clockwise,* as with angle β in figure 17.1**b,** the angle is said to be *negative.*

We now place the angle in the rectangular coordinate plane. An angle is said to be in **standard position** in the plane when:

1. its vertex is placed at the origin (intersection of the x and y axes)
2. its initial side lies on the positive x-axis.

See figure 17.2. The position of the terminal side in the plane determines the angle.

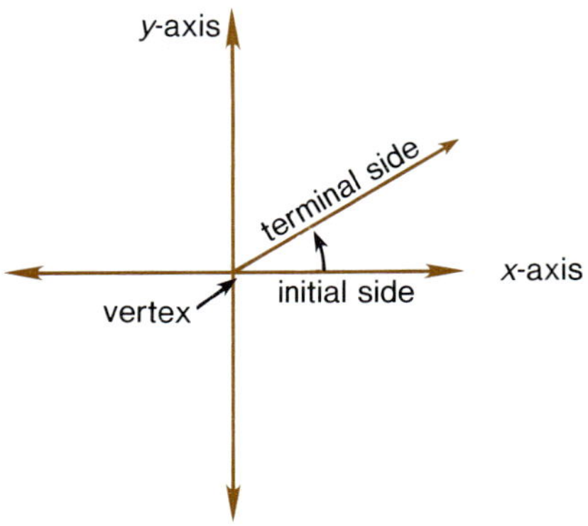

Figure 17.2

Recall that the x and y axes subdivide the plane into four equal parts called *quadrants*. The quadrants are named by their position, as shown in figure 17.3.

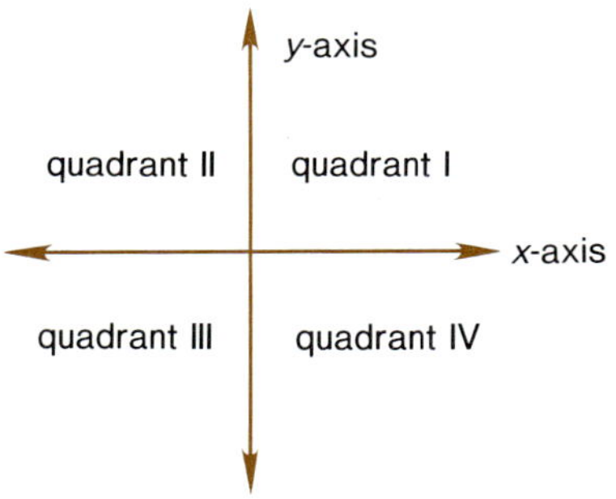

Figure 17.3

Angles in standard position are named by the quadrant in which their terminal side falls. That is, a **first-quadrant angle** has its terminal side in quadrant I, a **second-quadrant angle** has its terminal side in quadrant II, and so on. When the terminal side coincides with an axis, it is called a **quantrantal angle.**

Given an angle in standard position, any point in the plane, other than the origin, on the terminal side of the angle, uniquely determines the terminal side of the angle and the quadrant name of the angle. Consider figure 17.4. In this figure:

1. Point (3,2) determines first-quadrant angle α, which is positive.
2. Point $(-2,-4)$ determines third-quadrant angle θ, which is negative.
3. Point $(-4,0)$ determines quadrantal angle β.

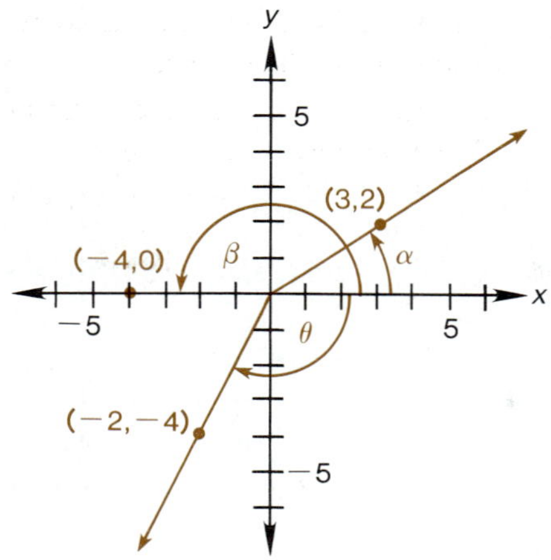

Figure 17.4

note

The terminal side of angle α also contains points (6,4) and (1,2/3) among others, so any one of these points could determine the terminal side of angle α.

Thus, for the terminal side of any given angle, there are many other angles, both positive and negative, that can share the same terminal side. See figure 17.5. For example, angle α has the same terminal side as negative angle β and positive angle γ. Many more angles have the same terminal side. Such angles are called *coterminal* angles. Thus, two or more angles in standard position whose terminal sides coincide are called **coterminal angles.** Angles α, β, and γ in figure 17.5 are coterminal angles.

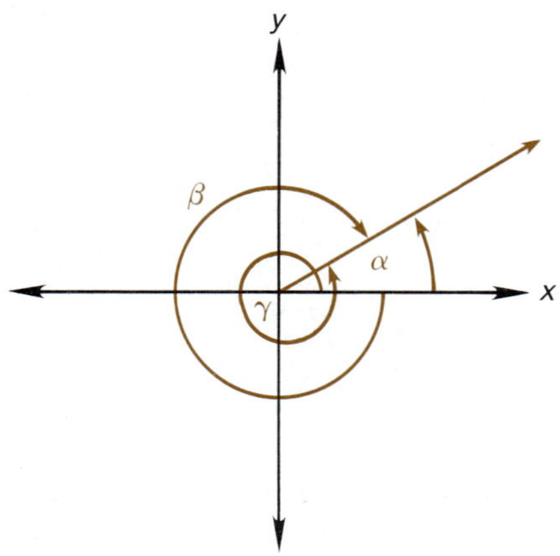

Figure 17.5

From this figure, we can see that to determine the measure of positive angles coterminal with an angle α, we add mutliples of 360°; to determine the measure of negative angles coterminal with an angle α, we add multiples of −360°.

note

360° is one full revolution and brings us back to the place where we started.

Examples

Find (a) two positive and (b) two negative angles coterminal with the given angle.

1. $65°$
   a. $65° + 360° = 425°$; $65° + 2(360°) = 65° + 720° = 785°$
   b. $65° + (-360°) = -295°$; $-295° + (-360°) = -655°$. Then $425°$, $785°$, $-295°$, and $-655°$ are coterminal with $65°$.
2. $218°$
   a. $218° + 360° = 578°$; $218° + 2(360°) = 218° + 720° = 938°$
   b. $218° + (-360°) = -142°$; $-142° + (-360°) = -502°$. Then, $578°$, $938°$, $-142°$, and $-502°$ are coterminal with $218°$.

Find the smallest positive angle that is coterminal with the given angle.

note

Add or subtract $360°$ as many times as necessary to get the smallest angle that is at least $0°$ but less than $360°$.

3. $587°$

   $587° - 360° = 227°$. Therefore, $227°$ is the smallest positive angle.
4. $1071°$

   $1071° - 2(360°) = 1071° - 720° = 351°$. Then, $351°$ is the smallest positive angle.
5. $-193°$

   $-193° + 360° = 167°$. Therefore, $167°$ is the angle we want.
6. $-830°$

   $-830° + 3 \cdot 360° = -830° + 1080° = 250°$, so $250°$ is the angle we want.

Consider now the ratios of the coordinates of points lying on the same terminal side. Recall that in figure 17.4, the terminal side of angle α contained the points (3,2), (6,4), and $\left(1, \frac{2}{3}\right)$. See figure 17.6. Since $\frac{3}{2} = \frac{6}{4} = \frac{1}{\frac{2}{3}}$, the ratios of the x-coordinate to the y-coordinate of any two points on the terminal side of the angle form a proportion (statement of equality between two ratios).

note

The same holds for the ratios of the y-coordinate to the x-coordinate of any two points on the terminal side.

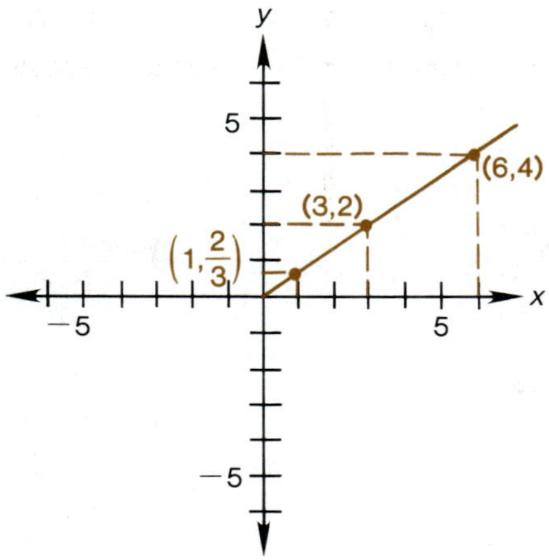

Figure 17.6

Using the x and y coordinates of a point on the terminal side of any angle α, together with the distance of that point from the origin, the point (0,0), we can define six ratios, which form the basis of trigonometry. See section 17.3.

Exercise Set 17.1

In exercises 1–9, (1) determine if the angle α is positive or negative and (2) name the angle by the quadrant in which its terminal side lies.

Example

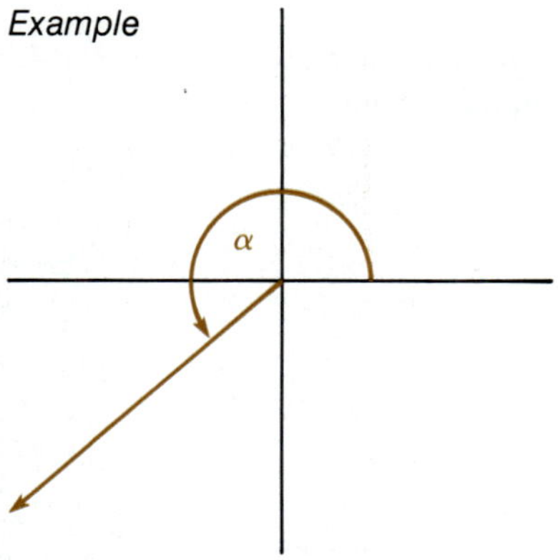

Since the rotation is counterclockwise, the angle is positive and it is a third-quadrant angle since the terminal side lies in quadrant III.

1.

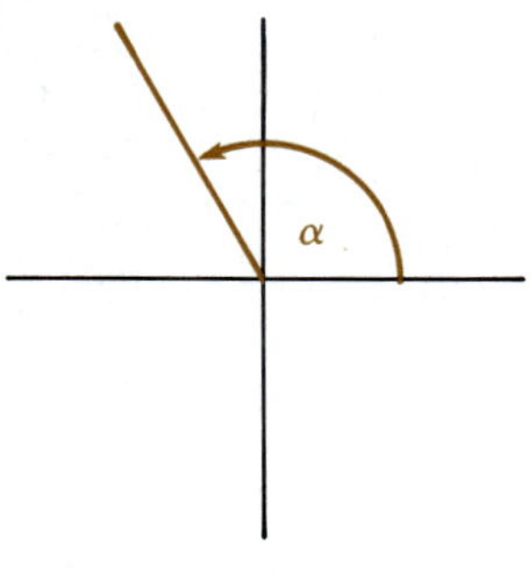

2.

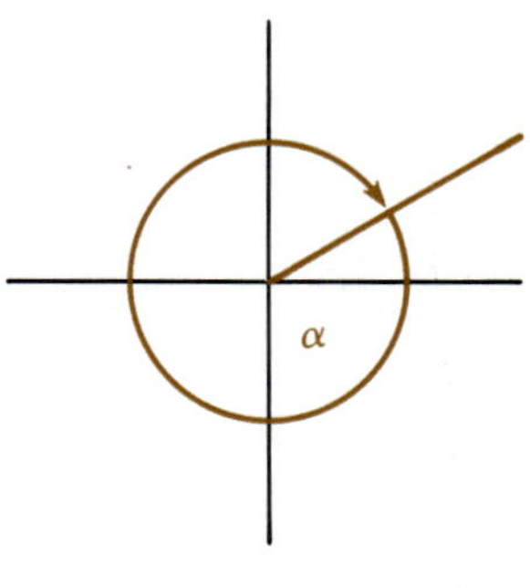

3.

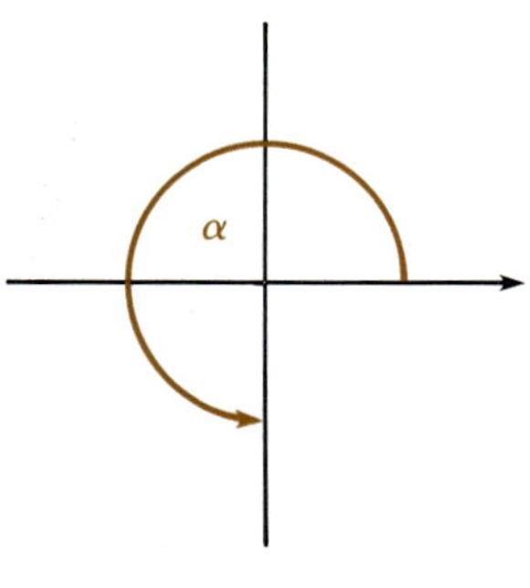

4.

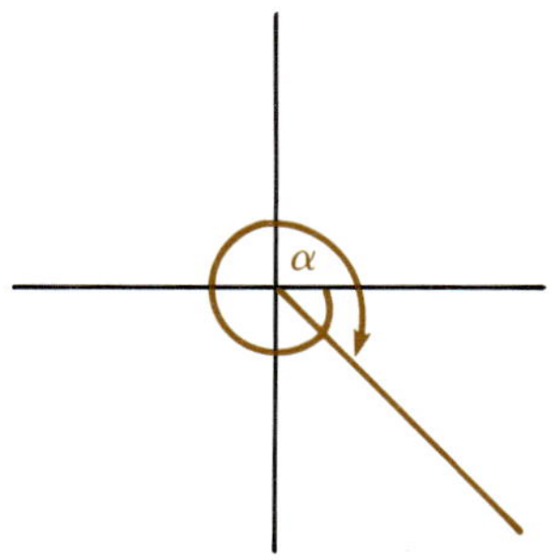

5.

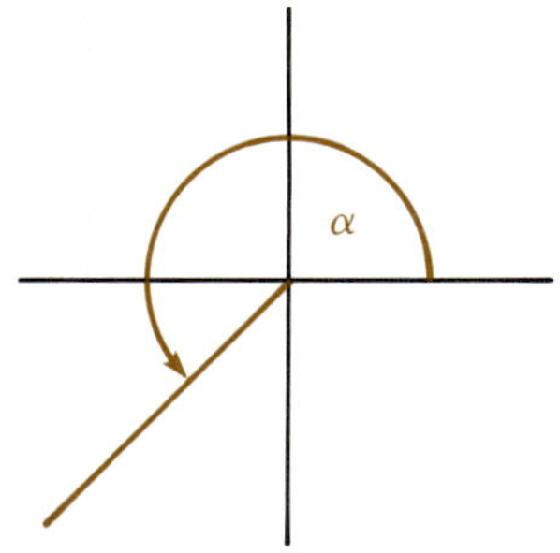

6.

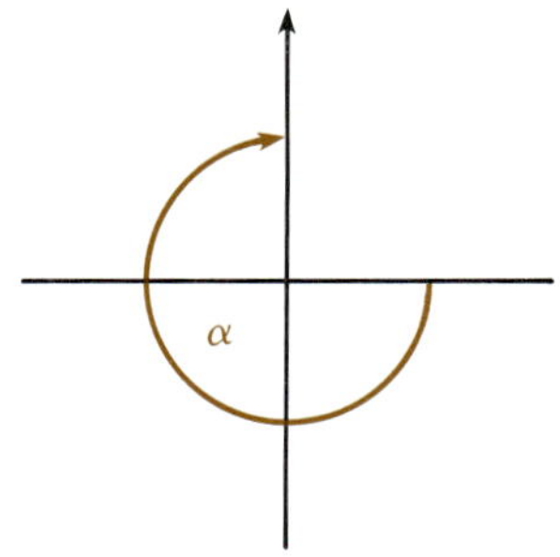

7.

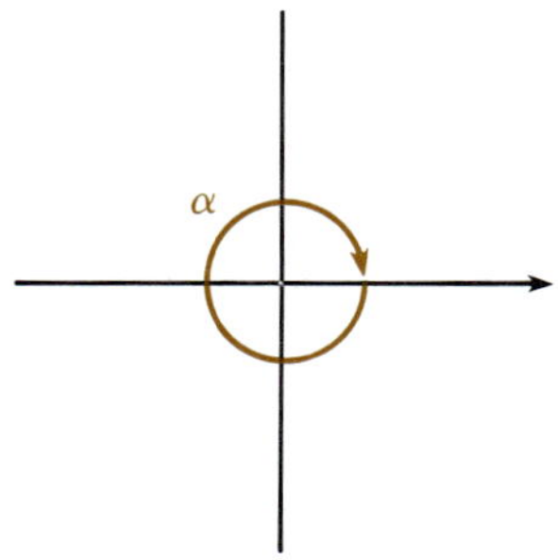

8.

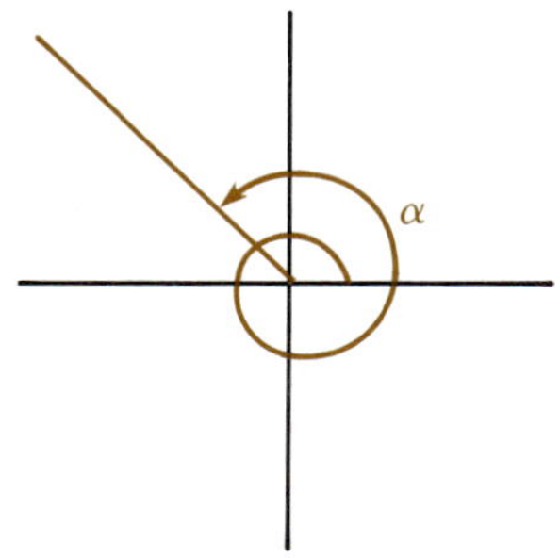

9.

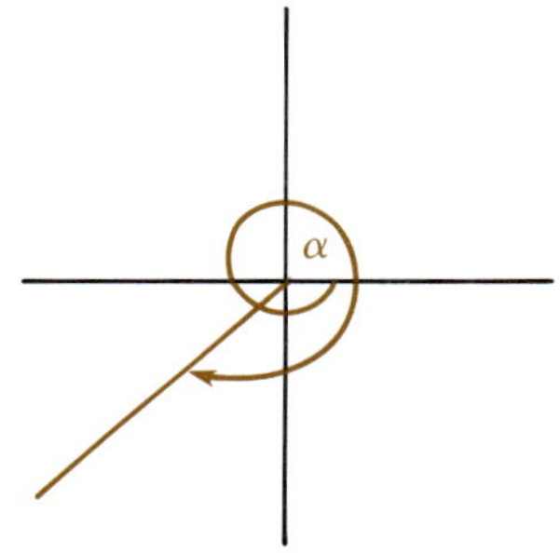

In exercises 10–13, given angle α, whose terminal side contains the given point, name three other points on the terminal side.

Example

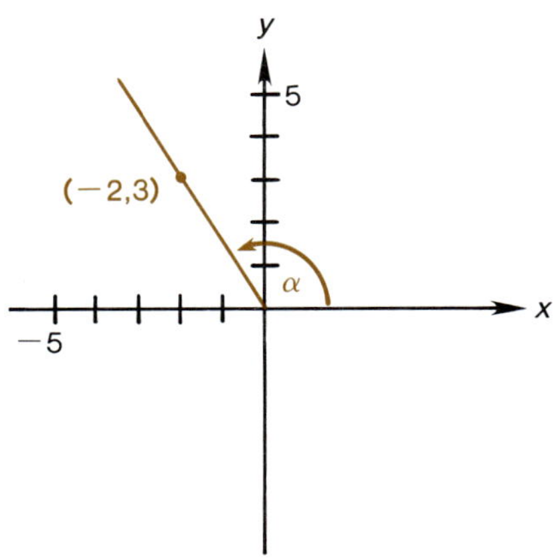

Other points:

1. (−4,6), found by multiplying the coordinates of (−2,3) by 2.
2. $\left(-1,\frac{3}{2}\right)$, found by multiplying the coordinates of (−2,3) by $\frac{1}{2}$.
3. (−6,9), found by multiplying the coordinates of (−2,3) by 3.

10.

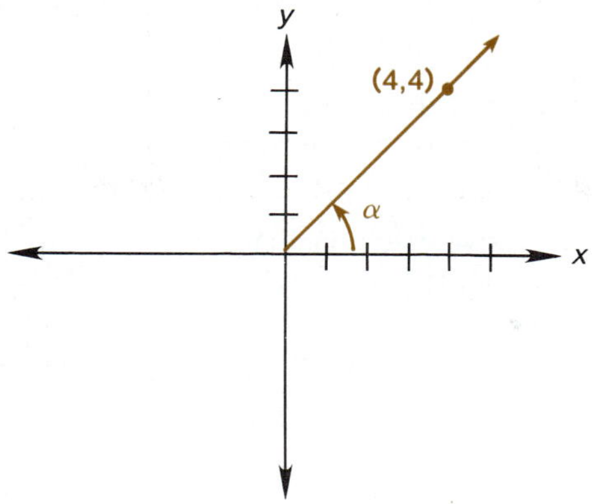

11.

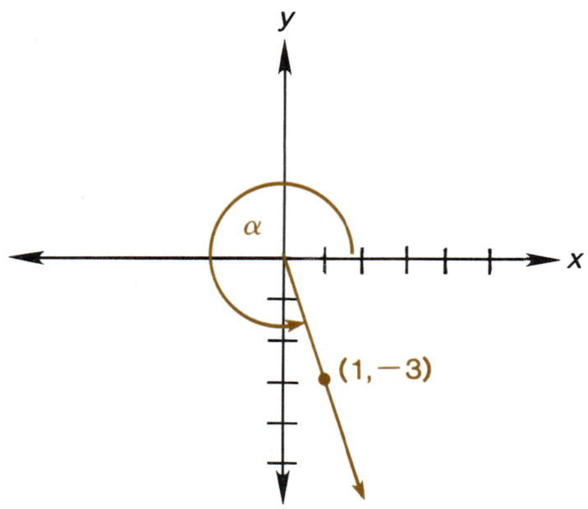

12.

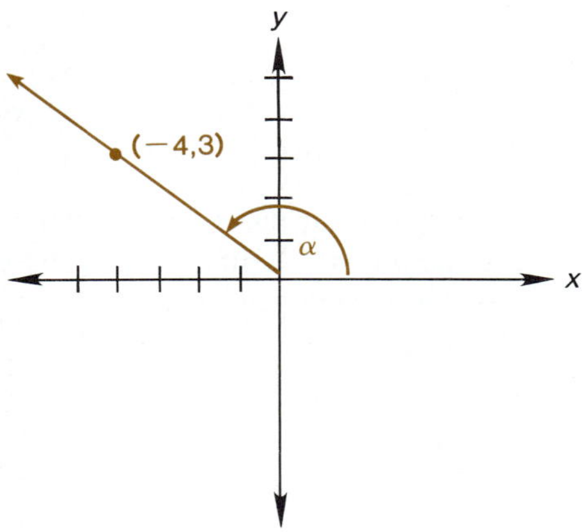

13.

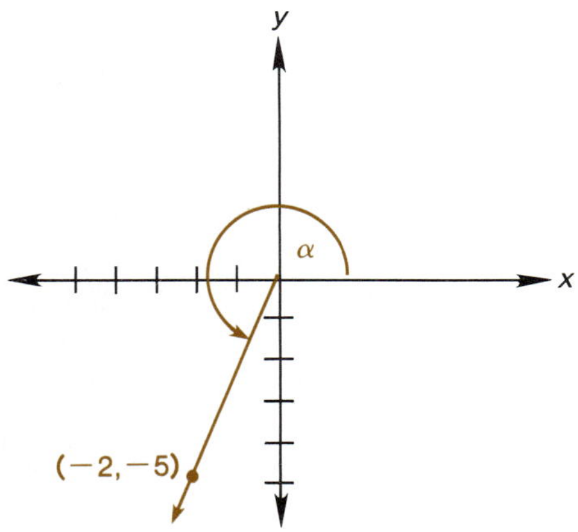

In exercises 14–20, determine two positive and two negative angles that are coterminal with the given angle.

•14. 83° 15. 127° 16. 251° 17. 329°
18. 90° 19. 180° 20. 270°

In exercises 21 through 30, find the measure of the smallest positive angle coterminal with the given angle.

21. 574° •22. 490° 23. 724° •24. 1,142°
25. −72° 26. −131° 27. −203° 28. −569°
29. −823.4° •30. −1,762.75°

Place the following angles in standard position. Draw an arrow representing the correct amount of rotation. Find two other angles, one positive and one negative, that are coterminal with the given angle.

Example Given an angle of $-140°$.

a. Positive coterminal angle:

$$-140° + 360° = 220°$$

b. Negative coterminal angle:

$$-140° + (-360°) = -500°$$

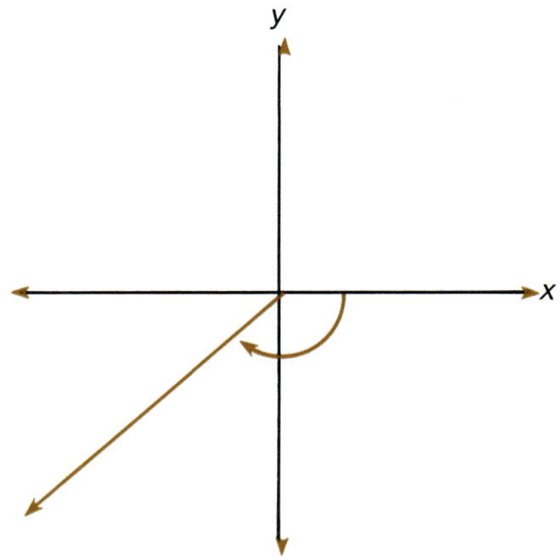

31. 76°	•32. 143°	33. 222°	34. 353°
35. $-62°$	36. $-157°$	37. $-253°$	38. $-307°$
39. 564°	•40. $-493°$		

In exercises 41 through 50, locate the following points in the plane. Draw a ray through the point starting at the origin. Use the ray and the positive *x*-axis to determine the angle and measure its size using your protractor.

41. (1,2)	•42. $(-1,3)$	43. $(-2,-2)$	44. $(5,-3)$
45. (4,4)	46. $(-7,-5)$	47. $(-3,3)$	48. $(2,-5)$
49. $(-4\sqrt{3},-4)$	•50. $(5,-5\sqrt{3})$		

17.3 The Trigonometric Ratios

Now we place a first-quadrant angle α in standard position in the plane. Recall from our previous discussion that given two points other than the origin, $A(a,b)$ and $C(c,d)$ on the terminal side, the ratios between the coordinates form a proportion. Now, from points A and C on the terminal side, drop a perpendicular line to the *x*-axis at points B and D, respectively. See figure 17.7.

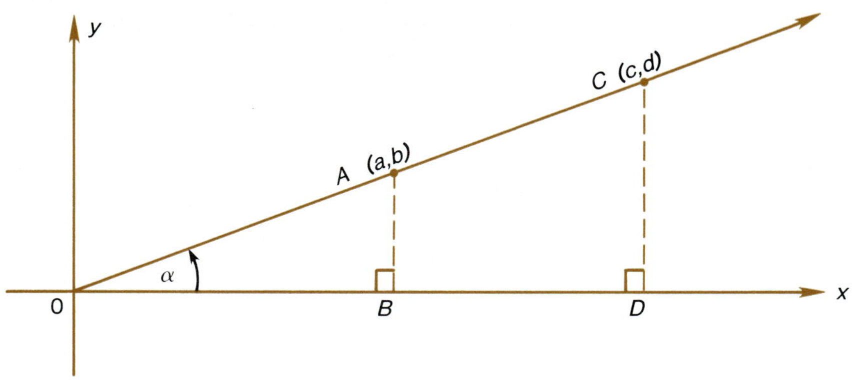

Figure 17.7

The right triangles *AOB* and *COD* thus formed are *similar triangles* since they have one acute angle, α, that is common to both triangles. Recall that corresponding sides of similar triangles are proportional, so equal ratios can be set up between corresponding sides of the triangles.

note

This situation is true for *any two* arbitrarily chosen points on the terminal side.

Thus, for any arbitrary point P on the terminal side of angle α, the ratio between any two sides of the triangle formed will *always* be the same. Consider figure 17.8.

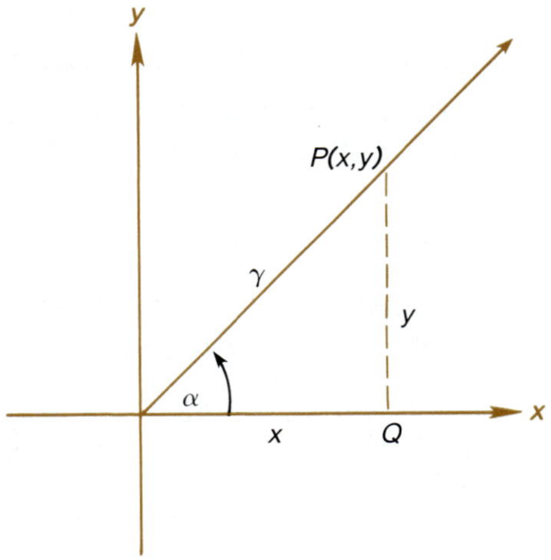

Figure 17.8

Let $P(x,y)$ be any point on the terminal side of angle α. We call the distance from the origin O to the point P the **radius vector,** denoted by r. The six trigonometric ratios of angle α are then defined by:

$$\text{sine of } \alpha = \sin(\alpha) = \frac{y}{r} \qquad \text{cosecant of } \alpha = \csc(\alpha) = \frac{r}{y}$$

$$\text{cosine of } \alpha = \cos(\alpha) = \frac{x}{r} \qquad \text{secant of } \alpha = \sec(\alpha) = \frac{r}{x}$$

$$\text{tangent of } \alpha = \tan(\alpha) = \frac{y}{x} \qquad \text{cotangent of } \alpha = \cot(\alpha) = \frac{x}{y}$$

The names of the six ratios are usually abbreviated, as indicated, for convenience in writing.

note

Since the terminal side of angle α can be in any quadrant, then x and y can be positive or negative. However, r is a distance, so r is *always positive*.

It is important to note that the cosecant, secant, and cotangent ratios are *reciprocals* of the sine, cosine, and tangent ratios, respectively. These three ratios are useful when working with triangles, as we will do later in the book.

We can now state the six trigonometric ratios for a given angle knowing a point on the terminal side.

Examples

Write the six trigonometric ratios of angle α placed in standard position, if the given point is on the terminal side.

1. (3,4)

   Note, in figure 17.9:

   a. The x-coordinate of the point is $x = 3$.

   b. The y-coordinate of the point is $y = 4$.

   c. Using the Pythagorean theorem, $r^2 = x^2 + y^2$. Then

$$r = \sqrt{x^2 + y^2} = \sqrt{3^2 + 4^2}$$
$$= \sqrt{9 + 16} = \sqrt{25} = 5 \qquad (r = 5)$$

By definition,

$$\sin(\alpha) = \frac{y}{r} = \frac{4}{5}$$

$$\cos(\alpha) = \frac{x}{r} = \frac{3}{5}$$

$$\tan(\alpha) = \frac{y}{x} = \frac{4}{3}$$

$$\csc(\alpha) = \frac{r}{y} = \frac{5}{4}$$

$$\sec(\alpha) = \frac{r}{x} = \frac{5}{3}$$

$$\cot(\alpha) = \frac{x}{y} = \frac{3}{4}$$

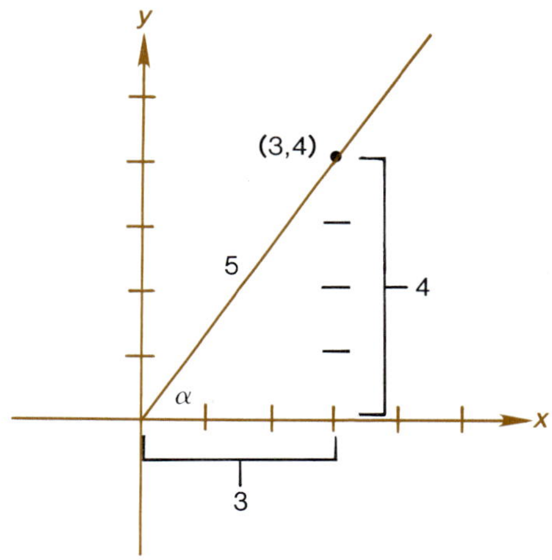

Figure 17.9

2. $(-2,1)$
   Note, in figure 17.10:
   a. $x = -2$
   b. $y = 1$
   c. $r = \sqrt{x^2 + y^2} = \sqrt{(-2)^2 + 1^2} = \sqrt{5}$. ($r = \sqrt{5}$)

   By definition,

$$\sin(\alpha) = \frac{1}{\sqrt{5}}$$

$$\cos(\alpha) = \frac{-2}{\sqrt{5}} = -\frac{2}{\sqrt{5}}$$

$$\tan(\alpha) = \frac{1}{-2} = -\frac{1}{2}$$

$$\csc(\alpha) = \frac{\sqrt{5}}{1} = \sqrt{5}$$

$$\sec(\alpha) = \frac{\sqrt{5}}{-2} = -\frac{\sqrt{5}}{2}$$

$$\cot(\alpha) = \frac{-2}{1} = -2$$

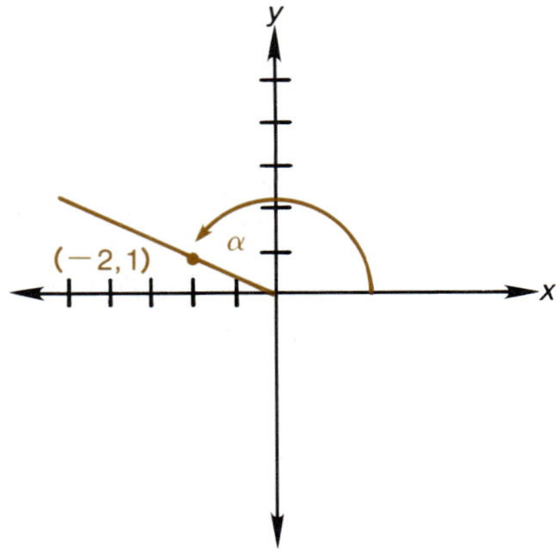

Figure 17.10

3. $(0,3)$
   Note, in figure 17.11:
   a. $x = 0$
   b. $y = 3$
   c. $r = \sqrt{3^2 + 0^2} = \sqrt{3^2} = 3$. ($r = 3$)

   Then,

$$\sin(\alpha) = \frac{3}{3} = 1$$

$$\cos(\alpha) = \frac{0}{3} = 0$$

$$\tan(\alpha) = \frac{3}{0} \text{ is undefined}$$

$$\csc(\alpha) = \frac{3}{3} = 1$$

$$\sec(\alpha) = \frac{3}{0} \text{ is undefined}$$

$$\cot(\alpha) = \frac{0}{3} = 0$$

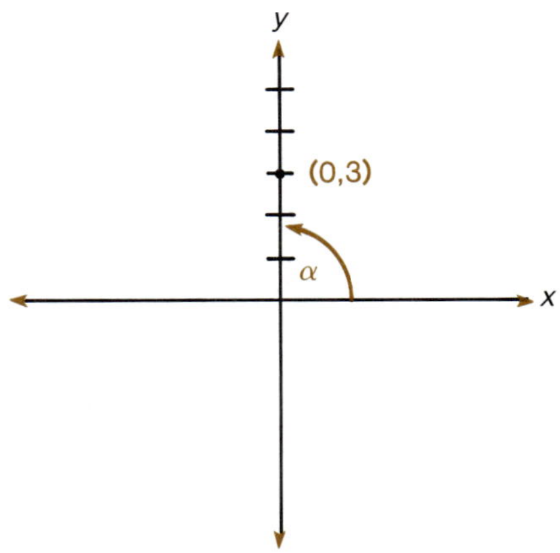

Figure 17.11

note

We say that $\tan(\alpha)$ and $\sec(\alpha)$ are undefined since division by zero is not defined. For certain angles, some ratios are undefined.

Knowing one of the ratios we can determine the values of the five remaining ratios for an angle in standard position.

Examples

1. Given $\tan(\alpha) = \frac{2}{3}$ and α is in quadrant I.

Note, in figure 17.12, that $x = 3$ and $y = 2$. Then

$$r = \sqrt{x^2 + y^2} = \sqrt{3^2 + 2^2} = \sqrt{9 + 4} = \sqrt{13}$$

$$\sin(\alpha) = \frac{2}{\sqrt{13}}$$

$$\cos(\alpha) = \frac{3}{\sqrt{13}}$$

$$\tan(\alpha) = \frac{2}{3}$$

$$\csc(\alpha) = \frac{\sqrt{13}}{2}$$

$$\sec(\alpha) = \frac{\sqrt{13}}{3}$$

$$\cot(\alpha) = \frac{3}{2}$$

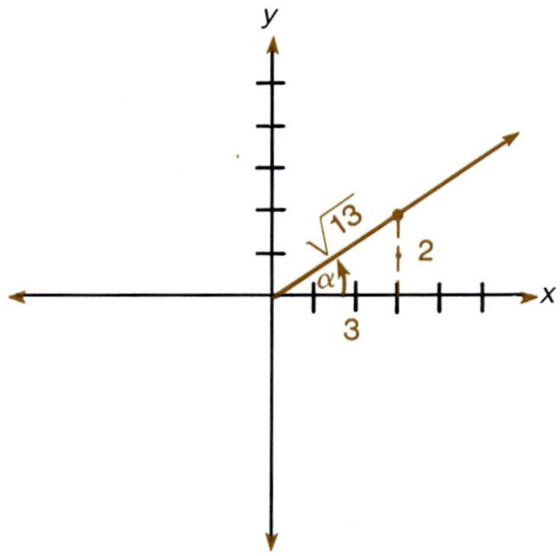

Figure 17.12

2. Given $\cos(\alpha) = -\frac{3}{4}$ and α is in quadrant II.

Note, in figure 17.13, that $x = -3$ and $r = 4$. Then

$$y = \pm\sqrt{4^2 - (-3)^2} = \pm\sqrt{16 - 9} = \pm\sqrt{7}$$

Since $y > 0$ in quadrant II, then $y = \sqrt{7}$.

$$\sin(\alpha) = \frac{\sqrt{7}}{4}$$

$$\cos(\alpha) = -\frac{3}{4}$$

$$\tan(\alpha) = -\frac{\sqrt{7}}{3}$$

$$\csc(\alpha) = \frac{4}{\sqrt{7}}$$

$$\sec(\alpha) = -\frac{4}{3}$$

$$\cot(\alpha) = -\frac{3}{\sqrt{7}}$$

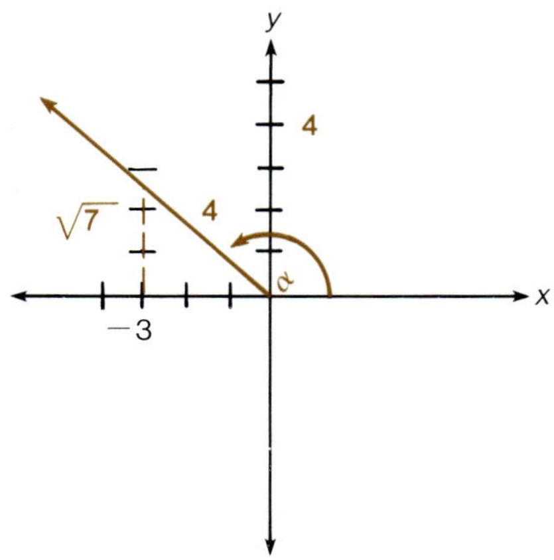

Figure 17.13

3. Given $\sin(\alpha) = -\dfrac{\sqrt{3}}{2}$ and α is in quadrant III.

Note, in figure 17.14, that $y = -\sqrt{3}$ and $r = 2$. Therefore,

$$x = \pm\sqrt{2^2 - (-\sqrt{3})^2} = \pm\sqrt{4 - 3} = \pm\sqrt{1} = \pm 1$$

Since $x < 0$ in quadrant III, then $x = -1$.

$$\sin(\alpha) = -\frac{\sqrt{3}}{2}$$

$$\cos(\alpha) = -\frac{1}{2}$$

$$\tan(\alpha) = \frac{-\sqrt{3}}{-1} = \sqrt{3}$$

$$\csc(\alpha) = -\frac{2}{\sqrt{3}}$$

$$\sec(\alpha) = -2$$

$$\cot(\alpha) = \frac{1}{\sqrt{3}}$$

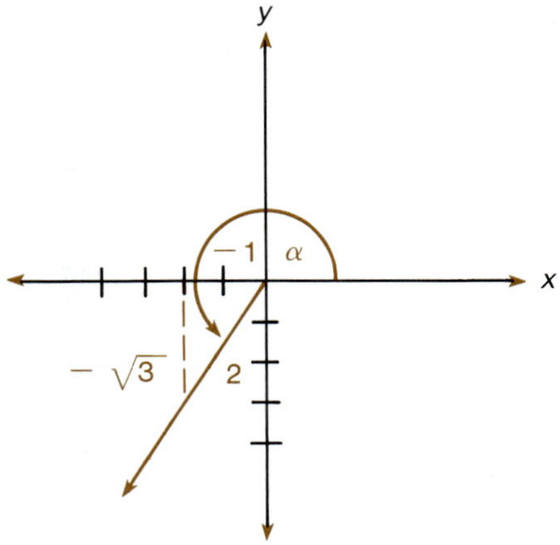

Figure 17.14

Exercise Set 17.2

In exercises 1 through 27, find the values of the six trigonometric ratios for the angle in standard position having the given point on the terminal side.

1. $(3,-4)$
•2. $(-3,-4)$
3. $(-3,4)$
4. $(5,-12)$
5. $(-12,-5)$
6. $(8,6)$
7. $(-6,-8)$
8. $(-9,12)$
9. $(-12,9)$
10. $(7,24)$
11. $(-24,-7)$
12. $(13,-12)$
13. $(0,2)$
•14. $(-3,0)$
15. $(0,-7)$
16. $(5,0)$
17. $(\sqrt{3},-1)$
18. $(1,-\sqrt{3})$
19. $(-2,2\sqrt{3})$
20. $(-2\sqrt{3},-2)$
21. $(4\sqrt{3},4)$
22. $(3\sqrt{2},3\sqrt{2})$
23. $(-2\sqrt{2},-2)$
24. $(-7,5)$
25. $(\sqrt{5},\sqrt{7})$
•26. $(-\sqrt{13},-\sqrt{3})$
27. $(-\sqrt{13},-\sqrt{12})$

Use your calculator in exercises 28 through 33 to find the necessary ratios.

28. $(4.73,-2.8)$
29. $(-5.893,-7.273)$
30. $(6.173,8.974)$
31. $(-252.7,753.6)$
•32. $(-122\sqrt{1.2},314\sqrt{18.7})$
33. $(-15\sqrt{143.7},-19.6\sqrt{814.0})$

In exercises 34 through 50, find the value of the five other trigonometric ratios for α in the given quadrant.

34. $\sin(\alpha) = \frac{3}{5}$; I

35. $\cos(\alpha) = -\frac{4}{5}$; II

36. $\tan(\alpha) = \frac{1}{3}$; I

37. $\sin(\alpha) = -\frac{1}{2}$; III

•38. $\cos(\alpha) = -\frac{\sqrt{3}}{2}$; II

39. $\csc(\alpha) = 2$; II

•40. $\sec(\alpha) = 2$; IV

41. $\cot(\alpha) = 1$; III

42. $\sin(\alpha) = \frac{1}{2}$; I

43. $\cos(\alpha) = -\frac{\sqrt{2}}{2}$; II

•44. $\tan(\alpha) = \frac{13}{12}$; III

45. $\cot(\alpha) = -\frac{7}{24}$; IV

46. $\cos(\alpha) = \frac{3}{5}$; IV

47. $\cot(\alpha) = \frac{7}{3}$; I

48. $\csc(\alpha) = -\frac{8}{3}$; III

49. $\sec(\alpha) = -\frac{3}{2}$; III

50. $\sin(\alpha) = \frac{4}{7}$; II

17.4 Values of the Trigonometric Ratios

Thus far we have considered the trigonometric ratios when we know (1) a point on the terminal side, or (2) one of the six ratios. Now we consider the measure of the size of an angle when we know one of the ratios or a point on the terminal side. There are certain special angles that we can determine because of the makeup of the ratio.

A protractor can help us determine the measure of an angle. First, we draw the angle in standard position using either a point on the terminal side or one of the known ratios. We then measure the size of the angle using the protractor.

Examples

1. Given that the point $(\sqrt{3}, 1)$ lies on the terminal side, then $x = \sqrt{3}$ and $y = 1$. Locate this point and draw the terminal side of the angle through this point. Using your protractor, measure the angle to be 30°.

note

We measure the distance from the origin to the point to get r. We find that $r = 2$ units.

From this information, we can determine (see figure 17.15):

$$\sin(30°) = \frac{1}{2}$$

$$\cos(30°) = \frac{\sqrt{3}}{2}$$

$$\tan(30°) = \frac{1}{\sqrt{3}}$$

$$\csc(30°) = 2$$

$$\sec(30°) = \frac{2}{\sqrt{3}}$$

$$\cot(30°) = \sqrt{3}$$

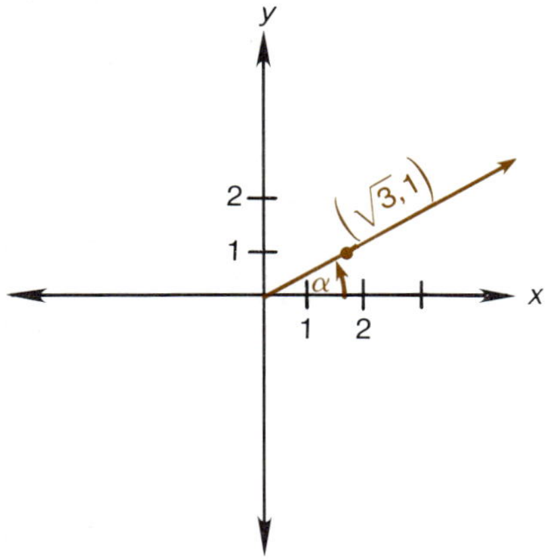

Figure 17.15

2. Given $\tan(\alpha) = 1$ and α lies in quadrant I, then $x = 1$ and $y = 1$. Measuring the size of the angle α, we get 45°.

note
Using the Pythagorean theorem, we get $r = \sqrt{2}$.

Then (see figure 17.16):

$$\sin(45°) = \frac{1}{\sqrt{2}}$$

$$\cos(45°) = \frac{1}{\sqrt{2}}$$

$$\tan(45°) = 1$$

$$\csc(45°) = \sqrt{2}$$

$$\sec(45°) = \sqrt{2}$$

$$\cot(45°) = 1$$

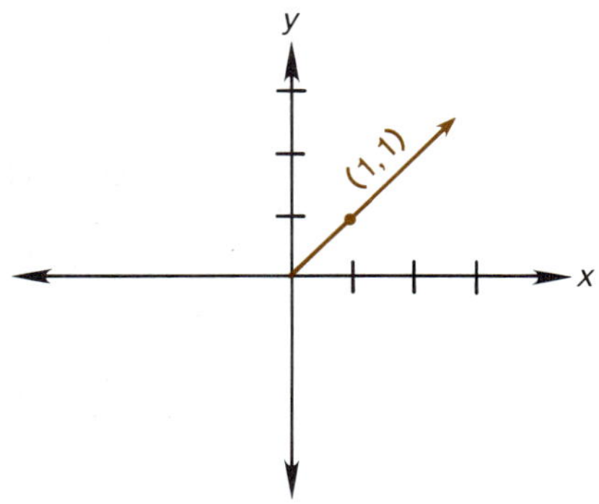

Figure 17.16

In like fashion, using the point $(1,\sqrt{3})$, we can determine

$$\sin(60°) = \frac{\sqrt{3}}{2} \qquad \csc(60)° = \frac{2}{\sqrt{3}}$$

$$\cos(60°) = \frac{1}{2} \qquad \sec(60°) = 2$$

$$\tan(60°) = \sqrt{3} \qquad \cot(60°) = \frac{1}{\sqrt{3}}$$

Using these examples and the definitions of the ratios, we can set up a table for the trigonometric ratios for angles 0°, 30°, 45°, 60°, and 90°, as given in table 17.1. These values should be learned since they are used frequently.

Table 17.1 Trigonometric Ratios for Selected Angles

Angle θ	$\sin\theta$	$\cos\theta$	$\tan\theta$	$\csc\theta$	$\sec\theta$	$\cot\theta$
0°	0	1	0	undef	1	undef
30°	$\frac{1}{2}$	$\frac{\sqrt{3}}{2}$	$\frac{1}{\sqrt{3}}$	2	$\frac{2}{\sqrt{3}}$	$\sqrt{3}$
45°	$\frac{1}{\sqrt{2}}$	$\frac{1}{\sqrt{2}}$	1	$\sqrt{2}$	$\sqrt{2}$	1
60°	$\frac{\sqrt{3}}{2}$	$\frac{1}{2}$	$\sqrt{3}$	$\frac{2}{\sqrt{3}}$	2	$\frac{1}{\sqrt{3}}$
90°	1	0	undef	1	undef	0

Obviously, to use the procedures just described for finding the values of the trigonometric ratios for any given angle would be tedious, time-consuming, and inaccurate. These values for angles, such as 27° and 59°, have been calculated by mathematicians and tabulated in table B in the appendix. The values can also be found on calculators if the necessary keys are on the keyboard.

Refer to table B to find the trigonometric value of any angle. The procedure is as follows. For "Angle α" whose measure is

1. from 0° to 45°, read *down* the *left-hand* column headed "Angle α" to the given angle. Read across this row to the column headed at the *top* by the given ratio.
2. from 45° to 90°, read *up* the extreme *right-hand* column headed "Angle α" *at the bottom*. Read across this row to the column headed at the *bottom* by the given ratio.

Examples

Find the values of the given ratios using table B.

1. sin (25°30′)

   Reading *down* the left-hand column headed "Angle α" and "Degrees" to 25°30′, we then move across the row to the column headed "sin (α)" at the top. Read the decimal number "0.0305". Thus,

   $$\sin(25°30') \approx 0.0305$$

2. sec (52°10′)

Reading *up* the right-hand column headed "Angle α" and "Degrees" at the bottom to 52°10′, we then move across this row to the column headed "sec (α)" *at the bottom*. Read the decimal number "0.0630". Thus,

$$\sec(52°10') \approx 0.0630$$

note

When reading up the right-hand column, be sure you read past the degree and to the "10" *above* 52°.

The table has been constructed to include the decimal representation for any of the trigonometric ratios of angles for every ten minutes from 0° to 90°. If the ratios are not given in the table, interpolation can be performed. We leave it to the student and the instructor to discuss the use of table B and interpolation. All computations in this text will be done on the calculator.

Often the measure of an angle is given in degrees and minutes. In such an event, to use the calculator we must state the angle in decimal degrees.

Examples

1. Find csc 35°12′. Since there are 60 minutes in a degree, we *divide* 12 by 60 to get $35°12' = 35\frac{12}{60}^{\circ} = 35\frac{1}{5}^{\circ} = 35.2°$. Then $\csc 35°12' = \csc 35.2° \approx 1.735$. So, $\csc 35°12' \approx 1.735$.
2. Find sin 80°42′. Dividing 42 by 60, we have $80°42' = 80\frac{42}{60}^{\circ} = 80\frac{7}{10}^{\circ} = 80.7°$. Therefore, $\sin 80°42' = \sin 80.7° \approx 0.9869$. So, $\sin 80°42' \approx 0.9869$.

When using the hand-held calculator, read its handbook to determine how the values are found. The trigonometric ratios are found by using the [sin], [cos], and [tan] keys. The angle whose ratio is given is found by using the "inverse" keys. These include the

[sin⁻¹], [arc|sin], or [inv] [sin]
[cos⁻¹], [arc|cos], or [inv] [cos]
[tan⁻¹], [arc|tan], or [inv] [tan]

keys. We read these keys as "the angle whose sine is," "the angle whose cosine is," and "the angle whose tangent is."

Examples

1. Find sin 42.3°. Enter [4] [2] [•] [3] [sin] and read **0.6730125135** on the display. Then, sin 42.3° ≈ 0.6730.
2. Find tan 63.4°. Enter [6] [3] [•] [4] [tan] and read **1.996953856** on the display. Then, tan 63.4° ≈ 1.9970.
3. Find sec 72.3°. Since the secant ratio is the reciprocal of the cosine ratio, then $\sec 72.3° = \frac{1}{\cos 72.3°}$. We find cos 72.3° and use the reciprocal key, [$\frac{1}{x}$]. Enter [7] [2] [•] [3] [cos] [$\frac{1}{x}$] and read **3.289115983** on the display. So, sec 72.3° ≈ 3.289.
4. Find csc 49°48′. Now, $49°48' = 49\frac{48}{60}° = 49\frac{4}{5}° = 49.8°$. Then, csc 49°48′ = csc 49.8° and we enter [4] [9] [•] [8] [sin] [$\frac{1}{x}$]. Read **1.309250065**. So, csc 49°48′ = csc 49.8° ≈ 1.3093.

   note

   We used the [sin] and [$\frac{1}{x}$] keys since the cosecant ratio is the reciprocal of the sine ratio. In like fashion, we would use the [tan] and [$\frac{1}{x}$] keys to find the cotangent of an angle.

Assume θ to be a first-quadrant angle in examples 5–8.

5. Find θ, given cos θ = 0.7513. If your calculator has a separate [arc] or [inv] key, press [•] [7] [5] [1] [3] [inv] [cos] and read **41.29688628** on the display. Thus, θ ≈ 41.3°.
6. Find θ, given sin θ = 0.8346. Press [•] [8] [3] [4] [6] [inv] [sin] and read **56.57421041** on the display. Thus, θ ≈ 56.6°.
7. Find θ, given cot θ = 0.5236. Since the cotangent ratio is the reciprocal of the tangent ratio, we use the [$\frac{1}{x}$] key following the entry of the number and then the [arc] [tan] keys. Press [•] [5] [2] [3] [6] [$\frac{1}{x}$] [inv] [tan] and read **62.36344561** on the display. Then θ ≈ 62.4°.
8. Find θ, given $\tan\theta = \frac{7}{8}$. Press [7] [÷] [8] [inv] [tan] and read **41.18592517** on the display. Then θ ≈ 41.2°.

Exercise Set 17.3

In exercises 1–15, determine the value of the indicated trigonometric ratio using a calculator.

1. sin 24.0°	2. cos 36.0°	3. tan 75.0°
•4. sin 52.3°	5. tan 75.3°	6. cos 47.8°
7. csc 18.3°	•8. sec 52.1°	9. cot 38.6°
10. sin 6.9°	11. sec 21.8°	12. cos 11.2°
13. tan 87.9°	•14. csc 79.4°	15. cot 60.8°

In exercises 16–27, write the angle to the nearest 0.01° and find the value of the given ratio using a calculator.

•16. sin 40°20′	17. cos 19°24′	18. tan 53°50′
19. sin 86°15′	•20. cos 68°12′	21. tan 72°23′
•22. csc 24°42′	23. sec 37°51′	24. cot 77°09′
25. sin 61°28′	26. sec 85°33′	27. cot 6°25′

In exercises 28–48, find the value of angle θ to the nearest 0.01° using your calculator, given that θ is a first-quadrant angle.

•28. $\sin \theta = 0.5505$	29. $\cos \theta = 0.6414$
•30. $\tan \theta = 1.423$	31. $\sin \theta = 0.0936$
32. $\cos \theta = 0.1143$	33. $\tan \theta = 0.2007$
34. $\sin \theta = 0.9813$	35. $\cos \theta = 0.8990$
36. $\tan \theta = 17.43$	37. $\csc \theta = 1.953$
38. $\sec \theta = 1.098$	39. $\cot \theta = 9.736$
•40. $\csc \theta = 4.876$	41. $\sec \theta = 1.037$
•42. $\cot \theta = 0.3111$	43. $\csc \theta = 1.214$
44. $\sin \theta = 0.5973$	•45. $\sec \theta = 27.43$
46. $\cos \theta = .0998$	47. $\cot \theta = 0.7499$
48. $\tan \theta = 124.3$	

In exercises 49–59, find angle θ using a calculator, given that θ lies in the first quadrant.

49. $\sin \theta = \frac{2}{3}$	•50. $\cos \theta = \frac{5}{6}$	51. $\tan \theta = \frac{8}{7}$
52. $\sin \theta = \frac{15}{16}$	53. $\cos \theta = \frac{7}{13}$	•54. $\sec \theta = \frac{9}{8}$
55. $\csc \theta = \frac{7}{5}$	56. $\cot \theta = \frac{2}{11}$	57. $\sin \theta = \frac{1}{9}$
58. $\cos \theta = \frac{7}{23}$	59. $\tan \theta = \frac{23}{47}$	

17.5 Right Triangle Trigonometry

In this section we investigate the applications of the trigonometric ratios to a right triangle. First we consider how to solve right triangles—how to find the unknown parts using the known parts.

Every triangle consists of six parts: three angles and three sides. If three of the six parts are known, one of which must be a side, we can determine the other three parts. With a right triangle we can solve the triangle if we know two parts other than the right angle. The two known parts can be (1) the measure of one of the acute angles and the length of a side, or (2) the lengths of two sides. Consider right triangle ABC, with sides a, b, and c. See figure 17.17.

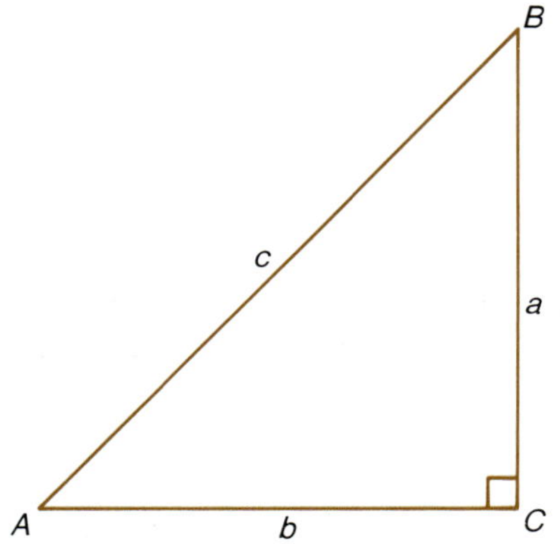

Figure 17.17

note

We always place vertex C at the right angle so side c is always the hypotenuse of the right triangle.

We define the side opposite a given angle to be the side that does not lie on the rays forming the angle. Then the side adjacent to a given acute angle of a right triangle is the side that, together with the hypotenuse, forms the angle. In figure 17.17:

1. Side a is *opposite* angle A and *adjacent* to angle B.
2. Side b is *opposite* angle B and *adjacent* to angle A.
3. Side c (the hypotenuse) is *opposite* right angle C.

Now place right $\triangle ABC$ in the plane so that angle A is in standard position (see figure 17.18). Using the definition for the six trigonometric ratios we then define:

$$\sin A = \frac{a}{c} \qquad \csc A = \frac{c}{a}$$

$$\cos A = \frac{b}{c} \qquad \sec A = \frac{c}{b}$$

$$\tan A = \frac{a}{b} \qquad \cot A = \frac{b}{a}$$

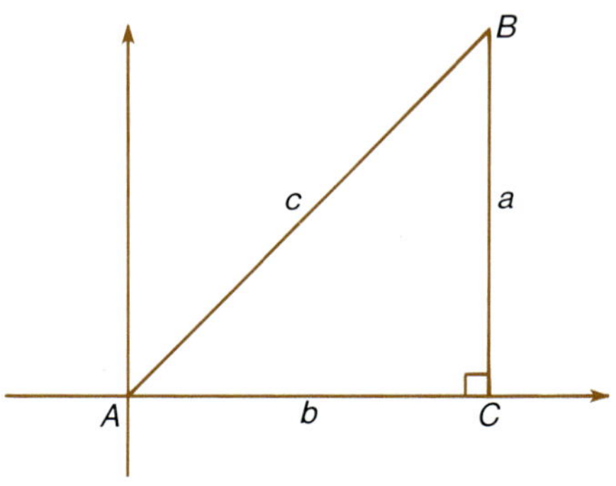

Figure 17.18

Example State the six trigonometric ratios for acute angle A of right $\triangle ABC$ (see figure 17.19).

$$\sin A = \frac{3}{5} \qquad \csc A = \frac{5}{3}$$

$$\cos A = \frac{4}{5} \qquad \sec A = \frac{5}{4}$$

$$\tan A = \frac{3}{4} \qquad \cot A = \frac{4}{3}$$

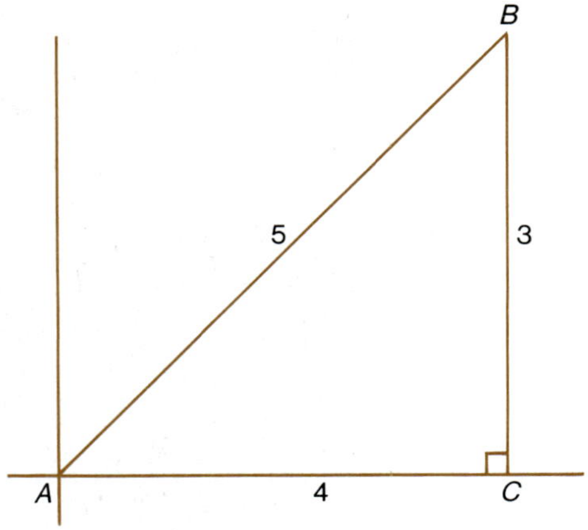

Figure 17.19

Now we use the preceding definitions to define the trigonometric ratios for arbitrary acute angle A in right $\triangle ABC$.

$$\sin A = \frac{a}{c} = \frac{\text{side opposite } A}{\text{hypotenuse}}$$

$$\cos A = \frac{b}{c} = \frac{\text{side adjacent to } A}{\text{hypotenuse}}$$

$$\tan A = \frac{a}{b} = \frac{\text{side opposite } A}{\text{side adjacent to } A}$$

$$\csc A = \frac{c}{a} = \frac{\text{hypotenuse}}{\text{side opposite } A}$$

$$\sec A = \frac{c}{b} = \frac{\text{hypotenuse}}{\text{side adjacent to } A}$$

$$\cot A = \frac{b}{a} = \frac{\text{side adjacent to } A}{\text{side opposite } A}$$

note

The same definitions apply to acute angle B, when B is substituted for A in the definitions. (The ratio of the sides will be different, however.)

We can now solve a given right $\triangle ABC$ using the derived definitions.

Examples

1. Given right $\triangle ABC$ with $a = 3$ and $m\,\angle A = 46°$, find the measure of the other acute angle B, the length of the hypotenuse, and the length of the other side. See figure 17.20.

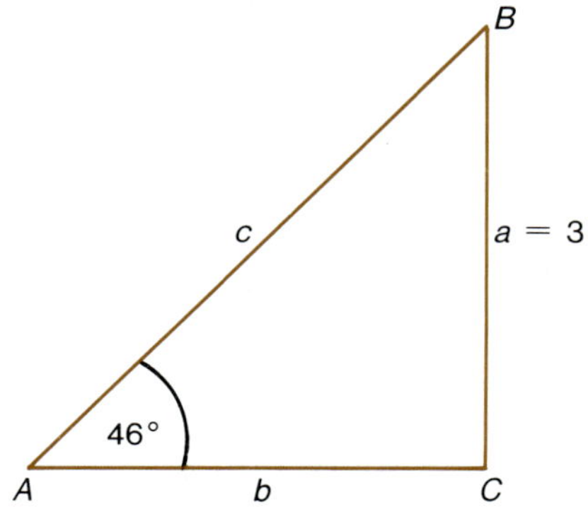

Figure 17.20

Now, since the measure of the angles of a triangle totals 180° and one angle contains 90°, the measure of the two acute angles must total 90°. Therefore,

$m\angle B = 90° - m\angle A = 90° - 46 = 44°$

Next, $m\,\angle A = 46°$ and *opposite side* $a = 3$ are known; to find the hypotenuse c, use $\sin 46° = \frac{3}{c}$ or $\csc 46° = \frac{c}{3}$. Since $c = \frac{3}{\sin 46°}$ or $c = 3 \csc 46°$, which is easier to calculate, use $c = 3 \csc 46° = 3(1.3902) \approx 4.17$.

Also, to find the *side adjacent* to angle A, side b, use $\tan 46° = \frac{3}{b}$ or $\cot 46° = \frac{b}{3}$. Since $b = \frac{3}{\tan 46°}$ or $b = 3 \cot 46°$ (again the easier to calculate), use $b = 3 \cot 46° = 3(.9657) \approx 2.897$. Then, $m\,\angle B = 44°$, $c \approx 4.2$, and $b \approx 2.9$.

2. Given right $\triangle ABC$ with side $a = 2.1$ and hypotenuse $c = 4.3$, solve the triangle. See figure 17.21. (*Note:* We want $m\angle A$, $m\angle B$, and the length of side b.)

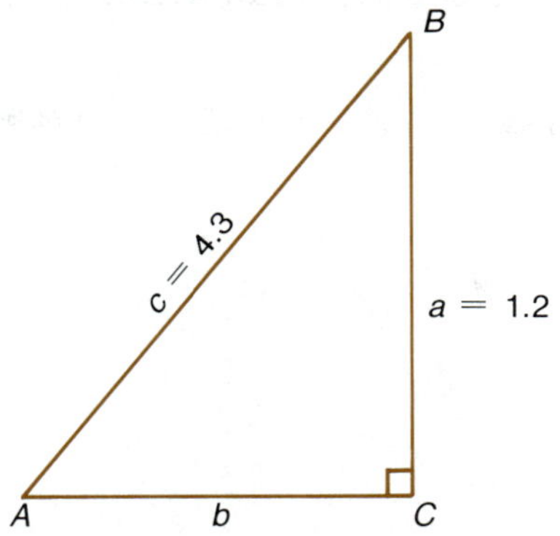

Figure 17.21

To find the measure of angle A, since the *hypotenuse* and the *side opposite* A are known, use $\sin A = \frac{a}{c} = \frac{2.1}{4.3} = 0.4884$. Then $m\angle A \approx 29.2°$. Now, $m\angle B = 90° - m\angle A = 90° - 29.2 \approx 60.8°$
Finally, using the Pythagorean theorem,

$$\begin{aligned} b &= \sqrt{c^2 - a^2} \\ &= \sqrt{(4.3)^2 - (2.1)^2} \\ &= \sqrt{18.49 - 4.41} \\ &= \sqrt{14.08}. \end{aligned}$$

So, $b \approx 3.75$. Therefore, $m\angle A \approx 29.2°$, $m\angle B \approx 60.8°$, and $b \approx 3.75$.

note

We used the Pythagorean theorem to find the side b since using the trigonometric ratios would have necessitated using a computed value.

A common application of right triangle trigonometry involves the **angle of elevation** and the **angle of depression** of an object in relation to a particular position. The angle of elevation is the angle between a horizontal sighting and an object above that sighting. The angle of depression is the angle between a horizontal sighting and an object below that sighting. See figure 17.22.

Figure 17.22

Examples

1. A boy is flying a kite. It is known that the angle of elevation of the kite with the ground is 58.2°. If the boy has let out 125 ft of string, at what height above the ground is the kite flying? See figure 17.23.

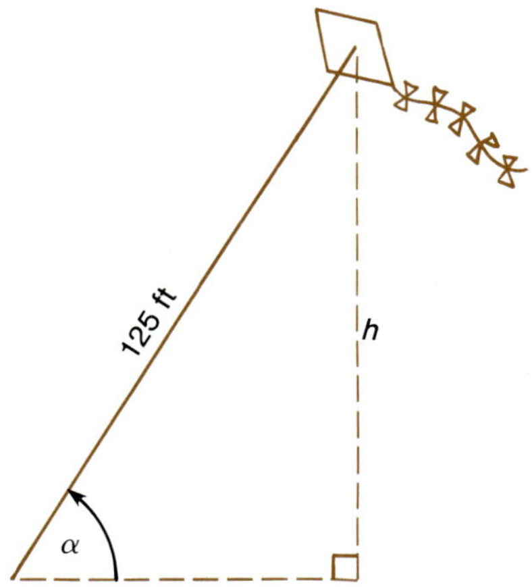

Figure 17.23

Since $m \angle(\alpha) = 58.2°$ and the hypotenuse c is 125 ft long are known, the *side opposite* the given angle α is the measure needed. Therefore, use

$$\sin 58.2° = \frac{h}{125} \text{ or } \csc 58.2° = \frac{125}{h}$$

Then, $h = 125 \sin(58.2°)$
$= 125(0.8499)$
≈ 106 ft.

Thus, the kite is flying approximately 106 ft above the ground.

note

In the study of plane geometry we learned that the alternate-interior angles formed by parallel lines cut by a transversal have the same measure. Using the same horizontal sighting, we can determine that their angle of elevation and angle of depression have the same measurement.

2. From the edge of a cliff that is 96 ft high, a ship is sighted anchored a distance of 425 ft from the base of the cliff. Find the angle of depression. See figure 17.24.

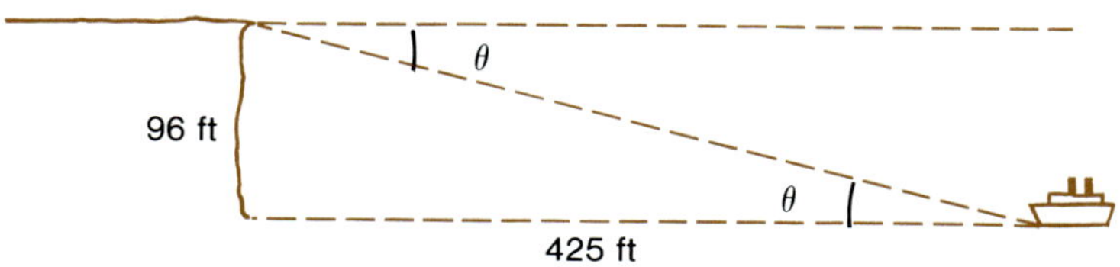

Figure 17.24

The *side opposite* and the *side adjacent* the angle θ are known; the angle measure is the one needed.
Therefore, use

$$\tan\theta = \frac{96}{425} \text{ or } \cot\theta = \frac{425}{96}$$

Then, $\tan\theta = \frac{96}{425}$

$= 0.2259,$

and $\theta \approx 12.7°$.

Thus, the angle of depression of the ship is approximately 12.7°.

3. A 25-ft ladder leans against a vertical wall. The foot of the ladder is at an angle of 35° with the ground. How far from the base of the wall is the foot of the ladder? See figure 17.25.

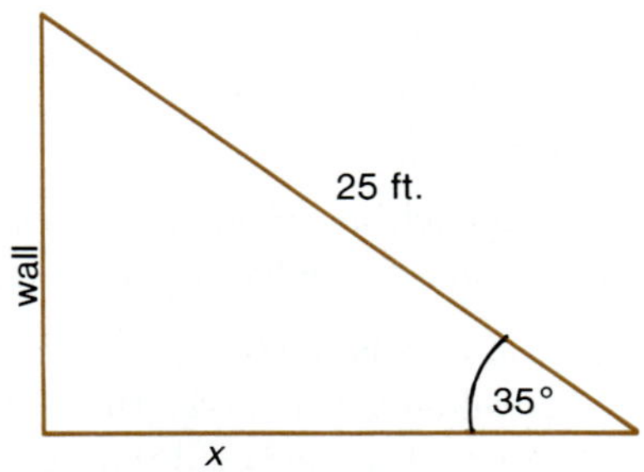

Figure 17.25

The angle of measure 35° and the length of the hypotenuse are given. Since the *side adjacent* the given angle is wanted, then use

$$\cos 35° = \frac{x}{25} \text{ or } \sec 35° = \frac{25}{x}$$

Then, $x = 25\cos 35°$

$= 25(0.8192)$

$\approx 20.$

Therefore, the foot of the ladder is approximately 20 ft from the base of the wall.

Exercise Set 17.4

In exercises 1–20, given right triangle *ABC*, solve the triangle with the given parts. Round the lengths of sides to three significant digits and the measure of angles to 0.1°.

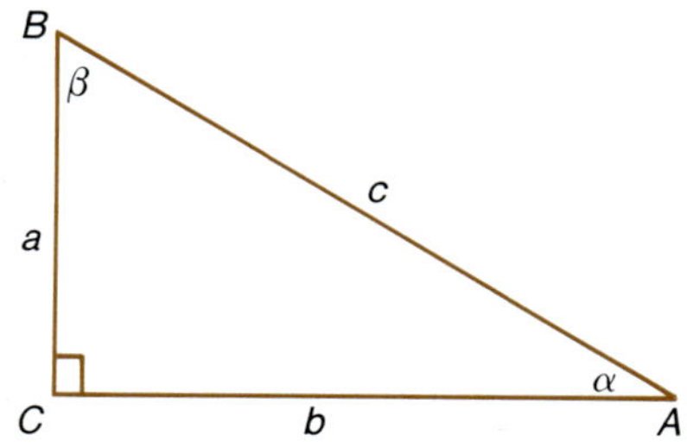

1. $a = 12$, $m\angle\alpha = 30°$
•2. $b = 8$, $m\angle\beta = 60°$
3. $a = 4$, $m\angle\beta = 45°$
4. $b = 1.3$, $m\angle\alpha = 34.5°$
5. $a = 10.3$, $m\angle\beta = 73.9°$
6. $b = 123.0$, $m\angle\beta = 62.8°$
7. $a = 376.0$, $m\angle\alpha = 52.1°$
8. $b = 4.76$, $m\angle\alpha = 18.4°$
9. $a = 5.0$, $b = 12.0$
•10. $a = 8.0$, $b = 6.0$
11. $a = 4.23$, $c = 8.73$
12. $b = 16.8$, $c = 23.4$
13. $a = 0.127$, $b = 0.346$
14. $c = 0.585$, $a = 0.279$
15. $b = 475$, $c = 763$
16. $a = 9.83$, $b = 5.21$
17. $m\angle\alpha = 83.1°$, $a = 2.93$
18. $m\angle\beta = 47.3°$, $b = 95.0$
19. $a = 97.6$, $b = 87.9$
20. $a = 7.83$, $c = 12.90$

In exercises 21 through 49, round distances and lengths correct to three significant digits and the measure of angles correct to the nearest 0.1°.

21. From the top of a building 153 ft above the street, the angle of depression as you look down on a man on the street measures 33.2°. Find, to the nearest tenth of a foot, how far the man is from the foot of the building at that instant.
•22. Find the angle of elevation of the top of a telephone pole that is 60.0 ft tall if it casts a shadow 72.3 ft long.
23. A tree casts a shadow 84.6 ft long when the angle of elevation of the sun is 39.2°. How tall is the tree?
•24. What is the angle of depression of a 6.00-ft man to the base of a tree that is 52.3 ft from the man?
25. From the lookout room of a lighthouse, a boat in the lake has an angle of depression of 18.1°. If the lookout is 195 ft above the water, how far is the boat from the lighthouse? Find the length of the line-of-sight from the lookout to the boat.
26. A 30-ft ladder leans against a building so that the foot of the ladder is 13.2 ft from the base of the building. Find the measure of the angle the ladder makes with the ground. How high on the building is the top of the ladder?
27. A ladder leans against a 9.70-ft high wall so that the top of the ladder is at the top of the wall and the foot of the ladder is 21.0 ft from the base of the wall. Find the length of the ladder. What is the measure of the angle the ladder makes with the wall?
28. Find the angle of elevation of a roadway that rises 175 ft for every 2,500 ft of road.
29. What is the angle of depression of an entrance ramp that is 2,050 ft long to a bridge that is 269.0 ft above ground level?

•30. What is the altitude of an isosceles triangle whose base angles have measure 51.0° if the base is 8.60 cm long? (*Hint:* The altitude to the base bisects the base.)

31. A guy wire to a telephone pole anchored to the ground at one end makes an angle of 48.9° with the ground. If the wire is anchored so that other end is 15.2 m above the ground, how long is the wire?

32. The altitude of the top of Pike's Peak outside of Colorado Springs, Colorado, is 14,010 ft. If the angle of depression from the top to the city is 8.1° and the horizontal distance from the base of Pike's Peak to the city is 5.2 mi, what is the altitude of Colorado Springs? (*Hint:* Change the miles to feet.)

33. A chisel is made from a steel rod with diameter 5.2 cm. If the tip of the rod is 4.8 cm long (see the accompanying figure), what is the angle of the tip?

34. A rectangular piece of land is 423.0 m by 673.2 m. Find the measure of the acute angle made by a diagonal of the rectangle with its longest side.

35. The length of a kite string is 750 ft and the angle of elevation of the kite is 31.4°. Find the vertical height of the kite.

36. If the distance from the north rim to the south rim of the Grand Canyon is measured to be 10.05 mi, how many feet above the south rim is the north rim if the angle of elevation from the south rim to the north rim is measured by a surveyor to be 1°24′? See the accompanying figure.

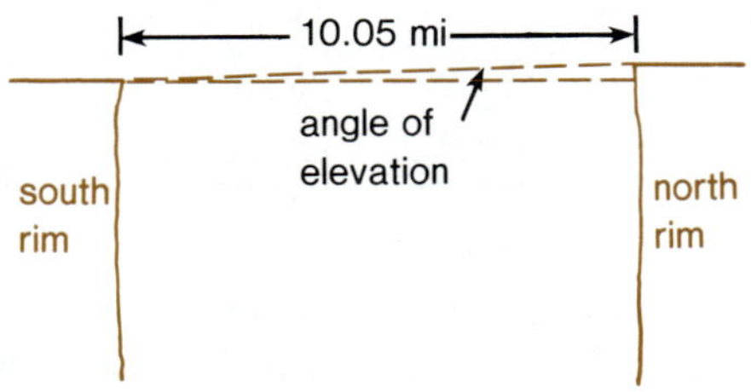

37. A road rises at an angle of elevation of 7.8°. In one-half mile of roadway, what is the increase in elevation of the road, measured in feet?

•38. When the sun is at an angle of elevation of 63°18′, what is the length of the shadow of the Washington Monument, which has height of 555 ft?

39. Find the rafter length in the accompanying figure.

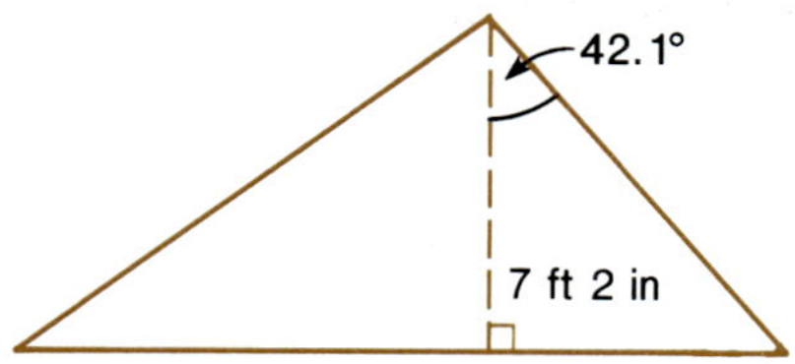

40. A salvage ship uses a sonar system to locate a sunken ship. If the slant distance from the salvage ship to the sunken ship is 1,670 m and the angle of depression is measured to be 21°42′, how deep is the water at the sunken ship's location? How far is the salvage ship from a point directly above the sunken ship?

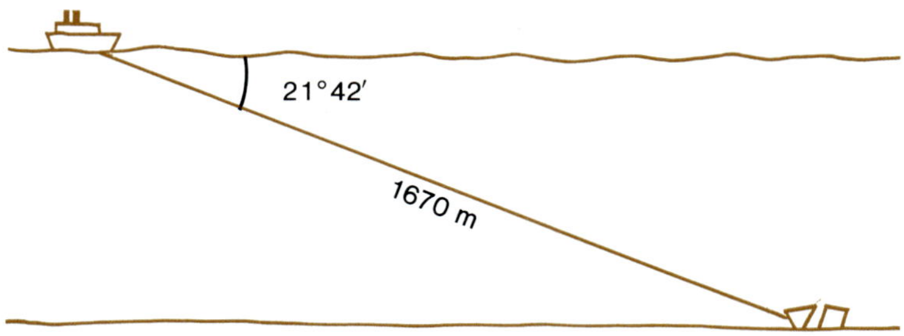

41. A landscape designer lays out the sprinkler system shown in the accompanying figure. How much pipe is necessary for the sprinkler system? (*Note:* Dotted lines do not represent pipe.)

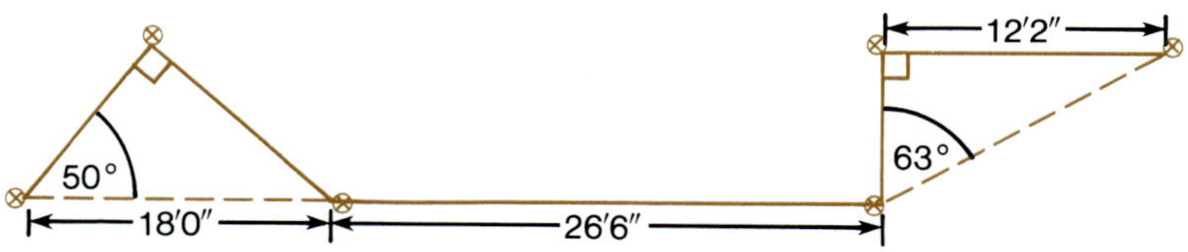

•42. The bases of an isosceles trapezoid are 8.0 cm and 12.0 cm. If the base angles measure 75°, find (a) the length of the equal sides x and (b) the altitude h of the trapezoid. See the accompanying figure.

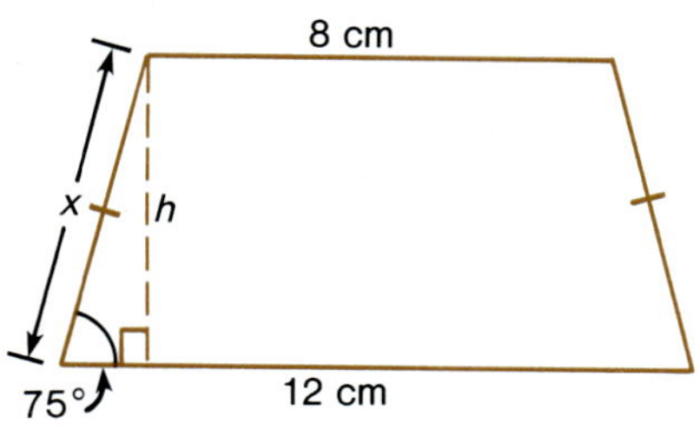

43. Holes are to be drilled in a steel plate as shown in the given figure. Find the lengths of $\overline{AC}$ and $\overline{AB}$.

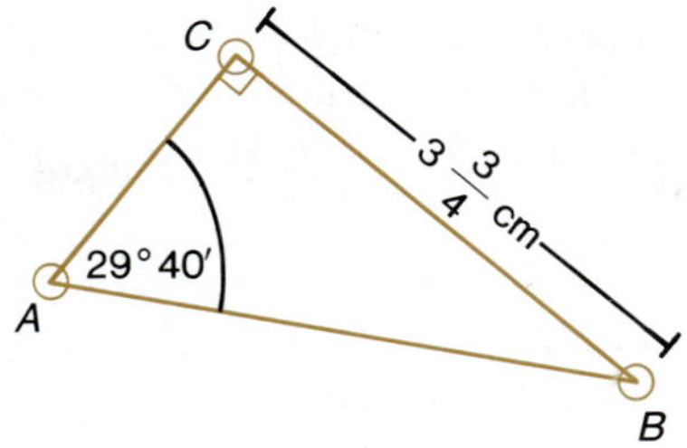

44. A chord 24 in long is 17.8 in from the center of a circle. See the accompanying figure.

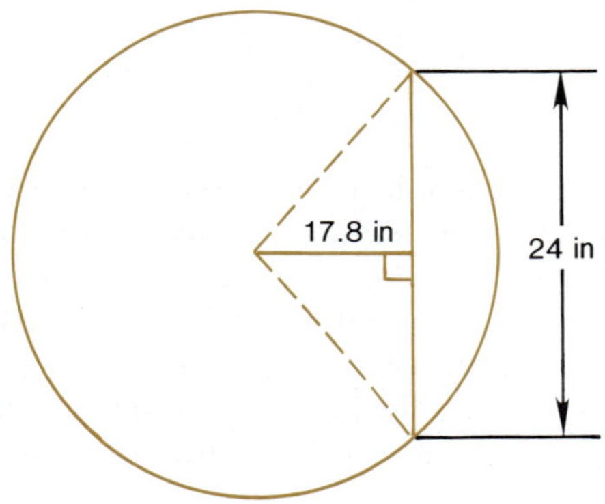

Find (a) the length of the radius of the circle, and (b) the central angle θ of the chord.

note

The line drawn to the chord from the center bisects the chord.

45. In an AC circuit, we plot the resistance R on the x-axis and the reactance x_i on the y-axis. The hypotenuse represents the impedance (z). Find z and $m\ \angle\theta$ when $x_i = 40$ ohms and $R = 25$ ohms. See the figure given.

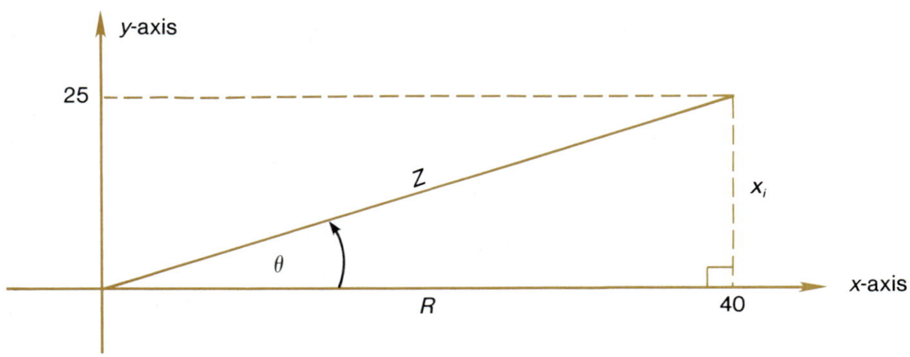

46. Using the diagram showing the relationship between the impedance (z), the resistance (R), and the reactance (x_i) of an electric circuit shown in the given figure, find:
   a. The impedance (z) when $R = 20$ ohms and $x_i = 30$ ohms
   b. The reactance (x_i) when $z = 20$ ohms and $R = 10$ ohms
   c. The resistance (R) if $z = 36$ ohms and $x_i = 25$ ohms

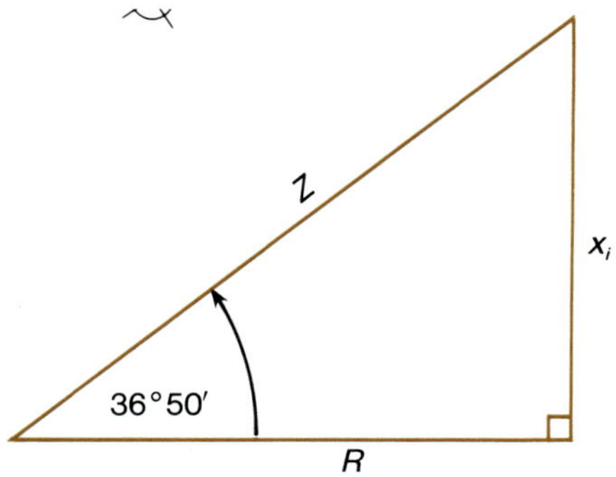

47. The tangent of the helix angle of a drill is equal to the circumference of the drill divided by the lead (the horizontal distance advanced by the helix in making a complete revolution). See the accompanying figure. Find:
   a. The helix angle if the pitch diameter is 4.70 in and the lead is 6.57 in. (*Note:* circumference = $\pi \times$ diameter.)
   b. The lead if the helix angle is 63° and the pitch diameter is 3.17 in.
   c. The circumference if the helix angle is 57.2° and the lead is 12.43 in.

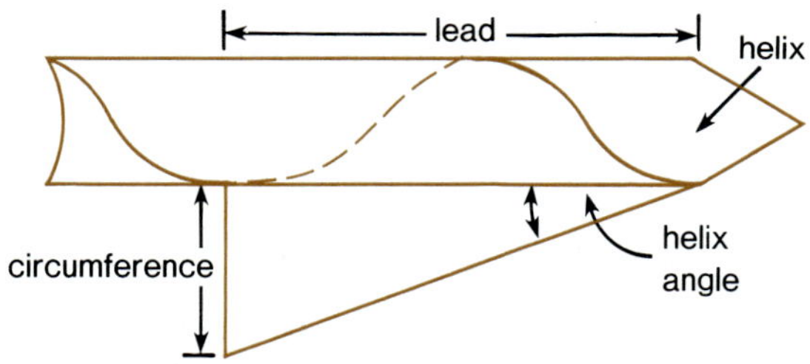

•48. Find the perpendicular distance from the center to the sides of a regular hexagon whose sides are 4.34 in long. See the accompanying figure. (*Hint:* Find the $m\ \angle\theta$ first.)

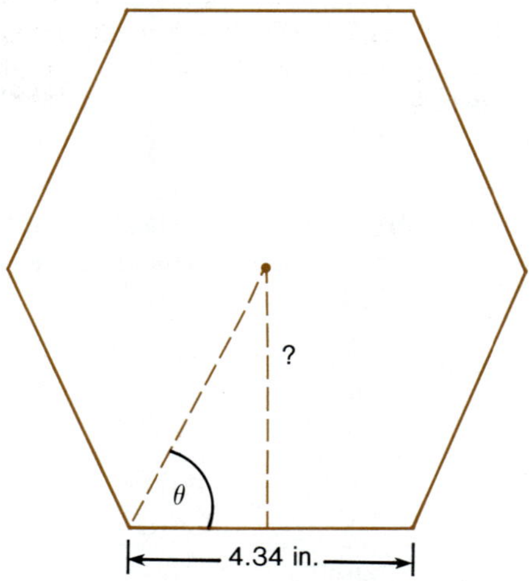

49. The accompanying figure is a sketch of a 700-ft radio tower with cables anchored at points A and B on a level with the base of the tower. Find:
    a. The length of cables AC and BC
    b. The distance AB between anchor points A and B

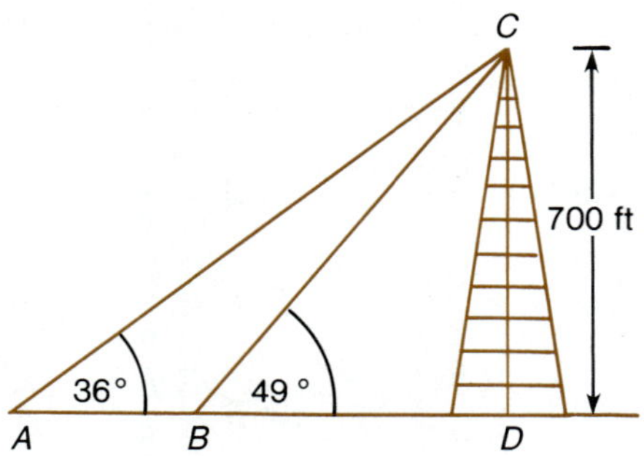

Chapter Summary

1. An angle in *standard position* has its vertex at the origin and one side, the *initial side,* on the positive x-axis and its other side, the *terminal side,* anywhere in the plane.
2. When the terminal side rotates away from the initial side in a *counterclockwise* direction, the angle is said to be **positive;** when it rotates away in a *clockwise* direction, the angle is said to be **negative.**
3. **Coterminal** angles are angles in standard position having the *same terminal side*.
4. Given angle θ in standard position, all angles that are coterminal with θ are of the form $\theta + 360\ k$ (k is an integer).

5. Given angle θ in standard position and point $P(x,y)$ on the terminal side of θ, by definition:

$$\sin\theta = \frac{y}{r} \qquad \csc\theta = \frac{r}{y}$$

$$\cos\theta = \frac{x}{r} \qquad \sec\theta = \frac{r}{x}$$

$$\tan\theta = \frac{y}{x} \qquad \cot\theta = \frac{x}{y}$$

where $r = \sqrt{x^2 + y^2}$, the distance from the vertex to point P.

6. The sine and cosecant, cosine and secant, tangent and cotangent ratios are *reciprocal ratios*. That is,

$$\csc\theta = \frac{1}{\sin\theta}$$

$$\sec\theta = \frac{1}{\cos\theta}$$

$$\cot\theta = \frac{1}{\tan\theta}$$

7. Given a right triangle ABC with the right angle at vertex C, we define:

$$\sin A = \frac{a}{c} \qquad \csc A = \frac{c}{a}$$

$$\cos A = \frac{b}{c} \qquad \sec A = \frac{c}{b}$$

$$\tan A = \frac{a}{b} \qquad \cot A = \frac{b}{a}$$

$$\sin B = \frac{b}{c} \qquad \csc B = \frac{c}{b}$$

$$\cos B = \frac{a}{c} \qquad \sec B = \frac{c}{a}$$

$$\tan B = \frac{b}{a} \qquad \cot B = \frac{a}{b}$$

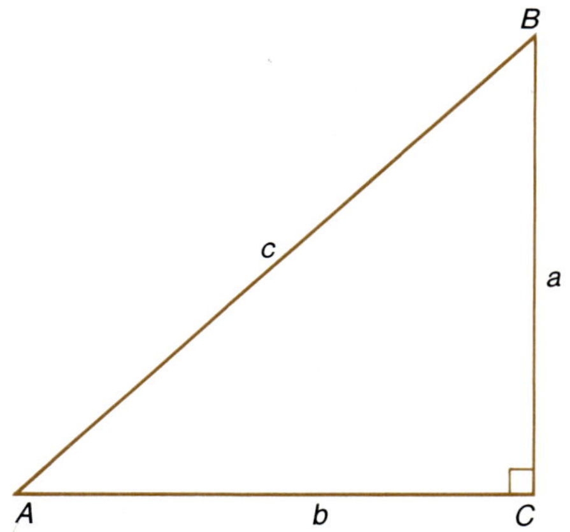

8. The sine and cosine, secant and cosecant, tangent and cotangent ratios are called **cofunctions** of one another.

9. Given right triangle ABC and arbitrary acute angle α of the triangle, we define:

$$\sin\alpha = \frac{\text{side opposite } \alpha}{\text{hypotenuse}} \qquad \csc\alpha = \frac{\text{hypotenuse}}{\text{side opposite } \alpha}$$

$$\cos\alpha = \frac{\text{side adjacent } \alpha}{\text{hypotenuse}} \qquad \sec\alpha = \frac{\text{hypotenuse}}{\text{side adjacent } \alpha}$$

$$\tan\alpha = \frac{\text{side opposite } \alpha}{\text{side adjacent } \alpha} \qquad \cot\alpha = \frac{\text{side adjacent } \alpha}{\text{side opposite } \alpha}$$

10. The **angle of elevation** of some object is the angle between the horizontal and the line of sight to the object, *upward* in relation to the observer.
11. The **angle of depression** of some object is the angle between the horizontal and the line of sight to the object, *downward* in relation to the observer.

Chapter Review

[17.2] Find two positive and two negative angles that are coterminal with the given angle.

1. $36°$ 2. $147°$ 3. $219°$ 4. $308°$

Find the smallest positive angle that is coterminal with the given angle.

5. $408°$ 6. $639°$ 7. $328°$ 8. $-73°$
9. $-149°$ 10. $-277°$ 11. -291 12. $-627°$

[17.3] Find the values of the six trigonometric ratios for the angle in standard position whose terminal side passes through the given point. Sketch the angle.

13. $(4,3)$ 14. $(12,-5)$ 15. $(-8,6)$ 16. $(-5,-12)$
17. $(2,-2\sqrt{3})$ 18. $(8,-8)$ 19. $(4,0)$ 20. $(9,-4)$

Find the values of the five other trigonometric ratios for angle α if α lies in the given quadrant.

21. $\sin(\alpha) = \frac{4}{5}, \quad 0 < \alpha < 90°$

22. $\cos(\alpha) = -\frac{5}{12}, \quad 90° < \alpha < 180°$

23. $\tan(\alpha) = \frac{7}{24}, \quad 180° < \alpha < 270°$

24. $\csc(\alpha) = -2, \quad 270° < \alpha < 360°$

25. $\sec(\alpha) = -\frac{8}{6}, \quad 90° < \alpha < 180°$

26. $\cot(\alpha) = \frac{1}{\sqrt{3}}, \quad 180° < \alpha < 270°$

[17.5] Solve right $\triangle ABC$ using the given conditions. Round off all lengths and distances correct to three significant digits and angles to the nearest $0.1°$.

27. $m\angle B = 15°48'$ and $c = 8.73$
28. $m\angle A = 63.2°$ and $b = 14.3$
29. $m\angle B = 57.6°$ and $b = 0.746$
30. $a = 630.1$ and $b = 259.6$
31. $a = 5.07$ and $c = 7.14$
32. $b = 14.7$ and $c = 23.9$

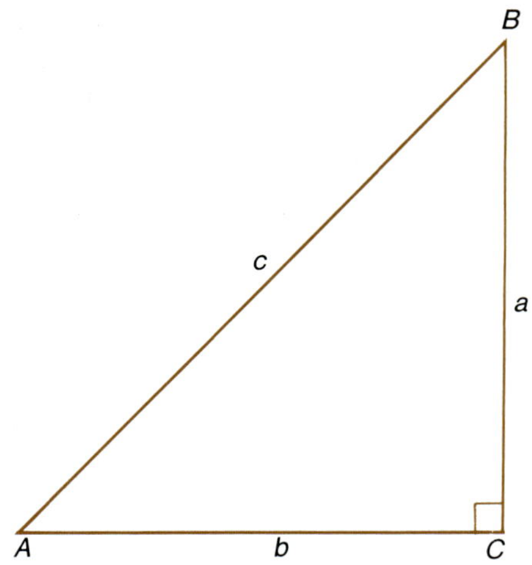

33. A guy wire attached to a tower makes an angle of $75°12'$ with the ground. If the wire is anchored 10.3 m from the base of the tower, how high on the tower is the wire attached?
34. At a certain time of the day a tree casts a shadow that is 16.0 m long. If the tree is 32.7 m tall, what is the angle of elevation of the sun?
35. From the top of his tower a forest ranger sights a fire at an angle of depression of $6.73°$. If the tower is 65.0 ft tall, how far from the base of the tower is the fire?
36. A balloon is held by a straight rope that is 202.3 m long. If the rope is inclined at an angle of $53°42'$, how high is the balloon?
37. A 20.0-ft ladder leans against the house, touching it 14 ft 3 in above the ground. What angle does the ladder make with the house? How far from the foot of the house is the other end of the ladder?
38. A beam of light is projected vertically making a spot on the clouds. The angle of elevation from a person standing 120.0 m from the light source to the light spot on the clouds is $82°30'$. How high is the cloud on which the light shines?
39. A 25-m swimming pool is 1.50 m deep at one end and 3.75 m deep at the other end. If the bottom has a constant slope, what is the angle of slope of the bottom from the horizontal? Find the length of the bottom from end to end.
40. If a shelf on a wall is supported by a straight brace 15.3 in long, and the brace is attached to the wall 10.5 in below the shelf, what angle does the brace make with the shelf?

Chapter Test

1. What are coterminal angles?
2. When is an angle in standard position?
3. When is an angle positive? negative?
4. Find the least positive angle that is coterminal with the following angles:
   a. $572°$ b. $-198.3°$ c. $-421°$

5. Given angle α in standard position, if the point $P(3, -5)$ is on the terminal side, state the six trigonometric ratios of angle α. In what quadrant does angle α lie?
6. Given third-quadrant angle θ such that $\sec \theta = -\frac{13}{5}$, find the values of the five other trigonometric ratios.
7. Using your calculator, find:
   a. $\sin 34°12'$ b. $\cos(-134°42')$
   c. $\csc 78.7°$ d. $\cot(324.7°)$
8. Using a calculator, find θ to the nearest 0.01°, given θ lies in the first quadrant and:
   a. $\tan \theta = 1.532$ b. $\sin \theta = 0.8176$
   c. $\sec \theta = 1.748$ d. $\cot \theta = 0.3146$
9. Using a calculator, find $m\angle\theta$ to the nearest 0.01°, given θ lies in the first quadrant and:
   a. $\sin \theta = \frac{7}{8}$ b. $\cos \theta = \frac{2}{9}$ c. $\csc \theta = \frac{34}{27}$

Round all lengths to the accuracy of the measured data and the size of the angle to the nearest 0.1°.

Given right $\triangle ABC$, find:

10. c when $m\angle A = 47.3°$ and $a = 5.6$
11. b when $m\angle B = 72.3°$ and $c = 14.9$ (Round b to three significant digits.)
12. $m\angle A$ to the nearest 0.01° when $a = 0.783$ and $b = 0.523$
13. $m\angle A$ to the nearest 0.1° and a when $m\angle B = 31°15'$ and $c = 425.6$

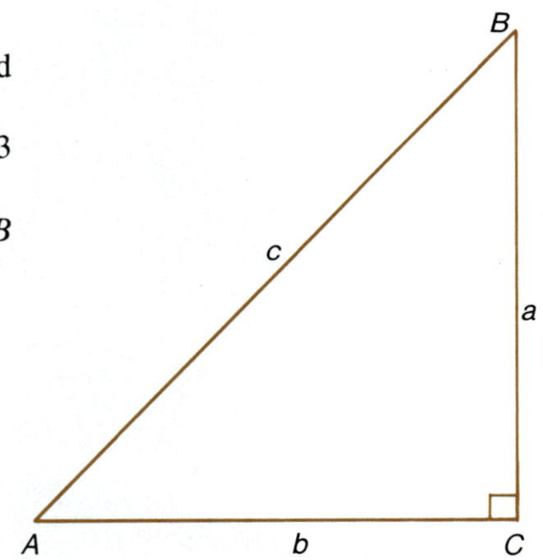

14. A ramp leading to a loading dock is 15.5 m long and makes an angle of 23.5° with the ground. How far above the ground is the loading dock?
15. An air traffic controller sitting in a tower that is 55.0 m high sights an airplane on the ground at an angle of depression of 32°14′. How far is the plane from the base of the tower?
16. A kite string makes an angle of elevation of 41.5°. If the girl flying the kite has paid out 143 ft of string, how far is it from the girl to a position on the ground directly below the kite?

18
Oblique Triangles and Vectors

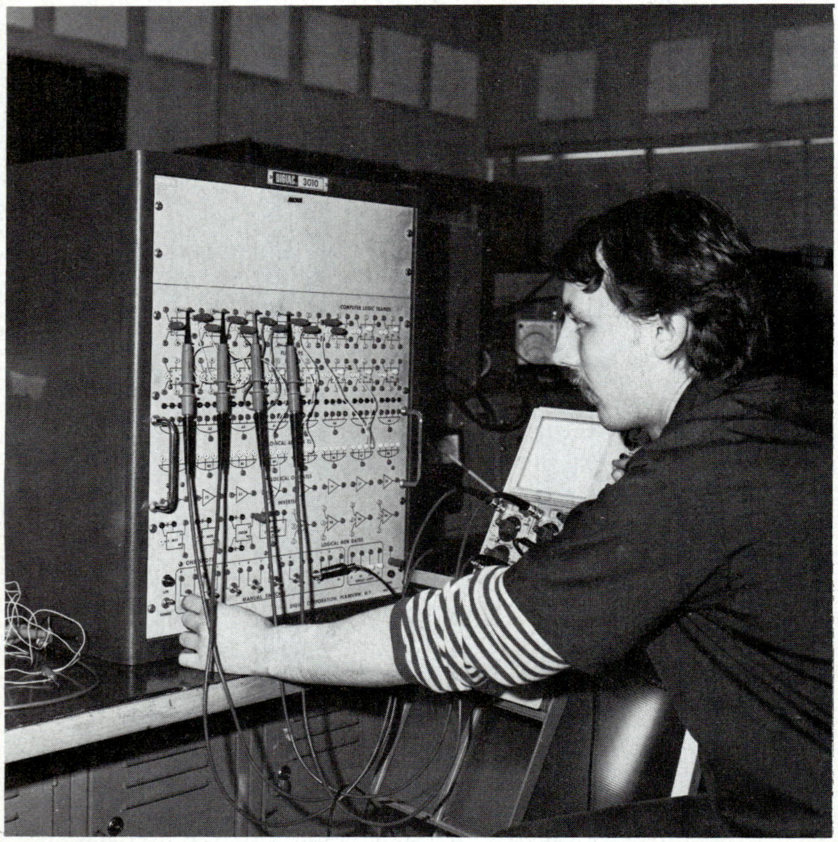

A technician studying a basic computer circuit on a shift register.
(Photograph by Kenneth Walters, courtesy of Henry Ford Community College)

18.1 Oblique Triangles

Thus far we have solved only right triangles for their unknown parts. Now we consider the class of all triangles that are *not right*—the **oblique triangles.** Recall that, when naming triangles by their angles, we have three kinds: right, acute, and obtuse. The class of oblique triangles thus contains the acute and obtuse triangles.

Using a protractor, compass, and ruler, a triangle can be constructed using any one of the following sets of conditions:

1. The measure of two angles and the length of a side
2. The lengths of two sides and the measure of an angle opposite one side
3. The lengths of two sides and the measure of an angle included by them
4. The lengths of the three sides

In chapter 17 we stated that a triangle can be solved if we know three parts of the triangle, one of which is the length of a side. Thus, we can solve for the unknown parts of *any* oblique triangle if we are given any one of the sets of conditions just listed.

18.2 The Sine Law

Consider oblique triangles ABC having sides a, b, and c in figure 18.1. The perpendicular dotted line dropped from vertex C to side $\overline{AB}$, or extended side $\overline{AB}$, in each triangle is the *altitude*, or *height*, denoted by h, of the triangle. From right triangle trigonometry, we can define

$$\sin B = \frac{h}{a} \qquad \text{and} \qquad \sin A = \frac{h}{b}$$

in figure 18.1**a.** In like fashion,

$$\sin B = \frac{h}{a} \qquad \text{and} \qquad \sin(180 - A) = \frac{h}{b} = \sin A$$

in figure 18.1**b.**

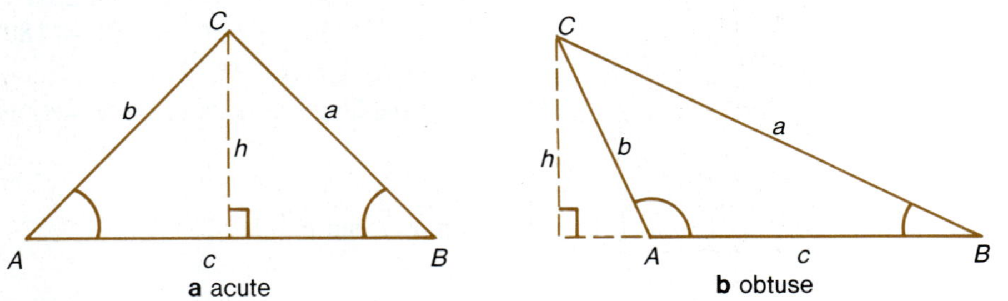

Figure 18.1

note

It can be shown that $\sin(180 - A) = \sin A$

Therefore, in either case

$$h = a \cdot \sin B \qquad \text{and} \qquad h = b \cdot \sin A$$

Since $a \cdot \sin B$ and $b \cdot \sin A$ are both equal to h, then

$$a \cdot \sin B = b \cdot \sin A$$

Dividing both sides of this equation by $\sin A \cdot \sin B$, we get

$$\frac{a}{\sin A} = \frac{b}{\sin B}$$

If we drop a perpendicular from vertex A to side $\overline{BC}$ and from vertex B to side $\overline{AC}$, we can show that each of the preceding ratios is equal to the ratio $\frac{c}{\sin C}$. Thus, the **Sine Law** (or **Law of Sines**) for oblique triangles states:

$$\frac{a}{\sin A} = \frac{b}{\sin B} = \frac{c}{\sin C}$$

That is, the Sine Law states: The ratio of a side of a triangle to the sine of the angle opposite this side is the same for all three angles and sides. Any two of these ratios forms a proportion that can be used to solve for the unknown parts of the triangle.

Recall that a proportion is a statement that two ratios are equal and that we can solve a proportion if we know three of the four parts. We can use the Sine Law to find the unknown parts of oblique triangles if we are given: (1) the measures of two angles and the length of one side, or (2) the lengths of two sides and the measure of an angle opposite one side.

Let us consider first an oblique triangle ABC, given the measure of two angles and the length of one side. We demonstrate this situation by the following examples.

Examples

1. Solve $\triangle ABC$ given $m\angle A = 43°$, $m\angle B = 62°$, and $a = 5$. (*Note:* We must find $m\angle C$ and the lengths of sides b and c.)
   a. To find $m\angle C$, use the fact that the sum of the measure of the angles in any triangle is 180°.
   Then,

   $$\begin{aligned} m\angle C &= 180° - (m\angle A + m\angle B) \\ &= 180° - (43° + 62°) = 180° - 105° \end{aligned}$$

   Therefore $m\angle C = 75°$.
   b. To find the length of side b, since we know a, $m\angle A$, and $m\angle B$, we use

   $$\frac{a}{\sin A} = \frac{b}{\sin B}.$$

   Then

   $$\frac{5}{\sin 43°} = \frac{b}{\sin 62°}$$

Taking the cross-products, we get

$$b \sin 43° = 5 \sin 62°$$

$$\text{and } b = \frac{5 \sin 62°}{\sin 43°}$$

$$= \frac{5(0.8829)}{(0.6820)}$$

$$= 6.4729$$

Therefore, $b \approx 6.47$, correct to three significant digits.

c. To find the length of side c, use a, $m\angle A$, $m\angle C$, and

$$\frac{a}{\sin A} = \frac{c}{\sin C}.$$

Then

$$\frac{5}{\sin 43°} = \frac{c}{\sin 75°}$$

and cross-multiplying, $c \sin 43° = 5 \sin 75°$.

Solving for c, we have

$$c = \frac{5 \sin 75°}{\sin 43°}$$

$$= \frac{5(0.9659)}{(0.6820)}$$

$$= 7.0814$$

Then, $c \approx 7.08$, correct to three significant digits.

note

To find the length of side c, we were forced to use a previously computed value, $m\angle C$. We do not want to do this unless we must, as in this case. Use *only* the given measurements unless the problem requires that you use a calculated value. If it is necessary to use a computed value, it is a good idea to compute this value to a *greater* accuracy than is necessary in order to help maintain the reliability of the answers that use this value. To illustrate, if $m\angle A \approx 33.4°$ and $m\angle B \approx 73.1°$, then $m\angle C \approx 73.5°$, which is less accurate than either angle A or angle B.

2. Solve $\triangle ABC$ given $m\angle B = 33°10'$, $m\angle C = 92°20'$, and $a = 6.4$. (*Note:* We must find the measure of $m\angle A$ and the lengths of sides b and c.)

a. $m\angle A = 180° - (m\angle B + m\angle C)$
$= 180° - (33°10' + 92°20')$
$= 180° - 125°30' = 54°30' = 54.5°$

b. To find the length of side c, use

$$\frac{a}{\sin A} = \frac{c}{\sin C}$$

Then

$$\frac{6.4}{\sin 54.5°} = \frac{c}{\sin 92.3°}$$

and $c \sin 54.5° = 6.4 \sin 92.3°$, so

$$\begin{aligned} c &= \frac{6.4 \sin 92.3°}{\sin 54.5°} \\ &= \frac{6.4(0.9992)}{(0.8141)} \\ &= 7.86 \end{aligned}$$

Therefore $c \approx 7.9$, correct to two significant digits.

c. To find the length of side b, use

$$\frac{a}{\sin A} = \frac{b}{\sin B}$$

Then

$$\frac{6.4}{\sin 54.5°} = \frac{b}{\sin 33.2°}$$

and $b \sin 54.5° = 6.4 \sin 33.2°$

Then

$$\begin{aligned} b &= \frac{6.4 \sin 33.2°}{\sin 54.5°} \\ &= \frac{6.4(0.5476)}{(0.8141)} \\ &\approx 4.30 \end{aligned}$$

Then, $b \approx 4.3$, correct to two significant digits.

Now we consider the **ambiguous case**—knowing the lengths of two sides and the measure of an angle opposite one side—so-called because of the possible triangles that can result from this set of conditions. The given information may produce one of four possible results:

1. two oblique triangles
2. one oblique triangle
3. one right triangle
4. no triangle

The following example illustrates the first result: two oblique triangles.

Example Solve oblique triangle ABC given $a = 4$, $b = 6$, and $m\angle A = 36.0°$.

In drawing the triangle, first construct angle $A = 36.0°$. Making one of the sides adjacent to angle A, side b, of length 6 units, set the compass to a distance of 4 units. Place the point of the compass at the end of side b, point C; strike an arc. The arc will cut the other side of angle A in two points, B and B'. Connect point C with these two points to form oblique triangles ABC and $AB'C$ in figure 18.2.

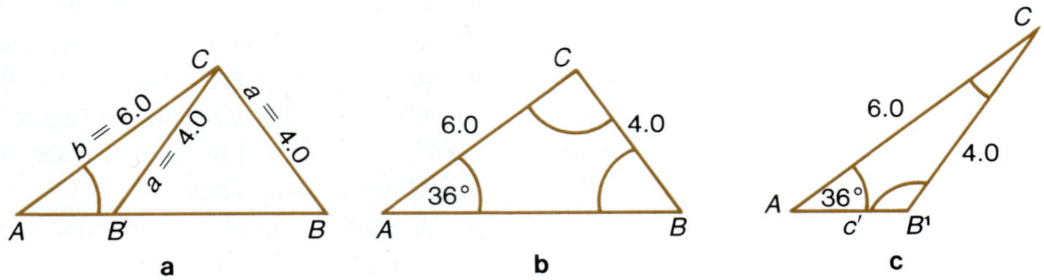

Figure 18.2

Use the Sine Law to find the measure of $m\angle B$ in $\triangle ABC$. Using

$$\frac{a}{\sin A} = \frac{b}{\sin B},$$

we substitute to obtain the proportion

$$\frac{4.0}{\sin 36.0°} = \frac{6.0}{\sin B}.$$

Cross-multiplying, we obtain $4.0 \sin B = 6.0 \sin 36.0°$.

$$\text{Then } \sin B = \frac{6.0 \sin 36.0°}{4.0}$$
$$= \frac{6.0(0.5878)}{4.0}$$
$$= 0.8817.$$

Thus $m\angle B \approx 61.8°$.

Now $m\angle C$ in $\triangle ABC$ is given by

$$\begin{aligned} m\angle C &= 180° - (m\angle A + m\angle B) \\ &= 180° - (36.0° + 61.8°) \\ &= 180° - 97.8° \\ &\approx 82.2°. \end{aligned}$$

Again use the Sine Law to find the length of side c of $\triangle ABC$:

$$\frac{a}{\sin A} = \frac{c}{\sin C}$$

and so

$$\frac{4.0}{\sin 36.0°} = \frac{c}{\sin 82.2°}$$

Therefore, $c \sin 36.0° = 4.0 \sin 82.2°$ and $c = \dfrac{4.0 \sin 82.2°}{\sin 36.0°}$.

Then

$$c = \frac{4.0(0.9907)}{(0.5878)}$$
$$= 6.742$$

and we get $c \approx 6.7$, correct to two significant digits.

Thus, in $\triangle ABC$, $m\ \angle B \approx 61.8°$, $m\ \angle C \approx 82.2°$, and $c \approx 6.7$.

Now consider the solution of $\triangle AB'C$ with angles A, B', and C. See figure 18.2**a.** Since $m\ \angle B = 61.8°$ and $\overline{BC} = \overline{B'C}$, then the measure of $m\ \angle AB'C = 61.8°$. Since this is supplementary to angle $AB'C$, then $m\ \angle AB'C = 180° - 61.8° = 118.2°$. Thus, the measure of angles A and $AB'C$ in $\triangle AB'C$ are known. Then

$$m\ \angle ACB' = 180° - (36.0° + 118.2°)$$
$$= 180° - 154.2° = 25.8°.$$

Find the length of c' by using

$$\frac{a}{\sin A} = \frac{c'}{\sin ACB'}.$$

Then

$$\frac{4.0}{\sin 36.0°} = \frac{c'}{\sin 25.8°},$$

so $c' \sin 36.0° = 4.0 \sin 25.8°$ and $c' = \dfrac{4.0 \sin 25.8°}{\sin 36.0°}$.

Thus

$$c' = \frac{4.0(0.4352)}{(0.5878)}$$
$$= 2.96$$
$$\approx 3.0, \text{ correct to two significant digits.}$$

Therefore, the solution of the second triangle formed by the conditions is $m\ \angle AB'C \approx 118.2°$, $m\ \angle ACB' \approx 25.8°$, and $c' \approx 3.0$.

The second result possible in the ambiguous case is one oblique triangle. One oblique triangle is obtained when the length of the side *opposite* the given angle is *longer* than the length of the side *adjacent* to the given angle. To illustrate, consider the following example.

Example

Let $b = 4.7$, $a = 6.3$, and $m\,\angle A = 52.3°$. From figure 18.3 note that when the arc is struck from point C using length 6.3, it intersects the side c at point B and an extension of side c at point A'. The angle at A' is not in the triangle, so there is only one triangle.

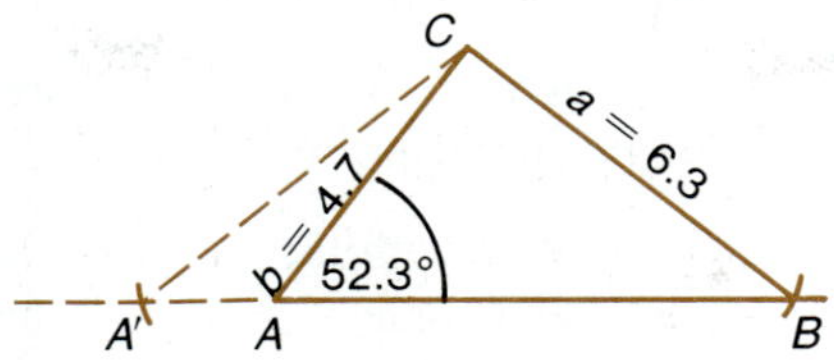

Figure 18.3

The last two possible results occur when:

1. The side opposite the given angle is *just long enough* to reach the other side when striking the arc, producing one right triangle.
2. The side opposite the given angle is *not long enough* to reach the other side when striking the arc, producing no triangle.

To summarize the ambiguous case, given acute angle A and side a (opposite angle A) and side b (one of those adjacent to angle A), the four possible results occur under the following conditions:

1. Two solutions if $b\sin(A) < a < b$
2. One solution if $a > b$
3. A right triangle solution if $a = b\sin(A)$
4. No solution if $a < b\sin(A)$

Since we are interested only in oblique triangles, we shall not consider either of the last two situations that can exist with this set of conditions.

Exercise Set 18.1

In exercises 1–16 solve the oblique triangles with the given parts using the Sine Law. Find the measure of the angles to the nearest 0.1° and round the lengths of the sides according to the accuracy of the measured data.

1. $a = 42$, $m\,\angle B = 22°$, $m\,\angle A = 73°$
•2. $b = 4.7$, $m\,\angle A = 27.3°$, $m\,\angle C = 89.2°$
3. $c = 14.7$, $m\,\angle C = 52.8°$, $m\,\angle B = 36.7°$
4. $a = 4.22$, $m\,\angle B = 75.0°$, $m\,\angle C = 47.8°$
5. $b = 345$, $m\,\angle B = 64.5°$, $m\,\angle A = 58.3°$
6. $c = 0.736$, $m\,\angle A = 22.3°$, $m\,\angle B = 114.7°$
7. $a = 78.0$, $m\,\angle B = 67°12'$, $m\,\angle C = 14°42'$
8. $b = 9.0$, $m\,\angle A = 72°30'$, $m\,\angle B = 39°50'$
9. $c = 0.217$, $m\,\angle C = 107°20'$, $m\,\angle A = 40°52'$
10. $a = 16.3$, $m\,\angle B = 136°36'$, $m\,\angle C = 14°15'$

11. $a = 92$, $b = 52$, $m\angle A = 37°$
•12. $b = 5.9$, $c = 7.02$, $m\angle C = 39.2°$
13. $a = 15.3$, $c = 32.5$, $m\angle C = 59°22'$
•14. $a = 8.41$, $b = 9.50$, $m\angle A = 54.5°$
15. $a = 8.84$, $c = 3.71$, $m\angle C = 19.8°$
16. $b = 24.3$, $m\angle A = 52.6°$, $m\angle B = 67.3°$

Use the Sine Law to solve the following exercises. Round off your lengths correct to three significant digits and angles to the nearest 0.1°.

17. A shelf is constructed with the supports shown in the accompanying figure. If angle $A = 29.3°$ and angle $B = 125.4°$, find the lengths of the supports.

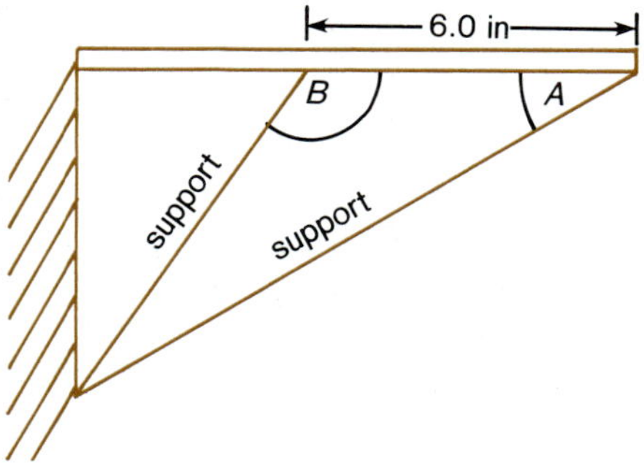

•18. A flagpole 55.0 ft high stands on top of a grassy knoll that slopes at an angle of 9.5° with the horizontal. How far from the flagpole is a person if the angle of elevation to the top of the flagpole is 23.4°? See the accompanying figure.

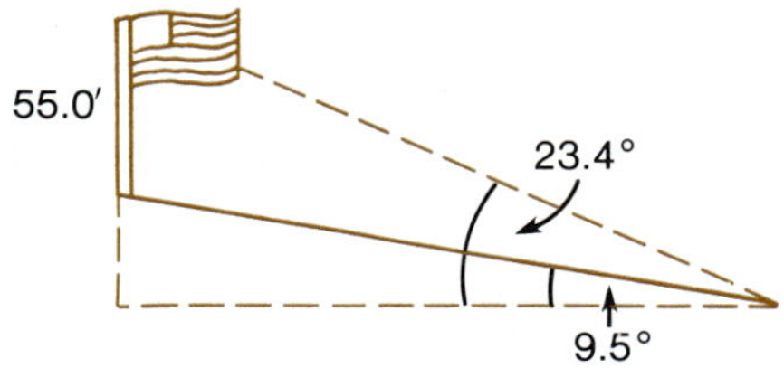

19. A metal plate contains three holes, as placed in the given figure. Find $m\angle A$.

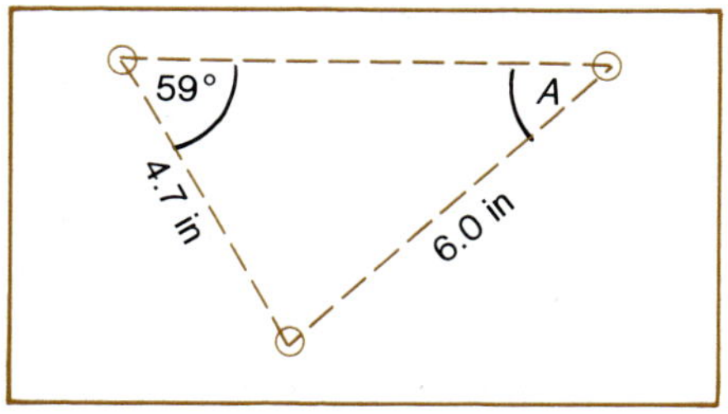

•20. A ship leaves port at San Francisco and travels due west for a time. At one point, the ship turns an angle of 112° from the original direction. It travels for 56.0 mi, at which time it is 79.5 mi from San Francisco. What angle does the direct line to San Francisco make with its original due west heading? How far from San Francisco did the ship make its turn?

21. One side of a triangular steel plate is 15.3 cm long and the angles this side makes with the other sides are 53.0° and 37.6°, respectively. Find the lengths of the other two sides.

22. The area of any oblique triangle *ABC* can be found by

$$\text{Area} = \frac{1}{2}bc \sin A$$

$$\text{Area} = \frac{1}{2}ac \sin B$$

$$\text{Area} = \frac{1}{2}ab \sin C$$

Using this formula, find the area of the following oblique triangles:

a. $m \angle B = 36.0°$, $m \angle C = 110.3°$, $a = 15.0$
b. $m \angle A = 37.6°$, $m \angle B = 63.1°$, $c = 5.80$
c. $m \angle A = 27.9°$, $m \angle C = 80.3°$, $b = 10.9$

(*Hint:* Use the sine law to find the missing side first.)

23. A woman 5.3 ft tall stands in a vertical position on a hillside having a slope of 22.0°. What is the length of the shadow she casts when the sun's elevation is 52.3° and the sun is behind her so that her shadow is *down* the hillside?

24. Two straight paths in a field make an angle of 142.0° with each other at an intersection, point *P*. John walks down one path at 6.50 km/hr and Joan walks down the other. After 2 h, they are 25.6 km apart. How far had Joan walked and how fast did she walk?

25. A ship is located 25.0° east of south from lighthouse L_1 and 43.2° west of south of lighthouse L_2. If the distance between lighthouses is 43.7 mi, how far is the ship from each lighthouse? (*Note:* Assume L_2 is due east of L_1.)

18.3 The Cosine Law

Suppose we are given (1) the length of two sides and the measure of the angle included between them, or (2) the lengths of the three sides of a triangle. If we try to use the Sine Law in either of these cases, we will soon discover that not enough information has been given. To illustrate, consider the following examples.

Examples

1. Given $a = 7$, $b = 4$, and $m \angle C = 52°$, solve the triangle.
   Substituting, we have

$$\frac{7}{\sin A} = \frac{4}{\sin B} = \frac{c}{\sin 52°}$$

   The given information contains only two of the four parts of any one proportion and we need three.

2. Given $a = 4$, $b = 5$, and $c = 8$, solve the triangle.
   Now

$$\frac{4}{\sin A} = \frac{5}{\sin B} = \frac{8}{\sin C}$$

   and we again know only *two parts* of any one proportion.

Another tool must be used to solve an oblique triangle with this information known. The **Cosine Law** (or **Law of Cosines**) is developed for this purpose.

Consider oblique triangle ABC with altitude h shown in figure 18.4.

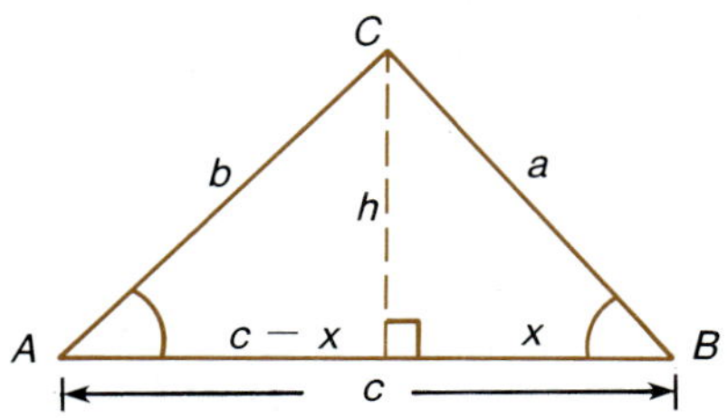

Figure 18.4

We can see that

$$\cos B = \frac{x}{a}.$$

Then $x = a \cos B$

By the Pythagorean theorem, $b^2 = h^2 + (c - x)^2$ and $a^2 = h^2 + x^2$, so $h^2 = a^2 - x^2$. Substituting $a^2 - x^2$ for h^2 in $b^2 = h^2 + (c - x)^2$ we obtain

$$\begin{aligned} b^2 &= (a^2 - x^2) + (c - x)^2 \\ &= a^2 - x^2 + c^2 - 2cx + x^2 \\ &= a^2 + c^2 - 2cx \end{aligned}$$

But $x = a \cos B$, and substituting $a \cos B$ for x in the preceding equation, we get

$$b^2 = a^2 + c^2 - 2ac \cos B$$

Similarly, we can determine that

$$\begin{aligned} a^2 &= b^2 + c^2 - 2bc \cos A \\ c^2 &= a^2 + b^2 - 2ab \cos C \end{aligned}$$

We can also show that the Cosine Law holds when the $\triangle ABC$ is obtuse with the obtuse angle at B.

For any oblique triangle (acute or obtuse), the Cosine Law states:

$$\begin{aligned} a^2 &= b^2 + c^2 - 2bc \cos A \\ b^2 &= a^2 + c^2 - 2ac \cos B \\ c^2 &= a^2 + b^2 - 2ab \cos C \end{aligned}$$

That is, the Cosine Law states: The square of the unknown side is equal to the sum of the squares of the other two sides minus two times their product times the cosine of the angle opposite the unknown side.

Examples

1. Solve oblique triangle ABC given $a = 7.2$, $c = 8.1$, and $m\angle B = 54°$. (*Note:* We are given the lengths of two sides and the measure of the angle included between them.)
   Solve first for the side opposite the given angle, side b. Use $b^2 = a^2 + c^2 - 2ac \cos B$:

   $$\begin{aligned} b^2 &= (7.2)^2 + (8.1)^2 - 2(7.2)(8.1)\cos 54° \\ &= 51.84 + 65.61 - 116.64(0.5878) \\ &= 117.45 - 68.56 \\ &= 48.89 \end{aligned}$$

   Therefore, $b = \sqrt{48.89} \approx 6.99 \approx 7.0$.
   Now use the Sine Law to find $m\angle A$.

   $$\frac{7.2}{\sin A} = \frac{6.99}{\sin 54°}$$

   $$6.99 \cdot \sin A = 7.2 \cdot \sin 54°$$

   $$\sin A = \frac{7.2 \sin 54°}{6.99} = \frac{7.2(0.8090)}{6.99} = 0.8333$$

   Then $\sin A = 0.8333$ and $m\angle A = 56.4°$.
   Since $m\angle C = 180° - (m\angle A + m\angle B)$, then

   $$\begin{aligned} m\angle C &= 180° - (56.4° + 54°) \\ &= 180° - 110.4° \\ &= 69.6° \end{aligned}$$

   Therefore, $b \approx 7.0$ $m\angle A \approx 56.4°$, and $m\angle C \approx 69.6°$.

   note

   We use the Cosine Law to find one missing value, and then use the Sine Law to find the remaining parts since it is simpler to use.

2. Solve oblique triangle ABC given $a = 18$, $b = 25$, and $c = 12$.

   note

   We are given the lengths of the three sides of the triangle. Solve for one of the angles first. When using the Cosine Law to find the measure of an angle, find the measure of the *largest* angle first—the one *opposite* the longest side.

   Solving for $m\angle B$, use $b^2 = a^2 + c^2 - 2ac \cos B$. Substitute to obtain:

   $$\begin{aligned} 25^2 &= 18^2 + 12^2 - 2(18)(12)\cos B \\ 625 &= 324 + 144 - 432 \cos B \\ 625 &= 468 - 432 \cos B \\ 157 &= -432 \cos B \end{aligned}$$

   Therefore, $\cos B = \dfrac{157}{-432} \approx -0.3634$.

note

The -0.3634 indicates that B is a second-quadrant angle, an angle between 90° and 180° (obtuse).

Using $\tilde{B}$ for the acute angle, we find $m\ \angle\tilde{B} = 68.7°$

So $m\ \angle\tilde{B} = 180 - m\ \angle B'$

$\approx 180 - 68.7°$

$\approx 111.3°$

To find $m\ \angle A$, use the Sine Law:

$$\frac{18}{\sin A} = \frac{25}{\sin 111.3°}$$

Then

$$\sin A = \frac{18 \sin (111.3°)}{25} = \frac{18(0.9317)}{25} = 0.6708$$

Therefore, $m\ \angle A \approx 42.1°$.

Finally, $m\ \angle C = 180° - (111.3° + 42.1°)$

$= 180° - 153.4°$

$\approx 26.6°$

Thus, $m\ \angle A \approx 42.1°$, $m\ \angle B \approx 111.3°$, and $m\ \angle C \approx 26.6°$.

note

The smallest angle should be opposite the shortest side and the middle-sized angle should be opposite the side whose length is between those of the other two. This is true in our example.

3. A man walks from point A to point B, a distance of 123 ft. He then turns 115° to the left and walks 98 ft to point C. How far is he from starting point A? See figure 18.5. (*Note:* $m\ \angle B = 180° - 115° = 65°$)

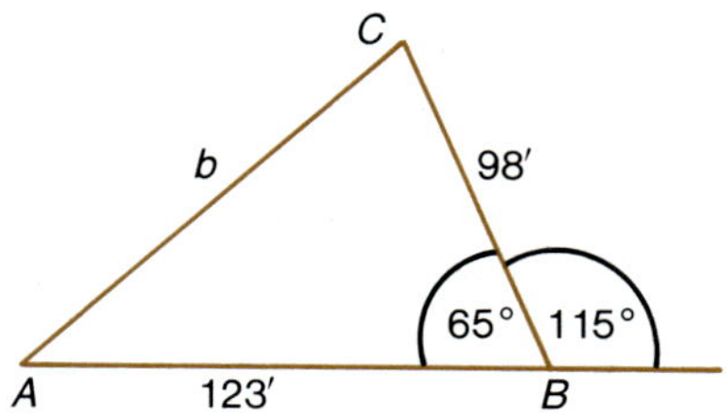

Figure 18.5

The measure required is the length of b, so use

$b^2 = a^2 + c^2 - 2ac \cos B$

$b^2 = (98)^2 + (123)^2 - 2(98)(123) \cos 65°$

$= 9{,}604 + 15{,}129 - 24{,}108(0.4226)$

$= 24{,}733 - 10{,}188.04$

$= 14{,}544.96$

Then $b = \sqrt{14{,}544.96} = 120.6 \approx 121$

The man is approximately 121 ft from starting point A.

4. In figure 18.6, if holes are drilled in a steel plate at points A, B, and C, find $m\ \angle B$.

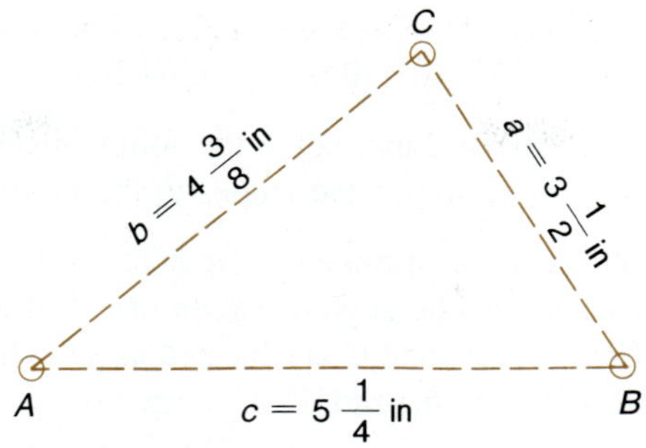

Figure 18.6

The measure required is that of angle B, so again use $b^2 = a^2 + c^2 - 2ac \cos B$. Then, solve again for $\cos B$:

$$\cos B = \frac{a^2 + c^2 - b^2}{2ac}$$

$$= \frac{(3.5)^2 + (5.25)^2 - (4.375)^2}{2(3.5)(5.25)}$$

$$= \frac{12.25 + 27.56 - 19.14}{36.75}$$

$$= \frac{20.67}{36.75}$$

$$= 0.5624$$

So, $m\ \angle B \approx 55.8°$.

Exercise Set 18.2

In exercises 1–15, solve oblique triangle ABC using first the Cosine Law. Round angles to the nearest 0.1° and lengths of sides to the same accuracy as the side measurements in the problem.

1. $a = 4$, $b = 9$, $m\ \angle C = 50°$
•2. $a = 3.7$, $c = 4.8$, $m\ \angle B = 43.9°$
3. $b = 14.9$, $c = 24.8$, $m\ \angle A = 63°12'$
4. $a = 4.7$, $b = 6.3$, $c = 5.4$
5. $a = 410$, $b = 563$, $c = 815$
•6. $a = 0.915$, $b = 0.207$, $c = 0.719$
7. $a = 36.4$, $b = 51.5$, $m\ \angle C = 81.3°$
8. $a = 172$, $c = 563$, $m\ \angle B = 101.2°$
9. $a = 21.9$, $b = 16.3$, $c = 24.5$
10. $a = 2.17$, $b = 3.76$, $c = 5.03$

11. $b = 0.046$, $c = 0.023$, $m\angle A = 10°30'$
12. $a = 43.6$, $b = 72.5$, $m\angle C = 112°06'$
13. $a = 58.0$, $b = 52.0$, $c = 8.1$
•14. $a = 11.7$, $b = 4.99$, $m\angle C = 76.9°$
15. $a = 0.165$, $b = 0.234$, $c = 0.309$

Using the Cosine Law, solve the following problems. (Round all lengths correct to three significant digits and angles to the nearest 0.1°.)

16. A triangular sprinkler system has sides of 75 ft, 50 ft, and 62.5 ft. Find the measure of the angle between the 75-ft and 50-ft sides.
17. Gears A, B, and C are located as shown in the accompanying figure, with radii of 4 cm, 6 cm, and 3 cm, respectively. Find AC, AB, BC, and angle α.

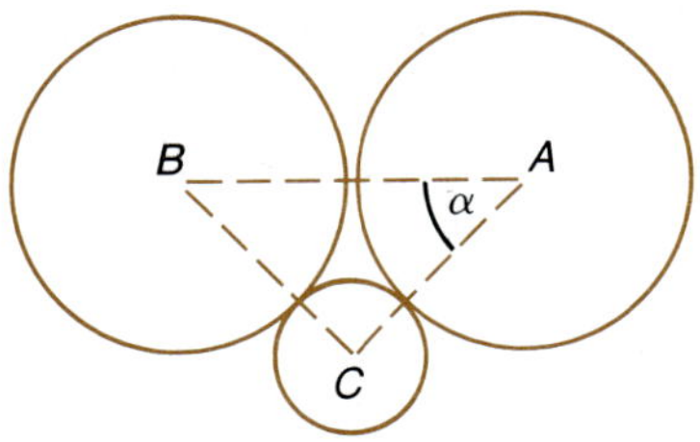

18. A steel block is to be finished to the dimensions shown in the given figure. Calculate angle α, the angle at which the sloping surfaces are to be planed.

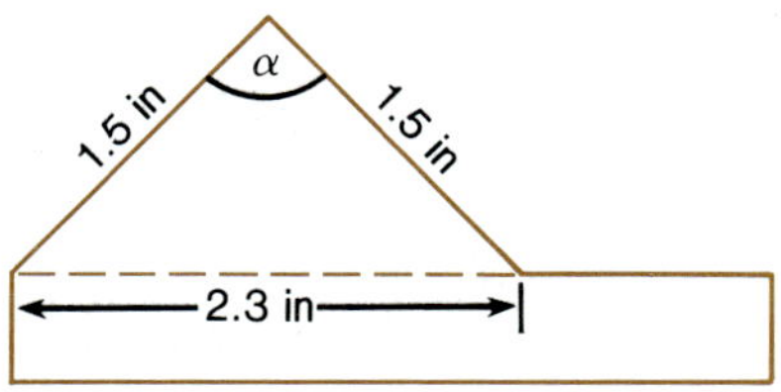

19. Given parallelogram $ABCD$ in the accompanying figure, find the length of the indicated diagonal $\overline{AC}$.

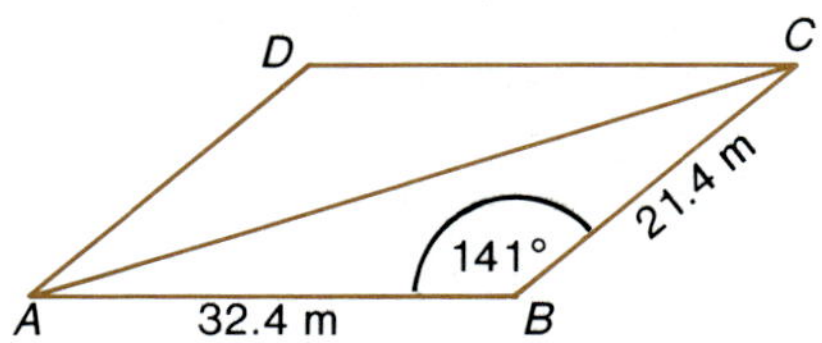

•20. A sheetmetal worker must make the triangular pattern shown in the accompanying figure. Find the angle of inclination α and the length of side x.

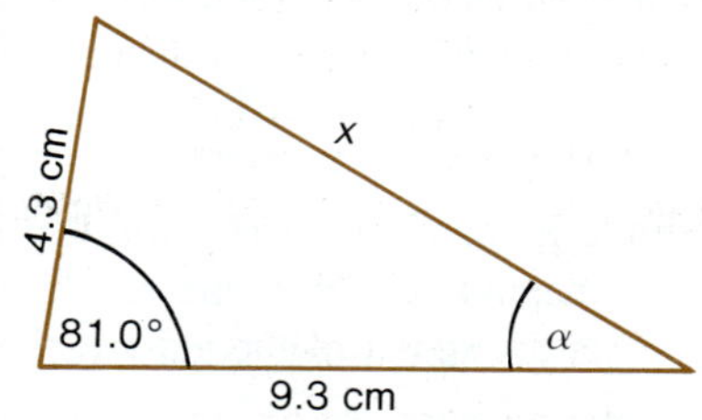

21. Two runners leave a given point A at the same time, running in straight lines. If the angle between their routes is 78° and if, after 5 min, the runners are 1.1 km and 0.98 km from the starting point, what is the straight line distance between runners after 5 min?
22. A surveyor wishes to calculate the distance between two trees across the river from where she is standing. If she is standing at a point C, the distance to the trees A and B is 98.9 m and 110 m, respectively, and the angle formed at C is 49.3°. Compute the distance between the trees.
23. The parallel sides of a trapezoid are 5.00 cm and 11.5 cm long. The oblique sides are 5.70 cm and 7.30 cm long. See the accompanying figure. If the measure of the diagonal shown is 8.30 cm, find the measures of angle α and β.

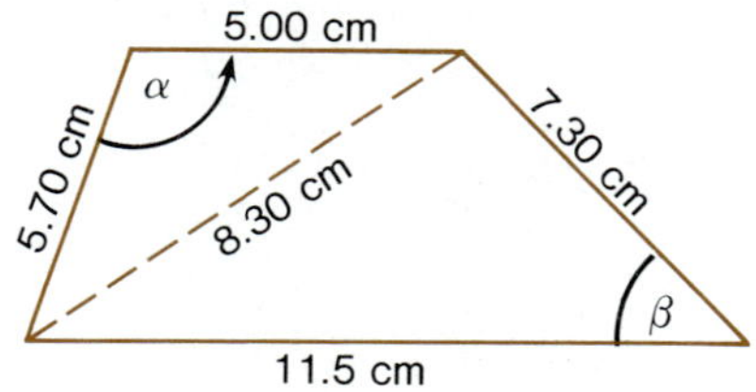

24. A triangular lot is to be enclosed by 450 m of fencing. If two sides of the lot are 150.0 m and 185.2 m long, find the measure of each angle of the lot.
25. A triangular sheet metal plate has sides 12.2 ft, 8.30 ft, and 7.01 ft. Find the measure of the angle included between the 12.2-ft and 8.30-ft sides.
26. A sheetmetal worker wishes to make a wedge whose side view is shown in the accompanying figure. Find the length of the inclined surface l and the angle of inclination of the wedge.

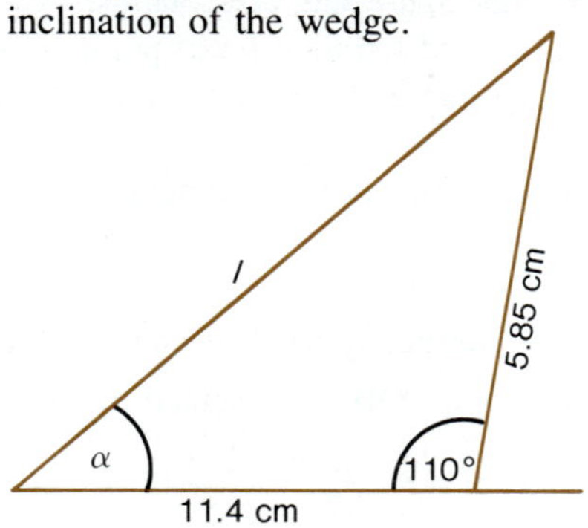

18.4 Vectors

The measurement of many physical quantities, such as length, weight, and area, is given by magnitude alone. The units of measure are specified and these quantities are represented by real numbers. Such numbers are called **scalars** since they relate to the numbers on a real number scale. To illustrate, consider the following example.

Example An automobile is traveling at 57 mi/h; this is a measure of the automobile's magnitude. Thus, the number 57 is a scalar quantity. The 57 mi/h represents the **speed** of the automobile.

Suppose we add in this example that the automobile is traveling at a speed of 57 mi/h *due east*. Now the *direction* of the automobile together with the *magnitude* is given and we call 57 mi/h the *velocity* of the automobile. This quantity is now called a *vector*. Physical quantities such as force, velocity, and acceleration, where direction as well as magnitude is required are called **vectors.**

The **magnitude** of a vector is represented by a line segment drawn to scale. The **direction** is represented by the line segment's orientation to a reference line. We use the rectangular coordinate system as the reference for our vectors. See figure 18.7.

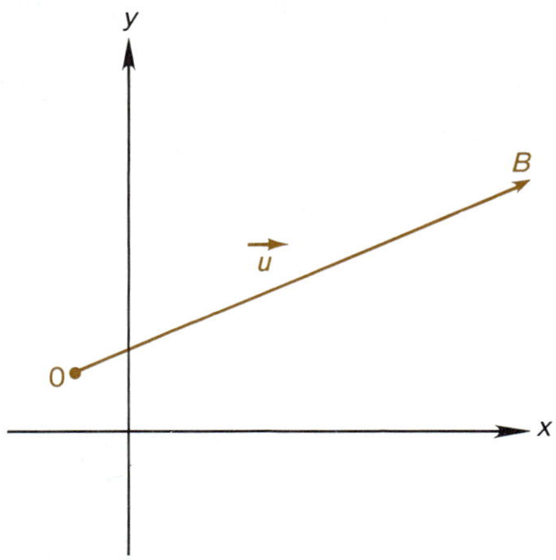

Figure 18.7

The vector in this figure has been named vector **u,** represented by $\vec{u}$, with initial (starting) point O and terminal (end) point B. We call O the **tail** and B the **head** of the vector. If we represent a quantity by A, this quantity is measured *only* by its magnitude.

A **vector** is a directed line segment.

note

We might alternately name vector **u** by its tail and head points. Thus, $\vec{u}$ can be called vector **OB,** represented by $\overrightarrow{OB}$.

The magnitude of vector **u,** or vector **OB,** is represented by the symbol u or OB. See figure 18.8.

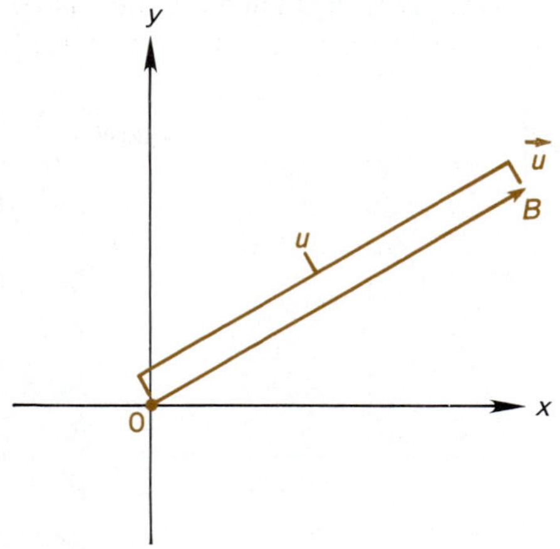

Figure 18.8

Vectors may be moved in the plane to different tail points and not change their value as long as they remain parallel to their original direction and maintain the same magnitude. That is: Vectors **u** and **v** are said to be equal if they have the same magnitude and the same direction. In figure 18.9, vectors **u** and **v** are equal even though they have different tail and head points.

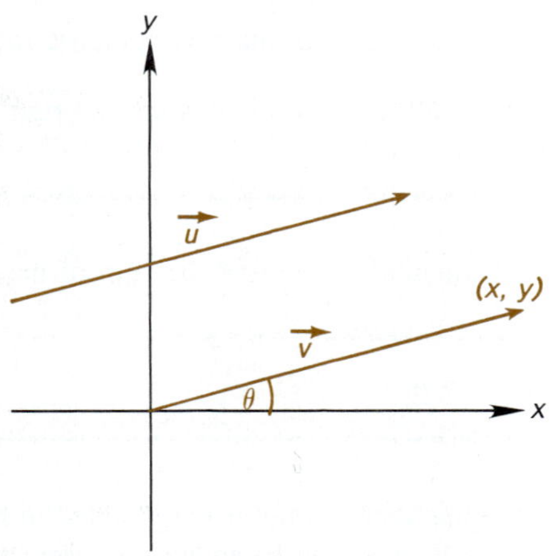

Figure 18.9

Notice that $\vec{v}$ has its tail point at the origin: thus, it is said to be in **standard position.** Numerical calculations are often simplified by translating vectors to standard position. We can then identify any vector by: (1) the length of the line segment and the measure of its angle θ made with the x-axis, or (2) the coordinates of the terminal point, x and y, as indicated in figure 18.9.

Example

Find the $m \angle \theta$ and magnitude of the vector in standard position with terminal point $(1, \sqrt{3})$. See figure 18.10.

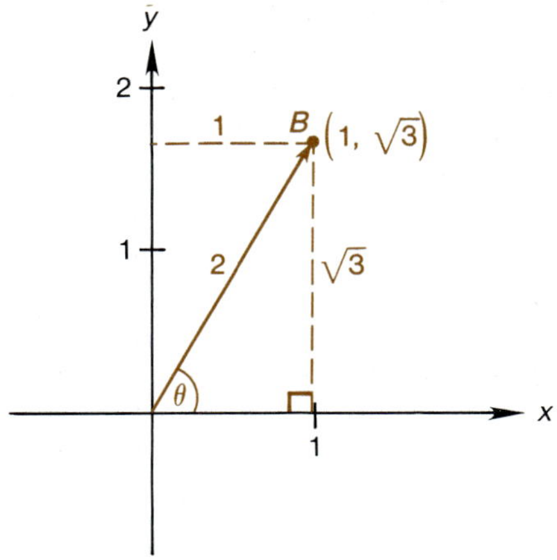

Figure 18.10

The hypotenuse of the right triangle formed by the point on the terminal side is

$$\sqrt{(\sqrt{3})^2 + 1^2} = \sqrt{4} = 2$$

Then $OB = 2$. Recall from the trigonometric ratios that $\sin 60° = \dfrac{\sqrt{3}}{2}$

Thus, $m \angle \theta = 60°$ and so this is a special right triangle: a 30°-60°-90° right triangle.

To add scalar quantities, we find the sum of their magnitudes.

Example

4 ft + 3 ft = 7 ft

However, vector quantities are not usually added in this manner since direction must also be considered. Vectors can be added, or subtracted, graphically or algebraically. Graphic addition gives a rough estimate of the sum while algebraic addition is more precise.

Example

Given a rectangular plot of ground that is 50 ft by 120 ft. Suppose a girl walks from one corner of the plot diagonally to the opposite corner. See figure 18.11. If she had walked along the sides, she would have walked 50 ft + 120 ft = 170 ft (a scalar sum). Using the Pythagorean theorem, the magnitude (length) of $\overrightarrow{R}$ is given by

$$R = \sqrt{(120)^2 + (50)^2} = \sqrt{16{,}900} = 130 \text{ ft}$$

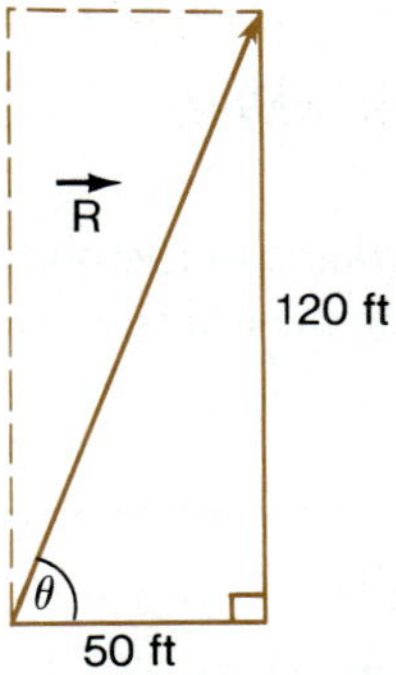

Figure 18.11

She actually walked 130 ft. What is the direction of her new position as a result of the walk?

Using trigonometric ratios, we obtain

$$\tan \theta = \frac{120}{50} = 2.4$$

Then $m\angle\theta = 67.4°$. Therefore, the vector whose tail is at the starting point and head is at the endpoint of her walk has

magnitude 130 ft and direction 67.4°

with the horizontal.

note

The vector that results from adding the vectors represented by the adjacent sides of the rectangle is called the **resultant,** denoted by $\vec{R}$.

In general, the sum, or resultant, $\vec{R}$, of two vectors, $\vec{u}$ and $\vec{v}$ is formed graphically by locating the tail of **v** at the head of **u** and then joining the tail of $\vec{u}$ and the head of $\vec{v}$. We call the vectors **u** and **v** the **components** of the resultant vector, $\vec{R}$. See figure 18.12.

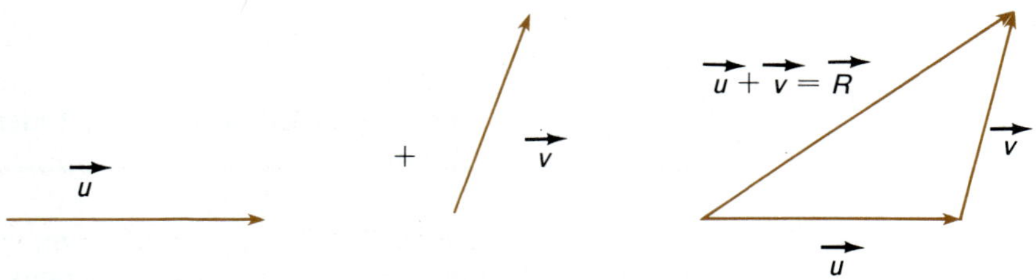

Figure 18.12

Three or more vectors may be added in the same general way. Consider figure 18.13, which adds $\vec{u} + \vec{v} + \vec{w} = \vec{R}$.

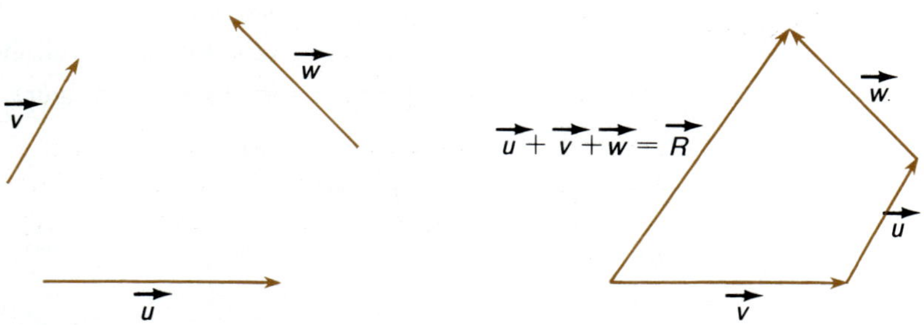

Figure 18.13

note

This is often called the **tail-to-head** (or **triangular**) **method** of adding vectors.

Another method used in adding *two* vectors is called the **parallelogram method.** Let the two vectors be the sides of a parallelogram formed by placing their tails at the same point. The *diagonal* of the parallelogram is the resultant vector, $\vec{R}$.

Example To add vectors $\vec{u}$ and $\vec{v}$, place the tails together as shown in the accompanying figure $\vec{R}$ is the diagonal of the parallelogram formed by drawing dotted lines parallel to and equal in length to vectors $\vec{u}$ and $\vec{v}$.

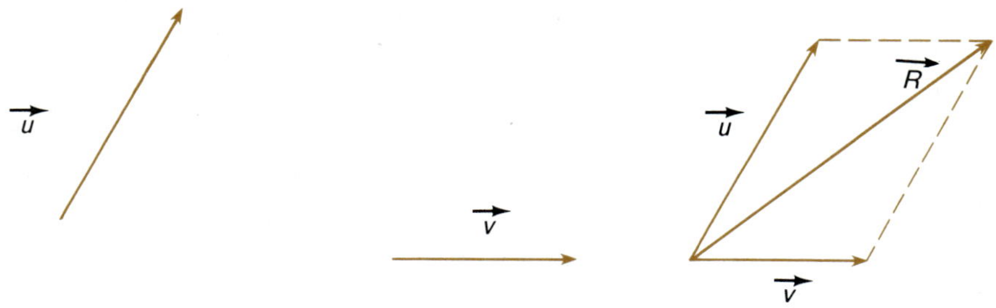

note

This is sometimes called the **tail-to-tail method** of adding vectors.

The tail-to-head method (triangular) of adding vectors is used more when adding two or more vectors. The tail-to-tail (parallelogram) method is used when *two forces* are acting on a single object.

Example Determine the resultant of two forces acting on an object:

1. a 30-lb force acting horizontally
2. a 40-lb force acting at a 40° angle with the horizontal

We construct a parallelogram by scale in figure 18.14.

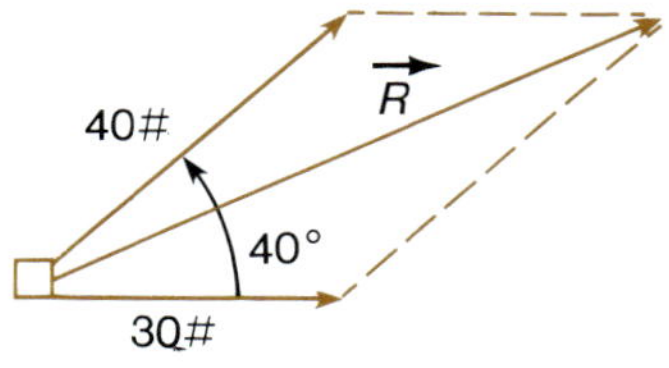

Figure 18.14

note

If two vectors have the *same* direction, the sum of the vectors is a resultant whose direction is the same as that of the two vectors and whose magnitude is the sum of the magnitudes of the vectors.

To subtract two vectors, we use the fact that given a vector **u,** then the negative (or opposite) of $\vec{u}$, denoted by $-\vec{u}$, is the vector having the same magnitude but *opposite direction*. See figure 18.15.

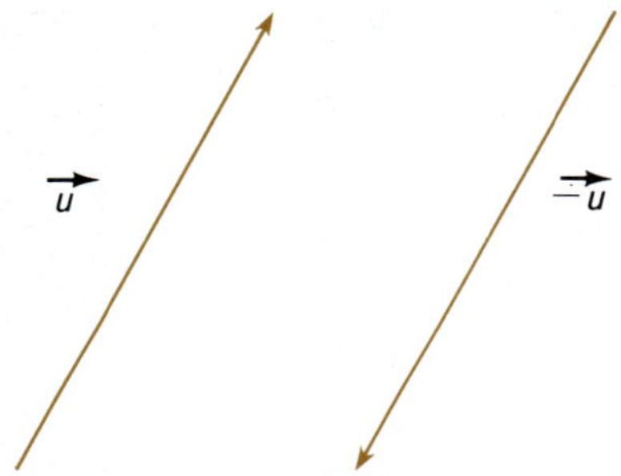

Figure 18.15

Using the definition for subtracting real numbers: $a - b = a + (-b)$, we define vector subtraction by

$$\vec{u} - \vec{v} = \vec{u} + (-\vec{v})$$

Given 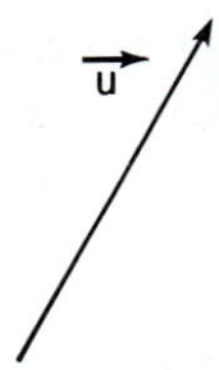

and 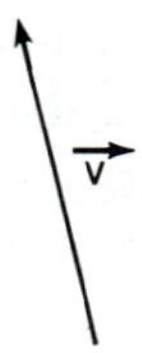

then 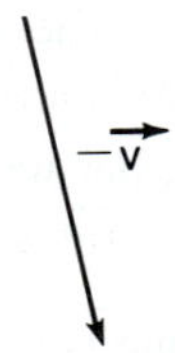

from which we get

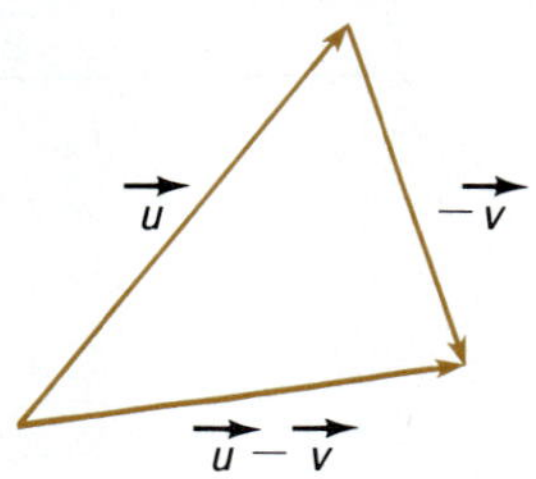

Example Given $\vec{u}$, $\vec{v}$, and $\vec{w}$ such that

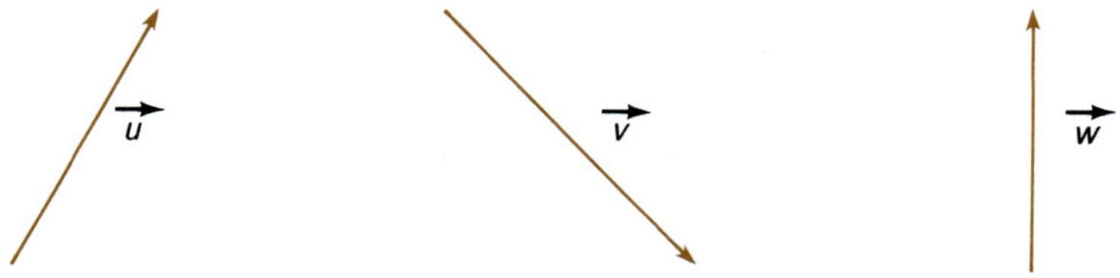

show the resultant, $\vec{R} = \vec{u} + \vec{v} - \vec{w}$. (*Note:* $\vec{u} + \vec{v} - \vec{w} = \vec{u} + \vec{v} + (-\vec{w})$)

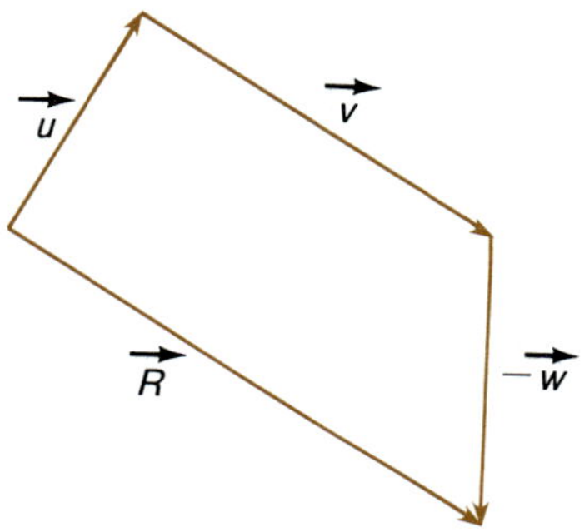

The product of two vectors is not discussed in this text, but we can define the product of a scalar c and a vector **u:** Given vector **v** with the same direction as vector **u,** but with magnitude c times that of $\vec{u}$, then

$$\vec{v} = c \cdot \vec{u}$$

That is, $3\vec{u}$ is a vector with the same direction as $\vec{u}$ but with magnitude three times that of $\vec{u}$. To illustrate:

note

$-2\vec{u}$ is represented by

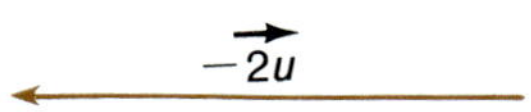

Examples

1. Given $\vec{u}$ and $\vec{v}$, determine $2\vec{u} + \vec{v}$.

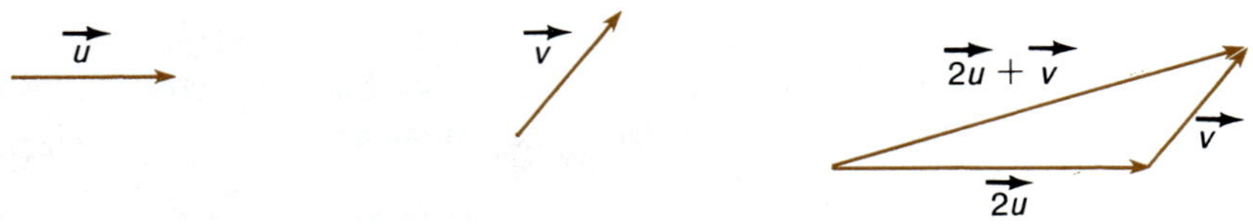

2. Using a parallelogram, add $2\vec{v} + 2\vec{u}$.

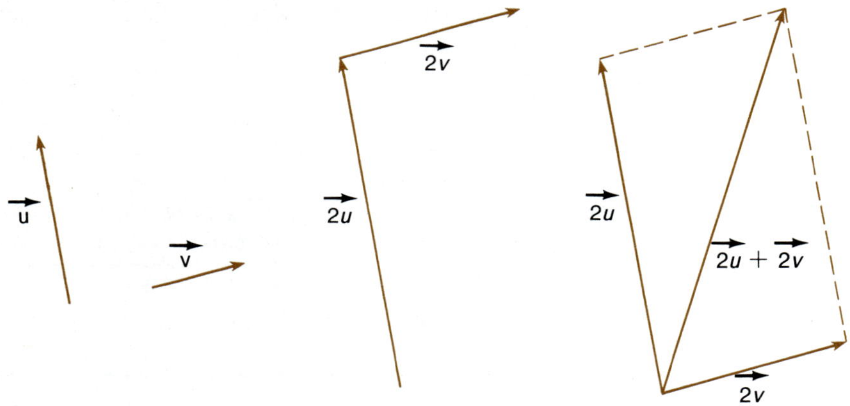

3. Subtract $2\vec{v}$ from $3\vec{u}$.

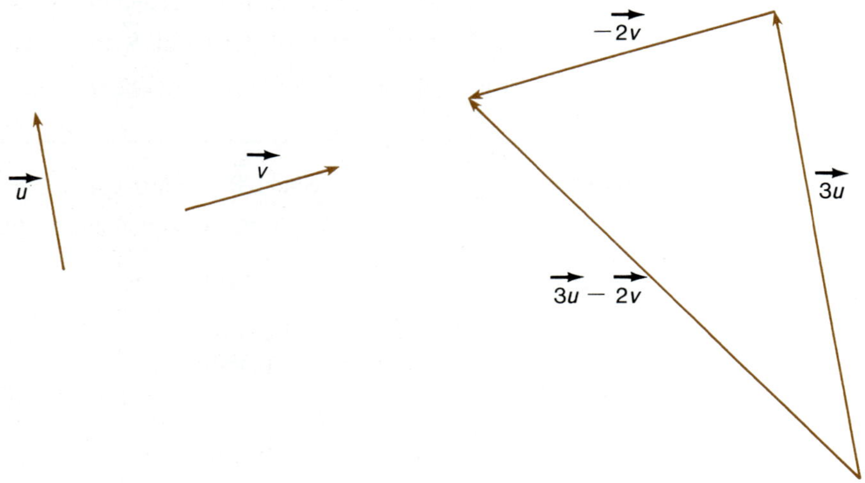

Recall, when adding two vectors, $\vec{u}$ and $\vec{v}$, to get the resultant vector, $\vec{R}$, that $\vec{u}$ and $\vec{v}$ are called the *components* of $\vec{R}$. Sometimes it is necessary to reverse the addition process and express a single vector, $\vec{R}$, as the sum of two vectors. This procedure is called **resolving the vector** into its components: a **vertical component** and a **horizontal component.**

Example The pilot of a plane knows the plane's air speed (magnitude) and angle of climb but wants to know the climbing velocity (vertical component) and the ground velocity (horizontal component). See figure 18.16.

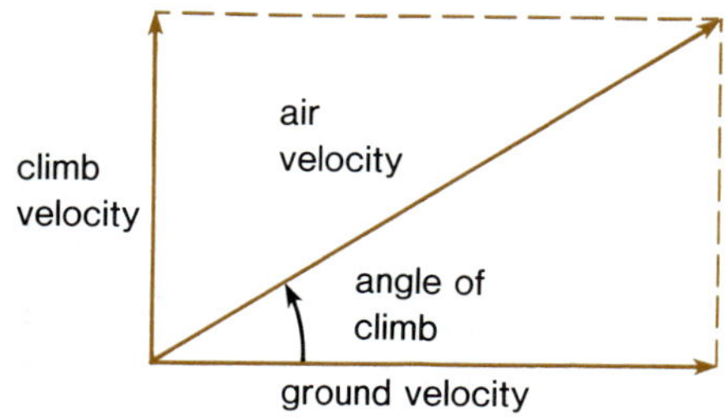

Figure 18.16

Resolving a vector into its components is usually done using the rectangular coordinate system. By placing the tail of the given vector at the origin, we relate the vertical component to the y-axis and the horizontal component to the x-axis. Then we call the horizontal component of $\vec{R}$ the ***x*-component,** denoted by R_x, and the vertical component of $\vec{R}$ the ***y*-component,** denoted by R_y.

Examples

1. Given a vector, $\vec{R}$, with magnitude $R = 5.0$ and direction $\theta = 63°$, resolve $\vec{R}$ into its horizontal and vertical components. See figure 18.17.

a. $\cos 63° = \dfrac{R_x}{5}$, so

$$\begin{aligned} R_x &= 5 \cos 63° \\ &= 5(0.4540) \\ &= 2.27 \end{aligned}$$

b. $\sin 63° = \dfrac{R_y}{5}$, so

$$\begin{aligned} R_y &= 5 \sin 63° \\ &= 5(0.8910) \\ &= 4.46 \end{aligned}$$

Therefore, vector $\vec{R}$ resolves into $R_x \approx 2.3$ and $R_y \approx 4.5$.

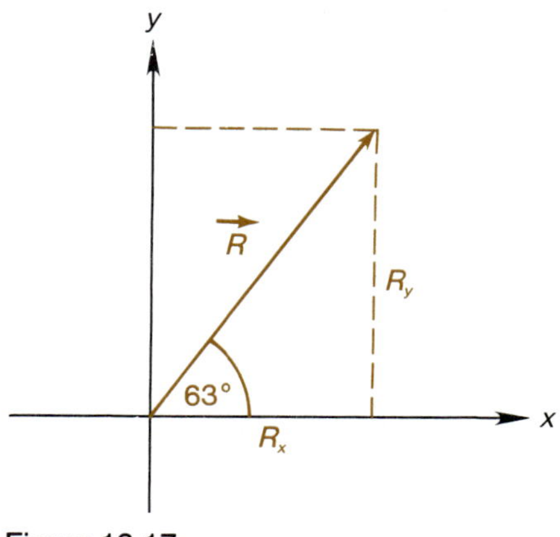

Figure 18.17

2. Resolve the vector, $\overrightarrow{R}$, with magnitude $R = 4.7$ and direction $\theta = 146°$ into its horizontal and vertical components. See figure 18.18.

   a. $\cos 146° = \dfrac{R_x}{4.7}$, so

   $$\begin{aligned} R_x &= 4.7 \cos 146° \\ &= 4.7(-0.8290) \\ &= -3.90 \end{aligned}$$

   b. $\sin 146° = \dfrac{R_y}{4.7}$, so

   $$\begin{aligned} R_y &= 4.7 \sin 146° \\ &= 4.7(0.5592) \\ &= 2.63 \end{aligned}$$

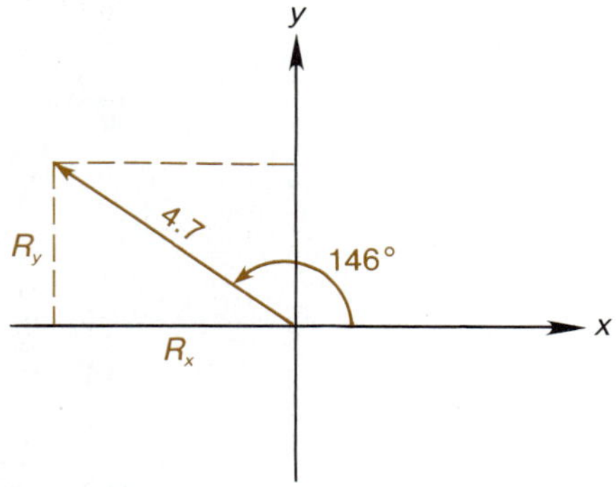

Figure 18.18

Therefore, the vector has been resolved into horizontal component 3.9 on the negative x-axis and 2.6 on the positive y-axis.

3. A ship is located 47° east of north from New York City at a distance of 313 mi. Find its horizontal (east) and vertical (north) components. See figure 18.19.
   (*Note:* $m\angle\theta = 90° - 47° = 43°$)

   a. $\cos 43° = \dfrac{R_x}{313}$, and so

   $$\begin{aligned} R_x &= 313 \cos 43° \\ &= 313(0.7314) \\ &= 228.9 \end{aligned}$$

   b. $\sin 43° = \dfrac{R_y}{313}$, and so

   $$\begin{aligned} R_y &= 313 \sin 43° \\ &= 313(0.6820) \\ &= 213.5 \end{aligned}$$

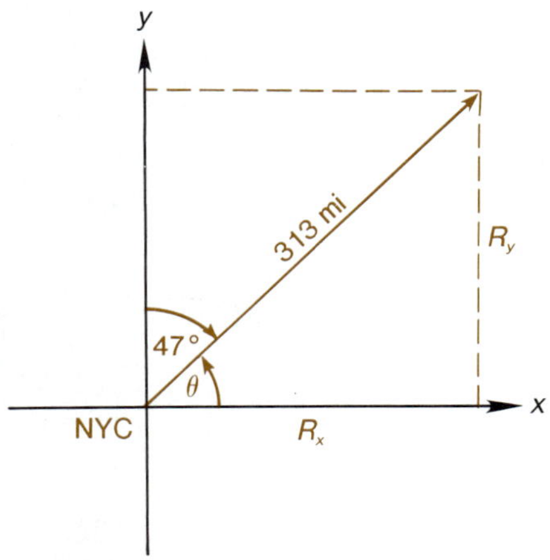

Figure 18.19

Therefore, the ship is approximately 228.9 mi east and 213.5 mi north of New York City.

Exercise Set 18.3

In exercises 1–17, find the indicated vector sums and differences using diagrams, given the vectors in the given figure.

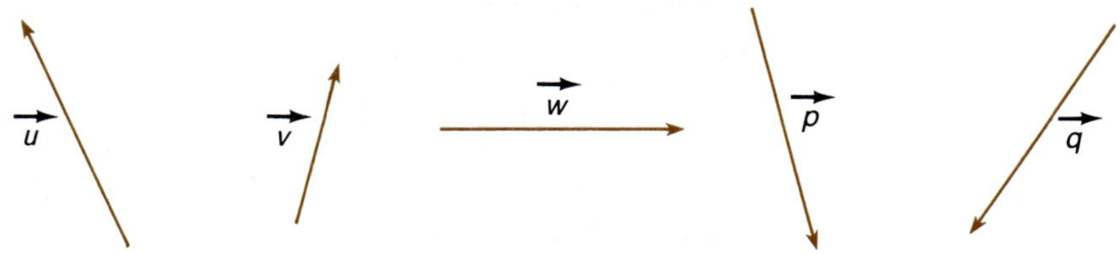

1. $\vec{u} + \vec{v}$
2. $\vec{w} + \vec{p}$
3. $\vec{p} + \vec{w}$
•4. $\vec{v} - \vec{u}$
5. $\vec{q} - \vec{p}$
6. $\vec{w} - \vec{u}$
7. $\vec{u} - \vec{w}$
•8. $\vec{u} + \vec{p} + \vec{q}$
9. $\vec{v} + \vec{w} + \vec{p}$
•10. $3\vec{v} + \vec{p}$
11. $2\vec{u} + \vec{q}$
•12. $3\vec{p} - 2\vec{v}$
13. $2\vec{q} - 3\vec{u}$
14. $\vec{u} + \vec{p} - \vec{v}$
15. $\vec{v} - \vec{w} + \vec{p}$
16. $2\vec{u} + \vec{v} - 2\vec{w}$
17. $\vec{q} - 2\vec{v} - \vec{w}$

In exercises 18–23, using the parallelogram (tail-to-tail) method, find $\vec{R}$ by means of a diagram given the following vectors.

18.

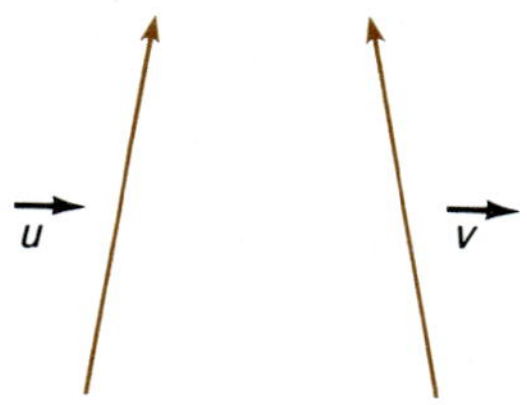

19.

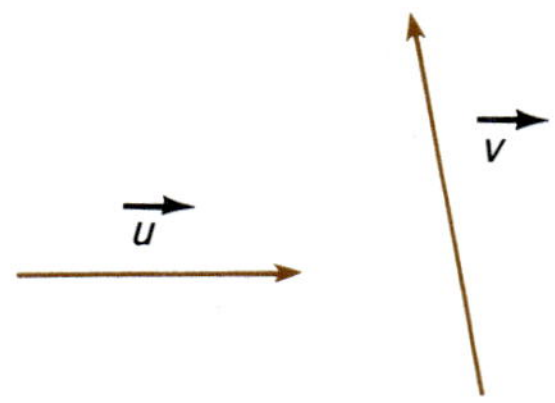

•20.

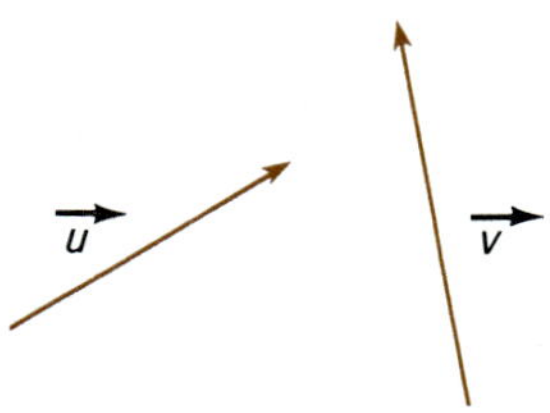

21.

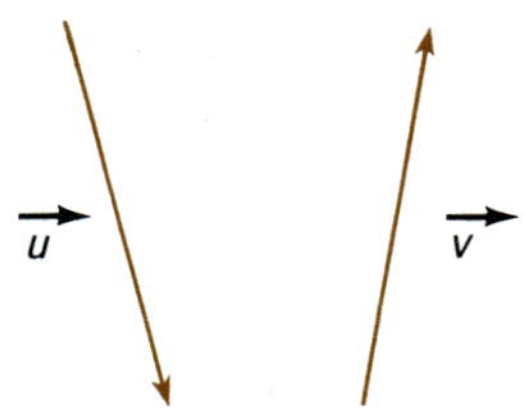

22.

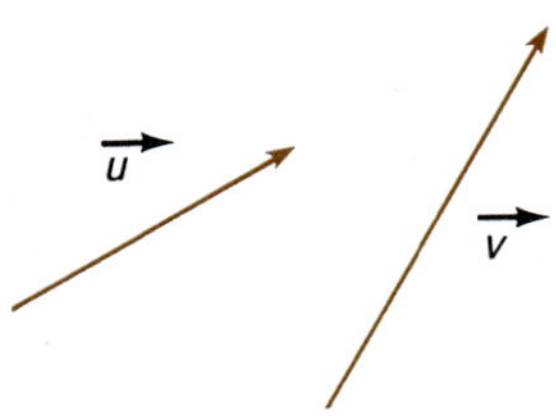

23.

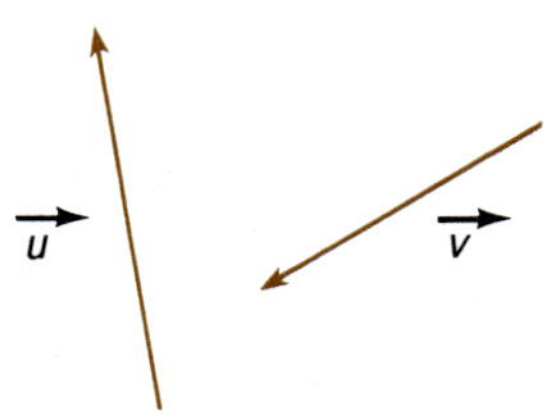

In exercises 24–35, resolve the given vector, $\vec{R}$, into its x-component, R_x, and y-component, R_y. Angle θ is with respect to the horizontal axis. (Round your answer to the nearest hundredth.)

24. $R = 5.2$, $\theta = 14.2°$
25. $R = 10.7$, $\theta = 62.3°$
•26. $R = 1{,}073$, $\theta = 110.0°$
27. $R = 0.745$, $\theta = 152.3°$
28. $R = 70.0$, $\theta = 90.0°$
29. $R = 82.3$, $\theta = 180.0°$
30. $R = 14.7$, $\theta = 0°$
31. $R = 84$, $\theta = 270°$
32. $R = 9.07$, $\theta = 2.3°$
33. $R = 245.0$, $\theta = 48.3°$
•34. $R = 1.15$, $\theta = -32.3°$
35. $R = 356.0$, $\theta = 236.3°$
36. A motorist travels a road that runs 64.0° north of east from Detroit. Find the horizontal and vertical components after he has traveled a distance of 75.3 mi.
37. A plane flies at 250 mi/h on an angle of inclination of 23.0°. Find its horizontal and vertical components.
•38. A balloon is rising at a rate of 6.0 ft/s, while a wind blowing at right angles with the balloon blows the balloon at an angle of 9.3° with the vertical. Find the horizontal and vertical components of the location of the balloon 1 min after it left the ground.
39. Two forces acting at right angles with one another have a resultant force of 250 lb acting at 25.3° with the horizontal. Find the forces.
40. Two holes are drilled in a rectangular steel plate so that hole B is 5.2 cm from hole A, and at an angle of 33.4° with the edge of the plate. Find the horizontal and vertical components of B with respect to A. See the accompanying figure.

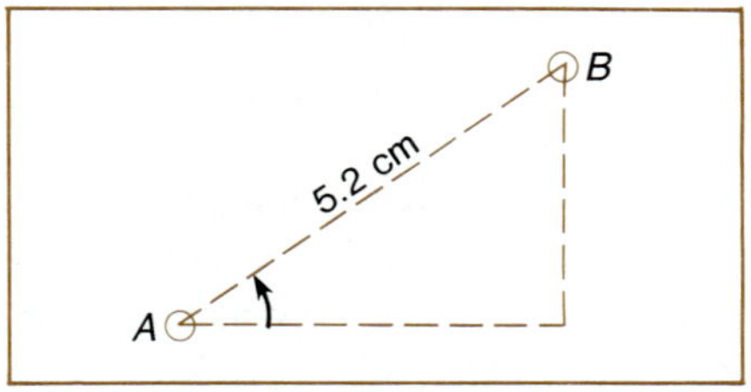

18.5 Analytic Vector Addition

In section 18.4 we added vectors by constructing diagrams, and we determined the resultant by these diagrams. Since a vector is not completely defined unless we know its magnitude and direction, we now find the magnitude and direction of the resultant vector, $\vec{R}$, more precisely.

If the two vectors are at right angles with one another, we can use the Pythagorean theorem and trigonometric ratios to define $\vec{R}$.

Examples

1. Given vectors, $\vec{u}$ and $\vec{v}$, at right angles with one another, with $u = 4.5$ and $v = 7.2$, find the magnitude and direction of the resultant vector, $\vec{R}$. See figure 18.20, with vectors **u, v,** and **R** placed in standard position.

   a. By the Pythagorean theorem

   $$\begin{aligned} R &= \sqrt{u^2 + v^2} \\ &= \sqrt{(4.5)^2 + (7.2)^2} \\ &= \sqrt{20.25 + 51.84} \\ &= \sqrt{72.09} \\ &\approx 8.49 \end{aligned}$$

   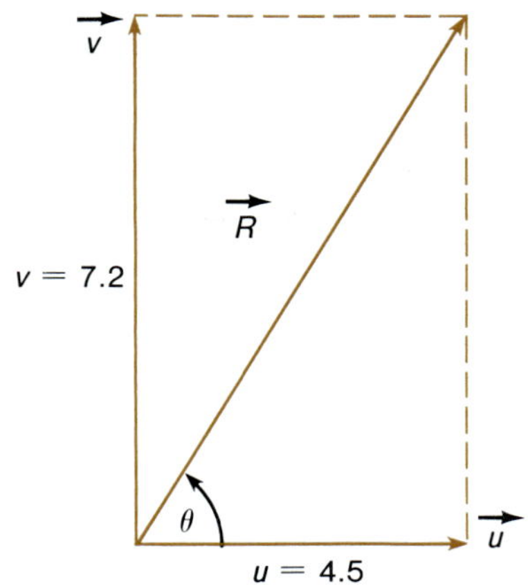

   Figure 18.20

   b. To find the direction of $\vec{R}$, find $m\,\angle\theta$. Since the side opposite and the side adjacent to θ are known, use

   $$\begin{aligned} \tan\theta &= \frac{v}{u} \\ &= \frac{7.2}{4.5} \\ &\approx 1.60. \end{aligned}$$

   Then $m\,\angle\theta \approx 57.99°$
   $\approx 58.0°$

   Therefore, $\vec{R}$ has magnitude 8.5 and direction 58.0° with the horizontal.

2. A 29-kg force acts horizontally on a hook while a 35-kg force acts vertically. Determine the resultant force on the hook. See figure 18.21. Since the forces act perpendicularly on one another, place the forces on the x and y axes. Now

   $$\begin{aligned} \mathrm{R} &= \sqrt{(35)^2 + (29)^2} \\ &= \sqrt{1225 + 841} \\ &= \sqrt{2066} \\ &= 45.5 \text{ kg} \end{aligned}$$

   Use $\tan\theta = \dfrac{35}{29}$
   ≈ 1.207

   Then $m\,\angle\theta \approx 50.4°$

   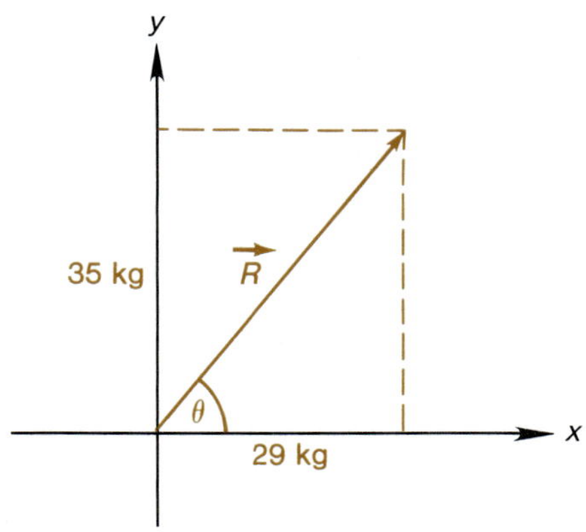

   Figure 18.21

   Therefore, the resultant has magnitude 45.5 kg and direction 50.4° with the horizontal.

When vectors to be added are not at right angles with one another, use the following procedure:

1. Place each vector with its tail at the origin.
2. Resolve each vector into its x and y components.
3. Determine the x and y components of R by *adding* the x and y components of each vector found in step 2.
4. Use the Pythagorean theorem and tangent ratio, as with vectors at right angles.

Examples

1. Find the resultant of two vectors, $\vec{u}$ and $\vec{v}$, given $u = 14.3$ with direction $\theta_u = 38.0°$ and $v = 9.2$ with the direction $\theta_v = 81.2°$. See figure 18.22.

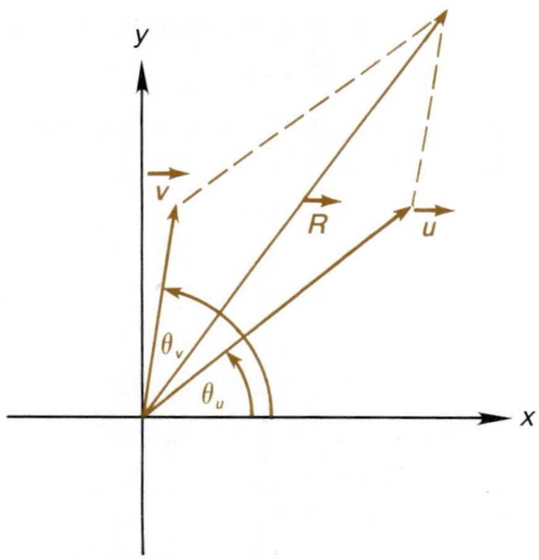

Figure 18.22

Find the x and y components of $\vec{u}$ and $\vec{v}$.

a. $\sin 38.0° = \dfrac{u_y}{14.3}$, and so $u_y = 14.3 \sin 38.0° = 8.8$

$\cos 38.0° = \dfrac{u_x}{14.3}$, and so $u_x = 14.3 \cos 38.0° = 11.3$

b. $\sin 81.2° = \dfrac{v_y}{9.2}$, and so $v_y = 9.2 \sin 81.2° = 9.1$

$\cos 81.2° = \dfrac{v_x}{9.2}$, and so $v_x = 9.2 \cos 81.2° = 1.4$

Now

$$R_x = u_x + v_x = 11.3 + 1.4 = 12.7$$
$$R_y = u_y + v_y = 8.8 + 9.1 = 17.9$$

Therefore

$$\begin{aligned} R &= \sqrt{(R_x)^2 + (R_y)^2} \\ &= \sqrt{(12.7)^2 + (17.9)^2} \\ &= \sqrt{161.29 + 320.41} \\ &= \sqrt{481.7} \\ &= 21.9 \end{aligned}$$

Also, $\tan \theta = \dfrac{R_y}{R_x} = \dfrac{17.9}{12.7} = 1.409$ and $\theta_R = 54.6°$. So the resultant vector has magnitude 21.9 and direction 54.6° with the horizontal.

note

In general, we can see that to find the

1. horizontal component, u_x, multiply the magnitude of the vector times $\cos \theta$.
2. vertical component, u_y, multiply the magnitude of vector times $\sin \theta$.

2. Find the resultant vector of a force system of four forces, shown in figure 18.23.

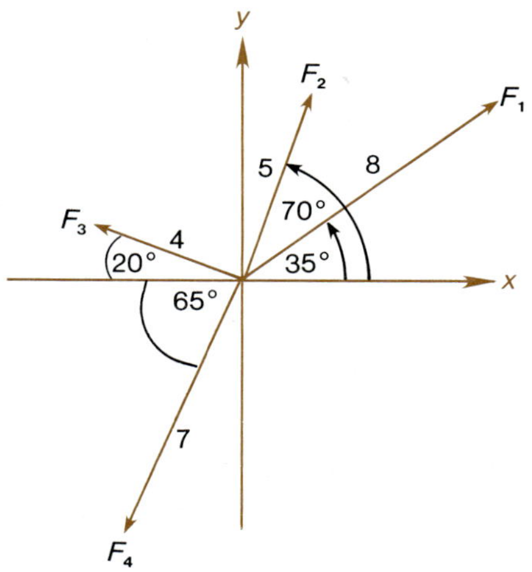

Figure 18.23

First find the sum of the x and y components of the four forces.

note

Components above the x-axis or to the right of the y-axis are considered *positive;* components to the left of the y-axis or below the x-axis are considered *negative*.

Then

$F_{1_x} = 8 \cos 35° = 6.55$ $\qquad F_{1_y} = 8 \sin 35° = 4.59$

$F_{2_x} = 5 \cos 70° = 1.71$ $\qquad F_{2_y} = 5 \sin 70° = 4.70$

$F_{3_x} = -4 \cos 20° = -3.76$ $\qquad F_{3_y} = 4 \sin 20° = 1.37$

$F_{4_x} = -7 \cos 65° = -2.96$ $\qquad F_{4_y} = -7 \sin 65° = -6.34$

Therefore,

$$R_x = 6.55 + 1.71 + (-3.76) + (-2.96) = 1.54$$
$$R_y = 4.59 + 4.70 + 1.37 + (-6.34) = 4.32$$

Then

$$\begin{aligned} R &= \sqrt{(1.54)^2 + (4.32)^2} \\ &= \sqrt{2.37 + 18.66} \\ &= \sqrt{21.03} \\ &\approx 4.59 \end{aligned}$$

Now, $\tan \theta_R = 4.32/1.54 = 2.81$. Then $\theta_R = 70.4°$. Therefore, the resultant vector has magnitude approximately 4.59 and direction 70.4° with the horizontal.

3. Three men are pulling an automobile with the forces indicated in figure 18.24. Find the resultant force.

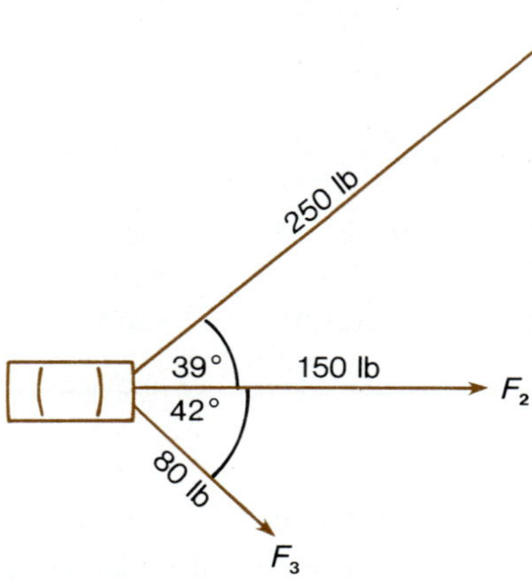

Figure 18.24

$$F_{1_x} = 250 \cos 39° = 194.3 \text{ lb} \qquad F_{1_y} = 250 \sin 39° = 157.3 \text{ lb}$$
$$F_{2_x} = 150 \cos 0° = 150 \text{ lb} \qquad F_{2_y} = 150 \sin 0° = 0 \text{ lb}$$
$$F_{3_x} = 80 \cos 42° = 59.5 \text{ lb} \qquad F_{3_y} = -80 \sin 42° = -53.5 \text{ lb}$$

Then

$$R_x = 194.3 + 150 + 59.5 = 403.8 \text{ lb}$$
$$R_y = 157.3 + 0 + (-53.5) = 103.8 \text{ lb}$$

Therefore

$$\begin{aligned} R &= \sqrt{(403.8)^2 + (103.8)^2} \\ &= \sqrt{163{,}054.4 + 10{,}774.4} \\ &= \sqrt{173{,}828.8} \\ &= 416.9 \text{ lb} \end{aligned}$$

Now $\tan \theta_R = \dfrac{103.8}{403.8} = 0.2571$ and $\theta_R = 14.4°$. Thus, the resultant force has magnitude 416.9 lb and direction 14.4° with the horizontal.

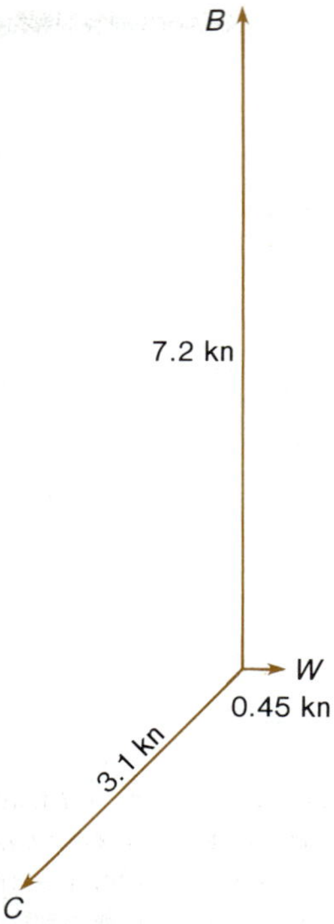

Figure 18.25

4. A sailboat is headed due north at a speed of 7.2 knots (a knot means nautical miles per hour and 1 knot is 1.15 mi/h). The pressure of the wind on the sails causes the boat to move sideways to the east at 0.45 knot. A tidal current is flowing to the southwest at 3.1 knots. What is the velocity of the sailboat relative to the earth's surface? See figure 18.25.

Now

$B_x = 7.2 \cos 90° = 0$ $\quad$ $B_y = 7.2 \sin 90° = 7.2$

$W_x = 0.45 \cos 0° = 0.45$ $\quad$ $W_y = 0.45 \sin 0° = 0$

$C_x = -3.1 \cos 45° = -2.2$ $\quad$ $C_y = -3.1 \sin 45° = -2.2$

Then $R_x = -1.75$ and $R_y = 5.0$.

Therefore,

$$\begin{aligned} R &= \sqrt{(-1.75)^2 + 5^2} \\ &= \sqrt{3.0625 + 25} \\ &= \sqrt{28.0625} \\ &\approx 5.30 \text{ knots} \end{aligned}$$

Now $\tan \theta_R = \dfrac{5.0}{-1.75} = -2.857$ and $\theta_R = 70.7°$. But since $R_x < 0$ and $R_y > 0$, then $\vec{R}$ is in the second quadrant. Thus, $\theta_R = 180° - 70.7° = 109.3°$. Hence, the resultant vector has magnitude 5.3 knots and direction 70.7° north of west.

Exercise Set 18.4

In exercises 1–5, determine the magnitude and direction of the resultant vector, $\vec{R}$, relative to the horizontal when the vectors, $\vec{u}$ and $\vec{v}$, are perpendicular at their tails, assuming $\vec{u}$ is horizontal. (Find all answers correct to the nearest tenth.)

1. $u = 3.0$ and $v = 4.0$
•2. $u = 483.0$ and $v = 214.0$
3. $u = 1070.0$ and $v = 2140.0$
4. $u = 14.3$ and $v = 10.4$
5. $u = 0.143$ and $v = 0.673$

In exercises 6–11, find the magnitude and direction of the resultant vector, $\vec{R} = \vec{u} + \vec{v}$. θ_u and θ_v are angles that the vectors, $\vec{u}$ and $\vec{v}$, make with the positive x-axis. (Find all answers correct to the nearest tenth.)

•6. $u = 31.0$ and $\theta_u = 16.0°$
$v = 43.2$ and $\theta_v = 76.3°$
7. $u = 1.3$ and $\theta_u = 40.7°$
$v = 3.4$ and $\theta_v = 110.1°$
8. $u = 13.1$ and $\theta_u = 0°$
$v = 15.6$ and $\theta_v = 39.2°$
9. $u = 14.85$ and $\theta_u = 90°$
$v = 6.73$ and $\theta_v = 136.5°$
10. $u = 0.443$ and $\theta_u = 71.0°$
$v = 0.714$ and $\theta_v = 14.9°$
11. $u = 145.0$ and $\theta_u = 87.6°$
$v = 93.6$ and $\theta_v = 180°$

In exercises 12–15, find the magnitude and direction of the resultant vector, $\vec{R}$, correct to the nearest tenth.

•12.

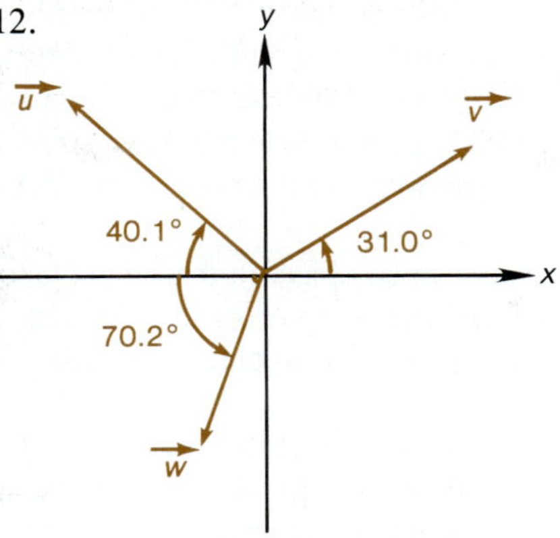

$u = 8.31$
$v = 10.10$
$w = 12.3$

13.

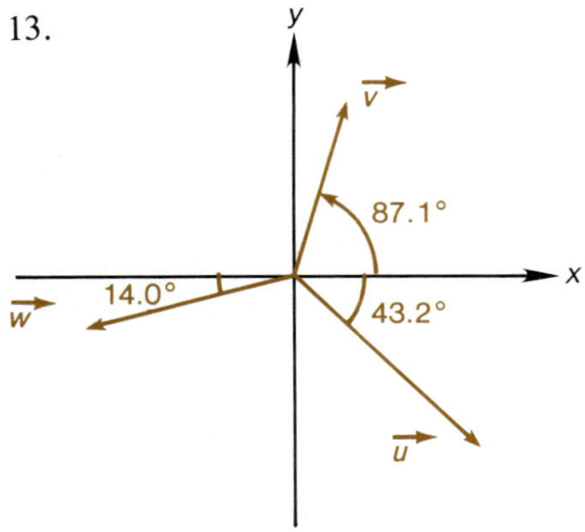

$u = 4.32$
$v = 2.83$
$w = 3.0$

14.

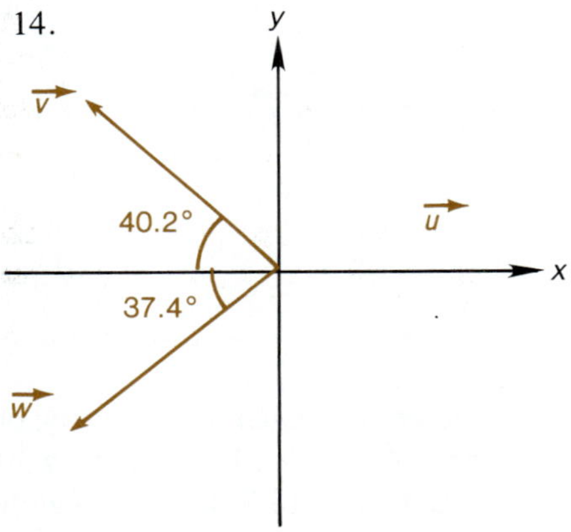

$u = 16.3$
$v = 14.2$
$w = 22.1$

15.

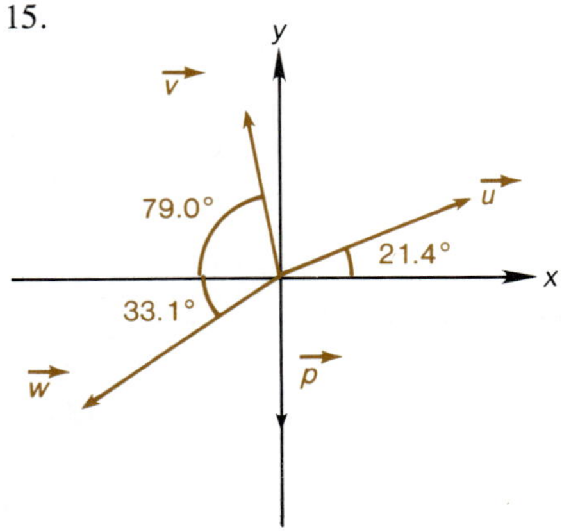

$u = 30.1$
$v = 21.4$
$w = 25.3$
$p = 33.4$

In exercises 16 through 29, round all answers to the nearest tenth.

16. Two forces act on an object at right angles to each other. If force F_1 is 150.0 lb and force F_2 is 240.0 lb, determine the magnitude and direction of the resultant force.
17. Determine the resultant of a 35-kg force acting vertically and a 40-kg force acting horizontally. If the forces are interchanged, do you get the same resultant?
•18. An airplane flying south at 430 mi/h is carried west by a wind having velocity 24.6 mi/h. Find the direction and ground speed of the plane.
19. Two forces of 104.0 lb and 225.0 lb act at right angles with one another. Find the magnitude of the resultant and the angle it makes with the larger force.

20. Forces of 45.6 lb and 32.9 lb act at right angles with one another. Find the magnitude of the resultant and the angle it makes with the larger force.
21. A jet flies 35° east of north at an airspeed of 473 mi/h. If a wind is blowing 59.2° west of north at 28.0 mi/h, find the direction and ground speed of the plane.
22. A ship is being towed by two tugs: one heading north of west at 34.7° with a force of 12,000 hp, and the other 57.6° south of west with a force of 18,000 hp. Find the horsepower and direction of the net force which the tugs are applying to the ship.
23. Two forces act on an object in the same plane. One force of 65.3 lb acts on the object, and the second force of 47.5 lb acts at an angle of 52.3° to the left of the first force. Find the resultant force. (*Hint:* place the first force on the positive x-axis.)
24. A motorboat travels at a still-water speed of 7.0 mi/h and heads across a river whose current flows at 2.30 mi/h. What is the resultant velocity of the motorboat? (See the accompanying figure.)

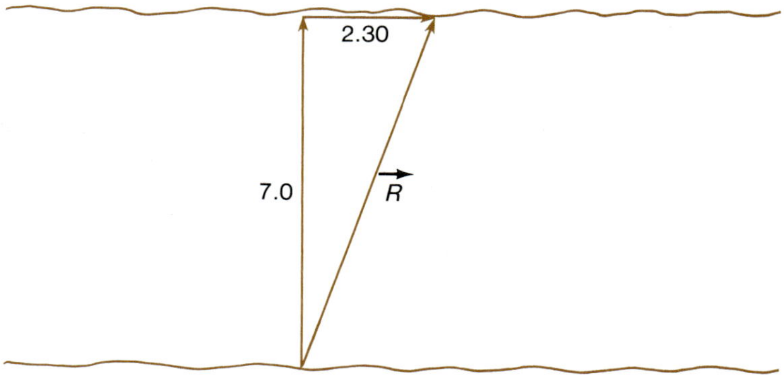

25. The following forces act on an object resting on a level, frictionless surface: F_1 is 15.0 lb to the north, F_2 is 21.2 lb to the west, F_3 is 14.6 lb at an angle 43.2° south of west, and F_4 is 18.7 lb at an angle of 73.2° south of east. Find the magnitude and direction of the resultant force acting on the object.
26. An airplane flies 300 mi east from city A to city B, and 250 mi south from city B to city C. It then flies 105 mi northwest to city D. How far is it from A to D?
27. A driver travels 15 mi south, 5.2 mi east, and 2.6 mi north looking for a farm. What is his resultant position from his starting place?
•28. Two alternators coaxially connected in series are out of phase by 23.0°. If the alternators have emf 125 v and 78 v, what will be the resultant emf? (See the accompanying figure.)

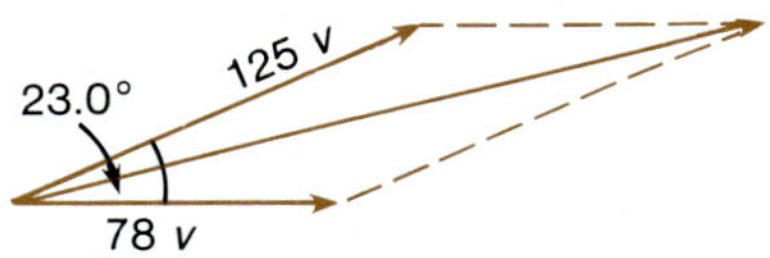

29. Two generators put out 105 v and 135 v, respectively. If the generators are out of phase by 15.6° and are coaxially connected in series, as in problem 28, find the resultant voltage of the generators.

Chapter Summary

1. The **Sine Law** states that in any oblique triangle ABC

$$\frac{a}{\sin A} = \frac{b}{\sin B} = \frac{c}{\sin C}$$

2. The *Sine Law* is used to solve oblique triangles given (a) two angles and a side or (b) two sides and an angle opposite one side.
3. Given two sides and an angle opposite one side, the *ambiguous case,* there may be:
   a. Two oblique triangles
   b. One oblique triangle
   c. One right triangle
   d. No triangle

   that satisfy the conditions.
4. The **Cosine Law** states that for any oblique triangle ABC

$$a^2 = b^2 + c^2 - 2bc \cos A$$
$$b^2 = a^2 + c^2 - 2ac \cos B$$
$$c^2 = a^2 + b^2 - 2ab \cos C$$

5. The *Cosine Law* is used to solve oblique triangles given (a) two sides and the included angle or (b) three sides.
6. The *area* of any oblique triangle ABC can be found by

$$A = \frac{1}{2}bc \sin A$$
$$= \frac{1}{2}ac \sin B$$
$$= \frac{1}{2}ab \sin C$$

7. A **vector** is a directed line segment with *magnitude* and *direction*.
8. The **magnitude** of a vector is measured by the *length* of the line segment.
9. The **direction** of a vector is determined by the angle θ it makes with a horizontal line (x-axis).
10. Vectors may be added using the (a) tail-to-tail (parallelogram) method or (b) tail-to-head (triangular) method.
11. Vectors are subtracted by $\vec{A} - \vec{B} = \vec{A} + (-\vec{B})$.
12. A vector, $\vec{A}$, may be resolved into its (a) vertical component, A_y and (b) horizontal component, A_x.
13. A vector, $\overrightarrow{(A + B)}$, is called the *resultant* vector, $\vec{R}$, of the vectors **A** and **B.**
14. The resultant vector, $\vec{R}$, may be resolved into its components by (a) $R_x = A_x + B_x$ and (b) $R_y = A_y + B_y$.
15. The direction of resultant vector, $\vec{R}$—θ_R—is given by

$$\tan \theta_R = \frac{R_y}{R_x}$$

Chapter Review

[18.2] Using the Sine Law, solve the oblique triangles ABC having the given conditions. (Round answers to the nearest tenth.)

1. $m\ \angle A = 52.3°$, $m\ \angle B = 71.3°$, and $c = 1.5$
2. $m\ \angle A = 21.5°$, $m\ \angle C = 108.3°$, and $c = 3.1$
3. $a = 21.4$, $b = 16.5$, and $m\ \angle A = 53.2°$
4. $b \doteq 0.343$, $c = 0.286$, and $m\ \angle C = 42.3°$
5. $a = 5.43$, $c = 8.13$, and $m\ \angle A = 69°54'$
6. $a = 72.2$, $b = 36.1$, and $m\ \angle B = 30.0°$

[18.3] Use the Cosine Law to solve $\triangle ABC$ given the following conditions. (Round answers to the nearest hundredth.)

7. $a = 5.73$, $b = 8.91$, and $m\ \angle C = 58.9°$
8. $a = 103.4$, $c = 201.6$, and $m\ \angle B = 98.1°$
9. $b = 73.3$, $c = 41.0$, and $m\ \angle A = 21°18'$
10. $a = 0.813$, $b = 0.724$, and $c = 0.216$
11. $a = 21.3$, $b = 10.7$, and $c = 11.0$

[18.3] Find the indicated part of oblique $\triangle ABC$ having the given conditions. (Correct to the nearest tenth.)
(*Note:* You might have to use either the Sine Law or the Cosine Law.)

12. Given $m\ \angle A = 33°12'$, $b = 14.3$, and $c = 10.9$. Find $m\ \angle C$.
13. Given $m\ \angle C = 101.3°$, $m\ \angle B = 42.1°$, and $a = 1.63$. Find side b.
14. Given $b = 0.119$, $c = 0.343$, and $a = 0.299$. Find $m\ \angle A$.
15. Given $a = 21.3$, $b = 32.6$, and $m\ \angle B = 50°54'$. Find $m\ \angle A$.
16. Given $a = 1.19$, $c = 2.79$, and $m\ \angle A = 25.12°$. Find $m\ \angle C$.
17. The angle of elevation of the top of a building from a point P is 32.3° and from another point Q is 52.9°. If point Q is 21.2 meters nearer to the building than point P on a line to the base of the building, find
    a. the straight line distance from point P to the top of the building.
    b. the height of the building.

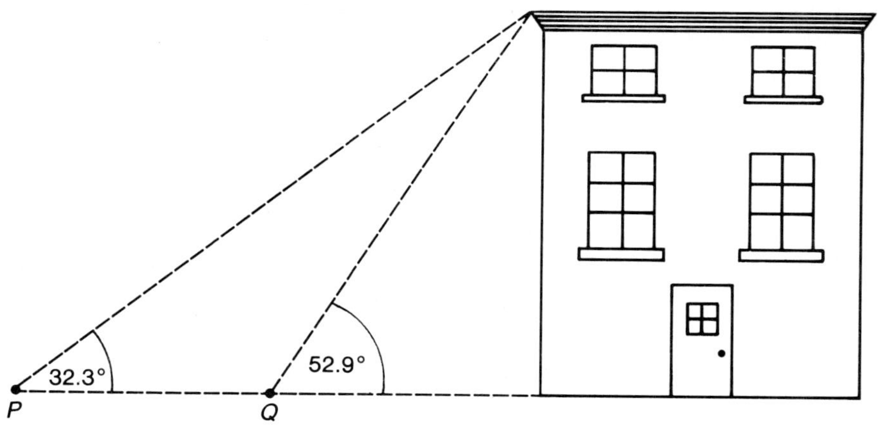

18. A steel plate in the shape of a rhombus has sides 12 cm long. If one angle of the rhombus is 48.0°, find the length of the diagonal opposite that angle. (*Note:* All sides have the same length.)
19. A machinist wishes to drill three holes A, B, and C in a rectangular plate. If the distances from center-to-center of each hole are as in the figure below, find the size of the angle at B.

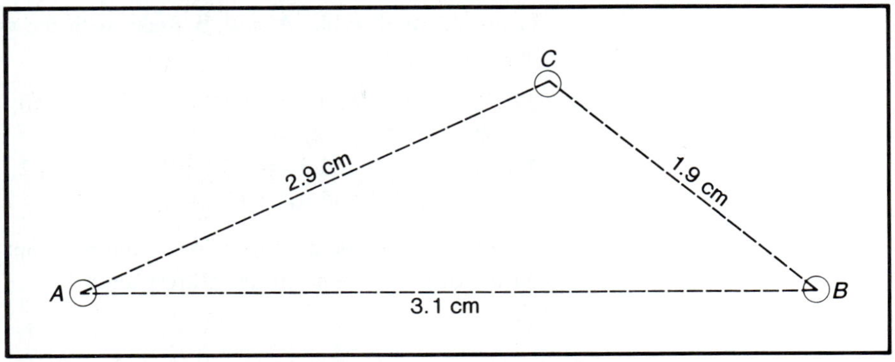

20. A ship in a fog is 30°15′ south of east of station A and 57°36′ south of west of station B. If station A is 62.3 km due west of station B, how far is the ship from A?
21. The diagonals of a parallelogram-shaped plot of ground are 52.0 ft and 63.4 ft long. If they intersect at an angle of 28.2°, find the sides of the parallelogram.

[18.4] Given the vectors

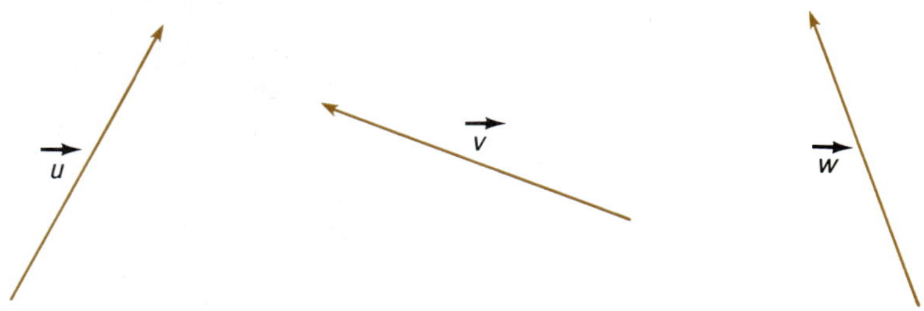

show the resultant, $\vec{R}$, using diagrams; use the tail-to-head (triangluar) method.

22. $\vec{u} + \vec{v} + \vec{w}$
23. $3\vec{u} + 2\vec{w}$
24. $2\vec{v} - \vec{u}$
25. $3\vec{w} - 2\vec{u} + \vec{v}$
26. $\vec{v} - 2\vec{u} - 3\vec{w}$

Using the preceding vectors, $\vec{u}$, $\vec{v}$, and $\vec{w}$, show the resultant vector, $\vec{R}$, using diagrams; use the tail-to-tail method.

27. $\vec{u} + \vec{w}$
28. $\vec{v} - \vec{w}$
29. $2\vec{v} + \vec{u}$
30. $3\vec{w} - \vec{v}$
31. $2\vec{u} + 3\vec{v}$
32. $2\vec{w} - 3\vec{u}$

[18.5] Resolve the vector, $\vec{R}$, into its components, R_x and R_y. Angle θ is with respect to the horizontal axis. (Round to the nearest tenth.)

33. $R = 4.2$, $\theta = 29.3°$
34. $R = 15.2$, $\theta = 72.4°$
35. $R = 522.0$, $\theta = 123°18'$
36. $R = 0.763$, $\theta = 271.8°$
37. $R = 12.73$, $\theta = 3°45'$
38. $R = 6.009$, $\theta = 180°$

Find the magnitude and direction of the resultant vector, $\vec{R} = \vec{A} + \vec{B}$, if θ_A and θ_B are the angles that $\vec{A}$ and $\vec{B}$ make with the positive x-axis. (Round to the nearest tenth.)

39. $A = 4.73$ and $\theta_A = 25.1°$
 $B = 5.06$ and $\theta_B = 0°$
40. $A = 0.837$ and $\theta_A = 52.7°$
 $B = 0.149$ and $\theta_B = 114.3°$
41. $A = 243.0$ and $\theta_A = 172°30'$
 $B = 141.3$ and $\theta_B = 0°$
42. $A = 17.3$ and $\theta_A = 180°$
 $B = 2.91$ and $\theta_B = 45°$

Find the magnitude and direction of the resultant vector, $\vec{R}$, given the forces acting on the origin. (Round to the nearest tenth.)

43.

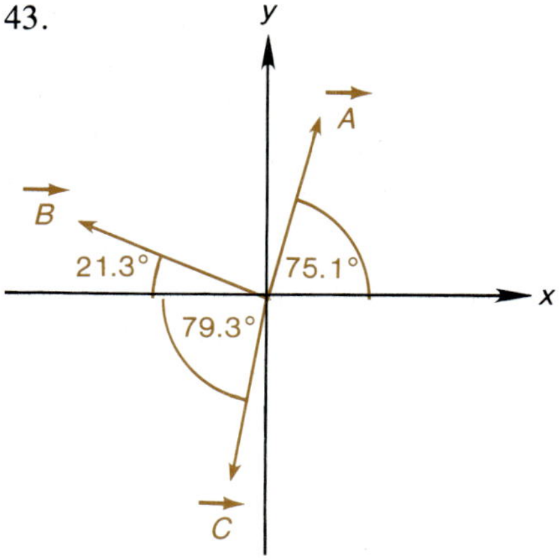

$A = 41.2$
$B = 29.3$
$C = 52.1$

44.

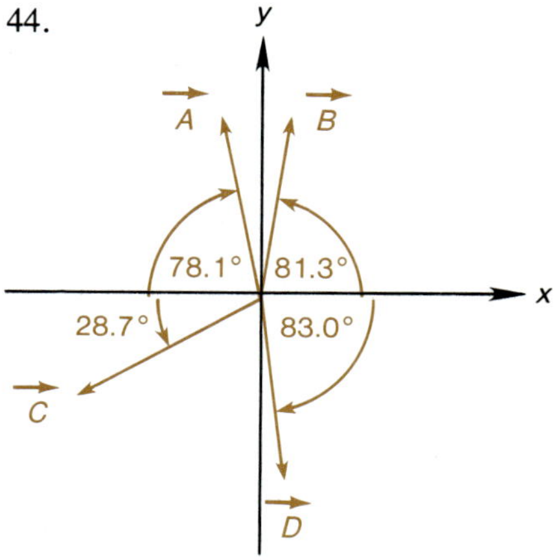

$A = 32.3$
$B = 28.4$
$C = 40.7$
$D = 43.5$

In exercises 45 through 50, round all answers to the nearest tenth.

45. Two forces acting at right angles with one another have a resultant force of 132.5 lb acting at 72°25′ with the horizontal. Find the forces.
46. A vector of magnitude 24.0 is inclined 52°16′ with the horizontal. Find its horizontal and vertical components.
47. An airplane heads in a direction of 10° north of west at an airspeed of 250 mi/h. If a wind is blowing at 25 mi/h from 52.0° south of west, find the ground speed (magnitude) and direction of the airplane.
48. Two forces of 24.0 lb and 15.2 lb act at right angles on an object. Find the magnitude of the resultant and the angle it makes with the smaller force.

49. A woman swims across a river whose current flows at 3.25 mi/h. If she is able to swim 1.70 mi/h in still water, how fast does she swim across the river?
50. Three forces F_1, F_2, and F_3 act on a motionless object lying on a level surface. If F_1 pulls west at 24.3 lb, F_2 pulls 48°42′ north of west at 41.2 lb, and F_3 pulls 81.3° south of west at 19.3 lb, find the magnitude and direction of the resultant force acting on the object. Assume no friction is involved.

Chapter Test

Round off all answers to the nearest tenth.

1. Given oblique $\triangle ABC$ with $m\angle A = 42.7°$, $m\angle B = 63.4°$, and $c = 18.1$, find angle C and side b.
2. Given oblique $\triangle ABC$ with $a = 4.7$, $b = 6.3$, and $c = 9.5$, find $m\angle A$.
3. Given oblique $\triangle ABC$ with $a = 14.7$, $b = 18.6$ and $m\angle A = 40.7°$, how many triangles will satisfy these conditions?
4. Two markers P and Q are on opposite ends of a lake. A third marker, A, is 210.0 m from P. If $m\angle P = 107.3°$ and $m\angle Q = 29.1°$, find the length of the lake.
5. A triangular tract of land is bordered on one side by 310.4 ft of fencing, by road frontage of 175.3 ft on another side, and by a tree line 207.3 ft long on the third side. Find the angle made by the tree line and the fence.
6. Find the area of a triangular steel plate if two sides of the plate are 16.2 cm and 10.7 cm long and the angle between these sides measures 52.3°.
7. Given the vectors shown, $\vec{u}$ and $\vec{v}$, draw diagrams showing:
   a. $2\vec{u} + 3\vec{v}$ (using tail-to-head method)
   b. $2\vec{v} - \vec{u}$ (using tail-to-tail method)

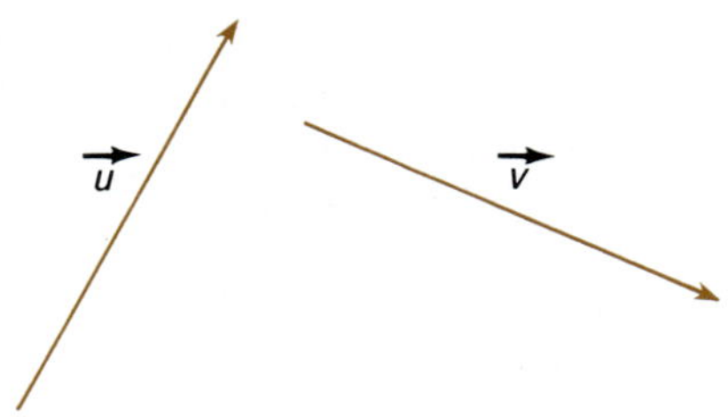

8. A 50-lb force acts on an object horizontally and a 40.2-lb force acts on the object at 36.3° with the horizontal. Determine the magnitude and direction of the resultant force.
9. An airplane flies at 320.0 mi/h on an incline of 31°54′. Determine ground velocity (horizontal component) and its climb velocity (vertical component).
10. Find the magnitude and direction of the resultant vector, $\vec{R} = \vec{u} + \vec{v}$, if $u = 15.6$, $\theta_u = 35.4°$, $v = 12.4$, and $\theta_v = 62.9°$.
11. A sailboat crosses a lake from south to north at a still-water speed of 15.0 knots. If an east to west wind blows at 20.6 knots, find the speed and direction of the sailboat.

12. A truck is being pulled by three men as shown in the accompanying figure. Determine the magnitude and direction of the resultant force.

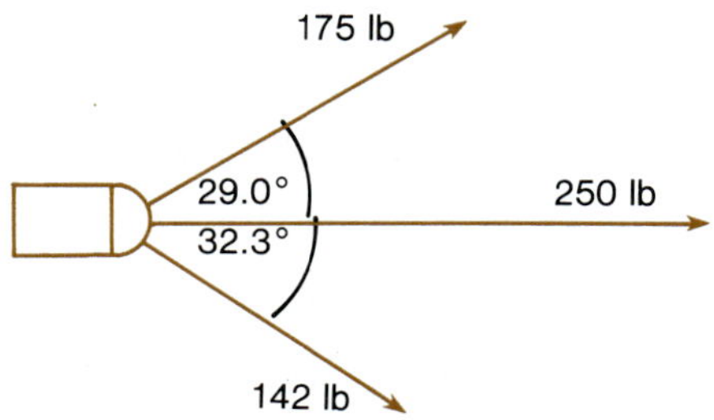

13. A woman standing at a point A, 950.5 ft from one end of a lake and 675.2 ft from the other end, makes an angle of 63.9° with the ends of the lake. How far is it from one end of the lake to the other? See the accompanying figure.

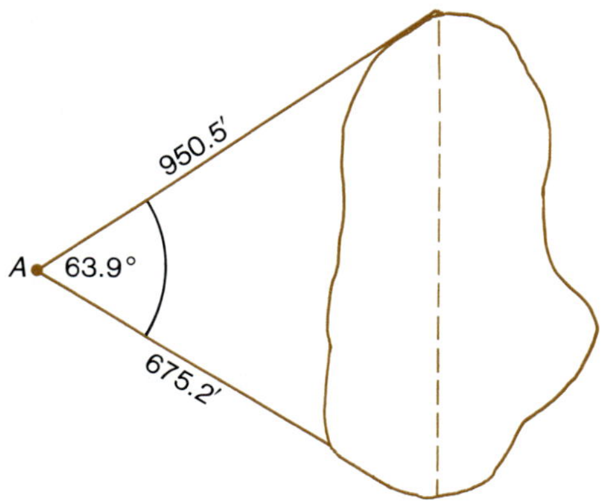

19

Radian Measure and Graphs of the Trigonometric Functions

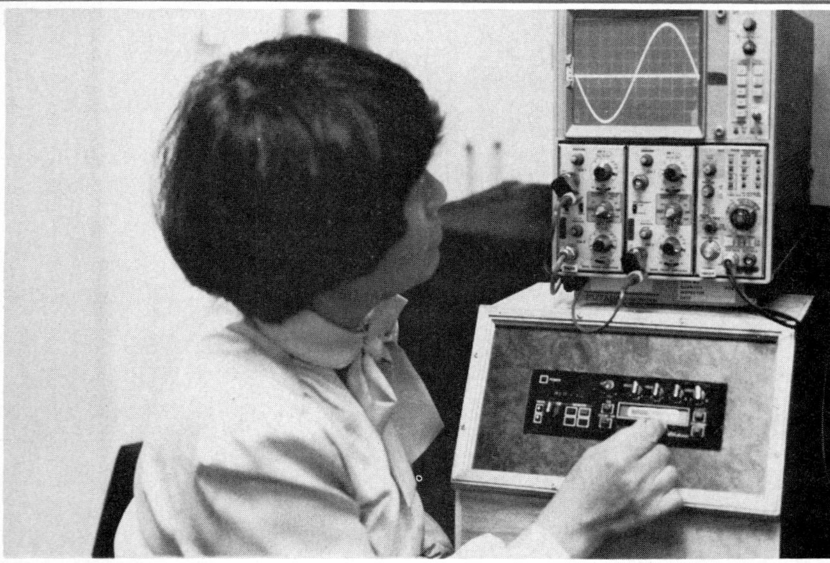

A technician evaluates a sine wave indicating local oscillator stability of an electronically tuned radio.
(Photograph by Mas Miyazaki, courtesy of Ford Motor Company)

19.1 Radians

Thus far, we have discussed the measure of the size of an angle in degrees. This measurement of angles has many applications, as we discovered in chapter 12. However, in some areas of science, where applications are involved, degree measurement is not convenient. A more convenient unit for measuring angles in these cases is the **radian (rad).** We define and discuss this measurement of an angle and consider some of its applications in this section.

The radian relates the central angle of a circle to the radius of that circle. Let a central angle θ cut an arc s of that circle. See figure 19.1. If the circle's radius is r, then the central angle θ is defined by

$$\theta \text{ in radians} = \frac{\text{arc length}}{\text{radius}} = \frac{s}{r}$$

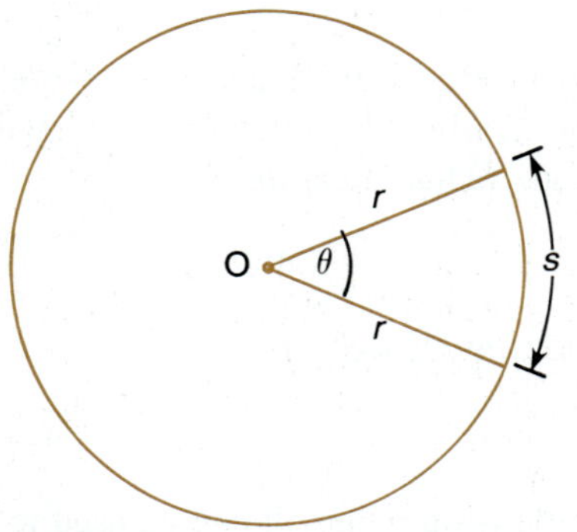

Figure 19.1

That is, the central angle θ between two radii of a circle, in radian measure, is defined to be the ratio of the arc s to the radius r. Obviously, when arc s has the same length as radius r, θ has measure *1 radian*.

note

s and r *must be* in the same unit of measurement.

Example Let $r = 4$ in and $s = 4$ in. Then θ in radians has measure $\frac{4 \text{ in}}{4 \text{ in}} = 1$ rad.

One radian is the measure of a central angle of a circle that intercepts an arc of the circle equal to the length of the radius of the circle. Then, in figure 19.2, the measure of angle θ in radians, denoted by θ^R, is 1 radian. Since the circumference of a circle is given by $C = 2\pi r$, the number of radians in one complete revolution is given by

$$\theta = \frac{s}{r} = \frac{2\pi r}{r} = 2\pi \text{ radians}$$

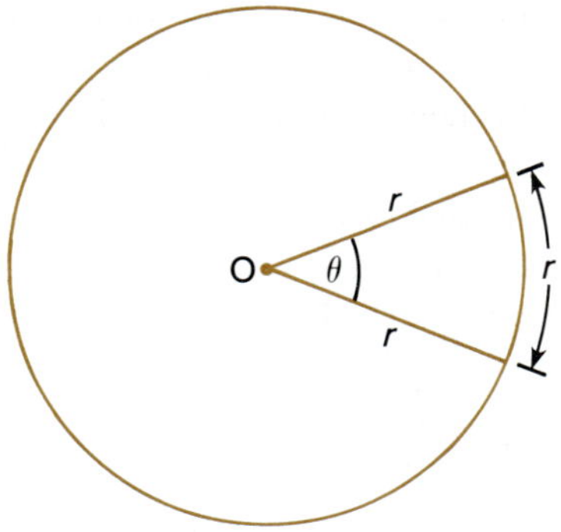

Figure 19.2

One revolution measures 360°, and one revolution measures 2π radians. Then 360° and 2π radians measure the same or equal angles. Given arbitrary angle θ having degree measure D and radian measure R,

$$\frac{D}{360°} = \frac{R}{2\pi}$$

Multiplying both sides by 2, we get

$$\frac{D}{180} = \frac{R}{\pi}$$

We call this the **angle proportion.** It can be used to convert from degree measure to radian measure and vice-versa.

note

Given the angle proportion $\frac{D}{180} = \frac{R}{\pi}$, then

1. $D = R \cdot \frac{180}{\pi} \approx R \cdot 57.3$
2. $R = D \cdot \frac{\pi}{180} \approx D \cdot (0.01745)$

From now on we shall indicate an angle θ as measured in degrees by $\theta°$, and in radians by θ^R. Then

$$\frac{\theta°}{180} = \frac{\theta^R}{\pi}$$

The radian is an unusual kind of unit since it has no dimension. An angle whose measure is expressed in radians is specified by a ratio of two lengths (arc and radius). It is, therefore, a *pure* number—not related to a unit. Thus, in our calculations, the unit rad *is always dropped* and, when no units are indicated, the radian is understood to be the unit of angular measure.

To convert the measure of an angle in degrees to radian measure, we multiply the number of degrees by $\frac{\pi}{180°}$ or 0.01745.

Examples

1. Given $\theta° = 30$, then $\theta^R = 30°\left(\frac{\pi}{180°}\right) = \frac{\pi}{6}$.

   So, 30° and $\frac{\pi}{6}$ are of equal measure.

2. Given $\theta° = 120$, then $\theta^R = 120°\left(\frac{\pi}{180°}\right) = \frac{2\pi}{3}$.

   So, 120° and $\frac{2\pi}{3}$ are of equal measure.

3. Given $\theta° = 86.4$, then $\theta^R = 86.4°\left(\frac{\pi}{180°}\right) = 0.48\pi \approx 1.51$.

   So, 86.4° and 1.51 have approximately equal measure.

4. Given $\theta° = 157.3$, then $\theta^R = 157.3°\left(\frac{\pi}{180°}\right) \approx 2.745$.

   So, 157.3° and 2.745 have approximately equal measure.

note

When expressing a measure in radians we often express the radians in terms of π, as in examples 1 and 2, since this is the most accurate measurement. We do this when the number of degrees in θ has factors common with 180: 30, 45, 60, 90, and so on.

To convert the measure of an angle in radians to degree measure, we multiply the number of degrees by $\frac{180}{\pi}$ or 57.3°.

Examples

1. Given $\theta^R = \frac{\pi}{4}$, then $\theta° = \frac{\pi}{4}\left(\frac{180°}{\pi}\right) = 45°$. So, $\frac{\pi}{4}$ and 45° are of equal measure.

2. Given $\theta^R = \frac{7\pi}{6}$, then $\theta° = \frac{7\pi}{6}\left(\frac{180°}{\pi}\right) = 210°$. So, $\frac{7\pi}{6}$ and 210° are of equal measure.

3. Given $\theta^R = 1.57$, then $\theta° = (1.57)\left(\frac{180°}{\pi}\right) \approx 90.0°$. Then, 1.57 and 90.0° have approximately equal measure.

4. Given $\theta^R = 5.09$, then $\theta° = (5.09)\left(\frac{180°}{\pi}\right) \approx 292°$. So, 5.09 and 292° have approximately equal measure.

Certain equivalences should be memorized. Since 360° = 2π rad, then 180° = π rad. From this we can obtain the following table.

θ in degrees	θ in radians
0	0
30	$\frac{\pi}{6}$
45	$\frac{\pi}{4}$
60	$\frac{\pi}{3}$
90	$\frac{\pi}{2}$
180	π
270	$3\frac{\pi}{2}$
360	2π

Exercise Set 19.1

In exercises 1–15 express the measure of the angle in radians using π. $\left(\textit{Note:}\text{ Convert using } \frac{\pi}{180}.\right)$

1. 20°
2. 12°
3. 36°
4. 75°
5. 80°
6. 110°
7. 150°
8. 210°
9. 225°
10. 260°
11. 300°
12. 315°
13. 320°
•14. 330°
15. 345°

In exercises 16–30 express the measure of the angle in radians to the nearest hundredth of a radian. (*Note:* Convert using 0.01745.)

16. 17°
17. 23.0°
•18. 31.9°
19. 63.7°
20. 109.4°
21. 167.0°
22. 214.7°
23. 256.3°
24. 294.1°
25. 313.2°
26. 25°18′
27. 142°18′
28. 214°42′
29. 301°48′
•30. 324°51′

In exercises 31–50 find the measure of the angle in degrees $\left(\text{use } \frac{180°}{\pi}\right)$ given the angle in radians.

31. $\frac{\pi}{10}$
32. $\frac{\pi}{12}$
33. $\frac{\pi}{9}$
34. $\frac{\pi}{15}$
35. $\frac{2\pi}{3}$
36. $\frac{5\pi}{6}$
37. $\frac{3\pi}{4}$
38. $\frac{2\pi}{5}$
39. $\frac{3\pi}{8}$
40. $\frac{7\pi}{8}$

41. $\frac{11\pi}{10}$ 42. $\frac{5\pi}{4}$ 43. $\frac{13\pi}{4}$ 44. $\frac{17\pi}{15}$ 45. $\frac{11\pi}{16}$

46. $\frac{23\pi}{12}$ 47. $\frac{5\pi}{3}$ •48. $\frac{19\pi}{10}$ 49. $\frac{17\pi}{9}$ 50. $\frac{9\pi}{5}$

In exercises 51–62 find the measure in degrees to the nearest tenth of a degree, given the angle in radians. (*Note:* Convert using 57.3°.)

51. 0.46 52. 0.93 53. 1.07 54. 1.41 55. 1.93
56. 3.09 57. 2.01 58. 3.85 59. 4.12 60. 4.63
61. 5.31 •62. 6.11

63. A flywheel turns 150 times in 1 min. Through how many radians does it turn in that minute?
•64. How many radians are there in 14.2 revolutions of the earth about its axis?
65. Through how many radians does the minute hand of a clock pass when moving from 12:00 to 12:20 P.M.?
•66. Through how many degrees and radians does the minute hand of a clock pass when the time changes from 1:00 P.M. to 3:15 P.M.?
67. A circle has radius 2 cm. How many radians are there in a central angle θ of a circle if the angle intercepts an arc s of length:

a. 4 cm b. 5 cm c. 10 cm d. $\frac{1}{2}$ cm?

(Remember: θ has measure 1 radian if $s = r$.)

19.2 Applications of Radian Measure

Recall that, by definition, given central angle θ, $\theta^R = \frac{s}{r}$, where s is the length of the arc of the circle the angle intercepts and r is the radius of the circle. Then, multiplying both sides of the equation by r, we determine that the length of the intercepted arc is given by $s = r \cdot \theta^R$.

Examples

1. Find the length of the intercepted arc s when
   a. $r = 5.1$ cm and $\theta^R = 2$
   Then, $s = (5.1)(2) = 10.2 \text{ cm} \approx 10 \text{ cm}$.
   b. $r = 3.6$ in and $\theta^\circ = 135°$
   Now we must find θ^R:
   $$\theta^R = (135)\left(\frac{\pi}{180}\right) = \frac{3\pi}{4}$$
   $$\text{So, } s = (3.6)\left(\frac{3\pi}{4}\right) = 8.48 \approx 8.5 \text{ in.}$$

2. A phonograph record having diameter 12.0 in turns through an angle of 150°. How far does a point on the rim travel? (*Note:* We want s when $r = 6.0$ in and $\theta° = 150°$.) Now

$$\theta^R = (150)\left(\frac{5\pi}{180}\right) = \frac{5\pi}{6},$$

$$\text{so, } s = 6.0\left(\frac{5\pi}{6}\right) = 5.0\pi \text{ in} \approx 15.7 \text{ in}$$

3. A wheel of an automobile has a 14-in radius. When the auto travels a distance of 72 ft, through what angle, measured in radians, does the wheel turn? In degrees?

$\left(\textit{Note:} \text{ We want } \theta^R = \frac{s}{r}.\right)$

Since r is measured in inches, we change this to feet; then $r = \frac{14}{12} = \frac{7}{6}$ ft, and

$$\theta^R = \frac{72}{\frac{7}{6}} = \frac{432}{7}$$

Therefore, $\theta^R \approx 61.7$ and $\theta° = (61.7)(57.30) \approx 3{,}500°$.

From the central angle of a circle we consider the portion of the circle enclosed by the angle—a "piece of pie" called a **sector.** See figure 19.3.

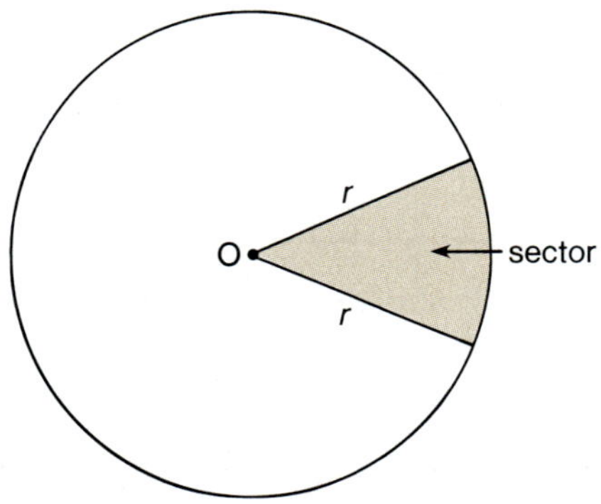

Figure 19.3

If the central angle has measure θ radians, the sector makes up the fraction $\frac{\theta^R}{2\pi}$ of a complete circle. Since the area of a circle is given by $A = \pi r^2$, then the area of the sector is given by $A = \frac{\theta^R}{2\pi} \cdot \pi r^2 = \frac{\theta^R}{2} \cdot r^2$. Thus, the area A of a sector of a circle is given by $A = \frac{\theta^R r^2}{2}$.

Example Find the area of a sector of a field in the shape of a circle if the central angle measures 15° and the circle has radius 294.0 m.

Now, $15° = (15)\left(\frac{\pi}{180}\right) = \frac{\pi}{12}$ rad, so

$$A = \frac{1}{2}\left(\frac{\pi}{12}\right)(294.0)^2 \approx 11{,}310 \text{ m}^2$$

Now we consider two kinds of velocity as they relate to an arc of a circle and to a central angle of a circle: linear velocity and angular velocity. We all are familiar with **linear velocity** (v), as it is defined by the distance s traveled divided by the time t. That is,

$$v = \frac{s}{t}$$

Example An automobile travels 200 mi in 4 hr. The linear velocity v is given by $v = 200$ mi/4 hr $= 50$ mi/hr.

When the object is not moving along a straight line, as in the example, but in a circular path (as on a circle) we must stipulate that the terminal side of the angle changes at a *constant rate*. Thus, the relationship that exists between velocity, time, and distance is the same that exists between the rate of change of the angle, the angle, and time. Dividing both sides of $s = r \cdot \theta^R$ by time t, we get $\frac{s}{t} = r \cdot \frac{\theta^R}{t}$. The ratio $\frac{\theta^R}{t}$ is defined to be the **angular velocity** (rate of change of the angle θ), and this velocity is denoted by ω (the Greek letter *omega*).

Thus, $\omega = \frac{\theta^R}{t}$

and since $v = \frac{s}{t}$

then $v = r \cdot \omega$

If θ is measured in radians, ω is measured in radians per unit of time and v is measured in linear measure (inches, feet, centimeters, meters) per unit of time.

note

ω is often measured by *revolutions* per unit of time. In order to use this unit in finding v we must change it to radians per unit of time by multiplying by 2π.

That is, we determine the linear velocity, v, of an object moving at a constant speed in a circular path by multiplying the radius of the circle by the angular velocity ω of the object.

note

Since $v = r \cdot \omega$, then the angular velocity $\omega = \frac{v}{r}$. When using the equation $v = r \cdot \omega$, it is helpful for dimensional soundness to express the radius r as a linear measure per radian.

Examples

1. Find the linear velocity of an object on a wheel whose radius $r = 7.0$ in if the wheel is turning at a rate of 150 revolutions per minute.
   Since $\omega = (150)(2\pi) = 300\pi$ rad/min, then:
   $$v = \left(7.0\,\frac{\text{in}}{\text{rad}}\right)\left(300\pi\,\frac{\text{rad}}{\text{min}}\right) \approx 6{,}600 \text{ in/min}$$
2. A rider is pedaling a bicycle with 26-in diameter wheels at a rate of 32 ft/sec. What is the wheel's angular velocity in (a) radians per second, (b) revolutions per minute?
   a. Given the diameter is 26 in, then the radius $r = 13$ in and $v = 32$ ft/sec. Change $r = 13$ in to $r = \frac{13 \text{ ft}}{12 \text{ rad}}$. Using $\omega = \frac{v}{r}$, then
   $$\omega = \frac{32\frac{\text{ft}}{\text{s}}}{\frac{13 \text{ ft}}{12 \text{ rad}}} = 32\frac{\text{ft}}{\text{s}} \cdot \frac{12 \text{ rad}}{13 \text{ ft}} = \frac{32 \cdot 12}{13} \approx 30 \text{ rad/s}$$
   b. To change to radians per minute multiply by 60. Then, $\omega \approx (30)(60) \approx 1{,}800$ rad/min. Therefore, $\omega \approx \frac{1{,}800}{2\pi}$ revolutions per minute, and
   $$\omega \approx 290 \text{ rev/min}$$
3. A giant C-194 Air Force cargo plane has propellers that make circles 16 ft in diameter at their tip. What is the speed of the propeller tip in feet per second when the shaft is turning at 1,200 rev/min?
   Given $\omega = 1{,}200$ rev/min, then converting to rad/sec
   $$\omega = \frac{1{,}200 \times 2\pi}{60} \text{ rad/sec} = 40\pi \approx 130 \text{ rad/sec}$$
   Since $d = 16$ ft, then $r = 8$ ft and
   $$v = 8 \times 130 \approx 1{,}000 \text{ ft/sec}$$

Exercise Set 19.2

Round all answers to three significant digits.

1. Find the length of the arc s on the circumference of a circle whose radius is 3.0 in subtended by a central angle of 45°.
2. Find the length of an arc s of a circle subtended by a central angle of 142° if the circle has diameter of 56 cm.
3. Find the measure of a central angle of a circle in (a) radians, (b) degrees, if the circle has radius 4.5 m and the intercepted arc is 5.4 m. (Find (b) to the nearest 0.1°.)
•4. A clock pendulum 1.5 ft long subtends an arc 10 in long as it oscillates through an angle. Find the measure of the angle in (a) radians, (b) degrees. (Find (b) to the nearest 0.1°.)
5. A clock pendulum having length 8.0 in swings through an angle of 10.2° each way from the vertical. Find the length of the arc through which the pendulum swings.
•6. Find the radius of a circle whose central angle of measure 3 radians intercepts an arc of 35 cm.
7. The tip of the minute hand of a clock subtends an arc 715 in long as it moves from 12 to 4 on the face of the clock. How long is the minute hand?
•8. Find the area of a sector of a circle having central angle θ and radius r such that

   a. $\theta^R = 21.1$ and $r = 17.3$ in
   b. $\theta^R = 3.05$ and $r = 4.15$ cm
   c. $\theta° = 32.4°$ and $r = 1.40$ m
   d. $\theta° = 52°15'$ and $r = 306.0$ mm

9. A wheel has 24 spokes, each equally spaced. If the wheel has radius 15 in, find the area of each sector formed by consecutive spokes.
•10. A standard industrial motor has speed 1,750 rev/min. How many radians per second is this?
11. The average distance of the earth from the sun is 93,000,000 mi. Taking the period of revolution around the sun (one year) as 365 days, what is the earth's linear velocity in miles per hour?
12. The shaft of a motor rotates at a constant angular velocity of 300 rev/min. How many radians will it have turned through in one-half minute?
13. A drill bit 8 mm in diameter is rotating at 300 rad/sec. What is the linear velocity of a point on its circumference?
•14. A grindstone 12 cm in radius is rotating at 1,650 rpm.
   a. What is its angular velocity?
   b. What is the linear velocity of a point on its rim?
15. A drill bit $\frac{1}{4}$ in in diameter is rotating at 1,200 rev/min.
   a. What is its angular velocity in rad/sec?
   b. What is the linear velocity of a point on its circumference?

•16. A wind-driven electric power station has a propeller 125 ft in diameter. When the propeller was turning at 32 rev/min, what was the linear velocity in miles per hour of one of the blade tips?

17. A steam turbine motor is 12 ft in diameter and turns at 1,500 rev/min. How many miles does a point on the rim travel in a 24-h day?

18. Find the linear velocity of a point on the circumference of a 10-in diameter record having a constant speed of $33\frac{1}{3}$ rev/min in inches per second. What is the angular velocity of the same point?

•19. An automobile having 13-in diameter wheels is traveling at 88 ft/sec. Find the angle through which the wheel rotates in 3 s. What is the angular velocity in radians per minute?

20. An automobile traveling at $\frac{220}{3}$ ft/sec has 15-in diameter wheels. Through what angle does the wheel rotate in 4 s? What is the angular velocity in degrees per minute? (Round to the nearest 0.1°.)

•21. A seesaw rotates through a 30° angle. If the end of the seesaw travels 6.2 ft, how long is the seesaw?

22. A shaft 6 in in diameter is being turned down on a lathe at 200 rev/min. The cutting tool takes off a continuous ribbon of metal. How many linear feet of metal ribbon will be cut off in 5 min?

•23. A satellite in circular orbit about the earth is 480 mi high. The satellite passes a point on the earth's surface once every 2.0 hr. What is its linear velocity? (Use 4,000 mi for the earth's radius.)

24. What is the linear velocity of a satellite in circular orbit about the earth if it makes a complete revolution every 75 min and is 600 mi above the earth's surface?

25. A fan blade rotates at a speed of 300 rev/min. If the blade has a radius of 12 cm, how fast is the outer tip of the blade moving in meters per second?

26. In a factory, thread is being wound on a 1.5-in diameter spool as the spool rotates at 2,000 rev/min. How fast is the thread being wound in feet per second?

27. A barrel 2 ft in diameter is rolling with an angular velocity of 5 rad/sec. What is the linear velocity in inches per minute of a point on its surface?

28. A truck whose tires have radius 60 cm travels at 20 m/s. Find the angular velocity of the tires in revolutions per second.

29. A truck whose tires have a 2-ft radius travels at 40 mi/h. What is the angular velocity of the tires?

19.3 The Sine and Cosine Waves

Thus far, trigonometry has been used to find the solutions of triangles and to determine velocities of objects going around in circles. We now consider the concept of trigonometric functions. Recall, from section 15.2, that a function is a special relationship that exists between two groups of numbers; that is, for each number in one group there exists *only one* corresponding number in the other group. The trigonometric ratios define functions.

When graphing algebraic equations, the groups of numbers were related by the equation $y = f(x)$. Arbitrary values for x were chosen and corresponding values for y were determined. These pairs of values were plotted as points in the rectangular coordinate system and connected by a smooth curve or line to get the graph of the equation.

We proceed in the same way to graph the equations of two of the trigonometric functions, defined by

$$f(x) = \sin x \qquad \text{and} \qquad g(x) = \cos x$$

which can be written as

$$y = f(x) = \sin (x) \qquad \text{or} \qquad y = \sin (x)$$

and the other

$$y = g(x) = \cos (x) \qquad \text{or} \qquad y = \cos(x)$$

where x is a number considered to be an angle measured in radians.

To graph the equation $y = \sin x$, set up a table of related values for x and y. To better relate the graph with angles in degrees, the degree measure for each number value of x is given in parentheses.

x	(0°) 0	(30°) $\frac{\pi}{6}$	(45°) $\frac{\pi}{4}$	(60°) $\frac{\pi}{3}$	(90°) $\frac{\pi}{2}$	(120°) $\frac{2\pi}{3}$	(135°) $\frac{3\pi}{4}$	(150°) $\frac{5\pi}{6}$	(180°) π
$y = \sin x$	0	0.5	0.71	0.86	1.0	0.86	0.71	0.5	0

x	(210°) $\frac{7\pi}{6}$	(225°) $\frac{5\pi}{4}$	(240°) $\frac{4\pi}{3}$	(270°) $\frac{3\pi}{2}$	(300°) $\frac{5\pi}{3}$	(315°) $\frac{7\pi}{4}$	(330°) $\frac{11\pi}{6}$	(360°) 2π
$y = \sin x$	−0.5	−0.71	−0.86	−1.0	−0.86	−0.71	−0.5	0

Plotting these points in the rectangular coordinate plane, we obtain the graph of $y = \sin x$ for values of $x = 0$ rads to $x = 2\pi$ rads. See figure 19.4.

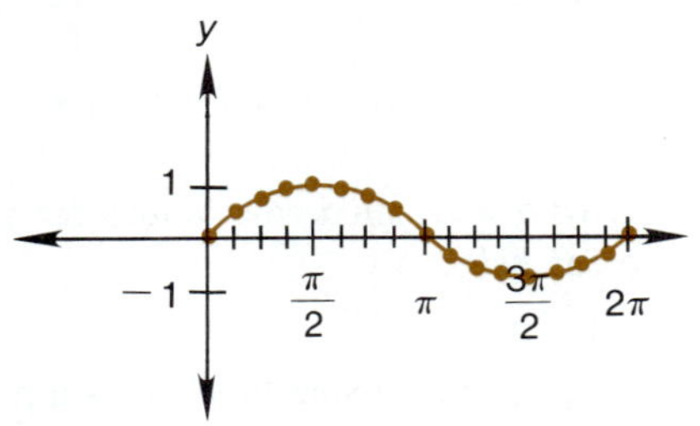

Figure 19.4

The graph of $y = \sin x$ is called a **sine wave** or **sinusoid.** Alternating current in the home as displayed on an oscilloscope is an example of this wave, where x corresponds to time and y corresponds to voltage.

Recall that any given angle has many coterminal angles. For the sine ratio, these angles were determined by adding (or subtracting) 360°, or $2\pi^R$, to the given angle. For this reason the values of the sine ratio begin repeating every 360°, or 2π, units. Then,

$$\sin x = \sin(x + 2\pi)$$

for any real number x. Because of this property, the sine function is called a **periodic function** with a period of 2π.

note

Given angle θ in degrees, $\sin \theta° = \sin(\theta° + 360°)$ and the *period* is then 360°. Graphically, this is one *cycle*.

Since this repetition of values continues indefinitely, the graph of $y = \sin x$, the sine wave, repeats every 2π units. See figure 19.5. Note that the range of values for y lies between and includes -1 and 1. That is,

$$-1 \leqslant y \leqslant 1 \qquad \text{or} \qquad -1 \leqslant \sin x \leqslant 1$$

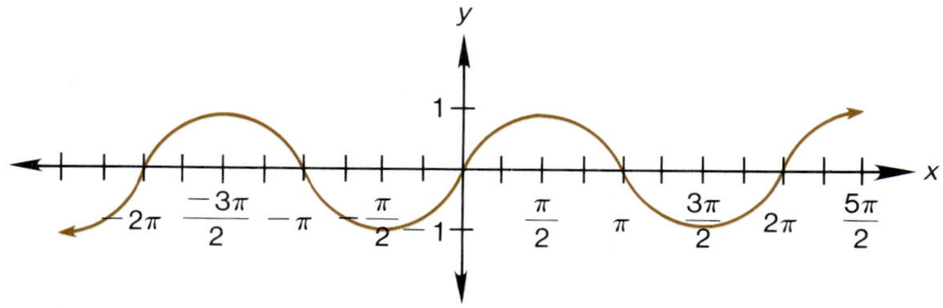

Figure 19.5

The graph of $y = \cos x$ is constructed in the same manner by choosing values for x and determining corresponding values for y.

x	0	$\frac{\pi}{6}$	$\frac{\pi}{4}$	$\frac{\pi}{3}$	$\frac{\pi}{2}$	$\frac{2\pi}{3}$	$\frac{3\pi}{4}$	$\frac{5\pi}{6}$	π	$\frac{7\pi}{6}$
$y = \cos x$	1.0	0.86	0.71	0.5	0	−0.5	−0.71	−0.86	−1.0	−0.86

x	$\frac{5\pi}{4}$	$\frac{4\pi}{3}$	$\frac{3\pi}{2}$	$\frac{5\pi}{3}$	$\frac{7\pi}{4}$	$\frac{11\pi}{6}$	2π
$y = \cos x$	−0.71	−0.5	0	0.5	0.71	0.86	1.0

See figure 19.6. As with $y = \sin x$, the values for $y = \cos x$ repeat every 360°, or 2π units. Thus, we can see

$$\cos x = \cos(x + 2\pi)$$

for any real number x, and the cosine function is a periodic function with period 2π. Likewise, as with the sine function,

$$-1 \leqslant y \leqslant 1 \qquad \text{or} \qquad -1 \leqslant \cos x \leqslant 1$$

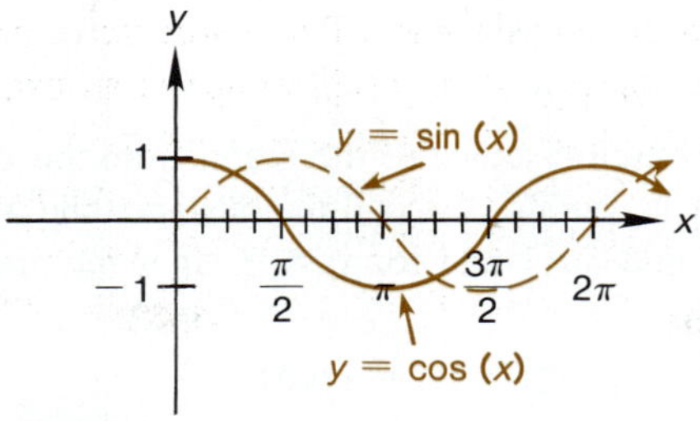

Figure 19.6

note

From the graphs of $y = \sin x$ and $y = \cos x$ we can see that the graph of $y = \sin x$ when *shifted* $\frac{\pi}{2}$ to the left gives us the graph of $y = \cos x$.

Since the basic shape of the curves of these graphs always remains the same, it is important to learn (1) the shape of the sine and cosine waves, (2) where the graphs cross the x-axis, and (3) where they reach their highest value (that is, 1) and their lowest value (that is, -1). In sketching variations of these functions, this information will be important. We don't want to have to plot numerous points, as we did previously, every time we wish to sketch the curve of such functions.

Consider now the graph of $y = 3 \sin x$. We again make a table of related values. The graph is given in figure 19.7.

x	0	$\frac{\pi}{4}$	$\frac{\pi}{2}$	$\frac{3\pi}{4}$	π	$\frac{5\pi}{4}$	$\frac{3\pi}{2}$	$\frac{7\pi}{4}$	2π
$y = 3 \sin x$	0	2.1	3	2.1	0	-2.1	-3	-2.1	0

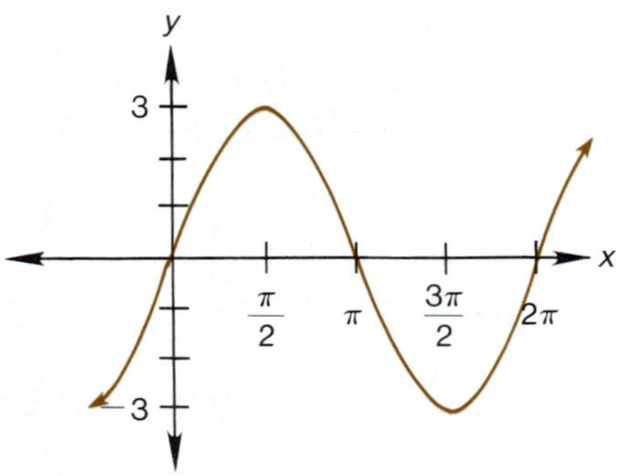

Figure 19.7

From figure 19.7 we can see

$$-3 \leq 3 \sin x \leq 3$$

Thus, the maximum distance of the graph from the x-axis of $y = \sin x$ is 1 unit, and of $y = 3 \sin x$ is 3 units. We call this distance the **amplitude** of the curve. For $y = \sin x$ the amplitude is 1 and for $y = 3 \sin x$ the amplitude is 3. In general, for the equations

$$y = a \sin x \qquad \text{or} \qquad y = a \cos x$$

the amplitude is given by $|a|$, the greatest value of y on the curve.

By noting where the values of the amplitude occur and where the graph crosses the axes we can sketch the graph of $y = a \sin x$ and $y = a \cos x$ with a minimum of points. By inspection we can see that the graph of

(1) $y = a \sin x$

(a) crosses the x-axis at both ends of a period and in the middle of a period, and (b) achieves a maximum value, $|a|$, one-fourth of the way through the period and minimum value three-fourths of the way through the period.

(2) $y = a \cos x$

(a) crosses the x-axis one-fourth and three-fourths of the way through the period, and (b) achieves maximum value at the ends of a period and minimum value one-half of the way through the period.

We use these characteristics to sketch the graph of any equation (function) involving sine or cosine.

Examples

Sketch the graph of the following functions between -2π and 2π.

1. $y = 2 \sin x$

   The amplitude is $|2| = 2$ and since the sine function (a) achieves amplitude values at $-\frac{3\pi}{2}$, $-\frac{\pi}{2}$, $\frac{\pi}{2}$, and $\frac{3\pi}{2}$, and (b) crosses the x-axis at -2π, $-\pi$, 0, π, and 2π, we sketch the graph in figure 19.8.

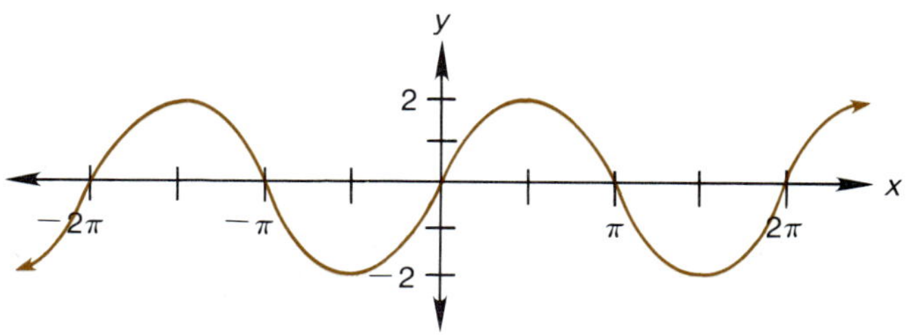

Figure 19.8

2. $y = \frac{1}{2} \cos x$

The amplitude is $\left|\frac{1}{2}\right| = \frac{1}{2}$. Since the cosine function (a) achieves amplitude values at -2π, $-\pi$, 0, π, 2π, and (b) crosses the x-axis at $-\frac{3\pi}{2}$, $-\frac{\pi}{2}$, $\frac{\pi}{2}$, and $\frac{3\pi}{2}$, then we sketch the graph in figure 19.9.

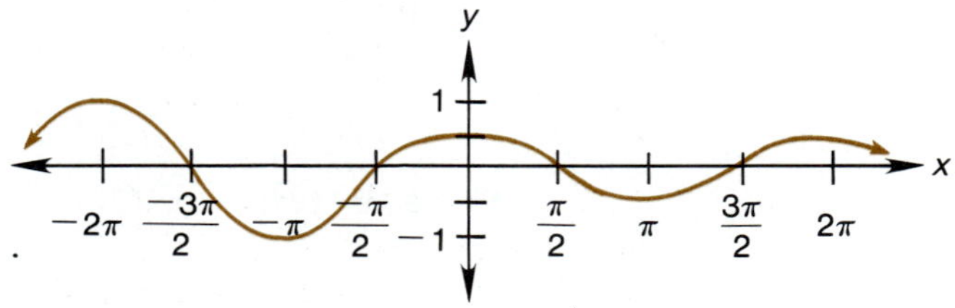

Figure 19.9

3. $y = -4 \sin x$
The amplitude is $|-4| = 4$. Since $a = -4$, consider the following table of values.

x	0	$\frac{\pi}{2}$	π	$\frac{3\pi}{2}$	2π
$y = 4 \sin x$	0	4	0	-4	0
$y = -4 \sin(x)$	0	-4	0	4	0

When $a < 0$, all previous values for $a \sin x$ are negated. Then the graph of $y = -4 \sin x$ is the graph of $y = 4 \sin x$ *turned over*. See figure 19.10.

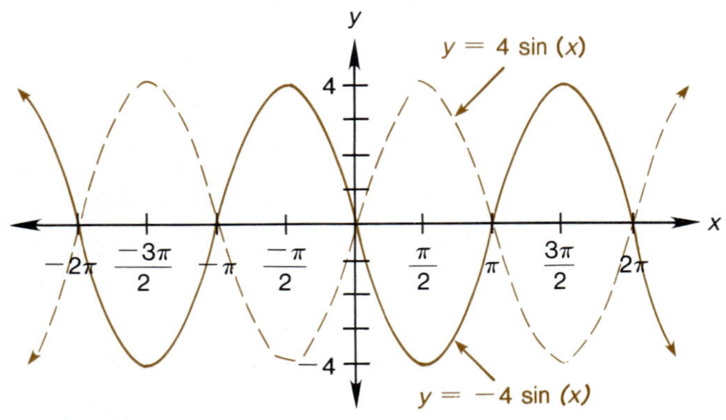

Figure 19.10

4. $y = -1.5 \cos x$

The amplitude is $|-1.5| = 1.5$. All values for $y = 1.5 \cos x$ are now negated. See figure 19.11.

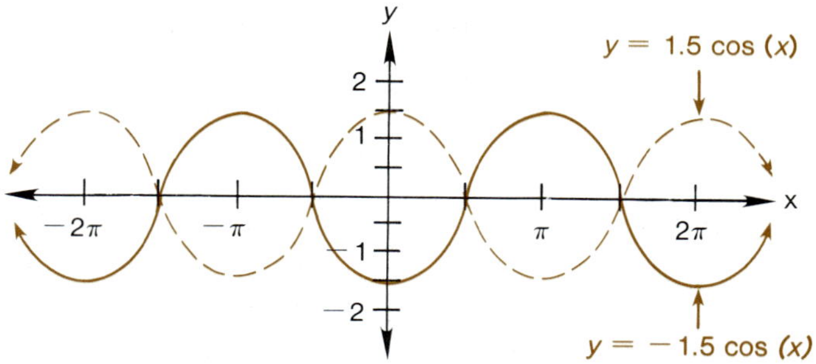

Figure 19.11

note

Again, $-1.5 < 0$ causes the graph to be $y = 1.5 \cos x$ *turned over*.

Now let us consider the graph of $y = \sin\left(\frac{1}{2}x\right)$. To find y for any value of x that we choose, we must multiply this value by $\frac{1}{2}$ and *then* find the value of the sine of this number. Consider the following table of values.

x	0	π	2π	3π	4π	5π	6π	7π	8π
$\frac{1}{2}x$	0	$\frac{\pi}{2}$	π	$\frac{3\pi}{2}$	2π	$\frac{5\pi}{2}$	3π	$\frac{7\pi}{2}$	4π
$y = \sin\left(\frac{1}{2}x\right)$	0	1	0	-1	0	1	0	-1	0

We can see from the table that the coefficient $\frac{1}{2}$ has caused the period to be 4π, two times the period of $y = \sin x$. Since

$$\frac{2\pi}{\frac{1}{2}} = 2\pi \cdot \frac{2}{1} = 4\pi$$

then the new period can be determined by dividing the fundamental period of 2π by $\frac{1}{2}$, in general, given

$$y = a \sin(bx) \qquad \text{or} \qquad y = a \cos(bx).$$

The period is determined by $\frac{2\pi}{b}$.

note

The number b is assumed positive in this chapter. If $b < 0$, there are properties of the trigonometric functions we could use to make $b > 0$. However, we will not do this in this text.

Examples Find the amplitude and period of the following.

1. $y = 3 \sin (2x)$

   $$\text{Amplitude} = |3| = 3$$

   $$\text{Period} = \frac{2\pi}{2} = \pi$$

2. $y = -2 \cos (3x)$

   $$\text{Amplitude} = |-2| = 2$$

   $$\text{Period} = \frac{2\pi}{3}$$

3. $y = -4 \sin \left(\frac{1}{3}x\right)$

   $$\text{Amplitude} = |-4| = 4$$

   $$\text{Period} = \frac{2\pi}{\frac{1}{3}} = \frac{2\pi}{1} \cdot \frac{3}{1} = 6\pi$$

4. $y = \frac{1}{2} \cos (\pi x)$

   $$\text{Amplitude} = \left|\frac{1}{2}\right| = \frac{1}{2}$$

   $$\text{Period} = \frac{2\pi}{\pi} = 2$$

The following are steps that can be used to graph $y = a \sin bx$ or $y = a \cos bx$ $(b > 0)$ through one period.

1. Determine the amplitude and mark off the appropriate scale on the y-axis.
2. Determine the period, $\frac{2\pi}{b}$, from the origin O. Lay off a distance $\frac{2\pi}{b}$ on the x-axis.
3. Divide the interval from O to $\frac{2\pi}{b}$ into *four* equal parts.

4. Locate the points where the graph crosses the x-axis.
   a. For $y = a \sin bx$ this occurs at the beginning, mid-point, and end of the interval.
   b. For $y = a \cos bx$ this occurs one-fourth and three-fourths of the way through the interval.
5. Locate the points where the maximum and minimum points are reached—these are the points *not found* in step 4, but are in step 3.
6. Connect these points with a smooth sine wave. If more accuracy is required, use a calculator to find additional points.

If more periods are required or desired, use this same procedure, extending the given period to the right or left as needed.

Examples Sketch the graph of the following equations.

1. $y = 2 \sin (3x)$ from $-\frac{2\pi}{3}$ to $\frac{2\pi}{3}$ (See figure 19.12.)
   a. The amplitude is $|2| = 2$.
   b. The period is $\frac{2\pi}{3}$.
   c. Mark off a convenient distance from 0 to $\frac{2\pi}{3}$. Subdivide this interval into points $\frac{\pi}{6}, \frac{\pi}{3}, \frac{\pi}{2}$.
   d. The graph crosses the x-axis at 0, $\frac{\pi}{3}$, and $\frac{2\pi}{3}$.
   e. Maximum and minimum values (2 and -2, respectively) are reached at $\frac{\pi}{6}$ and $\frac{\pi}{2}$.

note

To show two complete cycles, we have graphed one period to the left of the origin.

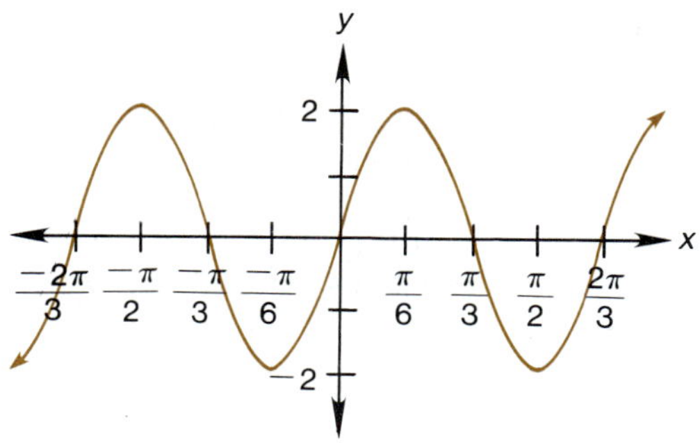

Figure 19.12

2. $y = -3\cos\left(\frac{1}{3}x\right)$ from -6π to 6π (See figure 19.13.)

a. The amplitude is $|-3| = 3$.

b. The period is $\frac{2\pi}{\frac{1}{3}} = 6\pi$.

c. Mark off a convenient distance from 0 to 6π. Divide this distance into four equal parts at points $\frac{3\pi}{2}$, 3π, and $\frac{9\pi}{2}$.

d. The graph crosses the x-axis at $\frac{3\pi}{2}$ and $\frac{9\pi}{2}$.

e. Minimum and maximum points of -3 and 3, respectively, are reached at 0, 3π, and 6π. (*Note:* The -3 means the graph goes to a minimum value first.)

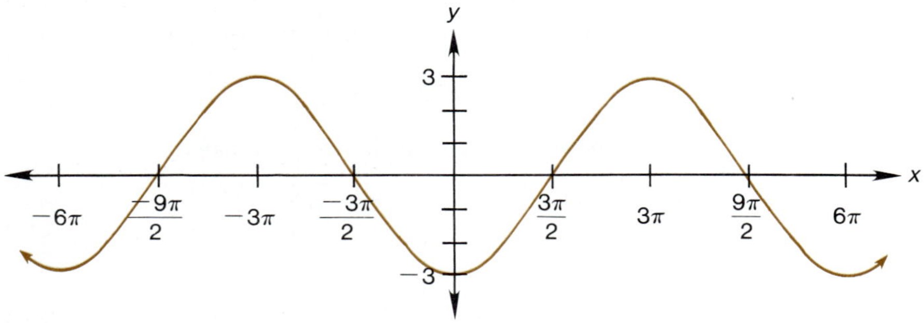

Figure 19.13

Finally, we consider the graph of $y = \sin\left(x + \frac{\pi}{2}\right)$. When setting up a table of values to find y we must *add* $\frac{\pi}{2}$ to the chosen values of x before taking the sine of the number.

x	$-\frac{\pi}{2}$	0	$\frac{\pi}{2}$	π	$\frac{3\pi}{2}$	2π
$x + \frac{\pi}{2}$	0	$\frac{\pi}{2}$	π	$\frac{3\pi}{2}$	2π	$\frac{5\pi}{2}$
$y = \sin\left(x + \frac{\pi}{2}\right)$	0	1	0	-1	0	1

In figure 19.14 we graph the curve of $y = \sin x$ (dotted line) and $y = \sin\left(x + \frac{\pi}{2}\right)$ (solid line) for comparison. Comparing the two graphs, we can see that the graph of

$y = \sin\left(x + \frac{\pi}{2}\right)$ is the graph of $y = \sin x$ *shifted* $\frac{\pi}{2}$ units *to the left*. We call $\frac{\pi}{2}$ the **phase shift** (sometimes this is called the **displacement**).

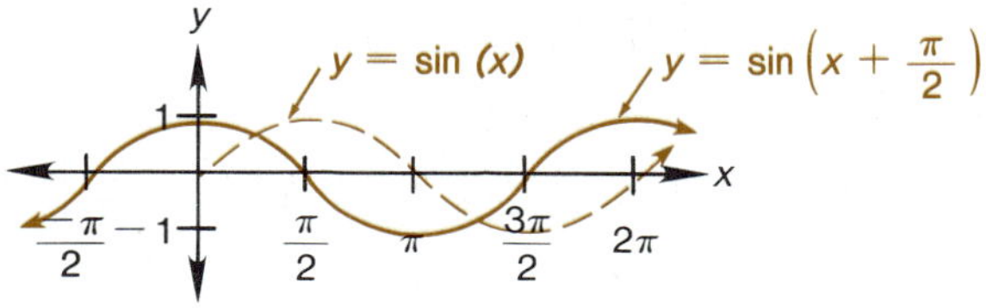

Figure 19.14

Sketching the graph of a trigonometric function involving phase shift usually presents difficulty. Consider again the function defined by $y = \sin\left(x + \frac{\pi}{2}\right)$. We call $x + \frac{\pi}{2}$ the **argument.** Since the period is 2π, it is important that the argument vary between 0 and 2π. Therefore, we consider

$$0 \leqslant x + \frac{\pi}{2} \leqslant 2\pi$$

Solving the inequality for x we have

$$0 - \frac{\pi}{2} \leqslant x \leqslant 2\pi - \frac{\pi}{2}$$

$$-\frac{\pi}{2} \leqslant x \leqslant \frac{3\pi}{2}$$

note

Refer again to figure 19.14 to see that the graph of $y = \sin\left(x + \frac{\pi}{2}\right)$ goes through one cycle (period) from $-\frac{\pi}{2}$ to $\frac{3\pi}{2}$.

From this example we can see the left-hand member of the inequality, $-\frac{\pi}{2}$, gives the phase shift, from 0 to $-\frac{\pi}{2}$, and the start of one cycle. The right-hand member shows the end of one cycle.

Consider now the general sine and cosine functions defined by

$$y = a \sin(bx + c) \qquad \text{and} \qquad y = a \cos(bx + c)$$

From the previous discussion we determine (1) the amplitude to be $|a|$, and (2) the argument to be $bx + c$. Since the argument varies from 0 to 2π, consider

$$0 \leqslant bx + c \leqslant 2\pi$$

and solving for x, adding $-c$ to all members:

$$0 - c \leqslant bx \leqslant 2\pi - c$$

Dividing through by b,

$$\frac{-c}{b} \leqslant x \leqslant \frac{2\pi - c}{b}$$

The phase shift is immediately determined by $-\frac{c}{b}$ and the period is then

$$\frac{2\pi - c}{b} - \left(\frac{-c}{b}\right) = \frac{2\pi - c + c}{b} = \frac{2\pi}{b} \qquad b > 0$$

After learning the basic sine and cosine waves (see figure 19.15), we can then sketch the graph of any sine or cosine function using a minimum of points.

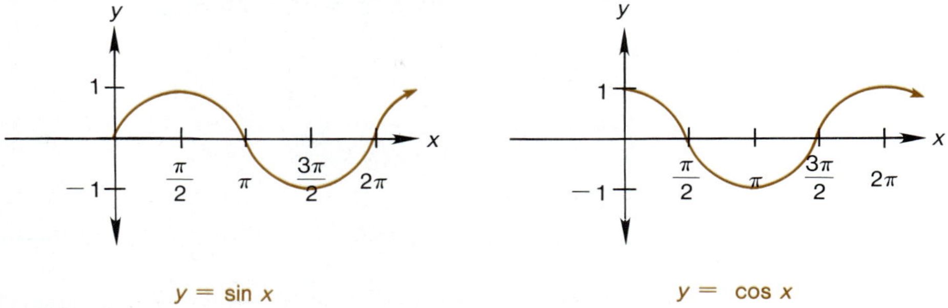

Figure 19.15

Examples Sketch the graph of the following functions through one period.

1. $y = 3\sin(2x + \pi)$ (See figure 19.16.)

   Now, amplitude $= |3| = 3$. The argument is $2x + \pi$. Consider

   $0 \leqslant 2x + \pi \leqslant 2\pi$

   Adding $-\pi$ to all members, $-\pi \leqslant 2x \leqslant \pi$. Dividing through by 2, $-\frac{\pi}{2} \leqslant x \leqslant \frac{\pi}{2}$. Then the phase shift is $-\frac{\pi}{2}$ and one cycle goes from $-\frac{\pi}{2}$ to $\frac{\pi}{2}$. Therefore, the period is π $\left(\text{or } \frac{2\pi}{2} = \pi\right)$.

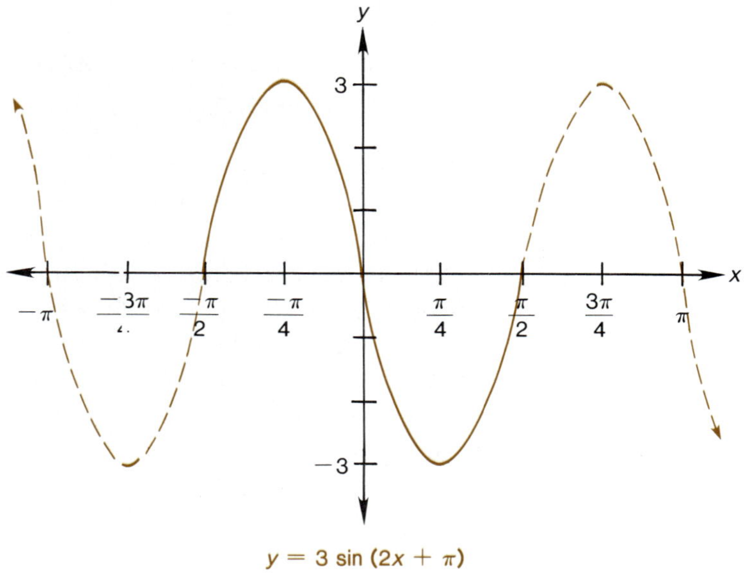

Figure 19.16

2. $y = 2\cos\left(3x - \frac{\pi}{2}\right)$ (See figure 19.17.)

The amplitude $= |2| = 2$. The argument is $3x - \frac{\pi}{2}$. Consider

$$0 \leqslant 3x - \frac{\pi}{2} \leqslant 2\pi$$

$$0 + \frac{\pi}{2} \leqslant 3x \leqslant 2\pi + \frac{\pi}{2}$$

$$\frac{\pi}{2} \leqslant 3x \leqslant \frac{5\pi}{2}$$

$$\frac{\pi}{6} \leqslant x \leqslant \frac{5\pi}{6}$$

The phase shift is then $\frac{\pi}{6}$ and the period is from $\frac{\pi}{6}$ to $\frac{5\pi}{6}$.

Since $\frac{5\pi}{6} - \frac{\pi}{6} = \frac{4\pi}{6} = \frac{2\pi}{3}$, the period is then $\frac{2\pi}{3}$.

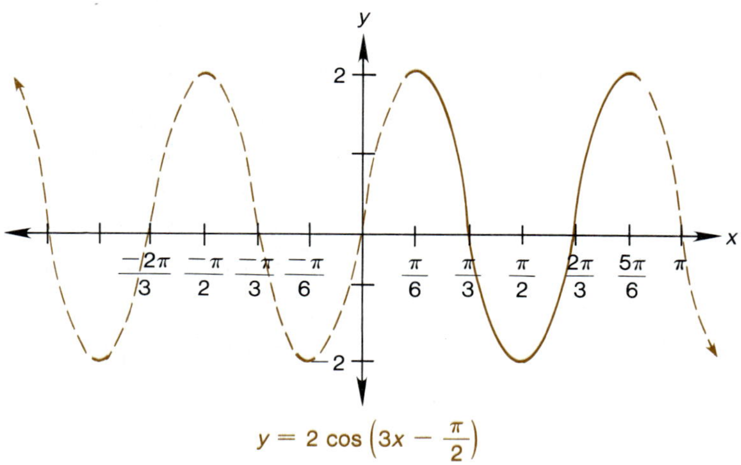

Figure 19.17

3. $y = -\frac{3}{2}\sin\left(\frac{1}{2}x + \frac{\pi}{4}\right)$ (See figure 19.18.)

The amplitude $= \left|-\frac{3}{2}\right| = \frac{3}{2}$. The argument is $\frac{1}{2}x + \frac{\pi}{4}$.

Then,

$$0 \leqslant \frac{1}{2}x + \frac{\pi}{4} \leqslant 2\pi$$

$$\frac{-\pi}{4} \leqslant \frac{1}{2}x \leqslant \frac{7\pi}{4}$$

$$\frac{-\pi}{2} \leqslant x \leqslant \frac{7\pi}{2}$$

The phase shift is $\frac{-\pi}{2}$ and the period is from $\frac{-\pi}{2}$ to $\frac{7\pi}{2}$.

Since $\frac{7\pi}{2} - \left(-\frac{\pi}{2}\right) = \frac{7\pi}{2} + \frac{\pi}{2} = 4\pi$, the period is 4π $\left(\text{or } \frac{2\pi}{\frac{1}{2}} = 2\pi \cdot 2 = 4\pi\right)$.

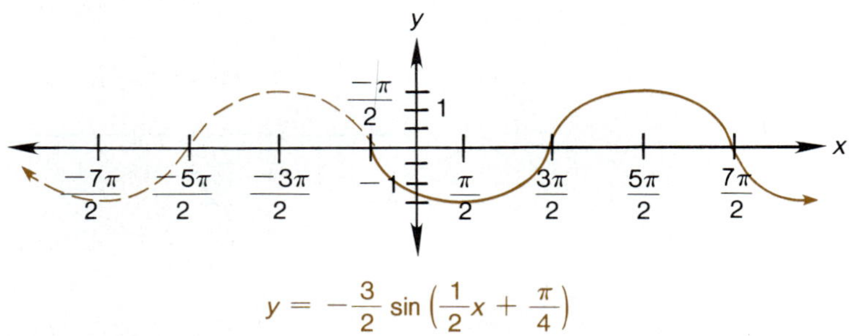

Figure 19.18

note

Since $a < 0$ in the equation, the graph went to a minimum value instead of a maximum value and is the graph of $y = \frac{3}{2}\sin\left(\frac{1}{2}x + \frac{\pi}{4}\right)$, "turned over."

4. $y = -\cos\left(\pi x - \frac{\pi}{2}\right)$ (See figure 19.19.)

The amplitude $= |-1| = 1$ and the argument is $\pi x - \frac{\pi}{2}$.

Then,

$$0 \leqslant \pi x - \frac{\pi}{2} \leqslant 2\pi$$

$$\frac{\pi}{2} \leqslant \pi x \leqslant \frac{5\pi}{2}$$

$$\frac{1}{2} \leqslant x \leqslant \frac{5}{2}$$

The phase shift is then $\frac{1}{2}$ and the period is $\frac{5}{2} - \frac{1}{2} = \frac{4}{2} = 2$ $\left(\text{or } \frac{2\pi}{\pi} = 2\right)$.

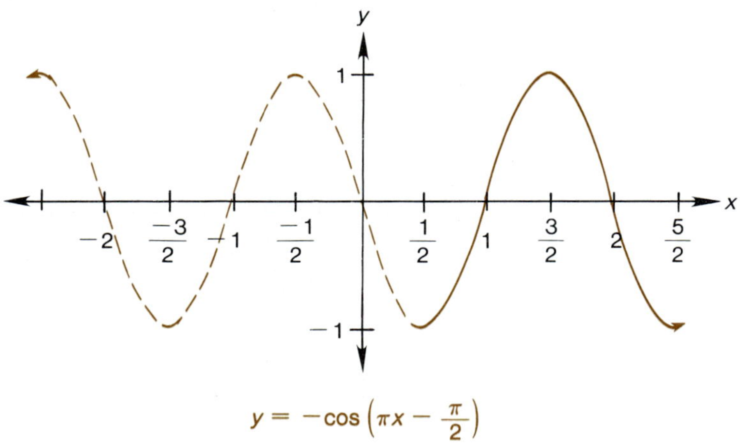

Figure 19.19

Exercise Set 19.3

In exercises 1–16, determine the amplitude of the given functions and sketch the graphs on the interval from -2π to 2π.

1. $y = 4 \sin x$
•2. $y = 5 \sin x$
3. $y = \frac{1}{3} \sin x$
4. $y = 0.8 \sin x$
5. $y = -\sin x$
6. $y = -4 \sin x$
7. $y = -\frac{3}{2} \sin x$
8. $y = -0.4 \sin x$
9. $y = 3 \cos x$
10. $y = 2 \cos x$
11. $y = \frac{1}{2} \cos x$
12. $y = 1.5 \cos x$
13. $y = -\cos x$
•14. $y = -5 \cos x$
15. $y = -\frac{5}{2} \cos x$
16. $y = -0.2 \cos x$

In exercises 17–32, determine the amplitude and period of the given functions. Sketch the graphs over one period.

17. $y = 3 \sin 2x$
•18. $y = 4 \sin 3x$
19. $y = -\sin 6x$
20. $y = -2 \sin 4x$
21. $y = \frac{1}{2} \sin \pi x$
22. $y = -0.6 \sin 2\pi x$
23. $y = 5 \sin \frac{2}{3}x$
24. $y = -3 \sin \frac{3}{2}\pi x$
25. $y = 2 \cos 3x$
26. $y = 3 \cos 4x$
27. $y = -\cos 2x$
•28. $y = -4 \cos 6x$
29. $y = \frac{3}{2} \cos 3\pi x$
•30. $y = -0.5 \cos \pi x$
31. $y = 6 \cos \frac{6}{5}x$
32. $y = -3 \cos \frac{1}{2}\pi x$

In exercises 33–42 use the argument to find the phase shift and the period.

33. $y = 3 \sin\left(x + \frac{\pi}{4}\right)$
34. $y = 2 \sin(3x - \pi)$
35. $y = -\cos\left(2x + \frac{\pi}{2}\right)$
36. $y = \cos\left(4x - \frac{\pi}{3}\right)$
37. $y = 4 \sin\left(\frac{1}{2}x - \frac{\pi}{2}\right)$
•38. $y = \cos\left(\frac{1}{4}x + \frac{\pi}{8}\right)$
39. $y = -\sin\left(\pi x - \frac{\pi}{4}\right)$
40. $y = \frac{5}{4} \cos(2\pi x + \pi)$
41. $y = -3 \sin\left(\frac{\pi}{2}x - \frac{\pi}{4}\right)$
•42. $y = -2 \cos(3\pi x + 2\pi)$

In exercises 43–62, determine the amplitude, period, and phase shift (if any) of the given functions. Sketch the graph through one period.

43. $y = 2 \sin x$

•44. $y = \frac{1}{2} \sin x$

45. $y = 3 \cos x$

46. $y = \frac{1}{4} \cos x$

47. $y = 3 \sin (2x)$

•48. $y = -2 \sin(4x)$

•49. $y = -2 \cos(2x)$

•50. $y = 4 \cos\left(\frac{1}{2}x\right)$

51. $y = \sin\left(x - \frac{\pi}{2}\right)$

52. $y = \sin(x + \pi)$

53. $y = \cos\left(x - \frac{\pi}{2}\right)$

•54. $y = \cos\left(x + \frac{\pi}{3}\right)$

55. $y = -2 \sin\left(x + \frac{\pi}{4}\right)$

56. $y = 3 \sin\left(x - \frac{\pi}{8}\right)$

57. $y = -4 \cos\left(x + \frac{\pi}{2}\right)$

58. $y = \frac{1}{2} \sin(3x + 2\pi)$

59. $y = \sin\left(2x - \frac{\pi}{4}\right)$

•60. $y = -3 \cos\left(\frac{2}{3}x - 3\pi\right)$

61. $y = 1.5 \sin(\pi x + \pi)$

•62. $y = 2 \cos(2\pi x - \pi)$

19.4 Graphs of the Other Trigonometric Functions

The graphs of the other four trigonometric functions—$y = \tan x$, $y = \csc x$, $y = \sec x$, and $y = \cot x$—can be graphed in much the same way as the sine and cosine functions. Their graphs will not be sine waves, however, even though each is a periodic function.

Recall that for some values of x, each of these functions is not defined. For example, the tangent ratio is not defined for $x = \frac{\pi}{2}$ and for $x = \frac{-\pi}{2}$. Thus, as values for x are chosen closer and closer to $\frac{\pi}{2}$ $\left(\text{but not including } \frac{-\pi}{2}\right)$, $\tan x$ takes on greater and greater values without bound (which means there is no greatest value). Likewise, if values for x are chosen approaching $\frac{-\pi}{2}$ $\left(\text{but not including } \frac{\pi}{2}\right)$, $\tan x$ gets more and more negative, again without bound. Thus, there is no least value for $\tan x$. Then, there is no point on the graph of $y = \tan x$ that corresponds to $x = \frac{\pi}{2}$ or $x = \frac{-\pi}{2}$.

This situation means that the graph of $y = \tan x$ will get closer and closer to, but never intersect, the lines $x = \frac{\pi}{2}$ and $x = \frac{-\pi}{2}$. These two lines are called *vertical asymptotes in the graph of* $y = \tan x$. A **vertical asymptote** is an imaginary line in the plane that the graph approaches but does not cross.

note

In the complete graph of $y = \tan x$ all vertical lines $y = \frac{\pi}{2} + \pi k$, where k is an integer, are vertical asymptotes.

To sketch the graph of $y = \tan x$, we determine some values of $\tan x$ where $\frac{-\pi}{2} < x < \frac{\pi}{2}$.

x	$\frac{-\pi}{3}$	$\frac{-\pi}{4}$	$\frac{-\pi}{6}$	0	$\frac{\pi}{6}$	$\frac{\pi}{4}$	$\frac{\pi}{3}$
$y = \tan x$	-1.7	-1	-0.6	0	0.6	1	1.7

Plotting these points and representing the vertical asymptotes $x = \frac{\pi}{2}$ and $x = \frac{-\pi}{2}$ by dotted lines, we show the graph of $y = \tan x$ in figure 19.20. To show more of the graph we repeat the curve as shown between $\frac{-\pi}{2}$ and $\frac{\pi}{2}$. Learn the basic curve for $y = \tan x$.

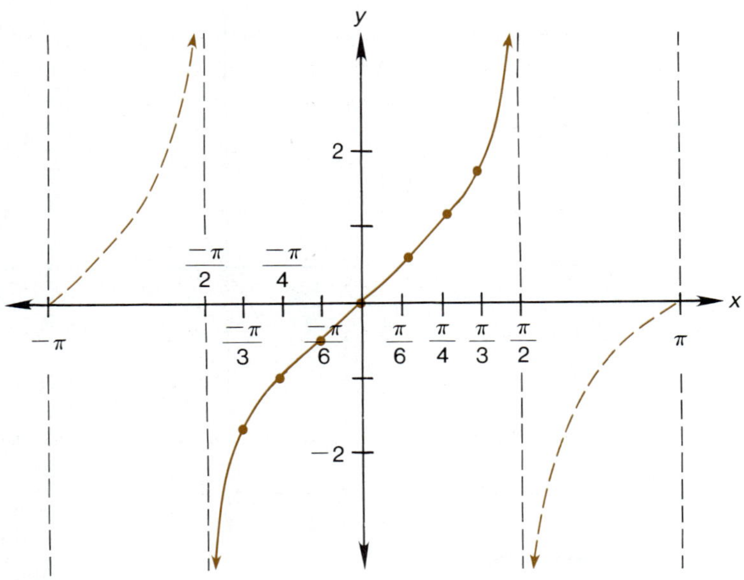

Figure 19.20

Since there is no greatest, or least, value for $y = \tan x$, the graph of the tangent function *does not have amplitude*.

note
The remaining three functions will not have amplitude either.

Consider then a table of values for $y = 2 \tan x$:

x	$\frac{-\pi}{3}$	$\frac{-\pi}{4}$	$\frac{-\pi}{6}$	0	$\frac{\pi}{6}$	$\frac{\pi}{4}$	$\frac{\pi}{3}$
$y = 2 \tan x$	-3.4	-2	-1.2	0	1.2	2	3.4

We note the values have just doubled and the points $\left(\frac{-\pi}{4}, -1\right)$ and $\left(\frac{\pi}{4}, 1\right)$ in the graph of $y = \tan x$ now are $\left(\frac{-\pi}{4}, -2\right)$ and $\left(\frac{\pi}{4}, 2\right)$ in the graph of $y = 2 \tan x$. We use these points as guide points in sketching the graph of the tangent function.

When graphing the sine and cosine functions, each period 2π was divided into four equal parts. Since the period of the tangent function is π, $\frac{-\pi}{2}$ to $\frac{\pi}{2}$, we again divide the period by the points $\frac{-\pi}{4}$, 0, and $\frac{\pi}{4}$ and get four equal parts. These points are then used to sketch the graph with a minimum number of points. See the graph of $y = 2 \tan x$ in figure 19.21.

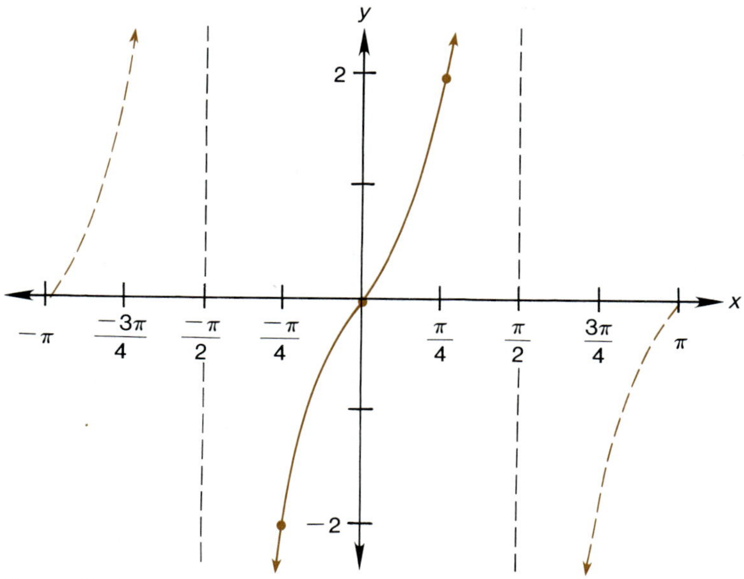

Figure 19.21

Thus, given $y = a \tan x$ $(a > 0)$, the points $\left(\frac{-\pi}{4}, -a\right)$ and $\left(\frac{\pi}{4}, a\right)$ will be important points in the graph of the tangent function.

note

As the period changes, the value of the first component in the above points changes. However, it is *always* the value of x that is one-fourth and three-fourths of the way through the period. The graph *always* crosses the x-axis halfway through the period.

Now we consider the graph of $y = a \tan(bx + c)$. The values of b and c have the same effect on the graph of this function as they did with the sine and cosine functions. That is, since the period of the tangent function is π, we consider the argument $bx + c$ as it varies from 0 to π:

$$0 < bx + c < \pi$$
$$0 - c < bx < \pi - c$$
$$\frac{0 - c}{b} < x < \frac{\pi - c}{b}$$

and so (1) the phase shift is $\frac{-c}{b}$ and (2) the period is

$$\frac{\pi - c}{b} - \frac{0 - c}{b} = \frac{\pi - c + 0 + c}{b} = \frac{\pi}{b} \qquad b > 0$$

Example

1. Determine the phase shift and period of $y = \tan\left(2x + \frac{\pi}{2}\right)$

Now,

$$0 < 2x + \frac{\pi}{2} < \pi$$
$$\frac{-\pi}{2} < 2x < \frac{\pi}{2}$$
$$\frac{-\pi}{4} < x < \frac{\pi}{4}$$

Thus, the phase shift is $\frac{-\pi}{4}$ and the period is $\frac{\pi}{4} - \left(\frac{-\pi}{4}\right) = \frac{\pi}{4} + \frac{\pi}{4}$ $= \frac{\pi}{2}$.

note

The period again is the period π divided by b. To say the phase shift is $\frac{-\pi}{4}$ is to shift the point (0,0) where the graph of $y = \tan x$ crosses the x-axis to the point $\left(\frac{-\pi}{4},0\right)$. This causes a corresponding shift of both asymptotes.

2. Sketch the graph of $y = 3\tan(3x - \pi)$. Consider:

$$0 < 3x - \pi < \pi$$
$$\pi < 3x < 2\pi$$
$$\frac{\pi}{3} < x < \frac{2\pi}{3}$$

The phase shift is $\frac{\pi}{3}$ and the period is $\frac{2\pi}{3} - \frac{\pi}{3} = \frac{\pi}{3}$. See figure 19.22.

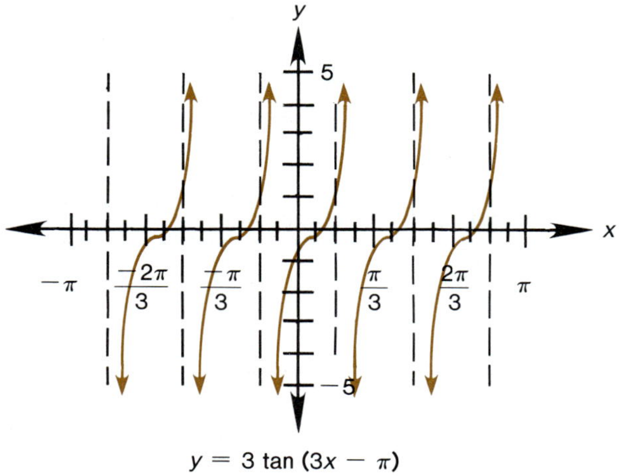

Figure 19.22

To sketch the graphs of the remaining three functions—$y = \cot x$, $y = \csc x$, and $y = \sec x$—we use the fact that they are reciprocals of $y = \tan x$, $y = \sin x$, and $y = \cos x$, respectively. Because of this relationship two very important conditions exist. The graphs of $y = \cot x$, $y = \csc x$, and $y = \sec x$:

(1) increase where their reciprocals decrease and decrease where their reciprocals increase, and

(2) have vertical asymptotes at values of x where the graphs of their reciprocals cross the x-axis.

Consider the graph of $y = \cot x$. The $\tan x$ increases (gets greater and greater) as x increases for all real number values of x. Then $\cot x$ decreases (gets lesser and lesser) as x increases, for all real number values of x. Also, since $\tan 0 = 0$ and $\tan \pi = 0$ and the graph of $y = \tan x$ crosses the x-axis at these points, then the graph of $y = \cot x$ has vertical asymptotes at these values of x. We graph $y = \cot x$ using these facts in figure 19.23.

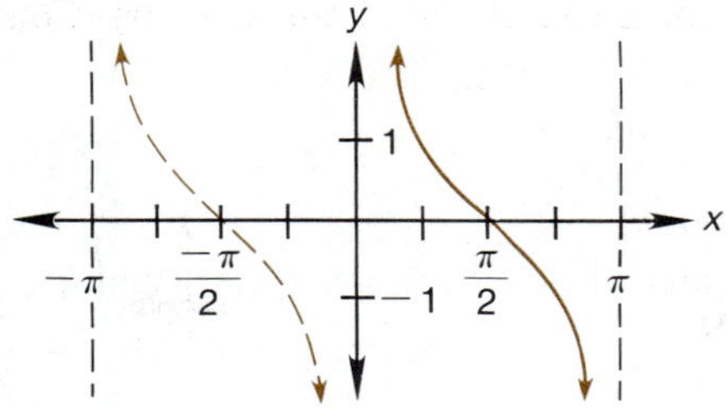

Figure 19.23

note

1. We again have guide points at $\left(\frac{\pi}{4}, 1\right)$ and $\left(\frac{3\pi}{4}, -1\right)$.
2. In the complete graph of $y = \cot x$, the lines $x = \pi k$, where k is an integer, are vertical asymptotes.
3. A cycle runs from 0 to π and the asymptotes are at the ends of the period.

Similarly, the following facts can be determined for the graphs of $y = \csc x$ and $y = \sec x$:

1. The lines $y = \pi k$, where k is an integer, are vertical asymptotes in the graph of $y = \csc x$.
2. The lines $y = \frac{\pi}{2} + \pi k$, where k is an integer, are vertical asymptotes in the graph of $y = \sec x$.
3. The period of both functions is 2π units.
4. Since $-1 \leqslant \sin x \leqslant 1$ and $-1 \leqslant \cos x \leqslant 1$, then
   a. $\csc x \leqslant -1$ or $\csc x \geqslant 1$
   b. $\sec x \leqslant -1$ or $\sec x \geqslant 1$

note

The reciprocal of a number between -1 and 1 is a number that is less than -1 or greater than 1.

Using these facts we graph $y = \csc x$ and $y = \sec x$ in figure 19.24. Notice that we have sketched first the graphs of $y = \sin x$ and $y = \sec x$ (in dotted lines). Using these graphs we then locate vertical asymptotes, where the graph crosses the x-axis, and then graph $y = \csc x$ and $y = \sec x$. In general, to graph $y = a \csc (bx + c)$ and $y = a \sec (bx + c)$, we graph first $y = a \sin (bx + c)$ and $y = a \cos (bx + c)$, respectively.

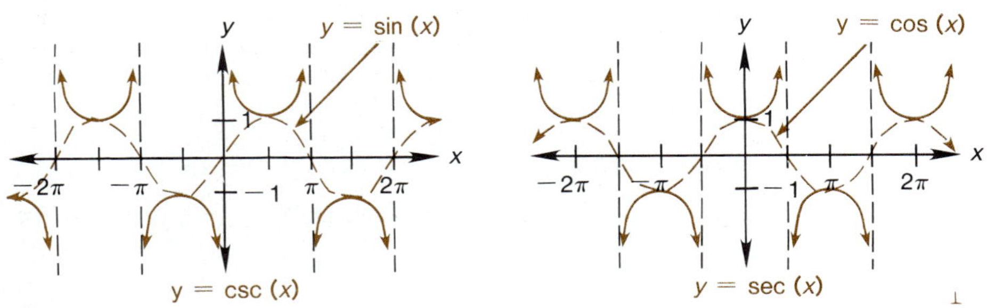

Figure 19.24

Examples Sketch the graphs of the following functions.

1. $y = \csc\left(2x - \frac{\pi}{2}\right)$ (See figure 19.25.)

We first sketch the graph of $y = \sin\left(2x - \frac{\pi}{2}\right)$.

Now,

$$0 \leqslant 2x - \frac{\pi}{2} \leqslant 2\pi$$

$$\frac{\pi}{2} \leqslant 2x \leqslant \frac{5\pi}{2}$$

$$\frac{\pi}{4} \leqslant x \leqslant \frac{5\pi}{4}$$

So the phase shift is $\frac{\pi}{4}$, amplitude = 1, and period = π.

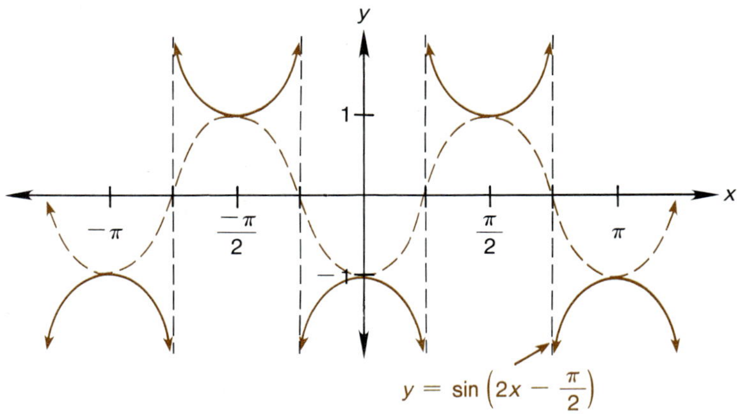

Figure 19.25

2. $y = 2 \sec\left(x + \frac{\pi}{2}\right)$ (See figure 19.26.)

We first sketch the graph of $y = 2 \cos\left(x + \frac{\pi}{2}\right)$.

Now,

$$0 \leqslant x + \frac{\pi}{2} \leqslant 2\pi$$

$$\frac{-\pi}{2} \leqslant x \leqslant \frac{3\pi}{2}$$

So the phase shift is $\frac{-\pi}{2}$, period = 2π, and amplitude = 2.

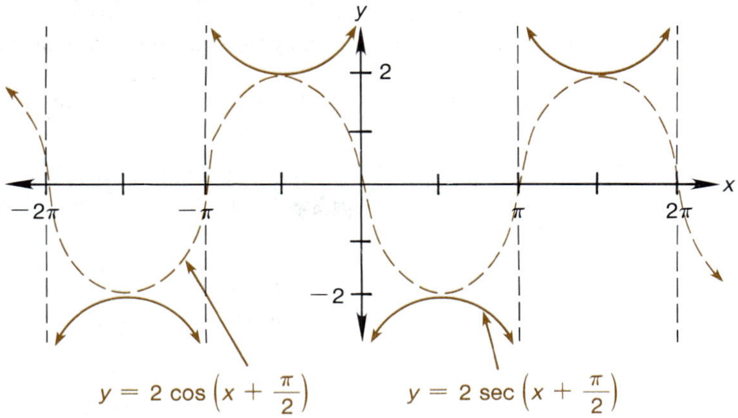

Figure 19.26

3. $y = 3\cot\left(\frac{1}{2}x\right)$ (See figure 19.27.)

Now,

$$0 \leqslant \frac{1}{2}x \leqslant \pi$$

$$0 \leqslant x \leqslant 2\pi$$

So the period $= 2\pi$ with no phase shift.
Notice, the asymptotes are now at -2π, 0, and 2π, endpoints of a period.

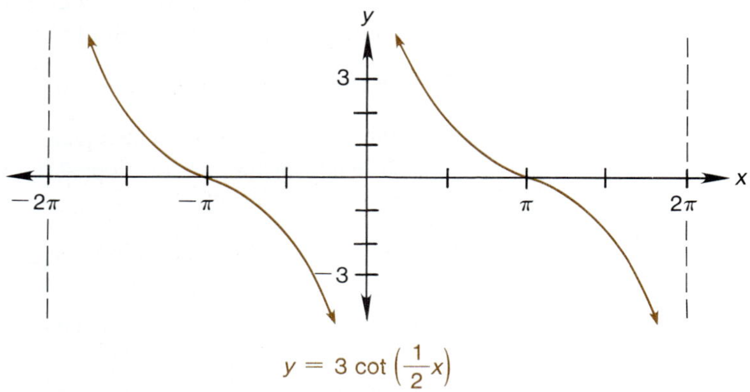

Figure 19.27

Exercise Set 19.4

Sketch the graph of each of the following functions over one period.

1. $y = \tan 3x$
•2. $y = \cot 2x$
3. $y = \csc \frac{1}{2}x$
4. $y = \sec \frac{1}{4}x$
5. $y = -\tan x$
6. $y = -\cot \frac{1}{2}x$
7. $y = 3 \csc x$
8. $y = -2 \sec x$
9. $y = \frac{1}{2} \cot(3x)$
•10. $y = 2 \tan 3x$
•11. $y = 2 \sec \frac{1}{3}x$
12. $y = -4 \csc 2x$
•13. $y = \csc\left(x + \frac{\pi}{4}\right)$
14. $y = \sec\left(x + \frac{\pi}{4}\right)$
15. $y = \tan\left(x - \frac{\pi}{4}\right)$
16. $y = \cot\left(x - \frac{\pi}{2}\right)$
17. $y = \sec(3x - \pi)$
18. $y = 3 \csc\left(x + \frac{\pi}{2}\right)$
19. $y = \cot\left(3x + \frac{\pi}{2}\right)$
•20. $y = 2 \tan(2x - \pi)$

Chapter Summary

1. Angles may be measured by real numbers called *radians*.
2. *One radian* is the measure of a central angle of a circle that intercepts an arc of the circle equal to the length of the radius of the circle.
3. A central angle, θ, of a circle has measure in radians given by $\theta^R = \frac{s}{r}$, where s is the length of the arc intercepted by the angle and r is the length of the radius.
4. The length of the arc intercepted by a central angle of a circle of measure θ^R is given by $s = r \cdot \theta^R$.
5. The area of a sector of a circle is given by $A = \frac{\theta^R \cdot r^2}{2}$.
6. $1° = 1° \cdot \frac{\pi}{180°} \approx 0.01745$ radians

   $1 \text{ rad} = 1 \text{ rad} \cdot \frac{180°}{\pi} \approx 57.3°$
7. To convert angle θ measured in radians to degrees, multiply by $\frac{180°}{\pi}$; in degrees to radians, multiply by $\frac{\pi}{180°}$.

8. The **angular velocity,** ω, of a point on a circle rotating at a constant rate is given by $\omega = \dfrac{\theta^R}{t}$, where θ^R is the angle through which the point rotates and t is a unit of time.
9. The **linear velocity,** v, of a point on a circle rotating at a constant rate is given by $v = r \cdot \omega$, where r is the radius of the circle expressed in linear measure per radian and ω is the angular velocity expressed in radians per unit of time.
10. The six trigonometric functions are periodic functions with (a) sine, cosine, secant and cosecant functions having period 2π units, and (b) tangent and cotangent functions having period π units.
11. The graphs of $y = \sin x$ and $y = \cos x$ produce the *sine wave*.
12. The graphs of $y = a \sin(bx + c)$ and $y = a \cos(bx + c)$ have amplitude $= |a|$, period $= \dfrac{2\pi}{b}$ $(b > 0)$, and phase shift $= -\dfrac{c}{b}$.
13. The graphs of $y = a \tan(bx + c)$ and $y = a \cot(bx + c)$ have period $\dfrac{\pi}{b}$ $(b > 0)$, and phase shift $= -\dfrac{c}{b}$, and no amplitude.

Chapter Review

[19.1] Convert the following angle measured in degrees to radians (to the nearest tenth).

1. 65°   2. 135°   3. 222.3°   4. 292°42′

Convert the following angles measured in radians to degrees (correct to the nearest tenth).

5. $\dfrac{7\pi}{8}$ 6. $\dfrac{13\pi}{4}$ 7. 1.72   8. 4.56

[19.2] Find the length of the arc intercepted by a central angle of a circle having: (Round to the accuracy of r.)

9. $r = 7.3$ cm and $\theta^R = 3.4$
10. $r = 10.9$ m and $\theta^\circ = 150$
11. $r = 0.17$ ft and $\theta^\circ = 81.7$

Find the area of a sector of a circle defined by a central angle of a circle given: (Round to the accuracy of r.)

12. $r = 4.2$ cm and $\theta^R = 3.7$
13. $r = 33.6$ mm and $\theta^\circ = 60$
14. $r = 6.73$ in and $\theta^\circ = 39.4$

Find the linear velocity, v, or angular velocity, ω, in the indicated units of measure. (Round to the accuracy of the measured data.)

15. Find v in centimeters per minute if $r = 16.2$ cm and $\omega = 20.0$ radians per minute.
16. Find ω in radians per second if $r = 1.700$ m and $v = 210.1$ m/s.
17. Find v in feet per second if the diameter $d = 14.2$ ft and $\omega = 301$ rev/s.
18. Find ω in degrees per second if $d = 240$ mm and $v = 31.2$ cm/s.
19. Find ω in revolutions per minute if $r = 7.90$ cm and $v = 61.2$ cm/s.
20. A belt runs a pulley of radius 5 in at 65 rev/min. Find the angular velocity of the pulley in radians per minute.
21. Suppose a point on a disk of radius 10 cm is rotating with an angular velocity of $\frac{\pi}{9}$ radians per second. Find the angle the point goes through in 5 s.
22. Find the linear velocity of a point on the edge of a flywheel with diameter 12 in if the flywheel is rotating 45 times per second.
23. An automobile wheel is rotating at 800 rev/min. Find:
    a. The angular velocity in radians per minute of a point on the tire tread 30 cm from the center
    b. The linear velocity of the tire tread in meters per minute
    c. The auto's speed in kilometers per hour
24. The film feeding onto the windup reel of a movie projector is moving at a constant linear velocity of 29.8 cm/s. When the radius of the film on the reel is 9.2 cm, find the angular velocity in (a) radians per second, (b) revolutions per second, (c) degrees per second. (Round to the nearest 0.1°.)

[19.3, 19.4] Determine the amplitude (where it exists), the period, and the phase shift, if there is any, for the graphs of the following functions.

25. $y = 3\sin\left(x - \frac{\pi}{4}\right)$
26. $y = -5\cos(2x + \pi)$
27. $y = \frac{1}{2}\tan\left(x - \frac{3\pi}{4}\right)$
28. $y = -\frac{1}{4}\sin(4x + 2\pi)$
29. $y = 1.5\cos\left(\pi x + \frac{\pi}{3}\right)$
30. $y = 4\tan(4x - \pi)$
31. $y = -\sin\left(2\pi x + \frac{3\pi}{4}\right)$
32. $y = -2\tan(3x)$
33. $y = 3\csc\left(3x - \frac{\pi}{2}\right)$
34. $y = \sec\left(\frac{\pi}{2}x\right)$
35. $y = 0.8\cot\left(\frac{1}{4}x + \pi\right)$
36. $y = 6\cos\left(\frac{\pi}{4}x - \pi\right)$

Sketch the graph of the following functions over two periods.

37. $y = 3 \sin x$

38. $y = -2 \cos x$

39. $y = \frac{1}{2} \cos 3x$

40. $y = -\sin \frac{1}{2}x$

41. $y = \sin\left(x - \frac{\pi}{4}\right)$

42. $y = \cos(x + \pi)$

43. $y = -2 \sin(2x + \pi)$

44. $y = 3 \cos\left(\frac{1}{3}x + \frac{\pi}{2}\right)$

45. $y = \frac{3}{2} \cos\left(4x - \frac{\pi}{2}\right)$

46. $y = 3 \tan\left(\frac{1}{2}x\right)$

47. $y = -\frac{1}{2} \sin\left(\frac{2}{3}x - \frac{\pi}{2}\right)$

48. $y = -4 \cos\left(\frac{3}{2}x - \frac{\pi}{2}\right)$

49. $y = \csc\left(x - \frac{\pi}{4}\right)$

50. $y = \sec\left(x - \frac{\pi}{2}\right)$

51. $y = \cot\left(x + \frac{\pi}{4}\right)$

Chapter Test

1. Convert 32.4° to radians correct to the nearest hundredth.
2. Convert $\frac{7\pi}{6}$ radians to degrees.

In exercises 3 through 7, round your answer to the accuracy of the given data.

3. Find the length of an arc of a circle having radius 2.4 in intercepted by a central angle of the circle of 70°.
4. Find the measure in degrees of a central angle of a circle that intercepts an arc 23.0 mm long if the circle has radius 15.6 mm.
5. Find the angular velocity ω in radians per second of a point on the circumference of a circle having radius 16.3 cm if the linear velocity v is 52.5 cm/s.
6. Mike Casey is riding a racing bicycle at a speed of 41.2 km/h. The wheels of the bicycle are 70.0 cm in diameter. What is the angular velocity of the wheels in radians per minute?

7. A machinist wishes to cut out the shaded portion of the metal disk in the accompanying figure. Find the area of the portion to be cut out.

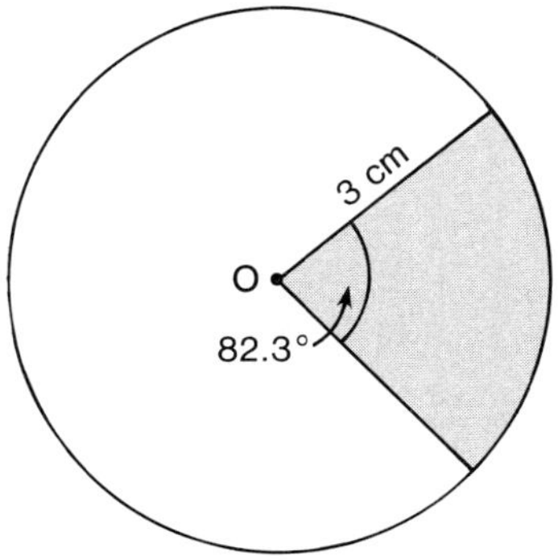

8. What is the amplitude and period length of the graph of $y = -3\sin(2x)$?
9. What is the period length and phase shift of the graph of
$$y = 2\tan\left(3x - \frac{\pi}{2}\right)?$$
10. What is the amplitude, phase shift, and period length of the graph of
$$y = 4\cos\left(\frac{3}{4}x + \pi\right)?$$

Sketch the graph of the following functions over one period.

11. $y = 2\sin\left(x + \frac{\pi}{4}\right)$
12. $y = -3\cos(2x - \pi)$
13. $y = 2\tan\left(\frac{1}{2}x\right)$
14. $y = \cot\left(x - \frac{\pi}{4}\right)$
15. $y = \csc\left(x + \frac{\pi}{2}\right)$

Appendix

Table A Squares, Square Roots, and Prime Factors

Number a	Square a^2	Square* Root $\sqrt{a}$	Prime Factors	Number a	Square a^2	Square* Root $\sqrt{a}$	Prime Factors
1	1	1.000		51	2,601	7.141	$3 \cdot 17$
2	4	1.414	2	52	2,704	7.211	$2^2 \cdot 13$
3	9	1.732	3	53	2,809	7.280	53
4	16	2.000	2^2	54	2,916	7.348	$2 \cdot 3^3$
5	25	2.236	5	55	3,025	7.416	$5 \cdot 11$
6	36	2.449	$2 \cdot 3$	56	3,136	7.483	$2^3 \cdot 7$
7	49	2.646	7	57	3,249	7.550	$3 \cdot 19$
8	64	2.828	2^3	58	3,364	7.616	$2 \cdot 29$
9	81	3.000	3^2	59	3,481	7.681	59
10	100	3.162	$2 \cdot 5$	60	3,600	7.746	$2^2 \cdot 3 \cdot 5$
11	121	3.317	11	61	3,721	7.810	61
12	144	3.464	$2^2 \cdot 3$	62	3,844	7.874	$2 \cdot 31$
13	169	3.606	13	63	3,969	7.937	$3^2 \cdot 7$
14	196	3.742	$2 \cdot 7$	64	4,096	8.000	2^6
15	225	3.873	$3 \cdot 5$	65	4,225	8.062	$5 \cdot 13$
16	256	4.000	2^4	66	4,356	8.124	$2 \cdot 3 \cdot 11$
17	289	4.123	17	67	4,489	8.185	67
18	324	4.243	$2 \cdot 3^2$	68	4,624	8.246	$2^2 \cdot 17$
19	361	4.359	19	69	4,761	8.307	$3 \cdot 23$
20	400	4.472	$2^2 \cdot 5$	70	4,900	8.367	$2 \cdot 5 \cdot 7$
21	441	4.583	$3 \cdot 7$	71	5,041	8.426	71
22	484	4.690	$2 \cdot 11$	72	5,184	8.485	$2^3 \cdot 3^2$
23	529	4.796	23	73	5,329	8.544	73
24	576	4.899	$2^3 \cdot 3$	74	5,476	8.602	$2 \cdot 37$
25	625	5.000	5^2	75	5,625	8.660	$3 \cdot 5^2$
26	676	5.099	$2 \cdot 13$	76	5,776	8.718	$2^2 \cdot 19$
27	729	5.196	3^3	77	5,929	8.775	$7 \cdot 11$
28	784	5.292	$2^2 \cdot 7$	78	6,084	8.832	$2 \cdot 3 \cdot 13$
29	841	5.385	29	79	6,241	8.888	79
30	900	5.477	$2 \cdot 3 \cdot 5$	80	6,400	8.944	$2^4 \cdot 5$
31	961	5.568	31	81	6,561	9.000	3^4
32	1,024	5.657	2^5	82	6,724	9.055	$2 \cdot 41$
33	1,089	5.745	$3 \cdot 11$	83	6,889	9.110	83
34	1,156	5.831	$2 \cdot 17$	84	7,056	9.165	$2^2 \cdot 3 \cdot 7$
35	1,225	5.916	$5 \cdot 7$	85	7,225	9.220	$5 \cdot 17$
36	1,296	6.000	$2^2 \cdot 3^2$	86	7,396	9.274	$2 \cdot 43$
37	1,369	6.083	37	87	7,569	9.327	$3 \cdot 29$
38	1,444	6.164	$2 \cdot 19$	88	7,744	9.381	$2^3 \cdot 11$
39	1,521	6.245	$3 \cdot 13$	89	7,921	9.434	89
40	1,600	6.325	$2^3 \cdot 5$	90	8,100	9.487	$2 \cdot 3^2 \cdot 5$
41	1,681	6.403	41	91	8,281	9.539	$7 \cdot 13$
42	1,764	6.481	$2 \cdot 3 \cdot 7$	92	8,464	9.592	$2^2 \cdot 23$
43	1,849	6.557	43	93	8,649	9.644	$3 \cdot 31$
44	1,936	6.633	$2^2 \cdot 11$	94	8,836	9.695	$2 \cdot 47$
45	2,025	6.708	$3^2 \cdot 5$	95	9,025	9.747	$5 \cdot 19$
46	2,116	6.782	$2 \cdot 23$	96	9,216	9.798	$2^5 \cdot 3$
47	2,209	6.856	47	97	9,409	9.849	97
48	2,304	6.928	$2^4 \cdot 3$	98	9,604	9.899	$2 \cdot 7^2$
49	2,401	7.000	7^2	99	9,801	9.950	$3^2 \cdot 11$
50	2,500	7.071	$2 \cdot 5^2$	100	10,000	10.000	$2^2 \cdot 5^2$

*All square roots have been rounded to three decimal places. $\pi = 3.142$**

**Rounded to three decimal places.

Table B The Trigonometric Functions for Radian and Degree Measure of Angle α

Angle α Degrees	Angle α Radians	sin α	cos α	tan α	csc α	sec α	cot α	Radians	Degrees
0° 00′	0.0000	0.0000	1.0000	0.0000	undef.	1.000	undef.	1.5708	90° 00′
10	0.0029	0.0029	1.0000	0.0029	343.8	1.000	343.8	1.5679	50
20	0.0058	0.0058	1.0000	0.0058	171.9	1.000	171.9	1.5650	40
30	0.0087	0.0087	1.0000	0.0087	114.6	1.000	114.6	1.5621	30
40	0.0116	0.0116	0.9999	0.0116	85.95	1.000	85.94	1.5592	20
50	0.0145	0.0145	0.9999	0.0145	68.76	1.000	68.75	1.5563	10
1° 00′	0.0175	0.0175	0.9998	0.0175	57.30	1.000	57.29	1.5533	89° 00′
10	0.0204	0.0204	0.9998	0.0204	49.11	1.000	49.10	1.5504	50
20	0.0233	0.0233	0.9997	0.0233	42.98	1.000	42.96	1.5475	40
30	0.0262	0.0262	0.9997	0.0262	38.20	1.000	38.19	1.5446	30
40	0.0291	0.0291	0.9996	0.0291	34.38	1.000	34.37	1.5417	20
50	0.0320	0.0320	0.9995	0.0320	31.26	1.001	31.24	1.5388	10
2° 00′	0.0349	0.0349	0.9994	0.0349	28.65	1.001	28.64	1.5359	88° 00′
10	0.0378	0.0378	0.9993	0.0378	26.45	1.001	26.43	1.5330	50
20	0.0407	0.0407	0.9992	0.0407	24.56	1.001	24.54	1.5301	40
30	0.0436	0.0436	0.9990	0.0437	22.93	1.001	22.90	1.5272	30
40	0.0465	0.0465	0.9989	0.0466	21.49	1.001	21.47	1.5243	20
50	0.0495	0.0494	0.9988	0.0495	20.23	1.001	20.21	1.5213	10
3° 00′	0.0524	0.0523	0.9986	0.0524	19.11	1.001	19.08	1.5184	87° 00′
10	0.0553	0.0552	0.9985	0.0553	18.10	1.002	18.07	1.5155	50
20	0.0582	0.0581	0.9983	0.0582	17.20	1.002	17.17	1.5126	40
30	0.0611	0.0610	0.9981	0.0612	16.38	1.002	16.35	1.5097	30
40	0.0640	0.0640	0.9980	0.0641	15.64	1.002	15.60	1.5068	20
50	0.0669	0.0669	0.9978	0.0670	14.96	1.002	14.92	1.5039	10
4° 00′	0.0698	0.0698	0.9976	0.0699	14.34	1.002	14.30	1.5010	86° 00′
10	0.0727	0.0727	0.9974	0.0729	13.76	1.003	13.73	1.4981	50
20	0.0765	0.0756	0.9971	0.0758	13.23	1.003	13.20	1.4952	40
30	0.0785	0.0785	0.9969	0.0787	12.75	1.003	12.71	1.4923	30
40	0.0814	0.0814	0.9967	0.0816	12.29	1.003	12.25	1.4893	20
50	0.0844	0.0843	0.9964	0.0846	11.87	1.004	11.83	1.4864	10
5° 00′	0.0873	0.0872	0.9962	0.0875	11.47	1.004	11.43	1.4835	85° 00′
10	0.0902	0.0901	0.9959	0.0904	11.10	1.004	11.06	1.4806	50
20	0.0931	0.0929	0.9957	0.0934	10.76	1.004	10.71	1.4777	40
30	0.0960	0.0958	0.9954	0.0963	10.43	1.005	10.39	1.4748	30
40	0.0989	0.0987	0.9951	0.0992	10.13	1.005	10.08	1.4719	20
50	0.1018	0.1016	0.9948	0.1022	9.839	1.005	9.788	1.4690	10
		cos α	sin α	cot α	sec α	csc α	tan α	Radians (Angle α)	Degrees (Angle α)

*Table values to four significant digits—figured on a TI-55 Texas Instrument calculator.

Table B (continued)

Angle α Degrees	Angle α Radians	sin α	cos α	tan α	csc α	sec α	cot α	Radians	Degrees
6° 00′	0.1047	0.1045	0.9945	0.1051	9.567	1.006	9.514	1.4661	84° 00′
10	0.1076	0.1074	0.9942	0.1080	9.309	1.006	9.255	1.4632	50
20	0.1105	0.1103	0.9939	0.1110	9.065	1.006	9.010	1.4603	40
30	0.1134	0.1132	0.9936	0.1139	8.834	1.006	8.777	1.4573	30
40	0.1164	0.1161	0.9932	0.1169	8.614	1.007	8.556	1.4544	20
50	0.1193	0.1190	0.9929	0.1198	8.405	1.007	8.345	1.4515	10
7° 00′	0.1222	0.1219	0.9925	0.1228	8.206	1.008	8.144	1.4486	83° 00′
10	0.1251	0.1248	0.9922	0.1257	8.016	1.008	7.953	1.4457	50
20	0.1280	0.1276	0.9918	0.1287	7.834	1.008	7.770	1.4428	40
30	0.1309	0.1305	0.9914	0.1317	7.661	1.009	7.596	1.4399	30
40	0.1338	0.1334	0.9911	0.1346	7.496	1.009	7.429	1.4370	20
50	0.1367	0.1363	0.9907	0.1376	7.337	1.009	7.269	1.4341	10
8° 00′	0.1396	0.1392	0.9903	0.1405	7.185	1.010	7.115	1.4312	82° 00′
10	0.1425	0.1421	0.9899	0.1435	7.040	1.010	6.968	1.4283	50
20	0.1454	0.1449	0.9894	0.1465	6.900	1.011	6.827	1.4254	40
30	0.1484	0.1478	0.9890	0.1495	6.765	1.011	6.691	1.4224	30
40	0.1513	0.1507	0.9886	0.1524	6.636	1.012	6.561	1.4195	20
50	0.1542	0.1536	0.9881	0.1554	6.512	1.012	6.435	1.4166	10
9° 00′	0.1571	0.1564	0.9877	0.1584	6.392	1.012	6.314	1.4137	81° 00′
10	0.1600	0.1593	0.9872	0.1614	6.277	1.013	6.197	1.4108	50
20	0.1629	0.1622	0.9868	0.1644	6.166	1.013	6.084	1.4079	40
30	0.1658	0.1650	0.9863	0.1673	6.059	1.014	5.976	1.4050	30
40	0.1687	0.1679	0.9858	0.1703	5.955	1.014	5.871	1.4021	20
50	0.1716	0.1708	0.9853	0.1733	5.855	1.015	5.769	1.3992	10
10° 00′	0.1745	0.1736	0.9848	0.1763	5.759	1.015	5.671	1.3963	80° 00′
10	0.1774	0.1765	0.9843	0.1793	5.665	1.016	5.576	1.3934	50
20	0.1804	0.1794	0.9838	0.1823	5.575	1.016	5.485	1.3904	40
30	0.1833	0.1822	0.9833	0.1853	5.487	1.017	5.396	1.3875	30
40	0.1862	0.1851	0.9827	0.1883	5.403	1.018	5.309	1.3846	20
50	0.1891	0.1880	0.9822	0.1914	5.320	1.018	5.226	1.3817	10
11° 00′	0.1920	0.1908	0.9816	0.1944	5.241	1.019	5.145	1.3788	79° 00′
10	0.1949	0.1937	0.9811	0.1974	5.164	1.019	5.066	1.3759	50
20	0.1978	0.1965	0.9805	0.2004	5.089	1.020	4.989	1.3730	40
30	0.2007	0.1994	0.9799	0.2035	5.016	1.020	4.915	1.3701	30
40	0.2036	0.2022	0.9793	0.2065	4.945	1.021	4.843	1.3672	20
50	0.2065	0.2051	0.9787	0.2095	4.876	1.022	4.773	1.3643	10
		cos α	sin α	cot α	sec α	csc α	tan α	Radians (Angle α)	Degrees (Angle α)

Table B (continued)

Angle α									
Degrees	Radians	sin α	cos α	tan α	csc α	sec α	cot α	Radians	Degrees
12° 00′	0.2094	0.2079	0.9781	0.2126	4.810	1.022	4.705	1.3614	78° 00′
10	0.2123	0.2108	0.9775	0.2156	4.745	1.023	4.638	1.3584	50
20	0.2153	0.2136	0.9769	0.2186	4.682	1.024	4.574	1.3555	40
30	0.2182	0.2164	0.9763	0.2217	4.620	1.024	4.511	1.3526	30
40	0.2211	0.2193	0.9757	0.2247	4.560	1.025	4.449	1.3497	20
50	0.2240	0.2221	0.9750	0.2278	4.502	1.026	4.390	1.3468	10
13° 00′	0.2269	0.2250	0.9744	0.2309	4.445	1.026	4.331	1.3439	77° 00′
10	0.2298	0.2278	0.9737	0.2339	4.390	1.027	4.275	1.3410	50
20	0.2327	0.2306	0.9730	0.2370	4.336	1.028	4.219	1.3381	40
30	0.2356	0.2334	0.9724	0.2401	4.284	1.028	4.165	1.3352	30
40	0.2385	0.2363	0.9717	0.2432	4.232	1.029	4.113	1.3323	20
50	0.2414	0.2391	0.9710	0.2462	4.182	1.030	4.061	1.3294	10
14° 00′	0.2443	0.2419	0.9703	0.2493	4.134	1.031	4.011	1.3265	76° 00′
10	0.2473	0.2447	0.9696	0.2524	4.086	1.031	3.962	1.3235	50
20	0.2502	0.2476	0.9689	0.2555	4.039	1.032	3.914	1.3206	40
30	0.2531	0.2504	0.9681	0.2586	3.994	1.033	3.867	1.3177	30
40	0.2560	0.2532	0.9674	0.2617	3.950	1.034	3.821	1.3148	20
50	0.2589	0.2560	0.9667	0.2648	3.906	1.034	3.776	1.3119	10
15° 00′	0.2618	0.2588	0.9659	0.2679	3.864	1.035	3.732	1.3090	75° 00′
10	0.2647	0.2616	0.9652	0.2711	3.822	1.036	3.689	1.3061	50
20	0.2676	0.2644	0.9644	0.2742	3.782	1.037	3.647	1.3032	40
30	0.2705	0.2672	0.9636	0.2773	3.742	1.038	3.606	1.3003	30
40	0.2734	0.2700	0.9628	0.2805	3.703	1.039	3.566	1.2974	20
50	0.2763	0.2728	0.9621	0.2836	3.665	1.039	3.526	1.2945	10
16° 00′	0.2793	0.2756	0.9613	0.2867	3.628	1.040	3.487	1.2915	74° 00′
10	0.2822	0.2784	0.9605	0.2899	3.592	1.041	3.450	1.2886	50
20	0.2851	0.2812	0.9596	0.2931	3.556	1.042	3.412	1.2857	40
30	0.2880	0.2840	0.9588	0.2962	3.521	1.043	3.376	1.2828	30
40	0.2909	0.2868	0.9580	0.2994	3.487	1.044	3.340	1.2799	20
50	0.2938	0.2896	0.9572	0.3026	3.453	1.045	3.305	1.2770	10
17° 00′	0.2967	0.2924	0.9563	0.3057	3.420	1.046	3.271	1.2741	73° 00′
10	0.2996	0.2952	0.9555	0.3089	3.388	1.047	3.237	1.2712	50
20	0.3025	0.2979	0.9546	0.3121	3.356	1.048	3.204	1.2683	40
30	0.3054	0.3007	0.9537	0.3153	3.326	1.048	3.172	1.2654	30
40	0.3083	0.3035	0.9528	0.3185	3.295	1.049	3.140	1.2625	20
50	0.3113	0.3062	0.9520	0.3217	3.265	1.050	3.108	1.2595	10
		cos α	sin α	cot α	sec α	csc α	tan α	Radians	Degrees
								Angle α	

Table B (continued)

Angle α									
Degrees	Radians	sin α	cos α	tan α	csc α	sec α	cot α	Radians	Degrees
18° 00′	0.3142	0.3090	0.9511	0.3249	3.236	1.051	3.078	1.2566	72° 00′
10	0.3171	0.3118	0.9502	0.3281	3.207	1.052	3.047	1.2537	50
20	0.3200	0.3145	0.9492	0.3314	3.179	1.053	3.018	1.2508	40
30	0.3229	0.3173	0.9483	0.3346	3.152	1.054	2.989	1.2479	30
40	0.3258	0.3201	0.9474	0.3378	3.124	1.056	2.960	1.2450	20
50	0.3287	0.3228	0.9465	0.3411	3.098	1.057	2.932	1.2421	10
19° 00′	0.3316	0.3256	0.9455	0.3443	3.072	1.058	2.904	1.2392	71° 00′
10	0.3345	0.3283	0.9446	0.3476	3.046	1.059	2.877	1.2363	50
20	0.3374	0.3311	0.9436	0.3508	3.021	1.060	2.850	1.2334	40
30	0.3403	0.3338	0.9426	0.3541	2.996	1.061	2.824	1.2305	30
40	0.3432	0.3365	0.9417	0.3574	2.971	1.062	2.798	1.2275	20
50	0.3462	0.3393	0.9407	0.3607	2.947	1.063	2.773	1.2246	10
20° 00′	0.3491	0.3420	0.9397	0.3640	2.924	1.064	2.747	1.2217	70° 00′
10	0.3520	0.3448	0.9387	0.3673	2.901	1.065	2.723	1.2188	50
20	0.3549	0.3475	0.9377	0.3706	2.878	1.066	2.699	1.2159	40
30	0.3578	0.3502	0.9367	0.3739	2.855	1.068	2.675	1.2130	30
40	0.3607	0.3529	0.9356	0.3772	2.833	1.069	2.651	1.2101	20
50	0.3636	0.3557	0.9346	0.3805	2.812	1.070	2.628	1.2072	10
21° 00′	0.3665	0.3584	0.9336	0.3839	2.790	1.071	2.605	1.2043	69° 00′
10	0.3694	0.3611	0.9325	0.3872	2.769	1.072	2.583	1.2014	50
20	0.3723	0.3638	0.9315	0.3906	2.749	1.074	2.560	1.1985	40
30	0.3752	0.3665	0.9304	0.3939	2.729	1.075	2.539	1.1956	30
40	0.3782	0.3692	0.9293	0.3973	2.709	1.076	2.517	1.1926	20
50	0.3811	0.3719	0.9283	0.4006	2.689	1.077	2.496	1.1897	10
22° 00′	0.3840	0.3746	0.9272	0.4040	2.669	1.079	2.475	1.1868	68° 00′
10	0.3869	0.3773	0.9261	0.4074	2.650	1.080	2.455	1.1839	50
20	0.3898	0.3800	0.9250	0.4108	2.632	1.081	2.434	1.1810	40
30	0.3927	0.3827	0.9239	0.4142	2.613	1.082	2.414	1.1781	30
40	0.3956	0.3854	0.9228	0.4176	2.595	1.084	2.394	1.1752	20
50	0.3985	0.3881	0.9216	0.4210	2.577	1.085	2.375	1.1723	10
23° 00′	0.4014	0.3907	0.9205	0.4245	2.559	1.086	2.356	1.1694	67° 00′
10	0.4043	0.3934	0.9194	0.4279	2.542	1.088	2.337	1.1665	50
20	0.4072	0.3961	0.9182	0.4314	2.525	1.089	2.318	1.1636	40
30	0.4102	0.3987	0.9171	0.4348	2.508	1.090	2.300	1.1606	30
40	0.4131	0.4014	0.9159	0.4383	2.491	1.092	2.282	1.1577	20
50	0.4160	0.4041	0.9147	0.4417	2.475	1.093	2.264	1.1548	10
		cos α	sin α	cot α	sec α	csc α	tan α	Radians	Degrees
								Angle α	

Table B (continued)

Angle α Degrees	Angle α Radians	sin α	cos α	tan α	csc α	sec α	cot α	Radians	Degrees
24° 00′	0.4189	0.4067	0.9135	0.4452	2.459	1.095	2.246	1.1519	66° 00′
10	0.4218	0.4094	0.9124	0.4487	2.443	1.096	2.229	1.1490	50
20	0.4247	0.4120	0.9112	0.4522	2.427	1.097	2.211	1.1461	40
30	0.4276	0.4147	0.9100	0.4557	2.411	1.099	2.194	1.1432	30
40	0.4305	0.4173	0.9088	0.4592	2.396	1.100	2.177	1.1403	20
50	0.4334	0.4200	0.9075	0.4628	2.381	1.102	2.161	1.1374	10
25° 00′	0.4363	0.4226	0.9063	0.4663	2.366	1.103	2.145	1.1345	65° 00′
10	0.4392	0.4253	0.9051	0.4699	2.352	1.105	2.128	1.1316	50
20	0.4422	0.4279	0.9038	0.4734	2.337	1.106	2.112	1.1286	40
30	0.4451	0.4305	0.9026	0.4770	2.323	1.108	2.097	1.1257	30
40	0.4480	0.4331	0.9013	0.4806	2.309	1.109	2.081	1.1228	20
50	0.4509	0.4358	0.9001	0.4841	2.295	1.111	2.066	1.1199	10
26° 00′	0.4538	0.4384	0.8988	0.4877	2.281	1.113	2.050	1.1170	64° 00′
10	0.4567	0.4410	0.8975	0.4913	2.268	1.114	2.035	1.1141	50
20	0.4596	0.4436	0.8962	0.4950	2.254	1.116	2.020	1.1112	40
30	0.4625	0.4462	0.8949	0.4986	2.241	1.117	2.006	1.1083	30
40	0.4654	0.4488	0.8936	0.5022	2.228	1.119	1.991	1.1054	20
50	0.4683	0.4514	0.8923	0.5059	2.215	1.121	1.977	1.1025	10
27° 00′	0.4712	0.4540	0.8910	0.5095	2.203	1.122	1.963	1.0996	63° 00′
10	0.4741	0.4566	0.8897	0.5132	2.190	1.124	1.949	1.0966	50
20	0.4771	0.4592	0.8884	0.5169	2.178	1.126	1.935	1.0937	40
30	0.4800	0.4617	0.8870	0.5206	2.166	1.127	1.921	1.0908	30
40	0.4829	0.4643	0.8857	0.5243	2.154	1.129	1.907	1.0879	20
50	0.4858	0.4669	0.8843	0.5280	2.142	1.131	1.894	1.0850	10
28° 00′	0.4887	0.4695	0.8829	0.5317	2.130	1.133	1.881	1.0821	62° 00′
10	0.4916	0.4720	0.8816	0.5354	2.118	1.134	1.868	1.0792	50
20	0.4945	0.4746	0.8802	0.5392	2.107	1.136	1.855	1.0763	40
30	0.4974	0.4772	0.8788	0.5430	2.096	1.138	1.842	1.0734	30
40	0.5503	0.4797	0.8774	0.5467	2.085	1.140	1.829	1.0705	20
50	0.5032	0.4823	0.8760	0.5505	2.074	1.142	1.816	1.0676	10
29° 00′	0.5061	0.4848	0.8746	0.5543	2.063	1.143	1.804	1.0647	61° 00′
10	0.5091	0.4874	0.8732	0.5581	2.052	1.145	1.792	1.0617	50
20	0.5120	0.4899	0.8718	0.5619	2.041	1.147	1.780	1.0588	40
30	0.5149	0.4924	0.8704	0.5658	2.031	1.149	1.767	1.0559	30
40	0.5178	0.4950	0.8689	0.5696	2.020	1.151	1.756	1.0530	20
50	0.5207	0.4975	0.8675	0.5735	2.010	1.153	1.744	1.0501	10
		cos α	sin α	cot α	sec α	csc α	tan α	Radians (Angle α)	Degrees (Angle α)

Table B (continued)

Angle α Degrees	Angle α Radians	sin α	cos α	tan α	csc α	sec α	cot α	Radians	Degrees
30° 00′	0.5236	0.5000	0.8660	0.5774	2.000	1.155	1.732	1.0472	60° 00′
10	0.5265	0.5025	0.8646	0.5812	1.990	1.157	1.720	1.0443	50
20	0.5294	0.5050	0.8631	0.5851	1.980	1.159	1.709	1.0414	40
30	0.5323	0.5075	0.8616	0.5890	1.970	1.161	1.698	1.0385	30
40	0.5352	0.5100	0.8601	0.5930	1.961	1.163	1.686	1.0356	20
50	0.5381	0.5125	0.8587	0.5969	1.951	1.165	1.675	1.0327	10
31° 00′	0.5411	0.5150	0.8572	0.6009	1.942	1.167	1.664	1.0297	59° 00′
10	0.5440	0.5175	0.8557	0.6048	1.932	1.169	1.653	1.0268	50
20	0.5469	0.5200	0.8542	0.6088	1.923	1.171	1.643	1.0239	40
30	0.5498	0.5225	0.8526	0.6128	1.914	1.173	1.632	1.0210	30
40	0.5527	0.5250	0.8511	0.6168	1.905	1.175	1.621	1.0181	20
50	0.5556	0.5275	0.8496	0.6208	1.896	1.177	1.611	1.0152	10
32° 00′	0.5585	0.5299	0.8480	0.6249	1.887	1.179	1.600	1.0123	58° 00′
10	0.5614	0.5324	0.8465	0.6289	1.878	1.181	1.590	1.0094	50
20	0.5643	0.5348	0.8450	0.6330	1.870	1.184	1.580	1.0065	40
30	0.5672	0.5373	0.8434	0.6371	1.861	1.186	1.570	1.0036	30
40	0.5701	0.5398	0.8418	0.6412	1.853	1.188	1.560	1.0007	20
50	0.5730	0.5422	0.8403	0.6453	1.844	1.190	1.550	0.9977	10
33° 00′	0.5760	0.5446	0.8387	0.6494	1.836	1.192	1.540	0.9948	57° 00′
10	0.5789	0.5471	0.8371	0.6536	1.828	1.195	1.530	0.9919	50
20	0.5818	0.5495	0.8355	0.6577	1.820	1.197	1.520	0.9890	40
30	0.5847	0.5519	0.8339	0.6619	1.812	1.199	1.511	0.9861	30
40	0.5876	0.5544	0.8323	0.6661	1.804	1.202	1.501	0.9832	20
50	0.5905	0.5568	0.8307	0.6703	1.796	1.204	1.492	0.9803	10
34° 00′	0.5934	0.5592	0.8290	0.6745	1.788	1.206	1.483	0.9774	56° 00′
10	0.5963	0.5616	0.8274	0.6787	1.781	1.209	1.473	0.9745	50
20	0.5992	0.5640	0.8258	0.6830	1.773	1.211	1.464	0.9716	40
30	0.6021	0.5664	0.8241	0.6873	1.766	1.213	1.455	0.9687	30
40	0.6050	0.5688	0.8225	0.6916	1.758	1.216	1.446	0.9657	20
50	0.6080	0.5712	0.8208	0.6959	1.751	1.218	1.437	0.9628	10
35° 00′	0.6109	0.5736	0.8192	0.7002	1.743	1.221	1.428	0.9599	55° 00′
10	0.6138	0.5760	0.8175	0.7046	1.736	1.223	1.419	0.9570	50
20	0.6167	0.5783	0.8158	0.7089	1.729	1.226	1.411	0.9541	40
30	0.6196	0.5807	0.8141	0.7133	1.722	1.228	1.402	0.9512	30
40	0.6225	0.5831	0.8124	0.7177	1.715	1.231	1.393	0.9483	20
50	0.6254	0.5854	0.8107	0.7221	1.708	1.233	1.385	0.9454	10
		cos α	sin α	cot α	sec α	csc α	tan α	Radians (Angle α)	Degrees (Angle α)

Table B (continued)

Angle α Degrees	Angle α Radians	sin α	cos α	tan α	csc α	sec α	cot α	Radians	Degrees
36° 00′	0.6283	0.5878	0.8090	0.7265	1.701	1.236	1.376	0.9425	54° 00′
10	0.6312	0.5901	0.8073	0.7310	1.695	1.239	1.368	0.9396	50
20	0.6341	0.5925	0.8056	0.7355	1.688	1.241	1.360	0.9367	40
30	0.6370	0.5948	0.8039	0.7400	1.681	1.244	1.351	0.9338	30
40	0.6400	0.5972	0.8021	0.7445	1.675	1.247	1.343	0.9308	20
50	0.6429	0.5995	0.8004	0.7490	1.668	1.249	1.335	0.9279	10
37° 00′	0.6458	0.6018	0.7986	0.7536	1.662	1.252	1.327	0.9250	53° 00′
10	0.6487	0.6041	0.7969	0.7581	1.655	1.255	1.319	0.9221	50
20	0.6516	0.6065	0.7951	0.7649	1.649	1.258	1.311	0.9192	40
30	0.6545	0.6088	0.7934	0.7673	1.643	1.260	1.303	0.9163	30
40	0.6574	0.6111	0.7916	0.7720	1.636	1.263	1.295	0.9134	20
50	0.6603	0.6134	0.7898	0.7766	1.630	1.266	1.288	0.9105	10
38° 00′	0.6632	0.6157	0.7880	0.7813	1.624	1.269	1.280	0.9076	52° 00′
10	0.6661	0.6180	0.7862	0.7860	1.618	1.272	1.272	0.9047	50
20	0.6690	0.6202	0.7844	0.7907	1.612	1.275	1.265	0.9018	40
30	0.6720	0.6225	0.7826	0.7954	1.606	1.278	1.257	0.8988	30
40	0.6749	0.6248	0.7808	0.8002	1.601	1.281	1.250	0.8959	20
50	0.6778	0.6271	0.7790	0.8050	1.595	1.284	1.242	0.8930	10
39° 00′	0.6807	0.6293	0.7771	0.8098	1.589	1.287	1.235	0.8901	51° 00′
10	0.6836	0.6316	0.7753	0.8146	1.583	1.290	1.228	0.8872	50
20	0.6865	0.6338	0.7735	0.8195	1.578	1.293	1.220	0.8843	40
30	0.6894	0.6361	0.7716	0.8243	1.572	1.296	1.213	0.8814	30
40	0.6923	0.6383	0.7698	0.8292	1.567	1.299	1.206	0.8785	20
50	0.6952	0.6406	0.7679	0.8342	1.561	1.302	1.199	0.8756	10
40° 00′	0.6981	0.6428	0.7660	0.8391	1.556	1.305	1.192	0.8727	50° 00′
10	0.7010	0.6450	0.7642	0.8441	1.550	1.309	1.185	0.8698	50
20	0.7039	0.6472	0.7623	0.8491	1.545	1.312	1.178	0.8668	40
30	0.7069	0.6494	0.7604	0.8541	1.540	1.315	1.171	0.8639	30
40	0.7098	0.6517	0.7585	0.8591	1.535	1.318	1.164	0.8610	20
50	0.7127	0.6539	0.7566	0.8642	1.529	1.322	1.157	0.8581	10
41° 00′	0.7156	0.6561	0.7547	0.8693	1.524	1.325	1.150	0.8552	49° 00′
10	0.7185	0.6583	0.7528	0.8744	1.519	1.328	1.144	0.8523	50
20	0.7214	0.6604	0.7509	0.8796	1.514	1.332	1.137	0.8494	40
30	0.7243	0.6626	0.7490	0.8847	1.509	1.335	1.130	0.8465	30
40	0.7272	0.6648	0.7470	0.8899	1.504	1.339	1.124	0.8436	20
50	0.7301	0.6670	0.7451	0.8952	1.499	1.342	1.117	0.8407	10
		cos α	sin α	cot α	sec α	csc α	tan α	Radians (Angle α)	Degrees (Angle α)

Table B (continued)

Angle α Degrees	Angle α Radians	sin α	cos α	tan α	csc α	sec α	cot α	Radians	Degrees
42° 00′	0.7330	0.6691	0.7431	0.9004	1.494	1.346	1.111	0.8378	48° 00′
10	0.7359	0.6713	0.7412	0.9057	1.490	1.349	1.104	0.8348	50
20	0.7389	0.6734	0.7392	0.9110	1.485	1.353	1.098	0.8319	40
30	0.7418	0.6756	0.7373	0.9163	1.480	1.356	1.091	0.8290	30
40	0.7447	0.6777	0.7353	0.9217	1.476	1.360	1.085	0.8261	20
50	0.7476	0.6799	0.7333	0.9271	1.471	1.364	1.079	0.8232	10
43° 00′	0.7505	0.6820	0.7314	0.9325	1.466	1.367	1.072	0.8203	47° 00′
10	0.7534	0.6841	0.7294	0.9380	1.462	1.371	1.066	0.8174	50
20	0.7563	0.6862	0.7274	0.9435	1.457	1.375	1.060	0.8145	40
30	0.7592	0.6884	0.7254	0.9490	1.453	1.379	1.054	0.8116	30
40	0.7621	0.6905	0.7234	0.9545	1.448	1.382	1.048	0.8087	20
50	0.7650	0.6926	0.7214	0.9601	1.444	1.386	1.042	0.8058	10
44° 00′	0.7679	0.6947	0.7193	0.9657	1.440	1.390	1.036	0.8029	46° 00′
10	0.7709	0.6967	0.7173	0.9713	1.435	1.394	1.030	0.7999	50
20	0.7738	0.6988	0.7153	0.9770	1.431	1.398	1.024	0.7970	40
30	0.7767	0.7009	0.7133	0.9827	1.427	1.402	1.018	0.7941	30
40	0.7796	0.7030	0.7112	0.9884	1.423	1.406	1.012	0.7912	20
50	0.7825	0.7050	0.7092	0.9942	1.418	1.410	1.006	0.7883	10
45° 00′	0.7854	0.7071	0.7071	1.000	1.414	1.414	1.000	0.7854	45° 00′
		cos α	sin α	cot α	sec α	csc α	tan α	Radians (Angle α)	Degrees (Angle α)

Table C The Common Logarithms to Base 10

N	0	1	2	3	4	5	6	7	8	9
1.0	0.0000	0.0043	0.0086	0.0128	0.0170	0.0212	0.0253	0.0294	0.0334	0.0374
1.1	0.0414	0.0453	0.0492	0.0531	0.0569	0.0607	0.0645	0.0682	0.0719	0.0755
1.2	0.0792	0.0828	0.0864	0.0899	0.0934	0.0969	0.1004	0.1038	0.1072	0.1106
1.3	0.1139	0.1173	0.1206	0.1239	0.1271	0.1303	0.1335	0.1367	0.1399	0.1430
1.4	0.1461	0.1492	0.1523	0.1553	0.1584	0.1614	0.1644	0.1673	0.1703	0.1732
1.5	0.1761	0.1790	0.1818	0.1847	0.1875	0.1903	0.1931	0.1959	0.1987	0.2014
1.6	0.2041	0.2068	0.2095	0.2122	0.2148	0.2175	0.2201	0.2227	0.2253	0.2279
1.7	0.2304	0.2330	0.2355	0.2380	0.2405	0.2430	0.2455	0.2480	0.2504	0.2529
1.8	0.2553	0.2577	0.2601	0.2625	0.2648	0.2672	0.2695	0.2718	0.2742	0.2765
1.9	0.2788	0.2810	0.2833	0.2856	0.2878	0.2900	0.2923	0.2945	0.2967	0.2989
2.0	0.3010	0.3032	0.3054	0.3075	0.3096	0.3118	0.3139	0.3160	0.3181	0.3201
2.1	0.3222	0.3243	0.3263	0.3284	0.3304	0.3324	0.3345	0.3365	0.3385	0.3404
2.2	0.3424	0.3444	0.3464	0.3483	0.3502	0.3522	0.3541	0.3560	0.3579	0.3598
2.3	0.3617	0.3636	0.3655	0.3674	0.3692	0.3711	0.3729	0.3747	0.3766	0.3784
2.4	0.3802	0.3820	0.3838	0.3856	0.3874	0.3892	0.3909	0.3927	0.3945	0.3962
2.5	0.3979	0.3997	0.4014	0.4031	0.4048	0.4065	0.4082	0.4099	0.4116	0.4133
2.6	0.4150	0.4166	0.4183	0.4200	0.4216	0.4232	0.4249	0.4265	0.4281	0.4298
2.7	0.4314	0.4330	0.4346	0.4362	0.4378	0.4393	0.4409	0.4425	0.4440	0.4456
2.8	0.4472	0.4487	0.4502	0.4518	0.4533	0.4548	0.4564	0.4579	0.4594	0.4609
2.9	0.4624	0.4639	0.4654	0.4669	0.4683	0.4698	0.4713	0.4728	0.4742	0.4757
3.0	0.4771	0.4786	0.4800	0.4814	0.4829	0.4843	0.4857	0.4871	0.4886	0.4900
3.1	0.4914	0.4928	0.4942	0.4955	0.4969	0.4983	0.4997	0.5011	0.5024	0.5038
3.2	0.5051	0.5065	0.5079	0.5092	0.5105	0.5119	0.5132	0.5145	0.5159	0.5172
3.3	0.5185	0.5198	0.5211	0.5224	0.5237	0.5250	0.5263	0.5276	0.5289	0.5302
3.4	0.5315	0.5328	0.5340	0.5353	0.5366	0.5378	0.5391	0.5403	0.5416	0.5428
3.5	0.5441	0.5453	0.5465	0.5478	0.5490	0.5502	0.5514	0.5527	0.5539	0.5551
3.6	0.5563	0.5575	0.5587	0.5599	0.5611	0.5623	0.5635	0.5647	0.5658	0.5670
3.7	0.5682	0.5694	0.5705	0.5717	0.5729	0.5740	0.5752	0.5763	0.5775	0.5786
3.8	0.5798	0.5809	0.5821	0.5832	0.5843	0.5855	0.5866	0.5877	0.5888	0.5899
3.9	0.5911	0.5922	0.5933	0.5944	0.5955	0.5966	0.5977	0.5988	0.5999	0.6010
4.0	0.6021	0.6031	0.6042	0.6053	0.6064	0.6075	0.6085	0.6096	0.6107	0.6117
4.1	0.6128	0.6138	0.6149	0.6160	0.6170	0.6180	0.6191	0.6201	0.6212	0.6222
4.2	0.6232	0.6243	0.6253	0.6263	0.6274	0.6284	0.6294	0.6304	0.6314	0.6325
4.3	0.6335	0.6345	0.6355	0.6365	0.6375	0.6385	0.6395	0.6405	0.6415	0.6425
4.4	0.6435	0.6444	0.6454	0.6464	0.6474	0.6484	0.6493	0.6503	0.6513	0.6522
4.5	0.6532	0.6542	0.6551	0.6561	0.6571	0.6580	0.6590	0.6599	0.6609	0.6618
4.6	0.6628	0.6637	0.6646	0.6656	0.6665	0.6675	0.6684	0.6693	0.6702	0.6712
4.7	0.6721	0.6730	0.6739	0.6749	0.6758	0.6767	0.6776	0.6785	0.6794	0.6803
4.8	0.6812	0.6821	0.6830	0.6839	0.6848	0.6857	0.6866	0.6875	0.6884	0.6893
4.9	0.6902	0.6911	0.6920	0.6928	0.6937	0.6946	0.6955	0.6964	0.6972	0.6981
5.0	0.6990	0.6998	0.7007	0.7016	0.7024	0.7033	0.7042	0.7050	0.7059	0.7067
5.1	0.7076	0.7084	0.7093	0.7101	0.7110	0.7118	0.7126	0.7135	0.7143	0.7152
5.2	0.7160	0.7168	0.7177	0.7185	0.7193	0.7202	0.7210	0.7218	0.7226	0.7235
5.3	0.7243	0.7251	0.7259	0.7267	0.7275	0.7284	0.7292	0.7300	0.7308	0.7316
5.4	0.7324	0.7332	0.7340	0.7348	0.7356	0.7364	0.7372	0.7380	0.7388	0.7396

Table C (continued)

N	0	1	2	3	4	5	6	7	8	9
5.5	0.7404	0.7412	0.7419	0.7427	0.7435	0.7443	0.7451	0.7459	0.7466	0.7474
5.6	0.7482	0.7490	0.7497	0.7505	0.7513	0.7520	0.7528	0.7536	0.7543	0.7551
5.7	0.7559	0.7566	0.7574	0.7582	0.7589	0.7597	0.7604	0.7612	0.7619	0.7627
5.8	0.7634	0.7642	0.7649	0.7657	0.7664	0.7672	0.7679	0.7686	0.7694	0.7701
5.9	0.7709	0.7716	0.7723	0.7731	0.7738	0.7745	0.7752	0.7760	0.7767	0.7774
6.0	0.7782	0.7789	0.7796	0.7803	0.7810	0.7818	0.7825	0.7832	0.7839	0.7846
6.1	0.7853	0.7860	0.7868	0.7875	0.7882	0.7889	0.7896	0.7903	0.7910	0.7917
6.2	0.7924	0.7931	0.7938	0.7945	0.7952	0.7959	0.7966	0.7973	0.7980	0.7987
6.3	0.7993	0.8000	0.8007	0.8014	0.8021	0.8028	0.8035	0.8041	0.8048	0.8055
6.4	0.8062	0.8069	0.8075	0.8082	0.8089	0.8096	0.8102	0.8109	0.8116	0.8122
6.5	0.8129	0.8136	0.8142	0.8149	0.8156	0.8162	0.8169	0.8176	0.8182	0.8189
6.6	0.8195	0.8202	0.8209	0.8215	0.8222	0.8228	0.8235	0.8241	0.8248	0.8254
6.7	0.8261	0.8267	0.8274	0.8280	0.8287	0.8293	0.8299	0.8306	0.8312	0.8319
6.8	0.8325	0.8331	0.8338	0.8344	0.8351	0.8357	0.8363	0.8370	0.8376	0.8382
6.9	0.8388	0.8395	0.8401	0.8407	0.8414	0.8420	0.8426	0.8432	0.8439	0.8445
7.0	0.8451	0.8457	0.8463	0.8470	0.8476	0.8482	0.8488	0.8494	0.8500	0.8506
7.1	0.8513	0.8519	0.8525	0.8531	0.8537	0.8543	0.8549	0.8555	0.8561	0.8567
7.2	0.8573	0.8579	0.8585	0.8591	0.8597	0.8603	0.8609	0.8615	0.8621	0.8627
7.3	0.8633	0.8639	0.8645	0.8651	0.8657	0.8663	0.8669	0.8675	0.8681	0.8686
7.4	0.8692	0.8698	0.8704	0.8710	0.8716	0.8722	0.8727	0.8733	0.8739	0.8745
7.5	0.8751	0.8756	0.8762	0.8768	0.8774	0.8779	0.8785	0.8791	0.8797	0.8802
7.6	0.8808	0.8814	0.8820	0.8825	0.8831	0.8837	0.8842	0.8848	0.8854	0.8859
7.7	0.8865	0.8871	0.8876	0.8882	0.8887	0.8893	0.8899	0.8904	0.8910	0.8915
7.8	0.8921	0.8927	0.8932	0.8938	0.8943	0.8949	0.8954	0.8960	0.8965	0.8971
7.9	0.8976	0.8982	0.8987	0.8993	0.8998	0.9004	0.9009	0.9015	0.9020	0.9025
8.0	0.9031	0.9036	0.9042	0.9047	0.9053	0.9058	0.9063	0.9069	0.9074	0.9079
8.1	0.9085	0.9090	0.9096	0.9101	0.9106	0.9112	0.9117	0.9122	0.9128	0.9133
8.2	0.9138	0.9143	0.9149	0.9154	0.9159	0.9165	0.9170	0.9175	0.9180	0.9186
8.3	0.9191	0.9196	0.9201	0.9206	0.9212	0.9217	0.9222	0.9227	0.9232	0.9238
8.4	0.9243	0.9248	0.9253	0.9258	0.9263	0.9269	0.9274	0.9279	0.9284	0.9289
8.5	0.9294	0.9299	0.9304	0.9309	0.9315	0.9320	0.9325	0.9330	0.9335	0.9340
8.6	0.9345	0.9350	0.9355	0.9360	0.9365	0.9370	0.9375	0.9380	0.9385	0.9390
8.7	0.9395	0.9400	0.9405	0.9410	0.9415	0.9420	0.9425	0.9430	0.9435	0.9440
8.8	0.9445	0.9450	0.9455	0.9460	0.9465	0.9469	0.9474	0.9479	0.9484	0.9489
8.9	0.9494	0.9499	0.9504	0.9509	0.9513	0.9518	0.9523	0.9528	0.9533	0.9538
9.0	0.9542	0.9547	0.9552	0.9557	0.9562	0.9566	0.9571	0.9576	0.9581	0.9586
9.1	0.9590	0.9595	0.9600	0.9605	0.9609	0.9614	0.9619	0.9624	0.9628	0.9633
9.2	0.9638	0.9643	0.9647	0.9652	0.9657	0.9661	0.9666	0.9671	0.9675	0.9680
9.3	0.9685	0.9689	0.9694	0.9699	0.9703	0.9708	0.9713	0.9717	0.9722	0.9727
9.4	0.9731	0.9736	0.9741	0.9745	0.9750	0.9754	0.9759	0.9763	0.9768	0.9773
9.5	0.9777	0.9782	0.9786	0.9791	0.9795	0.9800	0.9805	0.9809	0.9814	0.9818
9.6	0.9823	0.9827	0.9832	0.9836	0.9841	0.9845	0.9850	0.9854	0.9859	0.9863
9.7	0.9868	0.9872	0.9877	0.9881	0.9886	0.9890	0.9894	0.9899	0.9903	0.9908
9.8	0.9912	0.9917	0.9921	0.9926	0.9930	0.9934	0.9939	0.9943	0.9948	0.9952
9.9	0.9956	0.9961	0.9965	0.9969	0.9974	0.9978	0.9983	0.9987	0.9991	0.9996

Answers to Odd-Numbered Exercises, Chapter Reviews, and Chapter Tests

Chapter 1

Exercise 1.1

1. 633 3. 1,110 5. 1,465 7. 1,820 9. 1,665 11. 1,777 13. 2,301 15. 2,079 17. 16,535 19. 1,473,327 21. 253 23. 340 25. 17 27. 5,794 29. 2,510 31. 7,144 33. 31 35. 86 37. 13 39. 19 41. 6 ft 8 in 43. 15 ft 1 in 45. 19 h 5 min 47. 17 h 4 min 49. 9 h 51. 10 gal 53. 15 gal 2 qt 55. 3 ft 6 in 57. 4 ft 11 in 59. 2 h 33 min 61. 8 h 34 min 63. 12 gal 1 qt 65. 5 gal 2 qt 67. 12 ft 69. 13 ft 71. 19 ft 8 in 73. 2,818 kwh 75. 1,522 ohms 77. 6 ft 79. 13,168 81. 8,123

Exercise 1.2

1. $2\cdot2\cdot2 = 8$ 3. $4\cdot4\cdot4 = 64$ 5. $5\cdot5\cdot5\cdot5 = 625$ 7. $(4)(4) = 16$ 9. $3\cdot3\cdot3\cdot5\cdot5 = 675$ 11. $0\cdot0\cdot0 = 0$ 13. 3^3 15. 7^7 17. $3^2\cdot2^4$ 19. $9^2\cdot7^3\cdot5^2$ 21. 152 23. 160 25. 198 27. 266 29. 598 31. 500,486 33. 420,396 35. 1,342,192 37. 1,963,920 39. 17,258,274 41. 37,780,848 43. 8,121 45. 1,309*r*11 47. 706 49. 508*r*50 51. 75,512*r*41 53. (a) 8,640 ft, (b) 1,200 ft 55. 500 s or 8 min 20 s 57. $461\frac{1}{4}$ 59. 96 watts 128 watts 192 watts 61. 71 microfarads 63. \$63 65. 6 h 67. \$15,375

Exercise 1.3

1.

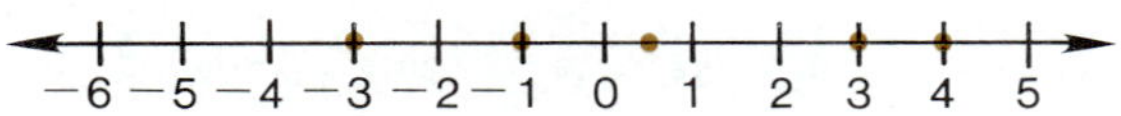

3.

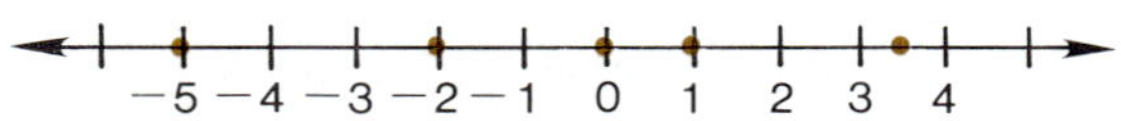

5.

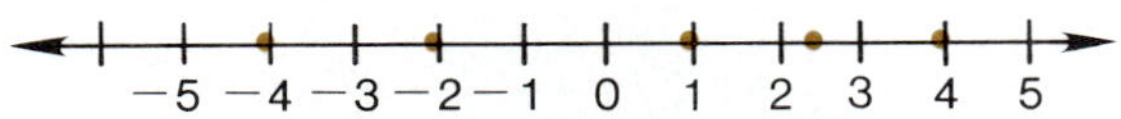

7. $-6, -4, 0, 3, 6$ 9. $-11, -7, -4, -1, 2$ 11. $-2, 3, 4, 5, 11$ 13. 4 15. 2 17. 5 19. 4 21. $\frac{1}{2}$ 23. $1\frac{1}{2}$ 25. $\frac{5}{8}$ 27. 6 29. $<$ 31. $>$ 33. $>$ 35. $<$ 37. $<$ 39. $>$ 41. $<$ 43. $<$ 45. $>$ 47. $>$

Exercise 1.4

1. 12 3. 5 5. -10 7. -5 9. -1 11. 6 13. -4 15. 3 17. 0 19. 12 21. 14 23. -6 25. -4 27. 9 29. -4 31. -8 33. -25 35. 0 37. -19 39. 14 41. -22 43. -10 45. -39 47. 128 49. 669 51. $>$ 53. $<$ 55. (a) 31°, (b) 55°, (c) $-9°$, (d) 57°, (e) $-54°$, (f) $-51°$ 57. $-\$372$ 59. (a) \$23.31, (b) \$23.59, (c) \$23.13, (d) \$23.59 61. 7°C

Exercise 1.5

1. 15 3. -28 5. -60 7. 0 9. 20 11. -105 13. 120 15. 2 17. -3 19. 0 21. 2 23. 9 25. -6 27. 16 29. 625 31. -32 33. 36 35. -4 37. 64 39. 8 41. Indeterminate 43. Undefined 45. $>$ 47. $<$ 49. $=$ 51. $4, -4$ 53. $11, -1$ 55. $3, -24$ 57. 4, 12 59. $-2, -2$ 61. $3, -3$ 63. $5, -3$ 65. 5, 5 67. $-14\frac{1}{3}$ points

69. If A = Assets, then $A - \$6 - \$6 - \$6 - \$6 - \$6 = A - \$30$. The change is $-\$30$.

Chapter Review

1. 1,780 2. 2,660 3. 3,186 4. 28,817 5. 51,564 6. 7,657 7. 2,085 8. 3,187 9. 3,446 10. 1,333 11. 9 12. 62 13. 17,204 14. 4,073,182 15. 9,269,589 16. 5,793*r*1 17. 742*r*2 18. 271 19. 11 20. $\frac{3}{4}$ 21. 4 22. -4 23. 3 24. -6 25. 1 26. -6 27. 15 28. -4 29. -9 30. 3 31. 15 32. -21 33. 12 34. 24 35. -144 36. 0 37. -7 38. 2 39. -6 40. Undefined 41. 0 42. $-9, -4$ 43. $=$ 44. (a) \$52,000 + ($-$\$3,000) + (\$$-$2,560) + ($-$\$3,300), (b) \$43,140 45. 40° 46. (a) $+9, +8, -5, -6$, (b) greater than, $+6$, (c) 58°, 69° + ($-11°$) = 58°

Chapter Test

1. -30 2. 12 3. 0 4. -13 5. 2 6. 20 7. 0 8. 36 9. 90,118 10. Indeterminate 11. 2,340 12. 1,052 13. 6 14. -3 15. 18,507 16. 18 17. 9 18. 7,159 19. -30 20. 16 21. 64 22. 426,321 23. -4 24. Undefined 25. -1 26. 1 27. -21 28. $226

Chapter 2

Exercise 2.1

1. $2\frac{4}{7}$ 3. $2\frac{3}{4}$ 5. $5\frac{7}{12}$ 7. $4\frac{3}{4}$ 9. $10\frac{5}{6}$ 11. $7\frac{1}{2}$ 13. $7\frac{4}{7}$ 15. $6\frac{2}{13}$ 17. 9 19. $7\frac{2}{8} = 7\frac{1}{4}$ 21. $3\frac{3}{6} = 3\frac{1}{2}$ 23. $7\frac{2}{4} = 7\frac{1}{2}$ 25. $2\frac{8}{16} = 2\frac{1}{2}$ 27. 2 29. $1\frac{16}{64} = 1\frac{1}{4}$ 31. $\frac{5}{2}$ 33. $\frac{20}{3}$ 35. $\frac{47}{16}$ 37. $\frac{103}{32}$ 39. $\frac{110}{21}$ 41. $\frac{41}{6}$ 43. $\frac{82}{17}$ 45. $\frac{97}{23}$ 47. $\frac{256}{32}$ 49. $\frac{60}{12}$ 51. $\frac{15}{4}$ 53. $\frac{39}{8}$ 55. $\frac{37}{4}$ 57. $\frac{177}{32}$ 59. $\frac{361}{4}$ 61. $\frac{176}{3}$ 63. 2^3 65. 2^6 67. $5 \cdot 13$ 69. $2^2 \cdot 5$ 71. $5 \cdot 41$ 73. $3 \cdot 5 \cdot 13$ 75. $2 \cdot 3 \cdot 17$ 77. $2 \cdot 3 \cdot 5 \cdot 7 \cdot 11$ 79. 15 81. 78 83. 35 85. 84 87. 228 89. 3 91. 5 93. 9 95. 9 97. 37 99. $-\frac{1}{3}$ 101. $\frac{1}{-2}$ 103. $-\frac{3}{16}$ 105. $\frac{-2}{9}$ 107. $-\frac{5}{9}$ 109. $\frac{2}{3}$ 111. $\frac{12}{5}$ 113. $\frac{40}{64}$ 115. $\frac{18}{32}$ 117. $\frac{20}{16}$ 119. $\frac{45}{60}$ 121. $\frac{92}{64}$ 123. $\frac{143}{6}$, improper fraction 125. $\frac{20}{64}$ in 127. $\frac{9}{10}$

Exercise 2.2

1. 120 3. 126 5. 144 7. 385 9. 36 11. 60 13. 126 15. 1,320 17. $\frac{7}{10}$ 19. $\frac{2}{3}$ 21. $\frac{11}{24}$ 23. $\frac{15}{8} = 1\frac{7}{8}$ 25. $\frac{23}{6} = 3\frac{5}{6}$ 27. $\frac{29}{24} = 1\frac{5}{24}$ 29. $\frac{16}{15} = 1\frac{1}{15}$ 31. $\frac{4}{5}$ 33. $\frac{5}{48}$ 35. $1\frac{1}{4}$ 37. $3\frac{9}{16}$ 39. $\frac{509}{720}$ 41. $\frac{149}{270}$ 43. $4\frac{5}{24}$ 45. $-14\frac{77}{720}$ 47. $\frac{9}{64}$ in 49. $\frac{5}{16}$ in 51. $11\frac{17}{96}$ in 53. $\frac{9}{64}$ in 55. $11\frac{15}{16}$ in 57. $5\frac{13}{20}$ amps

Exercise 2.3

1. $\frac{1}{2}$ 3. $\frac{49}{96}$ 5. $\frac{7}{12}$ 7. $\frac{15}{28}$ 9. $\frac{2}{7}$ 11. $\frac{25}{17} = 1\frac{8}{17}$ 13. $\frac{1}{5}$ 15. $\frac{51}{7} = 7\frac{2}{7}$ 17. $\frac{11}{130}$ 19. $\frac{615}{8} = 76\frac{7}{8}$ 21. 12 23. $\frac{264}{5} = 52\frac{4}{5}$ 25. $\frac{119}{87} = 1\frac{32}{87}$ 27. $\frac{68}{3} = 22\frac{2}{3}$ 29. $\frac{2}{17}$ 31. $\frac{1}{24}$ 33. $\frac{64}{45} = 1\frac{19}{45}$ 35. 5 37. (a) $187\frac{17}{48}$ cu. in, (b) $31\frac{39}{64}$ cu. in 39. $3\frac{379}{707}$ in 41. $76\frac{1}{8}$ in 43. 168 45. $10\frac{1}{2}$ v 47. $21\frac{1}{3}$

Exercise 2.4

1. 0.353 3. 40.7 5. 1,002.030 7. 23.015 9. 73.01 11. Thirty-six hundredths 13. Two and seven thousandths 15. Seven hundred three hundred thousandths 17. Fifty thousand seven and three hundred six ten thousandths 19. Forty-five and two thousand seven hundred sixty-three ten thousandths 21. 0.15 23. 0.438 25. 0.65 27. 0.76 29. 0.175 31. 0.615 33. 0.875 35. 7.167 37. 2.3 39. 5.643 41. $\frac{3}{8}$ 43. $\frac{1}{16}$ 45. $\frac{11}{16}$ 47. $\frac{23}{32}$ 49. $\frac{3}{4}$ 51. $\frac{1}{8}$ 53. $7\frac{3}{50}$ 55. $6\frac{17}{20}$ 57. $11\frac{3}{16}$

Exercise 2.5

1. 19.019 3. 540.2927 5. 13 5585 7. 156.9876 9. 1.06964 11. 9.699 13. 13.2227 15. 24.6828 17. 42,630 19. 56,076 21. 9.52816 23. 0.428412 25. 0.9100081809 27. 0.104 29. 0.57 31. 117.84 33. 1,533.64 35. 3.97 37. 0.138 in 39. 0.676 in 41. 3.466 in 43. 2,080.09 ohms 45. 22.54 v 47. 5.22 in 49. 31,643.75 lb 51. 148.8 kwh 53. 450 ma

Exercise 2.6

1. 4 3. 26 5. 3 7. 0 9. 3 11. $12\frac{1}{2}$ 13. 13 15. 6 17. 5 19. 121 21. 96 23. 3 25. 38 27. 4 29. $15\frac{5}{7}$ in^2 31. $7\frac{1}{41}$ in 33. $20\frac{28}{101}$ mi 35 $2\frac{52}{89}$ ohms 37. $23\frac{1}{3}$°C

Chapter Review

1. $\frac{5}{8}$ 2. $\frac{13}{24}$ 3. $\frac{19}{12} = 1\frac{7}{12}$ 4. 16 5. 60 6. $\frac{22}{35}$ 7. $\frac{1}{6}$ 8. $\frac{47}{36} = 1\frac{11}{36}$ 9. $\frac{239}{240}$ 10. $4\frac{1}{6}$ 11. $1\frac{1}{12}$ 12. $4\frac{9}{40}$ 13. $7\frac{37}{60}$ 14. $\frac{3}{7}$
15. $\frac{27}{14} = 1\frac{13}{14}$ 16. $\frac{16}{5} = 3\frac{1}{5}$ 17. $\frac{3}{10}$ 18. $\frac{9}{70}$ 19. $\frac{35}{36}$ 20. $3\frac{5}{33}$ 21. $\frac{2}{15}$ 22. $\frac{20}{3} = 6\frac{2}{3}$ 23. $\frac{3}{2} = 1\frac{1}{2}$ 24. 625.031 25. 4.0005
26. Three hundred ninety-four ten thousandths 27. Forty-five and forty-seven hundredths 28. 0.35 29. 0.188 30. 0.8 31. $\frac{23}{50}$
32. $\frac{73}{200}$ 33. $\frac{203}{400}$ 34. 430.8724 35. 2.7209 36. 0.01448 37. 7,600 38. $177.60 39. 0.275 in 40. $9\frac{11}{16}$ in 41. $2\frac{29}{32}$ in
42. $\frac{3}{20}$ v 43. $1\frac{19}{24}$ v 44. $12\frac{18}{101}$ ohms 45. 1,230.05 lb 46. 0.531 in 47. 23 48. 25 49. 11 50. 4 51. 49 52. 144 53. −1
54. −17 55. 49 56. 45

Chapter Test

1. $\frac{7}{12}$ 2. $\frac{4}{7}$ 3. $6\frac{3}{4}$ 4. $\frac{3}{38}$ 5. $\frac{16}{5} = 3\frac{1}{5}$ 6. 18 7. 20.64 8. 12.859 9. 28.392 10. 45 11. 19 12. $\frac{9}{35}$ 13. $1\frac{4}{9}$ 14. 42
15. $6\frac{7}{24}$ 16. 29 17. $1\frac{13}{22}$ 18. 0.00124 19. 38 20. $5\frac{19}{24}$ 21. 42,100 22. $\frac{2}{3}$ 23. −21 24. $\frac{9}{70}$ 25. $10\frac{1}{2}$ 26. 27
27. $0.025 = $2\frac{1}{2}$¢ 28. 13 29. $7\frac{3}{8}$ ft

Chapter 3

Exercise 3.1

1. $\frac{12}{7}$, 12:7 3. $\frac{8}{3}$, 8:3 5. $\frac{3}{1}$, 3:1 7. $\frac{8}{15}$, 8:15 9. $\frac{1}{12}$, 1:12 11. $\frac{1}{4}$, 1:4 13. $\frac{5}{4}$, 5:4 15. $\frac{39}{32}$, 39:32 17. $\frac{2}{1}$, 2:1 19. $\frac{3}{7}$, 3:7
21. 3:7 23. 5:2 25. 3:2 27. 5:12 29. 60:7 31. 1:26 33. $\frac{1}{6}$ lb/ft^3 35. 8 g/cm^2 37. 60 mi/h 39. 15:17 41. 5:12, 12:5
43. 8:3 45. 13:14 47. 21:2 49. 31:2 51. 7:50 53. 7 ft-lb/s 55. 3:16 57. 11:32

Exercise 3.2

1. $\frac{5}{4}$ 3. $\frac{100}{9}$ 5. $\frac{16}{5}$ 7. $\frac{12}{19}$ 9. 3 11. $\frac{50}{13}$ 13. 7 Weeks 15. 136 l 17. 44 pts 19. $3.15 21. 0.744 in 23. 75
25. $x = 4\frac{2}{7}$ in, $y = 5\frac{5}{7}$ in 27. 3,240 29. 4 h, 30 min 31. 2.5 qt 33. 67.19 yd^3 35. $40.28

Exercise 3.3

1. $s = kt$ 3. $I = kR$ 5. $V = kr^3$ 7. $V = khr^2$ 9. $E = \frac{k}{I}$ 11. $N = \frac{ks^3}{p^2}$ 13. $v = k\sqrt{ab}$ 15. $V = \frac{kx}{yz^2}$ 17. $M = kmv$
19. $F = \frac{kmv^2}{r}$ 21. 4 23. $\frac{22}{49}$ 25. 3 27. 120 29. $13\frac{1}{2}$ 31. 117 33. 16 35. $5\frac{1}{3}$ 37. $13\frac{1}{3}$ 39. $8\frac{1}{10}$
41. 120 lb 43. $3\frac{1}{3}$ amps 45. 57.6 ft/s 47. $\frac{5}{16}$ 49. $p = \frac{Tc}{V}$, $V = \frac{Tc}{p}$ 51. $117

Exercise 3.4

1. 0.52 3. 3.25 5. 0.008 7. 0.086 9. 0.0075 11. 37% 13. 473% 15. 0.8% 17. 1,500% 19. 0.056% 21. $\frac{14}{25}$
23. $\frac{31}{20}$ 25. $\frac{9}{1000}$ 27. $\frac{1}{400}$ 29. $\frac{3}{400}$ 31. 80% 33. 83.3% 35. 11.4% 37. 333.3% 39. 12%

Exercise 3.5

1. 7.2 3. 40 5. 70% 7. 256 9. 142.9% 11. 2.16 13. 79% 15. 25% 17. 50 19. 82.0 21. 46 is what % of 230? $\frac{46}{230} = \frac{P}{100}$, 20% 23. \$268 is 16% of what amount? $\frac{268}{B} = \frac{16}{100}$, \$1,675 25. 99.9% of 360 is what amount? $\frac{A}{360} = \frac{99.9}{100}$, 359.64 l 27. 450 is what % of 4,000? $\frac{450}{4{,}000} = \frac{P}{100}$, 11.25% 29. $2\frac{1}{2}$% of what amount is 33? $\frac{33}{B} = \frac{2\frac{1}{2}}{100}$, 1,320 31. 271.25 lb of lead, 22.75 lb of tin, 56 lb of antimony 33. 11.59 hp 35. 32.04 gal 37. 80% 39. 93.72 lb 41. \$17,850 labor, \$31.50 supervision, \$14,000 materials 43. \$5,011.76 45. \$140; \$156.80 47. \$461.54 49. \$322.56 51. 1,637.25 ft^2 53. 551.25 bd ft 55. \$71,092 57. \$108,170.91 59. \$237,333.33 61. \$28,341 63. \$65,649 65. 36 lb 67. 12.5%

Chapter Review

1. $\frac{2}{3}$, 2:3 2. $\frac{3}{1}$, 3:1 3. $\frac{3}{4}$, 3:4 4. $\frac{2}{5}$, 2:5 5. $\frac{3}{32}$, 3:32 6. $\frac{8}{21}$, 8:21 7. $\frac{21}{40}$, 21:40 8. $\frac{3}{1}$, 3:1 9. 4:7 10. 8:3 11. 2:3 12. 4:1 13. 5:32 14. 50 mi/gal 15. $3\frac{1}{3}$ 16. 40 17. $26\frac{2}{3}$ 18. $\frac{2}{3}$ 19. $\frac{1}{3}$ 20. $1\frac{1}{2}$ 21. $x = kp$ 22. $y = \frac{k}{z}$ 23. $a = kb^2$ 24. $R = \frac{k}{d^2}$ 25. $E = kIR$ 26. $x = \frac{ky}{z}$ 27. 15 28. 2 29. 2 30. $11\frac{1}{4}$ 31. 7:3 32. 9:14 33. 25 34. 1,650 35. $2\frac{1}{12}$ cm 36. (a) 0.46, (b) 1.45, (c) 0.006, (d) 0.0035 37. (a) 73%, (b) 4.7%, (c) 575%, (d) 0.15% 38. (a) $\frac{9}{25}$, (b) $\frac{41}{20}$, (c) $\frac{3}{2{,}500}$, (d) $\frac{3}{20{,}000}$ 39. (a) 80%, (b) 28%, (c) $18\frac{3}{4}$%, (d) $87\frac{1}{2}$% 40. 5.12 41. 25% 42. 175 43. 250 44. $666\frac{2}{3}$% 45. \$1102.50 46. 14.86 lb 47. 6.9% 48. 0.0013 49. 864 50. 15% 51. $112\frac{1}{2}$ lb 52. 1,680 53. 30.55 oz nickel, 0.975 oz manganese, 0.65 oz aluminum, 0.325 oz silicon 54. 12.11% lead, 87.89% tin 55. \$125.10; \$118.84 56. \$118.84 cash price

Chapter Test

1. 4:3 2. 6:1 3. 7:20 4. 1:2 5. 4:1 6. $14\frac{2}{5}$ 7. 30 gal 8. $16\frac{2}{3}$ qts 9. 21 10. 94.3 11. 27 ft; 12 s 12. (a) 0.13, (b) 4.20, (c) 0.0005, (d) 0.066 13. (a) 56%, (b) 7.2%, (c) 600%, (d) 0.14% 14. (a) $\frac{16}{25}$, (b) $\frac{7}{5}$, (c) $\frac{1}{200}$, (d) $\frac{1}{800}$ 15. (a) 80%, (b) 68%, (c) 76%, (d) $237\frac{1}{2}$% 16. $3\frac{1}{3}$% 17. 800 18. 27,300 19. 5.21% 20. \$832 21. 52.5 lb 22. 78% 23. 26.4 watts 24. 793.65 watts 25. 3.65%

Chapter 4

Exercise 4.1

1. 12 3. 22.13 5. 447.0393 7. 29.67 9. 36.797 11. 21.833 13. 2,282.209 15. 3,884.345

Exercise 4.2

1. 3,480 3. 239.42235 5. −10,690.065 7. 1,024.8722 9. 4,594.3118 11. 2,603.3767 13. −737,579.13

Exercise 4.3

1. 33.962264 3. 0.1133278 5. −2.7983294 7. 6.9781022 9. −0.0485788 11. −7.5631399

Exercise 4.4

1. 0.2 3. 0.02222 5. 0.212766 7. 219.29825 9. 81,225 11. 1,217.3121 13. 608.6089 15. 21 17. 2.078461 19. 62.64583 21. 2.3830443 23. 15,625 25. 0.039304 27. 531,441 29. 2 31. 0.2 33. 9 35. 2 37. 2.1147425 39. 0.3738218 41. 1.210101

Exercise 4.5

1. 87.5% 3. 283.33% 5. 125.33% 7. 20.688 9. 0.44184 11. 9.67346 13. \$252 15. \$14.23 17. \$1,354.95

Exercise 4.6

1. 34 3. 39.74 5. 41 7. 0.0606654 9. 8.9782609 11. 0.8055556 13. 0.4038835 15. 9.4762 17. 0.466129 19. 42.48 21. 20.063464 23. 0.6681373 25. 0.3622047 27. 0.4286624 29. −4.015 31. 2.6704099 33. 16.600512 35. *B; D* 37. *B* 39. 24.840699 41. 85.058935 43. 15.839192 45. 0.8210526

Exercise 4.7

1. 2.55×10^2 3. 1.2345×10^4 5. 1.55×10^5 7. 8.55076×10^2 9. 1.0076×10^6 11. 1.2×10^{-4} 13. 8.1×10^{-6} 15. 7×10^{-4} 17. 9.4×10^{-11} 19. -4.5×10^3 21. -5.85×10^6 23. -4.578×10 25. -2.985×10^{-8} 27. 49,900,000 29. 7.23 31. 0.0042 33. 0.00000147 35. 0.000789 37. −0.00000000482 39. −0.00000492

Exercise 4.8

1. 5.076 03 3. 4.7 06 5. 8.123 06 7. 7.63 −01 9. 2.3 −04 11. −5.6 04 13. 8.06 21 15. 5.787 −04 17. 2.2046 −03 19. 1.102 −03 21. 5.872 09

Exercise 4.9

1. 5.61179×10^{15} 3. 7.11732×10^0 5. 1.08726×10^{-2} 7. 1.3964286×10^{-7} 9. 1.879896×10 11. 1.2905×10^{-8} 13. 4.905892×10^{-12} 15. 4.0783089×10^{10} 17. 2.0089857×10^{-3}

Exercise 4.10

1. 42 3. 240 5. 14,000 7. 24 9. 4 11. 32 13. 3 15. 2 17. 6

Exercise 4.11

Refer to the answers of Exercises 4.1, 4.2, 4.3

Exercise 4.12

Refer to the answers of Exercise 4.4

Exercise 4.13

Refer to the answers of Exercise 4.5

Exercise 4.14

Refer to the answers of Exercise 4.6

Chapter Review

1. −2.452381 2. 1.2032258 3. 2.8365385 4. −1.5576923 5. 4.464684 6. 5.8622471 7. 2,097.64 8. 0.0000002 9. 4.100625×10^{18} 10. 6.8483575 11. 0.1529706 12. 272.205 13. 0.10486 14. 266.085 15. \$930.05 16. \$1,010.10 17. 56.63% 18. 0.84% 19. 150% 20. 1.904×10^8 21. 2.5424×10^{13} 22. 4.368×10^{-9} 23. 1.368×10^2 24. 3.6585366×10^7 25. 1.5755556×10^{-8} 26. 126,000 27. 20,000,000 28. 12 29. 0.12 30. 0.01

Chapter Test

1. 860.9004 2. 33.805106 3. −1.0648077 4. 6.8992754 5. 0.4857983 6. 5,246.1049 7. 0.0000004 8. 46.762712 9. 5,223.7767 10. 26.279236 11. 182.61% 12. 58.2504 13. \$15.95 14. 5.6×10^7 15. 7.32×10^{-3} 16. 143,000 17. 0.000563 18. 480 19. 3 20. 6.6216×10^{-11}

Chapter 5

Exercise 5.1

1. 2.294 lb 3. 4.732 long ton 5. 67.12 pt 7. 3.063 barrels 9. 774 in^2 11. 1,280 rd 13. 6.667 oz/in^2 15. 0.000185 yd^3/s 17. 39.6 t/mi 19. 737,600 ft-lb/s 21. 0.251 rd^2/s 23. 360 t/ft^2

Exercise 5.2

1. 0.362 m 3. 0.0803 kg 5. 43,070 g 7. 4.2 l 9. 0.0492 dm^2 11. 0.0000825 km^3 13. 3.833 m/s 15. 925,000 kg/m^3 17. 126,400 mg/min 19. 4,630 ml or 0.00463 kl

Exercise 5.3

1. 62.23 m 3. 12.39 lb 5. 102.52 l 7. 0.24 m^2 9. 90,610,835,000 cm^2 11. 77,906.88 cm/min 13. 0.30 lb/in^2 15. 3.08 g/cm^3 17. 0.26 cm^3/g 19. 93,895.2 m/h 21. 9.98 kg 23. 0.64 cm 25. 0.13 cm 27. 0.21 db/m 29. 59.06 ohm/m

Exercise 5.4

1. 3 significant digits, thousandths 3. 4 significant digits, thousandths 5. 4 significant digits, tens 7. 5 significant digits, hundred thousandths 9. 2 significant digits, millionths 11. 5 significant digits, tenths 13. 0.47 15. 9,007.1 17. 5.67 19. 73.2 21. 540 23. 10.9 25. Exact 27. Exact 29. Approximate 31. Approximate 33. Exact 35. Approximate 37. (1) 102,000,100,000; 82,730; 8,600,000. (2) 82,730; other two are the same. 39. (1) 6.031; 89.7; other two are the same. (2) 0.00073; 0.0056; 6.031; 89.7 41. $\pm .5$ 43. $\pm .05$ 45. ± 5 47. $\pm 50{,}000$ 49. 0.331 51. 250 53. 0.108 55. 70 57. 0.389 59. 595.87 61. 2.227 63. 1.129 65. 72.6 67. 413.2 69. 0.1709 71. 1.87 73. 3.0 75. 25.08 77. 16 79. 54.1 81. 16.4 in^2 83. 2.67 in 85. 225.7 ohm 87. 540 volts 89. 9.47 in

Chapter Review

1. 2.68 t 2. 45.04 pt 3. 66.24 ft^2 4. 2,560 rd 5. 0.239 m 6. 7.36 g 7. 5.3 l 8. 0.004 m^2 9. 41,280 oz/min 10. 0.094 dm^2 11. 6.167 m/s 12. 0.014 yd^3/min 13. 1 g/cm^2 14. 757,440 lb/ft^2 15. 0.025 cm^3 16. 3.80 km 17. 160.71 kg 18. 11.13 km^2 19. 158.59 l 20. 5,393.87 cm^2 21. 149,961.6 cm/min 22. 0.636 ft^3/h 23. 33.39 g/cm^2 24. 15.02 km/min 25. 2,658.13 kg/m^3 26. 117.31 lb 27. 0.95 cm 28. 0.18 in 29. 0.29 ohm 30. 96,000 kilocycles/h 31. 34.14 lb/in^2 32. 75.38 l 33. 3 34. 2 35. 4 36. 2 37. 5 38. 4 39. 4.04 40. 0.0074 41. 50.37 42. 463,430 43. $72.3 \pm .05$ 44. $82{,}300 \pm 50$ 45. $0.00093 \pm .000005$ 46. (a) Most 0.367; least 22,400 (b) Most 0.367 or 22,400; least 52 47. (a) Most 0.000873; least 79,000; (b) Most 702.6; least 79,000 48. 4.226 49. 27.04 50. 452.2 51. 3.77 52. 51.1 53. 131 lb 54. 2.01 in 55. $0.8\bar{0}$ in 56. $5{,}\bar{0}00$ ohms

Chapter Test

1. 4.53 l 2. 48 cups 3. 3,888 in^2 4. 47.35 cl 5. 48.27 km/h 6. 1.13 yd^2 7. 0.16 g/cm^3 8. 2.7 9. 3.4 10. 879.1 11. 10.88 12. 6.0 13. 14.5 14. 31.0 15. 6.54 in^3 16. 22 yd^2 17. 4.26 lb

Chapter 6

Exercise 6.1

1. 2 3. 3 5. 1 7. 3 9. 2 11. 1 13. 3 15. 8, 5, −9, −5 17. $a + b$ 19. $x - 7$ 21. $\frac{x + y}{z}$ 23. $a - 5$ 25. $\frac{1}{2}x - 2x$ 27. $x + 4$ 29. $\frac{x}{5}$ 31. $\frac{x - 6}{11}$

Exercise 6.2

1. 9 3. 5 5. 48 7. 5 9. 62 11. 288 13. 61 15. 0 17. −1 19. 1 21. −44 23. 20 25. 0 27. 43 29. (a) 10.91, (b) 105.6 31. (a) 5, (b) 6.67 33. (a) −156, (b) −1,016 35. 267.95 37. (a) 14,915, (b) 45,216 39. (a) 44, (b) 16, (c) 156.67 41. (a) 115.40, (b) 211.01 43. 2.81×10^{-26} 45. 17.5 47. 90 rpm 49. 0.327 51. 0.0280 53. 12.1 55. 427 rpm 57. (a) 529 watts, (b) 204 watts 59. (a) P is $\frac{1}{2}$ of the original value, (b) P is $\frac{1}{3}$ of the original value, (c) P is twice the original value 61. (a) P is 4 times as great, (b) P is 9 times as great, (c) P is $\frac{1}{4}$ as great 63. 18.2 65. 35.2 67. 2.1 69. 3.61 71. 5.15 73. −60.0 75. 60.4

Exercise 6.3

1. $14x$ 3. $-3ab$ 5. $7x^2 + 4z$ 7. $2a^3 + a^2b - 6ab^2 - b^3$ 9. $x^3 - x - 5$ 11. $-3i - 9I$ 13. $-3x^2y + 15xy$ 15. $4a^3 + a^2b + 6ab^2 - 5b^3$ 17. $70A + 9B$ 19. $-5eI + 14IR + 9I^2R$ 21. $-8I + 10$ 23. $80 - I$ 25. $25 + E$ 27. $E + 2E + E + 4E = 8E$ 29. $25R + (10R - 13R) = 22R$ 31. $C_1 = C - 180$ 33. $S + \frac{3}{4}S + 2(\frac{3}{4}S) = \frac{13}{4}S$ 35. $2(l - 36) + 2l = 4l - 72$ 37. $A \approx \frac{C}{2}$ 39. $\frac{4}{5}R + R + \frac{8}{5}R = \frac{17}{5}R$ 41. $\frac{5}{2}s + s = 50$ 43. $2.95x - 12.2y$ 45. $221.1R$ 47. $336.3A - 12.16B$ 49. $15.903y + 1.500z$ 51. $13.7x^2 + 72.81x - 7.3$ 53. $1.191x^2 + 88.77x + 6.382$

Exercise 6.4

1. $6^4 = 1{,}296$ 3. I^5 5. $(IR)^3 = I^3R^3$ 7. $4^5 = 1{,}024$ 9. I^6 11. $-15y^2$ 13. $20a^2b^2$ 15. $-x^3y^2z^2$ 17. $-30i^7$ 19. $x^2y(x + y)$ 21. x^3y^3 23. R^4 25. y^4 27. $4a^2b^2$ 29. $25x^4y^6$ 31. $-125i^6R^3$ 33. $-15a^{2n}$ 35. 10^{4z} 37. x^{5a} 39. $2^a a^{2a} b^a c^{3a}$ 41. $15x^2$ 43. $x^8 - 8x^6$ 45. x^{6a^2} 47. $2^p a^{np} b^{mp}$ 49. x^5y^6 51. $5a^4b^2$ 53. $-3\bar{0}$ 55. $19\bar{0}$ 57. $0.03x^9$ 59. $36a^{10}b^{11}$ 61. $0.096x^{10}y^{14}$ 63. $1{,}770x^3y^3z^{12}$

Exercise 6.5

1. $2a^3b - 2ab^2c + 2abc^2$ 3. $15RI^2 - 21Ri^2$ 5. $-15a^3b^2 + 5a^2b^3 - 20ab^4$ 7. $3i^3r - 6i^2r^2 - 3ir^3$ 9. $13e - 3$ 11. $29E + 3e$ 13. $2Ri^2 - 8Ri$ 15. $2E - 3e + R$ 17. $x^2 + 7x + 12$ 19. $y^2 - 13y + 36$ 21. $V^2 + 2V + 1$ 23. $R^2 - 6R + 9$ 25. $F^2 - 9$ 27. $6a^2 - 31a + 35$ 29. $4x^2 - 49$ 31. $3k^2 - 17kw - 6w^2$ 33. $E^2 + 12eE + 36e^2$ 35. $4e^2 - 9i^2$ 37. $a^3 + 2a^2b - 7ab^2 + 4b^3$ 39. $6I^3 + 21I^2 - 5I + 28$ 41. $x^4 - x^3 - x^2 - 11x - 12$ 43. $V^3 - 7V^2 + 4V + 12$ 45. $2R^3 - 3R^2r - 2Rr^2 + 3r^3$ 47. $a^3 - 3a^2b + 3ab^2 - b^3$ 49. $a^3 - 6a^2b + 12ab^2 - 8b^3$ 51. $1.2R^5 - 4.20R^4 + 2.24R^3$ 53. $-51a^3b^7 + 41.3a^4b^3 + 1.7a^3b^8$ 55. $a^2 - 4.82a + 5.20$ 57. $x^2 - 9.14x + 20.9$

Exercise 6.6

1. $4x^2$ 3. $-5x^3$ 5. $2ab$ 7. $4E^2$ 9. $3(a - b) = 3a - 3b$ 11. $2x - 3$ 13. $x - 1$ 15. $3x^2 - 2x + \frac{3}{4}$ 17. $3x^4 - 4x^3 - 2x + 1$ 19. $4\theta^2 + 2\theta$ 21. $x - y - 2y^2$ 23. $R - E$ 25. $I - i$ 27. $x + 3$ 29. $E - 9$ 31. $a + 3$ 33. $3a - 4$ 35. $x^2 + 2x + 4$ 37. $x + y$ 39. $5a + 6b$ 41. $x^2 + 2x + 3$ 43. $I^2 + 7I + 14 + \frac{6}{I - 1}$ 45. $-3a - 8$ 47. $x + 3$, 7, 42 units

Chapter Review

1. 3 terms trinomial 2. 1 term monomial 3. 2 terms binomial 4. 3 terms trinomial 5. 40 6. $\frac{73}{168}$ 7. 80 8. (a) 3, (b) $47\frac{1}{4}$ 9. 1,040 10. $6E$ 11. $8I^2 + 5I$ 12. $2P + 9R$ 13. $2V_0 - 2V_1$ 14. $2E - 6I - 2R + 1$ 15. x^7 16. x^{10} 17. $-6x^5$ 18. $16c^4d^8$ 19. $81a^8b^{12}c^{20}$ 20. $5x^3$ 21. a^{3x+4} 22. a^{5x} 23. x^{12a^2} 24. $2^ta^{rt}b^{st}$ 25. $12x^3 - 8x^2 + 16x$ 26. $12ER^3 - 24R^3 + 36E^2R^2$ 27. $6x^2 - x - 12$ 28. $14e^2 + ei - 3i^2$ 29. $4P^2 - 12P + 9$ 30. $9R^2 - 16$ 31. $E^3 - 3E^2 + 2E^2e - 6Ee - Ee^2 + 3e^2$ 32. $x^4 - 2x^3 - 10x^2 + 14x - 3$ 33. $6V^3 - 7V^2 - 7V + 6$ 34. $-8x^2$ 35. $2E - 3 + 5E^2$ 36. $4I + 1 - \frac{2}{2I - 1}$ 37. $5n^2 - n - 4$

Chapter Test

1. $2x^2 + 3x - 3$ 2. $6a^8b^7$ 3. $2x^2y + xy + 5xy^2 - 2x^2y^2$ 4. $4a^2 - 4ab + b^2$ 5. $4x^2 - 5xy$ 6. $24x^3 + 18x^2y$ 7. $3x^2 + 10xy - 8y^2$ 8. $27a^6b^4 + 6a^3b^4 - 21a^3b^6$ 9. $-8x^{14}y^3$ 10. $2ab^2$ 11. $x - 3$ 12. $4R^2 - 9S^2$ 13. $x - 6 + \frac{1}{x - 2}$ 14. $10x^7y^6 - 20x^6y^5 + 15x^5y^4$ 15. $9x^2 + 6xy + y^2$ 16. $-2x^3y^2$ 17. $3x^2 + 2x - 4 + \frac{1}{2x - 1}$ 18. $2a + 5b$ 19. $2x^4 - x^3 - 8x^2 - 5x - 4$ 20. $3x^2 + 8x + 5$ 21. $a^2 - b^2$ 22. $3a^4b^7c$ 23. $2x^2 + 3x - 4$ 24. $x^3 - 3x^2y + 3xy^2 - y^3$ 25. $2a^3 + a^2b - 2ab^2 - b^3$ 26. $5(2x - y) = 10x - 5y$ 27. $30a^3 - 56a^2b - 10ab^2 + 4b^3$ 28. 1 29. -36 30. 216

Chapter 7

Exercise 7.1

1. 2 3. 11 5. 22 7. 4 9. Identity 11. Identity 13. 12 15. Identity 17. Identity 19. 14 21. True 23. True 25. True 27. False 29. False 31. True 33. True 35. False 37. (a) 0.2, (b) 0.9 39. (a) 1, (b) 6, (c) 1.5 41. (a) 84, (b) 47 43. (a) 16, (b) -10 45. (a) 1,000, (b) 125

Exercise 7.2

1. 2 3. $\frac{11}{3}$ 5. -2 7. -4 9. 96 11. $\frac{54}{5}$ 13. 13 15. 1 17. 0 19. 3 21. 0 23. $\frac{5}{6}$ 25. 14 27. $\frac{11}{12}$ 29. 3 31. $\frac{10}{7}$ 33. -10 35. $\frac{16}{11}$ 37. $\frac{37}{10}$ 39. $\frac{2}{5}$ 41. 0.51 43. -1.87 45. 1.90 47. -0.092 49. -7.0

Exercise 7.3

1. 3 3. 54 5. 3 7. 3 9. 1 11. $\frac{5}{2}$ 13. 5 15. -1 17. $\frac{18}{31}$ 19. -6 21. $-\frac{5}{2}$ 23. 20 25. 0 27. $-\frac{7}{3}$ 29. -7 31. 3 33. 5 35. 2 37. $\frac{14}{5}$ 39. 10 41. 3 43. 0 45. 2.3 47. $1\tilde{0}$ 49. (a) -7.78°C, (b) -32.8°C, (c) -17°C 51. 0.42

Exercise 7.4

1. $\frac{3}{2y}$ 3. $-\frac{3}{2b}$ 5. $\frac{ab}{2}$ 7. $\frac{a+c}{b}$ 9. $3a$ 11. $\frac{K}{V}$ 13. $\frac{V}{lw}$ 15. $\frac{2s}{t^2}$ 17. $\frac{6M}{h^2}$ 19. $\frac{P_1V_1}{P_2}$ 21. $\frac{P-2W}{2}$ 23. $IR + Ir$
25. $\frac{5F-160}{9}$ 27. $\frac{Wv^2}{Fr}$ 29. $at + v_0$ 31. $\frac{b}{c+1}$ 33. $5a$ 35. $\frac{3a+6}{a+3}$ 37. $\frac{4b}{2a-2b+3}$ 39. $\frac{2a^2-8a}{5}$ 41. $\frac{S-2r^2}{2r}$
43. $\frac{A-2r^2}{2r}$ 45. $\frac{R_1R_2}{R_1+R_2}$ 47. $\frac{L_t}{1+at}$ 49. (a) 14.0°F, (b) 77.0°F, (c) 48°F 51. (a) $\frac{Z_2Z_1}{Z_2-Z_1}$, (b) $\frac{Z_1Z_2}{Z_1-Z_2}$ 53. (a) $\frac{pq}{q+p}$,
(b) $\frac{fp}{p-f}$ 55. $\frac{acx}{ac-cx-ax}$ 57. $\frac{rW}{W-2P}$ 59. $\frac{2A-hb_2}{h}$ 61. (a) $\frac{(0.3208)GPM}{A}$, (b) $\frac{AV}{0.3208}$ 63. $\frac{V^2}{2g}$ 65. (a) $\frac{NL}{7}$, (b) $\frac{7cs}{N}$
67. (a) $\frac{R_1L_2}{R_2}$, (b) $\frac{R_1L_2}{L_1}$

Exercise 7.5

1. 22, 40 3. 52, 35 5. 28 7. 7 9. 54 11. 15, 17, 19 13. 12, 108 15. 14, 7, 42 17. 9, 5 19. 19 ft by 16 ft
21. 4 ft, 8 ft 23. 28 amp, 52 amp 25. 38, 26 27. 47 V, 79 V 29. 19 ohm, 5 ohm 31. $13,000 at 8%, $7,000 at 6%
33. $8,000 at 10%, $10,000 at 8% 35. $17,000 at 18%, $8,000 at 11% 37. $14,000 39. $2\frac{16}{37}$ h
41. 400 kg of 68%, 1,600 kg of 83% 43. $955\frac{5}{9}$ l of 2.5%, $44\frac{4}{9}$ l of 25% 45. $5\frac{5}{7}$ cl of 38%, $94\frac{2}{7}$ cl of 3%

Exercise 7.6

1.
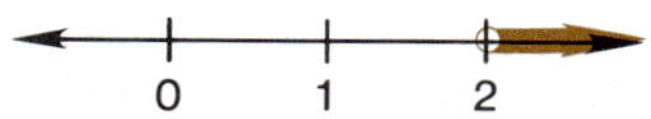
3.
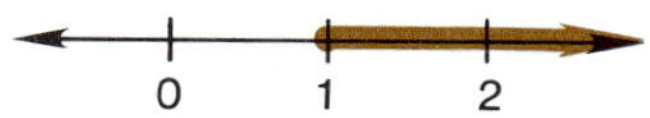
5.
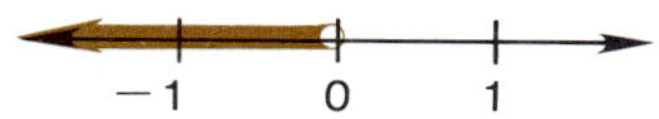
7.
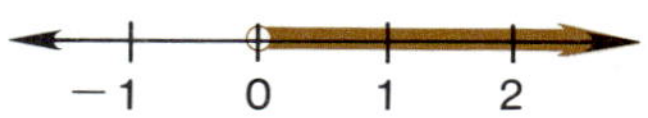
9.
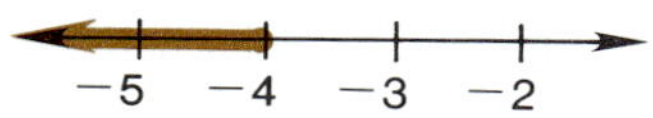
11.
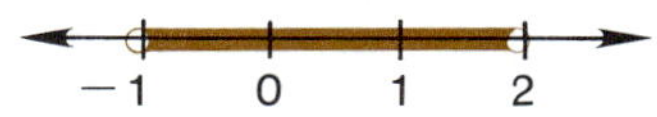
13.
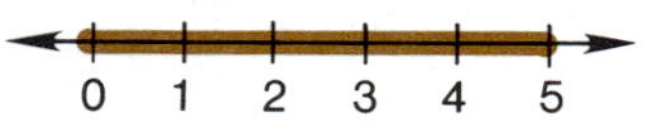
15.
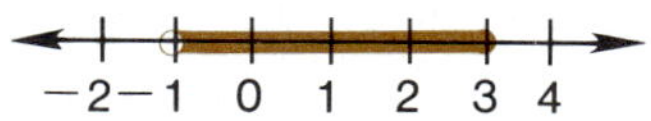
17. $x \leq \frac{5}{2}$
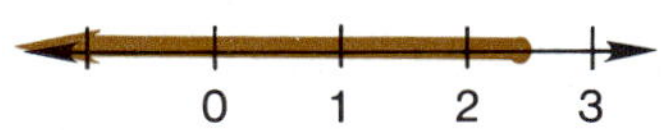
19. $x < 6$

21. $x \geq 18$
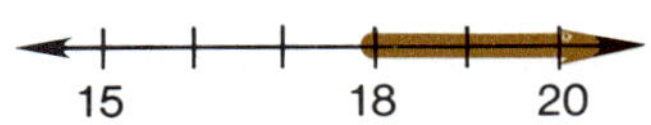
23. $x \geq -9$

25. $x > -5$

27. $x > -\frac{5}{2}$

29. $x > \frac{16}{3}$
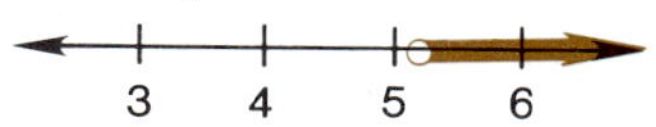
31. $x \leq 2$
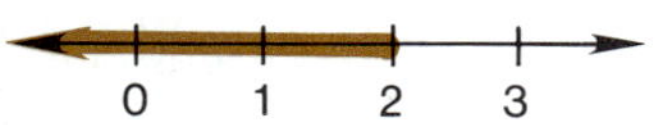
33. $x < 11$
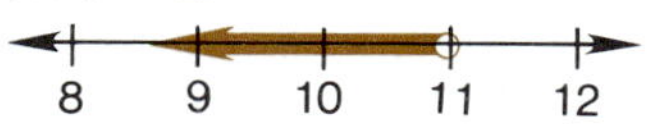
35. $x > 1$
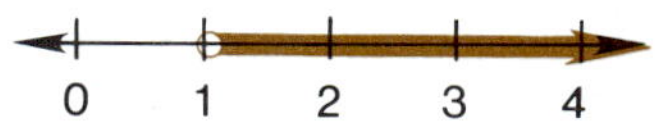

37. $x \geq 3$

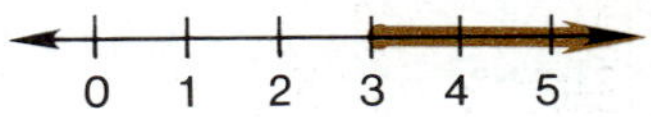

39. $x > \frac{10}{7}$

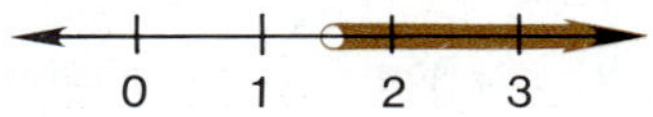

41. $x > -10$

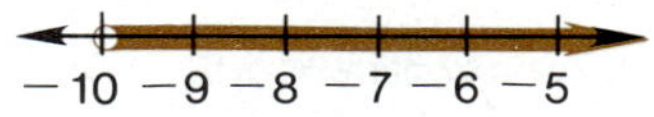

43. $x < \frac{16}{11}$

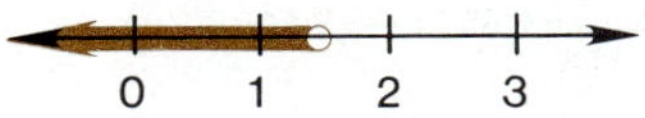

45. $x \leq 3\frac{7}{10}$

47. $x \geq \frac{5}{2}$

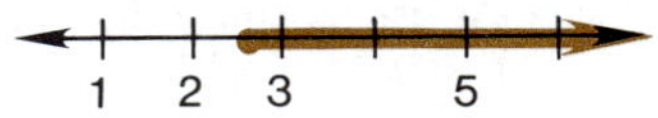

49. $x \geq \frac{2}{5}$

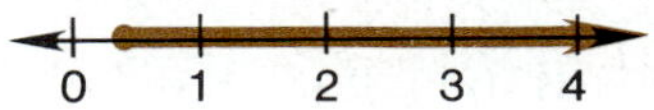

51. $-2 < x < \frac{1}{2}$

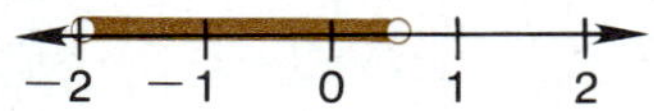

53. $\frac{-4}{5} \leq x \leq \frac{1}{5}$

55. $-2 < x \leq \frac{5}{4}$

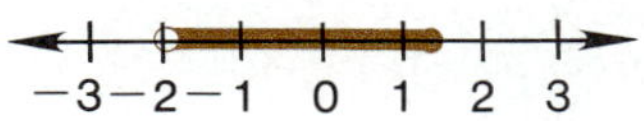

57. $-4 < x \leq 1$

59. $-1 \leq x < 7$

61. $-2 < x \leq \frac{1}{3}$

63. $x \geq 72$ 65. $x \geq 8$ 67. $P \geq 2C$ 69. $x < 6$ 71. $x \geq 7$ 73. $1 < x < 5$ 75. $5 < S < 27$

Chapter Review

1. True 2. True 3. False 4. False 5. False 6. True 7. True 8. True 9. 3 10. 3 11. -7 12. -7 13. 12 14. 14 15. 15 16. 21 17. 7 18. 21 19. 3 20. $\frac{14}{3}$ 21. $\frac{7}{3}$ 22. 3 23. $\frac{1}{2}$ 24. $\frac{-11}{4}$ 25. $\frac{-1}{4}$ 26. 9 27. $\frac{26}{3}$ 28. $\frac{64}{3}$ 29. -10 30. $\frac{8}{5}$ 31. $\frac{17}{2}$ 32. 5 33. 7 34. $\frac{7}{2}$ 35. $\frac{2}{5}$ 36. $\frac{-1}{5}$ 37. $\frac{4}{3a}$ 38. $\frac{7 + 3a}{2}$ 39. $-y$ 40. $\frac{-2by}{3a - 4b}$ 41. $3y$ 42. $\frac{-2a}{b}$ 43. $\frac{10b}{5ab - 1}$ 44. $\frac{b}{a}$ 45. $\frac{-6}{b}$ 46. $\frac{-ay - y}{a}$ 47. $\frac{-2ay - by}{a - 2b}$ 48. $\frac{by - 2ay}{4a - b}$ 49. 64, 41 50. 36 51. 45 52. \$11,000 at 8%, \$9,000 at 7% 53. \$12,000 at 12%, \$13,000 at 19% 54. $x > 4$ 55. $x \leq 3$ 56. $x > -7$ 57. $x < -4$ 58. $x < 2$ 59. $x > \frac{15}{7}$ 60. $x > \frac{9}{2}$ 61. $x \geq \frac{-6}{5}$ 62. $x \leq \frac{2}{9}$

Chapter Test

1. -4 2. $\frac{15}{4}$ 3. $-\frac{5}{2}$ 4. $\frac{5a + 2b + 1}{6}$ 5. $-\frac{8}{7}$ 6. $\frac{29}{5}$ 7. $\frac{5}{2}$ 8. $\frac{a}{a - b}$ 9. 1 10. $x \geq 2$ 11. $-\frac{4}{3}$ 12. $\frac{6a + b}{2a - 3b}$ 13. $\frac{-21}{2}$ 14. 4 15. $\frac{-4b}{b - 3a}$ 16. $x \geq \frac{-16}{7}$ 17. 8 18. $\frac{13}{15}$ 19. $\frac{-14}{3}$ 20. $x \geq 3$ 21. $\frac{7ab + b}{11a}$ 22. $\frac{7}{3}$ 23. $x > \frac{-9}{13}$ 24. $-\frac{3}{4}$ 25. \$6,000 at 10%, \$8,000 at 6% 26. 16, 23

Chapter 8

Exercise 8.1

1. $2(y + 3)$ 3. $4(x^2 + 2y)$ 5. $3(x^2y + 5z)$ 7. $7(e - 2i + 3r)$ 9. $3(5xy - 6z + x^2)$ 11. $42xy - 51y^2 + 7$ 13. $2(4x - 5y + 6z - 9w)$ 15. $5(4E^2 - 12ER + 9I^2)$ 17. $2x + 3$ 19. $3x - 4z - 6w$ 21. $4L - 5W + 2H$ 23. $-R^2 - 2R + 3E$ 25. $w - 2s + w^2 + w^3 + 3$ 27. $3xy(x + 2)$ 29. $2R^2(R^2 - 3)$ 31. $x(2x^2 - x + 1)$ 33. $3er(5 + 6r - e)$ 35. $xy(y + z + 1)$ 37. $2L(L^2 - 9 + L)$ 39. $5p(p + 2 - 3p^2)$ 41. $y + z$ 43. $-1 - e$ 45. $1 - x + x^2$ 47. $1 - x + y - z$ 49. $-2er + 3 - 4e^2r^2$ 51. $4x^2(x - 2), -4x^2(-x + 2)$ 53. $8a^2(a - 4b), -8a^2(-a + 4b)$ 55. $6a(4a^2 - 2 + a), -6a(-4a^2 + 2 - a)$ 57. $7R(R + 2 - 3R^2), -7R(-R - 2 + 3R^2)$ 59. $1(7x^2 + 9y^2), -1(-7x^2 - 9y^2)$ 61. $(a + b)(x + y)$ 63. $5(2a + b)(3x + 2y)$ 65. $9xy(a + 4b)$ 67. $(8a - 1)(b + 6)$ 69. $4xy(3x + 2y)(a - 2b)$ 71. $\frac{4W}{2W - 1}, \frac{52}{25}$ 73. $\frac{6}{P^2 - P - 1}, 6$ 75. $\frac{3}{R - 1}, \frac{-3}{4}$ 77. $\frac{-2D}{D - D_0}$ 79. $16t(35 - t)$ 81. $2\pi r(h + r)$

Exercise 8.2

1. $(x + 1)(x - 1)$ 3. $(a + 2)(a - 2)$ 5. $(3 + E)(3 - E)$ 7. $(1 + k)(1 - k)$ 9. $(4i + r)(4i - r)$ 11. $(x^2 + 1)(x + 1)(x - 1)$ 13. $(i + 4r)(i - 4r)$ 15. $(R + 2I)(R - 2I)$ 17. $(5 + W^2)(5 - W^2)$ 19. $3(a + 3b)(a - 3b)$ 21. $5(2 + b)(2 - b)$ 23. $(xy + 2z)(xy - 2z)$ 25. $(R^2a^2 + 4E^2)(Ra + 2E)(Ra - 2E)$ 27. $(4e^2 + 1)(2e + 1)(2e - 1)$ 29. $5(5p + 2v)(5p - 2v)$ 31. (a) $\frac{V}{8I}(h + 2v_1)(h - 2v_1)$, (b) $\frac{3V}{2A}\left(1 + \frac{2v_1}{h}\right)\left(1 - \frac{2v_1}{h}\right)$

Exercise 8.3

1. 3, 4 3. 6, 2 5. 4, 12 7. 2, 24 9. 6, −5 11. −7, −1 13. 12, 12 15. 18, 8 17. 8, −3 19. 8, −1 21. 12, −2 23. *DNE* 25. −4, −1 27. *DNE* 29. 4, 1 31. 4, 4 33. 7, 7 35. 12, 1 37. −15, −2 39. 2, 9 41. 10, −3 43. −2, 36 45. 5, 7 47. −18, 1 49. 6, −12 51. Yes; 3, 1 53. Yes; 6, 1 55. No 57. Yes; −5, 3 59. Yes; 6, 14 61. Yes; 3, −24 63. Yes; 4, 5 65. Yes; 14, 1 67. Yes; 14, −14 69. Yes; −9, −4 71. No 73. Yes; −7, 6 75. Yes; −16, −1 77. No 79. Yes; 5, 18

Exercise 8.4

1. $(2x + 1)(x + 3)$ 3. $2(2e - 1)(2e - 3)$ 5. $(8p - 1)(p + 6)$ 7. $(6E - R)(3E + 2R)$ 9. $(2x - 3)(x + 2)$ 11. $(2x + 1)(x + 1)$ 13. $(2R - 3)(R - 2)$ 15. $(5H - 3)(H + 2)$ 17. $(3I - 1)(3I - 1)$ 19. Will not factor 21. $(3a + 2)(2a + 3)$ 23. $(2x + 7)(2x + 3)$ 25. Will not factor 27. $(3y + 1)(3y - 8)$ 29. $(2V - 1)(5V + 6)$ 31. $(2H - 5)(H - 2)$ 33. $2(e + 2)(2e + 3)$ 35. $2(2I + 3)(I + 1)$ 37. $(3a - 1)(2a + 3)$ 39. $2(i - 2)(i + 5)$ 41. $(3x - 2)(2x + 3)$ 43. $(2x + 3)(2x + 3)$ 45. $(7a - 1)(a - 5)$ 47. $(5P - 1)(3P + 1)$ 49. $2a(a^2 - 4a - 10)$ 51. $I(2I + 1)(I + 7)$ 53. $(2a - 5)(4a + 3)$ 55. $-16(t - 1)(t - 1)$

Exercise 8.5

1. $(x + 12)(x - 1)$ 3. $(a - 12)(a - 2)$ 5. $(B + 6)(B + 2)$ 7. $(x - 4)(x + 3)$ 9. $3(e - 8)(e + 2)$ 11. Will not factor 13. $4(b - 3)(b + 2)$ 15. $(xy + 3)(xy - 6)$ 17. Will not factor 19. $(RS - 6)(RS + 5)$ 21. $4(LM - 7)(LM - 1)$ 23. $(x^2 - 6)(x + 1)(x - 1)$ 25. $(a^2 + 2)(a + 3)(a - 3)$ 27. $(x^2 + 2)(x + 5)(x - 5)$ 29. $3(x + 2)(x + 2)$ 31. $2a(a^2 - 4a - 10)$ 33. $I(2I + 1)(I + 7)$ 35. $(R^2 + 9)(R + 2)(R - 2)$ 37. $2(x + 3)(x - 3)(x^2 + 1)$ 39. $2(x^2 - 8)(x^2 + 1)$

Exercise 8.6

1. $(r + s)(r^2 - rs + s^2)$ 3. $(2x + y)(4x^2 - 2xy + y^2)$ 5. $(h - k)(h^2 + hk + k^2)$ 7. $(4x - y)(16x^2 + 4xy + y^2)$ 9. $(a + b)(c + d)$ 11. $(3x - 2y)(9x^2 + 6xy + 4y^2)$ 13. $(3x - 2y)(2a + b)$ 15. $(2a + 3b)(4a^2 - 6ab + 9b^2)$ 17. $2(a + 2)(a^2 - 2a + 4)$ 19. $2(x - 2)(x^2 + 2x + 4)$ 21. $x^2(x + 3y)(x^2 - 3xy + 9y^2)$ 23. $(5x - 3y)(4x + z)$ 25. $a^2(2b - a)(4b^2 + 2ab + a^2)$ 27. $2(3r + s)(9r^2 - 3rs + s^2)$ 29. $(b + 4)(b - 4)(x - y)$ 31. $(xy - z)(x^2y^2 + xyz + z^2)$ 33. $(a^5b^2 - 2c^3)(a^{10}b^4 + 2a^5b^2c^3 + 4c^6)$ 35. $(ab + 2)(a^2b^2 - 2ab + 4)$ 37. $(x + 3)(x - 3)(a + 1)(a - 1)$ 39. $(x^3y^4 + z^5)(x^6y^8 - x^3y^4z^5 + z^{10})$

Chapter Review

1. $x + 3$ 2. $x - 2$ 3. $y - 2$ 4. $e + 2i - 4r$ 5. $E + 3 - 6E^2$ 6. $3(E^2 - EI + I)$ 7. $xy(x + z + yz)$
8. $er(e + r)(e - r)$ 9. $R^2(5R - 2)(R + 1)$ 10. $4y(3y^2 + y + 2)$ 11. $x^2(x^2 + 3x + 9)$ 12. $4R^2S^2(4R - 3R^2S + 6)$
13. $5a^2b^2(2a^2b + 3 - 4a)$ 14. $(2 + x)(a + b)$ 15. $(x - 3z)(y + 4)$ 16. $(a + b)(3R + 1)$ 17. $(2a - 3b)(x - 3y)$
18. $(R + S)(R - S)(a - 2b)$ 19. $(x + 3)(x - 3)(R + S)$ 20. $(a + b)(a - b)(x - y)$ 21. $(x + 2)(x - 2)(3a + b)$
22. $(R + 4I)(R - 4I)$ 23. $(5 + E)(5 - E)$ 24. $(3x + y^2)(3x - y^2)$ 25. $(x^2 + 4y)(x^2 - 4y)$ 26. $(5A + 3B)(5A - 3B)$
27. $(6S + 7R)(6S - 7R)$ 28. $(x^2 + 4)(x + 2)(x - 2)$ 29. $(y^2 + 9)(y + 3)(y - 3)$ 30. $(2x + 1)(2x + 1)$ 31. $9(r - 2)(r - 2)$
32. $(4x - 1)(x - 1)$ 33. $(3i - 2)(3i + 5)$ 34. $(2a + 1)(4a - 3)$ 35. $(4x + 3)(6x + 1)$ 36. $(2a - 3)(4a - 3)$
37. $(2a + 3)(a + 6)$ 38. $(x - 7)(x - 2)$ 39. $2a(a - 5)(a + 1)$ 40. $(a + 12)(a + 2)$ 41. $(x - 8)(x + 4)$
42. $(a - 18)(a + 2)$ 43. $3(x - 5)(x + 2)$ 44. $x(x - 3)(x + 2)$ 45. $x(x - 7)(x + 3)$ 46. $(x + 2)(x - 2)(x + 1)(x - 1)$
47. $(x + 4)(x - 4)(x + 1)(x - 1)$ 48. $(x^2 + 1)(x + 1)(x - 1)$ 49. $(x^2 + 9)(x + 3)(x - 3)$ 50. $(ab + 3)(ab - 2)$
51. $(ab + 6)(ab + 4)$ 52. $(ab - 6)(ab - 3)$ 53. $(ab + 2)(ab - 10)$ 54. $(R + 2S)(R^2 - 2RS + 4S^2)$
55. $2(2x - 3)(4x^2 + 6x + 9)$ 56. $(3a + 5b)(9a^2 - 15ab + 25b^2)$ 57. $(xy - 1)(x^2y^2 + xy + 1)$ 58. $2(x^3 + 5)(x^6 - 5x^3 + 25)$
59. $(4x^4 - y^5)(16x^8 + 4x^4y^5 + y^{10})$ 60. $(ab^2 + c^3)(a^2b^4 - ab^2c^3 + c^6)$ 61. $(3a - b)(2x - y)$ 62. $(4a + 3b)(x + 2y)$
63. $(x + 2)(x - 2)(a + 3b)$ 64. $(x^2 + 4)(a - 2b)$

Chapter Test

1. $(R + 3)(R + 2)$ 2. $3a(a + 2b - 3)$ 3. $3(x + 5)(x - 2)$ 4. $(x + y)(m - n)$ 5. $(x + 5)(x + 2)$
6. $2xy(1 - 2xy - 4x^2y^2)$ 7. $(y - 4)(y + 2)$ 8. $(3a + 4)(a + 1)$ 9. $(x + 5y)(x - 5y)$ 10. Will not factor
11. $(a - 4b)(x + 2y)$ 12. Will not factor 13. $(4a + b)(4a - b)$ 14. $(5xy + 4z)(5xy - 4z)$
15. $(a - 5)(a - 2)$ 16. $(3R + 2)(2R + 3)$ 17. $2(a + 3b)(a - 3b)$ 18. $(ab + 8)(ab + 2)$ 19. $(b + 3)(b^2 - 3b + 9)$
20. $(2a + b)(3x - y)$ 21. Will not factor 22. Will not factor 23. $(x + 1)(x + 1)$ 24. $(y + 6)(y + 1)$ 25. Will not factor
26. $5(y + 3)(y + 1)$ 27. $(2z + 3)(2z - 3)$ 28. $(3x - 4)(x + 1)$ 29. $(s + 6)(s - 13)$ 30. $(4a^2 + 9b^2)(2a + 3b)(2a - 3b)$
31. $(s - 15)(s + 3)$ 32. $5x(4x^2 + 1)(2x + 1)(2x - 1)$ 33. $(2a + 1)(2a + 5)$ 34. $(R + 8)(R - 7)$
35. $2(x^2 + 4)(x + 2)(x - 2)$ 36. $(ab + 2)(ab - 4)$ 37. $(x^2 - 3)(x + 3)(x - 3)$ 38. $(2x - yz)(4x^2 + 2xyz + y^2z^2)$
39. $2(a - 2b^2)(a^2 + 2ab^2 + 4b^4)$ 40. $(5a + b)(x + 2)(x - 2)$

Chapter 9

Exercise 9.1

1. $\frac{15}{18}$ 3. $\frac{6a^2b}{9ab^2}$ 5. $\frac{10x}{2x^2 - 6x}$ 7. $\frac{4x - 12}{x^2 - 9}$ 9. $\frac{x^2 + 7x + 10}{x^2 - 4x - 12}$ 11. $\frac{b^2 - a^2}{-a^2 + 2ab - b^2}$ 13. $\frac{7}{8}$ 15. $\frac{1}{3a}$ 17. $\frac{1}{x + 3}$ 19. $\frac{1}{x + 4}$
21. $\frac{x + 7}{2x - 3}$ 23. $\frac{2(x + 5)}{x + 4}$ 25. $\frac{35x}{21}$ 27. $\frac{8a^2 - 6ab}{20ax}$ 29. $\frac{x^2 - 7x + 10}{x^2 - 4}$ 31. $\frac{4x - 3xy - 4y + 3y^2}{x^2 - y^2}$ 33. $\frac{3}{4}$ 35. $\frac{4a}{5b}$ 37. $\frac{5bc^2}{a}$
39. $\frac{x - 3}{x + 3}$ 41. $\frac{x - 5}{x - 3}$ 43. $\frac{2a - 3b}{2(2a + 3b)}$ 45. $\frac{m - n}{m + n}$ 47. $\frac{2y + 3}{4y - 1}$ 49. $\frac{4y + 1}{5 - 2y}$ 51. $\frac{16x - y}{y^2 - 16}$ 53. $\frac{3a - b}{6a}$
55. $-\frac{(x - 2y)^2}{2x + y}$ 57. $\frac{4(a + b)}{a - 2b}$ 59. $\frac{m(n - 4)}{n - 3}$ 61. 0 63. $\frac{5}{6}$ 65. $\frac{8}{13}$ 67. No

Exercise 9.2

1. $\frac{x}{4y}$ 3. $\frac{7c}{2b}$ 5. $\frac{16cx}{3az}$ 7. $\frac{2x^2 - 3xy + y^2}{2x^2 + 3xy + y^2}$ 9. $\frac{5x + 5y}{6}$ 11. $\frac{24x^2y - 96y}{25y + 50}$ 13. $\frac{3y - 15}{2}$ 15. $-4x - 12$
17. $\frac{3x + 3y}{4xy - 4y^2}$ 19. $\frac{10(m + 2n)(2x - 1)}{3(m - 3n)(m + 3n)}$ 21. $\frac{18x - 36}{x + 2}$ 23. $\frac{2x^2 - 3x + 1}{x^2 - x - 56}$ 25. $\frac{y^2 - 2y + 1}{3y + 2}$
27. $\frac{x + 5}{x - 4}$ 29. $\frac{x + 2}{5}$

Exercise 9.3

1. $\frac{3b}{2a}$ 3. $\frac{1}{2ab}$ 5. $\frac{4}{5x - 3y}$ 7. $\frac{x - 3}{x^2 - 2x + 1}$ 9. $4abc$ 11. $\frac{15a}{7b} \cdot \frac{14b^2}{a^2}$ 13. $\frac{24x - 8y}{x^2 - y^2} \cdot \frac{x + y}{3x - y}$
15. $\frac{a^2 - a - 42}{a^2 + a - 12} \cdot \frac{5s - 10}{2a^2 + a - 16}$ 17. $\frac{15ab}{7a - 4b} \cdot \frac{49a^2 - 16b^2}{25a^2b} \cdot \frac{ab}{7a + 4b}$ 19. $\frac{x^2 + 5x - 24}{x^2 + 10x + 25} \cdot \frac{1}{x - 3}$ 21. $\frac{64b}{35a}$ 23. $\frac{y}{6x}$ 25. $\frac{4}{5m}$
27. $\frac{5m^2}{3}$ 29. $\frac{12x^2 - 9x}{5x^2 - 8x - 4}$ 31. $\frac{r^3 + 4r^2 - 16r - 64}{r - 1}$ 33. $\frac{9m}{2m + 2}$ 35. $\frac{a - 3}{a - 5}$ 37. $\frac{3a + 3}{4a - 4}$ 39. $\frac{3x - 2}{x - 9}$ 41. $\frac{8a - 16}{a + 4}$
43. $\frac{1}{2m^2 - 3m - 2}$ 45. $\frac{a}{2a + 1}$

Exercise 9.4

1. $18x; \frac{3}{18x}, \frac{2}{18x}$ 3. $48x^2; \frac{15}{48x^2}, \frac{16x}{48x^2}$ 5. $42b^2; \frac{35a}{42b^2}, \frac{21ab}{42b^2}$ 7. $63z^4; \frac{91z^3}{63z^4}, \frac{135}{63z^4}$ 9. $54a^2bc^2; \frac{32ac}{54a^2bc^2}, \frac{3}{54a^2bc^2}$
11. $24x^3; \frac{18x}{24x^3}, \frac{40x^2}{24x^3}, \frac{27}{24x^3}$ 13. $140a^2b^3; \frac{24ab^2}{140a^2b^3}, \frac{70b^2}{140a^2b^3}, \frac{35a}{140a^2b^3}$ 15. $4(2x - 1); \frac{10}{4(2x - 1)}, \frac{3(2x - 1)}{4(2x - 1)}$
17. $a(a + 1)(a - 1); \frac{a - 1}{a(a + 1)(a - 1)}, \frac{a}{a(a + 1)(a - 1)}$ 19. $(a - 2)(a + 2)(a - 3); \frac{a(a + 2)}{(a - 2)(a + 2)(a - 3)}, \frac{(2a - 1)(a - 3)}{(a - 2)(a + 2)(a - 3)}$
21. $(R + 3)(R - 3)(R + 2); \frac{4R(R + 2)}{(R + 3)(R - 3)(R + 2)}, \frac{2R(R + 3)}{(R + 3)(R - 3)(R + 2)}$
23. $(E - 2)^2(E + 3)(E - 3); \frac{E(E - 2)^2}{(E - 2)^2(E + 3)(E - 3)}, \frac{2(E - 3)(E - 2)}{(E - 2)^2(E + 3)(E - 3)}, \frac{E(E^2 - 9)}{(E - 2)^2(E + 3)(E - 3)}$
25. $(a - 2)(a + 2)(a + 1)(a - 5); \frac{2a(a + 2)(a - 5)}{(a - 2)(a + 2)(a + 5)(a - 5)}, \frac{3a(a + 1)(a - 5)}{(a - 2)(a + 2)(a + 1)(a - 5)}, \frac{(a + 1)(a^2 - 4)}{(a - 2)(a + 2)(a + 1)(a - 5)}$
27. $2(x + 3)^2(x - 4); \frac{2(x - 3)(x - 4)}{2(x + 3)^2(x - 4)}, \frac{x(x + 3)(x - 4)}{2(x + 3)^2(x - 4)}, \frac{2(x - 1)(x + 3)}{2(x + 3)^2(x - 4)}$
29. $(x + 7)(x - 7)(x + 2); \frac{-5x(x + 2)}{(x + 7)(x - 7)(x + 2)}, \frac{3x(x + 7)}{(x + 7)(x - 7)(x + 2)}, \frac{7(x - 7)(x + 2)}{(x + 7)(x - 7)(x + 2)}$

Exercise 9.5

1. $\frac{23}{6x}$ 3. $\frac{1}{6a}$ 5. $\frac{13}{12z}$ 7. $\frac{52a}{45}$ 9. $\frac{8x + 1}{48}$ 11. $\frac{39b - 5}{36}$ 13. $\frac{11}{3}$ 15. $\frac{-3m^2 + 14m - 2}{4m}$ 17. $\frac{a + 5}{a^2b}$ 19. $\frac{33a + 14}{30a^2b}$
21. $\frac{10y + 7x - 1}{xy}$ 23. $\frac{9xy^2 + 10x^2y - 3}{12x^2y^2}$ 25. $\frac{9x + 7}{(x + 3)(x - 1)}$ 27. $\frac{45 - 9x}{2(2x - 3)(x + 3)}$ 29. $\frac{4a - 1}{2(a - 1)(2a + 1)}$
31. $\frac{11a + 6b}{a(a + b)(a - b)}$ 33. $\frac{4a - 3}{4a^2 - b^2}$ 35. $\frac{x^2 + 4x}{x^2 - 1}$ 37. $\frac{2xy + 4}{x^2(x - y)}$ 39. $\frac{a + b}{a - b}$ 41. $\frac{-3}{(y + 1)(y - 2)}$
43. $\frac{9x + 22}{(x + 3)(x - 3)(x + 2)}$ 45. $\frac{6 - a}{(a - 3)(a - 2)}$ 47. $\frac{-z^3 + 9z^2 - 13}{(z^2 - 4)(z^2 - 1)}$ 49. $\frac{3R^3 - 39R^2 + 133R + 15}{(R + 5)(R - 5)(R + 3)(R - 3)}$
51. $\frac{(u - 1)(R_2 + R_1)}{R_1R_2}$ 53. $\frac{bc + ac + ab}{abc}$ 55. $\frac{R_2R_3R_4 + R_1R_3R_4 + R_1R_2R_4 + R_1R_2R_3}{R_1R_2R_3R_4}$

Exercise 9.6

1. $\frac{8}{3}$ 3. $\frac{21}{2}$ 5. -22 7. 0 9. 15 11. $-\frac{5}{8}$ 13. $\frac{19}{24}$ 15. $-\frac{5}{27}$ 17. 1 19. $-\frac{13}{2}$ 21. $\frac{4}{37}$ 23. 36 25. $\frac{47}{9}$
27. $-\frac{11}{20}$ 29. $\frac{11}{5}$ 31. 40 33. $-\frac{19}{3}$ 35. $\frac{8}{3}$ 37. $x = \frac{2y}{3y - 1}$ 39. $c_1 = \frac{cc_2}{c_2 - c}$ 41. $a = \frac{8b}{b - 8}$ 43. $a = \frac{3b - 5}{4}$
45. $x = \frac{-(2y + 12)}{5}$ 47. $D = \frac{P}{pL}$ 49. $N = \frac{2D}{OD - D}$ 51. $f = \frac{Fa}{A}$ 53. $T_2 = \frac{P_2T_1V_2}{P_1V_1}$ 55. $L_1 = \frac{L_2}{d(t_2 - t_1) + 1}$
57. $R_1 = \frac{RR_2}{R_2 - R}$

Exercise 9.7

1. $\frac{5}{6}$ 3. $\frac{3}{2}$ 5. $\frac{8}{9}$ 7. $\frac{5}{2}$ 9. $\frac{23}{20}$ 11. $\frac{11}{2}$ 13. $\frac{4x+1}{4x-3}$ 15. $\frac{3a+1}{2-4a}$ 17. $\frac{4a^2+3}{5a^2-3a}$ 19. $\frac{ab-3}{ab+4}$ 21. $\frac{x+y}{y-x}$
23. xy 25. $\frac{5a^2+4b}{a^3b-a^2b^2}$ 27. $-\frac{y}{x}$ 29. $\frac{4xy}{(x^2-y^2)^2}$ 31. $\frac{2b+3}{a^3b^4}$ 33. $\frac{3a-1}{17-a}$ 35. $\frac{-5}{x+31}$ 37. $-\frac{3a^2+5a-8}{a+29}$
39. $\frac{(y-3)(y+6)}{(y+1)(y-4)}$ 41. $\frac{T_1}{T_2-T_1}$ 43. $\frac{L_1L_2L_3}{L_2L_3+L_1L_3+L_1L_2}$ 45. $f_s\sqrt{\frac{c+v}{c-v}}$

Chapter Review

1. $\frac{3b}{a}$ 2. $\frac{3xz}{2y^2}$ 3. $\frac{x-7}{x+7}$ 4. $\frac{x+3}{x+7}$ 5. $\frac{6}{5}$ 6. $-(x+y)$ 7. $\frac{3p-2}{5p+1}$ 8. $\frac{R-4}{3R-1}$ 9. $\frac{5-n}{2+n}$ 10. $\frac{9}{2}$ 11. $\frac{9a}{b}$
12. $\frac{2x-y}{2x+y}$ 13. $x-2$ 14. $\frac{mx-nx}{2}$ 15. $\frac{1}{(y-1)^2}$ 16. 2 17. $\frac{1-a}{3(2a-1)(x+5)}$ 18. $\frac{x-1}{x+1}$
19. $\frac{2a}{3}$ 20. $\frac{8b}{21}$ 21. $\frac{9}{2ab}$ 22. $\frac{14x-7}{3(3x+4)}$ 23. $\frac{1}{(x-2)(x+6)}$ 24. $\frac{10}{3}$ 25. $\frac{1}{x-8}$ 26. $\frac{3-b}{(b+3)(b-1)}$
27. $\frac{3a+3}{4(a-1)}$ 28. $36x^2$; $\frac{12}{36x^2}, \frac{15x}{36x^2}$ 29. $42a^2b^2$; $\frac{15ab}{42a^2b^2}, \frac{16}{42a^2b^2}$ 30. $2(x+3)(x-3)$; $\frac{2x}{2(x+3)(x-3)}, \frac{2x(x-3)}{2(x+3)(x-3)}$
31. $(y+5)(y-5)(y+3)$; $\frac{4y(y+5)}{(y+5)(y-5)(y+3)}, \frac{9y(y+3)}{(y+5)(y-5)(y+3)}$ 32. $\frac{68x}{105}$ 33. $\frac{23}{48a}$ 34. $\frac{47x-6}{(3x+1)(4x-3)}$
35. $\frac{4x^2-23x}{x^2-16}$ 36. $\frac{80a+48b-15ab}{20a^2b^2}$ 37. $\frac{4a^2+4a+6}{a+1}$ 38. $\frac{-(10x^2+3)}{x^2+1}$ 39. $\frac{16x^2+5x+15}{24x^2}$ 40. $\frac{13y^2-89y}{(y-9)(y^2-4)}$
41. $\frac{-2x^2-36x+2}{x^2-9}$ 42. $\frac{8x^2-40x-180}{x(x^2-x-20)}$ 43. $\frac{10x-5y}{(x+y)^2(x-2y)}$ 44. -96 45. $\frac{41}{48}$ 46. $\frac{13}{24}$ 47. $\frac{6}{5}$ 48. $-\frac{13}{5}$ 49. $-\frac{1}{14}$
50. $\frac{a+b}{3}$ 51. -3 52. $\frac{4a-3b}{a}$ 53. $-(4b+5c)$ 54. $\frac{Aey}{F}$ 55. $\frac{Wp-2E}{EF}$ 56. $\frac{TMAr}{TMA-2}$ 57. $\frac{Fr}{v^2+gr}$
58. $\frac{16}{63}$ 59. $\frac{1}{7}$ 60. 1 61. $\frac{y-x}{y+x}$ 62. $\frac{4+3x}{2-5x}$ 63. $\frac{a^2-b^2}{a^2+b^2}$ 64. $y-x$ 65. $\frac{xy}{y+x}$ 66. $\frac{a^3b-2a^2b^2-3ab^2}{ab-3b^2}$

Chapter Test

1. $\frac{6}{7}$ 2. $\frac{5b}{4a}$ 3. $\frac{x-5}{x-2}$ 4. $\frac{2a-b}{a^2+b^2}$ 5. $\frac{x-5}{x-3}$ 6. $\frac{a-3b}{3a+2b}$ 7. $\frac{3a^2}{4b^2}$ 8. $\frac{x^2-x-2}{x^2-9}$ 9. $\frac{a^2-2a+1}{a+2}$ 10. $\frac{x-6}{x+6}$
11. $\frac{7b}{a}$ 12. $\frac{(x-y)(x-1)}{x}$ 13. y^3 14. $\frac{x}{y}$ 15. $\frac{11}{3x}$ 16. $\frac{3b+2a}{ab}$ 17. $\frac{2x-3y}{x^2y^2}$ 18. $\frac{4a+2b}{(a+b)(a-b)}$ 19. $\frac{a^2-a+1}{a(a^2-1)}$
20. $\frac{7y+23}{(y+5)(y-5)(y+1)}$ 21. $\frac{x^2-9x}{(x+4)(x-3)(x-5)}$ 22. $\frac{15}{4}$ 23. 9 24. $\frac{2}{3}$ 25. $\frac{2b-a}{b+3}$ 26. $\frac{x^2y-x^2}{xy^2-y^2}$ 27. $\frac{4+3y}{2y-1}$
28. -18 29. 7 30. -1 31. $-\frac{2}{3}$ 32. 9 33. $\frac{20}{3}$ 34. $\frac{-y}{7y-2}$ 35. $\frac{fq}{q-f}$ 36. $\frac{RR_1R_3}{R_1R_3-RR_3-RR_1}$ 37. $v_2 = at + v_1$
38. $D = \frac{144p}{h}$ 39. $Q_2 = \frac{Q_1n}{1+n}$ 40. $D_0 = \frac{D_if}{D_i-f}$

Chapter 10

Exercise 10.1

1. 1 3. 5 5. 1 7. $\frac{1}{R^5}$ 9. $\frac{1}{9P^2}$ 11. $\frac{9}{C^4}$ 13. $5x^4$ 15. $\frac{x^2}{3}$ 17. 1 19. 1 21. 1 23. x^3 25. $\frac{1}{a^6}$ 27. $-\frac{1}{125}$ 29. $\frac{1}{8}$
31. $\frac{a^2}{b^7}$ 33. $\frac{S^5}{R^2}$ 35. $\frac{1}{R^2S^5}$ 37. $\frac{3}{a^5b^4}$ 39. $\frac{27}{x^6}$ 41. $\frac{y^{12}}{27x^3}$ 43. $\frac{R^5}{S^9}$ 45. $25R^5$ 47. $\frac{3z^4}{xy^5}$ 49. $\frac{4}{ab^9}$ 51. $\frac{3}{R^4}$ 53. $\frac{x^4}{9}$ 55. $\frac{1}{R^4}$
57. $\frac{b^3}{12a^4}$ 59. $\frac{1}{10}$ 61. $\frac{1}{10,000}$ 63. $\frac{1}{1,000,000}$ 65. 10^{-1} 67. 10^{-9}

Exercise 10.2

1. 10 3. 5 5. −12 7. −11 9. 2 11. 5 13. −3 15. 2 17. −3 19. 2 21. −3 23. 1 25. 1.2 27. −0.8 29. 0.3 31. −0.5 33. −0.4 35. 0.4 37. 9.00 cm 39. 24.0 amp 41. 13 cm 43. 1,550 ohms 45. 4.1 47. −1.2 49. 0.019 51. 1 53. −2.5

Exercise 10.3

1. $2\sqrt[3]{2}$ 3. $2\sqrt[3]{6}$ 5. $2\sqrt[5]{2}$ 7. $3\sqrt{7}$ 9. $5\sqrt{3}$ 11. $b^2\sqrt[3]{b^2}$ 13. y^3 15. $D^2\sqrt{D}$ 17. b^5 19. $3b^2c\sqrt{ac}$ 21. $2x^2z\sqrt{6xyz}$ 23. $2LM\sqrt[3]{2LM^2}$ 25. $S\sqrt[4]{S}$ 27. $3RS^3\sqrt[3]{3R^2S^2}$ 29. $9\sqrt{2}$ 31. 15 33. $2\sqrt{15}$ 35. $12\sqrt{3}$ 37. $3\sqrt{6}$ 39. 20 41. $5\sqrt{3}$ 43. $5x\sqrt{3}$ 45. $b\sqrt[3]{b}$ 47. a 49. $4b\sqrt{3a}$ 51. $3ab\sqrt[3]{2b}$ 53. $3S\sqrt[4]{3R^3}$ 55. $4a^5b^3\sqrt[3]{3b}$ 57. 2.0 59. 2.000

Exercise 10.4

1. $\frac{3}{5}$ 3. $\frac{5}{7}$ 5. $\frac{\sqrt{3}}{2}$ 7. $\frac{2}{3}$ 9. $\frac{2}{3}$ 11. $\frac{a}{b}$ 13. $\frac{y^2}{4}$ 15. $\frac{\sqrt[3]{x}}{y^4}$ 17. $\frac{a^3}{b}$ 19. $\frac{2x}{y^2}$ 21. $\frac{2\sqrt[5]{x^4}}{y}$ 23. $\frac{a^2b^3\sqrt[4]{ab}}{c^2}$ 25. $\frac{\sqrt{2}}{2}$ 27. $\frac{2\sqrt{7}}{7}$ 29. $\frac{\sqrt{15}}{15}$ 31. $\frac{2\sqrt{3}}{15}$ 33. $\sqrt{2}$ 35. $\frac{5\sqrt{2}}{2}$ 37. $\frac{2\sqrt[3]{3}}{3}$ 39. $\frac{2\sqrt[5]{3}}{3}$ 41. $\frac{2\sqrt[4]{5}}{5}$ 43. $\frac{x\sqrt{y}}{y}$ 45. $\frac{\sqrt{R}}{R}$ 47. $\frac{R\sqrt[3]{S}}{S}$ 49. $\frac{\sqrt[3]{ab}}{b}$ 51. $\frac{xy^2\sqrt[5]{z^2}}{z}$ 53. $x\sqrt[5]{y^2}$ 55. b 57. $\frac{2\sqrt[3]{x^2y}}{xy}$ 59. $\frac{\sqrt[5]{x^4y^2}}{xy}$ 61. $\frac{\sqrt[5]{a^3b}}{b}$ 63. $\sqrt[3]{a^2b}$ 65. $x\sqrt[7]{x^2y^5}$ 67. 3.0000 m

Exercise 10.5

1. $9\sqrt{3}$ 3. $10\sqrt{5}$ 5. $6\sqrt{3}$ 7. $3\sqrt{7}$ 9. $3\sqrt{a}$ 11. $8\sqrt{R}$ 13. $7\sqrt{xy}$ 15. $8\sqrt{R} + 2\sqrt{RS}$ 17. $4\sqrt{xy} + 3\sqrt{y}$ 19. $7\sqrt{2}$ 21. $\sqrt{3}$ 23. $17\sqrt{7}$ 25. $7\sqrt{2}$ 27. $2\sqrt{3} + 8\sqrt{2}$ 29. $7\sqrt{2a}$ 31. $-\sqrt{x}$ 33. $17\sqrt{2R}$ 35. $2\sqrt{2R} + 6\sqrt{3R}$ 37. $\sqrt[3]{4}$ 39. $17\sqrt[4]{3}$ 41. $5\sqrt[3]{2}$ 43. $3\sqrt[3]{3} + 10\sqrt[3]{2}$ 45. $5\sqrt[4]{x^3}$ 47. $\sqrt[3]{x^2y}$ 49. $2a\sqrt[3]{b^2}$ 51. 14 m 53. 8.10 cm by 8.10 cm

Exercise 10.6

1. $3\sqrt{2} + 3\sqrt{3}$ 3. $10\sqrt{6} + 5\sqrt{2}$ 5. $\sqrt{6} + \sqrt{14}$ 7. $2\sqrt{6} + \sqrt{15}$ 9. $2\sqrt{3} + 3\sqrt{2}$ 11. $7\sqrt{6} + 2\sqrt{35}$ 13. $9 + 18\sqrt{5}$ 15. $3R + \sqrt{RS}$ 17. $15\sqrt[3]{4} - 3$ 19. $4\sqrt[3]{3} + 6\sqrt[3]{4}$ 21. $R + R\sqrt[5]{R^2}$ 23. $3\sqrt[5]{y^4} + y\sqrt[5]{y}$ 25. $14 + 7\sqrt{2}$ 27. $15 + a + 8\sqrt{a}$ 29. $42 + 47\sqrt{y} + 10y$ 31. 2 33. 22 35. $16 - 6\sqrt{7}$ 37. $R + 3\sqrt{RS} + 2S$ 39. $a - 2\sqrt{ab} + b$ 41. $4a - b$ 43. $a^2b - c^2$ 45. $9R + 6\sqrt{RS} + S$ 47. $11 + \sqrt{3}$ 49. $5\sqrt{7} + \sqrt{2}$ 51. $\sqrt{R} - 3\sqrt{S}$ 53. $R\sqrt{S} + \sqrt{T}$ 55. $-\sqrt{3} - 2$ 57. $6 + 2\sqrt{6}$ 59. $\sqrt{6} + \sqrt{3}$ 61. $\frac{\sqrt{R} + \sqrt{S}}{R - S}$ 63. $\frac{a^2 + a\sqrt{b}}{a^2 - b}$ 65. $\frac{4\sqrt{3} + 2\sqrt{6}}{3}$ 67. $\frac{-5 + \sqrt{21}}{2}$ 69. 645

Exercise 10.7

1. $\sqrt{2}$ 3. $\sqrt[3]{6^2} = \sqrt[3]{36}$ 5. $\sqrt[4]{a}$ 7. $\sqrt[5]{X^3}$ 9. $\sqrt[5]{T^9} = T\sqrt[5]{T^4}$ 11. $7^{\frac{1}{2}}$ 13. $R^{\frac{1}{2}}$ 15. $x^{\frac{2}{3}}$ 17. $S^{\frac{3}{4}}$ 19. $x^{\frac{4}{5}}$ 21. 6 23. 5 25. a^2 27. S^2 29. 2 31. −3 33. 4 35. 8 37. 32 39. 64 41. 27 43. $\frac{1}{3}$ 45. $\frac{1}{9}$ 47. $-\frac{1}{3}$ 49. 4 51. R 53. $X^{\frac{7}{12}}$ 55. $B^{\frac{9}{10}}$ 57. $x^{\frac{1}{4}}$ 59. $S^{\frac{1}{2}}T^{\frac{1}{4}}$ 61. $R^{\frac{1}{2}}S^{\frac{1}{2}}$ 63. R^3 65. $T^{\frac{1}{6}}$ 67. $8a^3$ 69. xy^4 71. $4a^4b^2$ 73. $R^{\frac{13}{6}}$ 75. $y^{\frac{5}{4}}$

Exercise 10.8

1. $3j$ 3. $6j$ 5. $2\sqrt{3}j$ 7. $2\sqrt{6}j$ 9. $2\sqrt{5}j$ 11. $\sqrt{6}j$ 13. −9 15. −36 17. −3 19. −8 21. $-\sqrt{15}$ 23. −2 25. −5 25. −5 27. $-3\sqrt{2}$ 29. −4 31. −6 33. $-2\sqrt{7}$ 35. $-\sqrt{21}$ 37. −1 39. 1 41. $-j$ 43. $-j$ 45. 1 47. $6 + 2j$, complex 49. j, imaginary 51. $6 - 2j$, complex 53. −14, real 55. $4 - 4j$, complex

Chapter Review

1. 1 2. 1 3. 7 4. $\frac{4}{C^3}$ 5. $\frac{S^2}{3}$ 6. 5 7. $\frac{1}{X^3T^2}$ 8. $\frac{1}{8x^6y^3}$ 9. $\frac{2a^2b^2}{3}$ 10. $\frac{4}{x^2}$ 11. 9 12. 0.5 13. −3 14. −7 15. −3 16. −1 17. 3 18. −5 19. −2 20. 2 21. $2\sqrt[3]{4}$ 22. $3\sqrt{3}$ 23. $2\sqrt{10}$ 24. $a\sqrt[4]{a^3}$ 25. $3RS\sqrt{2S}$ 26. $3xy^2\sqrt[3]{3xz}$ 27. $2\sqrt{7}$ 28. $2\sqrt[3]{3}$ 29. $6\sqrt{5}$ 30. $2ab\sqrt[4]{ab^2}$ 31. $\frac{4\sqrt{17}}{17}$ 32. $\frac{\sqrt{14}}{6}$ 33. $\frac{\sqrt[3]{12}}{4}$ 34. $\frac{\sqrt{RS}}{S}$ 35. $\frac{\sqrt{xy}}{y^2}$ 36. $\frac{\sqrt{ab}}{b}$ 37. $\frac{2\sqrt{RS}}{S}$ 38. $\frac{\sqrt[3]{xy}}{y}$ 39. $\frac{\sqrt[3]{abc}}{c}$ 40. $\frac{\sqrt[7]{a^5b^4c^3}}{bc}$ 41. $7\sqrt{7}$ 42. $8\sqrt{2}$ 43. $3\sqrt{5}$ 44. $24\sqrt{3}$ 45. $\sqrt{2R}$ 46. $17\sqrt{x}$ 47. $16\sqrt[3]{4}$ 48. $12\sqrt[3]{2}$ 49. $x\sqrt[4]{y}$ 50. $5R\sqrt[3]{S}$ 51. $\sqrt{15} - \sqrt{21}$ 52. $2\sqrt{35} + 2\sqrt{15}$ 53. $R + R\sqrt[3]{R}$ 54. $8 - 2\sqrt{7}$ 55. $39 - 12\sqrt{3}$

56. -1 57. $8 + 2\sqrt{15}$ 58. $4a - b$ 59. $s - \sqrt{st} - 6t$ 60. $4R - 4\sqrt{RS} + S$ 61. $-\sqrt{3} - 2$ 62. $\dfrac{-\sqrt{6} + 4}{5}$

63. $\dfrac{\sqrt{R} - S}{R - S^2}$ 64. $\dfrac{\sqrt{xy} - x}{y - x}$ 65. $\dfrac{R^2 - R\sqrt{S}}{R^2 - S}$ 66. $\dfrac{-11 - 6\sqrt{2}}{7}$ 67. 6 68. 2 69. 4 70. 125 71. -2 72. $\dfrac{1}{4}$ 73. R

74. $S^{\frac{13}{12}}$ 75. $R^{\frac{1}{4}}$ 76. $a^{\frac{9}{8}}$ 77. x^2 78. $8a^3b^6$ 79. $R^{\frac{3}{2}}S^{\frac{1}{2}}$ 80. $x^{\frac{5}{6}}$ 81. $a^{\frac{5}{12}}$ 82. $B^{\frac{3}{5}}$ 83. $7j$ 84. $2\sqrt{7}j$ 85. -25 86. -5
87. -10 88. -12 89. $-\sqrt{21}$ 90. $-2\sqrt{3}$ 91. -6 92. -6

Chapter Test

1. $7\sqrt{3}$ 2. 12 3. -5 4. $6 + 3\sqrt{2}$ 5. $\dfrac{x\sqrt{y}}{y}$ 6. 4 7. 8 8. $B^{\frac{5}{4}}$ 9. $2RS^2\sqrt[5]{2R^2T^3}$ 10. $y^{\frac{1}{2}}$ 11. 29 12. $\dfrac{-\sqrt{5} - 3}{4}$

13. z 14. $14 - 6\sqrt{5}$ 15. $8j$ 16. -2 17. $\dfrac{1}{9}$ 18. $5\sqrt{R}$ 19. S^2 20. $\dfrac{\sqrt[5]{RS^3T^2}}{ST}$ 21. $x - 4y$ 22. $\dfrac{a\sqrt{a} - a\sqrt{b}}{a - b}$ 23. $3\sqrt{3}j$

24. $\dfrac{ab^3}{2}$ 25. -49 26. $\dfrac{\sqrt{6}}{3}$ 27. $\dfrac{9 - \sqrt{6}}{25}$ 28. $9R - 6\sqrt{RS} + S$ 29. -2 30. $x^{\frac{8}{3}}y^{\frac{3}{2}}$ 31. -4

32. $R^{\frac{5}{12}}$ 33. $-\sqrt{15}$ 34. 5.40 in by 7.20 in 35. 7.0 yd

Chapter 11

Exercise 11.1

$x^2 - 9x + 18 = 0, a = 1, b = -9, c = 18$ 3. $x^2 - x - 5 = 0, a = 1, b = -1, c = -5$
5. $x^2 - 6x = 0, a = 1, b = -6, c = 0$ 7. $12r^2 - 8r - 15 = 0, a = 12, b = -8, c = -15$
9. $12a^2 - 25a + 12 = 0, a = 12, b = -25, c = 12$ 11. $x^2 - 6x + 9 = 0, a = 1, b = -6, c = 9$
13. $6x^2 + x - 12 = 0, a = 6, b = 1, c = -12$ 15. $x^2 + 12x - 28 = 0, a = 1, b = 12, c = -28$
17. $25N^2 - 50N - 24 = 0, a = 25, b = -50, c = -24$ 19. $-12x^2 + 17x - 6 = 0, a = -12, b = 17, c = -6$
21. Not quadratic 23. $x^2 - x - 23 = 0, a = 1, b = -1, c = -23$ 25. Not quadratic 27. Not quadratic 29. $4x^2 - 9x + 1 = 0, a = 4, b = -9, c = 1$ 31. $6x^2 - 3 = 0, a = 6, b = 0, c = -3$ 33. $x^2 - 27 = 0, a = 1, b = 0, c = -27$
35. $x^2 + 9x + 27 = 0, a = 1, b = 9, c = 27$ 37. $x^2 - 4x - 8 = 0, a = 1, b = -4, c = -8$ 39. Not quadratic

Exercise 11.2

1. 4, 3 3. 4, -1 5. -7, 2 7. 0, -2 9. $\pm 2\sqrt{2}$ 11. $\pm\dfrac{5}{3}$ 13. $\pm\sqrt{2}$ 15. ± 3 17. ± 4 19. 0, -3 21. ± 3 23. 8

25. $\dfrac{5}{2}, -\dfrac{2}{3}$ 27. $3, \dfrac{-2}{3}$ 29. $\dfrac{-5}{6}, \dfrac{3}{2}$ 31. $-1, \dfrac{3}{2}$ 33. $\dfrac{4}{3}, \dfrac{-3}{2}$ 35. $\dfrac{\pm\sqrt{10}}{3}$ 37. $-1, \dfrac{9}{2}$ 39. $\dfrac{-5}{3}, \dfrac{-4}{3}$ 41. $-4, \dfrac{2}{3}$

43. (a) ± 1; solution is 1; (b) $\pm\dfrac{1}{2}$; solution is $\dfrac{1}{2}$ 45. -9, 3 47. (a) -14, 2; solution is 2; (b) -16, 4; solution is 4

Exercise 11.3

1. 2, 1 3. -3 5. ± 5 7. $\pm\sqrt{2}$ 9. $\dfrac{9 \pm \sqrt{65}}{2}$ 11. $-1 \pm \sqrt{7}$ 13. $4 \pm \sqrt{15}$ 15. $\dfrac{5 \pm \sqrt{97}}{6}$ 17. 0, 3 19. $0, \dfrac{9}{5}$

21. $\dfrac{-9 \pm \sqrt{57}}{6}$ 23. $2, \dfrac{-5}{3}$ 25. -4 27. $\dfrac{5}{2}$ 29. $\dfrac{-3}{2}$ 31. $\dfrac{1 \pm \sqrt{22}}{3}$ 33. $-1 \pm \sqrt{7}$ 35. $\dfrac{3 \pm \sqrt{87}j}{6}$

37. (a) $10 \pm 2\sqrt{21}$, (b) $\dfrac{50 \pm 2\sqrt{545}}{5}$ 39. (a) 10, 0, (b) $\dfrac{10 \pm 3\sqrt{10}}{2}$, (c) $5 \pm \sqrt{15}$ 41. 5.45, 0.55 43. 2.35, -0.85

45. 0.55, -1.22 47. ± 0.63 49. 2.89, -1.64

Exercise 11.4

1. 8, 3 3. -8, 7 5. -7, 4 7. 4 9. 4, 5 or -2, -1 11. -13, -11 or 11, 13 13. -14, -12 or 12, 14

15. 7, 13 or -13, -7 17. 4, 6 or -2, 0 19. -9, -4 21. 2, 7 or $-\dfrac{7}{3}$, -6 23. 9, 4 25. 9, 13 27. 4, 6 29. 2, 5

31. 6, 9 33. 15, 7 35. 8, 12 37. 18, 6 39. 8, 11 41. 2, 8 43. 1 45. 3 47. 8, 7 49. 4, 4, 8 51. 7, 8 53. 12, 4
55. 6, 36

Chapter Review

1. $3x^2 - 2x + 6 = 0, a = 3, b = -2, c = 6$ 2. $5e^2 - 4e = 0, a = 5, b = -4, c = 0$ 3. $7x^2 - 6 = 0, a = 7, b = 0, c = -6$ 4. $5x^2 + 4x - 8 = 0, a = 5, b = 4, c = -8$ 5. $x^2 - 3x + 7 = 0,$ $a = 1, b = -3, c = 7$ 6. $8y^2 + 4y + 3 = 0, a = 8, b = 4, c = 3$ 7. $-3, -5$ 8. $0, 5$ 9. ± 3 10. ± 3 11. 1 12. $7, -6$ 13. $-8, 1$ 14. $\frac{1}{2}, 1$ 15. $\frac{1}{3}, -\frac{1}{2}$ 16. $3, -1$ 17. $\frac{5}{2}, -1$ 18. $3, 2$ 19. $\frac{7}{2}, \frac{-3}{2}$ 20. $6, 1$ 21. $-\frac{9}{2}, 1$ 22. $6, 9$ 23. 8 or $-\frac{1}{4}$ 24. $8, 9$

Chapter Test

1. $5, -2$ 2. ± 9 3. $\frac{-3 \pm \sqrt{29}}{2}$ 4. $\pm\sqrt{10}$ 5. $1, -\frac{3}{2}$ 6. $0, 3$ 7. $\frac{1 \pm \sqrt{61}}{6}$ 8. ± 3 9. $1 \pm \sqrt{5}$ 10. $-2 \pm \sqrt{3}$ 11. $\pm\sqrt{3}$ 12. $0, 9$ 13. $\frac{3 \pm \sqrt{41}}{4}$ 14. $0, \frac{5}{2}$ 15. $-1, \frac{4}{5}$ 16. $\pm\sqrt{6}$ 17. $\pm\frac{4\sqrt{5}}{5}$ 18. $0, \frac{4}{5}$ 19. $-7, -2$ 20. $\frac{1 \pm \sqrt{2}j}{2}$ 21. $6, 8$ or $-6, -8$ 22. $\frac{4}{3}, \frac{3}{4}$ 23. $9, 5$ 24. $6, 16$

Chapter 12

Exercise 12.1

1. Line 3. Point 5. Ray 7. Broken line 9. Obtuse angle 11. Acute angle 13. Complementary angles 15. Supplementary angles 17. (a) $\angle ACE$ and $\angle ECD$, $\angle ECD$ and $\angle DCB$; (c) $\angle ACE$ and $\angle ECB$, $\angle ACD$ and $\angle DCB$; (e) $\angle ACE$ or $\angle ECD$; (g) $\angle ACB$ 19. $\angle\alpha$, $\angle B$, $\angle ABC$, $\angle CBA$ 21. 9.1° 23. 17.33° 25. 75.83° 27. 32°24′ 29. 10°3′ 31. 52°45′ 33. 1°7′12″ 35. 99°16′12″ 37. 125°, obtuse 39. 80°, acute

41. acute, 35°

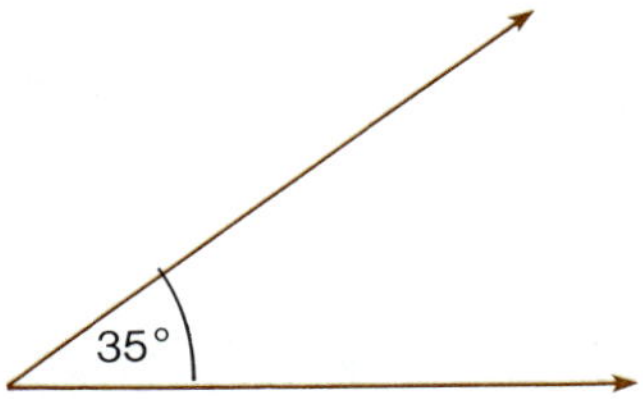

43. 115°, obtuse

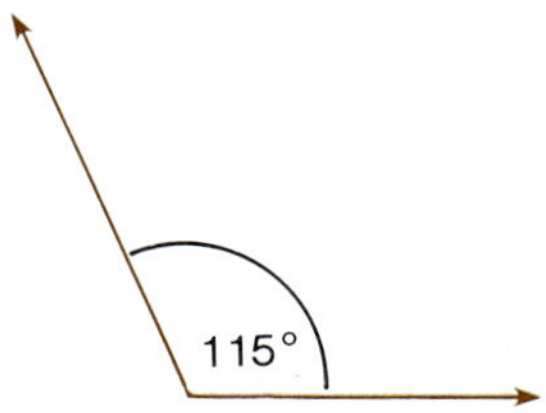

45. 123°, obtuse

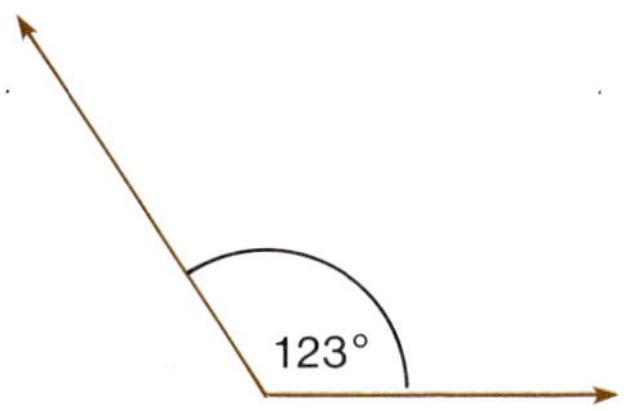

47. 175°, obtuse

49. 47°, acute

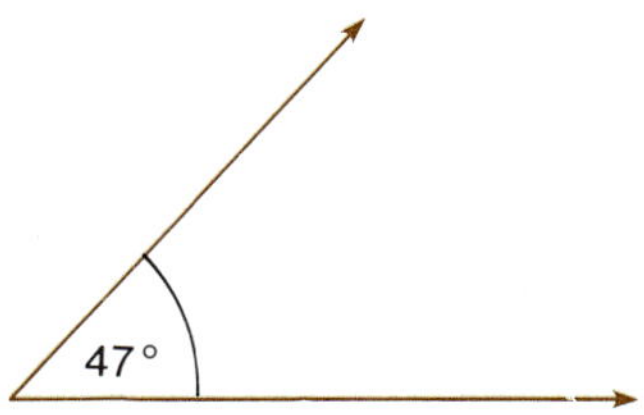

Exercise 12.2

1.

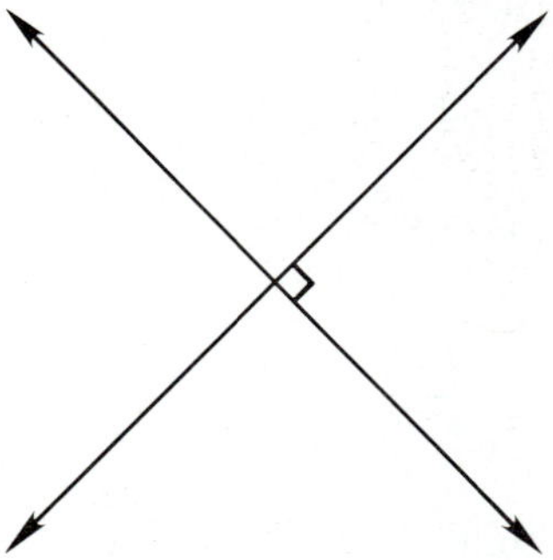

3. Alternate exterior 5. interior 7. Alternate interior 9. Corresponding 11. $m\angle E = m\angle H$, $m\angle E = m\angle A$, $m\angle E = m\angle D$, $m\angle C = m\angle F$, $m\angle C = m\angle G$, $m\angle C = m\angle B$, $m\angle D = m\angle H$, $m\angle D = m\angle A$, $m\angle F = m\angle G$, $m\angle F = m\angle B$, $m\angle A = m\angle H$, $m\angle B = m\angle G$ 13. 45°
15. Yes—the right angles are corresponding angles on two lines cut by a transversal.
17. $m\angle 1 = m\angle 3 = m\angle 5$, $m\angle 4 = m\angle 6 = m\angle 2$, $m\angle 2 = 27°$ 19. $m\angle 1 = 42°$, $m\angle 2 = 48°$, $BF \parallel CG$ Both lines form equal corresponding right angles with transversal $\overline{AD}$.

Exercise 12.3

1. Quadrilateral 3. Parallelogram 5. Octagon 7. Rectangle 9. Trapezoid 11. Equilateral triangle 13. Equilateral triangle 15. Scalene triangle 17. Acute triangle 19. Obtuse triangle 21. Acute triangle 23. RT and s, TS and r, RS and t 25. (a) 720°, (b) 1,080° 27. Square or rhombus 29. Parallelogram

31.

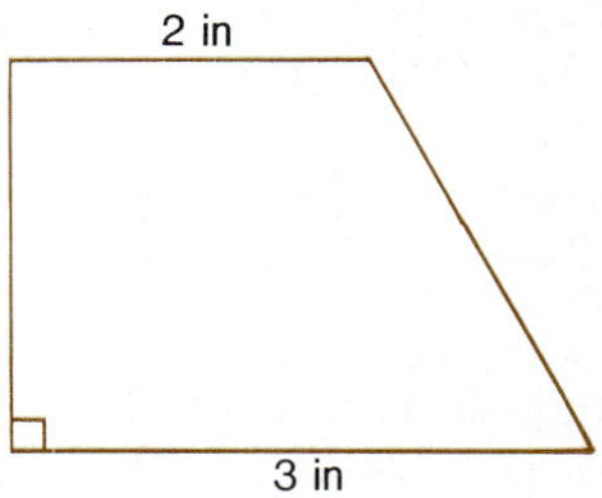

33. $\frac{7}{8}$ in; $m\angle 2 = m\angle 4$, $m\angle 1 = m\angle 3$

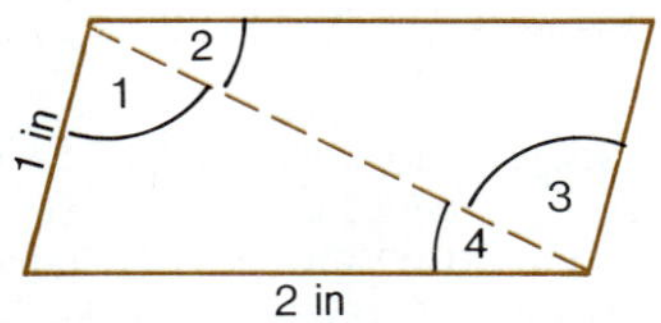

35. (a) 108°, (b) 120°, (c) 135° 37. 51° 39. 41°, 139°, 139° 41. $m\angle B = 52°$; $m\angle 1 = 52°$, $m\angle 2 = 76°$, $m\angle 3 = 52°$ 43. 106°

Exercise 12.4

1. 5.00 3. 13.0 5. 8.00 7. 18.0 9. 11.2 11. 7.00 13. 26.4 15. 4.83 17. 1.91 19. 0.302 21. (a) 5.66 cm, (b) 4.24 in, (c) 2.83 ft 23. 25.0 ft 25. 23.3 ft 27. 429.9 ft 29. 813.1 m 31. 2.6 in 33. 9.2 ft 35. 5.63 m 37. 302.9 ft 39. (a) 3.25 ohms, (b) 18.7 ohms

Exercise 12.5

1. $\overline{HD}$, $\overline{BF}$ 3. $\overleftrightarrow{CA}$, $\overleftrightarrow{AE}$ 5. $\angle DOF$, $\angle BOH$ 7. $\angle DBF$, $\angle BDH$ 9. $\overleftrightarrow{HD}$ 11. 120° 13. 60° 15. 30° 17. 72° 19. 105° 21. 60° 23. 35° 25. 140°

Exercise 12.6

1. $\overline{AC}$ and $\overline{DF}$, $\overline{CB}$ and $\overline{FE}$, $\overline{AB}$ and $\overline{DE}$, $\angle A$ and $\angle D$, $\angle B$ and $\angle E$, $\angle C$ and $\angle F$
3. $\overline{XZ}$ and $\overline{ST}$, $\overline{XY}$ and $\overline{RS}$, $\overline{ZY}$ and $\overline{RT}$, $\angle X$ and $\angle S$, $\angle Y$ and $\angle R$, $\angle Z$ and $\angle T$
5. $m\angle B = m\angle B$, $m\angle A = m\angle BDE$, $m\angle C = m\angle BED$; then $\triangle ABC \sim \triangle DBE$, $\overline{AB}$ and $\overline{DB}$, $\overline{AC}$ and $\overline{DE}$, $\overline{BC}$ and $\overline{BE}$
7. $\triangle PQR \sim \triangle QSR$ since $m\angle R = m\angle R$, $m\angle P = m\angle SQR$, $m\angle PQR = m\angle RSQ$; $\overline{PR}$ and $\overline{QR}$, $\overline{SQ}$ and $\overline{PQ}$, $\overline{RQ}$ and $\overline{SR}$.
$\triangle PQR \sim \triangle PSQ$ since $m\angle P = m\angle P$, $m\angle R = m\angle PQS$, $m\angle PQR = m\angle PSQ$; $\overline{PR}$ and $\overline{PQ}$, $\overline{PQ}$ and $\overline{PS}$, $\overline{SQ}$ and $\overline{RQ}$

9. $m\angle B = m\angle B$, $m\angle A = m\angle BED$, $m\angle C = m\angle BDE$, so $\triangle ABC \sim \triangle EBD$, $x \approx 4.2$
11. $m\angle A = m\angle E$, $m\angle B = m\angle D$, $m\angle BCA = m\angle ECD$, then $\triangle ABC \sim \triangle EDC$, $x = 13$, $y = 3.9$
13. $m\angle A = 80° = m\angle D$, $m\angle C = 70° = m\angle F$, $m\angle B = 30° = m\angle E$, then $\triangle ABC \sim \triangle EFD$, $x = 6.7$, $y = 5.4$
15. $m\angle CDB = 90° = m\angle ABC$, $m\angle C = m\angle C$, $m\angle A = m\angle CBD$, so $\triangle ABC \sim \triangle BDC$, $x = 4$
17. $m\angle C = m\angle C$, $m\angle A = m\angle CDE$, $m\angle B = m\angle CED$, so $\triangle ABC \sim \triangle DEC$, $x = 2.4$
19. $m\angle C = m\angle D$, $m\angle CBA = 53° = m\angle BED$, $m\angle A = 37° = m\angle EBD$, so $\triangle ABC \sim \triangle BED$, $x \approx 5.3$
21. $6\frac{3}{13}$ ft 23. 300 ft × 600 ft 25. 278 ft 27. $\overline{DE} \approx 9.3$ ft, $\overline{FG} \approx 4.7$ ft, $\overline{DC} \approx 15$ ft, $\overline{EF} \approx 13$ ft, $\overline{AB} \approx 39$ ft
29. 200 ft × 36 ft

Chapter Review

1. $\overrightarrow{AB}$ 2. $\overleftrightarrow{AB}$ 3. $\overline{AB}$ 4. Angle

5. (a) 47°, (b) 132° 6. Degrees, radians 7. (a) 19.3°, (b) 73.7°, (c) 27.9°

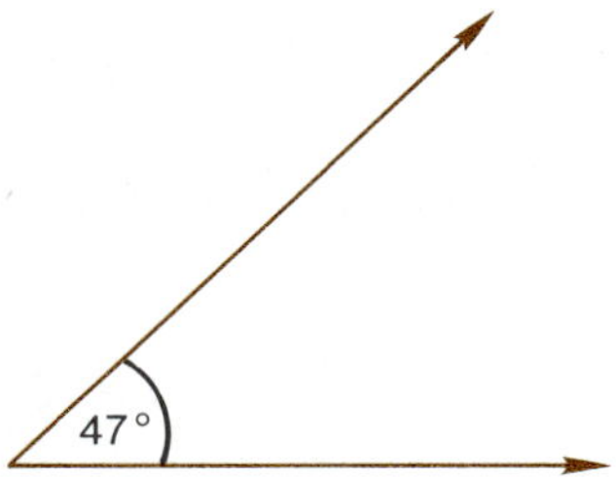

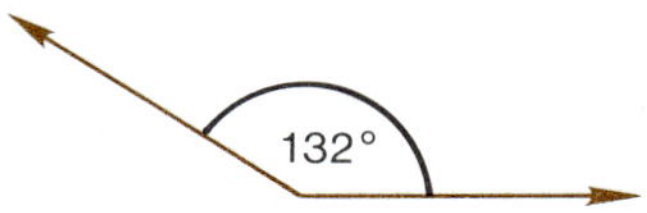

8. (a) 10°24′, (b) 37°15′, (c) 63°17′24″ 9. Obtuse, acute, right, straight 10. Adjacent 11. (a) Complementary, (b) Supplementary 12. Vertical, horizontal, oblique 13. Parallel 14. Perpendicular 15. (a) $\angle 3$ and $\angle 2$, $\angle 4$ and $\angle 1$; (b) $\angle 7$ and $\angle 6$, $\angle 5$ and $\angle 8$; (c) $\angle 4$ and $\angle 6$, $\angle 8$ and $\angle 2$, $\angle 3$ and $\angle 5$, $\angle 7$ and $\angle 1$ 16. (a) 105°, (b) 75°, (c) 105°, (d) 105° 17. (a) 54°, (b) 90°, (c) 54°, (d) 36° 18. (a) Pentagon, (b) Quadrilateral, (c) Triangle, (d) Hexagon, (e) Octagon, (f) Decagon 19. Regular 20. (a) Scalene, isosceles, equilateral; (b) Acute, obtuse, right equiangular 21. Altitude 22. Median 23. Diagonal 24. (a) Parallelogram; (b) Trapezoid; (c) Square, rhombus; (d) Rectangle, square; (e) Rhombus, square 25. 900° 26. Isosceles right △ 27. 35°, 110° 28. (a) $c \approx 8.1$, (b) $b \approx 5.7$, (c) $a \approx 8.0$ 29. 8 ft 1 in 30. 11.1 cm 31. 23 ft, no 32. Radius 33. Tangent, point of tangency, perpendicular 34. Secant, chord 35. Diameter 36. (a) 27° (b) 43°, (c) 90° 37. (a) =, (b) ~ 38. Similar 39. $m\angle AED = m\angle C$, $m\angle A = m\angle A$, $m\angle ADE = m\angle B$ 40. 29.0 ft 41. 15 ft and 20 ft 42. 13.8 in 43. congruent

Chapter Test

1. Point, plane, line 2. Straight, curved, broken 3. Line segment has length 4. Angle 5. One

6. (a) 70°, (b) 125° 7. 25.6°

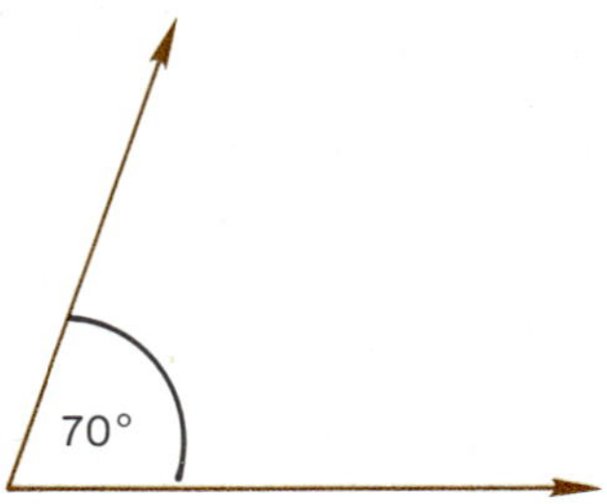

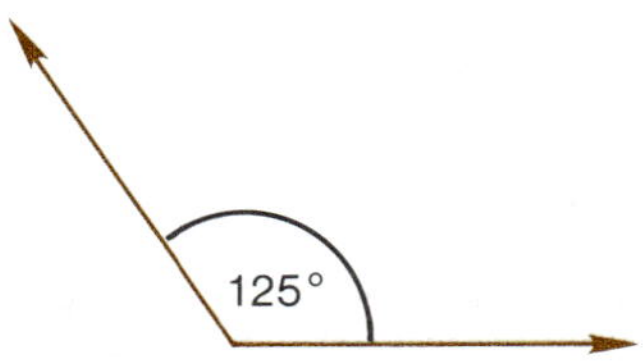

8. 122°48′ 9. (a) 90°, (b) $\angle DBE$ and $\angle EBA$, (c) Obtuse, (d) Complementary, (e) Straight, (f) Perpendicular 10. (a) 79°; (b) 98°; (c) 85°; (d) $\angle 8$, vertical; $\angle 1$, corresponding; $\angle 4$, alternate-interior 11. Regular 12. Altitude 13. 900° 14. (a) 60°, 120°, (b) 28°, (c) Congruent, (d) 180° 15. Square, Rhombus 16. (a) Obtuse, scalene; (b) Right, isosceles; (c) Equiangular, Equilateral 17. 10.8 in 18. 20.8 ft 19. (a) $\overline{AB}$, (b) 90°, (c) Inscribed angle, (d) Central angle, (e) 74°, (f) 115°, (g) $\overline{CB}$, $\overline{AC}$ 20. (a) Tangent, (b) Point of tangency, (c) 27° 21. (a) $m\angle ACB = m\angle BDC$, $m\angle B = m\angle B$, $m\angle A = m\angle DCB$; (b) 6.4 22. 35 ft by 45 ft

Exercise 13.1

1. 12.8 cm 3. 8.50 in 5. $2\frac{1}{12}$ in 7. 17.6 ft 9. 13.4 cm 11. 13.2 ft 13. 12.9 cm 15. 16.2 m 17. 14.9 cm 19. 19.00 min 21. 31.32 cm 23. 33.0 cm 25. 5.0 ft 27. Yes 29. 977 yd 31. $102.60 33. 18.5 in 35. 6.50 in 37. 653 in 39. 13.5 ft
39. 13.5 ft

Exercise 13.2

1. 400 ft^2 3. 256 cm^2 5. 10.4 yd^2 7. 7.40 ft^2 9. 225 in^2 11. 5 yd^2 13. 11.0 ft^2 15. 6.50 in^2 17. 8.45 in^2 19. $2\frac{1}{4}$ in^2 21. 23 ft^2 23. 5.5 ft^2 25. 50.5 in^2 27. 25 yd^2 29. 36.6 in^2 31. 15 ft 33. 7.97 in^2 35. 372,000 m^2 37. 298 ft 39. 49 in^2 41. 9 times greater 43. $100.80 45. $190.80 47. 5,360 in^2 49. 213 ft^2 51. 200.0 cm, 17,320 cm^2 53. 56 in 55. Doubles 57. 11.0 ft^2 59. 904 kg 61. 50.2 in^2 63. 4 times as great 65. 6.84 cm

Exercise 13.3

1. 1,050 ft^2, 1,570 ft^2, 4,250 ft^3 3. 110 in^2, 190 in^2, 170 in^3 5. 220 yd^2, 240 yd^2, 220 yd^3 7. 63 cm^2, 74 cm^2, 43 cm^3 9. 2,460 mm^2, 2,910 mm^2, 9,230 mm^3 11. 1.85 m^3 13. 3 gal, $32.85 15. 320 in^3 17. 0.6 lb, 23 lb, 19.3 lb 19. 1.60 cm

Exercise 13.4

1. 110 in^2, 140 in^2, 120 in^3 3. 24.2 cm^2, 28.4 cm^2, 9.84 cm^3 5. 250 cm^2, 280 cm^2, 250 cm^3 7. 44.0 in^3 9. 2.73 lb 11. 5.50 in 13. 58.4 ft^2 15. 55.0 lb 17. 21.2 ft^3 19. 0.030 in^3 21. 2,000 cm^3

Exercise 13.5

1. 64 in^3 3. 84.2 ft^3 5. 234,000 mm^2, 285,000 mm^2 7. 130 ft^2, 180 ft^2 9. 123,000 ft^3 11. 430 ft^3 13. (a) 308.4 ft^3, (b) 327.8 ft^2, (c) 133

Exercise 13.6

1. 81.8 cm^2, 203 cm^2 3. 7,210 cm^2, 13,600 cm^2 5. 390 cm^3 7. 139 in^3 9. 536 in^2, 690 in^2, 1,030 in^3 11. 1,720 in^3 not used 13. $118.43 15. 32.7 in^2 17. 15.9 ft, 5,390 ft^3

Exercise 13.7

1. 222 in^2, 310 in^3 3. 38.5 in^2, 22.4 in^3 5. 65 in^3 7. 65.4 in^3, 114 in^3 9. 7.00 in 11. 83.7 ft^2 13. 4.71 ft

Chapter Review

1. Perimeter area 2. 4.8 in, 1.4 in^2 3. 24.6 in, 34.4 in^2 4. 16.2 ft, 11.5 ft^2 5. 25.4 mm, 31.4 mm^2 6. 15.0 in, 10.8 in^2 7. 15.6 ft, 9.9 ft^2 8. 45 ft, 160 ft^2 9. 18.8 cm, 28.3 cm^2 10. 11.1 m, 7.3 m^2 11. 8.25 yd, 4.19 yd^2 12. 88.0 in^2 13. 63.8 ft, 3 gal 14. 528 ft, 7,960 ft^2 15. 40.0 in 16. 52.6 in^2 17. 150 in^2 18. 23.6 in^2 19. 9 ft^2, 1.93 ft^2 20. Volume 21. Trapezoidal prism 22. Rectangular solid 23. 87.8 cm^2, 130 cm^3 24. 50.2 in^2, 33.5 in^3 25. 55.0 in^2, 57.7 in^3 26. 61.3 ft^2, 41.2 ft^3 27. 82.4 cm^2, 87.1 cm^3 28. (a) 161 cm^2, (b) 82.7 in^2, (c) 132.6 cm^2 29. 9.69 in^3, 43.2 in^2 30. 472 lb 31. 696 in^2 32. 60.8 in^2, 10.8 lb 33. 2.55 lb, 57.0 in^2 34. 49.1 gal

Chapter Test

1. (a) 20.0 cm, 24.0 cm^2; (b) 9.00 in, 3.90 in^2; (c) 14.1 ft, 15.9 ft^2; (d) 22.1 m, 17.6 m^2; (e) 20.9 mm, 16.9 mm^2 2. 222 3. 9.06 in 4. 5.94 in^2 5. 2.31 in 6. 11.2 in^2 7. doubles 8. 0.859 lb 9. (a) 2,390 ft^3, 866 ft^2; (b) 38.7 in^3, 42.9 in^2; (c) 20.8 cm^3, 30.2 cm^2; (d) 42.9 m^3, 51.2 m^2 10. 7.55 lb 11. $13.73 12. 981 in^3, 0.693 lb 13. 28,500 in^2 14. $76.50 15. 34 16. (a) 10.6 in^3, (b) 0.883 in^3; (c) 11.5 in^3 17. 4.71 in 18. $95.67 19. 4.38 lb 20. 1,260 ft^2 21. 27 times greater 22. 6.83 cm

Chapter 14

Exercise 14.1

1. $\log_{10} 100 = 2$ 3. $\log_2 64 = 6$ 5. $\log_5 1 = 0$ 7. $\log_9 3 = \frac{1}{2}$ 9. $\log_{\frac{1}{2}} \frac{1}{4} = 2$ 11. $10^5 = 100{,}000$ 13. $5^4 = 625$ 15. $9^0 = 1$ 17. $8^3 = 512$ 19. $4^1 = 4$ 21. 5 23. 3 25. 2,401 27. $\frac{1}{256}$ 29. 3 31. 5

Exercise 14.2

1. $\log_b 3 + \log_b 5$ 3. $\log_3 2 + \log_3 13$ 5. $\log_5 7 - \log_5 11$ 7. $3 \log_b 4$ 9. $4 \log_{10} 5$ 11. $2 (\log_{\frac{1}{2}} 2 + \log_{\frac{1}{2}} 3)$

13. $3 \log_e 2 - 2 \log_e 3$ 15. $\log_5 3 + 3 \log_5 2 - 2 \log_5 5$ 17. $3 (\log_b x + \log_b y)$ 19. $\frac{1}{2} (\log_b x + \log_b y)$ 21. $\log_4 55$

23. $\log_{10} \left(\frac{25}{12}\right)$ 25. $\log_5 121$ 27. $\log_2 245$ 29. $\log_e \left(\frac{8}{5}\right)$ 31. $\log_3 \left(\frac{2}{3}\right)$ 33. $\log_e (18)$ 35. $\log_2 \left(\frac{567}{50}\right)$ 37. $\log_b \left(\frac{x^4}{y^z}\right)$

39. $\log_b \sqrt{xy}$ 41. $b^{1.3010} = 20$ 43. $b^{0.9030} = 8$ 45. $b^{2.6020} = 400$ 47. $b^{-0.1761} = \frac{2}{3}$ 49. $b^{0.8293} = \frac{27}{4}$

51. $b^{0.1590333} = \sqrt[3]{3}$ 53. $b^{-0.06245} = \sqrt{\frac{3}{4}}$ 55. 10

Exercise 14.3

1. 1.3617 3. 3.8669 5. 5.7185 7. 9.1959 − 10 9. 5.9736 − 10 11. 3.7586 13. 8.2539 15. 5.8825 − 10 17. 3.4131 − 10 19. 1.6721 21. 5.0973 23. 3.9415 25. 4.9571 27. 5.8825 29. 6.9325 31. 0.6776 33. 0.8414 35. 9.7582 − 10 37. 8.9196 − 10 39. 7.3802 − 10 41. 5.5551 − 10 43. 5.6075 45. 6.9777 47. 9.1501 − 10 49. (a) 306, (b) 10.9 51. 1.5534 53. 15.798

Exercise 14.4

1. 3.48 3. 31,600 5. 0.0566 7. 0.997 9. 0.000717 11. 0.00000000173 13. 682,300,000 15. 0.0000009764 17. 0.00005766 19. 501.19 21. 5069.91 23. 54,827,697 25. 5.75 27. 8.35 29. 0.0457 31. 0.000000015367 33. 0.00000543 35. 26.92 37. 15.81

Exercise 14.5

1. 3.58 3. 4.32 5. 0.030 7. 8.56 9. 10.1 11. −1.90 13. −2.54 15. −7.14 17. −7.11 19. 2.11 21. 1,620,000 23. 0.0534 25. 0.0000003 27. (a) −0.018, (b) 0.000000000215 29. (a) 4.5, (b) −1.6 31. 1,685,000,000,000 33. 4.72 35. 405 mg 37. 141 39. (a) \$5,637.48, (b) \$635.62, (c) \$2,073.83, (d) \$423.79, (e) \$4,586.67

Exercise 14.6

1. 1.47 3. 3.91 5. 0.5 7. 0.646 9. 1.54 11. −2.29 13. 1.33 15. 1.48

Exercise 14.7

1. 0.7925 3. 2.4650 5. 1.3023 7. −2.8076 9. 2.1236 11. 0.7996

Chapter Review

1. $3^2 = 9$ 2. $6^3 = 216$ 3. $25^{\frac{1}{2}} = 5$ 4. $p^s = q$ 5. $\log_5 125 = 3$ 6. $\log_{49} 7 = \frac{1}{2}$ 7. $\log_{\frac{1}{3}} \frac{1}{27} = 3$ 8. $\log_p s = q$ 9. 3

10. 2 11. 1,296 12. 8 13. 3 14. 4 15. $\log_5 2 + \log_5 3$ 16. $\log_4 7 + \log_4 3$ 17. $\log_b 5 - \log_b 7$ 18. $\log_e 2 + 2 \log_e 3$

19. $2 \log_{10} 4$ 20. $\log_6 2 + \log_6 7 - 2 \log_6 5$ 21. $2 \log_b m - 3 \log_b n$ 22. $\frac{1}{4} [\log_b 2 + \log_b 5]$ 23. $\frac{5}{3} \log_6 x$ 24. $\log_3 91$

25. $\log_b \left(\frac{16}{9}\right)$ 26. $\log_7 (81)$ 27. $\log_5 \left(\frac{49}{8}\right)$ 28. $\log_x 8$ 29. $\log_2 \left(\frac{10,125}{2}\right)$ 30. 2.708 31. −0.5108 32. 3.8066 33. 0.8047

34. 0.1703 35. 5.416 36. 0.7582 37. 1.9666 38. 2.5529 39. 3.9405 40. −0.1469 41. 7.1987 − 10 42. 1.21 43. 9.84 44. 218.78 45. 7.94×10^{30} 46. 1.711 47. 2.92×10^{35} 48. (a) 297.89, (b) 1.66 49. 5.6021 50. 2.1668 51. −1.1552 52. −4.6831 53. 9.2686 54. 13.1635 55. 2.27 56. 307.97 57. 6.52×10^{10} 58. 836.6 59. 1.02 60. 1.0008 61. (a) \$3,426.03, (b) \$50,116.21, (c) 28.8 62. (a) 18.66, (b) 21.39 63. (a) 0.839, (b) 26,364.56 64. 3.91 65. 0.48 66. −0.79 67. −32.89 68. 0.69 69. 1.86 70. 1.23 71. 0.87 72. +.16 73. −4.07 74. 0.2630 75. 1.43

Chapter Test

1. Common natural 2. $5^4 = 625$ 3. −2 4. 237.14 5. 1 6. 0 7. $\log_b \left(\frac{16}{5}\right)$ 8. (a) 2.4022, (b) 2.2618, (c) 4.1735, (d) 0.3691

9. 1.15 10. 1.56 11. 0.0000005 12. 35.6 13. 56.23 14. 0.2208 15. 3.55

Chapter 15

Exercise 15.1

1. (2,4) 3. (−1,3) 5. (−6,−1) 7. (0,4) 9. (5,0) 11. (−7,0) 13. $\left(\frac{1}{2},3\right)$ 15. $\left(\frac{3}{2},0\right)$

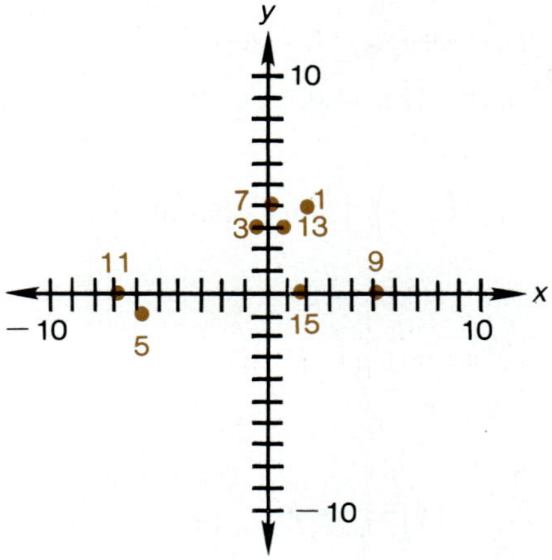

17. (2,−1) 19. (0,6) 21. (0,−6) 23. (−3,4) 25. (5,−5)

27.

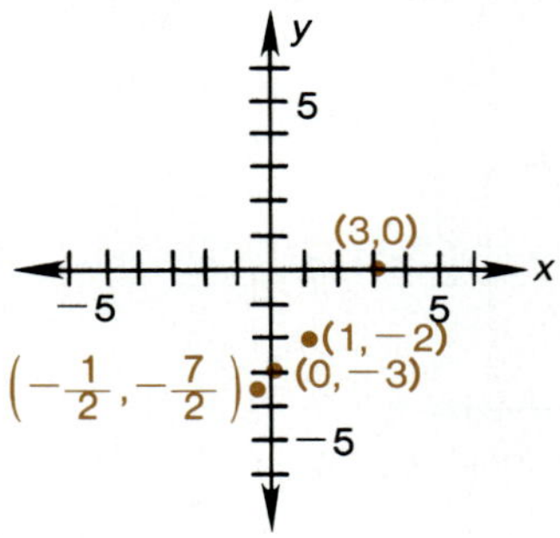

29.

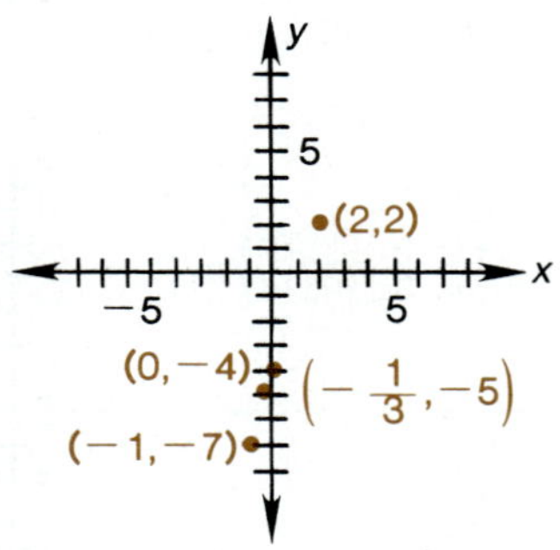

31.

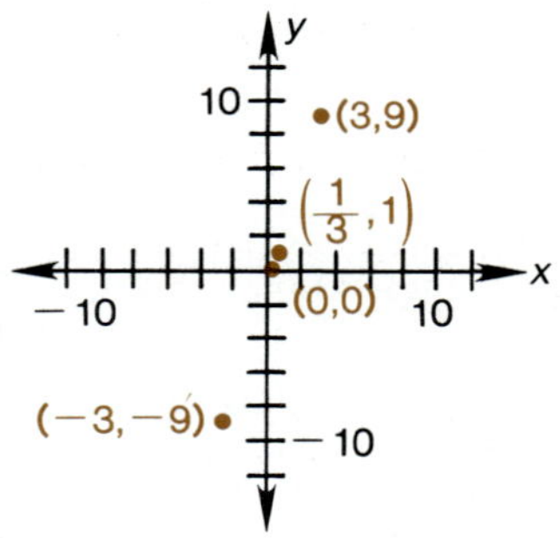

33.

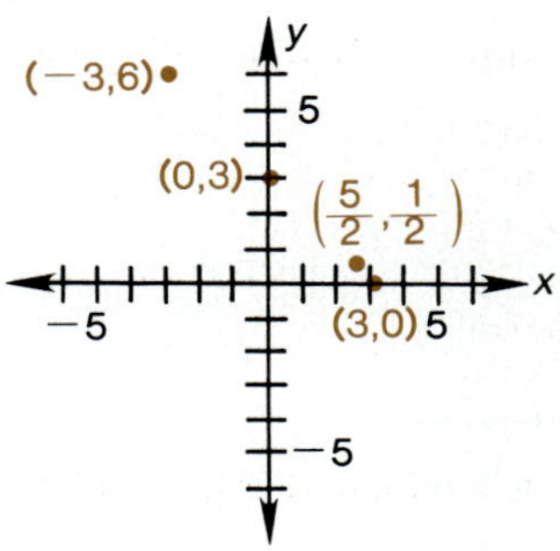

35.

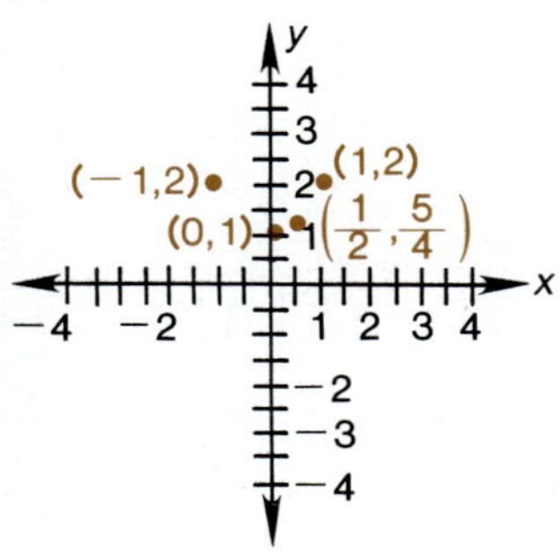

37.

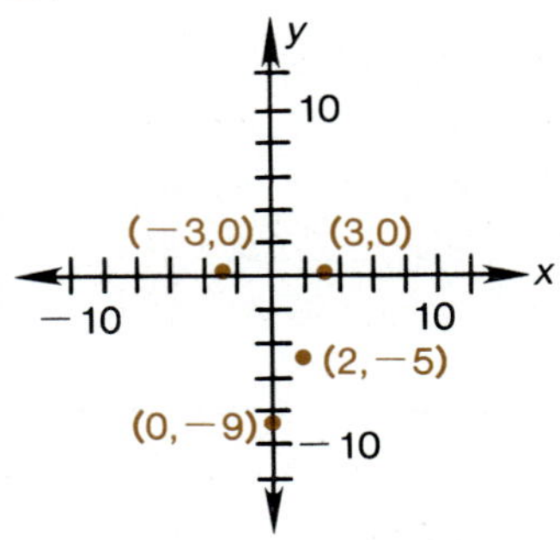

39. Third 41. First 43. Fourth 45. Third 47. Fourth 49. (a) Fourth, (b) Third, (c) Second, (d) First 51. 0

Exercise 15.2

1. 4, 0, −2   3. 10, 0, 5   5. −4, 0, 8   7. 4, 4, $\frac{1}{4}$ 9. 0, 3, 0   11. 0, 6, 0   13. $\frac{-5}{4}$, $\frac{5}{2}$, undefined

15. (1,4), $\left(\frac{1}{2},0\right)$, (0, −4), (−1.4, −15.2)   17. (2,0), (−2,0), (−3,5), (0.3, −3.91)   19. (1,4), (−3,0), (0,0), (−0.5, −1.25)

21. (−2,0), (0,8), (−4,0), (−3.9, −0.19)   23. $\left(1,\frac{1}{2}\right)$, (0,0), (−1.1,11), (−0.9, −9)   25. All real numbers   27. All real numbers

29. $x \geqslant -3$ 31. $x \leqslant 5$ 33. All real numbers except 6   35. 21.6, 29.4, 38.4   37. 172, 32.2, 302   39. 11.6, 54.4, 360

Exercise 15.3

1. (0,4), (−2,0)   3. (0,1), $\left(-\frac{1}{3},0\right)$ 5. (0,2), (3,0)   7. (0,2), (4,0)   9. (0,0)   11. (0,0)   13. $\left(0,\frac{11}{5}\right)$, $\left(\frac{11}{2},0\right)$ 15. $\left(0,-\frac{3}{2}\right)$,(3, 0)

17.

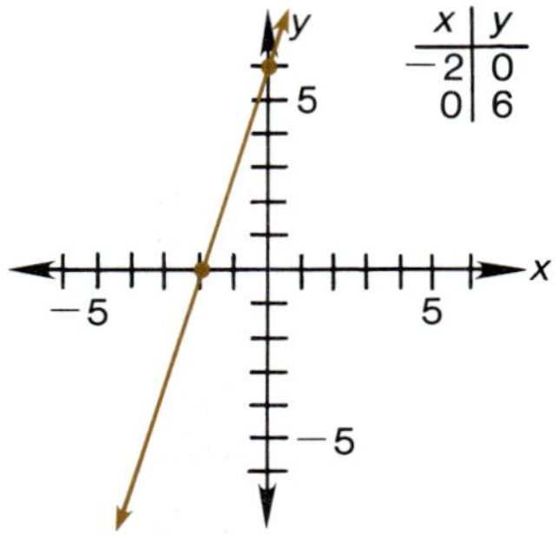

x	y
−2	0
0	6

19.

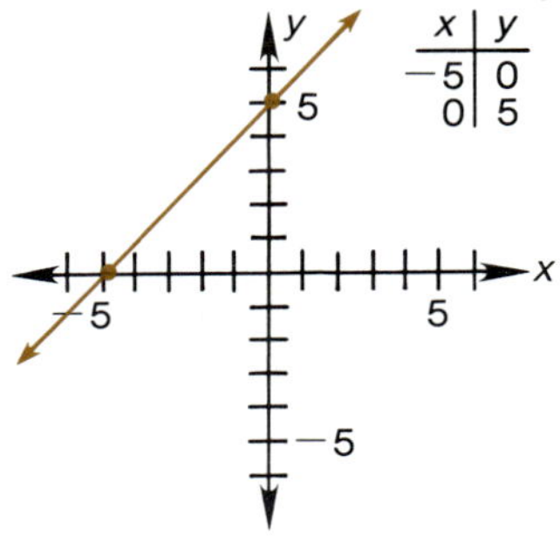

x	y
−5	0
0	5

21.

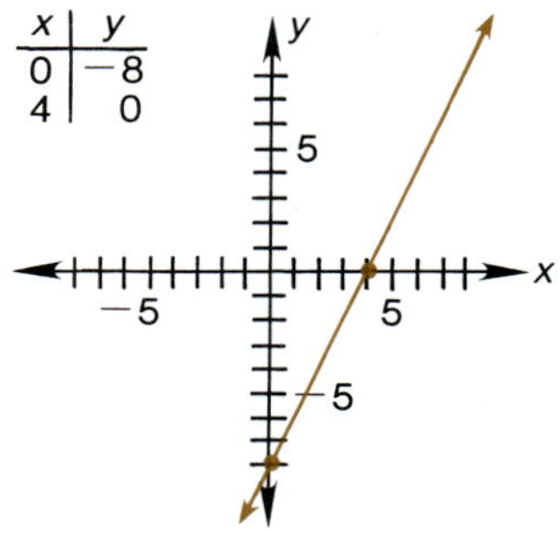

x	y
0	−8
4	0

23.

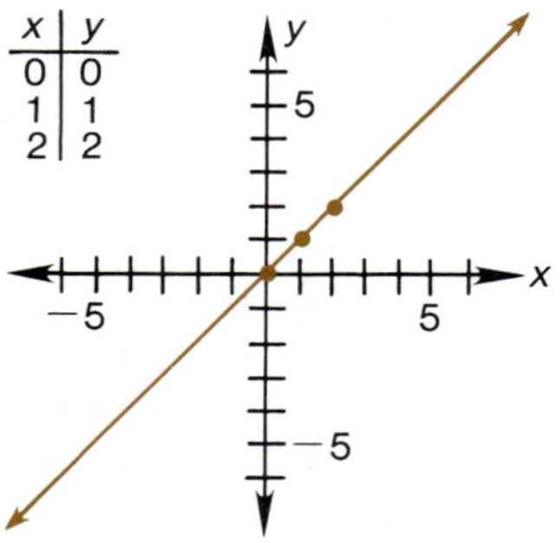

x	y
0	0
1	1
2	2

25.

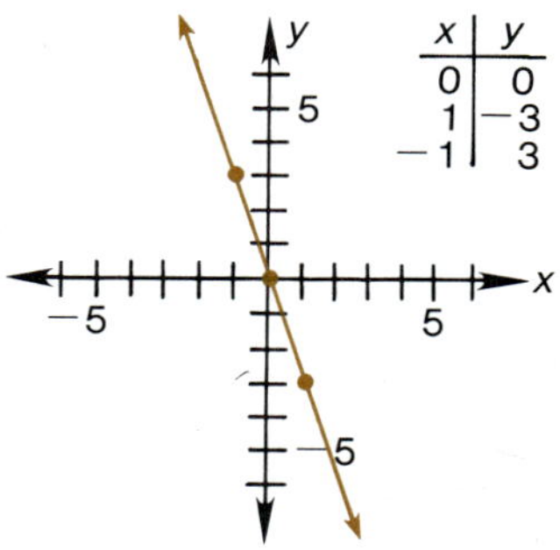

x	y
0	0
1	−3
−1	3

27.

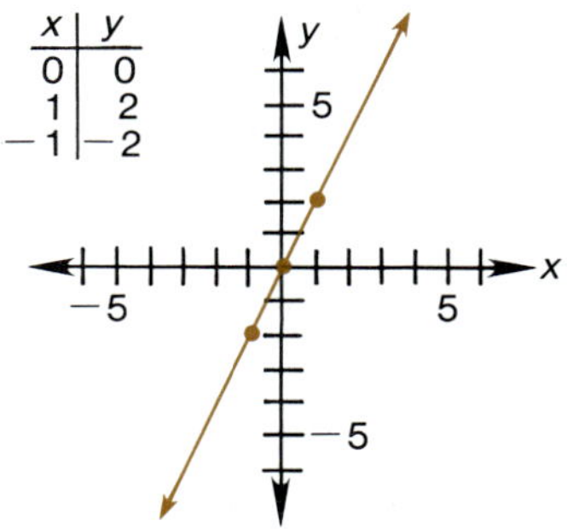

x	y
0	0
1	2
−1	−2

29.

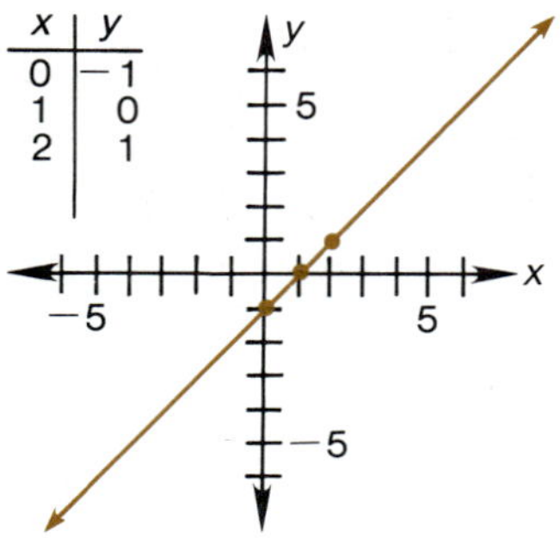

x	y
0	−1
1	0
2	1

31.

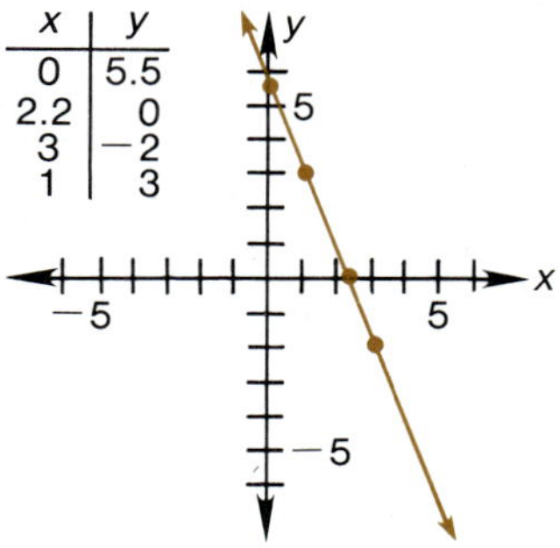

x	y
0	5.5
2.2	0
3	−2
1	3

33.

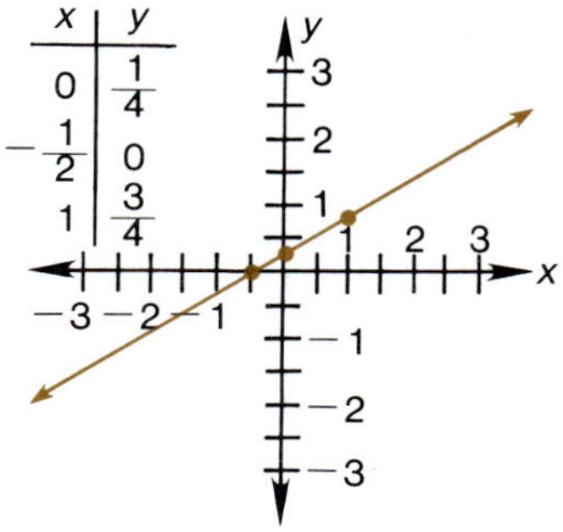

x	y
0	$\frac{1}{4}$
$-\frac{1}{2}$	0
1	$\frac{3}{4}$

35.

x	y
0	2
5	3
−10	0

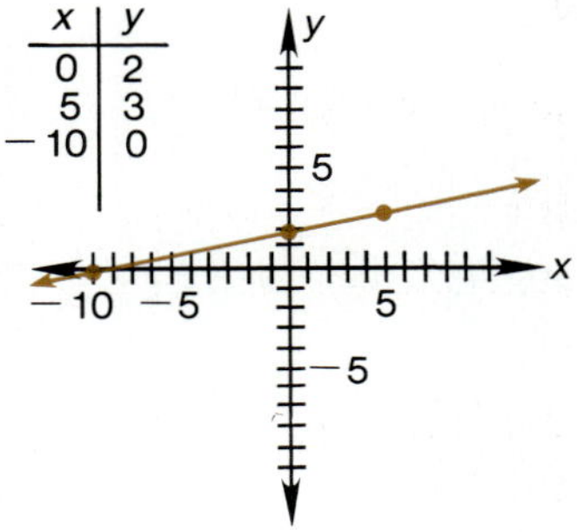

37. $y = 2x - 7$ 39. $y = \frac{4}{3}x + 3$ 41. $y = -\frac{7}{3}x + \frac{10}{3}$ 43. $y = \frac{1}{5}x + 1$ 45. $y = \frac{5}{8}x + \frac{9}{8}$

47.

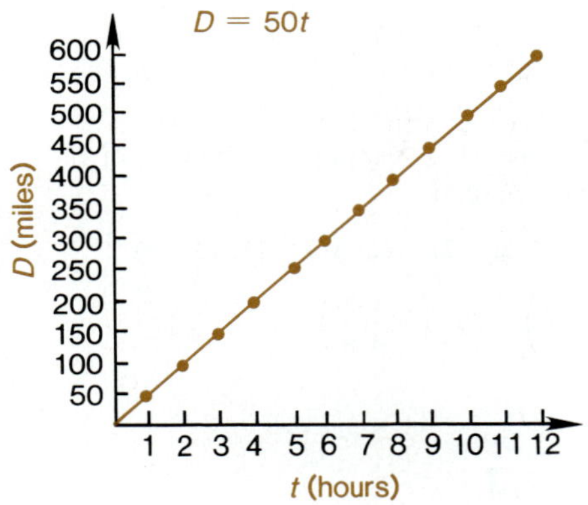

t	D
1	50
2	100
3	150
4	200
5	250
6	300
7	350
8	400
9	450
10	500
11	550
12	600

49.

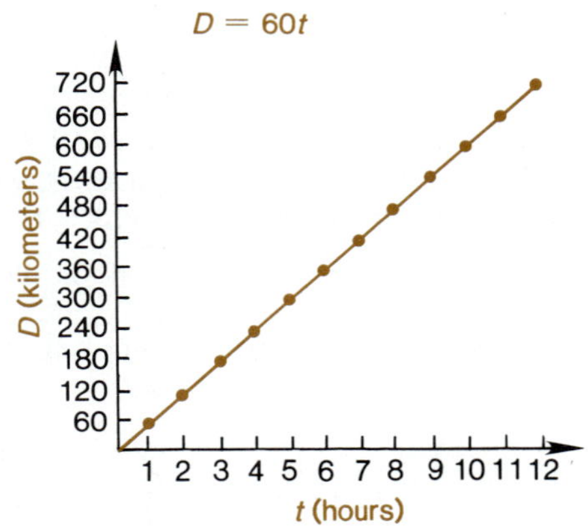

t	D
1	60
2	120
3	180
4	240
5	300
6	360
7	420
8	480
9	540
10	600
11	660
12	720

51.

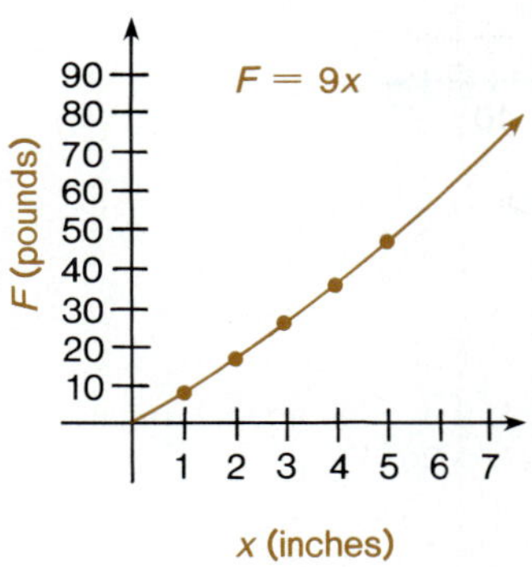

53.

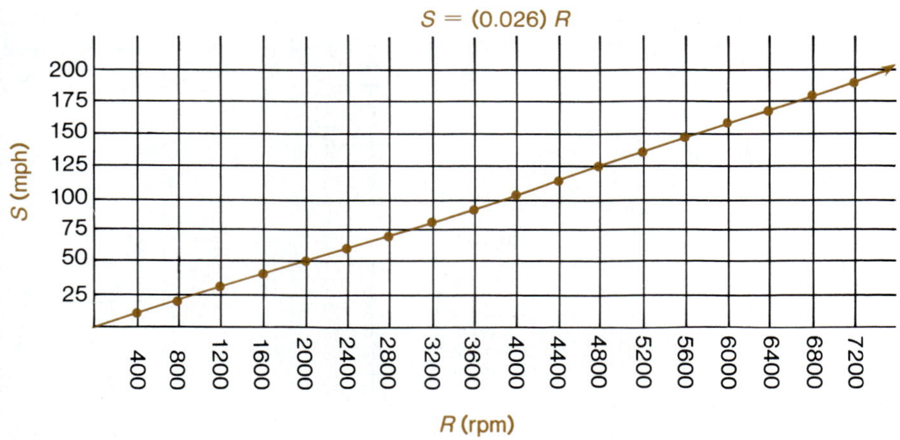

55.

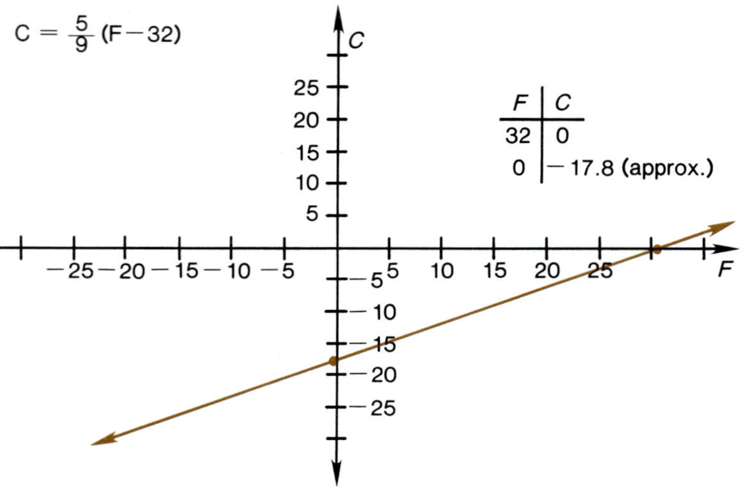

Exercise 15.4

1. $x: (4,0), (-4,0); y: (0,-16)$ 3. $x: (2,0), (4,0); y: (0,8)$ 5. $x: (-2,0), (-6,0); y: (0,12)$ 7. $x: (\sqrt{5},0), (-\sqrt{5},0); y: (0,5)$
9. $x: (-3,0); y: (0,9)$ 11. x intercept does not exist; $y: (0,5)$ 13. x intercept does not exist; $y: (0,6)$
15. $x: (4,0); y: (0,-16)$ 17. $x: \left(-\frac{1}{2},0\right), (-1,0);$ $y: (0,1)$ 19. $x: \left(\frac{3}{2},0\right), (-2,0); y: (0,6)$ 21. $(0,-16)$ 23. $(3,-1)$
25. $(-4,-4)$ 27. $(0,5)$ 29. $(-3,0)$ 31. $(0,5)$ 33. $(-2,2)$ 35. $(4,0)$ 37. $\left(-\frac{3}{4},-\frac{1}{8}\right)$ 39. $\left(-\frac{1}{4},\frac{49}{8}\right)$

41.

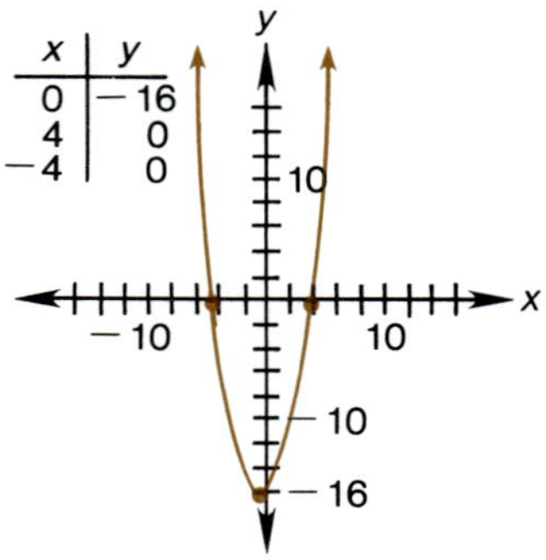

43.

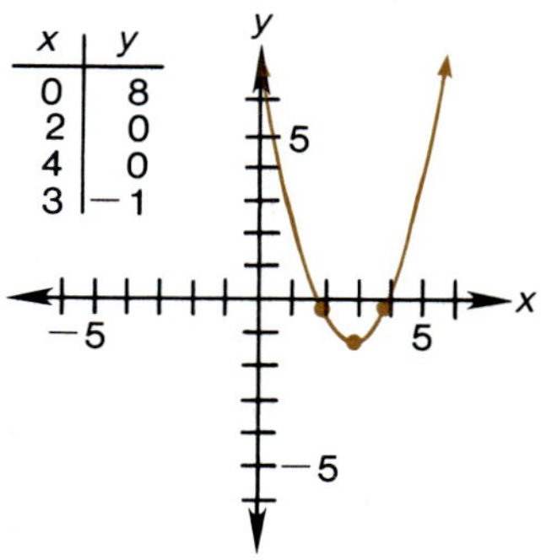

45.

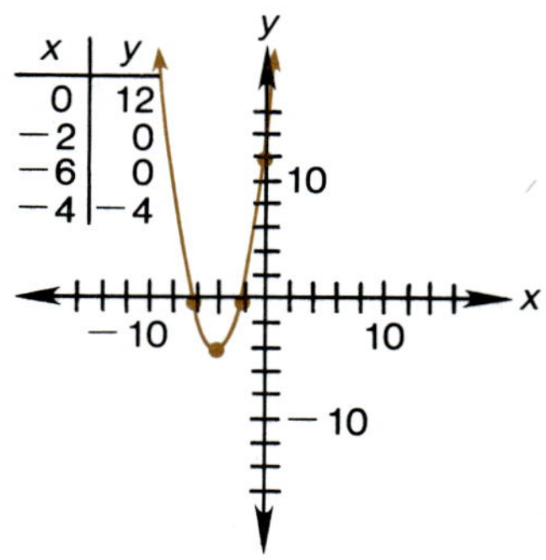

47.

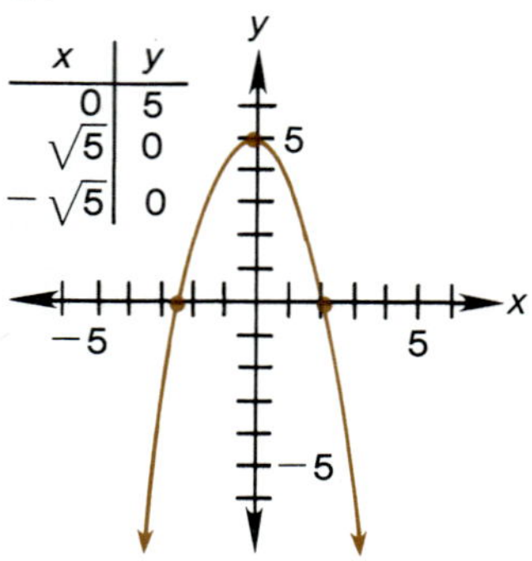

49.

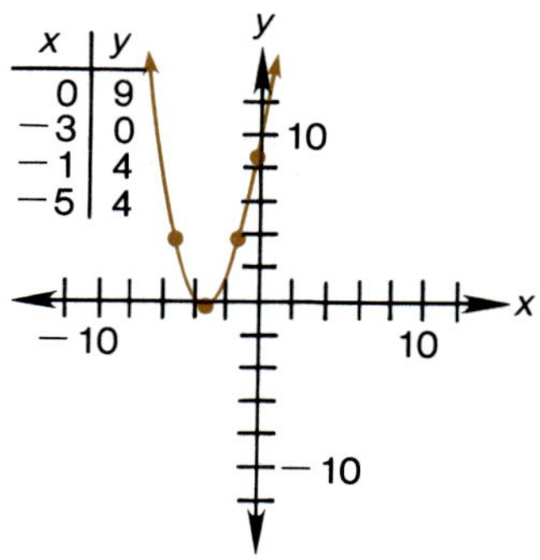

51.

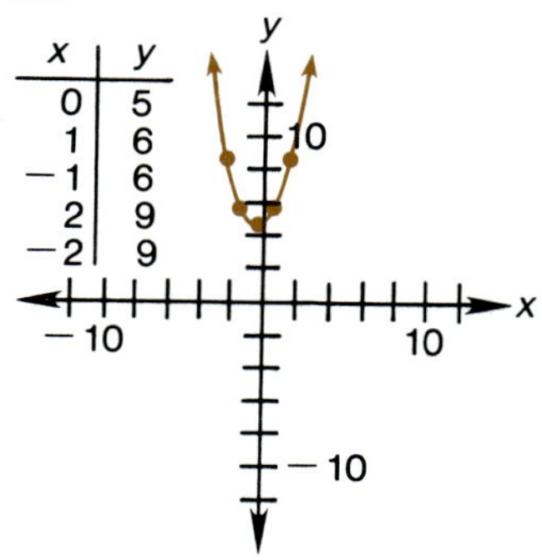

53.

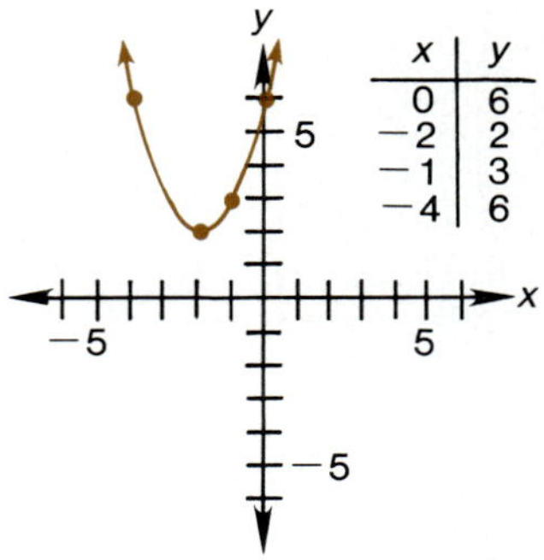

x	y
0	6
−2	2
−1	3
−4	6

55.

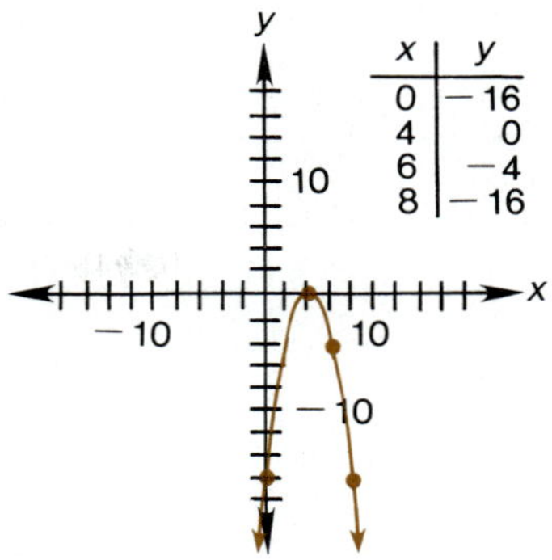

x	y
0	−16
4	0
6	−4
8	−16

57.

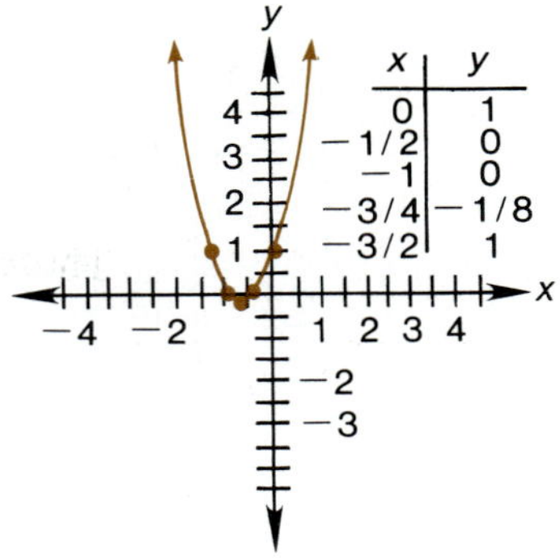

x	y
0	1
−1/2	0
−1	0
−3/4	−1/8
−3/2	1

59.

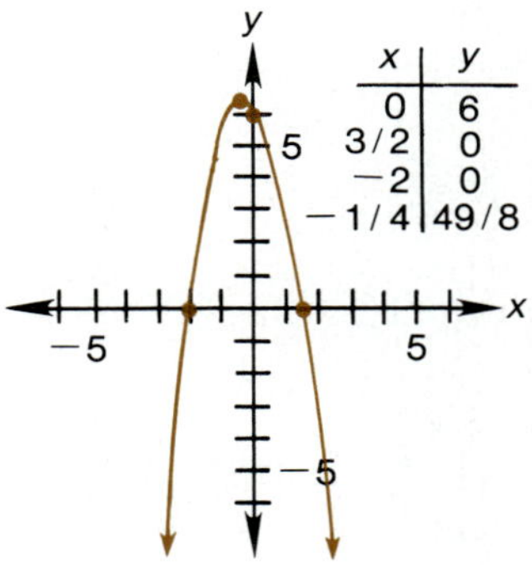

x	y
0	6
3/2	0
−2	0
−1/4	49/8

61.

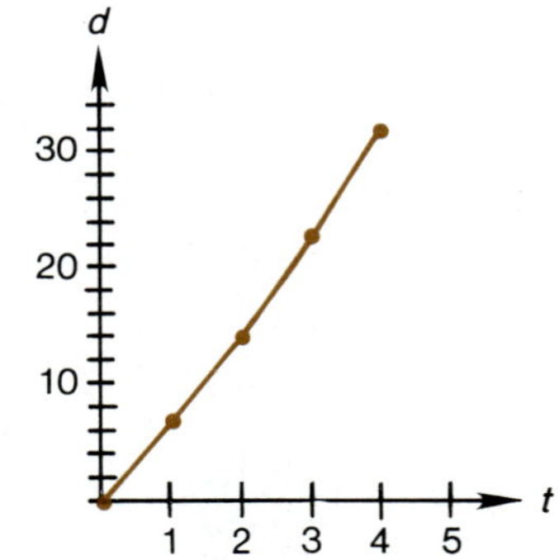

t	d
0	0
1	$6\frac{1}{2}$
2	14
3	$22\frac{1}{2}$
4	32

Distance is 14 feet when t is 2 seconds

63.

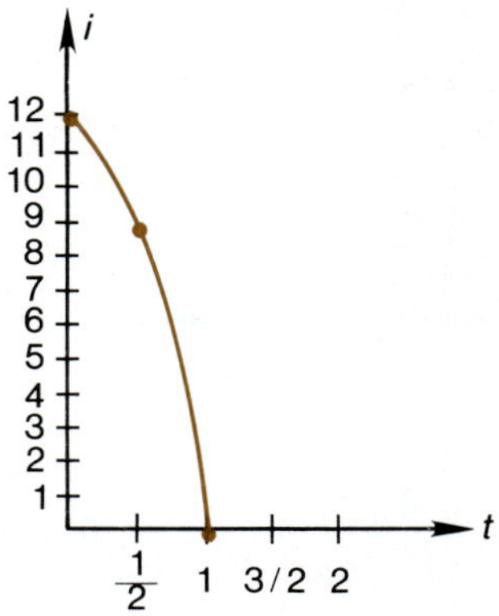

t	i
0	12
1	0
$\frac{1}{2}$	9

65.

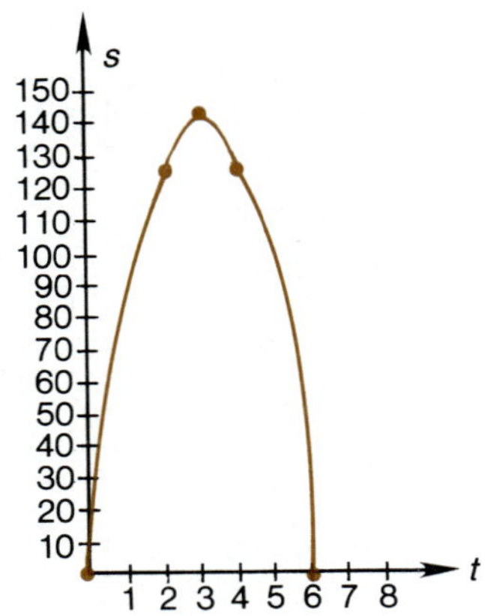

t	s
0	0
6	0
3	144
2	128
4	128

Maximum height is 144 feet when t is 3 seconds and the projectile will strike the ground when t is 6 seconds.

67.

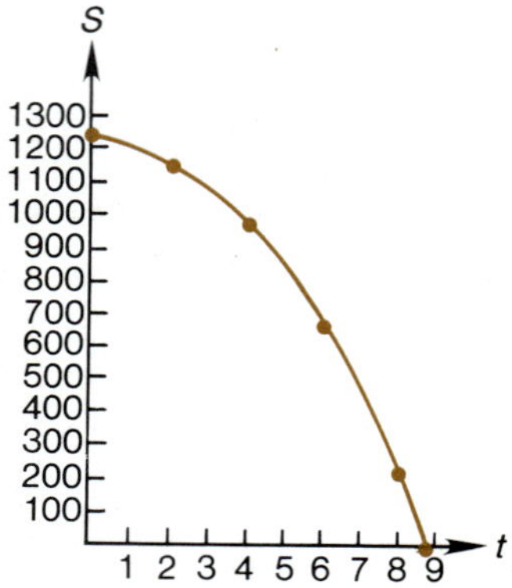

t	S
0	1250
≈ 8.8	0
2	1186
4	994
6	674
8	226

The object will strike the ground when t is $\frac{25\sqrt{2}}{4}$ seconds or approximately 8.8 seconds.

Exercise 15.5

1.

$y = x^3 + 1$

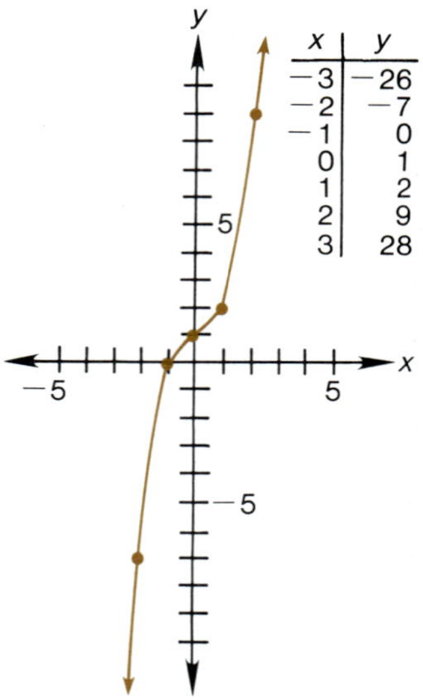

x	y	
−3	−26	
−2	−7	
−1	0	x int.
0	1	y int.
1	2	
2	9	
3	28	

3.

$y = x^3 - 4$

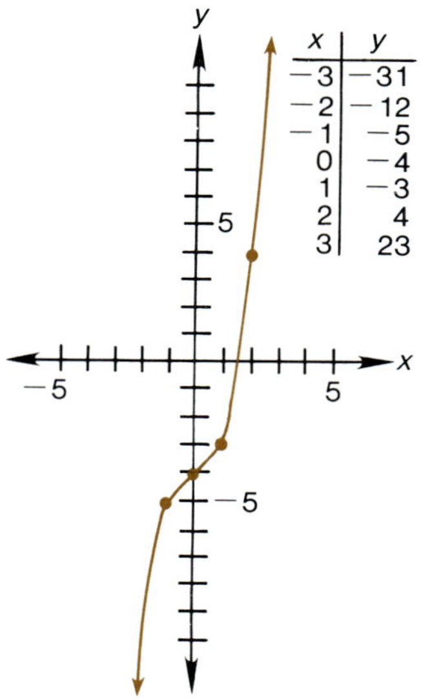

x	y	
−3	−31	
−2	−12	
−1	−5	
0	−4	y int.
1	−3	
2	4	
3	23	

5.

$y = x^4 - 1$

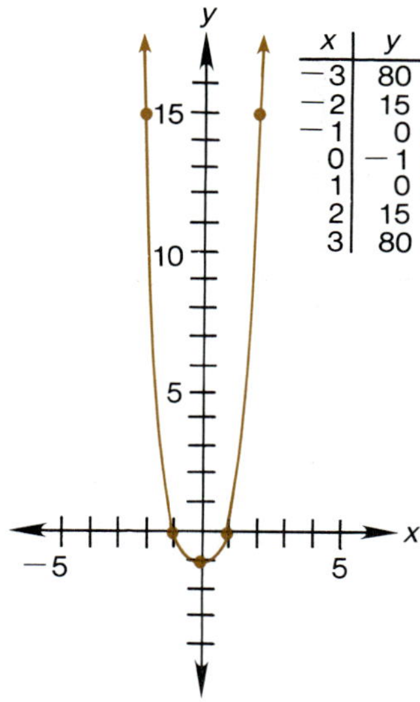

x	y	
−3	80	
−2	15	
−1	0	x int.
0	−1	y/int.
1	0	x int.
2	15	
3	80	

7.

$y = x^3 - 4x$

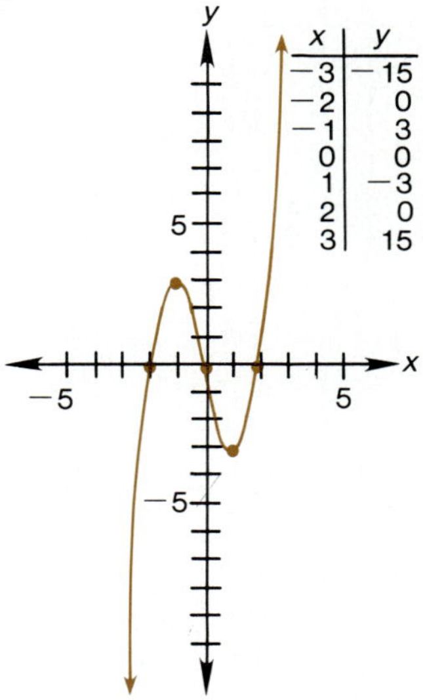

x	y	
-3	-15	
-2	0	x int.
-1	3	
0	0	x & y int.
1	-3	
2	0	x int.
3	15	

9.

$y = x^3 - 2x^2 - 8x$

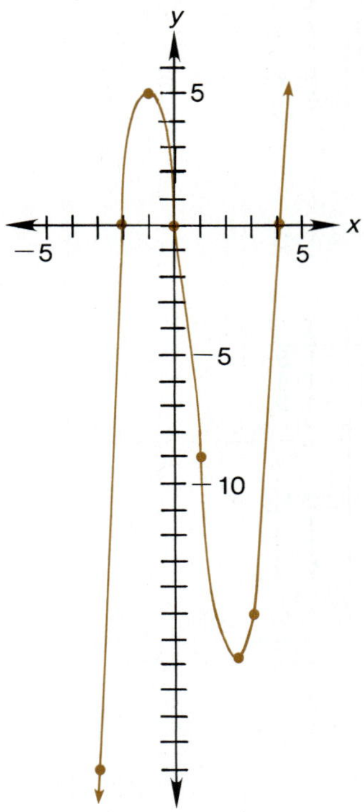

x	y	
-3	-21	
-2	0	x int.
-1	5	
0	0	x & y int.
1	-9	
2	-16	
3	-15	
4	0	x int.

11.

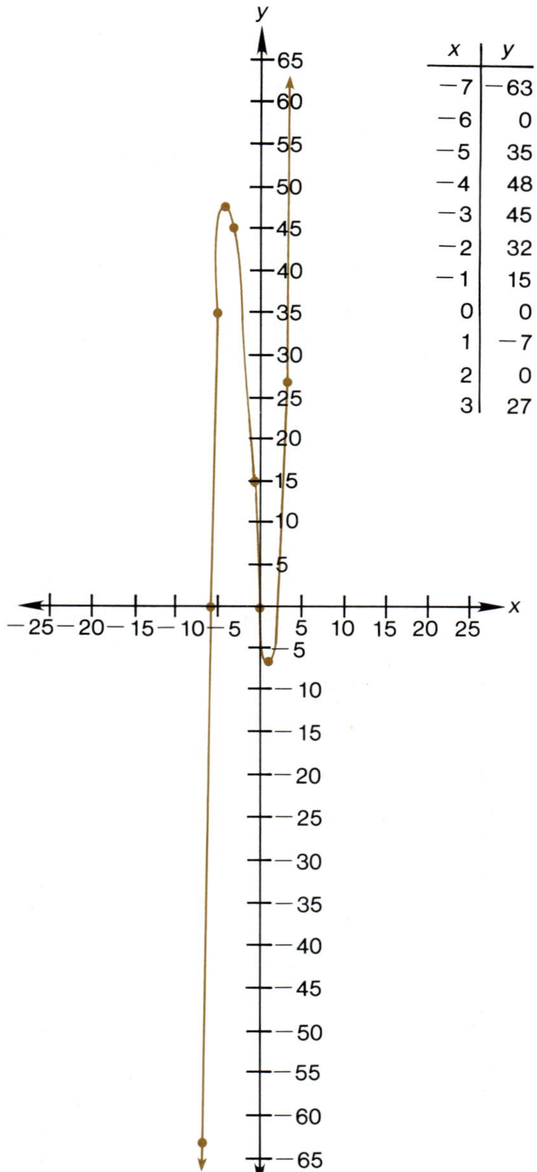

$y = x^3 + 4x^2 - 12x$

x	y	
-7	-63	
-6	0	x int.
-5	35	
-4	48	
-3	45	
-2	32	
-1	15	
0	0	x & y int.
1	-7	
2	0	x int.
3	27	

13.

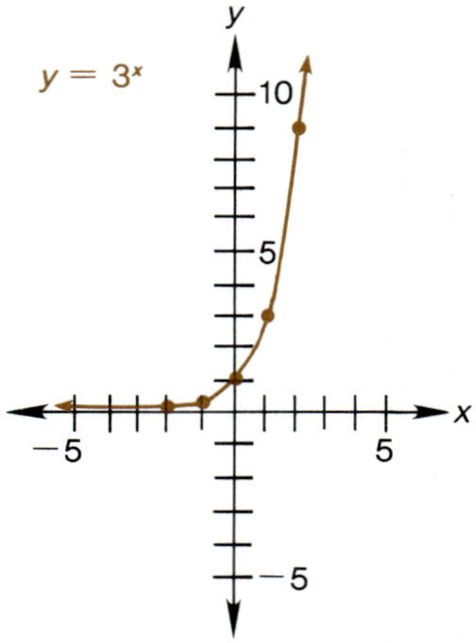

$y = x^3$

x	y
−2	$\frac{1}{9}$
−1	$\frac{1}{3}$
0	1 1 *y* int
1	3
2	9

15.

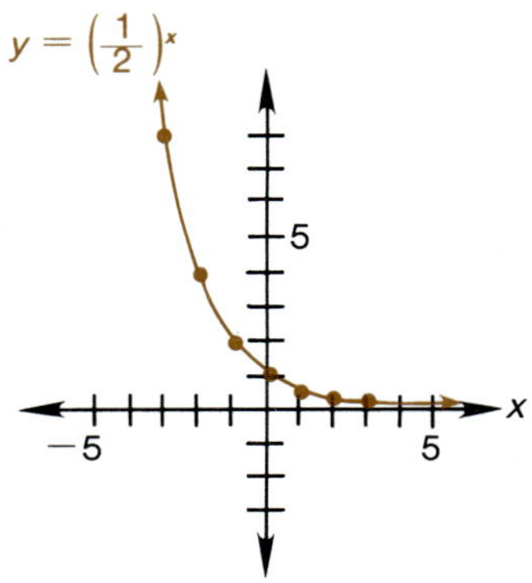

x	y
−3	8
−2	4
−1	2
0	1 *y* int.
1	$\frac{1}{2}$
2	1/4
3	1/8

17.

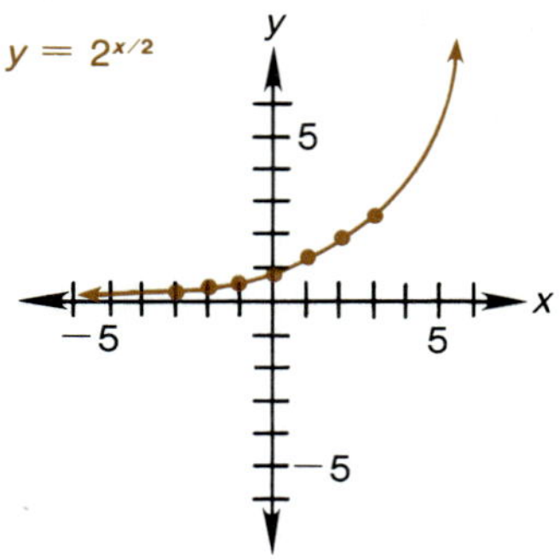

x	y
−3	$\frac{1}{2\sqrt{2}} \approx .35$
−2	$\frac{1}{2}$
−1	$1/\sqrt{2} \approx .707$
0	1 *y* int.
1	$\sqrt{2} \approx 1.414$
2	2
3	$2\sqrt{2} \approx 2.8$

19.

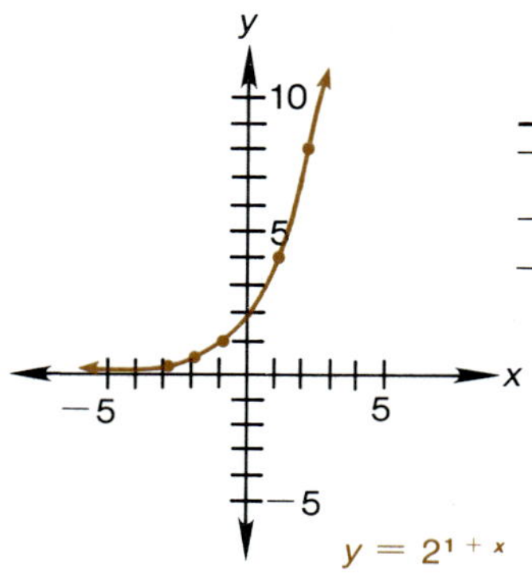

x	y
−3	$\frac{1}{4}$
−2	$\frac{1}{2}$
−1	1
0	2 *y* int.
1	4
2	8
3	16

21.

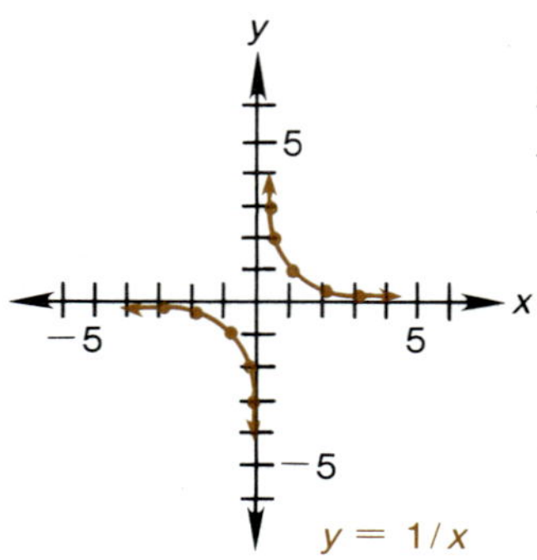

x	y
−3	−1/3
−2	$-\frac{1}{2}$
−1	−1
0	undef.
1	1
2	$\frac{1}{2}$
3	1/3
$\frac{1}{2}$	2
$\frac{1}{4}$	4

23.

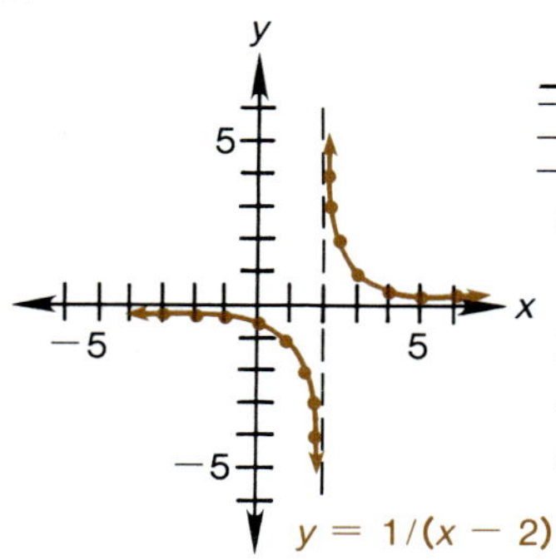

x	y
−3	−1/5
−2	−1/4
−1	−1/3
0	$-\frac{1}{2}$ *y* int.
1	−1
2	undef.
3	1
4	$\frac{1}{2}$
5	1/3

25.

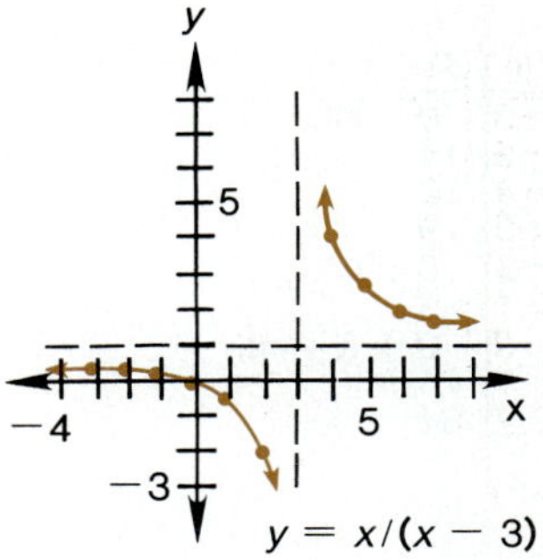

x	y	
−3	$\frac{1}{2}$	
−2	2/5	
−1	$\frac{1}{4}$	
0	0	x & y int.
1	$-\frac{1}{2}$	
2	−2	
3	undef.	
4	4	
5	5/2	
6	2	
7	7/4	

27.

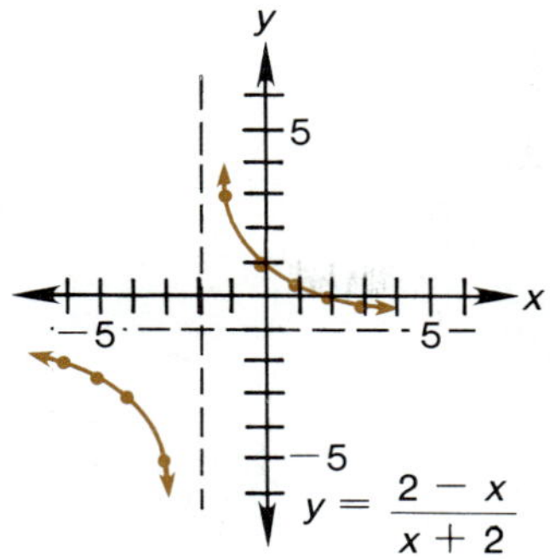

x	y	
−3	−5	
−2	undef.	
−1	3	
0	1	y int.
1	$\frac{1}{3}$	
2	0	x int.
3	$-\frac{1}{5}$	
−4	−3	
−5	$-\frac{7}{3}$	
−6	−2	

29.

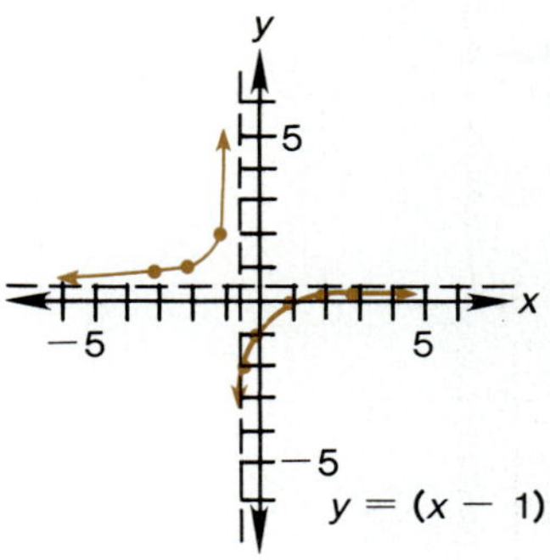

x	y	
−3	4/5	
−2	1	
−1	2	
0	−1	y int.
1	0	x int.
2	1/5	
3	2/7	
$-\frac{1}{2}$	undef.	

31.

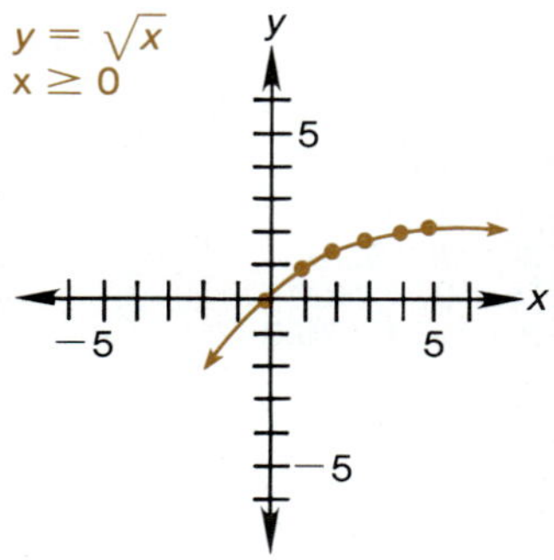

x	y	
0	0	x & y int.
1	1	
2	$\sqrt{2}$	
3	$\sqrt{3}$	
4	2	
5	$\sqrt{5}$	

33.

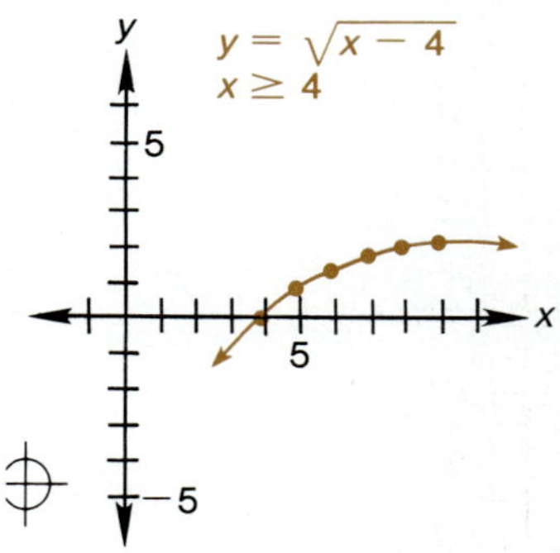

x	y	
4	0	x int.
5	1	
6	$\sqrt{2}$	
7	$\sqrt{3}$	
8	2	
9	$\sqrt{5}$	

35.

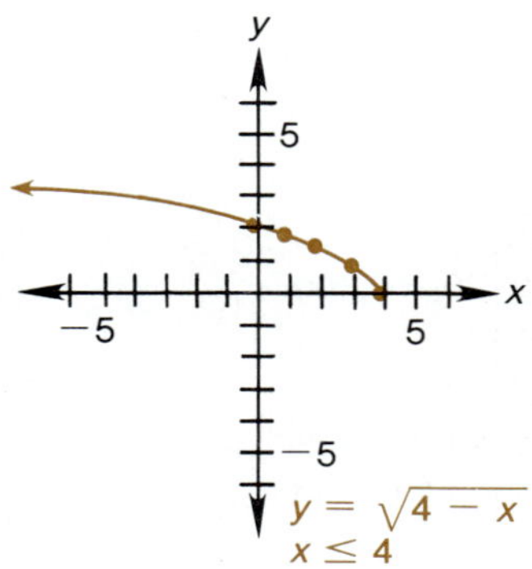

x	y	
4	0	x int.
3	1	
2	$\sqrt{2}$	
1	$\sqrt{3}$	
0	2	y int.
−1	$\sqrt{5}$	

37.

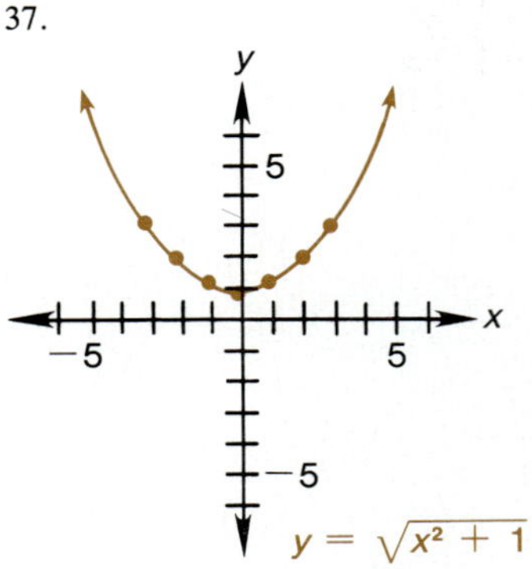

x	y	
−3	$\sqrt{10}$	
−2	$\sqrt{5}$	
−1	$\sqrt{2}$	
0	1	y int.
1	$\sqrt{2}$	
2	$\sqrt{5}$	
3	$\sqrt{10}$	

39.

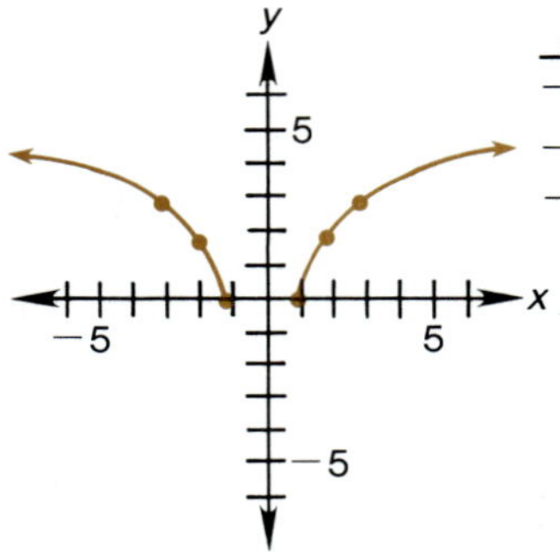

x	y	
-3	$\sqrt{8}$	
-2	$\sqrt{3}$	
-1	0	x int.
1	0	x int.
2	$\sqrt{3}$	
3	$\sqrt{8}$	

$y = \sqrt{x^2 - 1}$
$x \leq -1$ or $x \geq 1$

41.

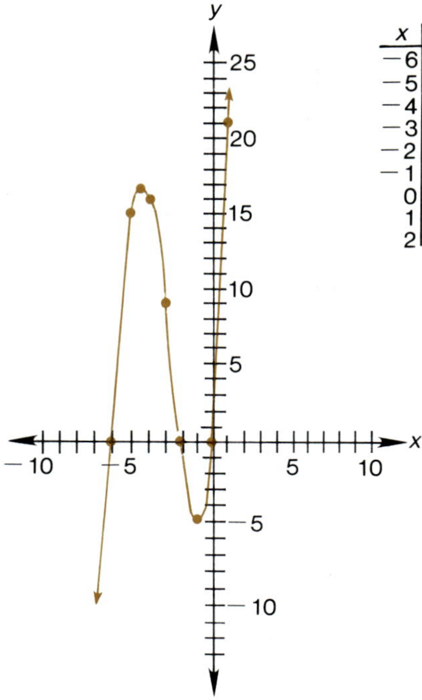

x	P	
-6	0	x int.
-5	15	
-4	16	
-3	9	
-2	0	x int.
-1	-5	
0	0	x & y int.
1	21	
2	64	

$P = x^3 + 8x^2 + 12x$

43.

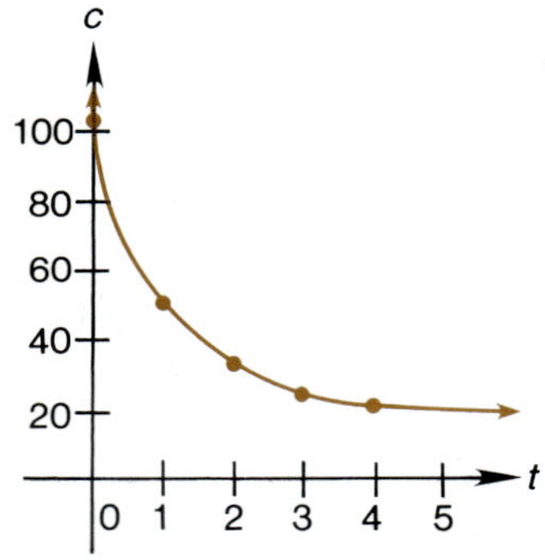

t	c
0	105
1	$155/3 \approx 51.6$
2	$305/9 \approx 33.9$
3	$755/27 \approx 28.0$
4	$2105/81 \approx 26.0$

$c = 25 + 80(1/3)^t$

45.

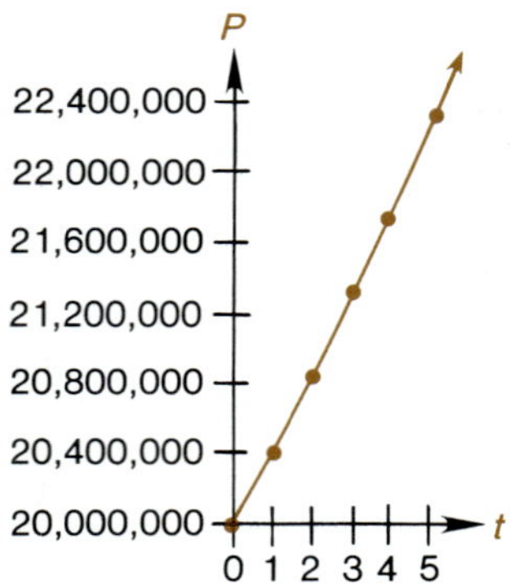

t	P
0	20,000,000
1	20,437,943
2	20,885,476
3	21,342,808
4	21,810,155
5	22,287,735

$P = I \cdot 2^{t/32}$
$(20{,}000{,}000)2^{t/32}$

47.

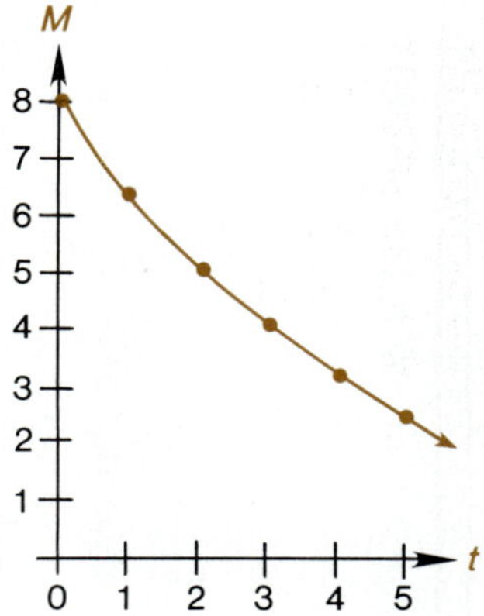

t	M
0	8
1	6.3
2	5.0
3	4
4	3.2
5	2.5
10	.79
20	.08

$M = A\left(\frac{1}{2}\right)^{t/3}$

$M = 8\left(\frac{1}{2}\right)^{t/3}$

49.

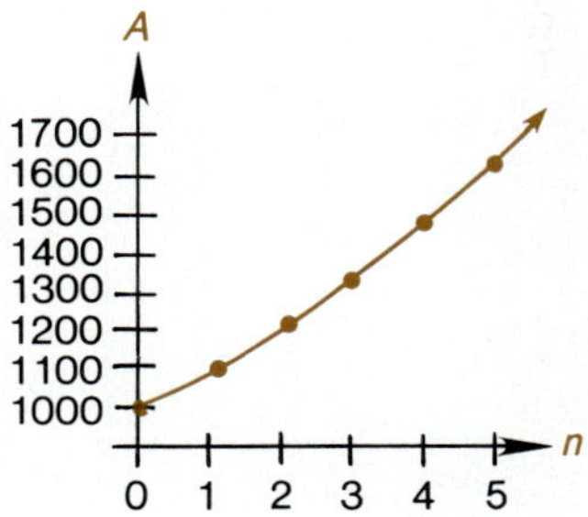

n	A
0	1000
1	1100
2	1210
3	1331
4	1464.1
5	1610.51

$A = P(1 + I)^n$

$A = 1000(1 + 10)^n$

51.

$S = \sqrt{A}$

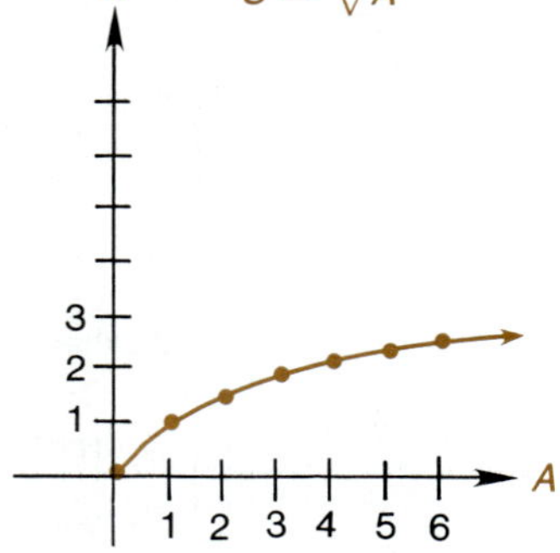

A	S
0	0
1	1
2	$\sqrt{2}$
3	$\sqrt{3}$
4	2
5	$\sqrt{5}$
6	$\sqrt{6}$

53.

$t = \frac{10}{7 - r}$

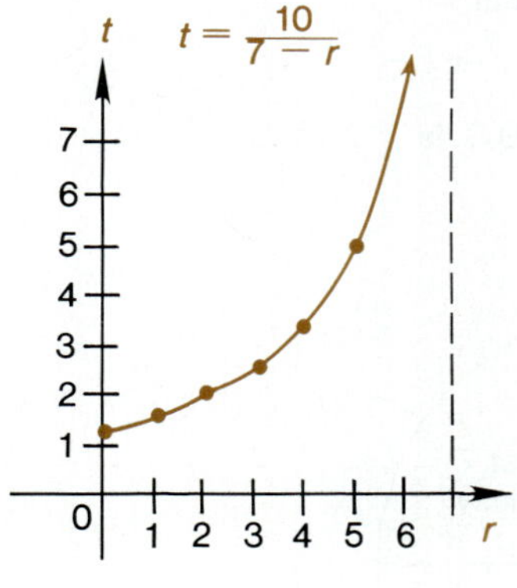

r	t
0	10/7
1	5/3
2	2
3	5/2
4	10/3
5	5
6	10

55.

$P = \frac{100}{1 + t}$

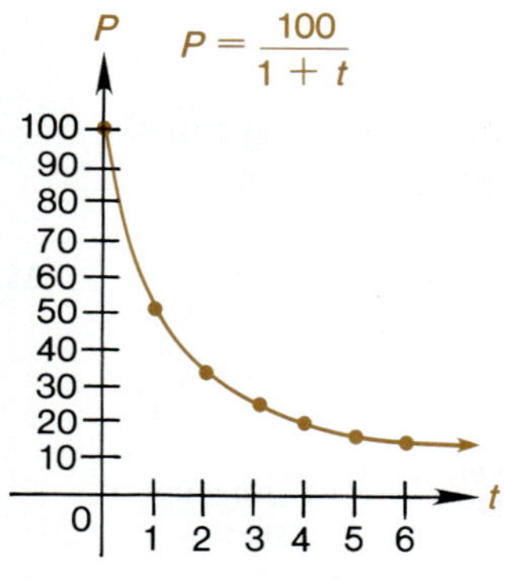

t	P
0	100
1	50
2	$33\frac{1}{3}$
3	25
4	20
5	50/3
6	100/7

Exercise 15.6

1.

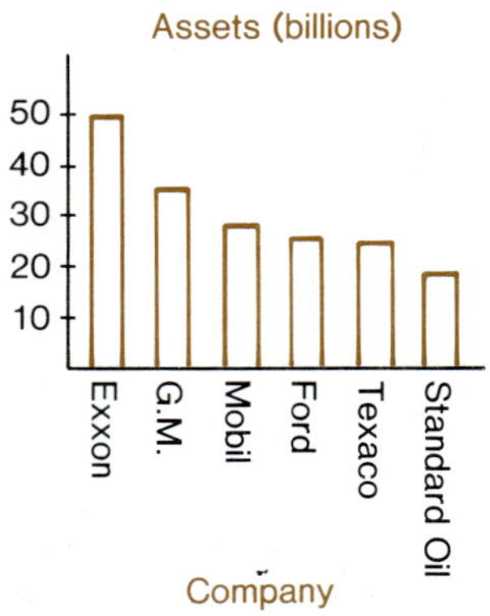

3.

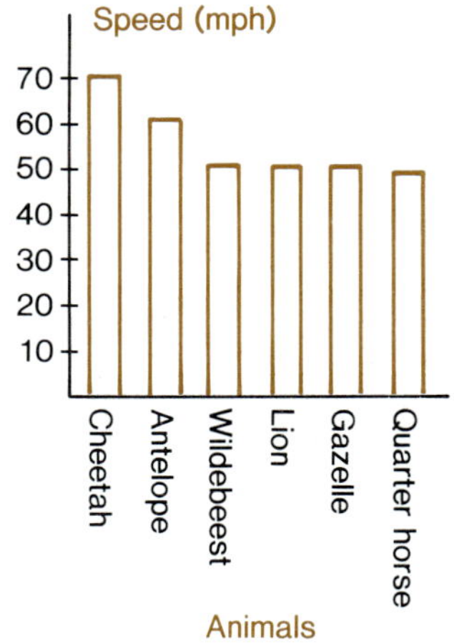

5.

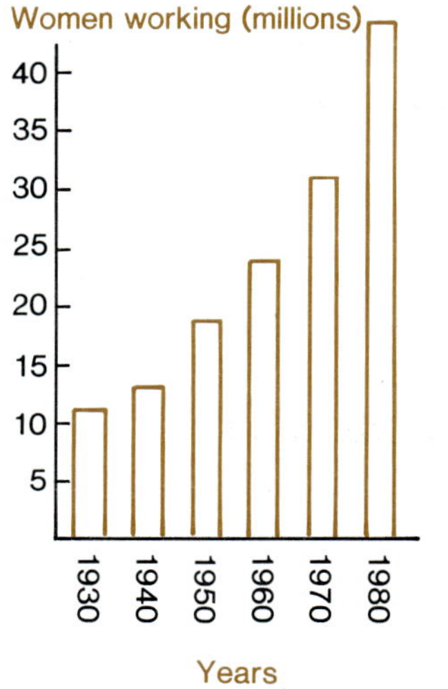

7.

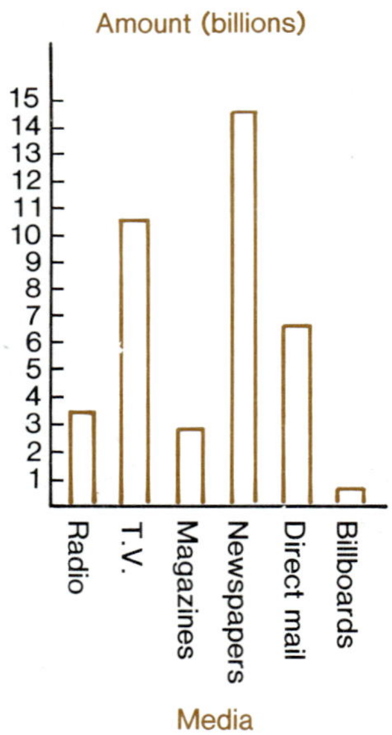

9.

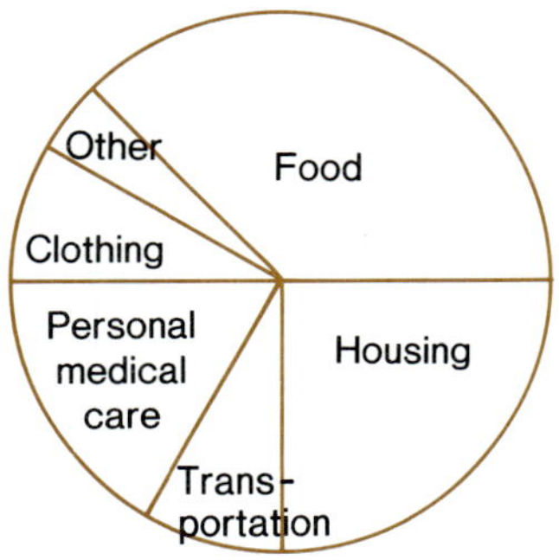

Breakdown of spending $10,642 by average family of four

11.

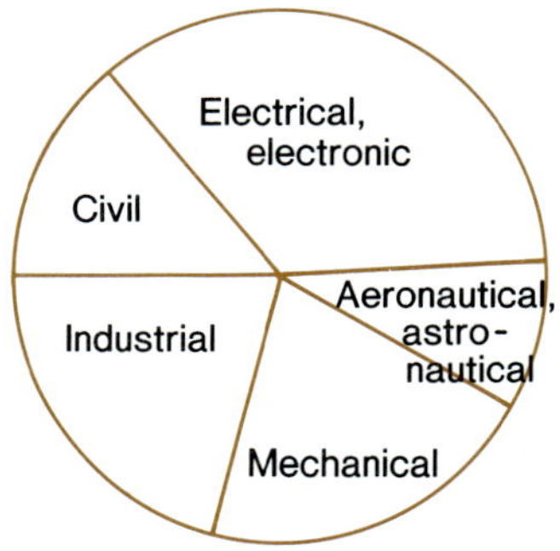

No. of engineers in U.S. is approx. 1,073,000.

13.

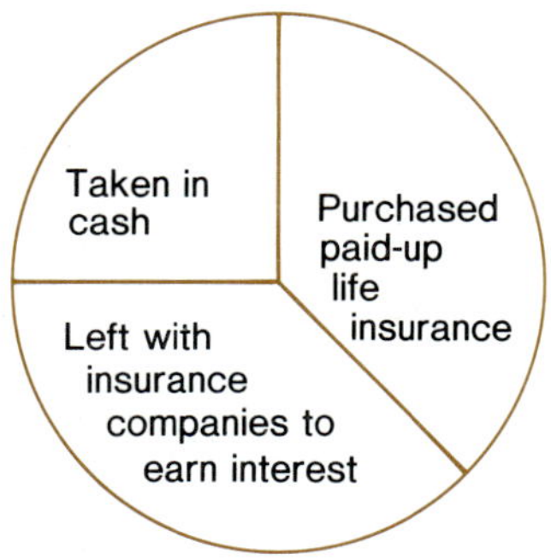

Dividend application of $7,120,224.

15.

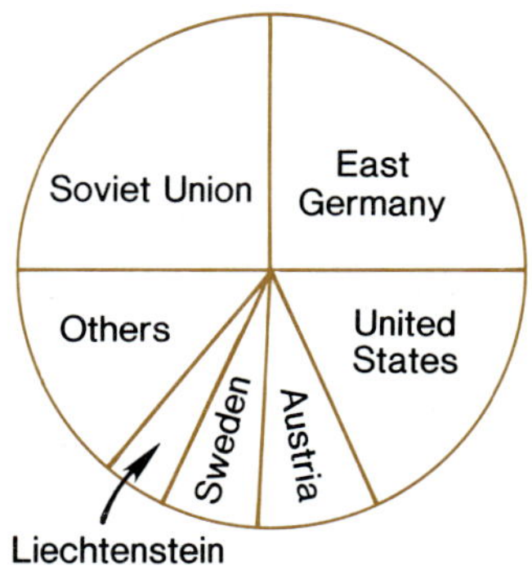

Gold Medals
1980 Winter Olympic Games

17.

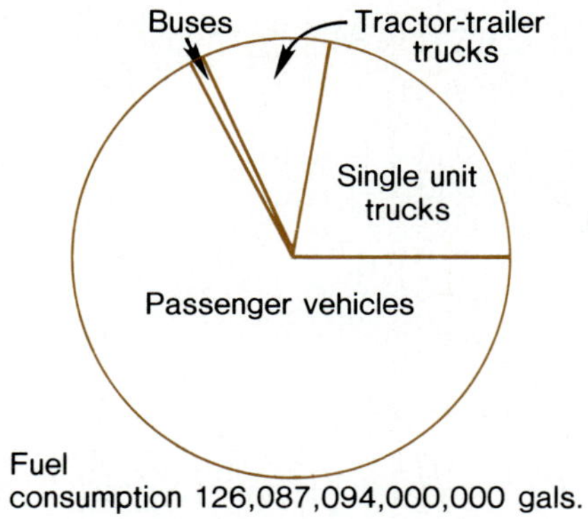

19.

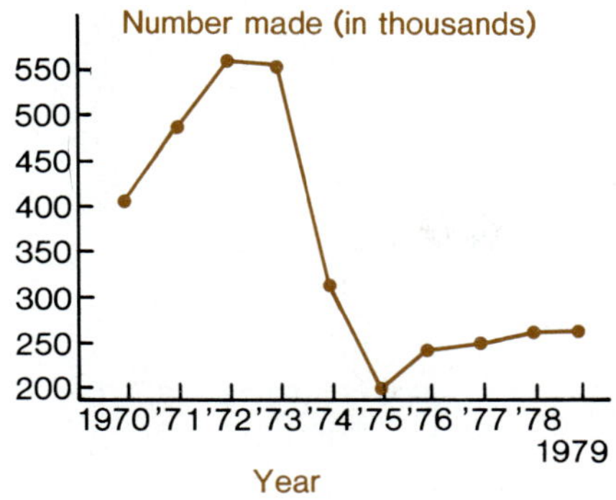

21.

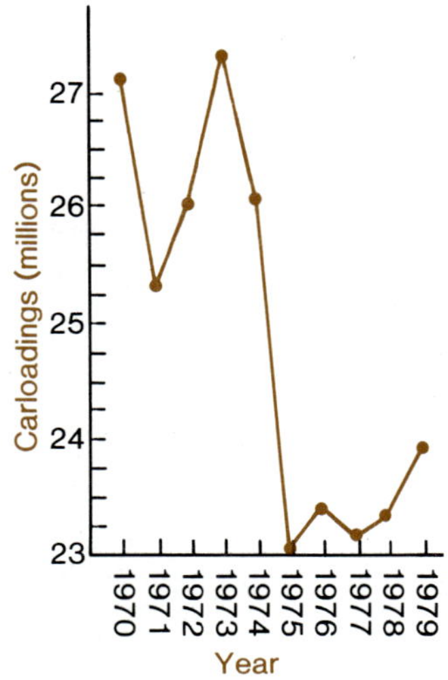

23.

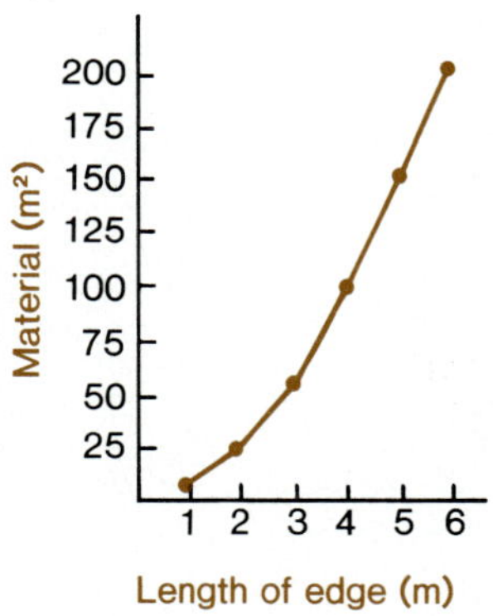

Chapter Review

7. $(-1, -2)$ 8. $(1, 1)$ 9. $(4, 1)$ 10. $(0, 3)$ 11. $(-3, 1)$ 12. $(3, -4)$

1. 2. 3. 4. 5. 6.

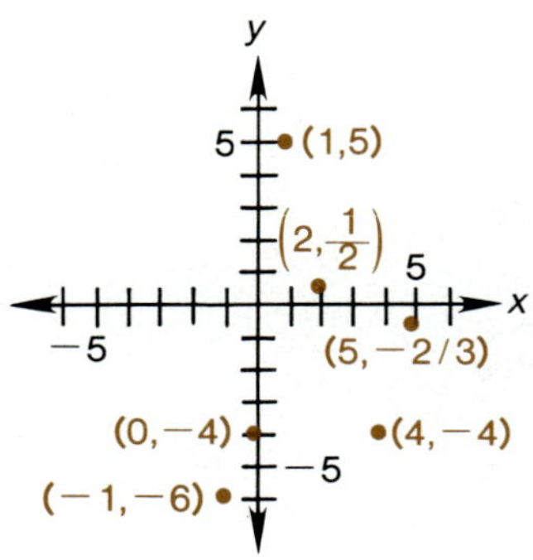

13.

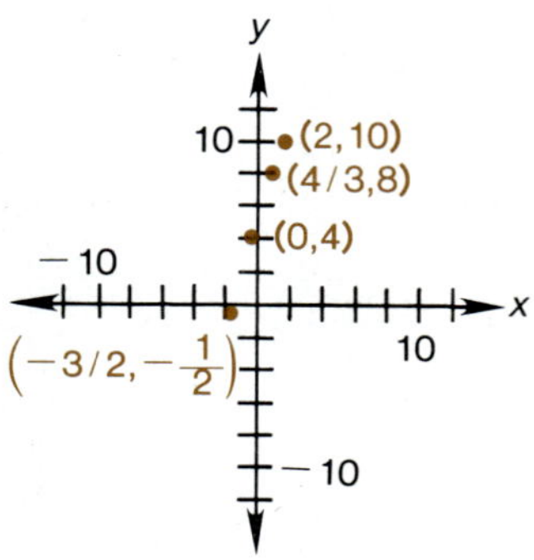

14.

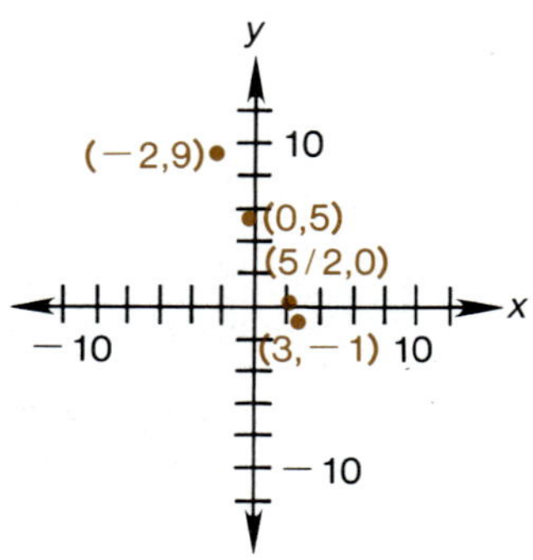

15.

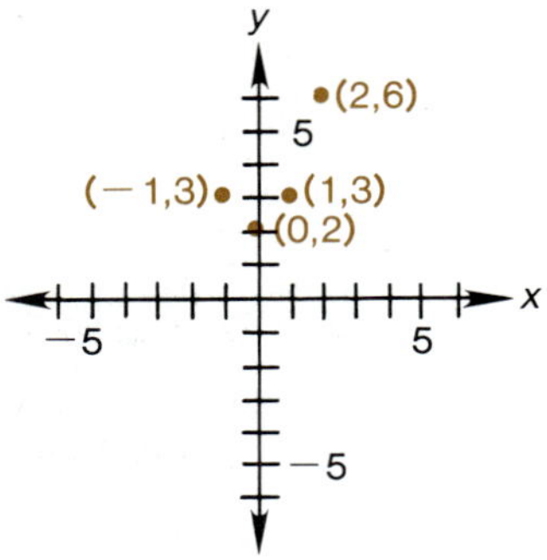

16.

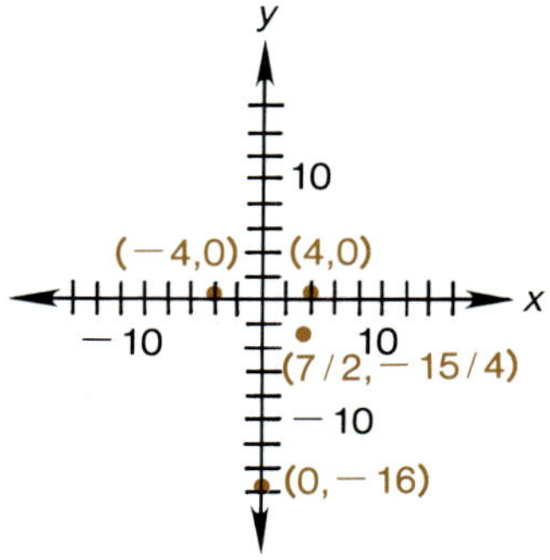

17. −9, 0, −3 18. 0, −2, 0 19. 7, −2, −1

20. $\frac{3}{4}, \frac{1}{2}, 0$ 21. $\frac{2}{5}, \frac{-2}{5}, 0$ 22. $x \geqslant -4$

23. All real numbers, $x \neq -1$

24.

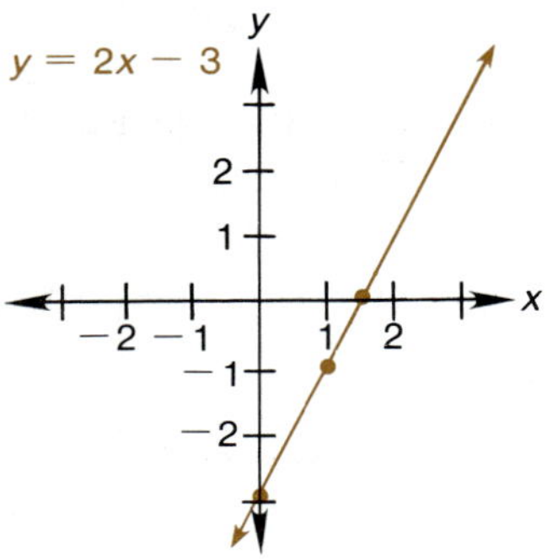

x	y	
0	−3	y int.
$\frac{3}{2}$	0	x int.
1	−1	

25.

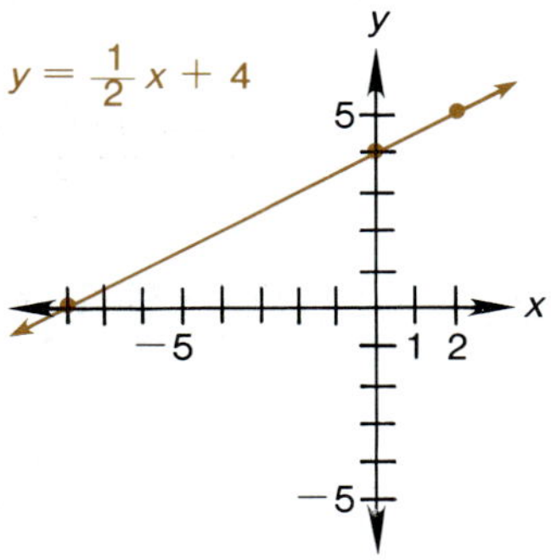

x	y	
0	4	y int.
−8	0	x int.
2	5	

26.

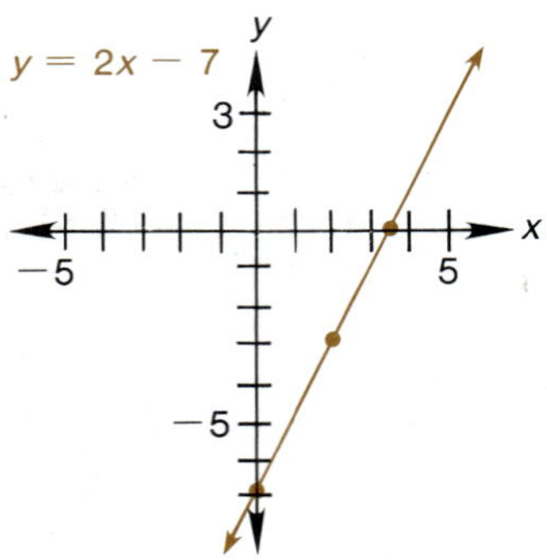

x	y	
0	−7	y int.
$\frac{7}{2}$	0	x int.
2	−3	

27.

$y = 5x$

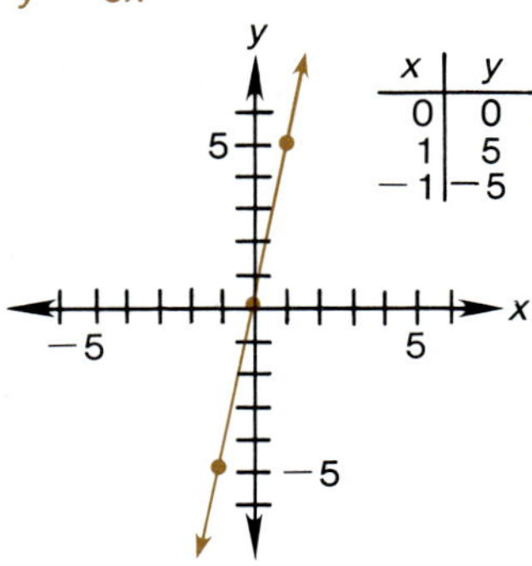

x	y	
0	0	x & y int.
1	5	
−1	−5	

28.

$y = \frac{5}{2}x + 2$

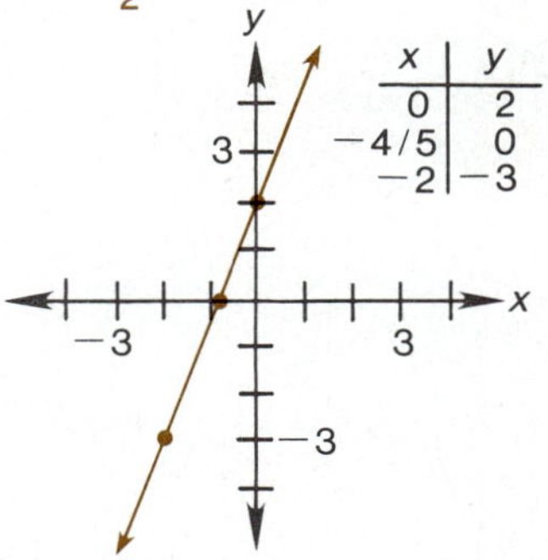

x	y	
0	2	y int.
$-4/5$	0	x int.
-2	-3	

29.

$y = \frac{1}{6}x + 1/3$

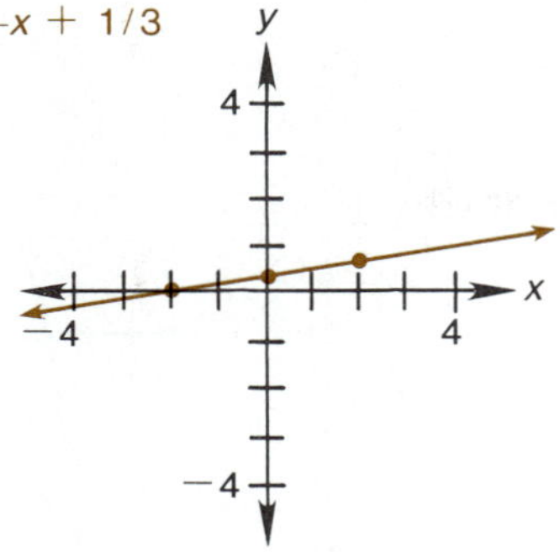

x	y	
0	$1/3$	y int.
-2	0	x int.
2	$2/3$	

30.

$\frac{1}{2}x - \frac{1}{4}y = 3/4$

$y = 2x - 3$

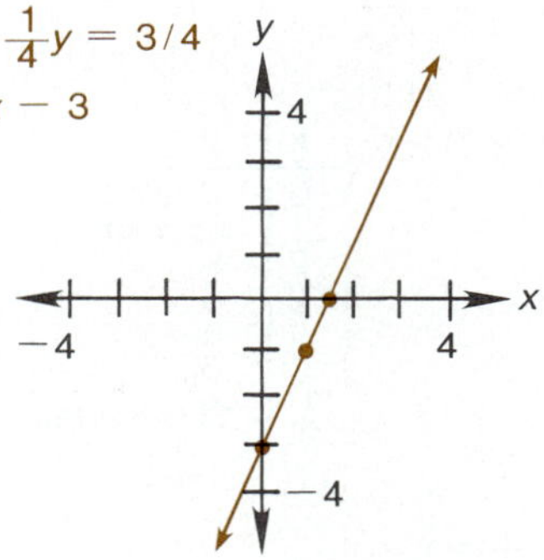

x	y	
0	-3	y int.
$3/2$	0	x int.
1	-1	

31.

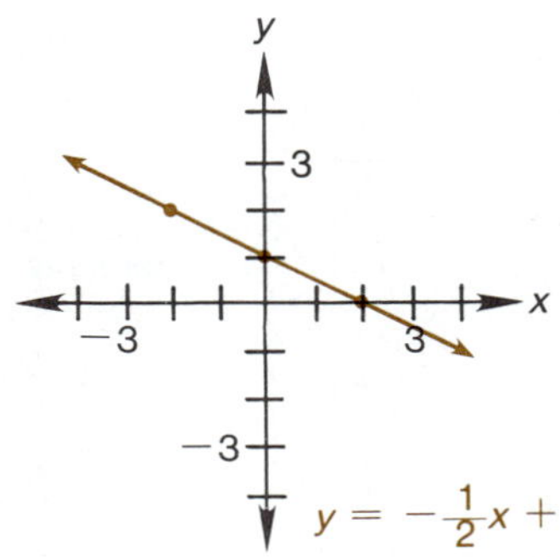

$y = -\frac{1}{2}x + 1$

x	y	
0	1	y int.
2	0	x int.
-2	2	

32.

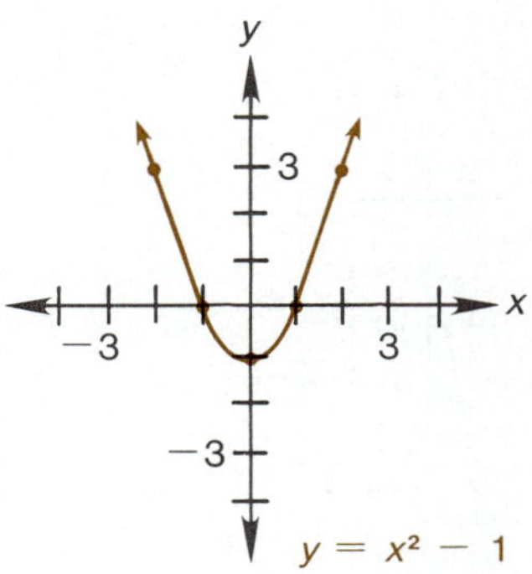

$y = x^2 - 1$

x	y	
0	-1	y int. (vertex also)
1	0	x int.
-1	0	x int.
2	3	
-2	3	

33.

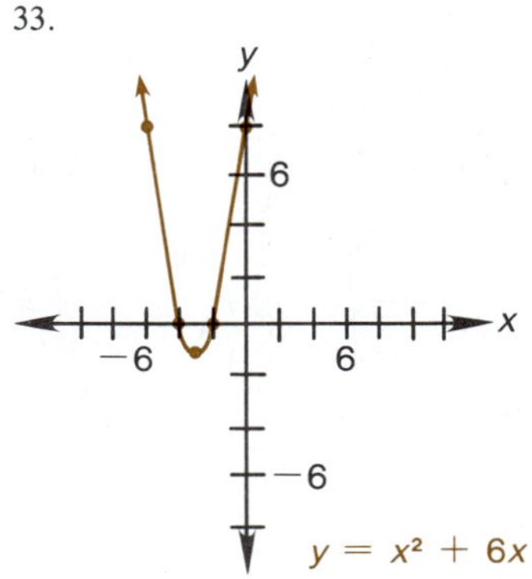

$y = x^2 + 6x + 8$

x	y	
0	8	y int.
-4	0	x int.
-2	0	x int.
-3	-1	vertex

34.

$y = x^2 - 8x + 12$

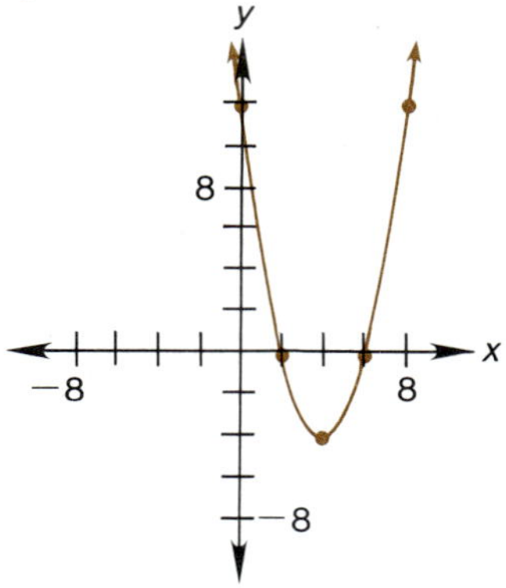

x	y	
0	12	y int.
6	0	x int.
2	0	x int.
4	−4	vertex

35.

$y = -x^2 + 4$

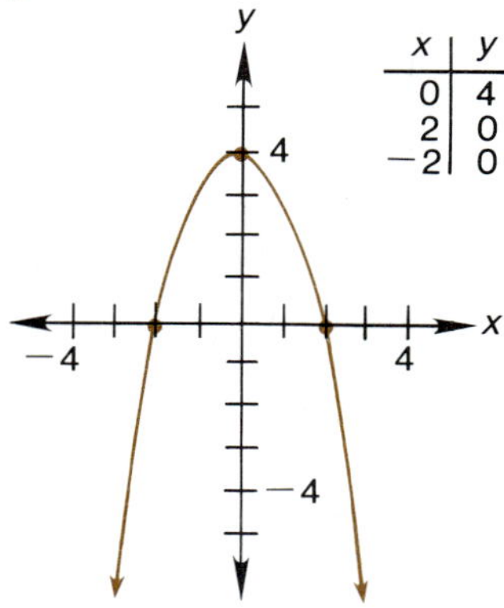

x	y	
0	4	y int. (vertex also)
2	0	x int.
−2	0	x int.

36.

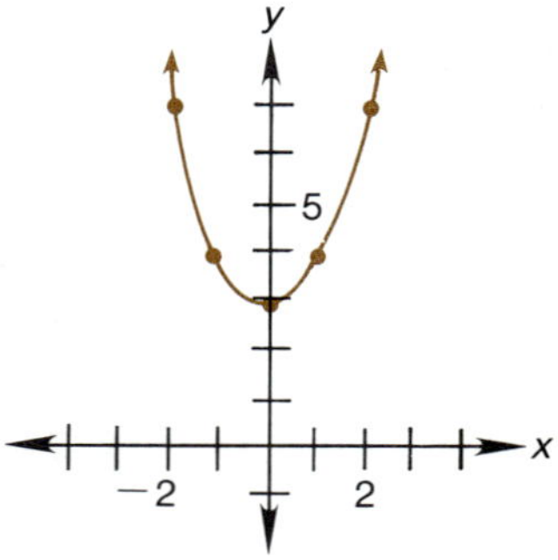

$y = x^2 + 3$

x	y	
0	3	y int. (vertex also) no x intercepts
1	4	
−1	4	
2	7	
−2	7	

37.

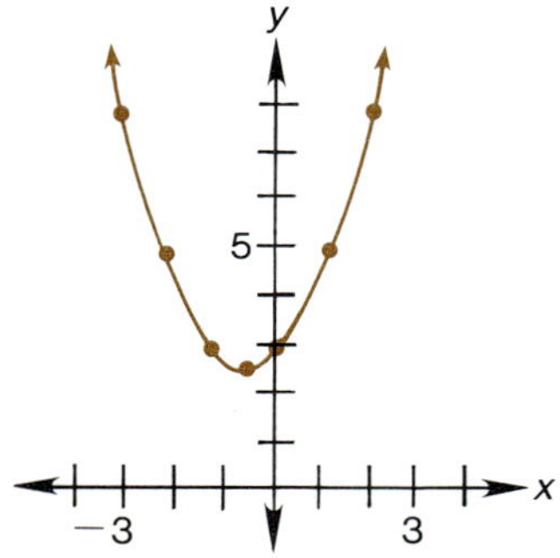

$y = x^2 + x + 3$

x	y	
0	3	y int no x int.
2	9	
−2	5	
1	5	
−1	3	
$-\frac{1}{2}$	11/4	vertex

38.

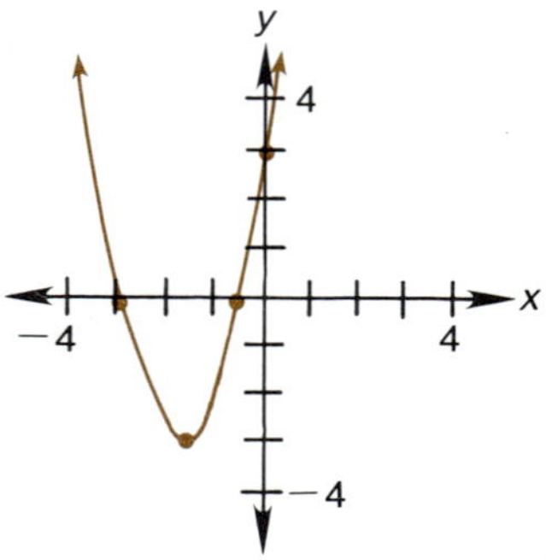

$y = 2x^2 + 7x + 3$

x	y	
0	3	y int.
$-\frac{1}{2}$	0	x int.
−3	0	x int.
−7/4	−25/8	vertex
−4	7	

39.

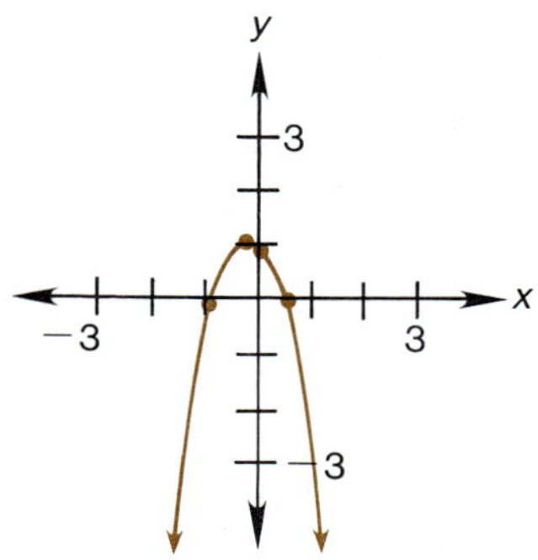

$y = -2x^2 - x + 1$

x	y	
0	1	y int.
$\frac{1}{2}$	0	x int.
−1	0	x int.
$-\frac{1}{4}$	9/8	vertex

40.

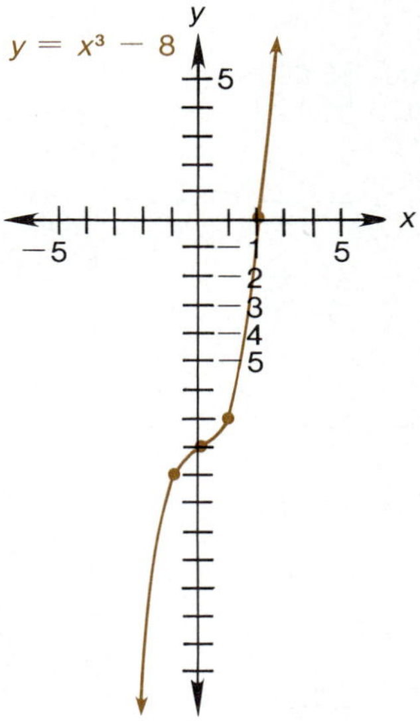

x	y	
0	-8	y int.
-2	-16	
-1	-9	
1	-7	
2	0	x int.
3	19	

41.

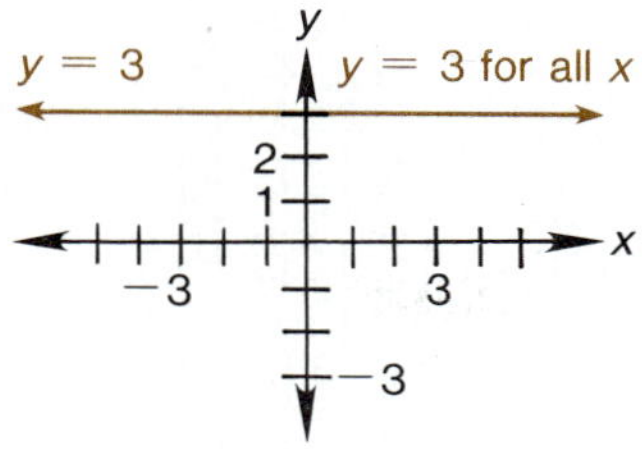

42.

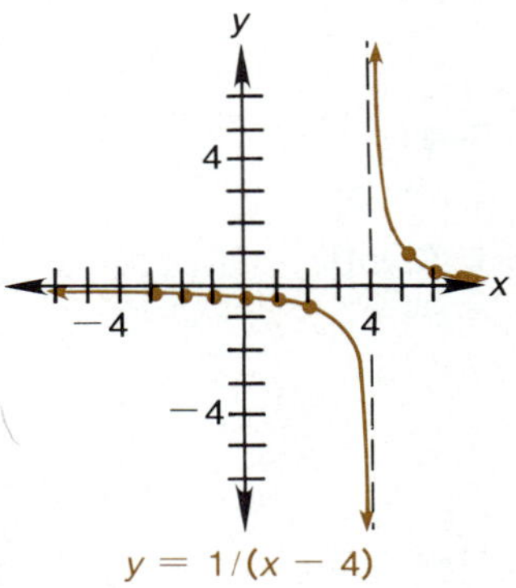

x	y	
-3	$-1/7$	
-2	$-1/6$	
-1	$-1/5$	
0	$-\frac{1}{4}$	y int.
1	$-1/3$	
2	$-\frac{1}{2}$	
3	-1	
4	undef.	
5	1	
6	$\frac{1}{2}$	
7	$1/3$	

no x intercept

43.

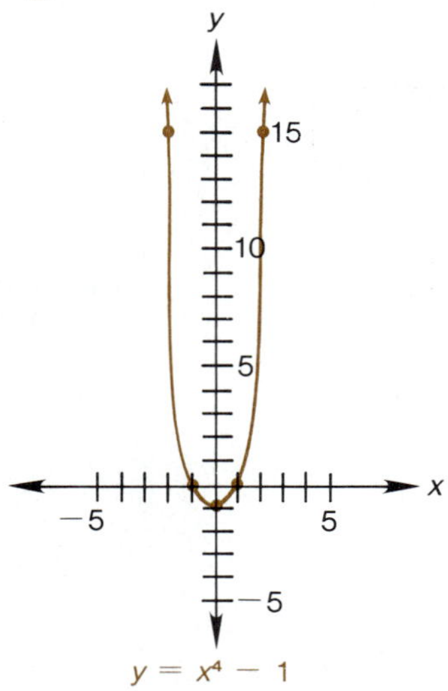

x	y	
-2	15	
-1	0	x int.
0	-1	y int.
1	0	x int.
2	15	

44.

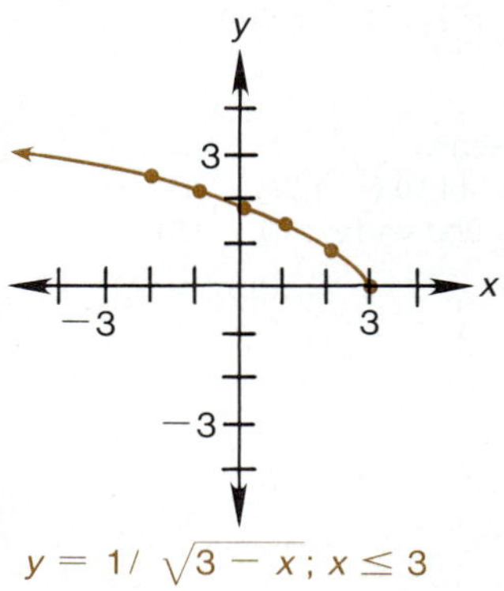

x	y	
3	0	X int.
2	1	
1	$\sqrt{2}$	
0	$\sqrt{3}$	Y int.
-1	2	
-2	$\sqrt{5}$	

45.

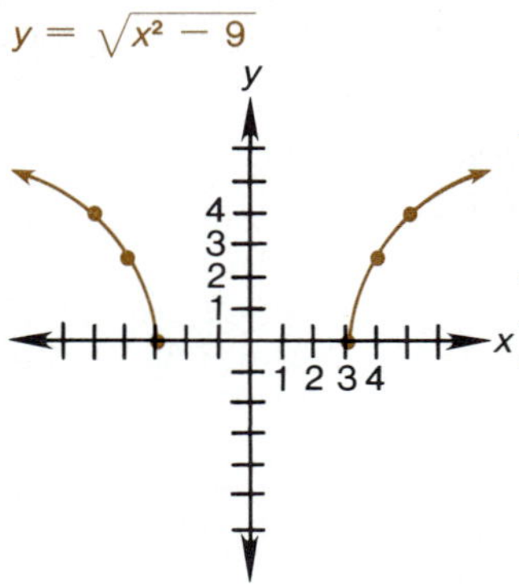

x	y	
3	0	x int.
4	$\sqrt{7}$	
5	4	
-4	$\sqrt{7}$	
-5	4	
-3	0	x int.

46.

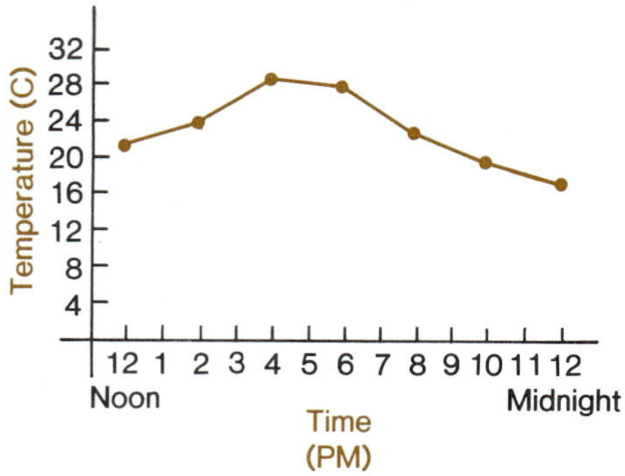

47.

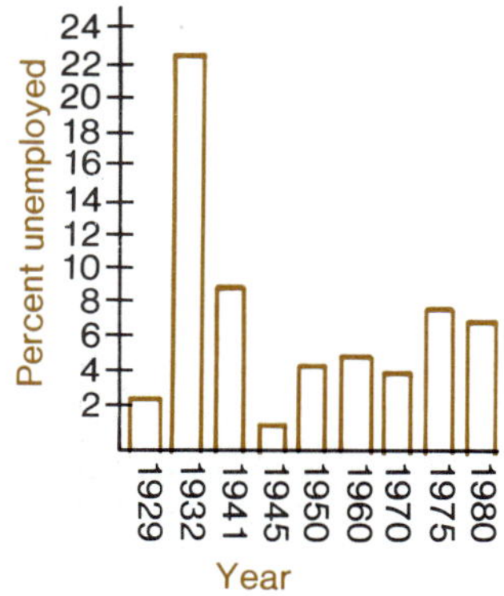

48.

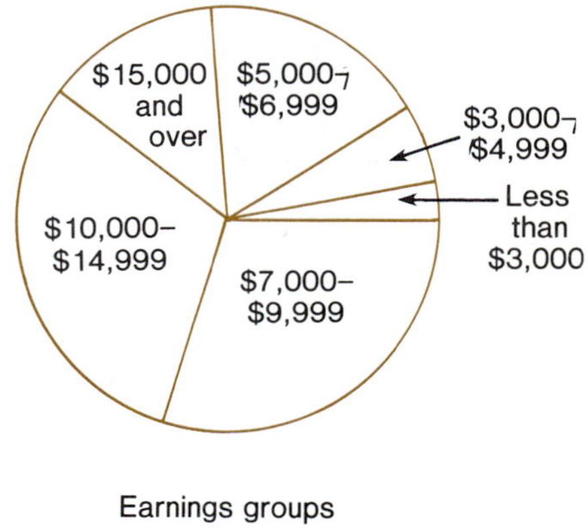

Chapter Test

1.

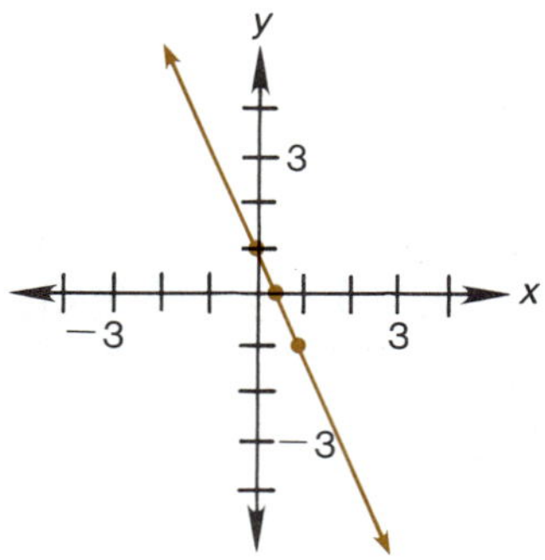

x	y	
0	1	y-int.
$\frac{1}{2}$	0	x-int.
1	-1	

Reals

x: $\left(\frac{1}{2}, 0\right)$

y: $(0, 1)$

2.

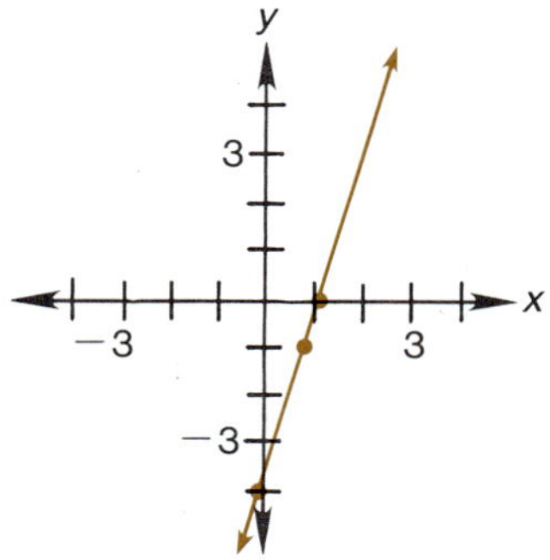

x	y	
0	-4	y int.
4/3	0	x int.
1	-1	

Reals

x: $\left(\frac{4}{3}, 0\right)$

y: $(0, -4)$

3.

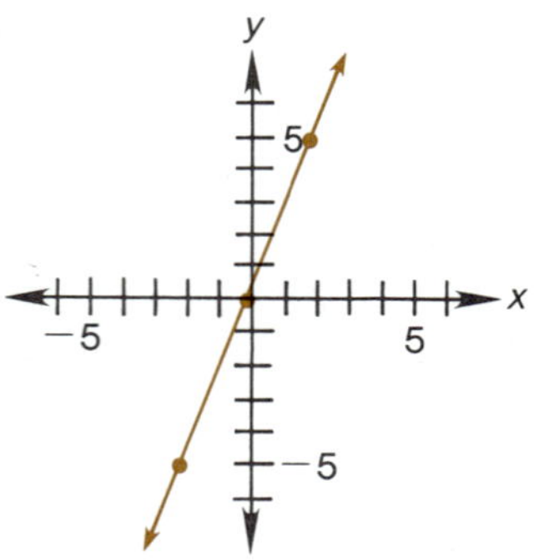

x	y	
0	0	x & y-int.
2	5	
-2	-5	

Reals

x and y: $(0, 0)$

4.

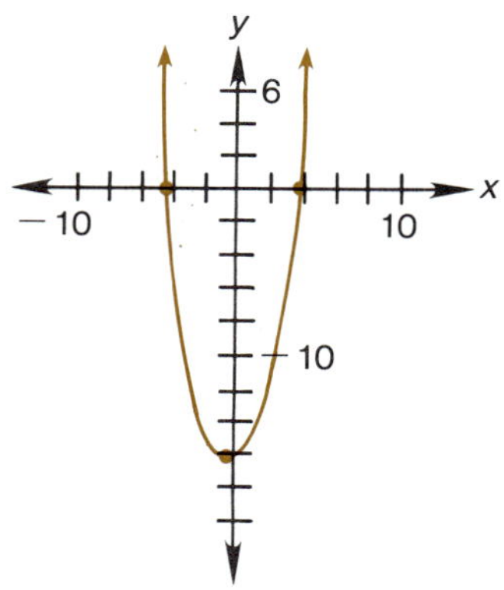

x	y	
0	-16	y int. (vertex also)
4	0	x int.
-4	0	x int.

Reals

x: $(4,0), (-4,0)$

y and vertex: $(0, -16)$

5.

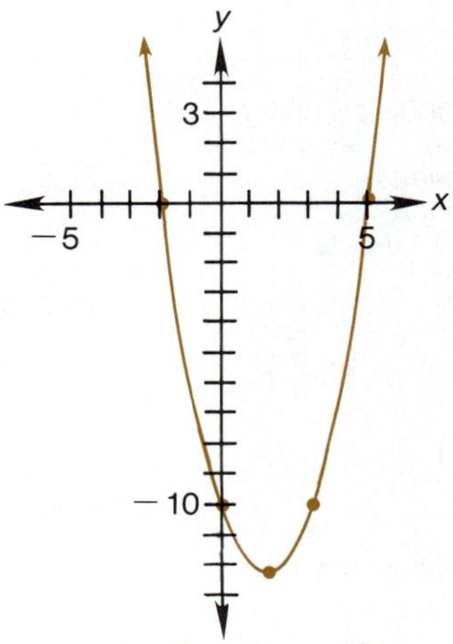

$y = x^2 - 3x - 10$

x	y	
0	−10	y int.
5	0	x int.
−2	0	x int.
$\frac{3}{2}$	$-12\frac{1}{4}$	vertex

Reals
x: (5,0), (−2,0)
y: (0,−10)
vertex: $\left(\frac{3}{2}, \frac{-49}{4}\right)$

6.

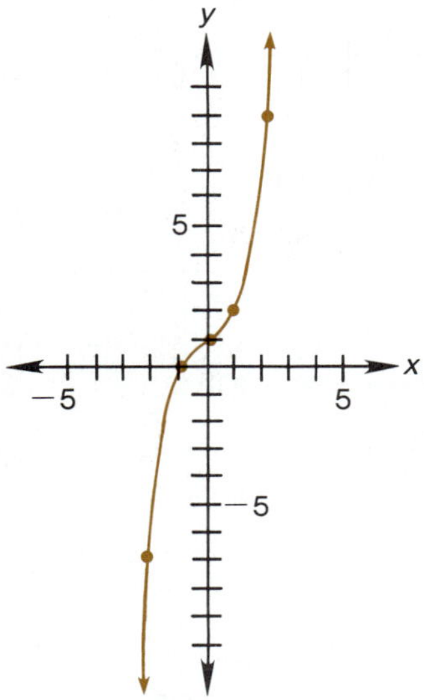

$y = x^3 + 1$

x	y	
0	1	y int.
−1	0	x int.
1	2	
2	9	
−2	−7	

7.

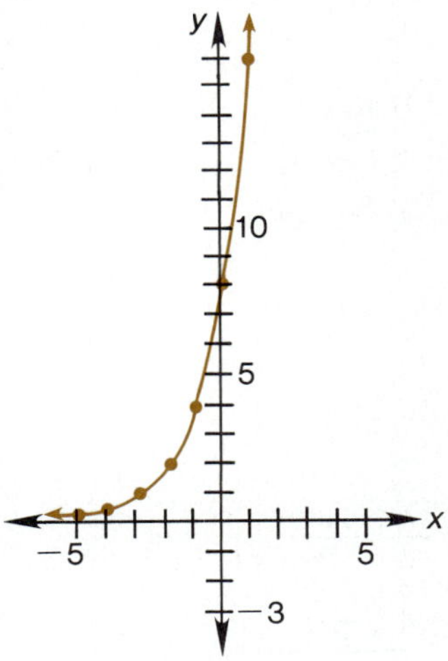

$y = 2^{x+3}$

x	y	
−5	$\frac{1}{4}$	
−4	$\frac{1}{2}$	
−3	1	
−2	2	
−1	4	
0	8	y int.
1	16	
2	32	

8.

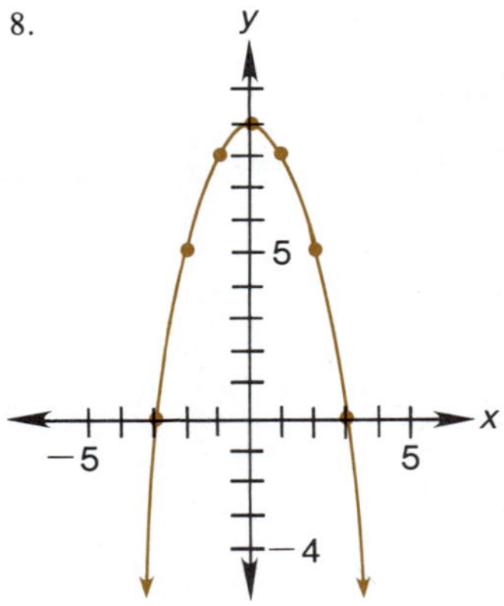

$y = 9 - x^2$

x	y	
0	9	y int.
3	0	x int.
−3	0	x int.
0	9	is vertex also
1	8	
2	5	
−1	8	
−2	5	

Reals
x: (3,0), (−3,0)
y and vertex: (0,9)

9.

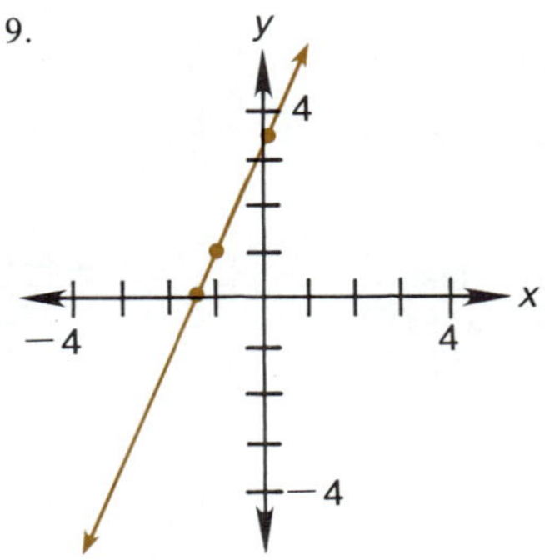

$5x - 2y + 7 = 0$
$y = \frac{5}{2}x + 7/2$

x	y	
0	7/2	y int.
−7/5	0	x int.
−1	1	

Reals
x: $\left(-\frac{7}{5}, 0\right)$
y: $\left(0, \frac{7}{2}\right)$

10.

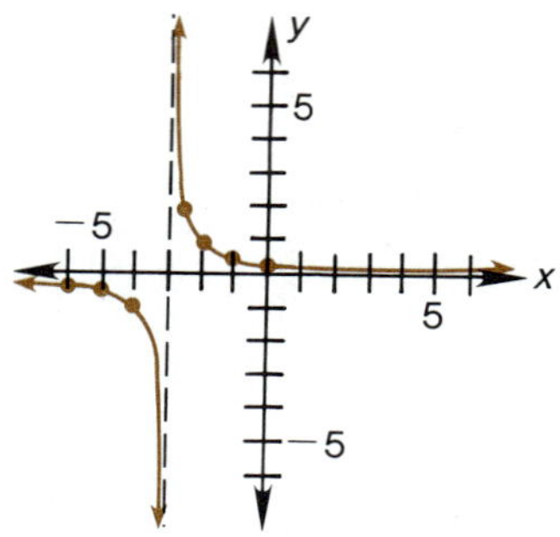

$y = 1/(x+3)\ x \neq -3$

x	y
-6	$-1/3$
-5	$-\frac{1}{2}$
-4	-1
-3	undef.
-2	1
-1	$\frac{1}{2}$
0	$1/3$ y int.
$-5/2$	2

Reals, $x \neq -3$

11.

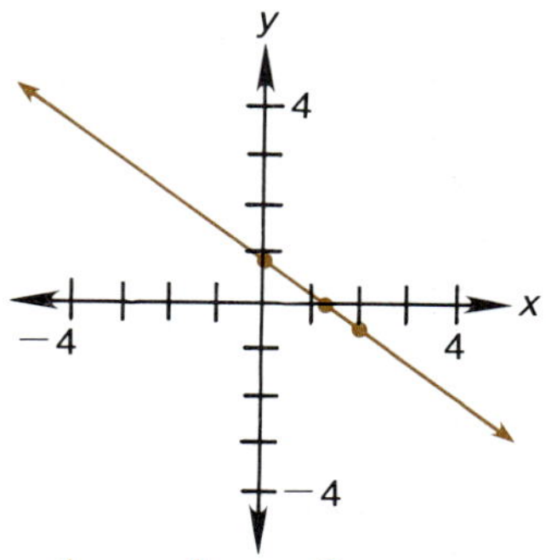

$\frac{1}{2}x + \frac{3}{4}y - \frac{2}{3} = 0$

$y = -\frac{2}{3}x + \frac{8}{9}$

x	y
0	$8/9$ y int.
$4/3$	0 x int.
2	$-4/9$

Reals

x: $\left(\frac{4}{3}, 0\right)$

y: $\left(0, \frac{8}{9}\right)$

12.

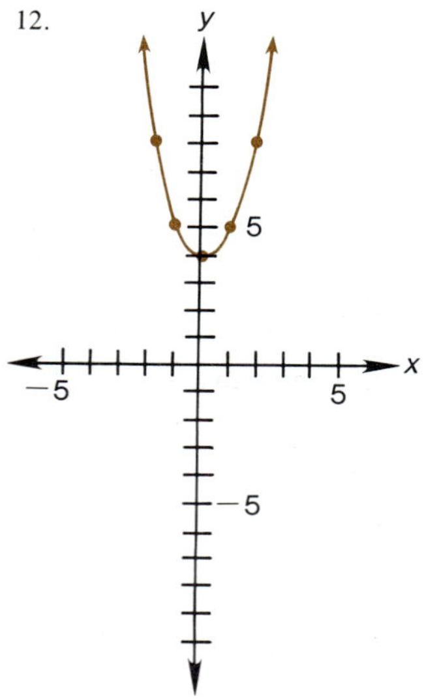

x	y
0	4 y int.
	no x intercepts
± 1	5
± 2	8

Reals

x: none

y and vertex: $(0,4)$

13.

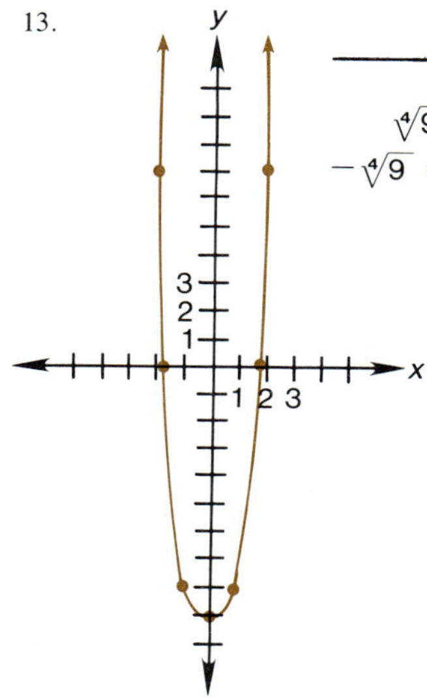

$y = x^4 - 9$

x	y	
0	-9	y int.
$\sqrt[4]{9} = \sqrt{3}$	0	x int.
$-\sqrt[4]{9} = -\sqrt{3}$	0	x int.
± 1	-8	
± 2	7	

14.

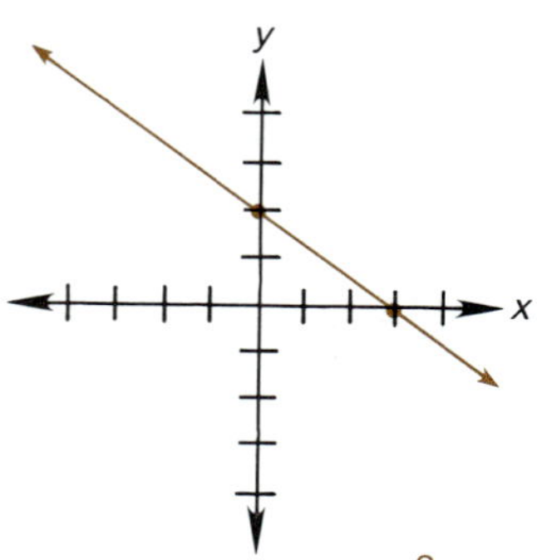

$x/3 + y/2 = 1$ or $y = -\frac{2}{3}x + 2$

x	y
0	2 y int.
3	0 x int.
-3	4

Reals

x: $(3,0)$

y: $(0,2)$

15.

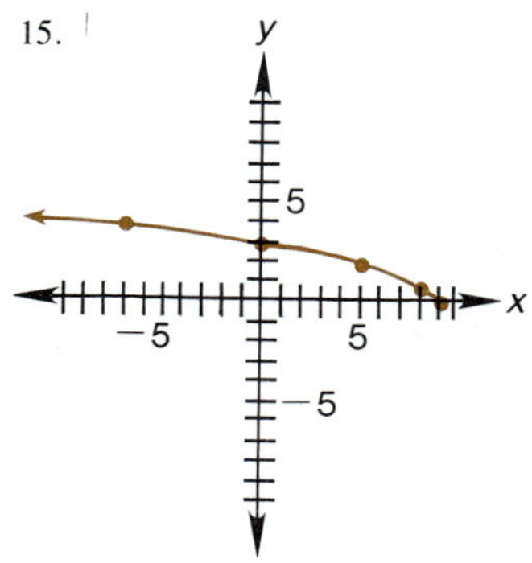

$y = \sqrt{9-x};\ x \leq 9$

x	y
-9	$3\sqrt{2}$
-8	$\sqrt{17}$
-7	4
-6	$\sqrt{15}$
-5	$\sqrt{14}$
-4	$\sqrt{13}$
0	3 y int.
5	2
8	1
9	0 x int.

16.

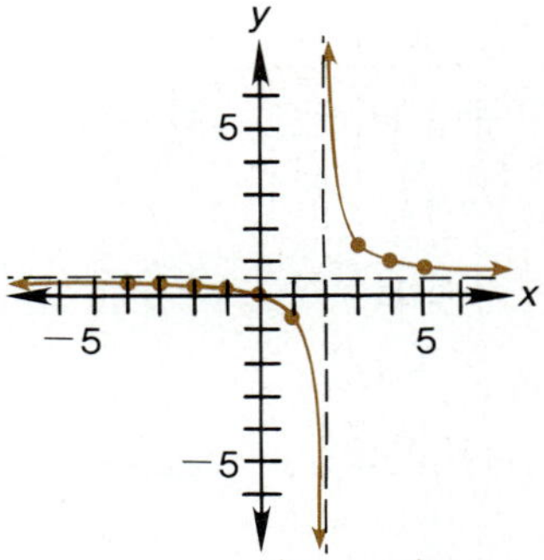

$y = x/(2x - 4)$

x	y	
-3	$3/10$	
-2	$\frac{1}{4}$	
-1	$1/6$	
0	0	x and y int.
1	$-\frac{1}{2}$	
2	undef.	
3	$3/2$	
4	1	
5	$5/6$	

Reals

$x \neq 2$

17.

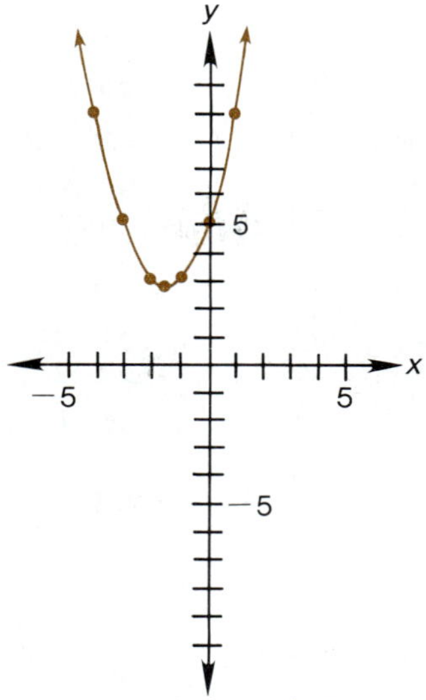

$y = x^2 + 3x + 5$

x	y	
0	5	y int.
1	9	
2	15	
$-3/2$	$11/4$	vertex
-1	3	
-2	3	

Reals

x: none

y: (0,5)

vertex: $\left(\frac{-3}{2}, \frac{11}{4}\right)$

18.

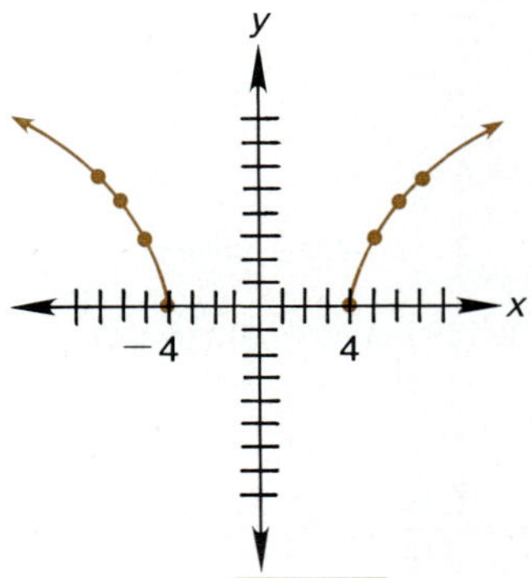

$y = \sqrt{x^2 - 16}$

x	y	
4	0	x int.
5	3	
6	$2\sqrt{5}$	
7	$\sqrt{33}$	
-4	0	y int.
-5	3	
-6	$2\sqrt{5}$	
-7	$\sqrt{33}$	

$x \geq 4$ or $x \leq -4$

19.

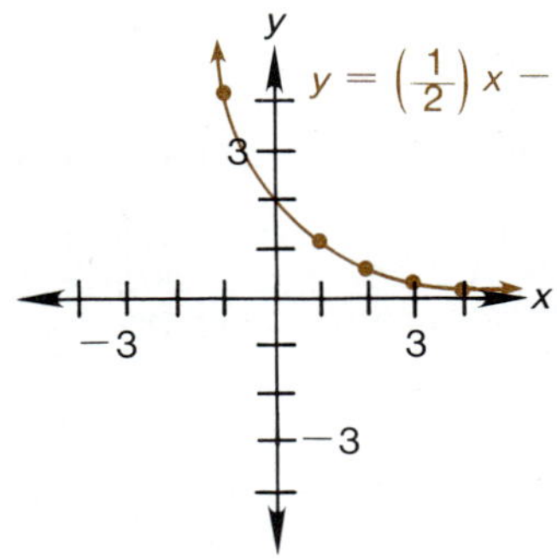

$y = \left(\frac{1}{2}\right)^{x-1}$

x	y	
1	1	
2	$\frac{1}{2}$	
3	$\frac{1}{4}$	
4	$1/8$	
-1	4	
-2	8	
0	2	y int.

Reals

20.

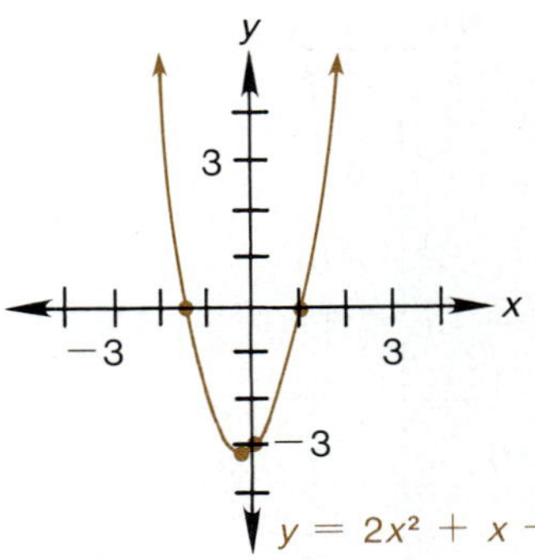

$y = 2x^2 + x - 3$

x	y	
0	-3	y int.
$-3/2$	0	x int.
1	0	x int.
$-\frac{1}{4}$	$-25/8$	vertex

Reals

x: $\left(-\frac{3}{2}, 0\right), (1,0)$

y: $(0, -3)$

vertex: $\left(-\frac{1}{4}, -\frac{25}{8}\right)$

Chapter 16

Exercise 16.1

1. One 3. One 5. Inconsistent 7. Dependent 9. Dependent 11. Yes 13. Yes 15. Yes 17. No 19. (3,2) 21. (4,1) 23. $\left(\frac{1}{2}, -2\right)$ 25. (2,3) 27. Inconsistent 29. (4, −2) 31. (0,1) 33. Dependent 35. (2,3) 37. $\left(\frac{11}{7}, \frac{6}{7}\right)$ 39. $\left(\frac{-15}{2}, \frac{95}{2}\right)$ 41. $\left(\frac{27}{13}, \frac{-4}{13}\right)$ 43. Inconsistent 45. $\left(-8, \frac{-40}{3}\right)$ 47. $\left(\frac{3}{2}, \frac{5}{2}\right)$ 49. $\left(\frac{21}{2}, \frac{-7}{4}\right)$

Exercise 16.2

1. (−2,3) 3. (3, −1) 5. (−1,3) 7. $\left(\frac{29}{10}, \frac{-3}{5}\right)$ 9. $\left(\frac{17}{21}, \frac{-11}{21}\right)$ 11. Inconsistent 13. (2, −1) 15. (−1,0) 17. Dependent 19. $\left(\frac{22}{21}, \frac{13}{21}\right)$ 21. $\left(\frac{4}{5}, \frac{9}{5}\right)$ 23. (7,3) 25. Inconsistent 27. (1, −5) 29. (−3,7) 31. Dependent 33. (5, −2)
35. $R = 4, S = 3$ 37. $a = -\frac{1}{15}, b = \frac{-19}{15}$ 39. $R = 4, S = 5$ 41. $a = 2, b = 3$ 43. Dependent 45. $N = 5, M = -2$
47. Inconsistent

Exercise 16.3

1. (3, −1) 3. (−6,2) 5. $\left(\frac{1}{2}, -2\right)$ 7. (4,5) 9. (3, −4) 11. $\left(\frac{7}{5}, \frac{1}{5}\right)$ 13. $\left(\frac{2}{13}, \frac{-14}{13}\right)$ 15. $R = 4, S = 1$ 17. $\left(\frac{17}{11}, \frac{-26}{11}\right)$ 19. $M = -1, N = 0$ 21. Inconsistent 23. $\left(\frac{15}{23}, \frac{13}{23}\right)$ 25. $\left(\frac{19}{22}, \frac{1}{11}\right)$ 27. $a = 4, b = -2$ 29. (−1,3) 31. Dependent 33. $\left(\frac{29}{15}, -\frac{3}{5}\right)$ 35. $\left(\frac{2}{3}, \frac{7}{3}\right)$ 37. Inconsistent 39. $\left(\frac{3}{5}, \frac{1}{5}\right)$ 41. $\left(\frac{92}{15}, \frac{-12}{5}\right)$ 43. (−5, −28) 45. $\left(\frac{5}{3}, 0\right)$

Exercise 16.4

1. 4 m, 14 m 3. 14 m, 36 m 5. 20 ft, 11 ft 7. \$14,000 at 9%, \$4,000 at $7\frac{1}{2}$% 9. \$22,000 at 7%, \$8,000 at 9%
11. \$13,300 at 9%, \$7,700 at 13% 13. 8 ft, 4 ft 15. 28 amps, 52 amps 17. 38, 26 19. 47 v, 79 v 21. 19 ohms, 5 ohms
23. 28 at \$125, 12 at \$185 25. 16 v, 11 v 27. 6, 17 29. $m = 3, b = -4$ 31. $m = -2, b = -1$
33. 40 ml of 10%, 80 ml of 4% 35. 9 oz of 80%, 3 oz of 60% 37. 100 of 2 grains, 100 of 3 grains 39. 10 l of 60%, 20 l of 30%
41. 12 mph, 4 mph

Chapter Review

1. (3,1) 2. (2,4) 3. (4,2) 4. (−2,6) 5. $\left(\frac{54}{5}, \frac{-2}{5}\right)$ 6. $\left(-\frac{1}{3}, 3\right)$ 7. $\left(\frac{-440}{249}, \frac{424}{83}\right)$ 8. $\left(\frac{1{,}667}{1{,}571}, \frac{-1{,}234}{1{,}571}\right)$ 9. (3,0) 10. (7,2)
11. (−1,1) 12. $a = 1, b = -1$ 13. $R = 0, S = 2$ 14. (5, −3) 15. (5,2) 16. Inconsistent
17. $L = 2, M = 1$ 18. Dependent 19. (5,3) 20. (5, −3) 21. Inconsistent 22. $s = \frac{24}{11}, t = \frac{5}{11}$ 23. $R = 3, S = 1$
24. $a = 7, b = 2$ 25. Dependent 26. $\left(\frac{19}{11}, \frac{-17}{11}\right)$ 27. $\left(\frac{-6}{5}, \frac{-13}{5}\right)$ 28. $\left(\frac{1}{2}, \frac{5}{2}\right)$ 29. $\left(\frac{-4}{3}, \frac{13}{3}\right)$ 30. (−42, −24)
31. $\left(\frac{-66}{35}, \frac{-73}{25}\right)$ 32. $\left(\frac{-11}{15}, \frac{-1}{25}\right)$ 33. 25 ft, 7 ft 34. \$7,000 at 23%, \$11,000 at 14% 35. 12, 54
36. 90 ml of 80%, 30 ml of 60%

Chapter Test

1. (2,1) 2. $a = 1, b = -2$ 3. $R = 1, S = 1$ 4. $y = 1, z = 2$ 5. $L = 4, M = 2$ 6. Dependent 7. $R = 1, S = -2$
8. $\left(\frac{61}{29}, \frac{-2}{29}\right)$ 9. Inconsistent 10. (4,0) 11. $r = \frac{41}{13}, s = \frac{7}{13}$ 12. $\left(\frac{-1}{11}, \frac{-52}{11}\right)$ 13. (−4,7) 14. $\left(\frac{31}{22}, \frac{29}{22}\right)$
15. $S = -1, R = 2$ 16. Inconsistent 17. \$23,000 at 8%, \$15,000 at 6% 18. 10 l of 60%, 20 l of 30%

Chapter 17

Exercise 17.1

1. Positive, II 3. Positive, Quadrantal 5. Positive, III 7. Negative, Quadrantal 9. Negative, III 11. $(2, -6)$, $\left(\frac{1}{2}, -\frac{3}{2}\right)$, $(3, -9)$

13. $\left(-1, -\frac{5}{2}\right)$, $(-4, -10)$, $\left(-\frac{1}{2}, -\frac{5}{4}\right)$ 15. 487, 847, −233, −593 17. 689°, 1,049°, −31°, −391°

19. 540°, 900°, −180°, −540° 21. 214° 23. 4° 25. 288° 27. 157° 29. 256.6°

31. 436°, −284°

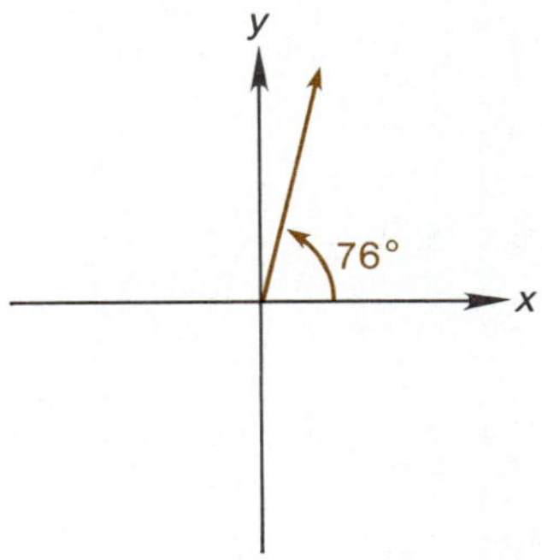

33. −138°, 582°

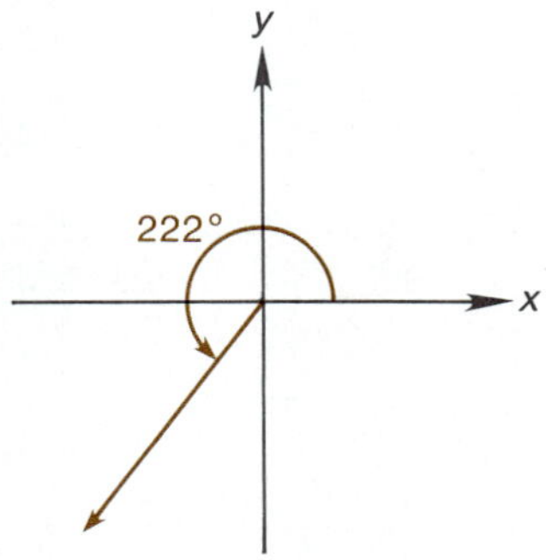

35. 298°, −422°

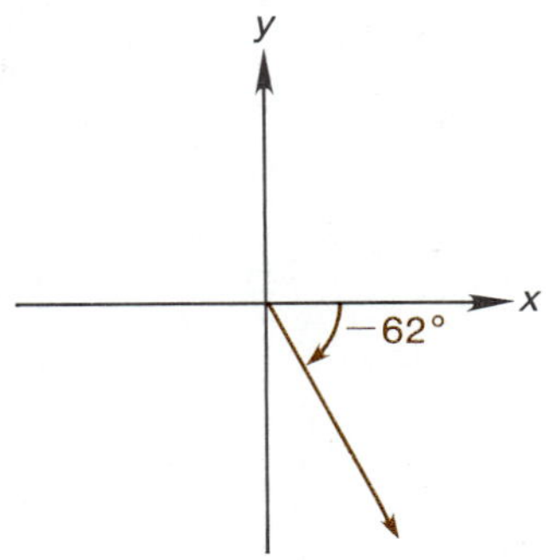

37. 107°, −613°

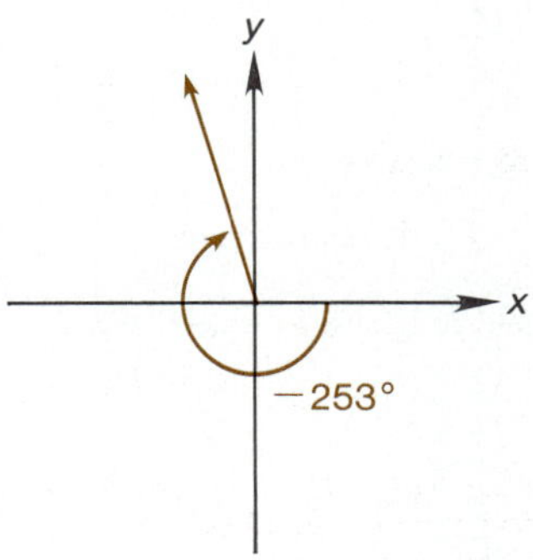

39. 204°, −156°

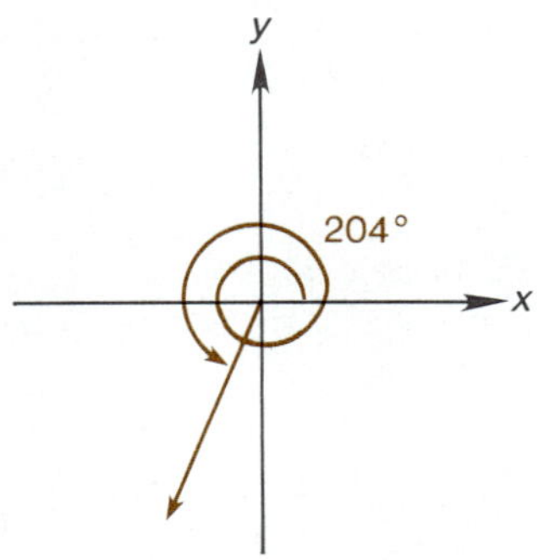

41.

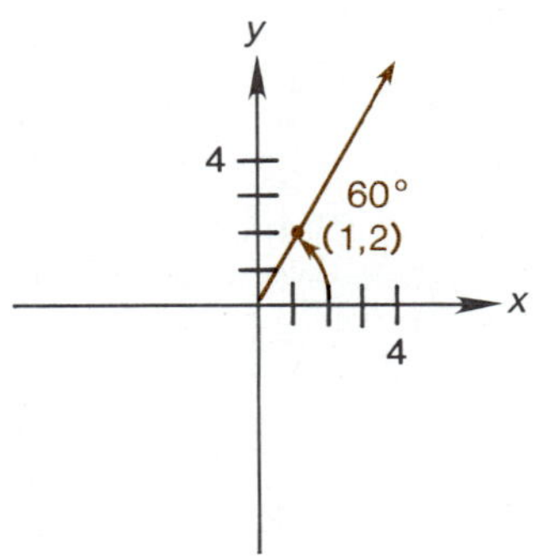

43.

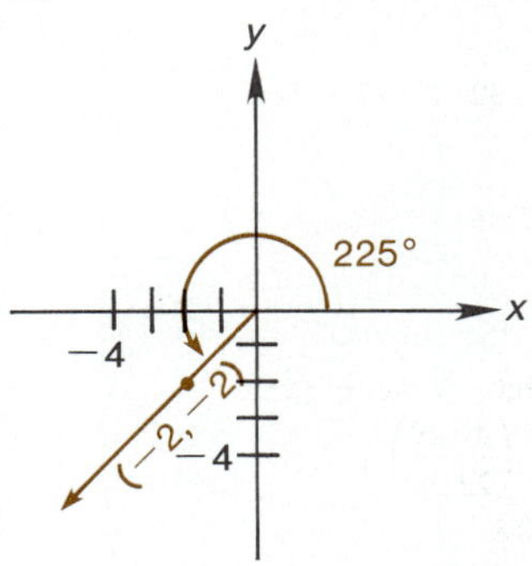

45.

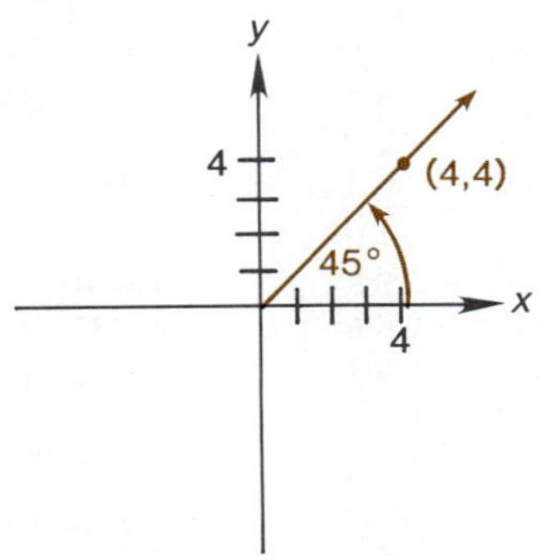

47.

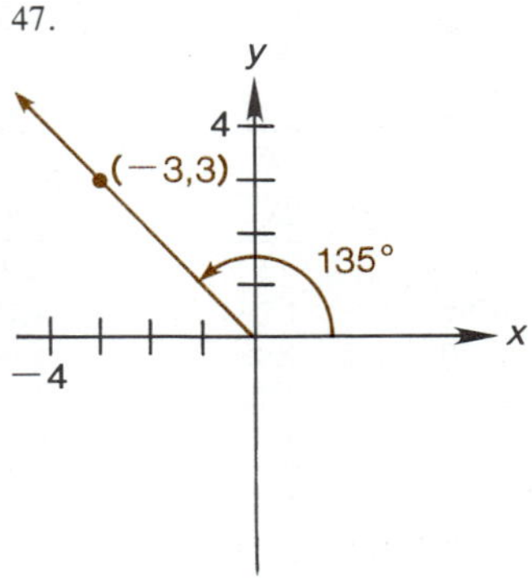

49.

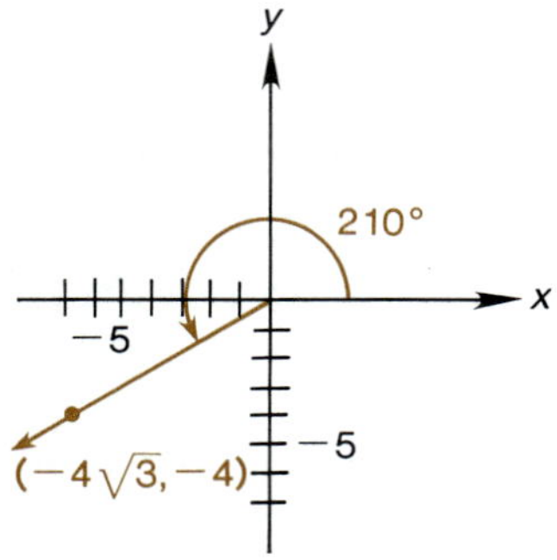

Exercise 17.2

	Sin	Cos	Tan	Csc	Sec	Cot
1.	$-\frac{4}{5}$	$\frac{3}{5}$	$-\frac{4}{3}$	$-\frac{5}{4}$	$\frac{5}{3}$	$-\frac{3}{4}$
3.	$\frac{4}{5}$	$-\frac{3}{5}$	$-\frac{4}{3}$	$\frac{5}{4}$	$-\frac{5}{3}$	$-\frac{3}{4}$
5.	$-\frac{5}{13}$	$-\frac{12}{13}$	$\frac{5}{12}$	$-\frac{13}{5}$	$-\frac{13}{12}$	$\frac{12}{5}$
7.	$-\frac{4}{5}$	$-\frac{3}{5}$	$\frac{4}{3}$	$-\frac{5}{4}$	$-\frac{5}{3}$	$\frac{3}{4}$
9.	$\frac{3}{5}$	$-\frac{4}{5}$	$-\frac{3}{4}$	$\frac{5}{3}$	$-\frac{5}{4}$	$-\frac{4}{3}$
11.	$-\frac{7}{25}$	$-\frac{24}{25}$	$\frac{7}{24}$	$-\frac{25}{7}$	$-\frac{25}{24}$	$\frac{24}{7}$
13.	1	0	Undef.	1	Undef.	0
15.	-1	0	Undef.	-1	Undef.	0
17.	$-\frac{1}{2}$	$\frac{\sqrt{3}}{2}$	$-\frac{1}{\sqrt{3}}$	-2	$\frac{2}{\sqrt{3}}$	$-\sqrt{3}$
19.	$\frac{\sqrt{3}}{2}$	$-\frac{1}{2}$	$-\sqrt{3}$	$\frac{2}{\sqrt{3}}$	-2	$-\frac{1}{\sqrt{3}}$
21.	$\frac{1}{2}$	$\frac{\sqrt{3}}{2}$	$\frac{1}{\sqrt{3}}$	2	$\frac{2}{\sqrt{3}}$	$\sqrt{3}$
23.	$-\frac{1}{\sqrt{3}}$	$-\frac{\sqrt{6}}{3}$	$\frac{1}{\sqrt{2}}$	$-\sqrt{3}$	$-\frac{\sqrt{6}}{2}$	$\sqrt{2}$
25.	$\frac{\sqrt{21}}{6}$	$\frac{\sqrt{15}}{6}$	$\frac{\sqrt{35}}{5}$	$\frac{2\sqrt{21}}{7}$	$\frac{2\sqrt{15}}{5}$	$\frac{\sqrt{35}}{7}$
27.	$-\frac{\sqrt{12}}{5}$	$-\frac{\sqrt{13}}{5}$	$\frac{2\sqrt{39}}{13}$	$-\frac{5}{\sqrt{12}}$	$-\frac{5}{\sqrt{13}}$	$\frac{\sqrt{39}}{6}$
29.	-0.7769	-0.6295	1.234	-1.287	-1.589	0.8104
31.	0.9482	-0.3179	-2.982	1.055	-3.146	-0.3353
33.	-0.9520	-0.3061	3.110	-1.050	-3.265	0.3215
35.	$\frac{3}{5}$	$-\frac{4}{5}$	$-\frac{3}{4}$	$\frac{5}{3}$	$-\frac{5}{4}$	$-\frac{4}{3}$
37.	$-\frac{1}{2}$	$-\frac{\sqrt{3}}{2}$	$\frac{1}{\sqrt{3}}$	-2	$-\frac{2}{\sqrt{3}}$	$\sqrt{3}$
39.	$\frac{1}{2}$	$-\frac{\sqrt{3}}{2}$	$-\frac{1}{\sqrt{3}}$	2	$-\frac{2}{\sqrt{3}}$	$-\sqrt{3}$

	Sin	Cos	Tan	Csc	Sec	Cot
41.	$-\frac{1}{\sqrt{2}}$	$-\frac{1}{\sqrt{2}}$	1	$-\sqrt{2}$	$-\sqrt{2}$	1
43.	$\frac{\sqrt{2}}{2}$	$-\frac{\sqrt{2}}{2}$	-1	$\sqrt{2}$	$-\sqrt{2}$	-1
45.	$-\frac{24}{25}$	$\frac{7}{25}$	$-\frac{24}{7}$	$-\frac{25}{24}$	$\frac{25}{7}$	$-\frac{7}{24}$
47.	$\frac{3}{\sqrt{58}}$	$\frac{7}{\sqrt{58}}$	$\frac{3}{7}$	$\frac{\sqrt{58}}{3}$	$\frac{\sqrt{58}}{7}$	$\frac{7}{3}$
49.	$-\frac{\sqrt{5}}{3}$	$-\frac{2}{3}$	$\frac{\sqrt{5}}{2}$	$-\frac{3}{\sqrt{5}}$	$-\frac{3}{2}$	$\frac{2}{\sqrt{5}}$

Exercise 17.3

1. 0.4067 3. 3.7321 5. 3.8118 7. 3.1848 9. 1.2527 11. 1.0770 13. 27.2715 15. 0.5589 17. 19.4°, 0.9432 19. 86.25°, 0.9979 21. 72.38°, 3.1486 23. 37.85°, 1.2664 25. 61.47°, 0.8786 27. 6.42°, 8.8872 29. 50.10° 31. 5.37° 33. 11.35° 35. 25.97° 37. 30.80° 39. 5.86° 41. 15.35° 43. 55.46° 45. 87.91° 47. 53.13° 49. 41.81° 51. 48.81° 53. 57.42° 55. 45.58° 57. 6.38° 59. 26.08°

Exercise 17.4

1. $b = 20.8$, $c = 24.0$, $m\angle B = 60.0°$ 3. $b = 4.00$, $c = 5.66$, $m\angle\alpha = 45.0°$ 5. $b = 35.7$, $c = 37.1$, $m\angle\alpha = 16.1°$ 7. $b = 293$, $c = 477$, $m\angle\beta = 37.9°$ 9. $c = 13.0$, $m\angle\alpha = 22.6°$, $m\angle\beta = 67.4°$ 11. $b = 7.64$, $m\angle\alpha = 29.0°$, $m\angle\beta = 61.0°$ 13. $c = 0.369$, $m\angle\alpha = 20.2°$, $m\angle\beta = 69.8°$ 15. $a = 597$, $m\angle\alpha = 51.5°$, $m\angle\beta = 38.5°$ 17. $b = 0.355$, $c = 2.95$, $m\angle\beta = 6.9°$ 19. $c = 131$, $m\angle\alpha = 48.0°$, $m\angle\beta = 42.0°$ 21. 233.8 ft 23. 69.0 ft 25. 597 ft, 628 ft 27. 23.1 ft, 65.2° 29. 7.5° 31. 20.2 m 33. 56.9° 35. 391 ft 37. 358 ft 39. 9.66 ft 41. 70.3 ft 43. $AC = 6.58$ cm, $AB = 7.58$ cm 45. $z \approx 47.2$, $m\angle\theta \approx 58.0°$, $m\angle Q = 58.0°$ 47. (a) 66°, (b) 5.07, (c) 19.3″ 49. (a) $AC = 1{,}190$ ft, $BC = 928$ ft; (b) 355 ft

Chapter Review

1. 396°, 756°, −324°, −684° 2. 507°, 867°, −213°, −573° 3. 579°, 939°, −141°, −501° 4. 668°, 1,028°, −52°, −412° 5. 48° 6. 279° 7. 688° 8. 287° 9. 211° 10. 83° 11. 69° 12. 93°

13.

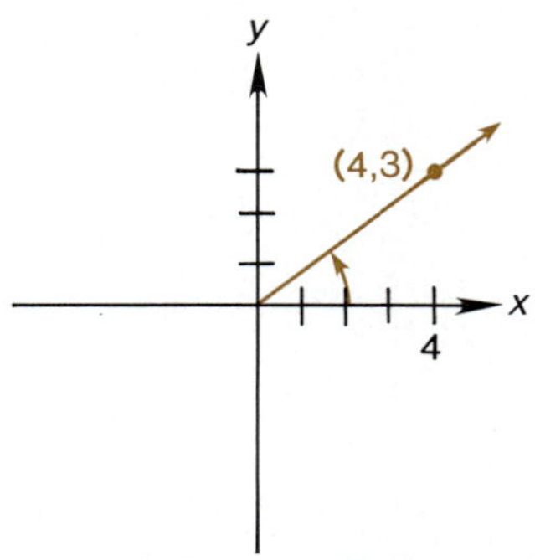

Sin (α) = 3/5 Csc (α) = 5/3
Cos (α) = 4/5 Sec (α) = 5/4
Tan (α) = 3/4 Cot (α) = 4/3

14.

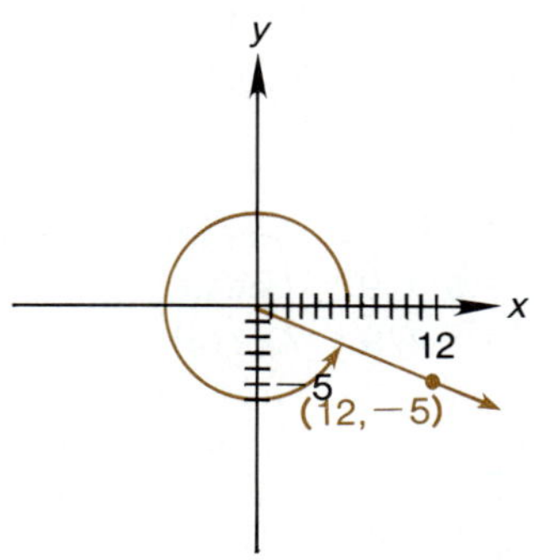

Sin (α) = −5/13 Csc (α) = −13/5
Cos (α) = 12/13 Sec (α) = 13/12
Tan (α) = −5/12 Cot (α) = −12/5

15.

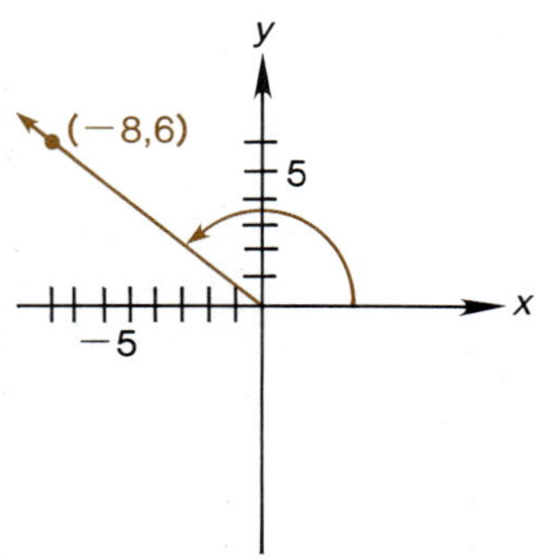

Sin (α) = 3/5 Csc (α) = 5/3
Cos (α) = −4/5 Sec (α) = −5/4
Tan (α) = −3/4 Cot (α) = −4/3

16.

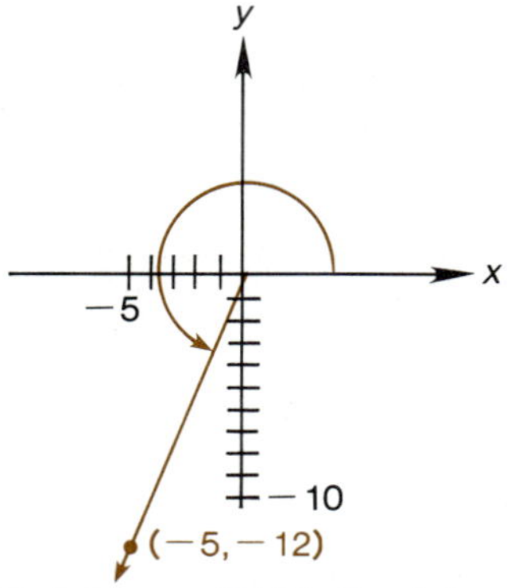

Sin $(\alpha) = -12/13$ Csc $(\alpha) = -13/12$
Cos $(\alpha) = -5/13$ Sec $(\alpha) = -13/5$
Tan $(\alpha) = 12/5$ Cot $(\alpha) = 5/12$

17.

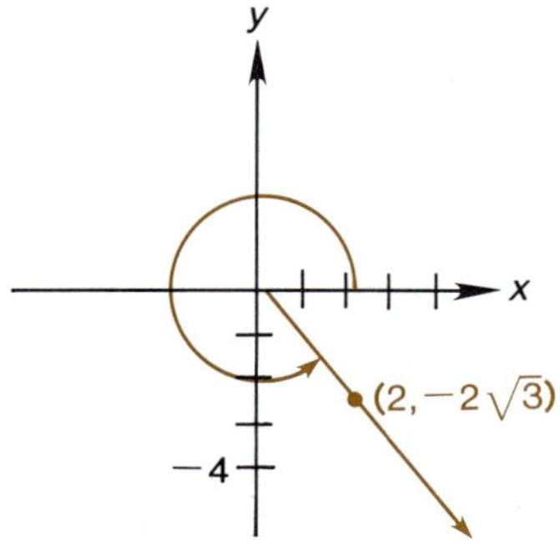

Sin $(\alpha) = -\sqrt{3/2}$ Csc $(\alpha) = \dfrac{-2}{\sqrt{3}}$
Cos $(\alpha) = \dfrac{1}{2}$ Sec $(\alpha) = 2$
Tan $(\alpha) = -\sqrt{3}$ Cot $(\alpha) = \dfrac{-1}{\sqrt{3}}$

18.

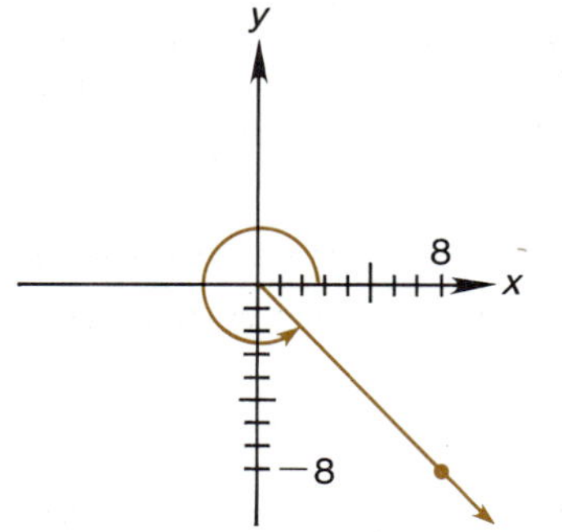

Sin $(\alpha) = -1/\sqrt{2}$ Csc $(\alpha) = -\sqrt{2}$
Cos $(\alpha) = 1/\sqrt{2}$ Sec $(\alpha) = \sqrt{2}$
Tan $(\alpha) = -1$ Cot $(\alpha) = -1$

19.

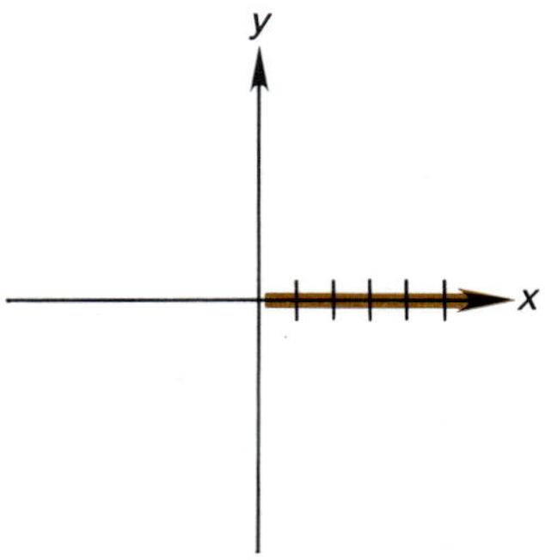

Sin $(\alpha) = 0$ Csc $(\alpha) =$ undef.
Cos $(\alpha) = 1$ Sec $(\alpha) = 1$
Tan $(\alpha) = 0$ Cot $(\alpha) =$ undef.

20.

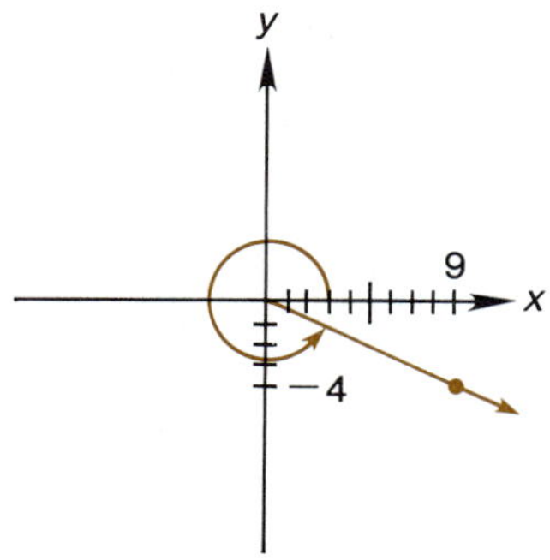

Sin $(\alpha) = -4/\sqrt{97}$ Csc $(\alpha) = -\sqrt{97}/4$
Cos $(\alpha) = 9/\sqrt{97}$ Sec $(\alpha) = \sqrt{97}/9$
Tan $(\alpha) = -4/9$ Cot $(\alpha) = -9/4$

	Sin	Cos	Tan	Csc	Sec	Cot
21.	$\frac{4}{5}$	$\frac{3}{5}$	$\frac{4}{3}$	$\frac{5}{4}$	$\frac{5}{3}$	$\frac{3}{4}$
22.	$\frac{\sqrt{119}}{12}$	$\frac{-5}{12}$	$-\frac{\sqrt{119}}{5}$	$\frac{12}{\sqrt{119}}$	$-\frac{12}{5}$	$-\frac{5}{\sqrt{119}}$
23.	$-\frac{7}{25}$	$-\frac{24}{25}$	$\frac{7}{24}$	$-\frac{25}{7}$	$-\frac{25}{24}$	$\frac{24}{7}$
24.	$-\frac{1}{2}$	$\frac{\sqrt{3}}{2}$	$-\frac{1}{\sqrt{3}}$	-2	$\frac{2}{\sqrt{3}}$	$-\sqrt{3}$
25.	$\frac{\sqrt{7}}{8}$	$-\frac{3}{4}$	$-\frac{\sqrt{7}}{6}$	$\frac{8}{\sqrt{7}}$	$-\frac{4}{3}$	$-\frac{6}{\sqrt{7}}$
26.	$-\frac{\sqrt{3}}{2}$	$-\frac{1}{2}$	$\sqrt{3}$	$-\frac{2}{\sqrt{3}}$	-2	$\frac{1}{\sqrt{3}}$

27. $a = 8.40$, $b = 2.38$, $m\angle A = 74.2°$ 28. $a = 28.3$, $c = 31.7$, $m\angle B = 26.8°$ 29. $a = 0.473$, $c = 0.884$, $m\angle A = 32.4°$
30. $c = 681$, $m\angle A = 67.6°$, $m\angle B = 22.4°$ 31. $b = 5.03$, $m\angle A = 45.2°$, $m\angle B = 44.8°$
32. $a = 18.8$, $m\angle A = 52.0°$, $m\angle B = 38.0°$ 33. 39.0 m 34. 63.9° 35. 551 ft 36. 163 m 37. 44.6°; 14.0 ft 38. 911 m
39. 5.1°; 25.1 m 40. 43.3°

Chapter Test

1. Angles in standard position having the same terminal sides. 2. When the vertex is at the origin and one side lies on the positive x-axis. 3. When the rotation of the terminal side is (a) counterclockwise, (b) clockwise. 4. (a) 212°, (b) 161.7°, (c) 299° 5. $\text{Sin}(\alpha) = \frac{-5}{\sqrt{34}}$, $\text{Cos}(\alpha) = \frac{3}{\sqrt{34}}$, $\text{Tan}(\alpha) = -\frac{5}{3}$, $\text{Csc}(\alpha) = -\frac{\sqrt{34}}{5}$, $\text{Sec}(\alpha) = \frac{\sqrt{34}}{3}$, $\text{Cot}(\alpha) = -\frac{3}{5}$, quad. IV 6. $\text{Sin}(\alpha) = -\frac{12}{13}$, $\text{Cos}(\alpha) = -\frac{5}{13}$, $\text{Tan}(\alpha) = \frac{12}{5}$, $\text{Csc}(\alpha) = -\frac{13}{12}$, $\text{Sec}(\alpha) = -\frac{13}{5}$, $\text{Cot}(\alpha) = \frac{5}{12}$ 7. (a) 0.5621, (b) −0.7034, (c) 1.0198, (d) −1.4124 8. (a) 56.87°, (b) 54.85°, (c) 55.1°, (d) 72.54° 9. (a) 61.04°, (b) 77.16°, (c) 52.57° 10. 7.6 11. 14.2 12. $m\angle A = 56.3°$ 13. $m\angle A = 58.45°$, $a = 363.9$ 14. 6.18 m 15. 87.2 m 16. 107 ft

Chapter 18

Exercise 18.1

1. $b = 16$, $m\angle C = 85°$, $c = 44$ 3. $b = 11.0$, $a = 18.5$, $m\angle A = 90.5°$ 5. $a = 325$, $c = 321$, $m\angle C = 57.2°$
7. $b = 72.6$, $c = 20.0$, $m\angle A = 98.1°$ 9. $a = 0.149$, $b = 0.120$, $m\angle B = 31.8°$
11. $c = 128$, $m\angle B = 19.9°$, $m\angle C = 123.1°$ 13. $b = 37.5$, $m\angle A = 23.9°$, $m\angle B = 96.7°$
15. $b = 10.5$, $m\angle A = 53.8°$, $m\angle B = 106.4°$; $b' = 6.12$, $m\angle CA'B = 126.2°$, $m\angle CBA = 34°$
17. $a = 6.87$ in, $b = 11.4$ in 19. $m\angle A = 42.2°$ 21. 9.34 cm, 12.2 cm 23. 6.1 ft 25. 34.3 m from L_1, 42.7 m from L_2

Exercise 18.2

1. $c = 7$, $m\angle A = 25.5°$, $m\angle B = 104.5°$ 3. $a = 22.4$, $m\angle B = 36.4°$, $m\angle C = 80.4°$
5. $m\angle A = 27.6°$, $m\angle B = 39.5°$, $m\angle C = 112.9°$ 7. $c = 58.4$, $m\angle A = 38.0°$, $m\angle B = 60.7°$
9. $m\angle A = 61.1°$, $m\angle B = 40.6°$, $m\angle C = 78.3°$ 11. $a = 0.024$, $m\angle B = 159.6°$, $m\angle C = 9.9°$
13. $m\angle A = 134.9°$, $m\angle B = 39.4°$, $m\angle C = 5.7°$ 15. $m\angle A = 31.7°$, $m\angle B = 48.2°$, $m\angle C = 100.1°$
17. $\overline{AC} = 7$, $\overline{AB} = 10$, $\overline{BC} = 9$, $m\angle\alpha = 60.9°$ 19. 50.8 m 21. 1.3 km 23. $m\angle\alpha = 101.5°$, $m\angle\beta = 46.0°$ 25. 33.6°

Exercise 18.3

1.

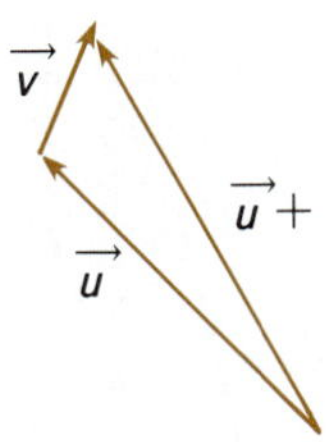

3.

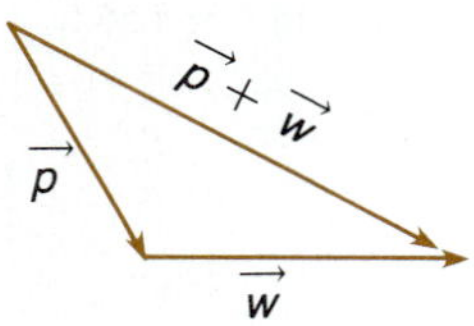

5.

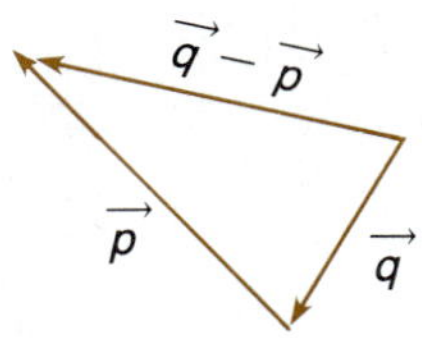

7.

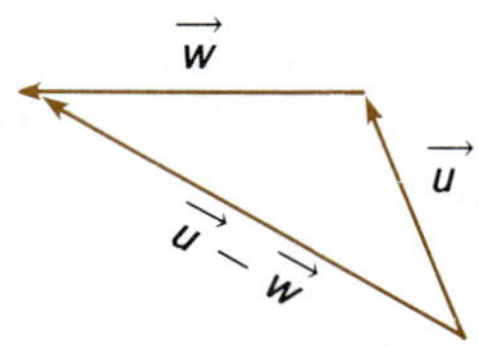

9.

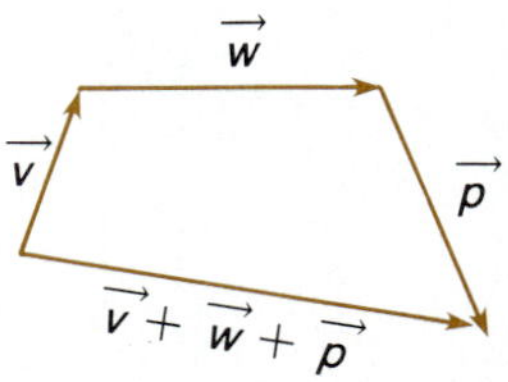

11.

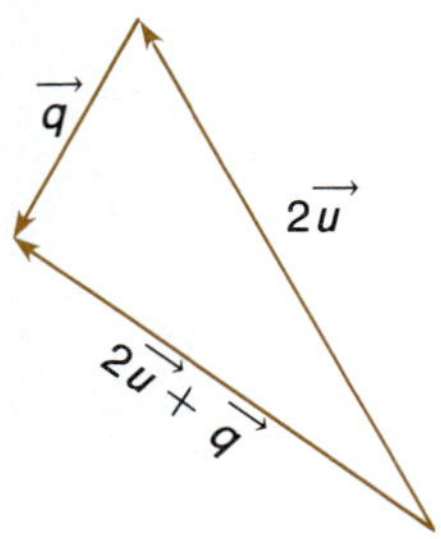

13.

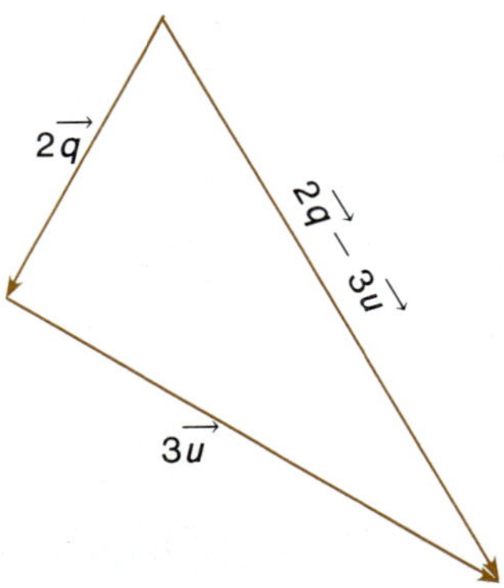

15.

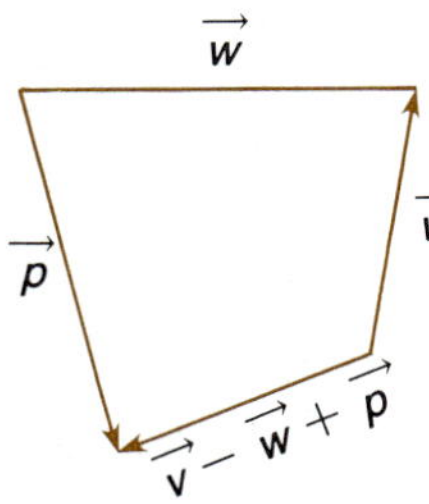

17.

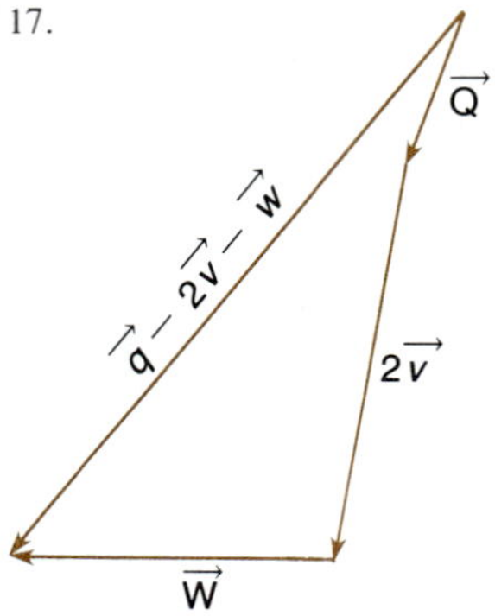

19.

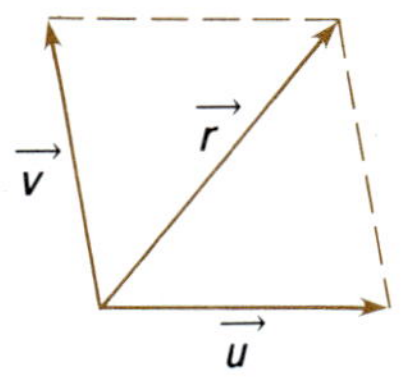

21.

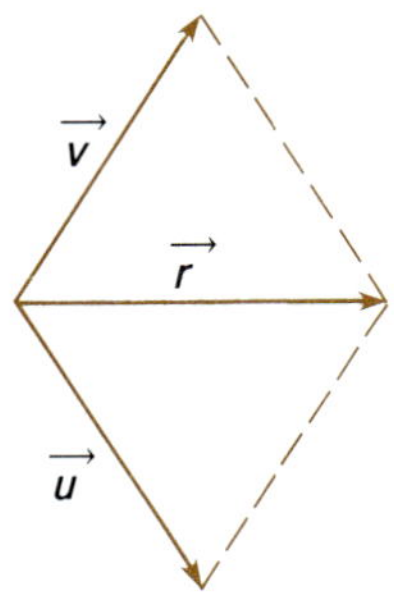

23.

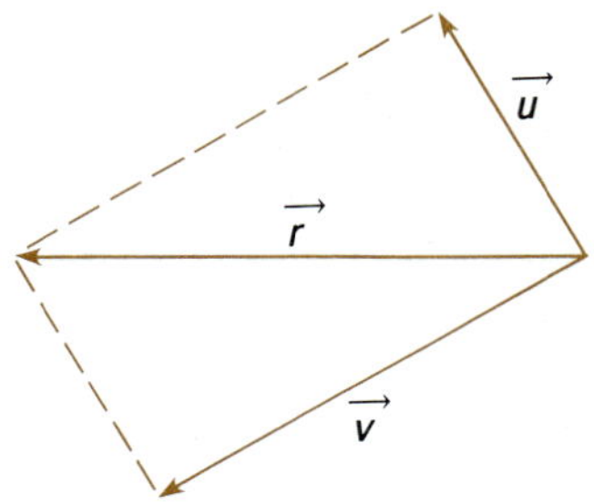

25. $\overrightarrow{R}_x \approx 4.97$, $\overrightarrow{R}_y \approx 9.47$ 27. $\overrightarrow{R}_x \approx -0.66$, $\overrightarrow{R}_y \approx 0.35$ 29. $\overrightarrow{R}_x \approx -82.3$, $\overrightarrow{R}_y = 0$ 31. $\overrightarrow{R}_x = 0$, $\overrightarrow{R}_y \approx -84$
33. $\overrightarrow{R}_x \approx 162.98$, $\overrightarrow{R}_y \approx 182.93$ 35. $\overrightarrow{R}_x \approx -197.52$, $\overrightarrow{R}_y \approx 296.18$ 37. $\overrightarrow{R}_x \approx 230$, $\overrightarrow{R}_y \approx 97.7$ 39. $\overrightarrow{R}_x \approx 226$, $R_y \approx 107$

Exercise 18.4

1. $\overrightarrow{R} = 5.0$, $\theta_R = 53.1°$ 3. $\overrightarrow{R} \approx 2{,}392.6$, $\theta_R \approx 63.4°$ 5. $\overrightarrow{R} \approx 0.7$, $\theta_R \approx 78.0°$ 7. $\overrightarrow{R} \approx 4.0$, $\theta_R \approx 92.6°$
9. $\overrightarrow{R} \approx 20.1$, $\theta_R \approx 104.1°$ 11. $\overrightarrow{R} \approx 169.3$, $\theta_R \approx 121.1°$ `13. $\overrightarrow{R} \approx 0.9$, $\theta/_R \approx 294.0°$ 15. $\overrightarrow{R} \approx 15.5$, $\theta_R \approx 208.1°$
17. $\overrightarrow{R} \approx 53.1$, $\theta_R \approx 41.2°$, magnitude same, angle changes to 48.8° 19. $\overrightarrow{R} \approx 247.9$, $\theta_R \approx 24.8°$ 21. $\overrightarrow{R} \approx 471.8$ mph, $\theta_R \approx 58.4°$ north of east 23. $\overrightarrow{R} \approx 101.5$, $\theta_R \approx 21.7°$ 25. $\overrightarrow{R} \approx 29.4$ lb, $\theta_R \approx 206°$ 27. 13.4 m, 292.8°
29. $\overrightarrow{R} \approx 237.8\ v$, $\theta_R \approx 8.8°$

Chapter Review

1. $a \approx 1.4$, $b \approx 1.7$, $m\angle C \approx 56.4°$ 2. $a \approx 1.2$, $b \approx 2.5$, $m\angle B \approx 50.2°$ 3. $c \approx 26.7$, $m\angle B \approx 38.1°$, $m\angle C \approx 88.7°$
4. $a \approx 0.423$, $m\angle B \approx 53.8°$, $m\angle A \approx 83.9°$, $\overline{CB} \approx 0.085$, $m\angle CB'A \approx 126.2°$, $m\angle CAB' \approx 11.5°$ 5. No solution
6. $m\angle A = 90°$, $m\angle C \approx 60°$, $c \approx 62.5$ 7. $c \approx 7.7$, $m\angle A \approx 39.5°$, $m\angle B \approx 81.6°$ 8. $b \approx 239.2$, $m\angle A \approx 25.3°$, $m\angle B \approx 56.6°$
9. $a \approx 38.1$, $m\angle B \approx 135.7°$, $m\angle C \approx 23°$ 10. $m\angle A \approx 106.7°$, $m\angle B \approx 58.5°$, $m\angle C \approx 14.8°$
11. $m\angle A \approx 158°$, $m\angle B \approx 10.8°$, $m\angle C \approx 11.2°$ 12. $m\angle C \approx 49.1°$ 13. $b \approx 1.83$ 14. $m\angle A \approx 58.7°$ 15. $m\angle A \approx 30.5°$
16. $m\angle C \approx 84.4°$ 17. (a) 40.6 m, (b) 25.7 m 18. 9.8 cm 19. 65.9° 20. 52.6 km 21. 15 ft and 56 ft

22.

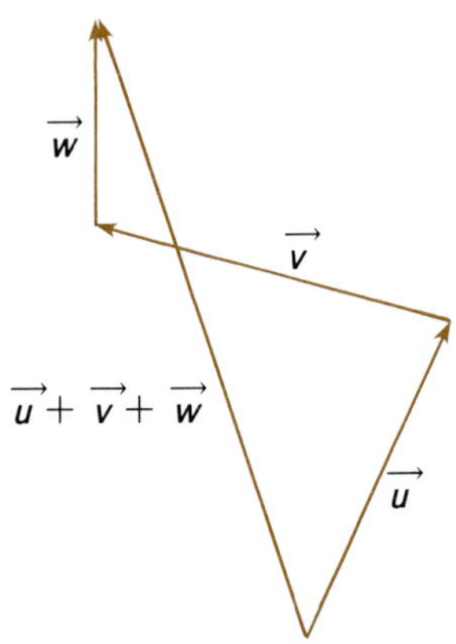

23.

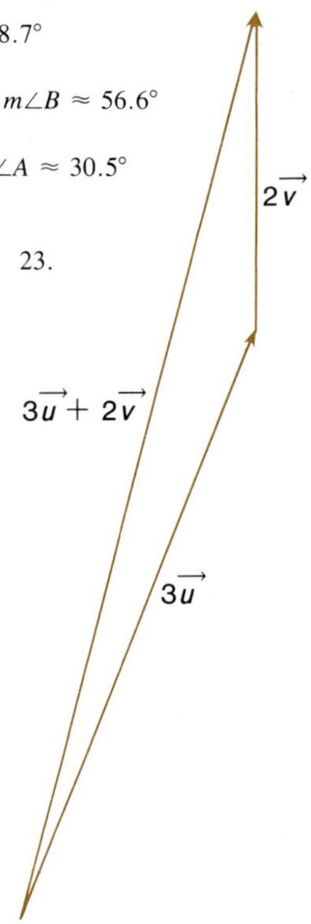

24.

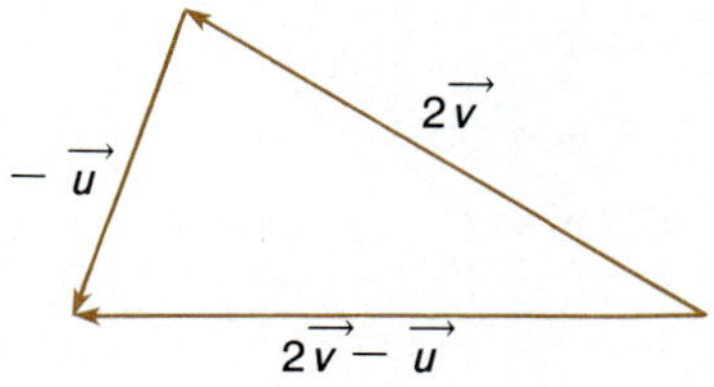

25.

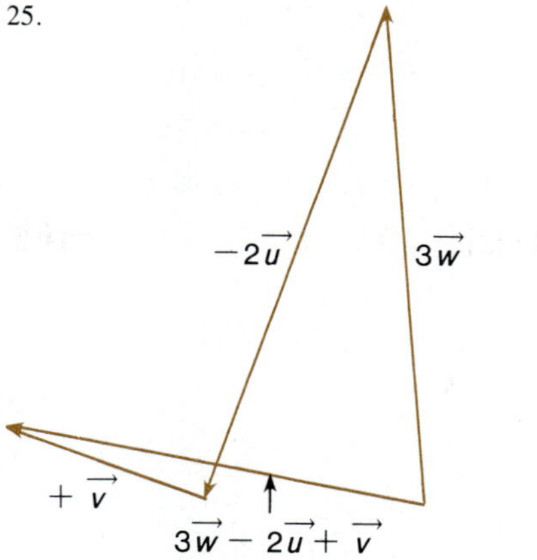

26.

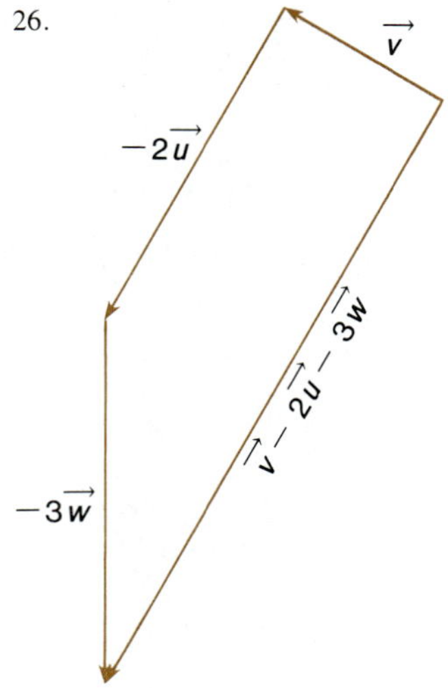

27.

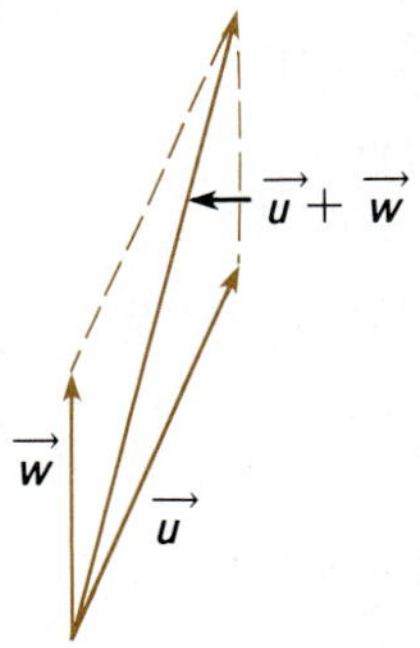

28.

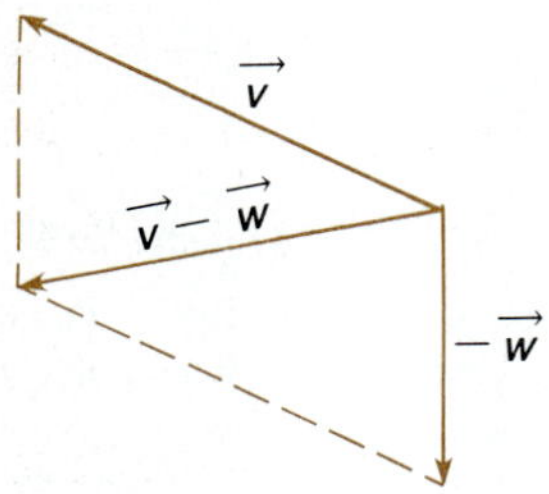

29.

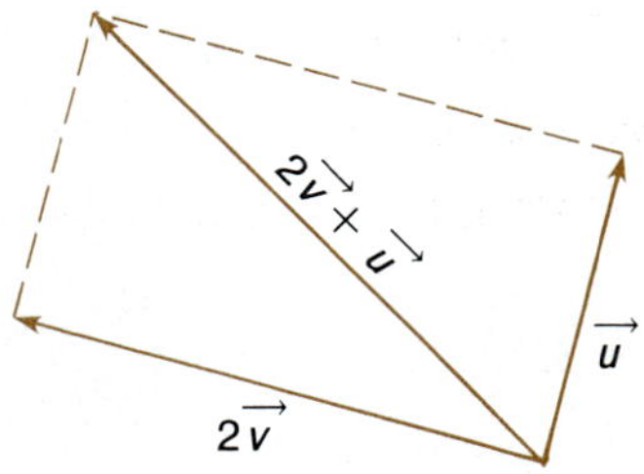

30.

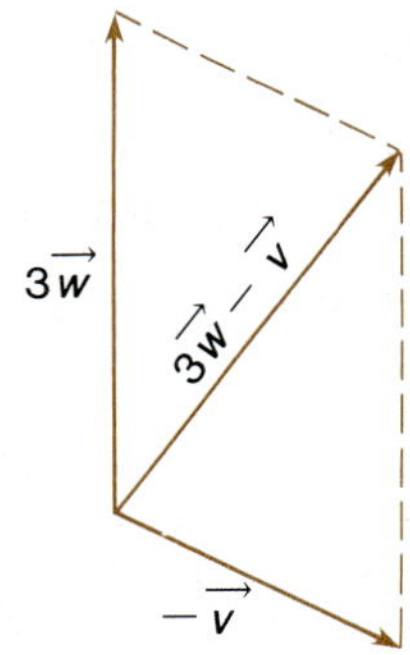

31.

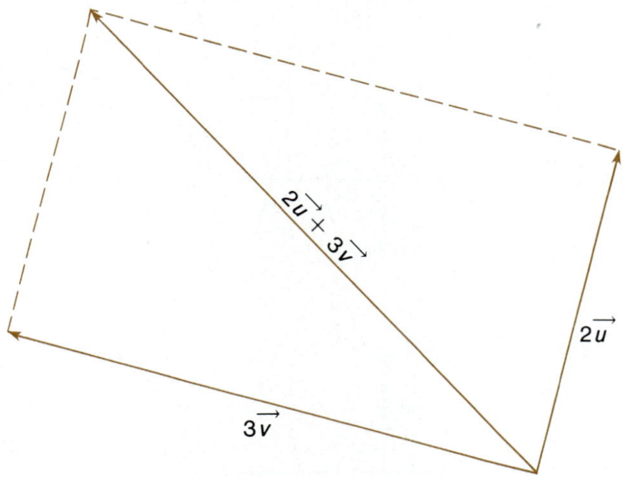

32.

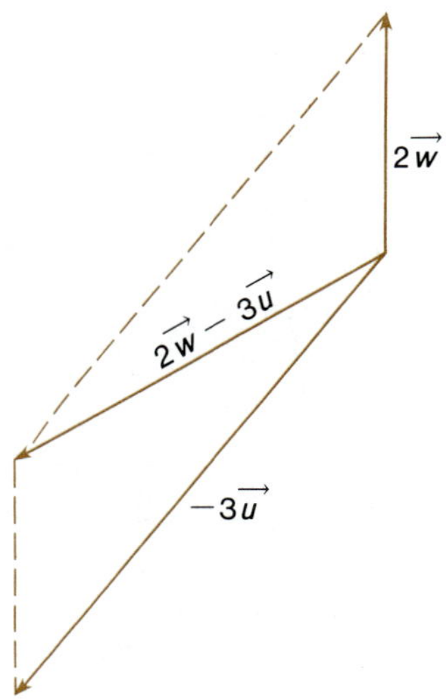

33. $\vec{R}_x \approx 3.7$, $\vec{R}_y \approx 2.1$ 34. $\vec{R}_x \approx 4.60$, $\vec{R}_y \approx 14.5$ 35. $\vec{R}_x \approx -286.6$, $\vec{R}_y \approx 436.3$ 36. $\vec{R}_x \approx 0.02$, $\vec{R}_y \approx -0.76$

37. $\vec{R}_x \approx 12.7$, $\vec{R}_y \approx 0.8$ 38. $\vec{R}_x \approx -6.0$, $\vec{R}_y \approx 0$ 39. $\vec{R} \approx 9.6$, $\theta_R \approx 12.1°$ 40. $\vec{R} \approx 0.9$, $\theta_R \approx 60.9°$

41. $\vec{R} \approx 104.5$, $\theta_R \approx 162.3°$ 42. $\vec{R} \approx 15.4$, $\theta_R \approx 172.2°$ 43. $\vec{R} \approx 26.4$, $\theta_R \approx 181.6°$ 44. $\vec{R} \approx 32.9$, $\theta_R \approx 185.3$

45. $\vec{R}_x \approx 40.0$, $\vec{R}_y \approx 126.3$ 46. $\vec{R}_x \approx 14.7$, $\vec{R}_y \approx 19.0$ 47. $\vec{R} \approx 262.7$ mph, $\theta_R \approx 174.8°$ or 5.2° north of west

48. $\vec{R} \approx 28.4$, $\theta_R \approx 57.7°$ 49. 3.7 mph 50. $\vec{R} \approx 55.7$, $\theta_R \approx 167.7°$

Chapter Test

1. $b \approx 16.8$, $m\angle C \approx 73.9°$ 2. $m\angle A \approx 25.7°$ 3. 2 4. 297.8 m 5. 32.5° 6. 68.6 cm²

7. (a)

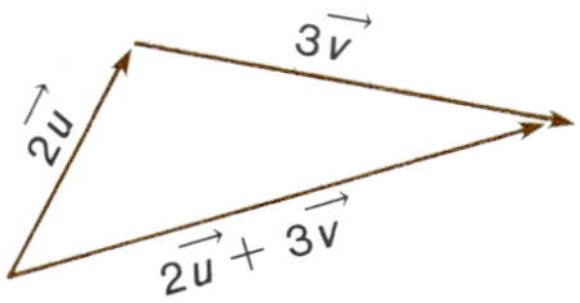

(b)

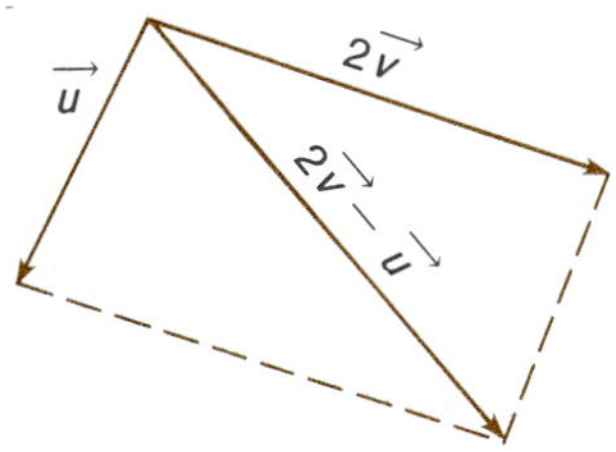

8. $\vec{R} \approx 85.8$, $\theta_R \approx 16.1°$ 9. $\vec{R}_x \approx 271.7$, $\vec{R}_y \approx 169.1$

10. $\vec{R} \approx 27.2$, $\theta_R \approx 47.5°$

11. $\vec{R} \approx 25.5$ knots, $\theta_R \approx 36.1°$ north of west

12. $\vec{R} \approx 523.2$ lb, $\theta_R \approx 1°$ 13. 891.5 ft

Chapter 19

Exercise 19.1

1. $\frac{\pi}{9}$ 3. $\frac{\pi}{5}$ 5. $\frac{4\pi}{9}$ 7. $\frac{5\pi}{6}$ 9. $\frac{5\pi}{4}$ 11. $\frac{5\pi}{3}$ 13. $\frac{16\pi}{9}$ 15. $\frac{23\pi}{12}$ 17. 0.40 19. 1.11 21. 2.91 23. 4.47
25. 5.47 27. 2.48 29. 5.26 31. 18° 33. 20° 35. 120° 37. 135° 39. 67.5° 41. 198° 43. 585° 45. 123.75° 47. 300°
49. 340° 51. 26.4° 53. 61.3° 55. 110.6° 57. 115.2° 59. 236.1° 61. 304.3° 63. 300π 65. 2.09
67. (a) 2, (b) 2.5, (c) 5, (d) $\frac{1}{4}$

Exercise 19.2

1. 2.36 in 3. (a) 1.20, (b) 68.8° 5. 2.85 in 7. 341 in 9. 29.5 in² 11. 66,700 13. 1,200 mm/s
15. (a) 126 rad/s or 40π rad/s, (b) 15.7 in/s or 5π rad/s 17. 15,400 mi 19. 487 rad, 9,740 rad/min 21. 23.7 ft
23. 14,100 mph 25. 3.77 m/s 27. 3,600 in/min 29. 106,000 rad/h

Exercise 19.3

1. amp = 4

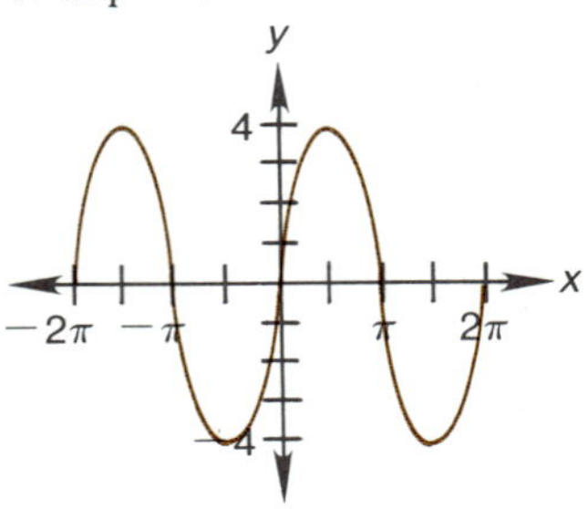

$y = 4$ Sin (x)

3. amp = $\frac{1}{3}$

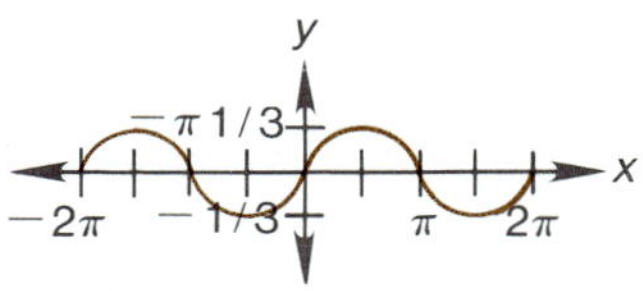

$y = 1/3$ Sin (x)

5. amp = 1

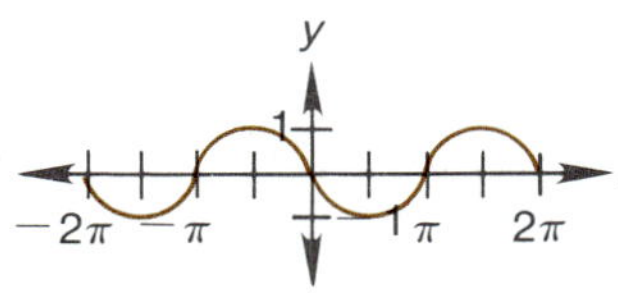

$y =$ Sin (x)

7. amp = $\frac{3}{2}$

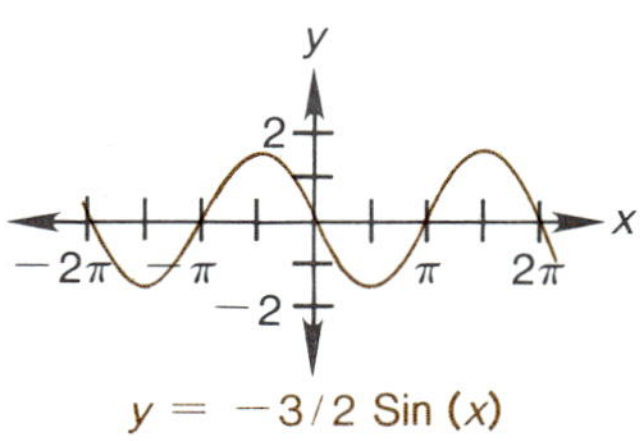

$y = -3/2$ Sin (x)

9. amp = 3

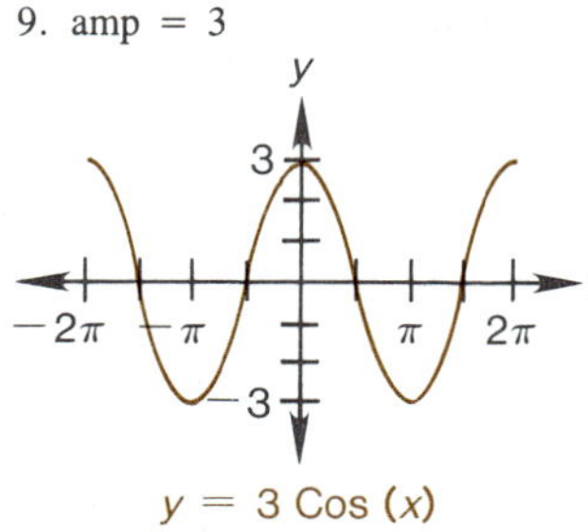

$y = 3$ Cos (x)

11. amp $= \frac{1}{2}$

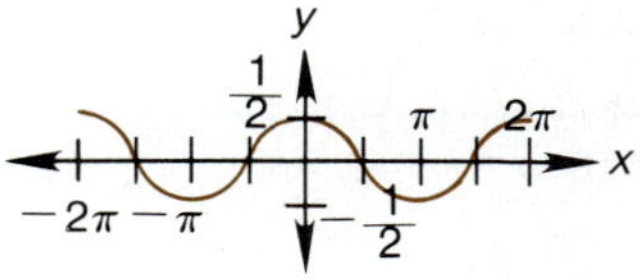

$y = \frac{1}{2}$ Cos (x)

13. amp $= 1$

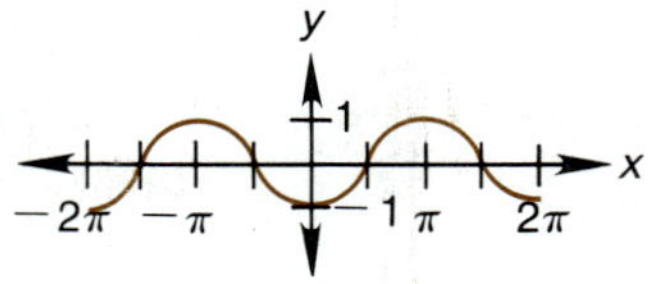

$y = -$Cos (x)

15. amp $= \frac{5}{2}$

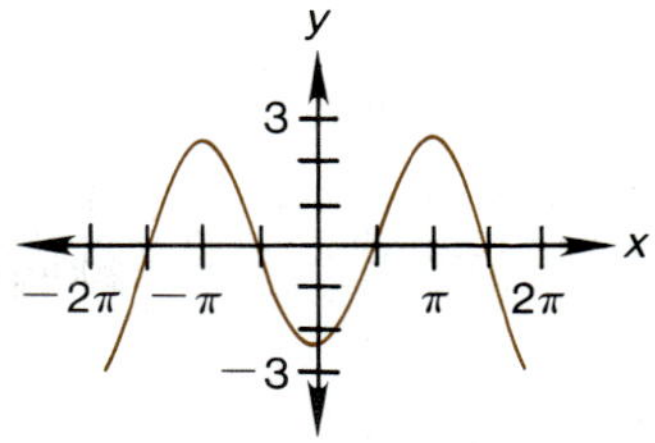

$y = -5/2$ Cos (x)

17. amp $= 3$, period $= \pi$

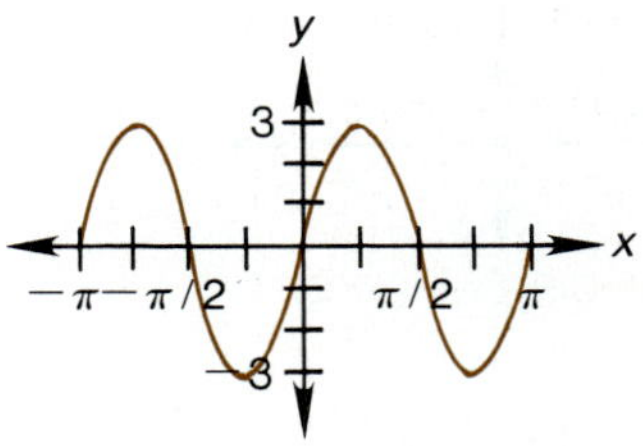

$y = 3$ Sin $(2x)$

19. amp $= 1$, period $= \frac{\pi}{3}$

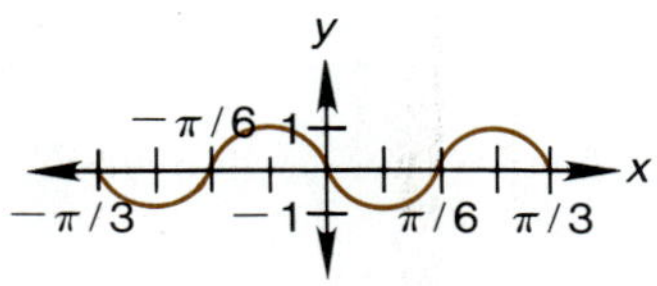

$y = -$Sin $(6x)$

21. amp $= \frac{1}{2}$, period $= 2$

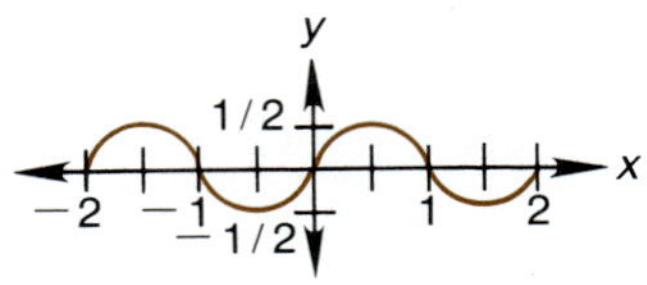

$y = \frac{1}{2}$ Sin (πx)

23. amp $= 5$, period $= 3\pi$

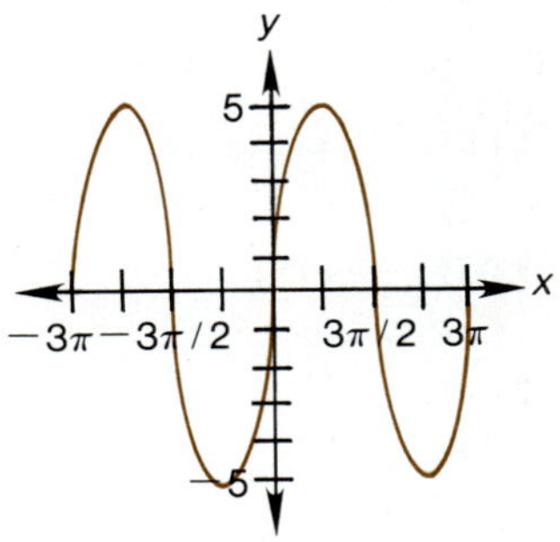

$y = 5$ Sin $(2/3x)$

25. amp $= 2$, period $= \frac{2\pi}{3}$

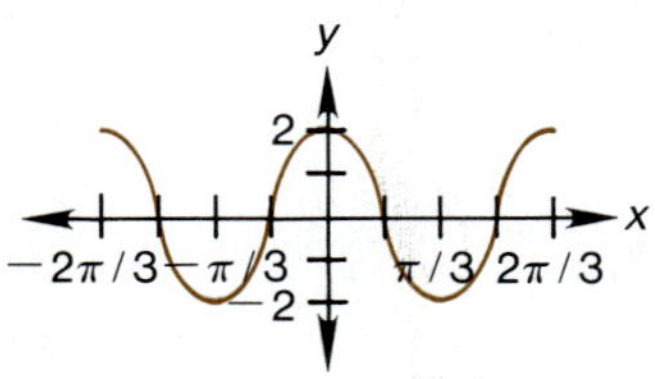

$y = 2$ Cos $(3x)$

27. amp $= 1$, period $= \pi$

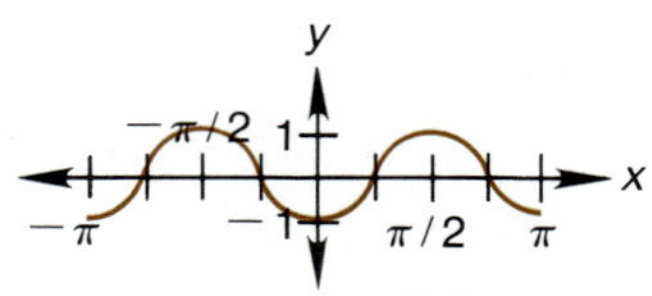

$y = -$Cos $(2x)$

29. amp $= \frac{3}{2}$, period $= \frac{2}{3}$

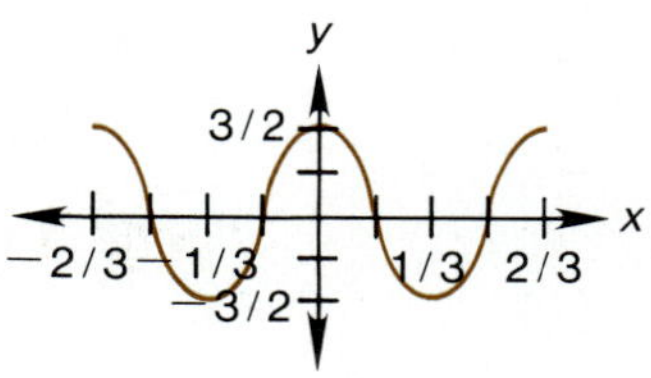

$y = 3/2$ Cos $(3\pi x)$

31. amp $= 6$, period $= \frac{5\pi}{3}$

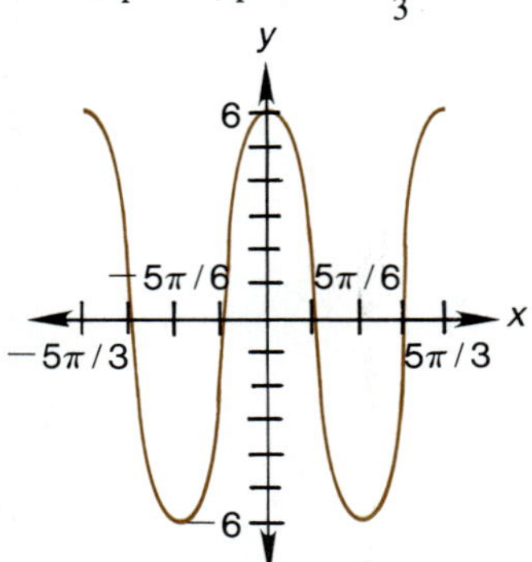

$y = 6$ Cos $\left(\frac{6}{5}x\right)$

33. shift $= -\frac{\pi}{4}$, period $= 2\pi$

35. shift $= -\frac{\pi}{4}$, period $= \pi$

37. shift $= \pi$, period $= 4\pi$

39. shift $= \frac{1}{4}$, period $= 2$

41. shift $= \frac{1}{2}$, period $= 4$

43. amp = 2, period = 2π

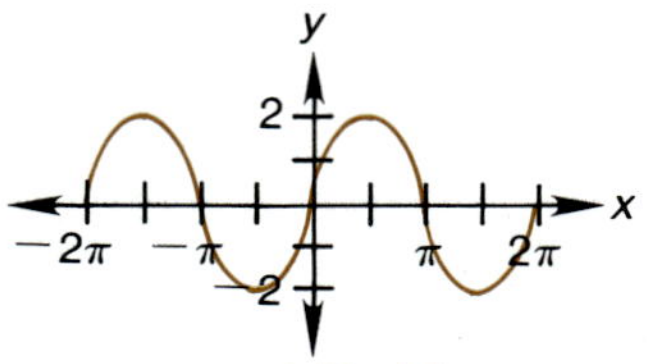

$y = 2 \text{ Sin } (x)$

45. amp = 3, period = 2π

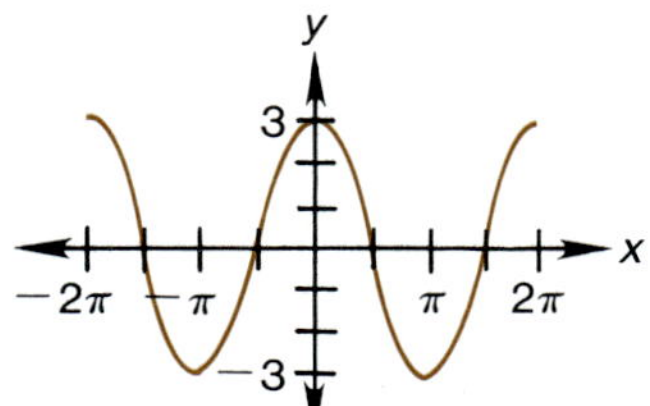

$y = 3 \text{ Cos } (x)$

47. amp = 3, period = π

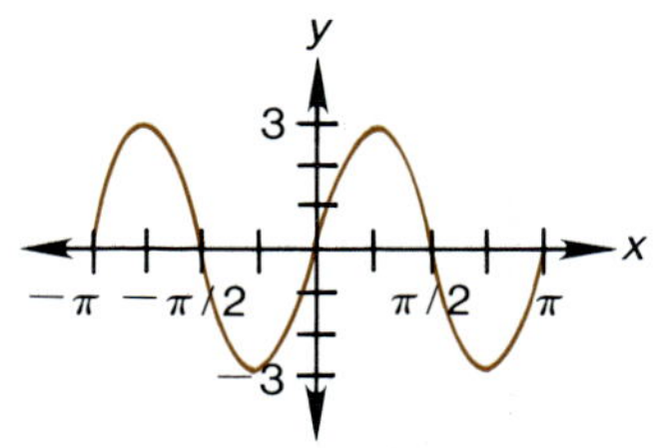

$y = 3 \text{ Sin } (2x)$

49. amp = 2, period = π

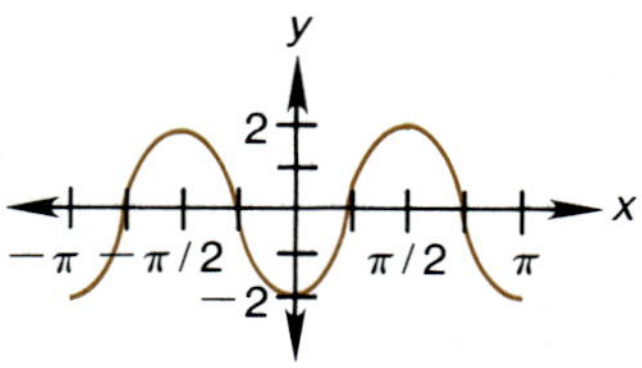

$y = -2 \text{ Cos } (2x)$

51. amp = 1, period = 2π, shift = $\frac{\pi}{2}$

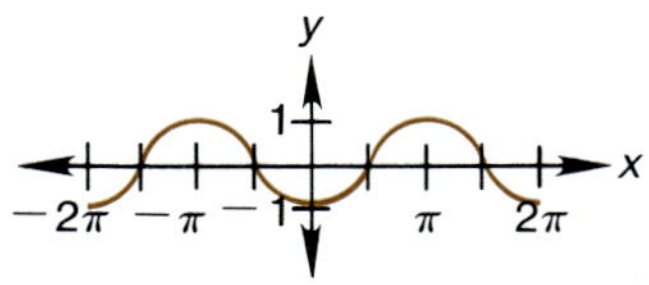

$y = \text{Sin } (x - \pi/2)$

53. amp = 1, period = 2π, shift = $\frac{\pi}{2}$

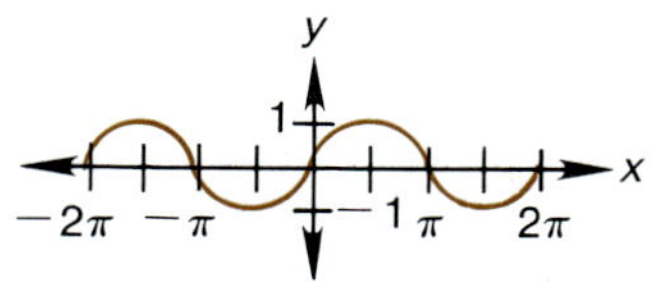

$y = \text{Cos } (x - \pi/2)$

55. amp = 2, period = 2π, shift = $\frac{-\pi}{4}$

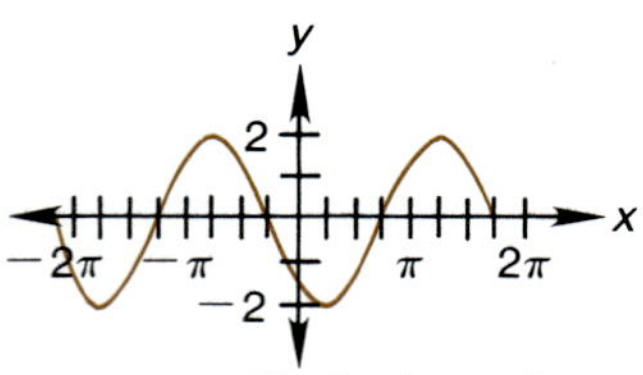

$y = -2 \text{ Sin } (x + \pi/4)$

57. amp = 4, period = 2π, shift = $-\frac{\pi}{2}$

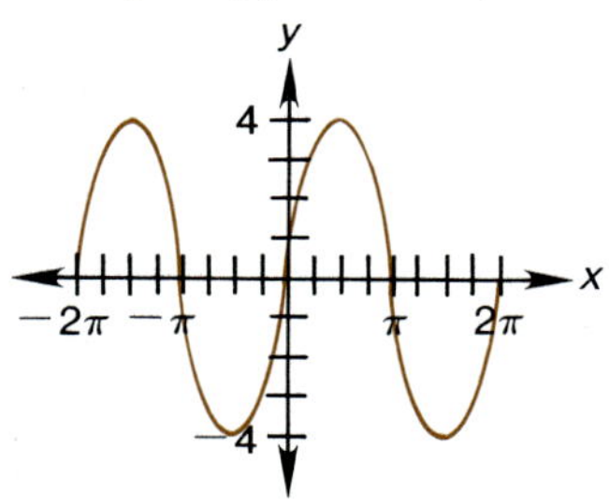

$y = -4 \text{ Cos } (x + \pi/2)$

59. amp = 1, period = π, shift = $\frac{\pi}{8}$

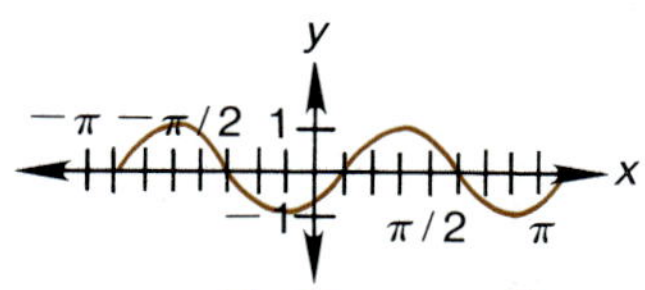

$y = \text{Sin } (2x - \pi/4)$

61. amp = 1.5, period = 2, shift = −1

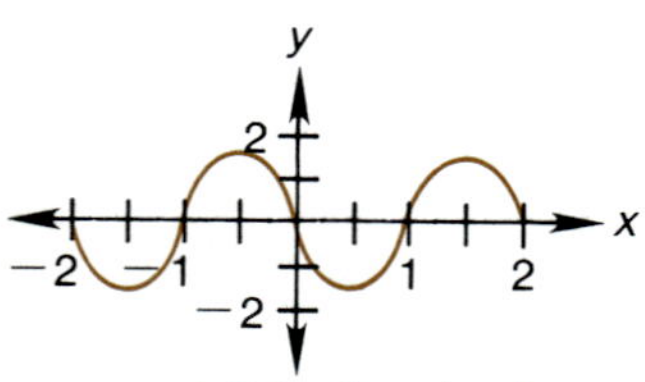

$y = 1.5 \text{ Sin } (\pi x + \pi)$

Exercise 19.4

1.

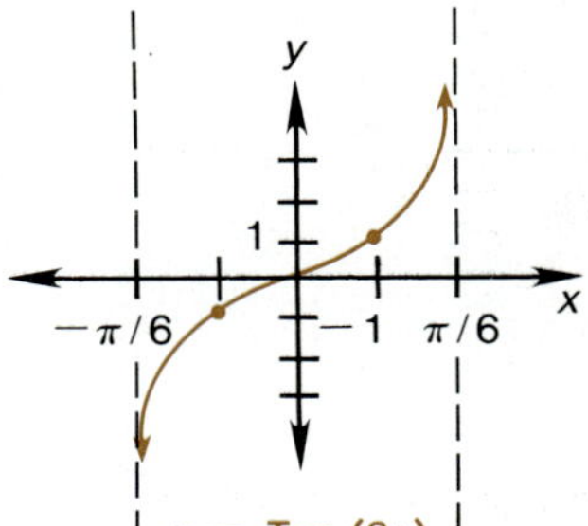

$y = \text{Tan}\,(3x)$

3.

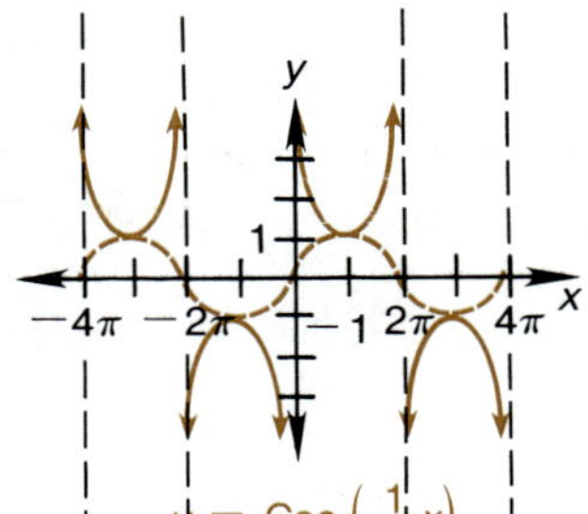

$y = \text{Csc}\left(\frac{1}{2}x\right)$

5.

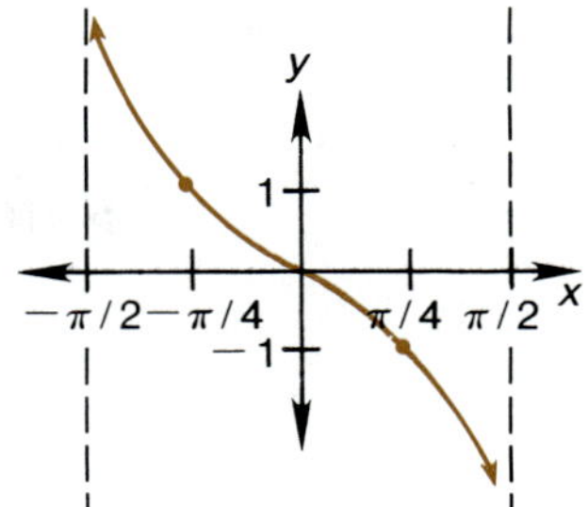

$y = -\text{Tan}\,(x)$

7.

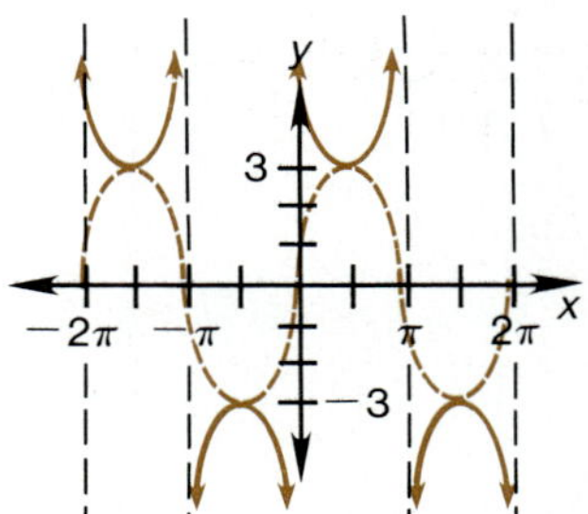

$y = 3\,\text{Csc}\,(x)$

9.

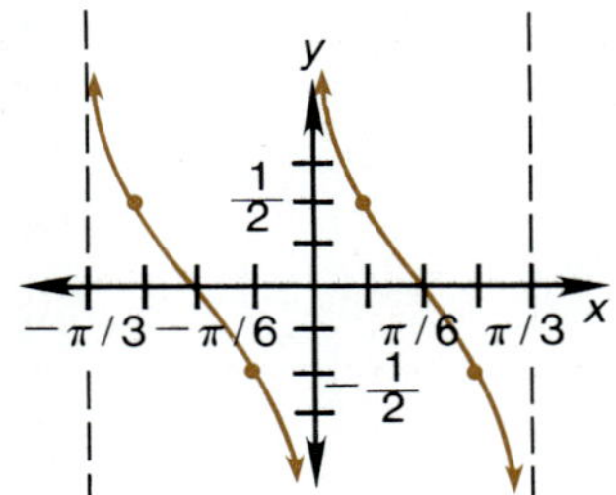

$y = \frac{1}{2}\,\text{Cot}\,(3x)$

11.

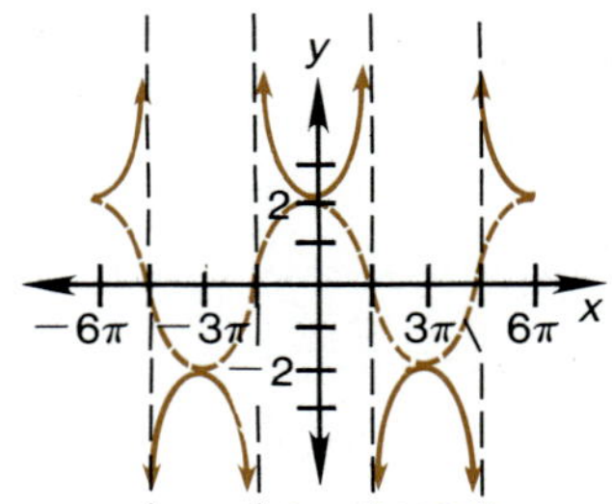

$y = 2\,\text{Sec}\,(1/3x)$

13.

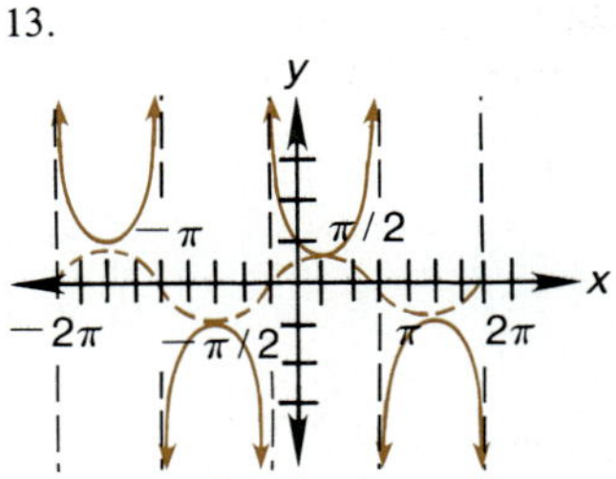

$y = \text{Csc}\,(x + \pi/4)$

15.

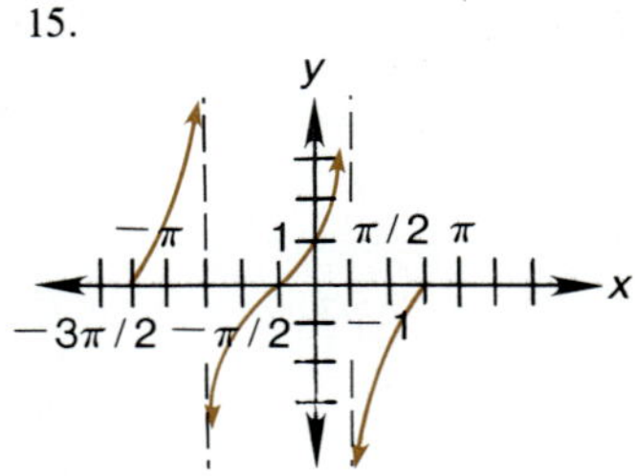

$y = \text{Tan}\,(x - \pi/4)$

17.

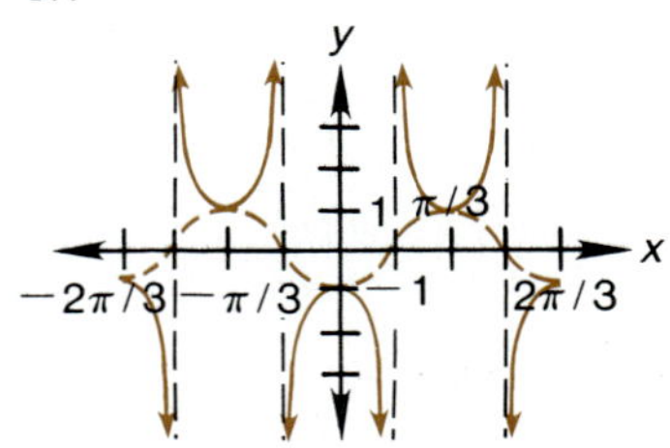

$y = \text{Sec}\,(3x - \pi)$

19.

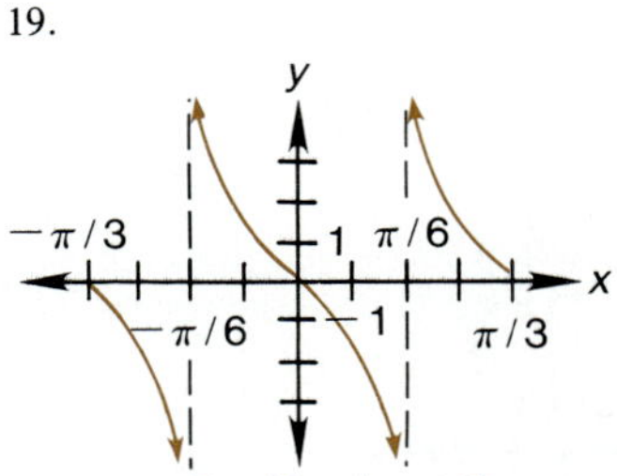

$y = \text{Cot}\,(3x + \pi/2)$

Chapter Review

1. 1.1 2. 2.4 3. 3.9 4. 5.1 5. 157.5° 6. 585° 7. 98.6° 8. 261.4° 9. 25 cm 10. 28.5 m 11. 0.24 ft 12. 33 cm^2 13. 591 mm^2 14. 15.6 in^2 15. 19,400 cm/min 16. 1.2 rad/s 17. 223 ft/s 18. 149 deg/s 19. 74 rpm 20. 6.8 rad/s 21. 100° 22. 1,700 in/s 23. (a) 84 rad/s, (b) 25 m/s, (c) 90 km/h 24. (a) 3.2, (b) 0.52, (c) 183

25. amp $= 3$, period $= 2\pi$, shift $= \frac{\pi}{4}$ 26. amp $= 5$, period $= \pi$, shift $= -\frac{\pi}{2}$ 27. period $= \pi$, shift $= \frac{3\pi}{4}$

28. amp $= \frac{1}{4}$, period $= \frac{\pi}{2}$, shift $= -\frac{\pi}{2}$ 29. amp $= 1.5$, period $= 2$, shift $= -\frac{1}{3}$ 30. period $= \frac{\pi}{4}$, shift $= \frac{\pi}{4}$

31. amp $= 1$, period $= 1$, shift $= -\frac{3}{8}$ 32. period $= \frac{\pi}{3}$ 33. period $= \frac{2\pi}{3}$, shift $= \frac{\pi}{6}$ 34. period $= 4$

35. period $= 8\pi$, shift $= -4\pi$ 36. amp $= 6$, period $= 8$, shift $= 4$

37.

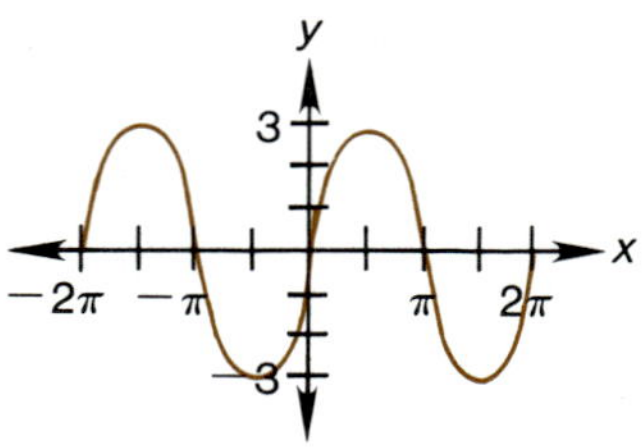

$y = 3 \text{ Sin } (x)$

38.

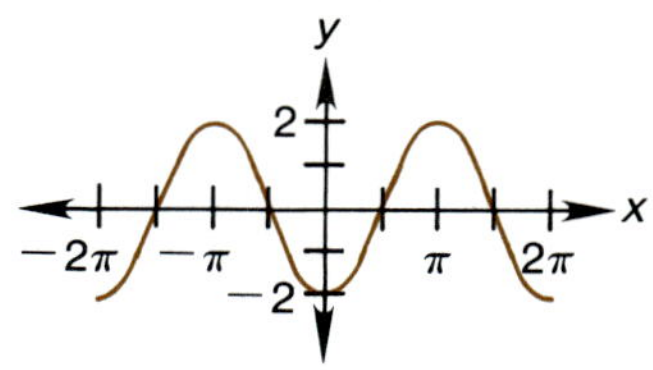

$y = -2 \text{ Cos } (x)$

39.

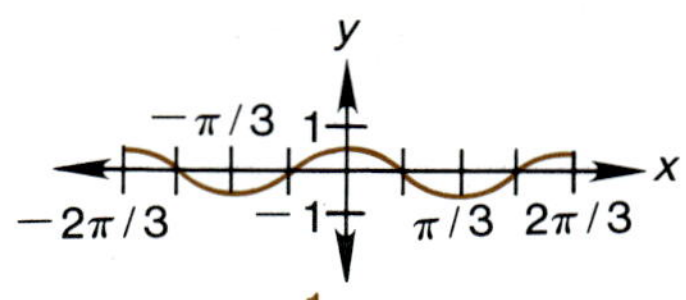

$y = \frac{1}{2} \text{ Cos } (3x)$

40.

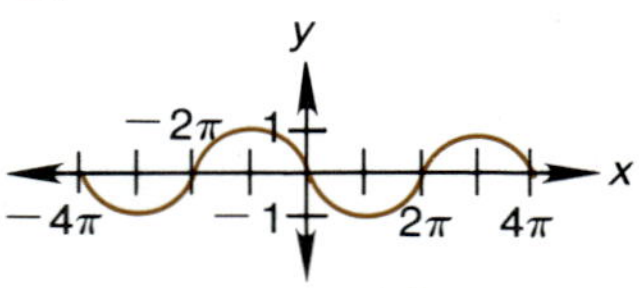

$y = -\text{Sin}\left(\frac{1}{2}x\right)$

41.

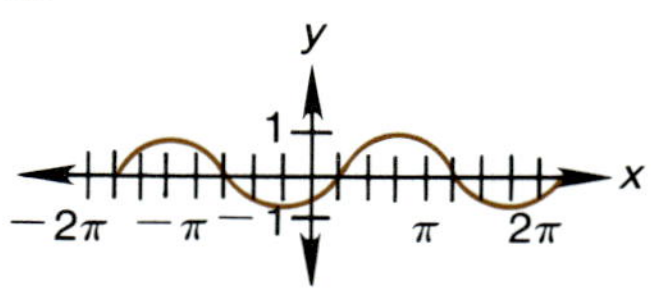

$y = \text{Sin } (x - \pi/4)$

42.

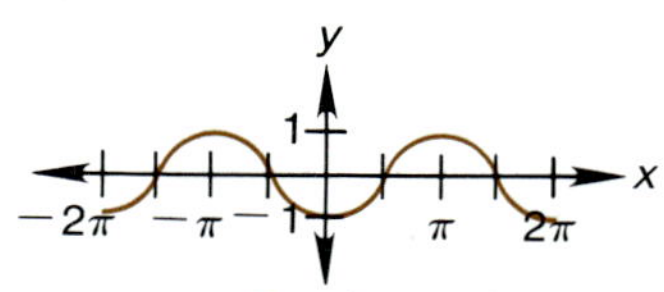

$y = \text{Cos } (x + \pi)$

43.

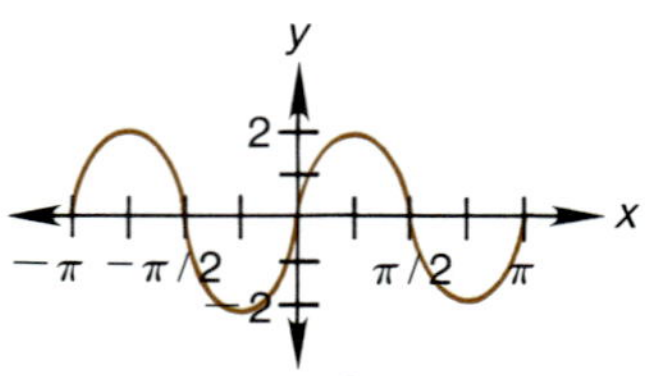

$y = -2 \text{ Sin } (2x + \pi)$

44.

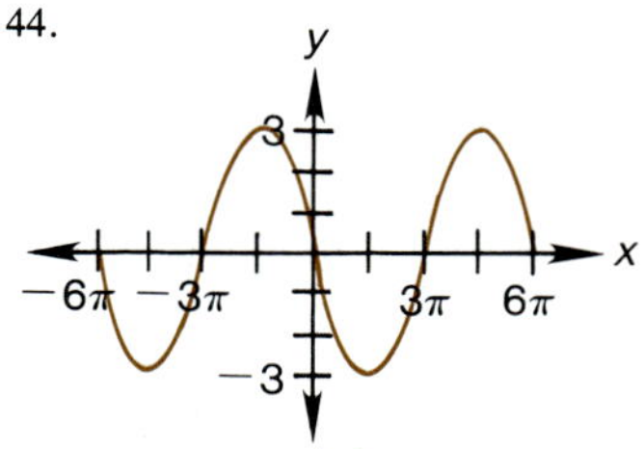

$y = 3 \text{ Cos}\left(\frac{1}{3}x + \frac{\pi}{2}\right)$

45.

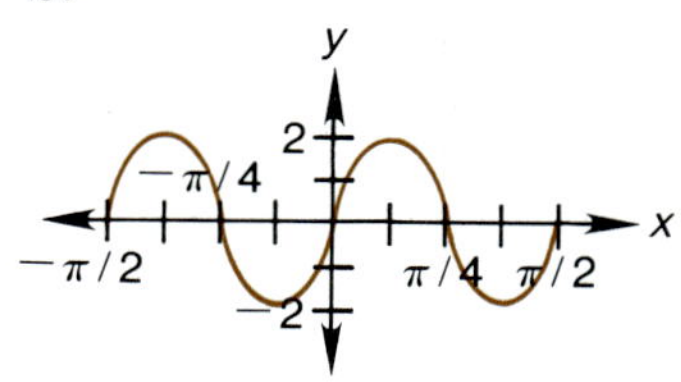

$y = 3/2 \text{ Cos } (4x - \pi/2)$

46.

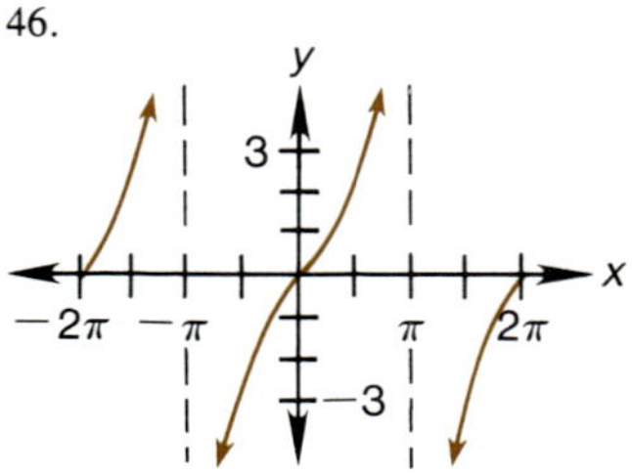

$y = 3 \text{ Tan}\left(\frac{1}{2}x\right)$

47.

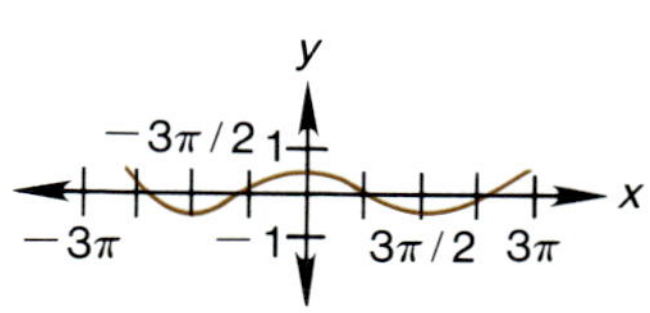

$y = -\frac{1}{2} \text{ Sin}\left(\frac{2}{3}x - \frac{\pi}{2}\right)$

48.

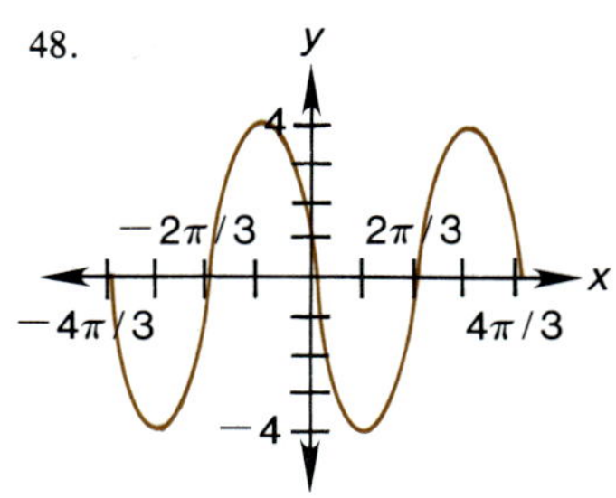

$y = -4 \text{ Cos } (3/2x - \pi/2)$

49.

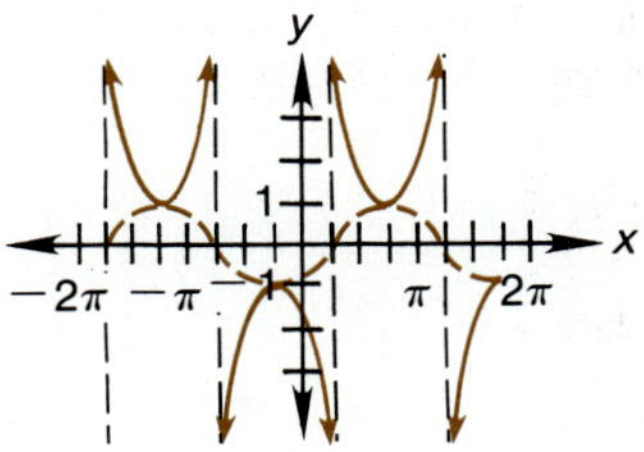

$y = \text{Csc}\,(x - \pi/4)$

50.

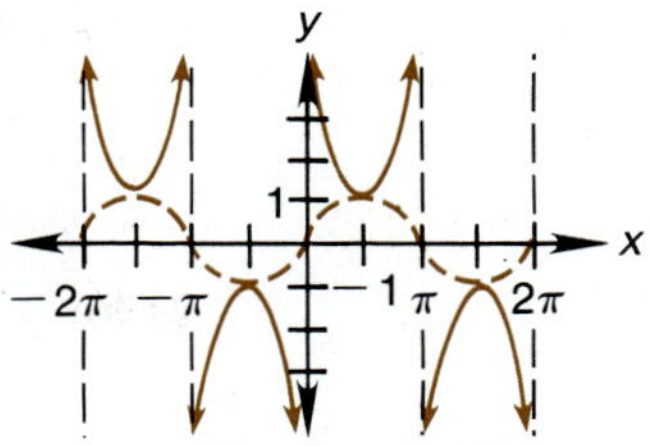

$y = \text{Sec}\,(x - \pi/2)$

51.

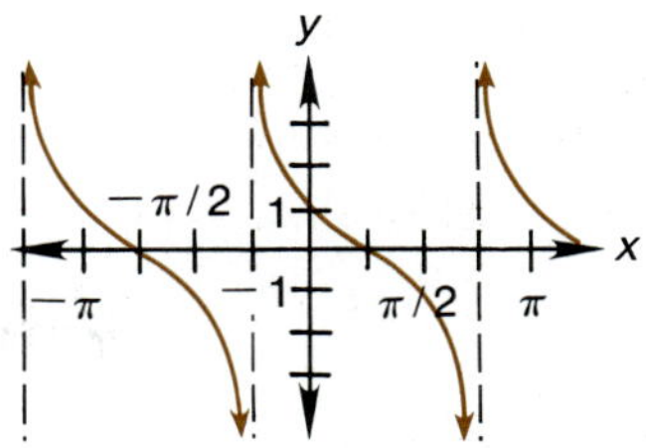

$y = \text{Cot}\,(x + \pi/4)$

Chapter Test

1. 0.57 rad 2. 210° 3. 2.9 in 4. 84.5° 5. 3.22 rad/s 6. 2,000 rad/min 7. 6 cm^2 8. amp = 3, period = π

9. period $= \frac{\pi}{3}$, shift $= \frac{\pi}{6}$ 10. amp = 4, period $= \frac{8\pi}{3}$, shift $= -\frac{4\pi}{3}$

11.

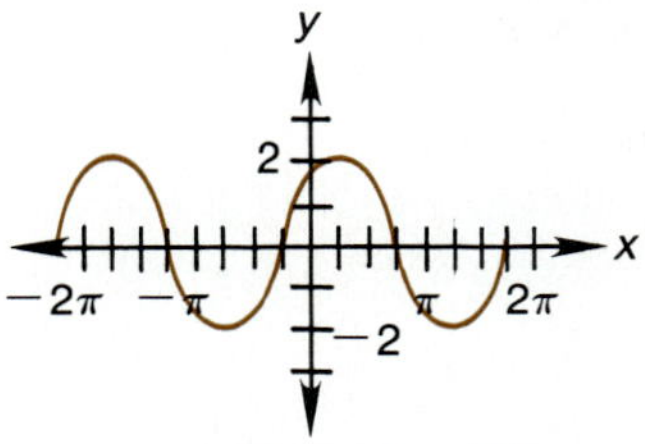

$y = 2\ \text{Sin}\,(x + \pi/4)$

12.

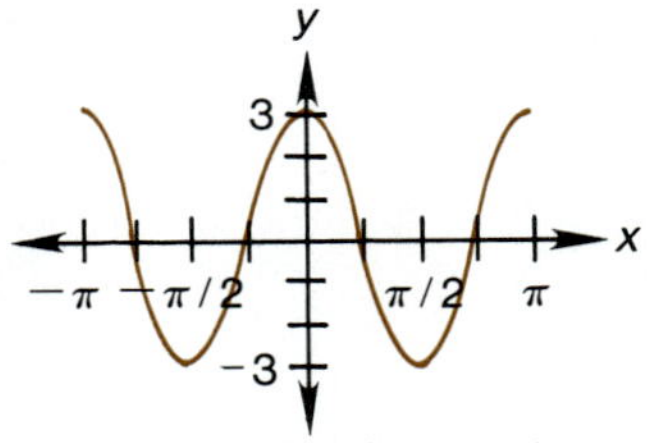

$y = -3\ \text{Cos}\,(2x - \pi)$

13.

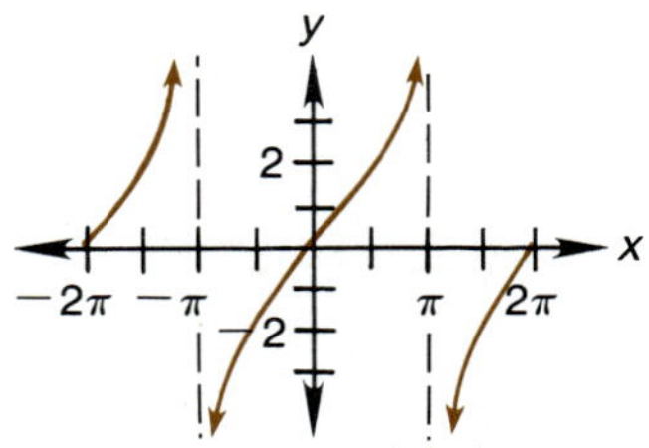

$y = 2\ \text{Tan}\left(\frac{1}{2}x\right)$

14.

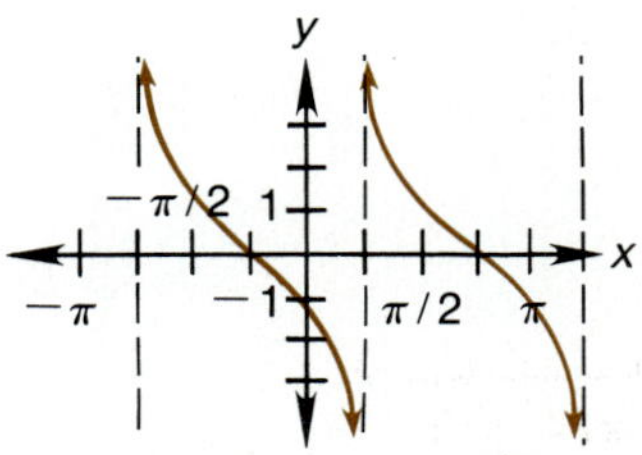

$y = \text{Cot}\,(x - \pi/4)$

15.

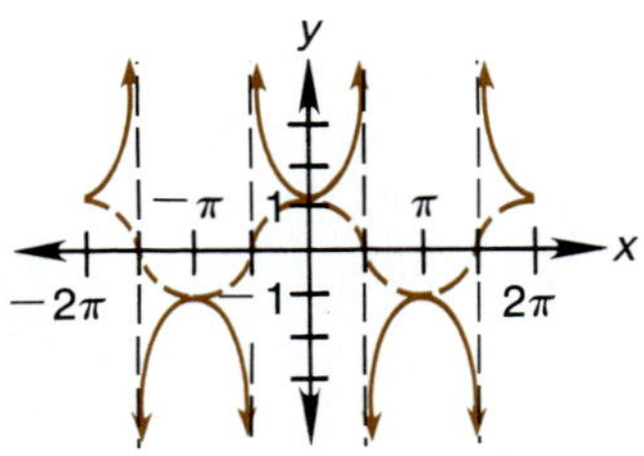

$y = \text{Csc}\,(x + \pi/2)$

Index

D

E

F

O

P

Q

R

British System of Measure

Units of Length

12 inches	=	1 foot
3 feet	=	1 yard
$16\frac{1}{2}$ feet	=	1 rod
5,280 feet	=	1 mile

Units of Area

144 in^2	=	1 ft^2
9 ft^2	=	1 yd^2
160 rd^2	=	1 acre
640 acres	=	1 mi^2

Dry Measure

2 pints	=	1 quart
8 quarts	=	1 peck
4 pecks	=	1 bushel

Energy

1 BTU	=	778 ft-lb
1 kilowatt-hour	=	3,413 BTU
1 BTU	=	252 calories

Power

1 horsepower (hp)	=	550 ft-lb/s
1 horsepower (hp)	=	0.75 kilowatt
1 kilowatt	=	3,413 BTU/h

Units of Mass or Weight

16 ounces	=	1 pound
2,000 pounds	=	1 short ton
2,240 pounds	=	1 long ton

Units of Volume

1,728 in^3	=	1 ft^3
27 ft^3	=	1 yd^3

Liquid Measure

2 cups	=	1 pint
2 pints	=	1 quart
4 quarts	=	1 gallon
$31\frac{1}{2}$ gallons	=	1 barrel
231 in^3	=	1 gallon

Symbolism

Unit	*Symbol*
inch	in
foot	ft
yard	yd
rod	rd
mile	mi
cup	c
pint	pt
quart	qt
gallon	gal
peck	pk
bushel	bu
ounce	oz
pound	lb
short ton	t
long ton	T
hour	h
minute	min
second	s
week	wk
year	yr